MOLECULAR MOTIONS IN LIQUIDS

MOLECULAR MOTIONS IN LIQUIDS

PROCEEDINGS OF THE 24TH ANNUAL MEETING
OF THE SOCIÉTÉ DE CHIMIE PHYSIQUE
PARIS-ORSAY, 2–6 JULY 1972

Jointly sponsored by the
ASSOCIAZIONE ITALIANA DI CHIMICA FISICA
DEUTSCHE BUNSENGESELLSCHAFT FÜR PHYSIKALISCHE CHEMIE
FARADAY DIVISION OF THE CHEMICAL SOCIETY

Edited by

JEAN LASCOMBE

Laboratoire d'infrarouge, Université de Bordeaux I, Bordeaux, France

D. REIDEL PUBLISHING COMPANY

DORDRECHT-HOLLAND / BOSTON-U.S.A.

Library of Congress Catalog Card Number 73-91947

ISBN-13:978-94-010-2179-1   e-ISBN-13:978-94-010-2177-7
DOI: 10.1007/978-94-010-2177-7

Published by D. Reidel Publishing Company,
P.O. Box 17, Dordrecht, Holland

Sold and distributed in the U.S.A., Canada, and Mexico
by D. Reidel Publishing Company, Inc.
306 Dartmouth Street, Boston,
Mass. 02116, U.S.A.

All Rights Reserved
Copyright © 1974 by D. Reidel Publishing Company, Dordrecht, Holland
Softcover reprint of the hardcover 1st edition 1974

No part of this book may be reproduced in any form, by print, photoprint, microfilm,
or any other means, without written permission from the publisher

TABLE OF CONTENTS

FOREWORD BY MICHEL MAGAT	IX
PREFACE	XI
ORGANIZING COMMITTEE	XIII
ACKNOWLEDGEMENTS	XIII
LIST OF PARTICIPANTS	XV
JACQUES YVON / Les problèmes de l'agitation moléculaire dans les liquides	1
PETER SCHOFIELD / Theoretical Aspects of Transport Phenomena	15
JOHN LAMB / Viscoelastic and Ultrasonic Relaxation Studies Related to Molecular Motions in Liquids	29
MANUEL G. VELARDE / On Cluster Expansions and Autocorrelation Functions: Further Remarks on the Divergence of the Density Expansion of Transport Coefficients	65
NARINDER K. AILAWADI / Evidence of Cooperative Phenomena in the Asymptotic Time Behavior of Correlation Functions: Self-Diffusion Coefficient of Methane	71
GENE F. MAZENKO and SIDNEY YIP / Fully Renormalized Kinetic Theory of Thermal Fluctuations in Liquids	79
HAROLD L. FRIEDMAN / Isotope Effects Upon Motions in Liquids in the Classical Limit	87
ROBERT H. COLE / Complex Permittivity and Dipole Correlations	97
CLAUDE BROT / Mouvements orientationnels: Fonctions de corrélation et illustrations expérimentales	107
H. KILP, G. KLAGES, and W. NOERPEL / The Influence of the FIR Absorption on the Dielectric Behaviour of Rigid Polar Molecules in Very Dilute Solutions	123
PIERRE DESPLANQUES, EUGÈNE CONSTANT, et RENAUD FAUQUEMBERGUE / Utilisation de la fonction mémoire dans l'interprétation des spectres d'absorption dipolaire des molécules toupies symétriques en phase liquide	133
JEAN-LOUIS GREFFE, JOSE GOULON, JEAN BRONDEAU, et JEAN-LOUIS RIVAIL / Champ de réaction dynamique en relaxation diélectrique, application au cas du chloroforme à 25 °C	151
JOSE GOULON, JEAN-LOUIS RIVAIL, JOHN CHAMBERLAIN, and GEORGE W. CHANTRY / Active Intramolecular Motion in Dielectric Relaxation of Pure Liquid Diacetyl	163
ROBERT S. WILSON / Dielectric Relaxation in Critical Polar Mixtures; Derivation of Glarum-Cole Equation by Projection Operator Method	171

JEAN-PIERRE BADIALI, HUBERT CACHET, ALAIN CYROT, et JEAN-CLAUDE LESTRADE / Etude, par relaxation diélectrique, du mouvement brownien des ions dans les liquides simples 179

SAVO BRATOS, YVES GUISSANI, and JEAN-CLAUDE LEICKNAM / Molecular Vibrations in Liquids 187

J. LASCOMBE, M. BESNARD, P. B. CALOINE, J. DEVAURE, et M. PERROT / Détermination par spectrométrie de vibration des fonctions de corrélation orientationnelles monomoléculaires dans les liquides 197

I. LAULICHT and S. MEIRMAN / Raman Spectroscopy and Molecular Reorientation in Liquids: $CDCl_3$ and CH_3Br 213

REINER ARNDT, R. MOORMANN, and A. SCHÄFFER / Raman Spectroscopic Studies on Temperature Dependence of Reorientational Motions of CH_3J and CD_3J Molecules 217

G. DÖGE / Raman Spectroscopic Observation of Collective Excitation of Molecular Vibrations in Liquid Methyl Iodide 225

P. C. M. VAN WOERKOM, J. DE BLEYSER, and J. C. LEYTE / Intermolecular Vibrational Relaxation in Liquids, Measured by the Isotopic Dilution Method 233

J. P. PERCHARD, C. PERCHARD, et D. LEGAY / Mouvements moléculaires du sulfure de carbonyle à l'état dissous par spectrométries Raman et infrarouge 235

B. BORŠTNIK, D. PUMPERNIK, and A. AŽMAN / Vibrational and Reorientational Relaxation of Hydrogen-Bonded Systems 241

WALTER G. ROTHSCHILD / Vibrational Relaxation in Liquid and Solid Quinoline 247

JEAN-CLAUDE LEICKNAM et YVES GUISSANI / Fonctions de corrélation tensorielles: formalisme général pour les molécules polyatomiques en solution et application au cas de la diffusion du rotateur asymétrique 257

K. MÜLLER, P. ETIQUE, and F. KNEUBÜHL / The Influence of Vibration-Rotation Coupling on Correlation Functions of Molecules 265

S. SUNDER and R. E. D. MCCLUNG / Raman Studies of Molecular Reorientation in Liquid Sulfur Hexafluoride 273

M. GILBERT et M. DRIFFORD / Etude dynamique par diffusion Raman-laser des hexafluorures de métaux et métalloïdes 279

J. P. MARSAULT, F. MARSAULT-HERAIL, J. L. SAULNIER, et G. LEVI / Réorientation de molécules toupie symétrique CH_3D et CD_3H dissoutes dans des fluides simples du point triple au point critique 287

JOSETTE VINCENT-GEISSE et CATHERINE DREYFUS / Moments de bandes et fonctions de corrélation rotationnelles en infrarouge pour une vibration perpendiculaire d'une molécule linéaire 295

JANINE SOUSSEN-JACOB, JOSETTE VINCENT-GEISSE, CATHERINE ALLIOT, ANNE-MARIE BIZE, JEAN-CLAUDE BRIQUET, ELISE DERVIL, JACQUELINE LOISEL, et JEAN-PAUL PINAN-LUCARRE / Etude des mouvements moléculaires dans les liquides au moyen du profil des bandes de vibration dans l'infrarouge 301

N. I. REZAEV / The Temperature Effect on the Width and Intensity of the Raman Spectra — 309

PATRICK DORVAL et PIERRE SAUMAGNE / Etude par spectroscopie infrarouge et par spectroscopie Raman des fluctuations orientationnelles de la molécule de furanne en solution — 319

JAMES E. GRIFFITHS / Anisotropy in Molecular Reorientational Motions in Simple Liquids — 327

H. G. HERTZ / Translational Motions as Studied by Nuclear Magnetic Resonance — 337

JIRI JONAS, J. DEZWAAN, and J. H. CAMPBELL / Nuclear Magnetic Resonance Relaxation Studies of Reorientational Motions in Liquids at High Pressure — 359

G. MARTINI, M. ROMANELLI, and L. BURLAMACCHI / Electron Spin Relaxation in the Study of the Motions of Paramagnetic Probes in Liquids — 371

R. ECKERT, G. LOOS, and H. SILLESCU / Association and Anisotropic Molecular Reorientation in Liquids as Studied by NMR — 385

R. MILLS / Self-Diffusion Measurements in Simple and Complex Liquids — 391

A. LOEWENSTEIN, E. GLASER, and R. ADER / Study of Molecular Motions in Liquid Methylamines by Nuclear Magnetic Resonance, II — 403

TASSO SPRINGER / Investigation of Molecular Liquids by Neutron Spectroscopy — 411

MANFRED D. ZEIDLER / Comparative Study of Molecular Motion in Liquids by NMR and Neutron Spectroscopy — 421

J. C. LASSEGUES and J. W. WHITE / Inelastic Neutron Scattering by Chemical Reactions. I: Structure and High Frequency Dynamics of Trifluoroacetic Acids — 439

K. CARNEIRO, M. NIELSEN, and J. P. MCTAGUE / Collective Excitations in Liquid Hydrogen Observed by Coherent Neutron Scattering — 461

LOUP VERLET / Molecular Dynamics: Computer 'Experiments' on Simple Liquids — 469

F. H. STILLINGER and A. RAHMAN / Molecular Dynamics Calculation of Neutron Inelastic Scattering from Water — 479

J. H. R. CLARKE, S. MILLER, and L. V. WOODCOCK / Computation and Interpretation of the Spectroscopic Properties of Liquids — 495

S. H. CHEN, Y. LEFEVRE, G. MAZENKO, and P. A. EGELSTAFF / Pressure Dependence of the Incoherent Scattering Law and the Time-Dependent Triplet Correlation Function in Dense Hydrogen Gas — 507

PIERRE LALLEMAND / Etude des mouvements moléculaires par diffusion de la lumière — 517

J. P. MUNCH et S. CANDAU / Etude des mécanismes de relaxation de vibration dans le tétrachlorure de carbone et le chloroforme liquides par analyse des spectres de diffusion Brillouin — 535

NAHUM D. GERSHON and IRWIN OPPENHEIM / Molecular Motion in Liquids from Low Frequency Depolarized Light Scattering — 553

B. KASPROWICZ-KIELICH, S. KIELICH, et J. R. LALANNE / Apport des processus de redistribution et de réorientation moléculaire dans les effets électro-optiques non linéaires — 563

C. DEMOULIN, C. J. MONTROSE, et N. OSTROWSKY / Etude des propriétés viscoélastiques d'un liquide par diffusion de la lumière — 575

J. V. CHAMPION and D. A. JACKSON / The Spectrum of Light Scattered from Liquid n-Alkanes — 585

B. Y. BAHARUDIN, D. A. JACKSON, and P. E. SCHOEN / Brillouin Scattering in the Inert-Gas Liquids: Argon and Krypton — 597

J. DILL and T. A. LITOVITZ / Rayleigh Scattering: Density Dependence of Orientational Motions in Acetone and Benzene — 605

MANSEL DAVIES / Far Infra-Red Absorptions in Non-Dipolar Liquids — 615

U. BUONTEMPO, S. CUNSOLO, P. DORE, and G. JACUCCI / Rototranslational Spectrum of H_2 in Liquid A and Comparison Between Intercollisional Interference in Para-H_2 and Rare Gases — 637

D. FRENKEL, R. M. VAN AALST, and J. VAN DER ELSKEN / Interpretation of Far Infrared Spectra in Terms of a Collision Distribution — 647

ULRICH STUMPER / Dielectric Absorption of Non-Polar Liquids in the Microwave and Far Infrared Region — 655

FOREWORD

When, in my capacity as President of the Société de Chimie physique, I opened the 24th Annual Meeting of this Society, devoted this year to 'molecular motions in liquids', I was stirred by a particular emotion. This had two reasons, one general and the other rather personal. I would like to give an explanation in the Foreword to this volume of communications to the Meeting and their ensuing discussions.

An essential characteristic of science is its international nature. It is like a symphony composed of contributions by all the countries playing together as an orchestra in unison. Just as a melody has different 'colours' when played by strings or woodwinds, so there exist similar 'colour' differences, subtle ones, between scientific contributions from different countries, rooted as they are in their own cultural history and liable to impoverish the ensemble if they should cease to participate. I have always had an impression of marked 'colour' differences prevailing among American, Russian, Japanese and European contributions, although within the latter group the timbre is very much the same. This is why I have dreamed of a European 'chamber orchestra' in addition to the great world orchestra.

I am very pleased, therefore, to announce that this European 'chamber orchestra' is in the formative stage and that it is happening in physical chemistry. Beginning in 1974, there will be a joint European meeting every other year, of physical chemists from the Bunsengesellschaft, the Faraday Division, the Société de Chimie Physique and, I hope, the recently formed Italian Association of Chemical Physics; each of these societies cancelled its own annual meeting this year.

The present meeting may be considered as the general rehearsal of forthcoming European meetings. In fact, once the subject, place and date of the present meeting had been decided on by the Société de Chimie Physique, we rapidly became aware of the interest aroused, and the number of contributions grew far beyond our capacity. We invited our three sister-organizations to assist us in the selection of contributions and the scientific organization of the meeting, and we can hardly be grateful enough for the assistance given to the Organizing Committee by their representatives, Messrs Hertz, Mansel Davies, and Paoloni.

The more personal reason for my desire to say a few words is that forty years of study and finding my way have gone by since I first read Schrödinger's article on the liquid state in the *Handbuch der Physik*. This article exposed the gap between our understanding of the crystalline and gaseous states on one hand, and our lack of understanding of the liquid state on the other. This was one of the inducements that led me into the field of physical chemistry with hopes of contributing something significant to our knowledge of the liquid state. I did not succeed in materializing this 'significant contribution', the dream of every young researcher, although in the course of forty years I devoted my efforts to the problem whenever an opportunity arose, and my interest has remained lively. So I have been very much aware of the progress of our knowledge, accumulated especially in the past thirty years, of the liquid state and particularly of the molecular motions in liquids. This progress owes as much to the development of new concepts, autocorrelations and molecular dynamics, as to the further perfection of more established methods such as dielectric dispersion, infrared and Raman spectroscopy and Rayleigh scattering, and to the appearance of new methods of investigation such as inelastic neutron scattering and ultrasonic sound propagation.

I believe that the contributions to this congress constitute a unique account of our knowledge to date. This knowledge has reached a level that justifies our hopes that the forthcoming decade may bring us a general theory of the liquid state connecting the molecular and macroscopic properties, unless it is impossible to derive such a theory. I hope that this will not be the case.

MICHEL MAGAT
President of the Société de Chimie physique

PREFACE

In the past forty years, the increase in knowledge on stationary states of molecules undoubtedly constitutes one of the most spectacular developments of molecular physical chemistry. This field is still the subject of active research, owing to the refinement of spectroscopic methods and to the possibilities of high speed calculations.

However, a new direction of research is currently being developed in which not only the stationary states are considered, but also the fluctuations of these states. This new approach holds interest to many aspects in physical chemistry, and especially so in the study of the liquid state. In a classical approach, only the average short range structures of the liquid molecules are formulated, but it is clearly evident that such an approach is essentially lacking, since the most fundamental properties of a liquid are related to transport phenomena.

To be able to study the liquid state on a microphysical scale as a function of time, it was necessary to develop the applications of statistical mechanics. Specialists in nuclear magnetic resonance (one of the most recent developments in spectroscopy) were quick to exploit these theories. Also, previous works such as those reported by Cabannes, Daure and Rousset on depolarized Rayleigh scattering in liquids have been reinterpreted in the light of these new theories. These theories are currently being applied by active research in the study of various spectral profiles, the determination of relaxation times and by computer simulation of molecular dynamics.

The purpose of the 'Molecular motions in liquids' conference in Orsay was to establish an up-to-date account of results obtained. We hope that this conference will enable physical chemists to have a general view of these problems and that it will stimulate the interest for further work in this field. It is also hoped that this will lead to a better understanding of the process of physicochemical transformations in the liquid state.

J. LASCOMBE

ORGANIZING COMMITTEE

J. Lascombe, *Président*

P. Barchewitz, S. Bratos, C. Brot, R. Cerf, M. Davies, G. Hertz, L. Volino,

P. Lallemand, J.-C. Lestrade, M. Magat, D. Massignon, L. Paoloni

J.-L. Rivail, L. Verlet, C. Troyanowsky, *Secrétaire général.*

ACKNOWLEDGEMENTS

The 24th Annual Meeting of the Société de Chimie physique was organized with the financial help of the following organizations:

- Ministère des Affaires étrangères
  Direction générale des relations culturelles
- Ministère de l'Education nationale:
  Direction de la Coopération universitaire internationale
- Délégation générale à la recherche scientifique et technique
- Direction des recherches et moyens d'essais
- Commissariat à l'Energie atomique

and the following firms and industrial groups:

- Compagnie française de raffinage
- Compagnie générale d'électricité
- Institut français du pétrole
- Société l'Air liquide
- Société Kodak Pathé
- Union des industries chimiques

We should like to renew our thanks to all the above.

LIST OF PARTICIPANTS

Ailawadi, N. K., Inst. f. Theoretische Physics, 1 Berlin 33, Arnimallee 3, R.F.A.
Arndt, R., Inst. f. Physikalische Chemie, D 33 Braunschweig, Göttingstrasse, R.F.A.
Badiali, J. P., Physique des liquides et électrochimie, 2 place Jussieu, Paris, F
Balcou, Y., Département de physique cristalline et chimie structurale, Univ. Rennes, F
Barbut, E. S. P. C. I., Résonance magnétique, 10 rue Vauquelin, Paris, F
Beguin, Cermo, B. P. 53, 38041 Grenoble Cedex, F
Bernard, J., Chimie physique, Faculté des sciences, 32 rue Mégevaud, 25000 Besançon, F
Bertoluzza, A., Istituto Chimico Ciamician, Via Selmi 2, Bologna, Italia
Besnard, Spectroscopie infrarouge, Université de Bordeaux I, 33405 Talence, F
Bonamy, L., Physique moléculaire, Faculté des sciences, La Bouloie, 25000 Besançon, F
Bordewijk, P., Gorlaeus Laboratoria der Rijksuniversiteit, Postbus 75, Leiden, Pays Bas
Borstnik, B., Kemijski Institut Borisa Kidrica, Ljubljana, Yugoslavia
Bottreau, Optique ultra-hertzienne, Université de Bordeaux I, 33405 Talence, F
Bourgou, Laboratoire de physique de l'Ecole Normale Supérieure de Tunis
Bourret, Chimie structurale USTL, Place E. Bataillon, 34060 Montpellier Cedex
Bovee, W., Univ. of Technology, Department of Physics, Lorentzweg 1, Delft, Pays Bas
Bratos, S., Physique théorique des liquides, 4 place Jussieu, 75005 Paris, F
Briant, J., I.F.P., 4 avenue de Bois Préau, 92502 Rueil Malmaison
Briguet, Spectroscopie et Luminescence, UER de physique, Université de Lyon, F
Brondeau, J., Chimie théorique, Case Officielle n° 140, 54037 Nancy Cedex, F
Brot, C., Physique Matière condensée, Faculté des sciences, 06 Nice, F
Brun, Interactions moléculaires, Université des sciences et techniques, Montpellier
Buback, M., Inst. f. Physikalische Chemie und Elektrochemie, D 7505, Ettlingen, R.F.A.
Buontempo, U., U.R. Fisica Molecolare e Superfluidi Istituto di Fisica G. Marconi, Roma, Italia
Burlamacchi, L., Istituto di Chimica Fisica, Via Gino Capponi, 9 Firenze, Italia
Cachet, H., Physique des liquides et électrochimie, 2 place Jussieu, 75005 Paris
Carneiro, K., AEC Risø, DK 4000 Roskilde, Danemark
Champion, J., Physics Department, City of London Polytechnic, Jewry Street, London EC3N Z EY
Chanu, J., Thermodynamique des milieux ioniques et biologiques, Univ. Paris VII
Charles, Physique appliquée, Univ. des sciences et techniques du Languedoc, 34 Montpellier

Chen, S. H., Nuclear Engineering Department, M.I.T., Cambridge, Mass. 02139, U.S.A.
Cole, R. H., Chemistry Department, Brown University, Providence, R.I. U.S.A.
Constant, E., Centre de recherches sur les propriétés hyperfréquences des milieux condensés, 59000 Lille
Cunsolo, S., Instituto di Fisica, P.delle Scienze 5, Roma
Cyrot, A., Physique des liquides et électrochimie, 2 place Jussieu, Paris, F
Dansas, P., Physique de la matière condensée. Parc Valrose, 06034 Nice Cedex, F
Davies, G. J., R 383, Post office research dept, Dollis Hill, London NW2 7 DT, G.B.
Davies, M., Chemistry Department, University of Wales, Aberystwyth, G.B.
Delmau, J., Spectroscopie et luminescence, Université Lyon I, F
Depeyre, D., Institut de chimie, Ecole centrale, 92290 Châtenay-Malabry
Desplanques, P., Centre de recherches sur les propriétés hyperfréquences des milieux condensés, B.P.36, 59 Villeneuve d'Ascq
Devaure, J., Spectroscopie infrarouge, Univ. de Bordeaux, 33405 Talence, F
Djabourov, M., EPCI. Résonance magnétique, 10 rue Vauquelin, 75005 Paris, F
Döge, G., Inst. f. Physikalische Chemie, Technische Universität, Braunschweig, R.F.A.
Dorfmuller, Institut f. Physikalische Chemie der T.U., 3 Hannover, Calinstrasse 46, R.F.A.
Dorval, P., Spectrochimie moléculaire, 74 rue A. France, 29200 Brest, F
Drifford, M., Chimie physique, CEN Saclay, 91190 Gif s/Yvette
Dupeyrat, R., Spectroscopie Raman, Tour 22, Université Paris VI, F
Duplan, J. C., Spectroscopie hertzienne, UER Physique, Lyon I, F
Dupuy, J., Physique des matériaux, Univ. Claude Bernard, Lyon Villeurbanne, F
Dworkin, A., Physico- chimie des rayonnements, Centre d'Orsay, 91405 Orsay, F
Fauquembergue, R., Centre de recherches sur les propriétés hyperfréquences des Milieux condensés, B.P. 36, 59 Villeneuve d'Ascq
Frenkel, D., Laboratorium voor Fysische Chemie, Nwe Prinsengracht 126, Amsterdam, P.B.
Friedman, H. L., Department of Chemistry, State University of New York, Stony Brook, N.Y. 11790, U.S.A.
Gerschel, A., Physico-chimie des rayonnements (Orsay), Physique de la matière condensée (Nice)
Gershon, N. D., The University of the Negev, Beer Sheva, Israël
Goulon, J., Laboratoire de chimie théorique, Université de Nancy
de Graaf, L. A., Interuniversitair Reactor Instituut, Berlageweg 15, Delft, Hollande
Greffe, J. L., Chimie théorique, Université Nancy 1, F
Griffiths, J. E., Bell Telephone Laboratories, Murray Hill, New Jersey, U.S.A.
Harris, K. R., Chemistry Dpt, Simon Fraser University, Burnaby 2, B.C., Canada
Hautecler, S., C.E.N./S.C.K., Boeretang 200, B-2400 MOL, Belgique
Hertz, G. H., Institut f. Physikalische Chemie, 75 Karlsruhe, Kaiserstrasse 12, R.F.A.
Hoare, M. R., Dpt. of Physics, Bedford College, Regent's Park, London NW1 4 NS, G.B.

Huron, M. J., Institut français du pétrole, B.P. 18, 92502 Rueil Malmaison, F
Hurwic, J., Chimie des diélectriques, Université de Provence, 3 place V. Hugo, 13031 Marseille
Jacucci, G., Instituto di Fisica, Roma, Italia
Jackson, D. A., Physics Dpt, University of Kent, Canterbury, G.B.
Janik, J., Instytut Fizyki Jadrowej, Krakow, 31342 Pologne
Jonas, J., Dept. of Chemistry, Université of Illinois, Urbana, Illinois, U.S.A.
Josien, M. L., Spectrochimie moléculaire, 8 rue Cuvier, Paris, F
Kilp, H., Institut f. Physik, Johannes-Gutenberg-Universität, Mainz, R.F.A.
Kneubühl, F., Physics Dept., ETH, Zürich, Suisse
Kranck, Laboratoire de biophysique, université de Nice, F
Laforgue, D., Electrochimie, Faculté des sciences, 51062 Reims Cedex
Lalanne, J. R., Centre de recherches Paul Pascal 33 Talence, F
Lallemand, P., Ecole Normale Supérieure, 24 rue Lhomond, 75005 Paris, F
Lamb, J., Dept. of Electronics & Electrical Engineering, The University, G12 8QQ, Glasgow, Scotland
Lantelme, F., Laboratoire d'électrochimie, Université de Paris VI
Lascombe, J., Laboratoire infrarouge, Université de Bordeaux I
Lassegues, J. C., Physical Chemistry Laboratory, South Parks Road, Oxford, G.B.
Laszlo, P., Chimie organique physique, Université de Liège, Belgique
Laulicht, I., Dept. of Physics, Bar-Ilan University, Ramat-Gan, Israël
Leicknam, J. C., Physique théorique des liquides, Paris VI, F
Lemanceau, B., Centre de recherches Paul Pascal, 33 Talence, F
Lestrade, J. C., Physique des liquides et électrochimie, 2 place Jussieu, 75005 Paris
Lindheimer, Interactions moléculaires, Faculté des sciences, 34 Montpellier, F
Litovitz, T. A., Physics Dpt., Catholic University of America, Washington, D.C., U.S.A.
Loewenstein, A., Chemistry Dept., Technion, Haïfa, Israël
McClung, R. E. D., Dept. of Chemistry, University of Alberta, Edmonton, Alberta, Canada
Magat, M., Physico-chimie des rayonnements, Orsay, F
Marsault, J. P., Physique expérimentale moléculaire, Université Paris VI
Martinez, A., Thermodynamique et physique moléculaire, 64016 Pau
Martini, G., Institute of Physical Chemistry, University of Florence, Via G. Capponi 9
Massignon, D., DPC/Centre d'études nucléaires, Saclay
Mathieu, J. P., Départment de recherches physiques, Tour 22, Université de Paris VI
Mazenko, G. F., Department of Nuclear Engineering, M.I.T., Harvard University, U.S.A.
Meirman, S., Dept. of Physics, Bar-Ilan University, Ramat-Gan, Israël
Michielsen, J., Laboratorium voor Elektrochemie, Plantage Muidergracht 14, Amsterdam, P.B.
Mills, R., Research school of physical sciences, Australia National University, Canberra, Australia

Misguich, J. H., STGI/CEA, B.P. N° 6, 92 Fontenay aux Roses
Monnerie, L., Physico-chimie structurale et moléculaire, E.S.P.C.I., 10 rue Vauquelin 75005 Paris
Morand, G., Physique générale, Tour 12, 9 quai Saint-Bernard, Paris
Morrow, J. C., Dept. of Chemistry, University of North Carolina, Chapel Hill, North Carolina, U.S.A.
Mueller, K., Physics Dept., ETH, Zürich, Switzerland
Munch, J. P., Acoustique moléculaire, 4 rue Blaise Pascal, 67070 Strasbourg
Natkaniec, I., Institute Laue-Langevin, B.P. 156, Centre de tri 38 Grenoble
Nguyen, Minh Hoang, Physique moléculaire, Besançon
Nguyen, Van Thanh, Laboratoire d'infrarouge, Bât 350, Campus d'Orsay, 91 Orsay
Noerpel, W., Institut f. Physik der Universität, 6202 Wiesbaden-Biebrich Riehlstr. 12, Mainz, R.F.A.
Nouchi, G., Optique moléculaire, Université de Bordeaux I, 33405 Talence
Oksengorn, B., Laboratoire des Interactions moléculaires et des Hautes Pression, 92 Bellevue
O'Reilly, P., Argonne National Laboratory, Argonne, Ill., U.S.A.
Ostrowsky, N., Ecole normale supérieure, 75005 Paris
Ouillon, Spectroscopie Raman, Université Paris VI, 4 place Jussieu, 75005 Paris
Papoular, CRTBT, BP 166, Cedex 38042 Grenoble Gare
Perrot, M., Spectroscopie infrarouge et Raman, Université de Bordeaux I, 33 Talence
Posch, H. A., Physikalisches Institut Strudlhofgasse 4, 1090 Wien, Autriche
Pourprix, B., Centre de recherches sur les propriétés hyperfréquences des milieux condensés, 59650 Villeneuve d'Ascq.
Quintard, P., Physique moléculaire, Faculté des sciences, 87100 Limoges
Rahman, A., Argonne National Lab., 9700 South Cass Avenue, Argonne, Ill. 60439, U.S.A.
Regnier, J., Chimie structurale, U.S.T., Montpellier, Cedex 34060
Rezaev, N. I., Physics Department, State University, 117234 Moscow, U.R.S.S.
Rivail, J. L., Chimie théorique, Université de Nancy 1,
Romero, F., Facultad de Ciencias Universidad, Sevilla, Espana
Rossi, I., Infrarouge, Université Paris VI, 91405 Orsay
Roth, E., Dept. de recherche et analyse, CEA, C.E.N. Saclay, B.P. N° 2, 91 Gif s/Yvette
Rothschild, W. G., Scientific Research Staff, Ford Motor Cy, Dearborn, Michigan 48121 U.S.A.
Rubalcava, H., Dept. of chemistry, University College, Dublin 4, Ireland
Rull, L. F., Facultad de Ciencias Universidad de Sevilla, Espana,
Saumagne, P., Spectrochimie moléculaire, Université de Bretagne Occidentale Brest
Schlaak, Spectroscopie infrarouge, Université de Bordeaux, 33 Talence
Schoen, P. E., Physics Lab., University of Kent, Canterbury, Kent, G.B.
Schofield, P., Theoretical Physics Division, AERE, Harwell, Didcot, Berks, G.B.
Sempere, R., Chimie structurale, Faculté des sciences, 34 Montpellier

Sillescu, H., Institut f. Physikalische Chemie, Robert Mayer Strasse 11, 6 Frankfurt, R.F.A.
Soussen-Jacob, J., Département de recherches physiques, Université Paris VI, 75230 Paris
Springer, T., Institut f. Festkörperforschung der Kernforschungsanlage, Jülich, R.F.A.
Stumper, U., III Physikalisches Institut der Universität, D 34 Göttingen, R.F.A.
Sundelof, L. O., Institute of Physical Chemistry, University of Uppsala, Suède
Szasz, G., General Electric, 37 Pelikanstrasse, CH 8001 Zürich, Switzerland
Taylor, A. D., Physical Chemistry Laboratory, University of Oxford, G.B.
Ter-Minassian, L., Physico-chimie des surfaces et des membranes, Paris
Tete, A., Thermodynamique des sels fondus, Université de Provence, 13013 Marseille
Tilenschi, S., Chimie physique des macromolécules et des colloïdes, Fac. de chimie, Bucarest
Totelin, C., Chimie organique physique, Université de Liège, Belgique
Troyanowsky, C., E.P.C.I., Chimie physique, 10 rue Vauquelin, 75005 Paris
Turq, P., Electrochimie, Paris VI, 8 rue Cuvier, 75005 Paris
Uebersfeld, J., Physique et métrologie des oscillateurs du C.N.R.S.
Vallauri, R., Laboratorio elettronica quantistica, via Panciatichi, 56/30 Firenze, Italia
Van Aalst, R. M., Laboratorium voor Fysische Chemie, Amsterdam, P.B.
Van Dael, Laboratorium voor Molekuulfysika, 3030 Heverlee, Belgique
Van Wechem, Laboratorium voor Electrochemie, Universiteit van Amsterdam, P.B.
Van Woerkom, P. C. M., Gorlaeus Laboratorium Afdeling Fysische Chemie III, Leiden, P.B.
Velarde, M., Universidad Autonoma de Madrid, Departamento de Física, Espagne
Verkerk, P., Interuniversitair Reactor Instituut, Berlageweg 15, Delft, P.B.
Verlet, L., Physique théorique et hautes énergies, 91405 Orsay
Vesely, F., Physikalisches Institut der Universität, Wien, Autriche
Vicq, G., Optique ultra-hertzienne, Université de Bordeaux 1, Talence
Vincent-Geisse, J., Département de recherches physiques, service infrarouge, univ. Paris VI
Viossat, Spectroscopie Raman, Département de physique, Tunis Belvédère
Volino, F., Institut Laue Langevin, 38042 Grenoble Cedex
Vorderwisch, P., HMI Reaktor, D 1000 Berlin, 39 Glienickerstrasse 100, R.F.A.
Weulersse, P., Département de physique, Faculté des sciences de Tunis
White, J. W., Physical Chemistry Lab., South Parks Road, Oxford, OXI, 3QZ, G.B.
Wilson, R. S., Michael Faraday Laboratories, Chemistry Dept., Northern Illinois Univ. Illinois, U.S.A.
Woodcock, L., School of Chemistry, Leeds University, Leeds LS2 9JT, G.B.
Yarwood, J., Chemistry Dept., University of Durham, Durham City, G.B.
Yip, S., M.I.T., Cambridge, Mass. 02138, U.S.A.
Yvon, J., Commissariat à l'Energie Atomique, 33 rue de la Fédération, 75015 Paris

Zeidler, M., Inst. Physikalische Chemie, Lehrstuhl II, 75 Karlsruhe, Kaiserstrasse 12, R.F.A.
Zmerli, A., Département de physique, Faculté des sciences El Menzah, Tunis
Zoppi, M., Laboratorio di Elettronica Quantistica del C.N.R., 50127 Firenze, Italia

LES PROBLEMES DE L'AGITATION MOLECULAIRE
DANS LES LIQUIDES

JACQUES YVON

Commissariat à l Energie Atomique, 29, rue de la Fédération, Paris 15e, France

Résumé. Tour d'horizon sur le programme de la conférence. Certains phénomènes peuvent s'interpréter dans le cadre de la mécanique classique, mais il ne faut pas même dans ce cas omettre de se référer à la théorie quantique sous-jacente. Les forces intermoléculaires et leur nature entièrement électromagnétique. L'impénétrabilité des molécules et le principe d'exclusion. Le liquide au repos. Surface libre, pesanteur et rôle du champ appliqué. La théorie des grands ensembles, le potentiel chimique, la pression, l'équation de Van der Waals et sa modernisation. Les états métastables, les pressions négatives. Les fluctuations: observables, moyennes simples, moyennes quadratiques et supérieures des fluctuations, leur permanence immuable dans l'équilibre. La diffusion des neutrons et de la lumière dans les liquides transparents: l'onde cohérente et l'organisation des phénomènes de diffusion.

Abstract. Survey of the conference's program. Some phenomena can be described within the framework of classical mechanics; however, even in these cases, one must not omit reference to the underlying Quantum Theory. Intermolecular forces and their electromagnetic character. Impenetrability of molecules and the exclusion principle. Liquids at rest. Free surfaces, gravity and the role of applied fields. The theory of grand ensembles, the chemical potential, the pressure, the Van der Waals equation and its modernization. Metastable states; negative pressures. Fluctuations: simple observable averages; quadratic and higher order fluctuations and their constancy at equilibrium. Neutron and light scattering in transparent liquids: coherent waves and organization of scattering phenomena.

1. Commentaire sur le programme de la conférence

Le programme de la conférence, décidé par le Comité d'organisation, était assez clair pour ne pas offrir de place à l'incompréhension.

Par contre l'ampleur de cette réunion a pris une tournure que je n'ai pas su prévoir lorsque j'ai accepté de tenir les premiers propos.

Comme me l'a écrit le distingué secrétaire général de la Société Française de Chimie Physique, cette dernière a un peu changé son fusil d'épaule. Ce qui veut dire qu'en cours de route, la Société Française s'est associée ses sœurs allemandes, anglaises et italiennes.

Non seulement le programme était clair, mais de plus il manifestait son actualité et son attrait. Il en est résulté l'extension que je viens de rappeler. Personnellement, j'y trouve un inconvénient car j'aurais préféré évoquer certaines de mes idées ou de mes interrogations devant un auditoire plus restreint avant d'affronter éventuellement une audience comme celle d'aujourd'hui.

Quoiqu'il en soit, les organisateurs ayant fixé les grandes lignes du programme, seule la publication in extenso des communications et des discussions pourra permettre de cerner l'ensemble des connaissances actuelles, les nouveautés et aussi les incertitudes et les points qui sont sujets à des contradictions. Aujourd'hui même je n'ai disposé, pour apprécier la situation, que des résumés préliminaires qui furent adressés au secrétariat, au nombre d'une cinquantaine il est vrai.

De quels liquides s'agit-il? Nous pouvons les identifier tout d'abord par les molécules constituantes. Les physico-chimistes peuvent être tentés par des produits exceptionnels, rares, susceptibles d'exalter tel ou tel phénomène. Mais en contrepartie, étant donné la difficulté pour un seul laboratoire d'aborder systématiquement tous les points de vue et étant donné les confrontations que permettent les produits classiques, il y a un intérêt évident à s'adresser à des molécules dont le contour est relativement compris. Je puis dire qu'en fait il s'agit de molécules dont la structure chimique a été convenablement auscultée, ce qui présuppose une certaine simplicité. Un caractère commun de la plupart de ces substances est qu'elles soient aisées à manipuler: en particulier, sauf erreur de ma part, je crois qu'elles sont toutes transparentes à la lumière du jour. Les communications portent en général sur des corps purs, mais un certain nombre examinent le cas de solutions. Grossièrement, ces molécules sont caractérisées par une géométrie approximative: molécules sphériques, allongées, aplaties. Une mention spéciale doit être accordée aux molécules électriquement polaires. Le carbone, l'hydrogène et les autres métalloïdes jouent un rôle privilégié. Naturellement, à l'état liquide, ces molécules peuvent donner lieu à des associations plus ou moins stables, plus ou moins complexes.

Il n'y a pas, dans un liquide, de phénomène statique ou évolutif, avec ou sans réaction chimique, qui ne soit conditionné par les mouvements de translation, de rotation, de vibration, les déformations de chaque molécule constituante, chacune de ces molécules étant impliquée dans le concert général.

Du point de vue expérimental, l'accent sera mis sur des phénomènes fins. En général le liquide en équilibre thermodynamique est soumis à de faibles sollicitations périodiques dont on observe les effets. Il s'agit d'ultra-sons, de rayonnements infrarouges ou millimétriques observés en transmission, de diffusion de la lumière ou des neutrons, d'effet Raman, de résonance magnétique. Les théoriciens, les calculateurs peuvent se permettre plus impunément d'aller au cœur des problèmes, d'où des études de dynamique moléculaire, avec une nouveauté qui est l'étude des milieux polaires et des considérations générales sur les phénomènes de transport.

Je termine cette introduction en mentionnant deux points de vue extrêmes: le premier consiste à s'intéresser essentiellement aux molécules en essayant de dégager leurs propriétés particulières au sein de leur environnement. Le second consiste à comprendre pourquoi malgré leur diversité les liquides présentent une certaine unité. C'est sur ce second point de vue que je désire insister aujourd'hui.

2. Première excursion théorique

Les organisateurs ont décidé que les liquides quantiques seraient exclus du cadre de la réunion. Ce que j'interprète plus exactement comme une exclusion des diverses variétés d'hydrogène et d'hélium liquide.

En fait tous les liquides sont quantiques, pour la raison majeure que personne n'a jamais vu un système non quantique, ni Archimède, ni Newton, ni Lavoisier. Si la constante de Planck était nulle, il n'y aurait ni molécules, ni atomes, ni corps solides, ni levier, ni pommes. Newton n'a jamais pu avoir l'idée qu'il créait la cinématique et la dynamique classique : le mot classique comme nous le comprenons aujourd'hui lui était aussi étranger que le mot quantique. La théorie classique doit être considérée aujourd'hui comme une approximation de la théorie quantique, approximation qui est valable lorsque la constante de Planck est négligeable.

Cette filiation, qui va à l'inverse de la filiation historique, a quelques conséquences qui peuvent heureusement s'énoncer assez simplement.

En théorie quantique, la particule ne suit pas une trajectoire. Son mouvement est décrit par une fonction d'onde. Ce qui nous importe pour les phénomènes que nous étudions c'est l'interprétation probabiliste, proposée par Max Born, du module carré de cette fonction. Qui dit probabilité dit aussi calcul de moyennes, travaux de statistique. La description statistique des systèmes moléculaires s'introduit beaucoup plus naturellement en théorie quantique qu'en théorie classique [8].

Un autre point concerne le suivi des particules. Dans un système classique, que les particules soient ou non de même nature, il est indispensable de les numéroter pour décrire les équations du mouvement. Cette numérotation se retrouve lorsqu'on veut suivre l'évolution statistique du système. La quantité qui évolue est la densité en phase et l'équation d'évolution est l'équation de Liouville. A première vue il n'y a rien de particulier à ajouter lorsque, par exemple, toutes les particules sont de même nature. La théorie quantique est plus impérative : elle exige que les particules de même nature soient indiscernables. Cette proposition a pour conséquence que la numérotation n'a aucun sens. Elle est pourtant bien commode. On s'entire dans la description classique en imposant à la densité en phase d'être invariante quand on y permute les coordonnées des particules d'une manière quelconque.

Une dernière remarque concerne une question dont le développement demande beaucoup de soins, mais dont les grandes lignes permettent d'introduire une notion d'importance dans l'étude des systèmes moléculaires. Je viens de dire que les particules ne suivent pas une trajectoire. Cette affirmation contredit à première vue l'observation de trajectoires de particules chargées, par exemple un proton, dans une chambre de Wilson ou dans une chambre à bulles. Il s'agit, bien entendu, de particules dont l'énergie cinétique est très supérieure à l'énergie de liaison des électrons dans les molécules du milieu détecteur. En fait l'explication de cette contradiction a été donnée il y a longtemps par Heisenberg dans le langage de la mécanique ondulatoire. Le mémoire de Heisenberg était passablement compact. Aujourd'hui, en profitant des progrès de la physique mathématique et de la terminologie, on pourrait reprendre la question d'une manière plus décontractée, si j'ose dire. Quoiqu'il en soit, et c'est là où je voulais en venir, dans le langage plus sophistiqué que nous utilisons aujourd'hui, nous dirons qu'il s'agit d'un phénomène d'autocorrélation : il y a, dans les conditions expérimentales précitées, une très forte corrélation entre les positions de la même particule à des instants consécutifs.

3. Les forces

Les systèmes moléculaires que nous étudions sont constitués de molécules bien définies.

On ne peut toutefois exclure l'existence à petite dose d'ions des deux signes, de radicaux libres, d'associations moléculaires plus ou moins stables.

En pénétrant plus avant dans la structure, disons que nous avons affaire à des systèmes constitués de noyaux atomiques stables et d'électrons. Cette considération est utile parce qu'elle nous permet de ramener toutes les interactions qui peuvent intervenir à un seul type : ce sont des interactions entre particules chargées. Ce sont des interactions électromagnétiques.

Pour être plus complet, il faut mentionner aussi que noyaux, souvent, et électrons, toujours, sont porteurs d'un moment magnétique. Il en résulte des interactions supplémentaires, secondaires du point de vue quantitatif, mais primordiales du point de vue de l'exploration des milieux moléculaires par les méthodes de résonance magnétique. Toutefois nous restons dans le cadre précité : les interactions supplémentaires qui apparaissent ainsi sont encore de caractère électromagnétique.

Il n'était sans doute pas inutile de souligner que l'électromagnétisme domine, en arrière fond peut-être mais très fondamentalement, la chimie et, consécutivement, la biologie.

Parmi les techniques qui permettent d'explorer les milieux moléculaires, il en est une toutefois qui fait appel à un autre type de phénomènes : je veux parler de la diffusion des neutrons lents. L'interaction de ces particules avec les noyaux atomiques relève d'un autre secteur de la physique. Cette remarque est marginale, bien entendu, pour notre propos.

Si nous n'envisageons que des phénomènes lents, et à cet égard l'agitation moléculaire aux températures qui nous intéressent est un phénomène lent, la considération d'une doctrine aussi vaste que l'électromagnétisme est superflue. Il s'agit, plus modestement, d'électrostatique et plus précisément, de la loi de Coulomb.

La théorie quantique, mariée avec l'attraction coulombienne exercée par les noyaux sur les électrons, explique l'existence de systèmes relativement stables comme les atomes, les molécules. On sait qu'un atome contient un nombre déterminé d'électrons. Ce nombre est tel que l'atome est électriquement neutre. Cette situation paraît tellement naturelle qu'on pourrait négliger d'en évoquer un facteur profond. Sans ce facteur, l'atome ne serait neutre que statistiquement, qu'en moyenne. Ce facteur, c'est le principe d'exclusion : deux électrons ne peuvent se trouver simultanément dans le même état. En conséquence l'atome a tendance à expulser les intrus. Le principe d'exclusion précise, notons le, le principe de l'indiscernabilité des particules de même nature.

Ce qui vaut pour les atomes vaut aussi pour les molécules. Ce que nous venons de mentionner explique la relative impénétrabilité de ces objets les uns pour les autres. Ils ont, au moins en première approximation, un volume propre. Pour les molécules qui ont des articulations ce volume suit naturellement les déformations.

Quand deux molécules sont proches l'une de l'autre sans être au contact, il se produit un effet contraire. L'existence de la tension superficielle, la coexistence d'un liquide et sa vapeur, l'existence de solides cristallins formés par des particules neutres, suppose des attractions. C'est ce que Van der Waals avait compris dès l'an 1873 et ce que F. London a mis en forme en 1928. Il est remarquable que ces attractions ne rappellent guère la loi de Coulomb: elles diminuent très vite avec la distance.

Un calcul précis de l'énergie potentielle intermoléculaire, résultant de l'effet combiné des répulsions et des attractions, à partir des premiers principes, présente trop de difficultés pour être mené à bien. On se contente de déterminations empiriques.

Dans une première démarche on admet que les molécules interagissent deux à deux et que l'effet global des interactions est la somme des interactions binaires. De plus, pour les molécules sphériques ou approximativement sphériques, ces interactions ne dépendent que de la distance. Moyennant ces limitations, la connaissance du potentiel intermoléculaire résulte de la détermination expérimentale du réseau d'isotherme du milieu gazeux. Il reste à ce stade assez d'arbitraire pour qu'il soit nécessaire d'utiliser des formules empiriques: la plus populaire, à juste titre, est celle de Lennard-Jones, qui comporte finalement seulement deux paramètres arbitraires. Ceux-ci restent à ajuster aux observations. Le fait que deux paramètres seulement suffisent est conforme à la loi des états correspondants.

Il faut dire que de nombreux mémoires théoriques conduisent à penser que l'hypothèse de seuls effets binaires n'est qu'une approximation.

Un mémoire récent [1] attire de son côte l'attention sur l'idée que les effets ne sont pas instantanés.

Dans le cas des molécules polaires, il faut compléter le schéma par l'existence d'interactions entre les dipôles: ici cet élément de l'interaction est exactement connu. Rendre compte de ses effets est assez délicat: ce sujet intéresse particulièrement la transmission du rayonnement hertzien. Les mémoires de Debye ont ouvert le sujet qui continue à faire l'objet de travaux très intéressants.

Une nouveauté due aux progrès des techniques est l'observation directe des interactions moléculaires par l'envoi d'un jet de molécules à travers une cible gazeuse. Je signale ici, sans aucune prétention à être exhaustif, des mesures relatives à l'hélium 4 [2] et d'autres relatives à la molécule H_2O [3]. Les premières ont permis de préciser la partie répulsive de l'interaction.

4. Surface libre et homogénéité

Je vous propose de contempler en imagination un verre d'eau posé sur une table. Vous constatez l'existence d'une surface libre, plane et horizontale, sauf au voisinage des parois où l'eau mouille le verre. Si l'eau reste dans le verre, cela est dû à l'action de la pesanteur. C'est aussi la pesanteur qui impose à la surface libre d'être un plan horizontal.

Tout cela est familier et très simple. Mais en compensation le liquide n'est pas homogène. La pression est plus élevée au fond du verre que près de la surface, il en va de même de la densité. Heureusement beaucoup d'observations du physico-chimiste n'ont guère à tenir compte de ces complications. Mais je dois insister sur le fait qu'il n'en va pas de même pour le théoricien scrupuleux. Tout d'abord parce qu'il doit être capable d'évaluer à partir des premiers principes les moindres caractéristiques de la densité et de la pression et ensuite parce que c'est sa mission de prévoir quels seraient les phénomènes en l'absence de pesanteur.

Par prudence toutefois, nous consulterons d'abord des témoignages d'observateurs. Ce que nous devons à la munificence de la NASA [4]. Celle-ci a procédé à des expériences en chute libre dans le vide sur une hauteur d'une quarantaine de mètres. Dans un ballon sphérique et transparent on dispose soit du mercure, soit du méthanol. Le ballon est libéré en haut de la tour. On observe la position de la surface libre du liquide un peu avant l'impact final. La chute libre soustrait le système à l'action de la pesanteur. Dans le cas du mercure, on voit la masse liquide décoller en partie de son assise initiale, cette masse prend la forme qui est habituellement celle d'une gouttelette. Là où la pesanteur disparaît, la tension superficielle commande. Le théoricien peut maintenant reprendre la parole et imaginer ce qui se passerait si la tour était plus haute : le mercure finirait par former une grosse boule et par décoller de la paroi. Avec le méthanol la situation est inversée : avant l'impact le liquide s'est répandu à la surface du ballon. Cette fois-ci la surface libre, qu'on imagine plus qu'on ne la voit, est concave.

Tenons-nous en au mercure, qui présente une situation plus intéressante, et imaginons de traiter par les voies savantes et recommandables de la mécanique statistique l'équilibre thermodynamique d'une masse de mercure liquide en présence de sa vapeur et en l'absence de pesanteur. Parmi d'autres complications que je ne désire pas énumérer, nous devons penser au phénomène suivant : la position de la boule n'est pas déterminée. Le théoricien va devoir sortir de son formalisme deux sortes de mouvements, d'une part l'agitation des atomes de mercure dans la masse liquide, dans la vapeur et en allant d'un milieu à l'autre et d'autre part les lents mouvements browniens de la boule dans son ensemble, mouvements qui ne sont pas indépendants les uns des autres. S'il ne le fait pas, c'est qu'il a triché.

On est bien libre d'exprimer l'opinion que le traitement théorique correct de cette situation acrobatique ne présente aucun intérêt. C'est tout au moins ce que je pense personnellement. Mais nous sommes alors ramenés à notre point de départ : imposer au milieu une situation plus stable en ne réduisant pas la pesanteur au néant.

Entre les deux descriptions, apesanteur et pesanteur, les géomètres consciencieux font une distinction importante : en situation d'apesanteur la symétrie est plus grande. La terminologie appropriée devant une telle situation consiste à dire qu'en présence de pesanteur on a affaire à une symétrie brisée.

L'hamiltonien du système moléculaire comprend donc trois sortes de termes :

(1) la somme des hamiltoniens individuels des molécules constituantes, qui comprend notamment l'énergie cinétique de translation ;

(2) la somme, également molécule par molécule, des termes d'énergie potentielle dus au champ appliqué ;

(3) la somme des énergies potentielles intermoléculaires, plus complexe.

Il ne reste plus, comme chacun sait, qu'à introduire cet hamiltonien – ou cette fonction de Hamilton – si le traitement est classique, dans l'exponentielle de Boltzmann-Gibbs et à procéder principalement à des calculs de moyennes, c'est-à-dire à des intégrations. C'est tout simple comme on voit, mais nous allons y revenir.

5. L'équilibre thermodynamique des systèmes simples

Suivre les mouvements individuels des molécules sur de longues périodes de temps est une tentative héroïque, aussi bien du point de vue théorique que du point de vue expérimental. Assurément, l'observation du mouvement brownien dans les suspensions colloïdales en donne une bonne idée. On ne saurait en surestimer l'importance historique. La possibilité de mélanger au milieu moléculaire étudié des molécules d'une nature très voisine, particulièrement par l'emploi d'isotopes radioactifs permet de cerner de plus près l'effet des mouvements désordonnés de translation. Enfin les moyens modernes de calcul, qui peuvent être orientés de diverses manières, permettent de traiter des échantillons constitués de quelques milliers de molécules. Les modèles moléculaires sont actuellement assez simples, mais des progrès de toutes sortes interviendront.

Toutefois l'observation courante met en évidence des phénomènes moins fantasques que le profond désordre sous-jacent. C'est que, quand plusieurs molécules viennent au contact, les péripéties de leur vie antérieure n'ont guère d'importance. Cette proposition est particulièrement valable quand l'équilibre thermique se trouve réalisé. Cet équilibre impose une discipline inattendue. Les grands énoncés de la thermodynamique en arrivent à ignorer la structure moléculaire. Les notions de température, de flux de chaleur, le principe de Carnot en sont les meilleurs témoins. Ces notions, ce principe ont affronté avec une superbe indifférence l'avènement de la théorie quantique: il incombe aux nouvelles idées de s'y conformer. L'exploration des basses températures n'a pas non plus apporté de contestation au moule initial.

La théorie moléculaire de l'équilibre thermodynamique ne repose pas uniquement sur l'exponentielle de Boltzmann-Gibbs. L'emploi de celle-ci repose sur l'idée que le système moléculaire étudié contient un nombre déterminé de molécules. Il est préférable d'introduire une variable aléatoire supplémentaire, qui est le nombre de molécules présentes. On passe ainsi, suivant la terminologie de Gibbs, de la théorie des ensembles de systèmes, à la théorie des grands ensembles. A première vue, c'est une complication. A l'épreuve on constate que l'analyse mathématique est plus claire. En particulier la théorie introduit maintenant très directement la notion de potentiel chimique.

La théorie des systèmes simples a fait l'objet de développements approfondis. Par systèmes simples il faut entendre des systèmes dont les molécules sont aussi simples que possible. Le modèle physique est fourni par les gaz rares. La molécule est monoatomique. Ses seuls mouvements sont des mouvements de translation. On suppose, ce qui est peut-être moins réaliste, que les forces sont de type binaire et de plus qu'elles ne dépendent que de la distance: ce sont des forces instantanées.

La conception suivante de laquelle le nombre des molécules présentes est aléatoire mafeste ici son avantage: elle a pour conséquence que les corrélations spatiales entre les molécules s'estompent rapidement avec la distance. Au cas contraire, elles ne s'estompent jamais rigoureusement. Reconnaissons que cet avantage est plus technique que physique.

Il est légitime de se demander si la théorie des systèmes simples justifie des développements aussi étudiés, alors que la liste des milieux intéressés est plutôt mince. D'une part, on peut répondre qu'il faut bien commencer par là. Mais il faut ajouter aussi que cette théorie sert également, convenablement adaptée, manipulée aux très basses températures ou même au zéro absolu, à aborder des sujets qui sont loin d'être épuisés, liquides quantiques, systèmes électroniques, atomes eux-mêmes et systèmes nucléaires, à commencer par les noyaux atomiques. Dans ces extrapolations, il convient de faire appel à la théorie quantique, ce qui ne s'impose pas pour les liquides que nous avons en vue aujourd'hui.

La notion de pression mérite un commentaire particulier. Précisons tout d'abord qu'elle déborde le cas de l'équilibre thermique. Sa définition est entièrement mécanique. Dans sa généralité, sa définition relève de la statistique, mais elle ne fait aucune allusion à la notion de température. Elle intervient en hydrodynamique, dans l'équation du mouvement – qui ne fait jamais que traduire, dans un cadre multimoléculaire, la relation élémentaire entre l'accélération d'une molécule et les forces qui lui sont appliquées. Sa structure est complexe puisqu'elle comprend un terme cinétique, qui résulte d'effets d'inertie et un terme intermoléculaire qui s'exprime à l'aide des corrélations binaires et de la force. C'est en général un tenseur symétrique, ayant donc six composantes indépendantes. Dans l'équilibre thermique, le terme cinétique se réduit à sa diagonale principale et il est isotrope. Sa valeur serait la même si le liquide pouvait s'assimiler à un gaz parfait. Le terme intermoléculaire est isotrope seulement lorsque le liquide est homogène et lui-même isotrope dans la région intéressée. Cette situation n'est jamais réalisée exactement au voisinage des frontières. Elle constitue en général une très bonne approximation au sein de la masse liquide.

A noter que la pression est envisagée ici du point de vue interne : la pression exercée sur la paroi supposerait une description moléculaire de celle-ci, qui nous entraînerait trop loin.

Lorsque le liquide est pratiquement homogène et isotrope – sauf peut-être aux frontières, la pression, qui ne comporte plus qu'une seule composante distincte, jouit d'une propriété remarquable. L'une des grandeurs fondamentales de la théorie est la grande fonction de partition. Cette grandeur a une interprétation moins pompeuse. Dans ce système où le nombre de molécules présentes est aléatoire, il y a une certaine probabilité π_0 très faible à la vérité, pour que le système n'en contienne aucune. C'est la probabilité du vide. La grande fonction de partition est l'inverse de π_0. Ceci dit, le logarithme naturel de la grande fonction de partition est égal au produit de la pression par le volume divisé par le produit de la constante de Boltzmann par la température absolue :

$$-\log \pi_0 = \frac{PV}{kT} \quad (P > 0).$$

Certains auteurs sont tentés de prendre cette formule comme une définition de la pression. Cette manière de faire est regrettable, car elle revient à dépouiller la mécanique statistique d'un de ses succès, étant donné que la formule ci-dessus s'accorde avec le principe de Carnot.

Quoiqu'il en soit, la pression dans le milieu homogène en équilibre thermodynamique peut s'exprimer par un développement en série vis-à-vis de la densité numérique, les coefficients de ce développement dépendant de la température et de la loi de force. Ce sont les coefficients du viriel et l'équation est l'équation d'Ursell.

Il semble – en attendant mieux – que la meilleure technique pour obtenir, à propos des liquides relativement denses qui nous intéressent, de bons résultats, consiste à traiter en première approximation les molécules comme des boules dures. On décrit ainsi un milieu sans existence réelle car sans cohésion mais qui manifeste la tendance des molécules à s'encager les unes les autres. Un calcul de perturbation permet ensuite de corriger cette approximation pour tenir compte des attractions, lesquelles sont beaucoup moins intenses que les répulsions [5]. Ce n'est jamais là que la modernisation du traitement séculaire de Van der Waals. Si on laisse de côté la région critique, le type d'équation obtenu permet de comprendre l'allure générale du réseau d'isothermes et les phénomènes de condensation.

6. Les états métastables

Parmi les retards au changement de phase, nous nous intéressons aujourd'hui au retard à la condensation et même plus spécialement au maintien à l'état liquide d'un liquide déprimé.

Le phénomène s'observe aisément lorsqu'on désire répéter l'expérience de Torricelli. Avec un peu de soin on peut soulever le tube barométrique sans que le mercure décolle du sommet. Il suffit ensuite d'un léger choc pour décoller le liquide et ramener son niveau supérieur à une situation plus orthodoxe.

Avant le décollement, la région haute du mercure se trouve dans un état métastable. Qualitativement le phénomène est prévu par l'équation de Van der Waals – ou par ses émules plus modernes – si l'on suppose que ces équations peuvent s'extrapoler hors du domaine de stabilité de manière continue. C'est prendre au sérieux la boucle de Van der Waals ou du moins le tracé de ses branches au voisinage de l'équilibre.

Je voudrais – sans prétendre résoudre un problème qui demanderait avant tout de meilleures observations – expliquer pourquoi l'équation d'état est susceptible de déborder le domaine des états d'équilibre. Il existe en effet deux techniques pour obtenir l'équation d'état. La première consiste à éliminer le potentiel chimique au profit de la densité numérique. La seconde part de la suite des équations de récurrence connues sous le sigle BBGKY. On spécifie que le système est stationnaire, qu'il n'y a pas de corrélation entre les vitesses et les coordonnées spatiales et que la loi de distribution des vitesses la loi de Maxwell est ramenée à une température unique. La suite des opérations conduit à la même équation d'état que l'autre méthode sans qu'on ait jamais fait appel aux propriétés extrêmales de la grande fonction de partition, lesquelles sont nécessaires pour assurer que l'équilibre thermodynamique est stable. En définitive l'équation d'état aurait un domaine d'application plus large que le domaine thermodynamique. Un point discutable, entre autres, est qu'il paraît difficile de généraliser le raisonnement au cas quantique.

L'expérience de dépression peut être poussée beaucoup plus loin que ci-dessus. L'expérience suivante est récente [6]. De l'eau est disposée dans un tube horizontal muni de cornes comme l'indique la figure 1 :

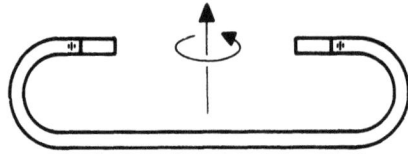

Fig. 1.

Ce tube est mis en rotation rapide autour d'un axe vertical. La pression du liquide dans la région centrale atteint aisément des valeurs négatives. A 18 000 tours par minute la pression de l'eau dans la région médiane peut atteindre − 100 atmosphères. Les auteurs estiment qu'il y a continuité entre les pressions positives et les pressions négatives. Ces essais mériteraient d'être poursuivis.

Il est d'ailleurs possible de contrôler l'existence de pressions négatives sans se fatiguer : elles sont révélées par l'existence de la tension superficielle. Il s'agit cette fois d'états stables.

7. Fluctuations

La terminologie quantique définit les observables. Ce sont des grandeurs microscopiques. Elles ne doivent donc impliquer aucune grandeur macroscopique, aucune probabilité. Les plus simples concernent la position des particules, leur quantité de mouvement, l'énergie intermoléculaire, le potentiel appliqué. Pour simplifier je supposerai ici que celui-ci ne dépend pas du temps.

Les observables élémentaires ne dépendent pas explicitement du temps. Mais à toute observable élémentaire on peut faire correspondre une infinité d'observables qui se déduisent de la première en suivant l'évolution du système au cours du temps suivant les lois de la dynamique :

G ou $G(0)$ engendre la famille $G(t)$,

t de signe quelconque. Seules les grandeurs conservatives correspondent à une observable indépendante du temps.

En vertu du principe d'indiscernabilité, une observable fait jouer le même rôle à toutes les particules de même nature. Dans le cas des grands systèmes moléculaires, les observables deviennent de ce fait très complexes, et, il faut bien le constater, particulièrement inobservables. La première information que l'on puisse souhaiter sur une observable est sa valeur moyenne − au sens de Max Born.

Si une plus grande ambition est possible, on considère l'écart entre l'observable en question et sa valeur moyenne. C'est la fluctuation. On en calcule la moyenne quadratique. On peut ultérieurement calculer des moyennes mixtes et des moyennes de degré plus élevé.

Ces observables ont leur équivalent en théorie classique. Leur définition y présente plus de souplesse parce qu'il n'est plus gênant dans ce cas de localiser exactement les particules. De plus la famille $G(t)$ s'interprète aisément par le suivi des trajectoires.

Ce qui donne, dans le cas des gros systèmes, une importance primordiale aux moyennes, c'est que les écarts quadratiques sont généralement très petits.

Les évaluations des moyennes peuvent concerner aussi bien un système en évolution qu'un régime permanent ou que l'équilibre thermodynamique. Dans ce dernier cas tous les $G(t)$ ont la même moyenne.

Un exemple d'un examen relativement aisé concerne le nombre de molécules présentes dans un volume donné, du moins dans le cas de molécules simples et si le volume est grand devant la portée des corrélations. Si le liquide est homogène dans la région envisagée, les moyennes de tous ordres sont chacune proportionnelles au volume. Leur ensemble n'est pas gaussien. Il est à noter qu'au voisinage du point critique, la proportionalité au volume cesse soit parce que la pesanteur s'oppose à l'homogénéité soit parce que les corrélations acquièrent une grande portée. En conséquence elles restent toujours finies.

Il convient de retenir de cet exemple que les fluctuations des observables ne sont généralement pas gaussiennes.

Des grandeurs comme la température, l'entropie, la résistance électrique, la constante diélectrique ne sont pas des observables, contrairement à la densité, l'énergie interne, la pression, le potentiel électrique, le flux de chaleur. Par exemple la résistance électrique est le quotient de la moyenne du potentiel électrique par la moyenne du courant électrique. Il n'est question de diviser quoi que ce soit par l'observable courant électrique: les observables n'ont généralement pas d'inverse.

Il n'est pas interdit de considérer des fluctuations de température ou d'entropie, mais ces notions doivent être manipulées avec circonspection.

Même dans un système en équilibre de température, les fluctuations ne disparaissent jamais. Croire le contraire serait supposer que l'agitation thermique serait susceptible de s'arrêter. Il est clair que les grandes fluctuations régressent rapidement: il se pose là le problème de la dynamique des fluctuations. Toutefois ce traitement ne saurait être unilatéral. Dans un système en équilibre leur probabilité est immanente. Leur apparition n'est pas plus surnaturelle que leur disparition.

La dynamique des fluctuations ne peut s'étudier dans le détail. On est amené à considérer de nouvelles observables comme:

$$G(t)\,G(t+\tau)$$

qui se prêtent à nouveau à des calculs de moyennes. Celles-ci déterminent des corrélations temporelles. Dans les cas complexes on se limitera à des valeurs infinitésimales de τ. C'est ainsi que l'on peut procéder à l'étude des ondes d'agitation thermique.

Il convient de ne pas mélanger deux notions bien distinctes: les fluctuations, qui sont spontanées et inévitables avec les perturbations qui résultent, si par exemple le liquide est initialement au repos, de l'action d'un agent extérieur au système. A la suite d'une perturbation les moyennes des fluctuations sont naturellement modifiées, mais il en est de même des moyennes simples et des autres grandeurs macroscopiques, qui présentent un intérêt plus immédiat.

8. Phénomènes de diffusion optique

La nécessité d'étudier d'abord les moyennes macroscopiques – qui ont le mérite d'être mesurables et si bien mesurables que les mécaniciens, les thermiciens en avaient une bonne connaissance avant la découverte de l'électron – ne doit pas nous détourner de la recherche des situations où les fluctuations viennent s'installer à l'avant-scène.

C'est ce qui se produit en particulier lors des phénomènes de diffusion. Je précise: diffusion des neutrons, diffusion de la lumière. Les neutrons donnent lieu à des calculs plus simples, mais les principes requis sont les mêmes. La diffusion de la lumière a l'intérêt de présenter des traits plus accusés et l'avantage de poser plus sérieusement les problèmes.

Le liquide est transparent vis-à-vis du rayonnement incident. Celui-ci n'émet pas non plus spontanément un rayonnement de même espèce: dans le cas des neutrons il est aisé de choisir des molécules convenables. Dans le cas de la lumière, il est impossible d'éviter de toute manière l'existence du rayonnement thermique: aux températures qui nous intéressent son rôle ne semble pas important et nous le négligerons.

Que ce liquide soit en repos ou en mouvement, les phénomènes de la propagation et de la diffusion du rayonnement ne changent guère. Toutefois le cas d'un liquide initialement en équilibre thermique permet d'énoncer plus simplement la théorie. C'est pourquoi je m'en tiendrai là.

Il est commode de se placer dans le cadre de la théorie quantique. Avant que le rayonnement ne pénètre le milieu diffusant, l'équilibre thermique des états quantiques est réalisé. La probabilité de chacun d'eux est donnée par l'exponentielle de Boltzmann. Le rayonnement incident provoque des transitions, ce qui a pour effet d'altérer à la fois le rayonnement et le liquide. Les transitions qui amènent le liquide d'un état dans un autre altèrent la fréquence du rayonnement, à moins toutefois que le niveau d'énergie du liquide soit changé. Mais il existe des transitions virtuelles qui laissent le liquide dans son état quantique initial: la fréquence du rayonnement ne subit aucune altération, mais sa propagation est modifiée [7].

Dans les liquides transparents, l'effet des transitions virtuelles l'emporte de beaucoup sur celui des autres transitions. C'est cet effet qui commande les phénomènes de réfraction et réflexion, le changement de longueur d'onde dans le milieu traversé et qui, malgré le désordre moléculaire et l'agitation thermique, assure la cohérence de l'essentiel du rayonnement transmis.

L'onde cohérente est tellement dominante que les transitions non virtuelles ne prennent naissance que sur le passage de l'onde cohérente. Ces transitions correspondent à une rupture de l'équilibre thermique et à une diffusion de la radiation. Dans les conditions expérimentales normales la rupture de l'équilibre thermique n'influe pas sur la diffusion: ce sont deux phénomènes concomitants. Pour observer des interférences entre ces deux phénomènes il serait nécessaire d'employer des sources intenses. Alors il n'y aurait plus proportionnalité entre la radiation incidente et la radiation diffusée. Si on néglige les phénomènes de diffusion multiple, on peut dire que la radiation diffusée se propage à son tour suivant les lois de l'optique cohérente.

La diffusion des radiations est assurément liée à des fluctuations dans le milieu diffusant. Ces fluctuations sont les fluctuations propres du milieu diffusant : elles ne sont pas provoquées par le rayonnement, celui-ci se contente de les révéler. L'observation de la diffusion est donc un moyen d'analyser l'agitation thermique dans un liquide. On pourrait certes objecter que dans des expériences prolongées, l'équilibre thermique pourrait être rompu ; en fait le liquide se charge lui-même de revenir à l'équilibre. Ce retour à l'équilibre ne peut pas toutefois initier des effets du premier ordre. Dans le cas de la lumière, les progrès de l'optique permettent une analyse de plus en plus fine de la raie Rayleigh et des raies Brillouin. Il convient de ne pas négliger parallèlement les mesures de dépolarisation. Mais, c'est du moins mon avis, l'interprétation qui doit permettre d'évaluer les fluctuations en fonction des observations, nécessite encore des développements théoriques.

9. Conclusion

En profitant des impératifs de quelques phénomènes simples et familiers, j'ai essayé de montrer comment leur observation peut s'accorder avec l'existence des populations moléculaires et de leur agitation désordonnée.

Il me faut reconnaître qu'ayant laissé complètement de côté les comparaisons quantitatives, j'ai pu donner l'impression d'une relative facilité. En fait, même les principes ont sans doute encore des côtés superficiels ou même incertains et c'est aux études minutieuses, dont le programme n'est certes pas épuisé et dont la situation présente est l'objet même de l'actuelle conférence, de mettre les choses en meilleure place. J'ajoute que j'ai laissé presque entièrement de côté, comme il convenait, le secteur essentiel et difficile des phénomènes irréversibles.

Bibliographie

1. Schram, K.: *Phys. Letters* **43A**, 282 (1973).
2. Feltgen, R., Pauly, H., Torello, F., et Vehmeyer, H.: *Phys. Rev. Letters* **30**, 829 (1973).
3. Snow, W. R., Dowell, J. T., Chevrenak, J. G., et Berek, H. E.: *J. Chem. Phys.* **58**, 2517 (1973).
4. NASA: note technique D. 2075 (U.S.A. 1963).
5. Rocard, Y.: *Thermodynamique* **340**, Masson, Paris, 1967.
6. Winniek, J. et Cho, S. J.: *J. Chem. Phys.* **55**, 2092 (1971).
7. Yvon, J.: *Nuclear Phys.* **5**, 150 (1958).
8. London, F. et Bauer, E.: *La théorie de l'observation en mécanique quantique*, Hermann, Paris, 1939.

THEORETICAL ASPECTS OF TRANSPORT PHENOMENA

PETER SCHOFIELD

Theoretical Physics Division, AERE Harwell, Didcot, Berks., England

Abstract. In this paper the structure of the time correlation functions which determine the transport coefficients in a simple liquid are discussed. It is pointed out that the success of the rigid sphere model requires a modification of the usual memory function approach based on Mori's continued fraction method. The role of binary collisions is treated in some detail and it is shown that their effect can be represented by a function of time which scales with a parameter representing the 'hardness' of the interaction, but which gives a contribution to the transport coefficient approximately independent of 'hardness'.

Résumé. Dans cette communication, on considère la structure des fonctions de corrélations dans le temps qui déterminent les coéfficients de transport d'un liquide simple. Le succès du modéle 'sphère rigide' demande une modification du traitement usuel fondé sur la méthode de fraction continue de Mori.

Le rôle des collisions binaires reçoit un traitement détaillé et on démontre qu'il est possible de représenter leur effet par une fonction du temps avec un paramètre qui représente la 'dureté' de l'interaction, mais qui donne une contribution aux coefficients presqu' indépendante de cette dureté.

1. Introduction

In recent years considerable progress has been made in the calculation of the free energy, and hence the equation of state, of simple fluids by a double perturbation theory [1, 2, 3]. The procedure is to consider the interatomic potential as consisting of two parts, a strongly repulsive short range component plus a weaker longer range attraction. The theory is then based on the calculation of the repulsive potential system as a perturbation about a rigid sphere model, using either theoretical or computer generated distribution functions for this system, and the calculation of the attractive contribution also by perturbation theory, by expanding in powers of the attractive potential. There are indications that a similar approach will work for the calculation of transport coefficients – some of the evidence is summarised in Section 3 of this paper. One may also cite a recent review by Hanley, McCarty and Cohen [4], where the modified Enskog theory is assessed in its application to simple dense fluids. One of the aims of the present paper is to outline a programme by which this double perturbation might be carried through and to present some preliminary results.

The transport coefficients may be calculated in principle in one of two ways, either by calculating the changes in molecular distribution functions from their values in a uniform system in the presence of gradients of velocity, temperature, concentration, etc., or equivalently by the Kubo formulae in which the transport coefficients are expressed as time integrals of certain time correlation functions evaluated in equilibrium.

Up to about ten years ago all calculations of transport coefficients (including the Enskog theory, of course) used the first method, based on the Boltzmann equation.

From the point of view adopted in this paper it is more appropriate to approach the theory by the Kubo method. The reason for this is that one may study the time dependence of equilibrium correlations directly either by computer experiment or by radiation scattering such as inelastic neutron scattering, whereas the non-equilibrium molecular distribution functions are less tangible. Without the results of computer experiments much of the progress made in the understanding of the liquid state during the past decade would not have been possible – not only in the sense that the computer has produced well-defined results from a given intermolecular potential for a quantitative test of theory, but also in assessing the validity of hypotheses used in theoretical treatments.

In Section 3, we review some computer experiment results in relation to self-diffusion, (from the theoretical point of view the simplest transport coefficient) which will be used to illustrate the theory in this paper.

In Section 4, we outline the 'memory function' approach to the calculation of correlation functions, based on the generalised Langevin equation of Mori. In Section 5, we consider generally the time evolution of a classical many body system. Returning to the main theme of the paper, we then discuss how the rigid sphere system may be regarded as a limit of an inverse power law repulsion and evaluate the memory function associated with a binary collision, in the limit of strong repulsion.

2. Transport Coefficients in Terms of Correlation Functions

The Kubo formulae for transport coefficients are most simply derived by comparing the form of correlation function one obtains from the macroscopic equations in terms of which it is defined [5]. That is, if one has a macroscopic conserved variable $\bar{x}(R, t)$, satisfying a continuity equation

$$\dot{\bar{x}}(R, t) + \nabla \cdot \bar{y}(R, t) = 0 \tag{1}$$

then the linear transport coefficient relates $y(R, t)$ to $\bar{x}(R, t)$ by

$$y(R, t) = - D \cdot \nabla \bar{x}(R, t) \tag{2}$$

D is the 'diffusivity' (having dimensions (length)^2/time).

Equations (1) and (2) may be Fourier transformed and solved, to give the correlation function:

$$\langle \bar{x}_\kappa(t) \, \bar{x}_{-\kappa}(0) \rangle = \langle \bar{x}_\kappa \bar{x}_{-\kappa} \rangle \exp\{- D\kappa^2 |t|\}, \tag{3}$$

where $\langle \bar{x}_\kappa \bar{x}_{-\kappa} \rangle$ represents the mean square fluctuation of the κ the Fourier component of $\bar{x}(R, t)$. The microscopic expression for D is then obtained by finding the term in the correlation function for the microscopic equivalent of $\bar{x}(\kappa, t)$, $x(R, t)$ which has the asymptotic behaviour given by (3) and one deduces that, if the limit exists,

$$D = \int_0^\infty dt \lim_{\kappa \to 0} \{\langle y_\kappa(t) \, y_{-\kappa}(0) \rangle / \langle x_\kappa(0) \, x_\kappa(0) \rangle\} \tag{4}$$

and the Equation (2) is valid. Here $y_\kappa(t)$, defined by

$$\dot{x}_\kappa(t) = i\kappa y_\kappa(t). \tag{5}$$

Because of the relation between x and y, (4) may be expressed in a slightly different way in terms of the time Fourier transform of the time dependent correlation function $\langle x_\kappa(t) x_{-\kappa}(0) \rangle$. If

$$X(\kappa, \omega) = \frac{1}{2\pi} \int_{-\infty}^{\infty} dt \, e^{i\omega t} \langle x_\kappa(t) x_{-\kappa}(0) \rangle \tag{6}$$

then

$$D = \lim_{\omega \to 0} \omega^2 \lim_{\kappa \to 0} \kappa^{-2} X(\kappa, \omega)/X(\kappa) \tag{7}$$

with $X(\kappa)$ written for $\langle x_\kappa(0) x_{-\kappa}(0) \rangle$.

Thus the problem of the evaluation of transport coefficients is part of the more general problem of the microscopic evaluation of time correlation functions for equilibrium fluctuations.

In particular, we have for a simple fluid the shear and bulk viscosity coefficients η, ϕ given by fluctuations in the momentum density,

$$J_\kappa^\alpha = \sum_i p_i^\alpha \exp\{i\kappa \cdot r_i\}. \tag{8}$$

Rotational symmetry determines the corresponding correlation functions $C^{\alpha\beta}(\kappa, \omega)$ to be expressible in terms of longitudinal, and transverse functions $C_L(\kappa, \omega)$, $C_T(\kappa, \omega)$ as

$$C^{\alpha\beta}(\kappa, \omega) = \hat{\kappa}^\alpha \hat{\kappa}^\beta C_L(\kappa, \omega) + (\delta_\alpha^{\alpha\beta} - \hat{\kappa}^\alpha \hat{\kappa}^\beta) C_T(\kappa, \omega). \tag{9}$$

($\hat{\kappa}$ is a unit vector parallel to κ). The diffusivity associated with C_T and C_L via (7) are then the transverse and longitudinal kinematic viscosities η/ρ, and $(\phi + \frac{4}{3}\eta)/\rho$ with ρ the density.

Likewise the thermal conductivity, λ, is related to the entropy fluctuation [5], via the thermal diffusivity $\lambda/\rho C_p$ (C_p specific heat). It is interesting to note, however, that since in the hydrodynamic limit, the entropy fluctuation forms part of the density fluctuation (the Rayleigh peak in the light scattering spectrum), one can relate λ to a limiting expression of the density-density correlation function, $S(\kappa, \omega)$.

Finally, we shall consider the self-diffusion of a particle in a liquid, related to the single density correlation function $S_S(\kappa, \omega)$.

3. Experimental Results

In this section, we call attention to some of the findings of the computer experiments concerning the velocity autocorrelation function.

Figures 1 and 2 show, respectively, the mean square displacements and the velocity autocorrelation functions of atoms interacting with (a) a Lennard-Jones 6–12 potential and (b) just the repulsive part of this potential.

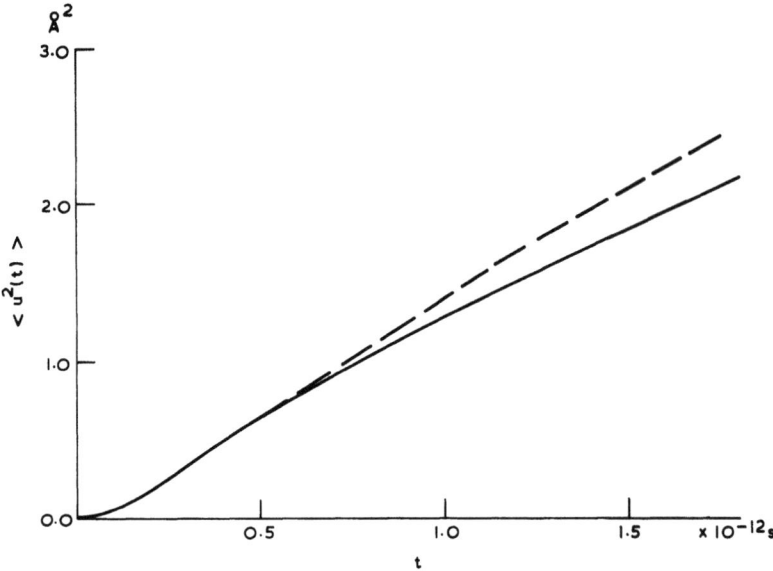

Fig. 1. Mean square displacement of atom in Lennard-Jones fluid in neighbourhood of triple point ($\varrho = 1.430$ gm cm^{-3}, T = 87.5 K). The full line is that for the full potential, the broken line, for just the repulsive part of the potential.

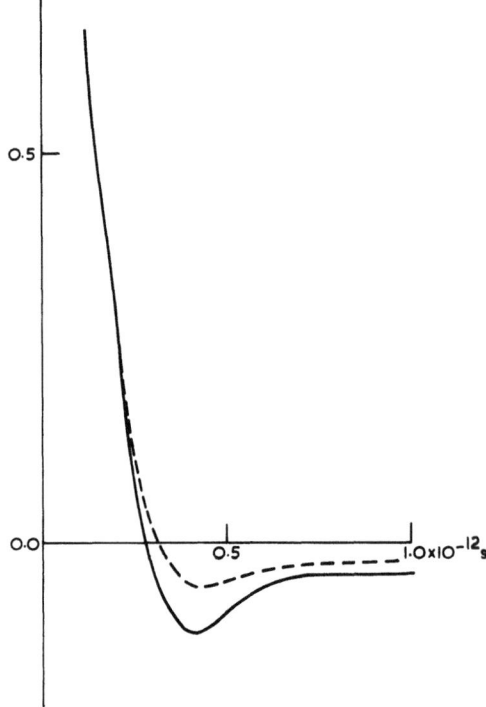

Fig. 2. The velocity autocorrelation function of an atom in Lennard-Jones fluid for the two cases of Figure 1.

The following points should be noted:

(I) At short times the curves are very similar, which seems to indicate that the repulsive interaction between a pair of atoms determines the initial behaviour of the velocity autocorrelation function.

At longer times, *both* autocorrelation functions show the negative region with a plateau first pointed out by Rahman [6]. It is important to note that this plateau is *not* a consequence of an attractive potential tail as implied in many of the theoretical models.

At the state point shown (near the triple point) the diffusion coefficient is reduced by about 25% by the attractive potential. However on the coexistence curve at a reduced temperature of $T^* = T/T_c = 0.87$, the difference is less than the experimental uncertainty [7]. Thus as in the calculation of the static distribution functions, one may say the attractive potential plays an important role only in the liquid region.

(II) At *liquid* densities, as shown here, the diffussion coefficient, given by the asymptotic slope of the mean square displacement, is pretty well determined after a short time of about 20×10^{-13} s, and, more significantly, by which time the atom has moved a root mean square distance of only 1.5Å – less than half the interparticle spacing.

(II) Turning now to the rigid sphere system, Figure 3 shows the diffusion coefficient as a function of density, as calculated by Alder, Gaas and Wainwright [8]. This is plotted relative to the value given by Enskog theory, D_E. It can be seen that this theory is an underestimate by as much as 30% at intermediate densities and an overestimate by 40% at high densities.

(IV) Finally, we compare the rigid sphere system with results for an inverse power repulsive potential, $\varepsilon(\sigma/R)^v$, with $v=12$ [9]. Table I gives values of the diffusion

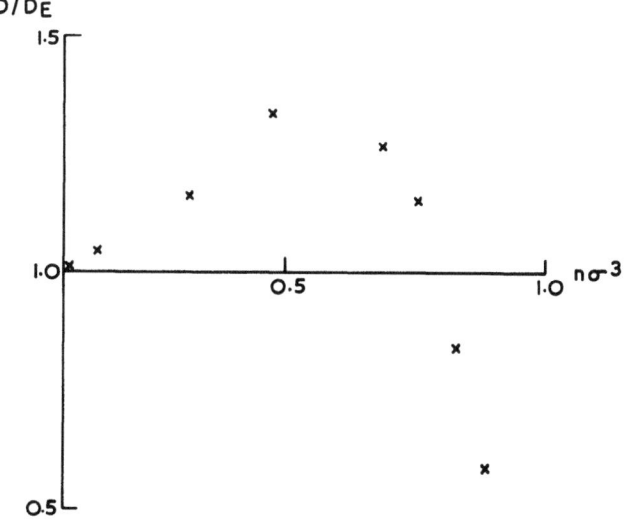

Fig. 3. Results of Alder *et al.* [8] for self-diffusion coefficient of rigid sphere fluid relative to Enskog value.

TABLE I

Self-diffusion for inverse twelfth power potential

Reduced density	D (Dimensionless units)	Rigid sphere value
0.470	0.0271	0.0278
0.556	0.0189	0.0194
0.631	0.0133	0.0140
0.710	0.0090	0.0096
0.815	0.0045	0.0054
0.862	0.0032	0.0039

constant obtained by the molecular dynamics method compared with the rigid sphere model, where the sphere diameter is chosen to fit the free energy. The first column gives the molecular dynamics results and the second the Alder, Gass and Wainwright result for the appropriate rigid sphere diameter. It can be seen that the agreement is reasonably good.

Table II shows a similar comparison at a single density as a function of temperature given by Levesque and Verlet [10]. Although on physical grounds it might not be too surprising that a rigid sphere model works, when one comes to consider the time

TABLE II

Self-diffusion of Lennard-Jones fluid at reduced density $\tilde{n} = 0.85$ (Levesque and Verlet [10])

T^*	D	Rigid sphere
0.76	0.0048 (0.0064)[a]	0.0059
1.08	0.0078	0.0090
1.27	0.0095	0.0112
2.14	0.0182	0.0195
2.81	0.0245	0.0254
4.70	0.0400	0.0460

[a] Value obtained for repulsive part of potential alone (Schofield, unpublished).

correlation function which determines the diffusion coefficient, one finds apparently quite different time behaviour at short times – for a rigid core, collisions are instantaneous and this results in the velocity autocorrelation function having a finite slope at the origin. With a differentiable potential, on the other hand the slope is zero. We examine this point, in the next section.

4. The Velocity Autocorrelation Function

The self-diffusion coefficient D is given by the time integral of the velocity autocorrelation function [5]:

$$D = \tfrac{1}{3} \int_0^\infty \langle v(0) \cdot v(t) \rangle \, dt \qquad (10)$$

$$= \frac{\kappa T}{M} V(t), \qquad (11)$$

where M is the particle mass and $V(t)$ the autocorrelation function

$$V(t) = \langle v(0) \cdot v(t) \rangle \, dt / \langle v^2 \rangle. \qquad (12)$$

If the interparticle potential $\Phi(\{r_i\})$ is a differentiable function of the atomic positions, then at short times

$$V(t) = 1 - \tfrac{1}{2}\omega_e^2 t^2 + \cdots, \qquad (13)$$

where ω_e^2 (the square of the Einstein frequency) is given by

$$\omega_e^2 = \langle a^2 \rangle / \langle v^2 \rangle, \qquad (14)$$

where a is the particle acceleration. Labelling the atom under consideration 0,

$$a^\alpha = \frac{1}{M} \frac{\partial \Phi}{\partial r_0^\alpha}. \qquad (15)$$

For simplicity, we consider the case where Φ is a sum of pair potentials, $\phi(r_{ij})$. We then have the following equivalent expressions for ω_e^2:

$$\omega_E^2 = \frac{1}{3M\kappa T} \langle (\nabla \Phi)^2 \rangle \qquad (16a)$$

$$= \frac{1}{3M} \langle \nabla^2 \Phi \rangle. \qquad (16b)$$

(16a) yields

$$\omega_E^2 = \frac{1}{3M\kappa T} \left\{ \left\langle \sum_i (\nabla \phi(r_{0i}))^2 \right\rangle + \left\langle \sum_i \nabla \phi(r_{0i}) \sum_{j \neq i} \nabla \phi(r_{0j}) \right\rangle \right\} =$$

$$= \frac{1}{3M\kappa T} \left[4\pi n \int_0^\infty g(R) [\phi'(R)]^2 R^2 \, dR - \right.$$

$$\left. - n^2 \int d^3R \, d^3S \, g(R, S) \frac{R \cdot S}{RS} \phi'(R) \phi'(S) \right], \qquad (17)$$

where $g(R)$ and $g(R, S)$ are the two and three particle distribution functions, and n the density.

On the other hand (16b) yields:

$$\omega_E^2 = \frac{4}{3M} \int d^3R\, g(R)\, \nabla^2 \phi(R) \tag{18}$$

which, after some manipulation can be written

$$\omega_E^2 = \frac{4\pi n \kappa T}{3M} \int y'(R) f'(R) R^2\, dR - \frac{4\pi n}{3M} \int y(R) f'(R) \phi'(R) R^2\, dR, \tag{19}$$

where $f(R)$ is the Mayer function $[\exp(-\beta\phi(R))-1]$ and

$$y(R) = g(R) \exp(\beta\phi(R)). \tag{20}$$

Note that the second term in (19) is the same as the first in (17), therefore the other two terms are equivalent.

ω_E^2 has been written in this way, (19), in order to bring out what happens as the strength of the repulsive potential increases and tends to the rigid sphere limit. Consider the inverse power potential $\varepsilon(\sigma/R)^\nu$. For large ν the Mayer function rapidly approaches -1 for $R < \sigma$ and 0 for $R > \sigma$. Hence $f'(R)$ which enters (19) approaches the delta function $\delta(R-\sigma)$, while $y(R)$ remains differentiable at $R=\sigma$. Hence

$$\omega_E^2 \sim \tfrac{4}{3}\pi\, \frac{\kappa T}{M\sigma^2} \cdot n\sigma^3 \left[\frac{\nu\varepsilon}{\kappa T} y(\sigma) + \sigma y'(\sigma) \right]. \tag{21}$$

Thus as ν increases the 'pair' contribution to ω_E^2 diverges as ν, while the 'triplet' contribution remains finite.

However, although for a strongly repulsive potential, the mean square force is very large, the correlation will fall very rapidly to zero with time. We show below, that this 'binary collision time', τ, goes to zero as $(M\sigma^2/\kappa T)^{1/2}\,\nu^{-1}$ as ν becomes large, so that the product $\omega_E^2\,\tau$ approaches a finite value Ω_0.

Consequently, one may show, in this limit

$$V(t) = 1 - \Omega_0 |t| + O(t^2). \tag{22}$$

To make this reasoning more precise, we consider now the generalised Langevin equation of Mori [11], and the so-called memory function formalism.

The equations defining the memory function, $M(t)$ associated with a variable $x(t)$ are

$$\dot{x}(t) + i\omega_0 x(t) + \int_0^t M(t') x(t-t')\, dt' = \sigma(t) \tag{23}$$

with

$$i\omega_0 = -\langle \dot{x}x \rangle / \langle xx \rangle \tag{24}$$

and

$$\langle \sigma(t) x(0) \rangle = 0. \tag{25}$$

These equations are sufficient to define $M(t)$ as a correlation function of $\sigma(t)$ with a modified time evolution operator. The details need not concern us here. A good account has been given by Berne and Harp [12]. ω_0 is a zero order propogation frequency. If x represents a diffusive mode, as in the cases considered in this paper, ω_0 vanishes. From (23) and (25) it follows that the correlation function satisfies

$$\dot{X}(t) + \int_0^t M(t') X(t-t') \, dt' = 0. \tag{26}$$

By differentiation, one obtains an equation for $M(t)$:

$$X(0) M(t) = -\ddot{X}(t) - \int_0^t M(t') \dot{X}(t-t') \, dt'. \tag{27}$$

The Laplace transform of Equation (26) gives

$$\tilde{X}(z) = X(0)(z + \tilde{M}(z))^{-1} \tag{28}$$

($\tilde{f}(z) = \int_0^\infty f(t) \exp\{-zt\} \, dt$). Note that if $X(t)$ is differentiable, (27) enables one to obtain the Taylor expansion of $M(t)$ in terms of the derivatives of $\dot{x}(t)$ at $t=0$. In particular

$$M(0) = -\ddot{x}(0). \tag{29}$$

Further, if the integral of $X(t)$ gives a transport coefficient, then the integral of $M(t)$ gives its inverse. Thus for example the self-diffusion coefficient is given by

$$D = \frac{\kappa T}{M\Omega}, \tag{30}$$

where Ω is the time integral of the memory function of the velocity autocorrelaton function.

In the continued fraction representation of Mori [11], a generalised Langevin equation is written for the memory function and so on, introducing a hierarchy of equations

$$M_n(0) M_{n+1}(t) = -\ddot{M}_n(t) - \int_0^t M_{n+1}(t') M_n(t-t') \, dt' \tag{31}$$

so that the Laplace transform of $X(t)$ is expressed as a continued fraction

$$\tilde{X}(z) = X(0)/(z + M_1(0)/(z + M_2(0)/(z + \cdots))), \tag{32}$$

where the values of $M_n(0)$ are determinable from the Taylor expansion of $X(t)$.

Clearly such a representation is only possible if the Taylor expansion exists — thus excluding the rigid sphere system. Since we now believe that the rigid sphere system is a reasonable zeroth order approximation, it follows that the continued fraction representation in its simple form is not acceptable. However, a slight modification is possible, in which the rigid core limit is obtained naturally. We can see from Equation (22) that for the rigid core system

$$\ddot{V}(t) = -2\Omega_0 \delta(t) + V'(t),$$

where $V'(0)$ exists — thus the general $\ddot{V}(t)$ can be split into two parts, one of which is well-behaved and the other not at $t=0$ in the rigid core limit. In fact, this is the separation of $\langle a(0) \cdot a(t) \rangle$ into a 'pair' part $\langle F_{0i}(0) \cdot F_{0i}(t) \rangle$ and a 'triplet' part $\langle F_{0i}(0) \times F_{0j}(t) \rangle$. A similar distinction can be made in the time derivative of the second term.

Thus we replace Equation (31) by

$$M'_n(0) M_{n+1}(t) = -\ddot{M}_n^1(t) - \int_0^t M_{n+1}(t') M_n^1(t-t') dt$$

$$M_n(t) = M_n^c(t) + M_n^1(t), \tag{33}$$

where $M_n^c(t)$ is a hard core part, which in the rigid core limit gives a $\delta(t)$ behaviour. Thus (32) is replaced by

$$\tilde{X}(z) = X(0)/(z + \tilde{M}_1^c(z) + M_1^1(0)/(z + \tilde{M}_2^c(z) + M_2^1(0)/(z + \cdots), \tag{34}$$

where in the rigid core limit $\tilde{M}_n^c(z) \to \Omega_n$, independent of z.

In the next section, we consider the behaviour of $M^c(t)$.

5. The Time Evolution of Correlation Functions

The time evolution of a classical many body system is described by an evolution operator, $S(t)$, which may be written in terms of the Liouville operator L:

$$S(t) = \exp(Lt)$$

$$L = \sum_i \left\{ \frac{p_i^\alpha}{M} \frac{\partial}{\partial r_i^\alpha} - \frac{\partial \Phi}{\partial r_i^\alpha} \frac{\partial}{\partial p_i^\alpha} \right\}. \tag{34}$$

It may be shown [13] that within a correlation function $\langle A(0) B(t) \rangle$, $S(t)$ may be written as the exponential of another operator $\hat{L}(t)$ which contains differential operators acting on both A and B. This follows from the type of integration by parts already indicated in the previous section.

$$\hat{L}(t) = L_0(t) + L_1(t) + L_2(t) + \cdots, \tag{35}$$

where $L_0(t)$ corresponds to the free particle motion and $L_n(t)$ is an nth order term in the potential Φ. We quote here the explicit form of the first few terms for the case where A and B are functions of position only:

$$L_0(t) = -\tfrac{1}{2}\mathbf{L}_0 \cdot \mathbf{R}_0 t^2 \tag{36}$$

$$L_1(t) = \frac{d}{dt}\int dt_1 (g_{01}(t-t_1)-1)\Phi^{(1)}(g_{10}(t_1)-1)$$

$$L_2(t) = \int_0^t dt_1 \dot{g}_{01}(t-t_1)\Phi^{(1)}g_{10}(t_1) \times$$

$$\times \int_0^{t_1} dt_2 g_{02}(t-t_2)(g_{12}(t_1-t_2)-1)\Phi^{(2)}\dot{g}_{20}(t_2), \tag{37}$$

where

$$g_{ij}(t) = \exp[-\tfrac{1}{2}\mathbf{L}_i \cdot \mathbf{R}_j t^2] \tag{38}$$

and $\mathbf{L}_0$ represents $\partial/\partial r$ acting on A, $\mathbf{R}_0$, $\partial/\partial r$ on B and $\mathbf{L}_i$, $\mathbf{R}_i$ differential operators acting on $\Phi^{(i)}$.

With this representation we have calculated [14] the asymptotic form, for large ν, of the pair force correlation function $\langle \mathbf{F}_{0i}(0) \mathbf{F}_{0i}(t) \rangle$ for an inverse power potential $\varepsilon(\sigma/R)^\nu$. For large ν: (i) the leading term of an L or R operator is equivalent to $\nu(\mathbf{r}/r^2)$; (ii) in configurational averages the function is strongly peaked around $r=\sigma$. We find, therefore, that

$$\langle \mathbf{F}_{0i}(0) \cdot \mathbf{F}_{0i}(t) \rangle \approx \langle F_{0i}^2 \rangle Y\left(\nu\left(\frac{\kappa T}{M\sigma^2}\right)^{1/2} t\right). \tag{39}$$

Since $\langle F_{0i}^2 \rangle$ is to leading order proportional to ν it follows that the time integral of (39) is independent of ν, so that the contribution $\tilde{M}_c(0)$ to the diffusion coefficient is independent of ν – supporting our interpretation of the experimental results.

The form of $Y(\tau)$ has been evaluated in two approximations, shown in Figure 4. The solid line shows a 'ladder' approximation in which $S(t)$ is evaluated as $\exp\{L_0 + L_1\}(1+L_2)$ and the broken line shows the expansion in Φ, $\exp L_0(1+L_1+\tfrac{1}{2}L_1^2+L_2)$. At short times these are very close. The value of $\int_0^\infty Y(t)\,dt$ in these approximations may be compared to the exact Chapman-Enskog value for rigid sphere binary collisions. The first approximation above gives 1.07 times the correct result, and the second gives 0.96 times.

Also shown in Figure 4 (crosses) is the experimental form of the pair force correlation function obtained by molecular dynamics for the Lennard-Jones potential. The result has been normalized to the same curvature at $t=0$ by choosing an effective repulsive power ν^{eff}, which is in good agreement with the estimate from the position of the peak in the integrand of the mean square force. For values of $\tau > 2$, the experimental function has a long plateau arising from repeated collisions between the two particles due to the cage effect of their neighbours, which is not accounted for in the low order terms in the expansion of $L(t)$. ν^{eff} is found to be 16, in good agreement with

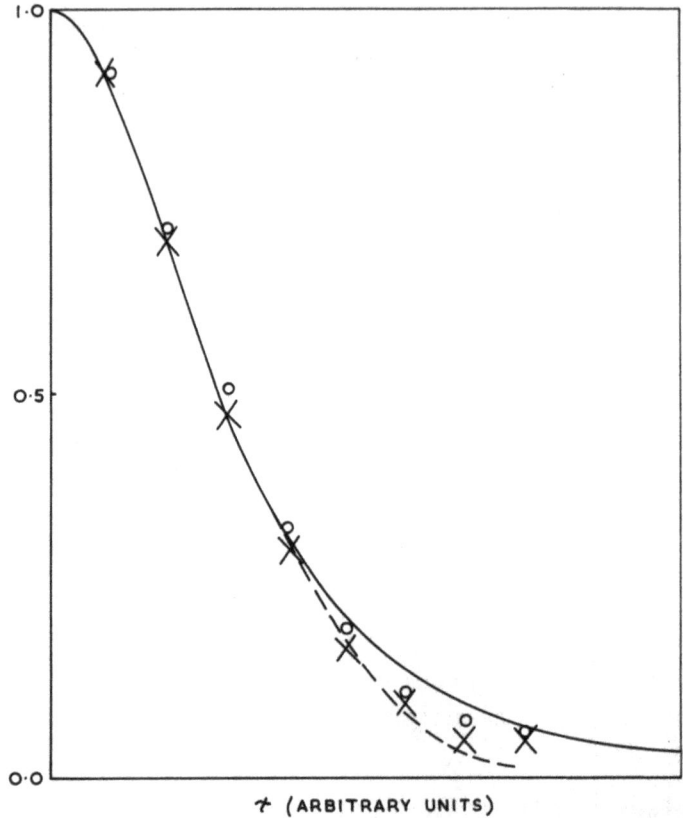

Fig. 4. Binary collision contribution to the memory function for Lennard-Jones potential. The solid line is the 'ladder' approximation, the dashed line the contribution of terms up to second order in the potential. Crosses represent the computer results for self-diffusion and circles the binary collision contribution to the stress-stress correlation functions.

the theoretical value, 16.5, obtained from the logarithmic derivative of $\phi(R)$ taken where $\phi(R) = \kappa T$ [14].

One may also attempt to evaluate other transport coefficients using this formalism. Again, we concentrate on the binary collision contribution. For the viscosity and thermal conductivity, there are terms already occurring in the correlation functions $\langle y(0) y(t) \rangle$ which are singular in the hard core limit [5].

For viscosity these are of the form $\langle r^\alpha_{0i}(0) F^\beta_{0i}(0) r^\gamma_{0i}(t) F^\delta_{0i}(t) \rangle$, and for thermal conductivity $\langle \mathbf{r}_{0i}(0) \cdot \mathbf{F}_{0i}(0) \mathbf{r}_{0i}(t) \cdot \mathbf{F}_{0i}(t) \rangle$. In Figure 4, the circles give the computed normalised correlation function for viscosity – again it follows very closely the theoretical form given by $L(t)$ (the same value of v^{eff} is used).

The similarity of these functions shows that the binary collision contributions, denoted by suffix c are related by symmetry:

$$\Omega_0 = \frac{\kappa T}{MD_c} = \frac{10\eta_c}{\varrho\sigma^2} = \frac{50\phi_c}{3\varrho\sigma^2} = \frac{4\lambda_c}{c_v\varrho\sigma^2}. \tag{40}$$

6. Conclusions

In this paper we have presented some preliminary work aimed at understanding the time behaviour of correlation functions which determine the transport coefficients in simple fluids. The observation that the rigid sphere fluid is a reasonable first approximation leads to a proposed modification of the usual continued fraction type approach, based on sum rules. It has been demonstrated that the time behaviour of binary collisions can be expressed in terms of a function which scales with a parameter representing the 'hardness' of the potential, but which gives a contribution to the transport coefficient independent of the hardness.

Other contributions to the transport coefficients, arising from collective motion may contribute as much as 50% at liquid densities. These will be considered in later work, using the methods of Sections 4 and 5. In particular the modified continued fraction approach will be applied in the calculation of the neutron inelastic scattering cross-section via the density-density correlation function $S(\kappa, \omega)$.

References

1. Weeks, J. B., Chandler, D., and Anderson, H. C.: *J. Chem. Phys.* **54**, 5237 (1971).
2. Gubbins, K. E., Smith, W. R., Tham, M. K., and Tiepel, E. W.: *Mol. Phys.* **22**, 1089 (1971).
3. Verlet, L. and Weis, J. J.: *Phys. Rev.* **A5**, 939 (1972).
4. Hanley, H. J. M., McCarty, R. D., and Cohen, E. G. D.: *Physica* **60**, 322 (1972).
5. Schofield, P.: in H. N. V. Temperley, J. S. Rowlinson and G. S. Rushbrooke (eds.), *Physics of Simple Liquids*, North Holland, 1969, Chapter XIII.
6. Rahman, A.: *Phys. Rev.* **136**, A405 (1964).
7. Schofield, P.: *Computer Phys. Comm.* **5**, 17 (1973).
8. Alder, B. J., Gass, B. M., and Wainwright, T. E.: *J. Chem. Phys.* **53**, 3813, (1970).
9. Ross, M. and Schofield, P.: *J. Phys.* **A4**, L305, (1971).
10. Levesque, D. and Verlet, R.: *Phys. Rev.* **A2**, 2514 (1970).
11. Mori, H.: *Prog. Theor. Phys.* **33**, 423 (1965); **34**, 399 (1965).
12. Berne, B. J. and Harp, G. D. *Adv. Chem. Phys.* **XVII**, 63 (1970).
13. Schofield, P.: 1974, *J. Phys.* **A6**.
14. Schofield, P. and Trainin, J. E.: to be published.

DISCUSSION

Yip: Regarding the ratio of diffusion coefficients D/D_{Enskog} that Alder et al. obtained for hard spheres, can you say what are the effects of the non-hard sphere modification?

Schofield: Not yet.

Mazenko: Is this perturbation scheme on the modified Liouville operator introduced by Mori? How does one go from an operator to a function of time?

Schofield: The operator L (t) defined in the paper results from averaging out the velocities in the correlation function $\langle A \rangle$. It thus contains position operators only. If A and B are expressed as Fourier transforms then L(t) is a function of κ and the coordinates.

Janik: Is it possible and worthwhile to generalize the theory for more complicated molecules? I am thinking about representing a complicated molecule by hard spheres connected by hard rods, and allowing rotation about some bonds. Perhaps it is possible to obtain in this way a realistic connection between diffusion coefficients and viscosities.

Schofield: Eventually we hope that such a programme will be possible.

Ailawadi: In one of the slides you had what you call Ladder approximation as well as Potential approximation. How many terms you have to sum to get the results you mentioned?

Schofield: I think these approximations are made clear in written version of the paper.

Comment by M. R. Hoare

While considerable progress has clearly been achieved using linear-response and Langevin-type formulations, I would hope that this will not divert too much attention from the study of transport theories involving explicit scattering models, where a great deal remains to be done and where the mathematical difficulties may not be quite as forbidding as is commonly thought.

There has, for example, been very encouraging progress in our deeper understanding of the one-dimensional hard-particle gas. Percus and Lebowitz (*Phys. Rev.* **155**, 122 (1967)) have obtained exact expressions for time-dependent distribution and autocorrelation functions and very recently Levitt (*J. Stat. Phys.* **7**, 329 (1973)) has rederived these by informal physical arguments.

Mizan Rahman and I have recently discovered the exact eigenfunction solution for the scattering kernel of the problem, which underlies the previous results. (Hoare, M. R. and Rahman, M.: *J. Phys.* **A10**, 1461 (1973)). These eigenfunctions are of quite unfamiliar type and expose the singular character of the problem in its essentials. (They are singular distributions combining delta-function terms with Hadamard 'pseudo-functions' of a kind hitherto unknown in statistical physics. The eigenvalue spectrum is continuous from a threshold point, there being no discrete relaxation times other than the equilibrium value. The entire continuum set are correctly orthogonal and complete.)

The importance of these solutions could be said to be that they demonstrate that the singularities in linear transport theory are not simply an awkward consequence of using non-differentiable potentials, such as the hard-sphere case, but are an essential feature of the underlying kinetic equations. While it seems that these singularities can be overcome by various approximations, it is conceivable that something of crucial importance is lost in the process.

The moral of this is perhaps that further insights should be sought through the study of generalized functions and the integral scattering operators which give rise to them. The popularity of the linear-response formalism which, it should at least be mentioned, is still open to some fundamental doubts (See e.g., Van Kampen, N. G.: *Physica Norvegica* **5**, 279 (1971)), to some extent obscures this aspect.

Yip: The exact solutions for $S(\kappa, \omega)$ in the case of gas of one-dimensional hard points can be obtained by solving the appropriate linearized Boltzmann equation.

Mazenko: The problem of one-dimensional hard rods has been solved in the case of the velocity auto-correlation by Lebowitz(?), Percus and Sykes. They find an exponential decay with a decay rate given by the Enskog theory. It includes 'static' corrections in the decay rate.

Misguich: The correction you find to the Enskog theory depends on $\gamma(\sigma)$ and on the derivative of γ at σ. More general corrections, of the 'non-Enskogian' type have been obtained in the PNM theory [1, 2] and in the generalized Rice Allnatt theory [3], which depend not only on the contact value $\gamma(\sigma)$ of the radial correlation function, but also on the value of this function at each point $\int dr\, f(r)\, Y(r)$. This form expresses a high density effect, which cannot be found in Enskog's theory of dense gases, and which can be interpreted as the indirect interaction of two hard spheres at a distance larger than σ, only by multiple interactions with other hard spheres of the dense liquid. These contributions in the PNM theory have been shown to arise from at least triple correlations.

References

1. Prigogine, I., Nicolis, G., and Misguich, J.: *J. Chem. Phys.* **43**, 4516 (1965).
2. Misguich, J.: *J. Phys.* **11**, 221 (1969).
3. Misguich, J. and Nicolis, G.: *Mol. Phys.* **11**, 309 (1972).

VISCOELASTIC AND ULTRASONIC RELAXATION STUDIES RELATED TO MOLECULAR MOTIONS IN LIQUIDS

JOHN LAMB

Dept. of Electronics and Electrical Engineering, The University, Glasgow, G12 8QQ, U.K.

Abstract. A review is given of recent work in which the rotational and translational motions of molecules in non-polymeric liquids are studied by linear perturbations involving externally applied time-varying stress. The investigations described are mainly concerned with the use of cyclic shear waves at selected frequencies of shear alternation covering a wide range of temperature and, for a small number of liquids, pressures up to 1.4 GN m^{-2}. Reference is made to associated measurements of shear creep and compressional wave propagation. Results obtained are described in terms of a relaxational (Maxwell) and a retardational (Davidson-Cole) contribution to the total shear compliance of the liquid.

Résumé. Revue de travaux récents dans lesquels les mouvements de translation et de rotation de molécules dans des liquides non polymériques sont étudiés par des perturbations linéaires qui mettent en jeu une contrainte externe variant avec le temps. Les études poursuivies ont porté principalement sur l'emploi d'ondes de cisaillement cycliques à des fréquences choisies de variation du cisaillement couvrant un large domaine de température et, pour quelques liquides, de pressions allant jusqu'à 1,4 GN m^{-2}. On mentionne les mesures associées de réactivité de cisaillement et de propagation de l'onde de compression. Les résultats obtenus sont décrits en fonction de contributions relaxationnelle (Maxwell) et retardationnelle (Davidson-Cole) à la compliance totale au cisaillement du liquide.

1. Introduction

The most characteristic property of a liquid is its viscosity which reflects the relative difficulty with which translational movement of the molecules can occur. At temperatures which are sufficiently high for a liquid to have more than about 15% of free volume the molecules rapidly attain equilibrium with nearest neighbours following a disturbance and the viscosity is determined principally by the energy required for a molecule to jump from one site to an adjacent available one. The dependence of viscosity upon temperature is then governed by the Arrhenius equation:

$$\ln \eta = A + E/RT. \tag{1}$$

On cooling, the majority of liquids crystallise at or slightly below their melting points, where their viscosities are only about 1 P, so that the whole extent of their viscosity-temperature behaviour is, in general, covered by the Arrhenius equation. There are, however, many liquids which can be readily supercooled and it is with such liquids that this review is principally concerned. At high temperatures their dependence of viscosity on temperature follows the Arrhenius equation but at lower temperatures this no longer applies, as is demonstrated by Figure 1.

It is convenient to refer to an Arrhenius temperature, T_A, below which the availability of free volume is the chief factor determining viscous flow. Longuet-Higgins and Widom [1] have suggested that in this region it is the short range repulsion forces which determine the structure while the longer range attractive forces merely provide

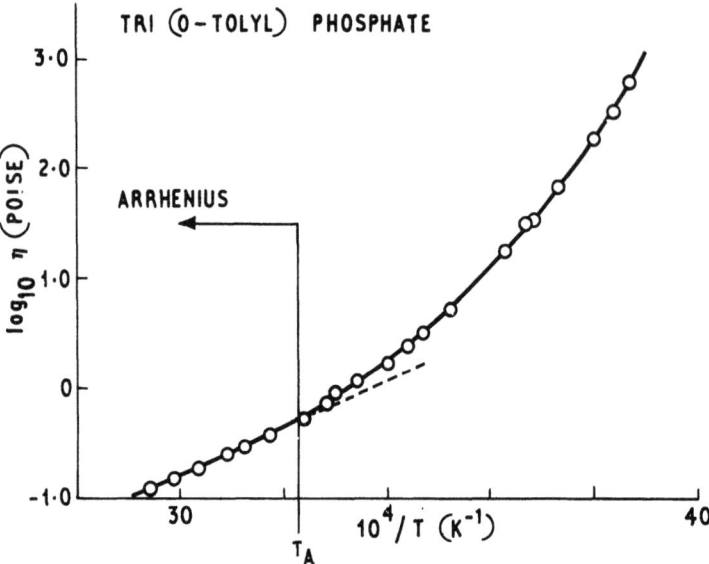

Fig. 1. $\log_{10}\eta$ plotted vs $1/T$ for tri(o-tolyl) phosphate.

a general background potential which holds the liquid together. This is compatible with the view that the transport properties of such liquids are mainly controlled by density (or free volume) rather than by energy. Thus Cohen and Turnbull [2] assume that a molecule can take part in translational motion only if a critical amount of free volume is available. The dependence of viscosity on free volume which their theory predicts is of the form:

$$\ln \eta = A + B/(V - V_0).$$

V is the specific volume and V_0 is interpreted as the volume of a hypothetical system in which no free volume would be available for flow. The parameter A contains a weakly temperature-dependent term, $\frac{1}{2} \ln T$. If either the volume or the density varies linearly with temperature, then this equation reduces to that of the empirical Vogel [3] equation:

$$\ln \eta = A + B/(T - T_0). \tag{2}$$

As shown by Barlow et al. [23], this equation successfully describes the dependence of viscosity upon temperature of supercooled liquids, as illustrated by the results shown in Figure 2.

The approximate temperature, T_A, marks the change from behaviour represented by that of Equation (1) to that of Equation (2). At T_A, the viscosity has a value of about 1 P and the relative free volume is typically 12 to 15%. With decreasing temperature the free-volume can become as low as 1% close to the glass transition temperature, T_g.

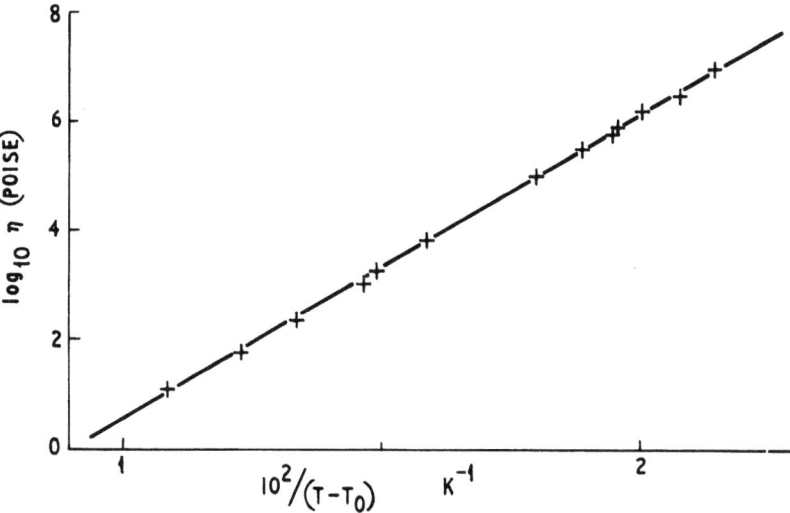

Fig. 2. $\log_{10}\eta$ plotted vs $1/(T-T_0)$ for *bis*(*m*-(*m*-phenoxy phenoxy) phenyl) ether. $T_0 = 224.1$ K.

If the liquid is subjected to a constant externally applied shear stress, the preferred direction for translational motion of the molecules is parallel to the direction of shear. Moreover, if the magnitude of the applied stress is sufficiently small, it merely perturbs the existing equilibrium but does not sensibly affect the time constants governing the processes of molecular flow or local reorientation, which are statistically determined by the pressure and temperature of the liquid. For present purposes, we take the Maxwell relaxation time, τ_m, of molecular translational equilibrium as the reference time constant, although subsequently it will be argued that the retardation time of local molecular adjustment can also be derived from the experiments to be described.

A similar situation pertains if the applied shear stress is alternating sinusoidally in time, provided that the periodic time greatly exceeds the time constant, τ_m. ($\omega\tau_m \ll 1$). If, however, the variable stress changes so rapidly that its periodic time is much less than τ_m ($\omega\tau_m \gg 1$), then the translational flow mechanism does not have time to operate and the response of the liquid to the external stress is simply that of elastic deformation. We therefore expect to observe a shear rigidity modulus of the order of that of a typical solid but with some reduction to allow for the presence of the structure-weakening holes. At intermediate frequencies the behaviour changes from purely viscous to purely elastic with increasing value of the parameter ($\omega\tau_m$). This is the simple physical basis of viscoelastic relaxation as observed in alternating shear. Under appropriate conditions, this could in principle be studied also by suitably conducted experiments of shear creep or stress relaxation. Any analytical or model representation of this phenomenon must yield the behaviour of a purely viscous liquid at low frequencies (or long times) and that of an elastic solid at sufficiently high frequencies. The simplest linear representation of viscoelastic response in accordance with these restrictions is that of the Maxwell model, in which the complex compliance at angular frequency ω is expressed by:

$$J^*(j\omega) = J_\infty + 1/j\omega\eta = J_\infty[1 + 1/j\omega\tau_m]. \tag{3}$$

J_∞ is the limiting compliance at very high frequencies, $\tau_m = \eta J_\infty$ and $J^*(j\omega)$ is equal to the strain divided by the stress. The corresponding creep function would be $J(t) = J_\infty + t/\eta$ and for stress relaxation, $G(t) = G_\infty \exp(-t/\tau_m)$ with $G_\infty = 1/J_\infty$.

Measurements show that the value of J_∞ is between 10^{-10} and 5×10^{-10} cm^2 dyn^{-1}: taking a typical value of 2×10^{-10} cm^2 dyn^{-1} ($= 2 \times 10^{-9}$ m^2 N^{-1}) for J_∞ gives the value of the characteristic frequency, f_c, corresponding to $\omega_c \tau_m = 1$ as $10^{10}/4\pi\eta$ Hz $\simeq$ $\simeq 0.8/\eta$ GHz, where η is in poise. Thus for $\eta = 1$ cP, 10 cP, 1 P, 10 P and 100 P the corresponding values of f_c are 80 GHz, 8 GHz, 800 MHz, 80 MHz and 8 MHz respectively. It transpires that, for experimental reasons, alternating shear wave measurements on liquids are sensibly confined to frequencies below 1 GHz and hence, in order to determine the viscoelastic as distinct from the purely viscous response, it is necessary to work with supercooled liquids in the region where the steady-flow viscosity is greater than 1 P. On this account, the viscoelastic behaviour of liquids in the Arrhenius region ($\eta < 1$ P) can only be deduced from measurements made using other experimental techniques, e.g., by light scattering. In practice, however, many liquids do supercool readily and for these there is available a viscosity range extending from about 1 P at T_A to 10^{13} P at T_g. Similar considerations apply to conditions where hydrostatic pressure is also employed as an experimental variable, the only practical restriction being that the liquid shall not crystallise as the temperature is lowered and/or the pressure increased.

2. Shear Wave Propagation and Viscoelastic Equations

Consider the propagation of a transverse shear wave parallel to the Z-direction through a fluid medium. Let the oscillatory shear displacement, u, be in the X-direction due to the shear stress T. We define the complex compliance $J^*(j\omega) = J' - jJ''$ as the ratio strain/stress, so that $J^*(j\omega) = (\partial u/\partial z)/T$. The equation of motion for an elemental volume of fluid is then

$$\varrho \frac{\partial^2 u}{\partial t^2} = \frac{\partial T}{\partial z} = \frac{1}{J^*}\frac{\partial^2 u}{\partial z^2}.$$

The solution of this wave equation can be represented in terms of a complex velocity, $C^* = 1/(\varrho J^*)^{1/2}$, by $u = u_0 \exp[j\omega(t - z/C^*)]$.

The quantity which is measured directly in the experimental systems with which we are here concerned is the complex shear mechanical impedance of the liquid $Z_L = R_L + jX_L = -T/(\partial u/\partial t)$. It follows from the above relationships that

$$J' - jJ'' = \varrho/Z_L^2.$$

Hence,

$$J' = \frac{\varrho(R_L^2 - X_L^2)}{(R_L^2 + X_L^2)^2} \quad \text{and} \quad J'' = \frac{2\varrho R_L X_L}{(R_L^2 + X_L^2)^2}. \tag{4}$$

The equation originally introduced by Maxwell to account for viscoelastic relaxation can be written:

$$T + \tau_m \frac{\partial T}{\partial t} = \eta \frac{\partial}{\partial t}\left(\frac{\partial u}{\partial z}\right), \quad \text{where } \tau_m \text{ is a time constant.}$$

Replacing the time derivative by $j\omega$ for a sinusoidal wave gives

$$J^*(j\omega) = (1 + j\omega\tau_m)/j\omega\eta. \tag{5}$$

For viscous flow at low frequencies ($\omega\tau_m \ll 1$) $J^* \simeq 1/j\omega\eta$, whereas at sufficiently high frequencies $J^* \to \tau_m/\eta$. Hence the high frequency elastic compliance $J_\infty = \tau_m/\eta$ or, alternatively, the limiting shear modulus $G_\infty = 1/J_\infty = \eta/\tau_m$. The Maxwell relaxation time is therefore simply the product of the steady-flow viscosity and the high frequency

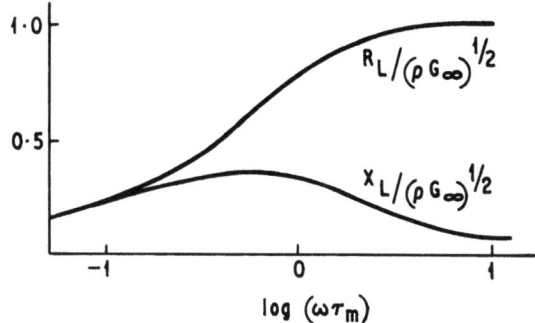

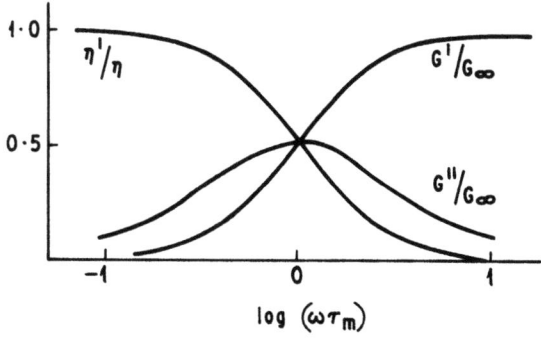

Fig. 3a.

Fig. 3a–b. Calculated curves for the components of the shear mechanical impedance and of the complex elastic modulus plotted in normalised form vs log($\omega\tau_m$). (a) For the Maxwell model (Equation (3)). (b) For the Barlow et al. [4] Equation (6).

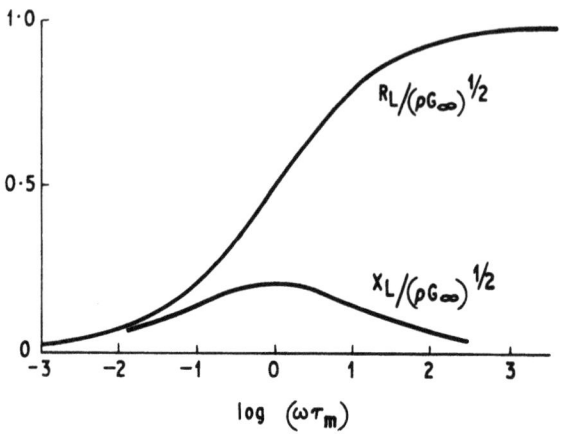

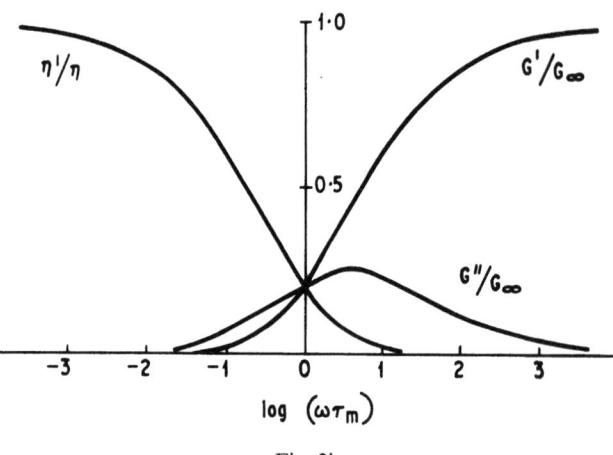

Fig. 3b.

compliance, $\tau_m = \eta J_\infty = \eta/G_\infty$. The expression for $J^*(j\omega)$ given in Equation (5) now becomes that quoted previously in Equation (3):

$$J^*(j\omega) = J_\infty [1 + 1/j\omega\tau_m]. \tag{3}$$

Corresponding expressions for the components of the complex shear modulus, $G^*(j\omega) = G' + jG''$ are:

$$G' = G_\infty \frac{\omega^2 \tau_m^2}{(1 + \omega^2 \tau_m^2)} \quad \text{and} \quad G'' = G_\infty \frac{\omega \tau_m}{(1 + \omega^2 \tau_m^2)}.$$

The behaviour represented by this single relaxation formulation is shown in Figure 3(a): corresponding curves are shown in Figure 3(b) derived from the Barlow et al. [4] equation as described below.

This simple model does not provide a good description for experimental results obtained on a large number of liquids. These demonstrate that although normalised curves can be fitted to the results obtained for different liquids, the relaxation region is much broader than is predicted by the Maxwell equation, extending over several decades of effective frequency. Barlow et al. [4] found that by adding the reciprocal impedances for a Newtonian liquid and a Hookian solid it was possible to provide an adequate description for experimental results. The resulting BEL equation can be written:

$$J^*(j\omega) = J_\infty [1 + (j\omega\tau_m)^{1/2}]^2/j\omega\tau_m$$
$$= J_\infty [1 + 1/j\omega\tau_m] + 2J_\infty/(j\omega\tau_m)^{1/2}. \quad (6)$$

The first term is recognised as the Maxwell relaxation whilst the $\omega^{-1/2}$ dependence of the second term suggests a diffusion controlled mechanism of the type first put forward by Glarum [5] to describe the results of dielectric relaxation. Glarum's model is concerned with the diffusion of defects in one dimension and considers only the defect nearest to a given site. If D is the self-diffusion coefficient of defects and ϱ the average defect density, then the defect diffusion time $\tau_D = 1/4\varrho^2 D$. Glarum found that the relaxation function derived from this model fitted the experimental results at times shorter than τ_D but failed at long times. To circumvent this, he introduced a second independent relaxation process which could be ascribed physically to the relaxation of a site in the absence of a defect arrival. This was assigned an exponentially-decaying relaxation function with a time constant τ_e and the resulting equation was found to give a fit to experimental results, provided τ_e was arbitrarily chosen to be approximately equal to τ_D.

Hunt and Powles [6] followed an essentially similar argument in fitting their nuclear magnetic relaxation results to the defect-diffusion model and, like Glarum, they found it necessary to introduce a second independent process in order to obtain agreement with experimental results. The fact that this second process was required to have a characteristic time similar to τ_D for different molecular systems appeared physically unrealistic, since it would have to be a chance occurrence that the characteristic times of two independent processes were roughly equal. Noting that the failure of the defect-diffusion model occurs at times longer than τ_D, Phillips has pursued this analysis to take into account the effect of the second-nearest defects. Thus, on average, the distance of a second-nearest defect to a site is twice the distance of the nearest defect so that the second nearest defect will take about four times as long to arrive at the site and will arrive after an interval of about $4\tau_D$. Predictions of the nearest-neighbour defect-diffusion model of Glarum can therefore be expected to fail at times of the order of $4\tau_D$ and longer. At long times, therefore, the effects of the non-nearest defects plays an important role and this is the basis of the arguments presented by Phillips et al. [7]. The corrected analysis proves to be mathematically intractable but a series of mathematical approximations is developed, giving a physically acceptable long-time response together with the original satisfactory short-time response of the Glarum treatment and without the introduction of an arbitrary second independent process.

This leads to theoretical derivation of the BEL equation on this one-dimensional model basis including second-nearest defects.

$$J^*(j\omega) = J_\infty [1 + (j\omega\tau_m)^{1/2}]^2 / j\omega\tau_m,$$

with the Maxwell relaxation time $(\tau_m = \eta J_\infty)$ equal to the defect diffusion time τ_D.

In the light of the experimental results to be described subsequently in Section 5, it should be noted that inherent in the defect-diffusion model are the concepts of a 'site' and of a 'defect'. In the one-dimensional treatment outlined above a site may be taken as a point. In three dimensions, for which an extended analysis has not been attempted, the site would require to be dimensionally specified. Thus, the fact that the site might be identified as a molecule having a degree of orientational equilibrium within the force field of neighbouring molecules has not been taken into account in the derivation of Equation (6). On the other hand, a defect is generally considered to be a hole of molecular dimensions or less which moves through the system by a random walk process. It is assumed that relaxation at a site cannot occur until a defect or hole arrives at that site and it is thus the arrival of the first defect that causes relaxation independent of the state of other sites. Sites are thus considered to be sensibly immobile while the motion of the defects is described by the continuum diffusion equation. The underlying physical reasoning for these mathematical statements is essentially that the relative concentration of defects to sites is much less than unity so that, with a sufficiently small step length for the random walk of a defect, the number of steps taken by a defect before it arrives at a given site enables the continuum diffusion equation to be used to describe what is in effect a large number of discrete motions. Identifying a defect with a hole and a site with a molecule, or a group of molecules, justifies the application of these assumptions and hence the equations based upon them, to molecular liquids in the supercooled region. They would not, in general, be expected to apply to liquids in the Arrhenius region nor to ionic, network or polymeric liquids. However, at sufficiently low values of relative free volume even ionic or network liquids might be expected to conform to the description afforded by the defect-diffusion model.

Equation (6), developed empirically by Barlow et al. [4], has now been given further substance by the subsequent theoretical study of Phillips et al. [7]. The results of this analysis are given above in the form appropriate to the experiments involving alternating shear. The corresponding creep function is:

$$J(t) = J_\gamma + t/\eta + 2J_\infty (t/\pi\tau_m)^{1/2} \tag{7}$$

and the stress autocorrelation function is:

$$J_\infty \varphi(t) = [1 + 2(t/\tau_m)] \exp(t/\tau_m) \operatorname{erfc}(t/\tau_m) - 2(t/\pi\tau_m)^{1/2}. \tag{8}$$

3. Shear Compliance and Elastic Modulus of Liquids

At sufficiently high values of the 'effective frequency' $(\omega\tau_m)$ the results of experiment and the theoretical predictions show that the reactive part, X_L, of the shear mechanical impedance tends asymptotically to zero with increasing value of $(\omega\tau_m)$. Thus, under these conditions, $J'' \to 0$ and $J' \to \varrho/R_L^2$, from Equation (4), so that corresponding measurements of ϱ/R_L^2 give the limiting (high-frequency) shear compliance, J_∞. In order to measure the value of J_∞ it is therefore necessary to work at high frequencies and, in addition, to select the conditions of temperature and/or pressure so that $\tau_m(=\eta J_\infty)$ is much greater than the value it has at room temperature and atmospheric pressure. If the behaviour followed the predictions of the single relaxation-time (Maxwell) model it would be sufficient to arrange that $\omega\tau_m > 10$ in order to ensure operating in the elastic region. However, because of the broad spread of the relaxation region found in practice and in accordance with the BEL equation, it is necessary to arrange that $\omega\tau_m > 10^4$ for measurements of J_∞ to be possible. It has already been stated that measurements of reasonable accuracy are confined to frequencies below 1 GHz: a highest operating frequency of 450 MHz is chosen as a compromise between the conflicting requirements of high frequency and acceptable accuracy of measurement. In addition, it is advisable to make measurements at a second frequency in order to ensure that the resulting value of $\varrho/R_L^2 = J_\infty$ is independent of frequency, i.e., that the measured behaviour is truly elastic within the limits of experimental accuracy. The second frequency is chosen to be 30 MHz. Thus for $\omega\tau_m > 10^4$ the steady flow viscosity must exceed about 2.5×10^5 P at a frequency of 30 MHz and about 2×10^4 P at 450 MHz.

The elastic properties of a large number of liquids have been determined. In all cases, following the initial work of Barlow et al. [8], it has been found that J_∞ depends linearly upon temperature: for convenience, this is cast in the form

$$J_\infty = J_0 + C(T - T_0), \tag{9}$$

where T_0 is the reference temperature involved in the modified free-volume equation for viscosity. Tabulated values of the constants J_0, C and T_0 are to be found in the paper by Lamb [9]. At T_g the value of J_∞ for different liquids ranges from 0.5×10^9 to 0.9×10^9 N m^{-2}. At a given frequency of measurement the extent of the linear part of the plot of ϱ/R_L^2 vs T, which yields J_∞ in Equation (9), depends upon the liquid in question but generally covers some 20 °C or more. As temperature increases, the liquid no longer behaves elastically and extrapolations are then made using Equation (9) into the temperature range of the relaxation measurements: sufficient evidence has been accumulated to confirm the justification of this procedure. Representative plots are shown in Figure 4.

Again, when pressure is employed as an experimental variable it is necessary to work in the corresponding viscosity range in order to determine the elastic behaviour. In addition, however, the atmospheric pressure value of the elastic modulus, $G_\infty(0)$, (or its inverse the compliance, J_∞) is known at the particular temperature used for the variable pressure measurements from the results of measurements made at atmospheric pressure as a function of temperature. Relatively few liquids have been studied at high

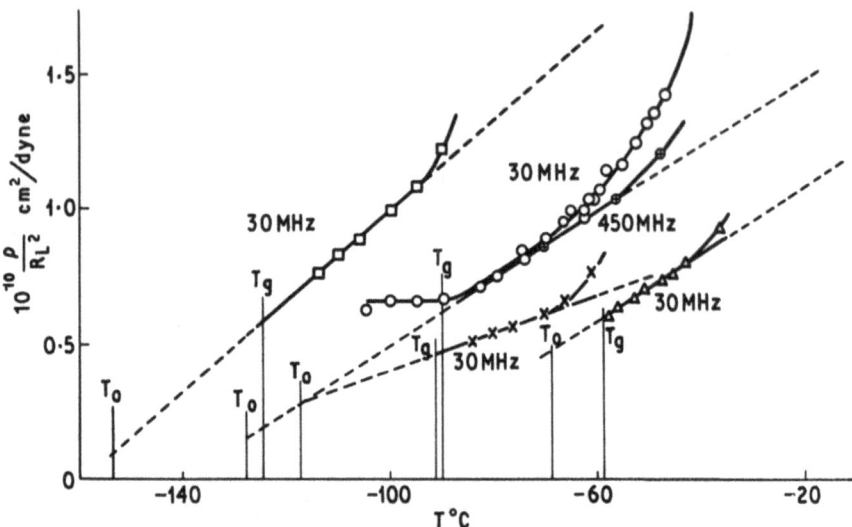

Fig. 4. Measured values of (ϱ/R_L^2) vs temperature showing the linear variation of J_∞ with temperature in the elastic region. □, *tetra* (2-ethylhexyl) silicate, 30 MHz; ○, *di(iso*-butyl) phthalate, 30 MHz; ⊕, *di(iso*-butyl) phthalate, 450 MHz; ×, *tri(β*-chloroethyl) phosphate, 30 MHz; △, *tri(m*-tolyl) phosphate, 30 MHz.

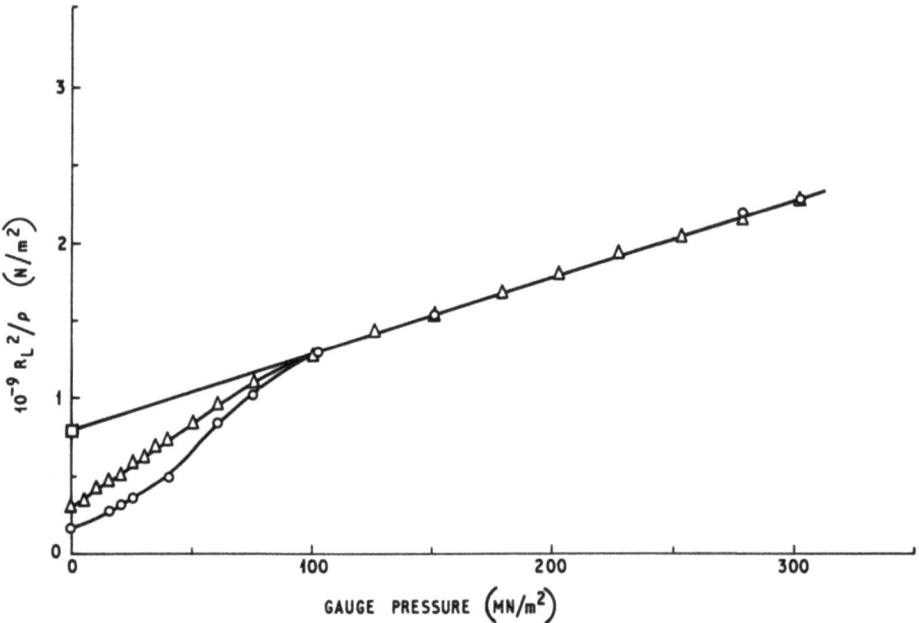

Fig. 5. Measured values of R_L^2/ϱ at 30 °C as a function of hydrostatic pressure for bis(*m*-(*m*-phenoxy phenoxy) phenyl) ether, showing the linear variation of G_∞ with pressure in the elastic region. △, 30 MHz; ○, 10 MHz; □, extrapolated value at 30 °C obtained from the measured linear variation of J_∞ with temperature at atmospheric pressure. Taken from Barlow *et al.* [11], 1972. By permission of the Royal Society.

pressures but available results show unambiguously that the limiting shear modulus varies linearly with pressure [10-13].

$$G_\infty(p) = G_\infty(0) + Dp. \qquad (10)$$

In the elastic region ($\omega\tau_m \gg 1$), $R_L^2/\varrho \to G_\infty$ and verification of Equation (10) is shown in Figure 5 for the six-ring meta-linked polyphenyl ether, $bis(m\text{-}(m\text{-phenoxy phenoxy})$ phenyl) ether.

4. Viscoelastic Relaxation

The BEL equation is most conveniently expressed in terms of the components of the complex compliance by Equation (6). Corresponding equations can be obtained for the components of the complex shear modulus or for those of the shear mechanical impedance. The latter are of particular interest, since these are the quantities which are measured directly along with the density, steady-flow viscosity and the limiting high frequency compliance.

Since $Z_L^2 = \varrho/J^*(j\omega)$, we have for the BEL model:

$$\frac{Z_L}{(\varrho G_\infty)^{1/2}} = \frac{R_L + jX_L}{(\varrho G_\infty)^{1/2}} = \frac{(\omega\tau_m/2)^{1/2}[1 + (2\omega\tau_m)^{1/2} + j]}{[1 + (\omega\tau_m/2)^{1/2}]^2 + (\omega\tau_m/2)}.$$

The components on the right-hand side of this expression involve a single variable and hence to test the predictions of the model it serves to plot $R_L/(\varrho G_\infty)^{1/2}$ and $X_L/(\varrho G_\infty)^{1/2}$ vs $\log(\omega\tau_m)$ in the form of 'universal curves' and to compare these with experimental results plotted similarly. This is illustrated by the curves of Figure 6 where, to avoid confusion, experimental values are given for only a few of the liquids measured, these being representative of a much larger group comprising liquids of widely differing molecular type. These results were obtained at frequencies from 6 to 450 MHz and as a function of temperature from T_A to T_g, at atmospheric pressure. There is good agreement within the limits of experimental error; moreover, as additional confirmation, the results of Figure 7 show that frequency and temperature are equivalent and interchangeable variables. At high effective frequencies ($\omega\tau_m \gg 1$), R_L tends to the value $(\varrho G_\infty)^{1/2}$ which is the basis of the method described in Section 3 for measuring the elastic properties. At low values of $\omega\tau_m (\ll 1)$ the curves for R_L and X_L merge: this is the purely viscous or Newtonian region where in the low frequency limit, $R_L = X_L = (\pi f \eta \varrho)^{1/2}$. It is on this account that preference is given to plotting directly the measured values of R_L and X_L, in normalised form, rather than the components of the complex modulus or compliance, since the quantity $(R_L^2 - X_L^2)$ appears in the numerator of the expressions for G' or for J' (c.f. Equation (4)) with the resulting enhancement of error particularly at low frequencies.

Measurements have also been made on some liquids as a function of pressure at frequencies of 10 and 30 MHz. Experimental difficulties preclude the measurement of the reactive component of the shear mechanical impedance and only the resistive component has been determined, together with the density and viscosity. The ex-

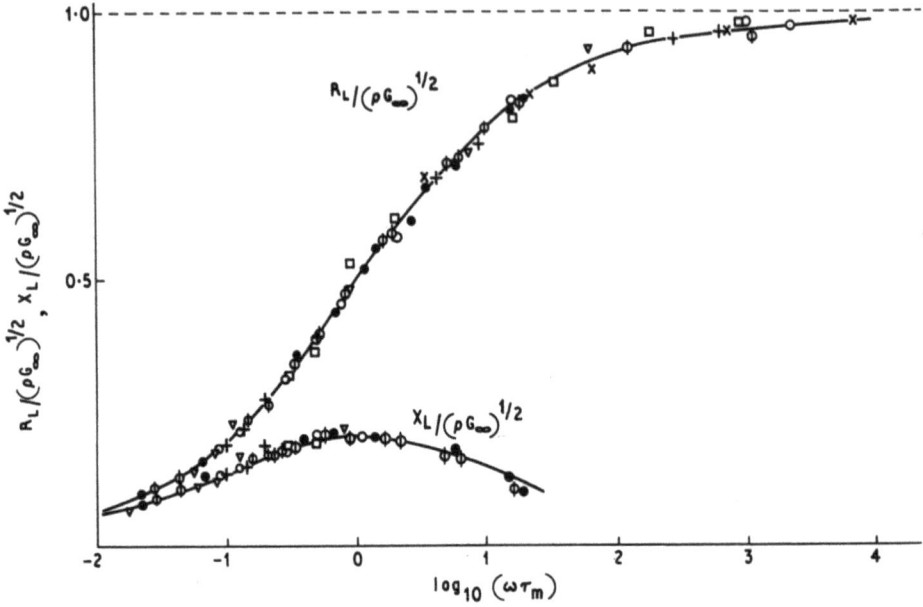

Fig. 6. Comparison of experimental results for a number of liquids with the curves predicted from the Barlow et al. [4] Equation (6). ▽, squalane; +, squalene; □, tri(β-chloroethyl) phosphate; ○, tri(m-tolyl) phosphate; X, tris(2-ethylhexyl) phosphate; ⊕, di(iso-butyl) phthalate, below 278 K (see Section 7); ●, di(n-butyl) phthalate. Results obtained at atmospheric pressure.

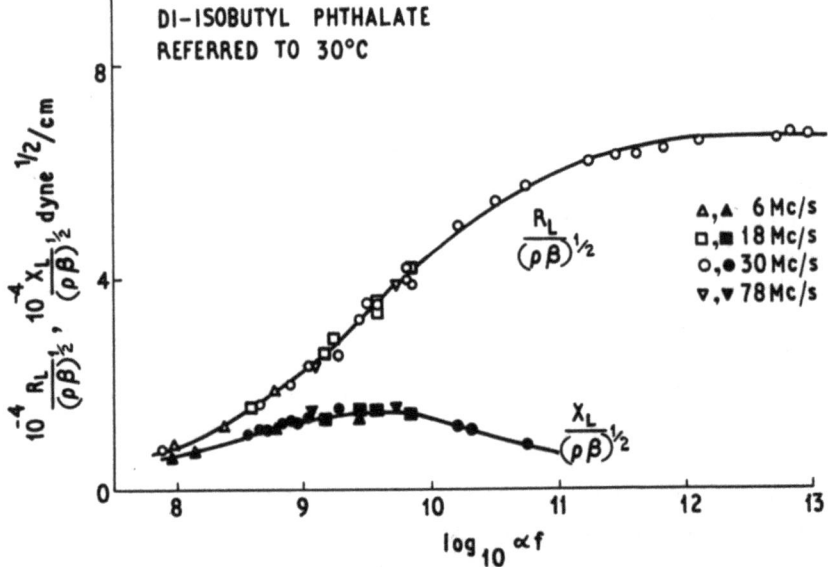

Fig. 7. Measured values of the resistive and reactive components of the shear mechanical impedance of di(isobutyl) phthalate plotted in reduced form and referred to 30°C, $\alpha = (\tau_m)_T/(\tau_m)_{30°C}$; $\beta = (G_\infty)_T/(G_\infty)_{30°C}$; $\alpha\beta = (\eta)_T/(\eta)_{30°C}$. △, ▲, 6 MHz; □, ■, 18 MHz; ○, ●, 30 MHz; ▽, ▼, 78 MHz. Taken from Barlow et al. [8], Paper I. By permission of the Royal Society.

perimental error is greater than is the case for measurements made at atmospheric pressure but, when the results are normalised and plotted, they again conform to the prediction of the BEL equation as shown by Figure 8 [11].

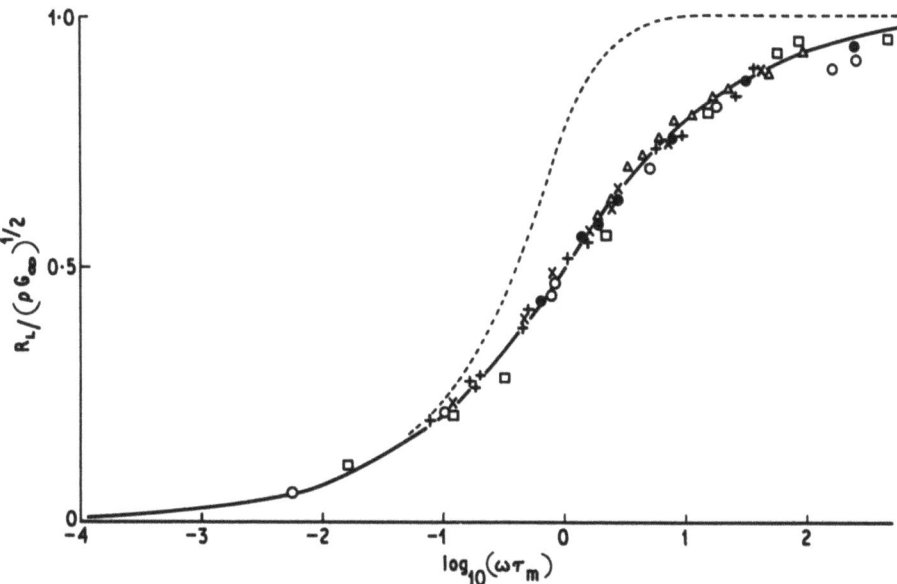

Fig. 8. Normalised values of the resistive component of the shear mechanical impedance. Obtained at frequencies in the range 6 to 450 MHz and as a function of temperature at atmospheric pressure: +, bis(m-(m-phenoxy phenoxy) phenyl) ether; X, di(2-ethylhexyl) phthalate. Obtained at 30.0°C as a function of pressure; ▽, bis(m-(m-phenoxy phenoxy) phenyl) ether, 10MHz; △, bis(m-(m-phenoxy phenoxy) phenyl) ether, 30MHz; ○, di(2-ethylhexyl) phthalate, 10MHz; □, di(2-ethylhexyl) phthalate, 30MHz. The solid curve represents the behaviour calculated from the Barlow et al. [4] Equation (6). The broken line represents the Maxwell model for a single relaxation time, Equation (3). Taken from Barlow et al. [11]. By permission of the Royal Society.

5. Viscoelastic Relaxation and Retardation

Despite the good agreement found between experimental results and calculated values based upon the BEL equation, it was realised that this equation can only be an approximate representation [14]. The reason for this is that, when transformed into the time domain, it leads to the prediction of unlimited recoverable creep strain, as is evident from the term in $t^{1/2}$ in Equation (7), and this is unacceptable on physical grounds. The basis of the explanation for this is to be found in the earlier theoretical work of Gross [15], who showed that, provided experimental conditions are restricted to the amplitude region where linear viscoelastic theory applies, the complex compliance can be described by the equation:

$$J^*(j\omega) = J_\infty + 1/j\omega\eta + J_r \int_0^\infty \frac{N(\tau)\,d\tau}{1 + j\omega\tau}, \tag{11}$$

where $N(\tau)$ is the spectrum of retardation times normalised so that the integral of $N(\tau)\,\mathrm{d}\tau$ is unity

$$J_r^*(j\omega) = J_1 - jJ_2 = J_r \int_0^\infty \frac{N(\tau)\,\mathrm{d}\tau}{1+j\omega\tau} = J_r\chi(j\omega); \quad (12)$$

is the complex retardational compliance. The real part of $J^*(j\omega)$ in Equation (11) tends to $(J_\infty + J_r)$ at low frequencies and to J_∞ at high frequencies. Nevertheless, in accordance with the conditions set out in Section 1, the term $(1/j\omega\eta)$ predominates at low frequencies, so that the behaviour is essentially viscous under these conditions.

The corresponding time variation of strain $\varepsilon(t)$ in response to a constant stress γ, imposed at $t=0$ is given by:

$$\frac{\varepsilon(t)}{\gamma} = J_\infty + t/\eta + J_r \int_0^\infty N(\tau)[1-\exp(-t/\tau)]\,\mathrm{d}\tau =$$
$$= J_\infty + t/\eta + J_r\psi(t), \quad (13)$$

where $\psi(t)$ is the normalised retardation function. Comparison of Equations (11) and (6) shows that the amplitude $N(\tau)$ of the retardation spectrum appropriate to the BEL equation is proportional to $\tau^{1/2}$ and therefore the corresponding recoverable creep strain increases without limit as $t^{1/2}$ in accordance with Equation (7). The solution of this problem must clearly involve some termination of the spectrum at long times in order to yield physically acceptable predictions. Moreover, since experimental results in alternating shear are in such good agreement with the predictions of the BEL equation, it follows that any correction to the defect-diffusion model to account for this inconsistency can only have a slight influence on the calculated curves for R_L and X_L which must be within the experimental error of previous extensive measurements. The problem has been identified with the long time or low frequency behaviour, so that its effect is most likely to be found in the experimentally ill-defined region between the viscous Newtonian region $(R_L = X_L)$ and the onset of viscoelastic relaxation. In an attempt to resolve this question at issue, a new experimental system was constructed and designed to give results at a single frequency of 30 MHz of an order of magnitude greater accuracy than that of previous techniques. Measurements of R_L and X_L were made as a function of temperature within the supercooled region and the results of this investigation are reported by Barlow and Erginsav [16]. These show that the retardational compliance (Equation (2)) can be fitted within experimental error by the form of the empirical equation used by Davidson and Cole [17] to fit their dielectric relaxation data:

$$J_r^*(j\omega) = J_1 - jJ_2 = J_r/(1+j\omega\tau_r)^\beta. \quad (14)$$

τ_r is a characteristic retardation time which is the longest time constant defining the cut-off in the retardation spectrum and β is a constant for a given temperature,

$0<\beta<1$. For supercooled liquids β increases slightly with increasing temperature but is often found to be approximately 0.5.

The total compliance is therefore expressed by:

$$\frac{J^*(j\omega)}{J_\infty} = [1 + 1/j\omega\tau_m] + \frac{J_r/J_\infty}{(1 + j\omega\tau_r)^\beta}. \tag{15}$$

Taking $\beta=0.5$, and in the light of the above comments, it is noted that for $\omega\tau_r \gg 1$ the second term of Equation (15) becomes approximately $(J_r/J_\infty)/(j\omega\tau_r)^{1/2}$ which, in a formal sense, is indistinguishable from the corresponding term in the BEL equation, namely, $2J_\infty/(j\omega\tau_m)^{1/2}$, if J_r/J_∞ and τ_r/τ_m are constants independent of temperature. These conditions are found to apply in the region $\omega\tau_m > 1$.

The analysis of experimental results proceeds as follows. Measurements of density and of steady-flow viscosity are made over the temperature range of the viscoelastic measurements and extrapolated values of J_∞ are obtained in this range by the proce-

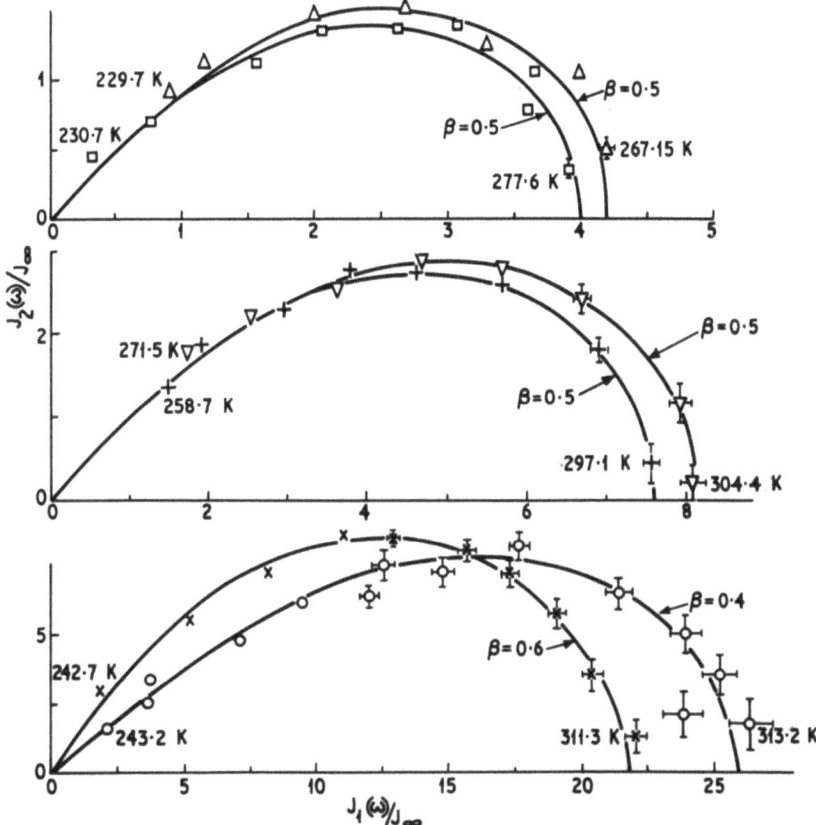

Fig. 9. Relative variation of the components of the retardational compliance at 30 MHz. ○, Tri-(β-chloroethyl) phosphate; X, squalane; +, tri(m-tolyl) phosphate; ▽, tri(o-tolyl) phosphate; △, di(n-butyl) phthalate; □, di(iso-butyl) phthalate. The experimental accuracy decreases as $J_1(\omega)/J_\infty$ increases, the error bars define the estimated semi-inter-quartile range of the error distributions for $J_1(\omega)/J_\infty$ and $J_2(\omega)/J_\infty$. Taken from Barlow and Erginsav [16]. By permission of the Royal Society.

dure outlined in Section 3 using Equation (9). The components of the shear mechanical impedance are measured at selected temperatures and the corresponding values of J' and J'' calculated from Equation (4). Using the known value of $\tau_m = J_\infty \cdot \eta$, the components J_1 and J_2 of the complex retardational compliance are calculated from Equation (15). For the BEL model $J_1 = J_2 = J_\infty/(\omega\tau_m/2)^{1/2}$ so that a plot of J_2/J_∞ vs J_1/J_∞ would be a straight line of slope equal to 45°. In practice, the Davidson-Cole 'skewed arc' plots of Figure 9 are obtained and for cases where $\beta = \frac{1}{2}$ the slope at the origin is indeed 45°. The intercept of each curve on the J_1/J_∞ axis gives directly the value of J_r/J_∞ at the corresponding temperature. The temperature dependence of J_r is, however, unknown but there is no indication of any appreciable change in the value of this quantity over the 30 to 40 K range covering the results of Figure 9. In the subsequent analysis it has been assumed that J_r has the same dependence upon temperature as J_∞ over this restricted range, this being about 15% overall according to the linear variation for J_∞ given by Equation (9). Variation of τ_r with temperature is obscured by the results plotted in the manner of Figure 9. In Figures 10 and 11 the components J_1/J_∞ and J_2/J_∞ are plotted separately as a function of $\log(\omega\tau_m)$ for tri(m-tolyl) phosphate and for di(iso-butyl) phthalate, respectively, the value of β for each liquid being equal to 0.5. The broken line on each of these figures is the prediction of the BEL equation and it can be seen that the measured values converge to this curve beyond $\omega\tau_m = 1$. Reference to Figure 6 shows that the major part of the relaxation region falls above $\omega\tau_m = 1$ where the predictions of the Davidson-Cole formula [15] are indistinguishable from those of the BEL Equation (6), and τ_r has the same dependence upon temperatures as τ_m. The limiting value of J_1 at low frequencies is simply the retardational compliance, J_r. Values of the ratio J_r/J_∞ and of the parameter β are listed in Table I for the liquids which have been studied in detail.

In the region where $\omega\tau_m$ is less than unity the change of J_1/J_∞ and J_2/J_1 with $\omega\tau_m$ cannot be accounted for by assuming that the characteristic retardation time is a constant multiple of τ_m as the temperature is varied. However, assuming the validity

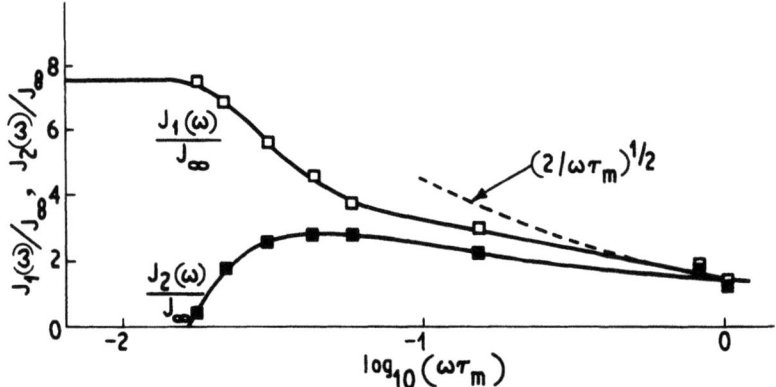

Fig. 10. Measured values of $J_1(\omega)/J_\infty$ and $J_2(\omega)/J_\infty$ for tri(m-tolyl) phosphate plotted vs $\log_{10}(\omega\tau_m)$. The broken line represents the prediction of the Barlow et al. [4] equation, for which $J_1 = J_2 = J_\infty/(\omega\tau_m/2)^{1/2}$.

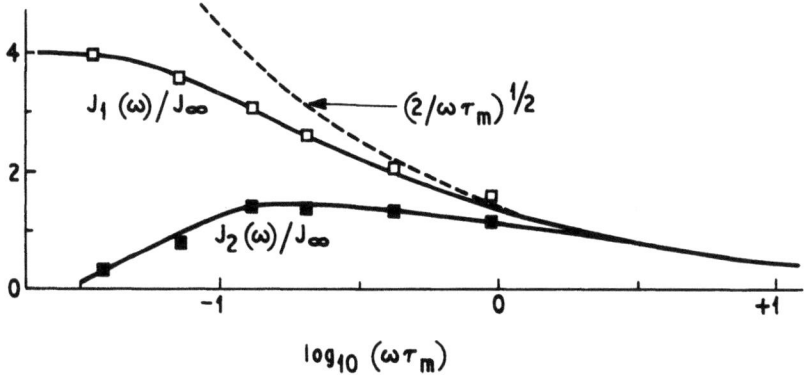

Fig. 11. Measured values of $J_1(\omega)/J_\infty$ and $J_2(\omega)/J_\infty$ for di(iso-butyl) phthalate plotted vs $\log_{10}(\omega\tau_m)$. The broken line represents the prediction of the Barlow et al. [4] equation for which $J_1 = J_2 = J_\infty/(\omega\tau_m/2)^{1/2}$. Taken from Barlow and Erginsav [16]. By permission of the Royal Society.

TABLE I

Values of the ratio J_r/J_∞ and the parameter β in the Davidson-Cole Equation (15)

Liquid	J_r/J_∞	β
Bis(m-(m-phenoxy phenoxy) phenyl) ether	> 50	?
Squalane	21.8	0.6[a]
Tri(β-chloroethyl) phosphate	26.0	0.4[a]
Tri(o-tolyl) phosphate	8.1	0.5[b]
Tri(m-tolyl) phosphate	7.6	0.5[b]
Di(n-butyl) phthalate	4.2	0.5[b]
Di(iso-butyl) phthalate	4.0	0.5[c]
1,3,5-tri(α-naphthyl) benzene	2.2	0.3[d]

[a] Measured from T_A to $T_A - 70\,°C$, approximately.
[b] Measured from T_A to $T_A - 40\,°C$, approximately.
[c] Measured from T_K to $T_K - 40\,°C$, approximately.
[d] Measured close to T_g (at $T_g + 4\,°C$).

of Equations (14) and (15), it is possible to calculate the values of τ_r at any given temperature using the experimental values of J_1 and J_2 plotted against $(\omega\tau_m)$ as in Figures 10 and 11. The variation of $\log(\tau_r/\tau_m)$ with $(1/T)$ is shown in Figure 12 for tri(m-tolyl) phosphate ($\beta = 0.5$) together with the plot of $\log\eta$ against $(1/T)$. In the low temperature region below about 260 K ($10^4/T = 38.5\,K^{-1}$) to the glass transition temperature of 203 K ($10^4/T = 49\,K^{-1}$) (τ_r/τ_m) is constant and equal to the value of $(J_r/2J_\infty)^2$, confirming the agreement in this temperature range between the Davidson-Cole and the BEL expressions. However, in the higher temperature range from 260 K to the Arrhenius temperature of approximately 300 K, τ_r does not have the same dependence upon temperature as does τ_m and the ratio (τ_r/τ_m) decreases rapidly with increasing temperature as the Arrhenius temperature, T_A, is approached. An observation which may be of some significance is that $\tau_r/\tau_m \simeq J_r/J_\infty$ at T_A and this seems to be the case for all the six liquids studied so far.

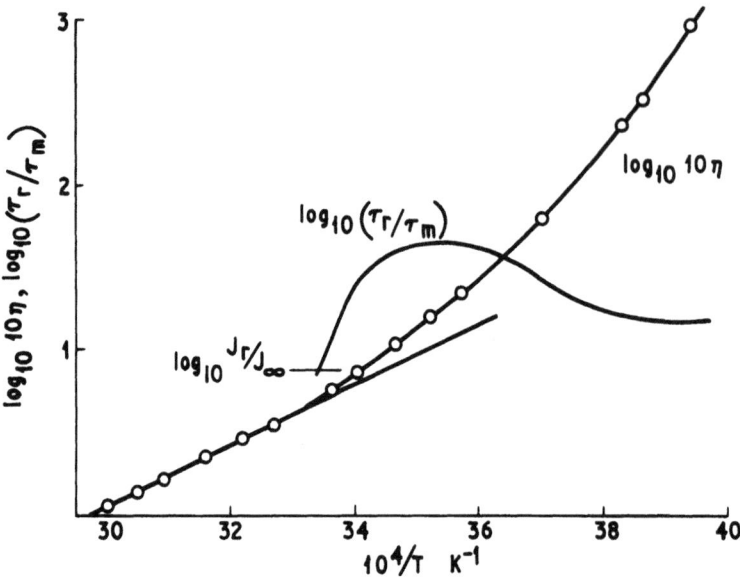

Fig. 12. The variation of $\log_{10}(\tau_r/\tau_m)$ and $\log_{10}(10\eta)$ with $1/T$ for *tri(m-tolyl) phosphate*: η is measured in Poise. Taken from Barlow and Erginsav [16]. By permission of the Royal Society.

Comparisons between the Davidson-Cole and BEL equations are not readily made for liquids in which the value of β is significantly different from 0.5. Nevertheless, the pattern of behaviour in such cases would appear to be similar to that for liquids with $\beta = 0.5$. The plot of $\log(\tau_r/\tau_m)$ vs $1/T$ is shown in Figure 13 for *tri(β-chloroethyl) phosphate*, for which $\beta = 0.6$ and $J_r/J_\infty = 26$. Again, τ_r/τ_m is approximately constant though apparently decreasing slightly with decreasing temperature at lower temperatures but falling rapidly with increasing temperature towards T_A. At T_A the ratio is again approximately equal to J_r/J_∞.

Unfortunately, experimental considerations do not permit reliable measurements to be made in the higher temperature, lower viscosity, region above T_A and viscoelastic behaviour in the Arrhenius region remains a matter of speculation.

In principle, the retardational properties of liquids in the linear viscoelastic region can be investigated either by methods using alternating shear or by the measurement of shear creep and recoverable compliance. The recoverable compliance of several polymer melts has been determined but the only available data for a non-polymeric liquid would appear to be those of Plazek and Magill [18] for *1,3,5-tri(α-napthyl) benzene* at temperatures near to T_g. They find that recoverable compliance data obtained at different temperatures can be reduced to a single curve, which is a much more stringent test than reduction of relaxational data owing to the fact that the latter is dominated by the steady-flow viscosity.

The total recoverable compliance, $J_r\psi(t)$ (Equation (13)) of 1,3,5-*tri*(α-naphthyl) benzene reduced to 64.2°C is shown in Figure 14, which is taken from Plazek and Magill. At short times the curve tends to a value corresponding to $J_\infty = 0.81 \times$

$\times 10^{-10}$ cm^2 dyn^{-1} and to $J_0 = J_\infty + J_r = 2.58 \times 10^{-10}$ cm^2 dyn^{-1} at long times. Hence $J_r = 1.77 \times 10^{-10}$ cm^2 dyn^{-1} and the ratio $J_r/J_\infty = 2.2$. By transforming this time response into a frequency domain it is a straightforward matter to calculate the frequency-dependent compliance which is found to fit a Davidson-Cole equation with $\beta = 0.3$ and $\tau_r = 7.8\, \tau_m$. The measured viscosity at 64.2°C is $10^{12.35}$ P. A recalculation of the total recoverable compliance using these values gives a line coincident with that of Figure 14, showing that the empirical Davidson-Cole equation is also consistent with creep data. The value of $\beta = 0.3$ is somewhat less than that found for the liquids studied in alternating shear, but the temperature here is close to T_g and from dielectric studies β would be expected to be lower in value than at higher temperatures well above T_g.

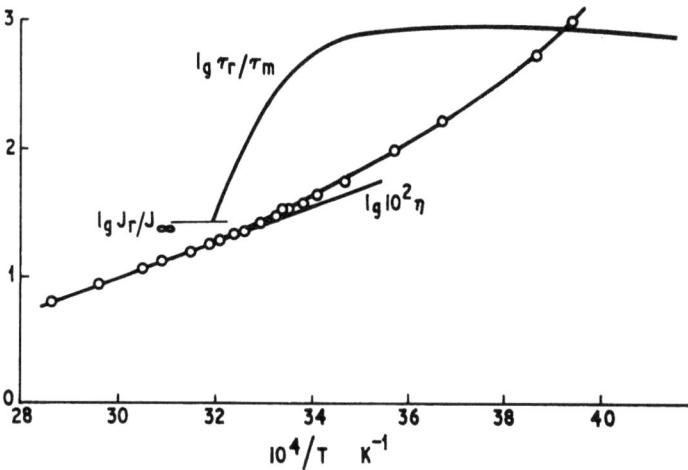

Fig. 13. The variation of $\log_{10}(\tau_r/\tau_m)$ and $\log(100\eta)$ with $1/T$ for *tri(β-chloroethyl) phosphate*: η is measured in Poise. Taken from Barlow and Erginsav [16]. By permission of the Royal Society.

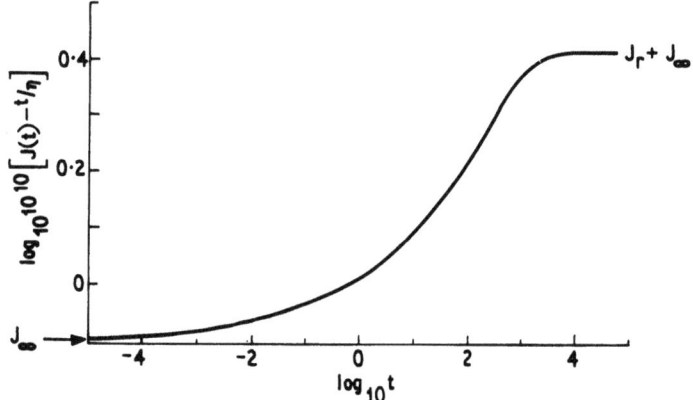

Fig. 14. Recoverable compliance of 1,3,5-*tri*(α-naphthyl) benzene at 64.2°C. The curve is drawn through the temperature-reduced data of Plazek and Magill [18]. The corresponding variation with frequency is given by: $J^*(j\omega) = J_\infty + 1/j\omega\eta + J_r/(1 + j\omega\tau_r)^\beta$, with $J_\infty = 0.81 \times 10^{-10}$ cm^2 dyn^{-1}, $J_r = 1.77 \times 10^{-10}$ cm^2 dyn^{-1}, $\beta = 0.3$, $\eta = 10^{12.35}$ P and $\tau_r = 7.8\tau_m$. Taken from Barlow and Erginsav [16]. By permission of the Royal Society.

6. Interpretation of Viscoelastic Retardation and Dielectric Relaxation

The terms $[J_\infty + 1/j\omega\eta]$ and $[J_\infty + t/\eta]$ in the respective expressions for the complex compliance and the creep compliance involve the combination of the steady-flow viscosity and the high frequency, or instantaneous time, compliance J_∞, according to the Maxwell model. Both η and J_∞ are essentially macroscopic quantities governed by the interactions between very many molecules. In the supercooled region J_∞ reflects the influence of a typical solid compliance enhanced by the presence of structure weakening holes and is approximately an order of magnitude greater than the value of the shear compliance of a solid. The viscosity is governed by the Brownian motion of defects or holes and their diffusion through the liquid and again requires structural co-operative rearrangements of large groups of molecules. On the other hand, the retardational compliance, J_r, represents the delayed storage of elastic energy under shear stress and it is reasonable to presume that this is determined by local fluctuations in the immediate neighbourhood of a molecule and hence by the intermolecular forces which control reorientation. Independent evidence for this point of view is provided by the results of experiments on dielectric relaxation. With the helpful collaboration of Dr G. Williams, it is now possible to compare the results of dielectric and viscoelastic experiments on the same liquids. Thus, Shears and Williams [19] have measured the dipole relaxation in two of the liquids studied in this viscoelastic work, namely, *tri*(*o*-tolyl) phosphate and *di-n*-butyl phthalate, each of which is described in viscoelastic relaxation by $\beta = 0.5$ in the Davidson-Cole expression (c.f. Figure 9). The dielectric measurements were made at frequencies in the range 100 to 10^5 Hz and hence at somewhat lower temperatures than the viscoelastic measurements made at 30 MHz. A marginally lower value of β might therefore be expected in the dielectric case.

A comparison is shown in Figure 15 for the 'skewed-arc' plots of the viscoelastic retardation and the dipole relaxation in these two liquids. The dielectric results are plotted as $\varepsilon''/(\varepsilon_0 - \varepsilon_\infty)$ vs $(\varepsilon' - \varepsilon_\infty)/(\varepsilon_0 - \varepsilon_\infty)$ whilst the viscoelastic data are plotted as J_2/J_r vs J_1/J_r. The agreement found between these two quite independent studies on the same liquids is remarkable, bearing in mind that β might be expected to be somewhat below 0.5 for the lower temperatures employed in the dielectric work. The conclusion drawn from this is that, since the dipole relaxation is essentially a reorientation of the molecule in the force field of surrounding molecules, the retardational viscoelastic process must likewise be one of local reorientation adjustment. On this account the value of J_r will be strongly influenced by the type of molecule. Larger values of J_r are to be expected as the flexibility of the molecule increases and, with the limited evidence available, this is found to be the case.

Previous arguments of Litovitz and McDuffie [20] and Kono *et al.* [21] have compared the dielectric relaxation time τ_D with the Maxwell relaxation time τ_m. It is clear from the results described above that a close relation exists between τ_D and the retardation time, τ_r, and it is on this basis that future comparisons should be made.

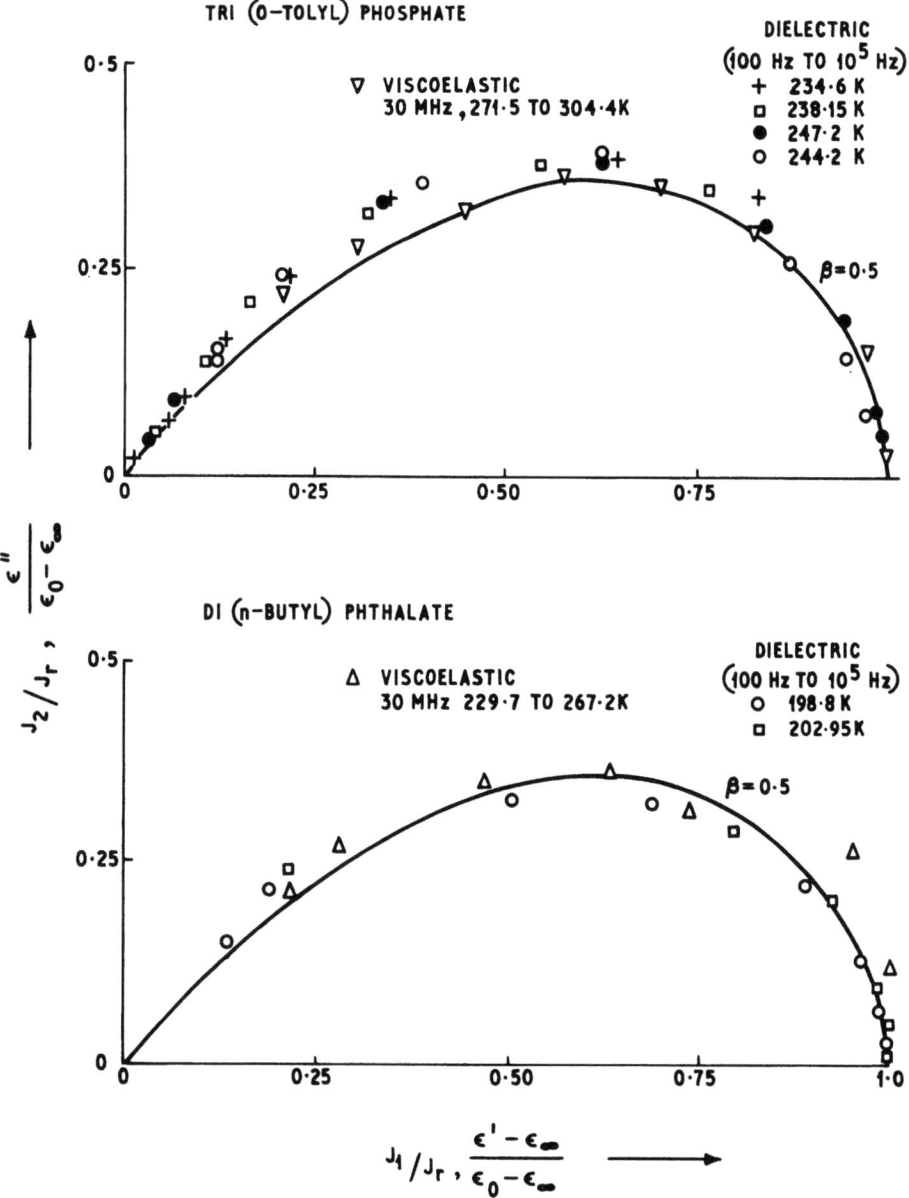

Fig. 15. Comparison of the dielectric relaxation data of Shears and Williams [19], and the retardational compliance. Dielectric results are plotted as $\varepsilon''/(\varepsilon_0 - \varepsilon_\infty)$ vs $(\varepsilon' - \varepsilon_\infty)/(\varepsilon_0 - \varepsilon_\infty)$ and the retardational compliance as $J_2(\omega)/J_r$ vs $J_1(\omega)/J_r$. The curves are drawn in accordance with the Davidson-Cole equation with $\beta = 0.5$. Tri(o-tolyl) phosphate: ▽, viscoelastic data, 30 MGz, 271.5 to 304.4 K. Dielectric data, 100 Hz to 10^5 Hz; +, 234.6 K; □, 238.15 K; ●, 247.2 K; ○, 244.2 K. Di(n-butyl) phthalate: △, viscoelastic data, 30 MHz, 229.7 K to 267.2 K. Dielectric data, 100 Hz to 10^5 Hz; ○, 198.8 K; □, 202.95 K.

The BEL equation is found to fit experimental results above $\omega\tau_m = 1$ and indeed this is an acceptable approximation over the whole of the viscoelastic region (c.f. Figure 6). Small deviations from the behaviour calculated from this equation have been noted by Davies et al. [22], but it is only when extreme care is taken to achieve high accuracy that the retardational effects brought out in the work of Barlow and Erginsav [16] can be delineated. Above $\omega\tau_m = 1$ the form of the Davidson-Cole equation is indistinguishable from that of the BEL equation. The BEL equation has, however, been derived theoretically from the defect diffusion model, in which it is visualised that flow takes place by movement of holes rather than directly by movement of molecules, whereas the retardation time process is identified above with molecular readjustments within a local environment. It seems reasonable, therefore, to assume that the same processes which give rise to the diffusion of defects are responsible for the local readjustments in the neighbourhood of a molecular site. Hence, at temperatures well below T_A, the concept is that of diffusional motion of defects through the system which, on the appropriate time scale (τ_m), permit readjustment of relatively large molecular groups to occur, which we identify as translational flow. However, local reorientation of a molecule within a group is also dependent upon the arrival of a defect at the site in question by the same process of diffusional motion. The result that τ_r is greater than τ_m can be attributed to the fact that the elastic constraints on molecular rotation are considerable weaker than those which operate on the translational motion of defects, since J_r is greater than J_∞ (Table I). Moreover, the relaxation time $\tau_m = \eta J_\infty$ and, by analogy, we can define a retardational viscosity, η_r, given by $\tau_r = \eta_r J_r$. In the low temperature region under consideration we have shown that $\tau_r/\tau_m = (J_r/2J_\infty)^2$ and hence $\eta_r/\eta = J_r/4J_\infty$. Taking for illustration a typical example of tri(o-tolyl) phosphate, we have for this liquid, $J_r/J_\infty = 8.1$ so that $\tau_r/\tau_m \simeq 16$ and $\eta_r \simeq 2$. The retardational viscosity, therefore, also exceeds the macroscopic steady flow viscosity by a factor of 2 in this case. The fact that τ_m and τ_r have the same temperature dependence can be attributed to the assumption that both translational flow and local reorientation are controlled by the same process of defect diffusion.

At higher temperatures approaching T_A the ratio τ_r/τ_m decreases markedly with increasing temperature, as is evident from Figures 12 and 13. At T_A we find approximately that $\tau_r/\tau_m = J_r/J_\infty$ implying that $\eta_r = \eta$. Considering the fact that the available relative free volume has increased by an order of magnitude from T_g to T_A, the relative changes in τ_r and τ_m are comparatively small. For tri(o-tolyl) phosphate the ratio was 16 at low temperatures and this has fallen to 8 at T_A. However, in the Arrhenius region above T_A, where the relative free volume exceeds 15%, the generally accepted picture is that translational motion takes place by molecular jumps, so that the viscosity is here determined principally by the energy required for a molecule to surmount the barrier due to its nearest neighbours and not by the process of defect diffusion assumed to apply at lower temperatures. At high temperatures molecules rapidly attain equilibrium with respect to nearest neighbours, so that it could be reasonably assumed that as temperature increases above T_A, τ_r would become progressively less than τ_m. Experimental evidence is not available to test this hypothesis, but the trends apparent below T_A are entirely in accord. The free volume equation for the dependence of viscosity upon temperature applies only a few degrees K below T_A but a reduction of some 30 K below T_A is required before τ_r reaches a constant value with respect to τ_m. In this temperature interval molecules presumably have some freedom for local orientation irrespective of the diffusion of holes through the body of the liquid. At lower temperatures the local readjustment is considered to be entirely governed by the availability of free volume in the neighbourhood of the molecule, through the process of defect-diffusion.

7. Viscosity Discontinuities in the Supercooled Region

Although the modified free volume Equation (2) gives an excellent description of the viscosity of most supercooled liquids in the temperature range below T_A, there are some liquids in which it is necessary to apply this equation in two separate temperature regions with two different sets of values of the constants A, B and T_0, as shown in Figure 16 for sec butyl benzene [23]. Similar behaviour has subsequently been found in other liquids by Davies and Matheson [24] who, by calculating the volume required for free rotation, have argued that the transition from one type of dependence of viscosity on temperature to another is due to imposed restrictions on molecular rotation as the temperature is lowered. Specifically, they suggest that in the Arrhenius region molecules can rotate many times about at least two axes during the time between translational jumps. In the non-Arrhenius (supercooled) region, no more than one rotational mode is free and in those liquids which show a discontinuity in viscosity behaviour at a temperature T_K in the non-Arrhenius region, this occurs when the additional rotational mode is restricted. Translational motion would be presumed to cease at the lower value of T_0, but the glass transition at a higher temperature prevents this from being observed experimentally.

The retardation time, τ_r, has been associated with reorientation of the molecule in its local environment and hence should reflect any significant changes which occur in the degree of rotational freedom. One of the liquids which exhibits a discontinuity in the non-Arrhenius viscosity region is $di(iso\text{-butyl})$ phthalate, for which the viscosity plots for the two regions are shown in Figure 17 together with its isomer $di(n\text{-butyl})$ phthalate, which shows no such discontinuity. The data shown in Figure 9 for $di(iso\text{-butyl})$ phthalate was obtained below the transition temperature T_K, the behaviour there being described by a Davidson-Cole plot with $\beta=0.5$. However, when the measurements were extended above T_K, a second retardation process was observed in the higher temperature region as shown in Figure 18 and this is described by a single retardation time, $\tau_{r2}(\beta=1)$. Although, as mentioned by Barlow and Erginsav [16], it would be unwise to place too much reliance upon these measurements because of the low viscosity in this region and the consequent deterioration in experimental accuracy, there is no doubt that a second retardational process exists and that the change from one regime to the other coincides closely with the temperature of the viscosity discontinuity at $T_K(\simeq 277\text{K})$. A time-dependent process is therefore directly associated

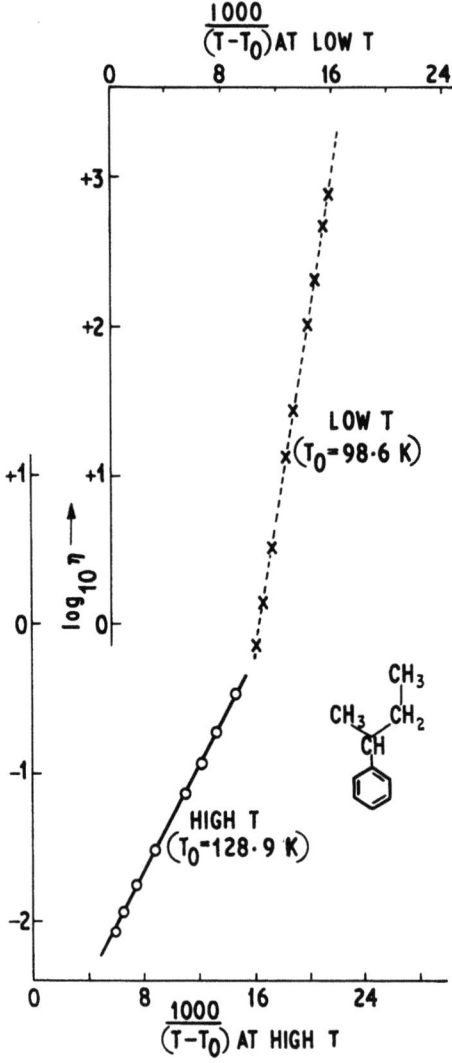

Fig. 16. Viscosity of *sec*-butyl benzene in the supercooled region plotted as $\log_{10}\eta$(P) vs $1/(T-T_0)$ for the 'high' and 'low' temperature regions. Taken from Barlow *et al.* [23]. By permission of the Royal Society.

with a region of the viscous behaviour even though the retardation mechanism does not contribute to the steady flow viscosity. The retardation time, τ_{r2}, of the higher temperature process is longer than the time τ_{r1} which characterises the behaviour below T_K. Each contributes a delayed compliance of approximately $4J_\infty$. The ratios of the retardational viscosities $\eta_{r1}(=\tau_{r1}/J_{r1})$ and $\eta_{r2}(=\tau_{r2}/J_{r2})$ to the steady-flow viscosity, η, are shown in Figure 19, together with the plot of $\log\eta$ vs $1/T$. At T_K, η_{r1} becomes equal to η, and an extrapolation of the curve for η_{r2}/η indicates that η_{r2} is also approximately equal to η at T_A.

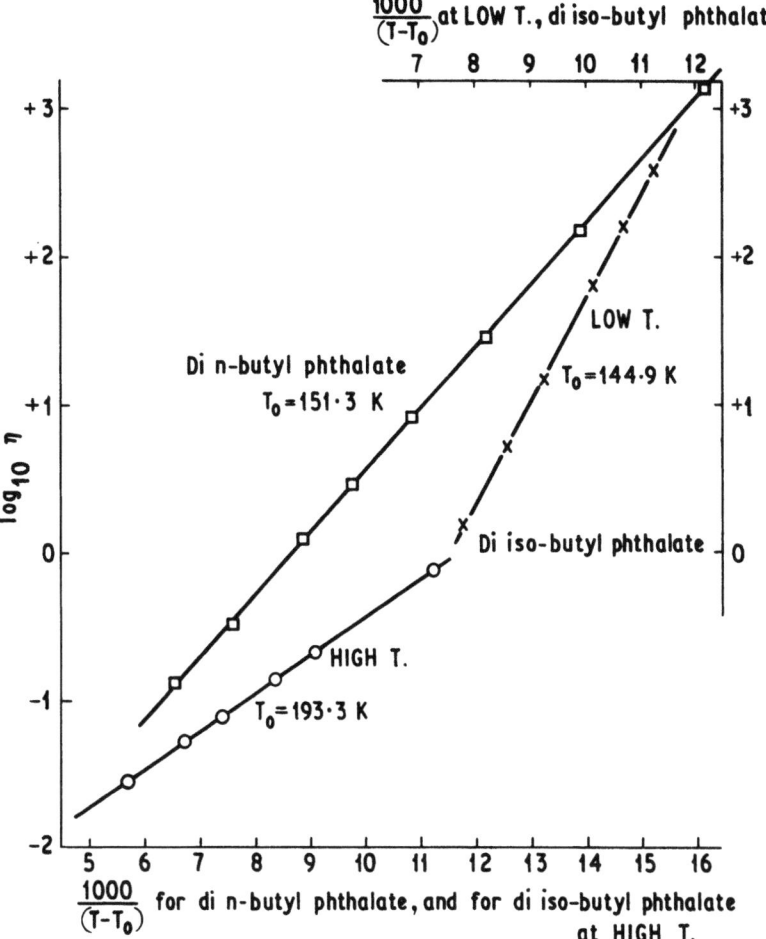

Fig. 17. Viscosity of *di*(*n*-butyl) phthalate and *di*(*iso*-butyl) phthalate plotted as $\log_{10}\eta(P)$ vs $1/(T-T_0)$. Taken from Barlow *et al.* [23]. By permission of the Royal Society.

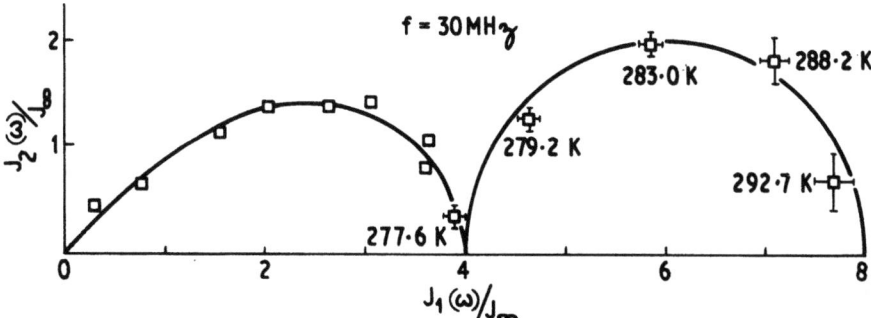

Fig. 18. Components of the retardational compliance of *di*(*iso*-butyl) phthalate above and below the viscosity discontinuity in the supercooled region. Error bars show the estimated semi-interquartile range of error distributions. Taken from Barlow and Erginsav [16]. By permission of the Royal Society.

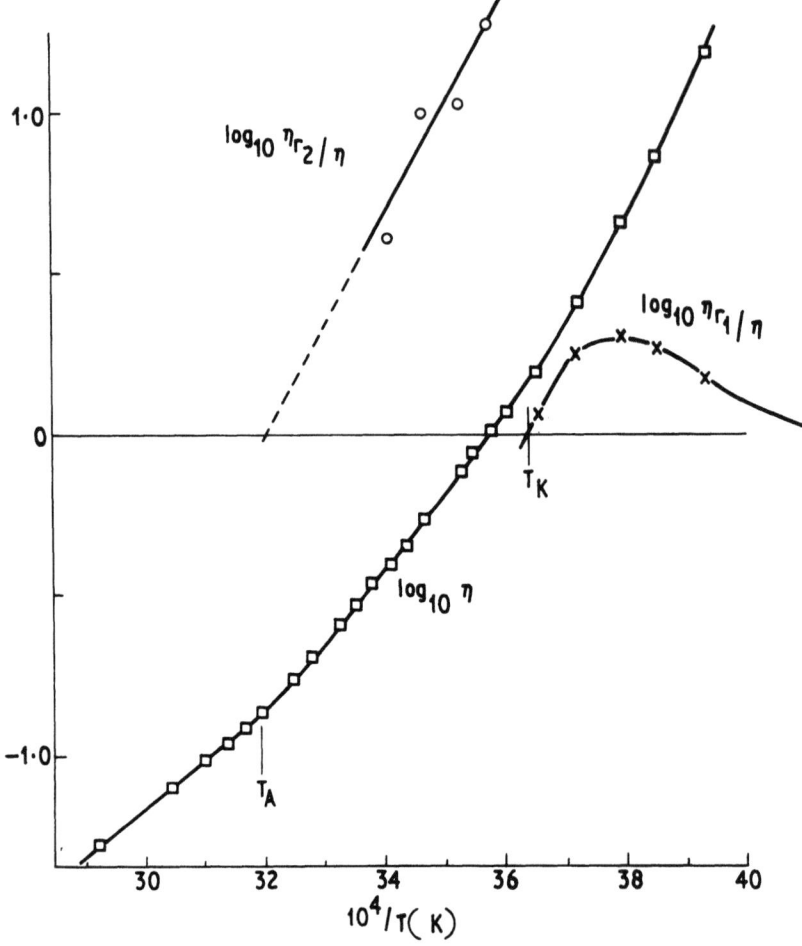

Fig. 19. Variation of $\log_{10}\eta$, $\log_{10}\eta_{r1}/\eta$ and $\log_{10}\eta_{r2}/\eta$ with $1/T$ for di(iso-butyl) phthalate. Taken from Barlow and Erginsav. [16]. By permission of the Royal Society.

These findings support the general ideas put forward in Section 6 relating the retardation time process to rotational adjustment of the molecule in the force field of its nearest neighbours.

8. Modifications of the BEL Equation for Mixtures

In some of the earlier experiments reported [4, 8], deviations were found from the predictions of the BEL equation for a few of the liquids studied. It was suspected that the cause might be due to impurity, as was later confirmed, and this led to a study of the behaviour of liquid mixtures [14]. These measurements were not of sufficient accuracy to permit observation of the retardational process and it was found that all results obtained on binary and other mixtures could be fitted within experimental

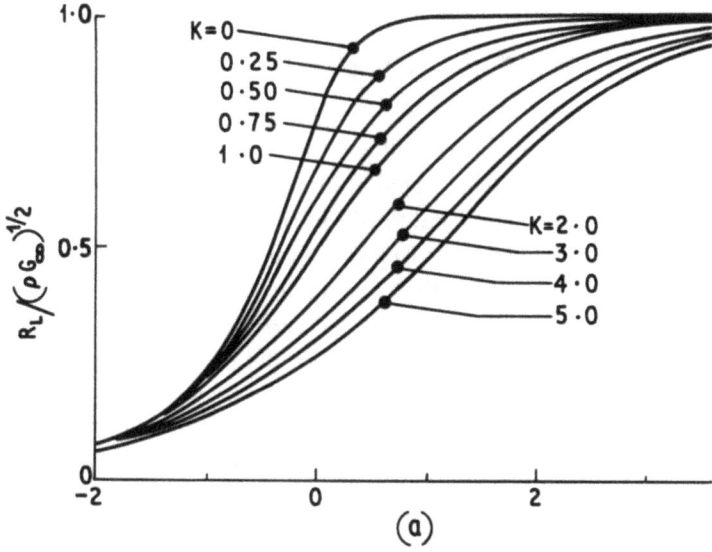

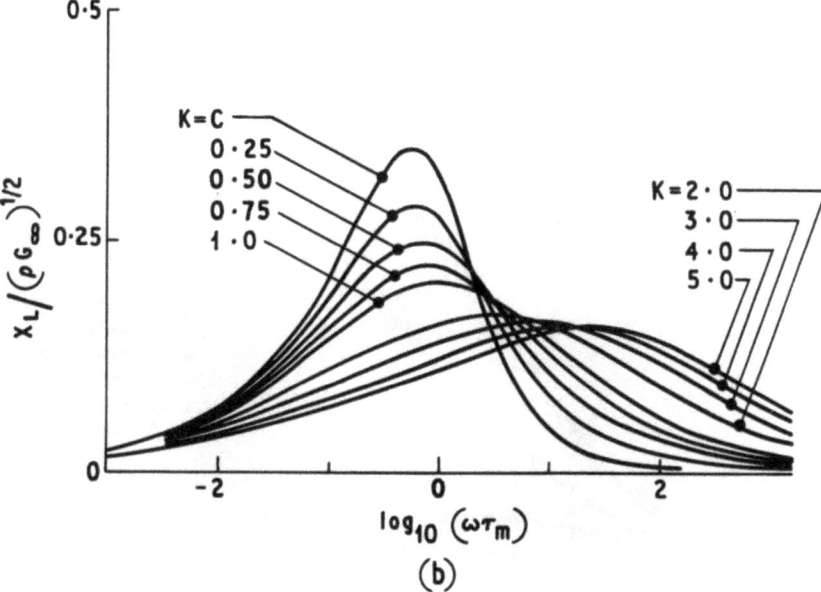

Fig. 20. Calculated relaxation curves for the normalised components of the shear mechanical impedance for different values of the factor K in the modified Barlow et al. [4] Equation (16). Taken from Barlow et al. [14]. By permission of the Royal Society. [Note that the ordinate axes on Figures 1 and 2 of the original publication should be interchanged].

accuracy by a more general form of the BEL equation:

$$J^*(j\omega) = J_\infty [1 + 1/j\omega\tau_m] + 2KJ_\infty/(j\omega\tau_m)^{1/2}. \tag{16}$$

The value of K in each case is obtained by comparing the experimental results for $R_L/(\varrho G_\infty)^{1/2}$ and $X_L/(\varrho G_\infty)^{1/2}$ with a series of computed curves as shown in Figure 20. $K=1$ corresponds to the original BEL equation and this value was found for all the binary mixtures at equal mole fraction and for the pure components. For other mixing ratios a minimum value of $K=0.25$ was found to apply at 0.2 mole fraction of the lower molecular weight component with a maximum value of $K=1.8$ at 0.8 mole fraction. The highest value, $K \simeq 2.9$ was found for castor oil (Figure 21). Since this behaviour differs markedly from that of pure liquids and of binary mixtures and because of interest in castor oil as a lubricant, the previous work has recently been extended by investigating the behaviour of this liquid as a function of pressure and temperature [13]. Results obtained at a series of pressures up to 300 MN m^{-2} showed that in the elastic region J_∞ is proportional to temperature at each pressure, as given in Figure 22. In addition, for a series of temperatures in the range -10 to $-30\,°C$, measurements of R_L^2/ϱ in the elastic range establish that G_∞ is proportional to pressure at each selected temperature as shown in Figure 23. Taking all the results at different pressures and temperatures and at frequencies of 10 and 30 MHz gives the normalised

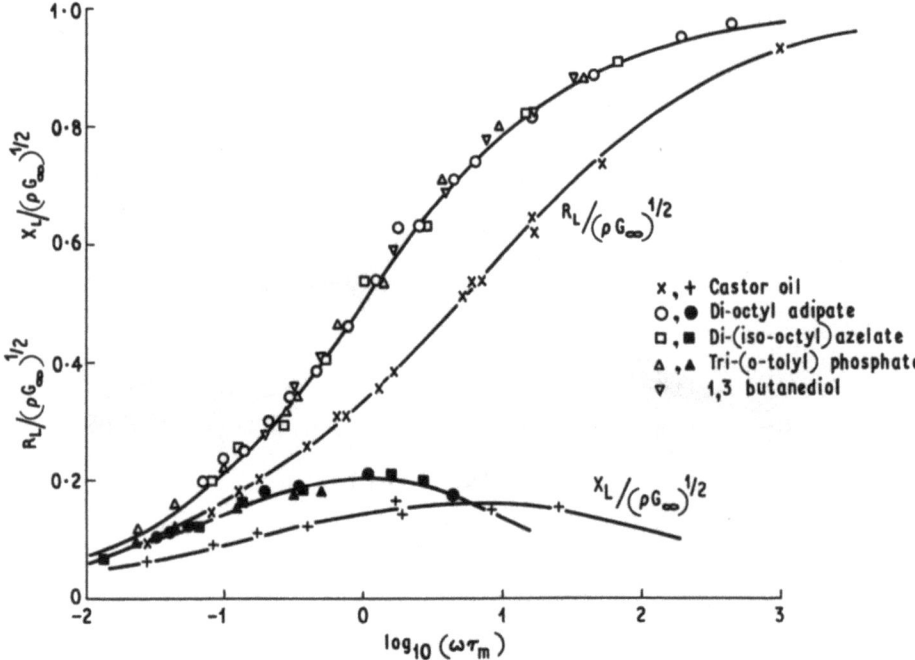

Fig. 21. Normalised components of the shear mechanical impedance plotted vs $\log_{10}(\omega\tau_m)$ for four pure liquids and for castor oil. Curves are calculated from Equation (16) with $K = 2.9$ for castor oil and $K = 1$ for the other liquids. Results obtained at atmospheric pressure. Taken from Barlow et al. [13]. By permission of the Chemical Society.

plot shown in Figure 24 for $R_1/(\varrho G_\infty)^{1/2}$ vs $(\omega \tau_m)$. These are fitted by a curve derived from Equation (16) with $K=2.6$. The difference between this value and the previous figure of 2.9 obtained from results taken at only atmospheric pressure merely reflects the consequency of fitting the computed curves of Figure 20 to extended data and the change is not significant. The conclusion drawn is that the modified form of the BEL equation agrees with experimental results on a wide range of liquid mixtures including high pressure measurements. The constant K is determined by fitting to computed curves but its significance in physical terms remains obscure at the present time.

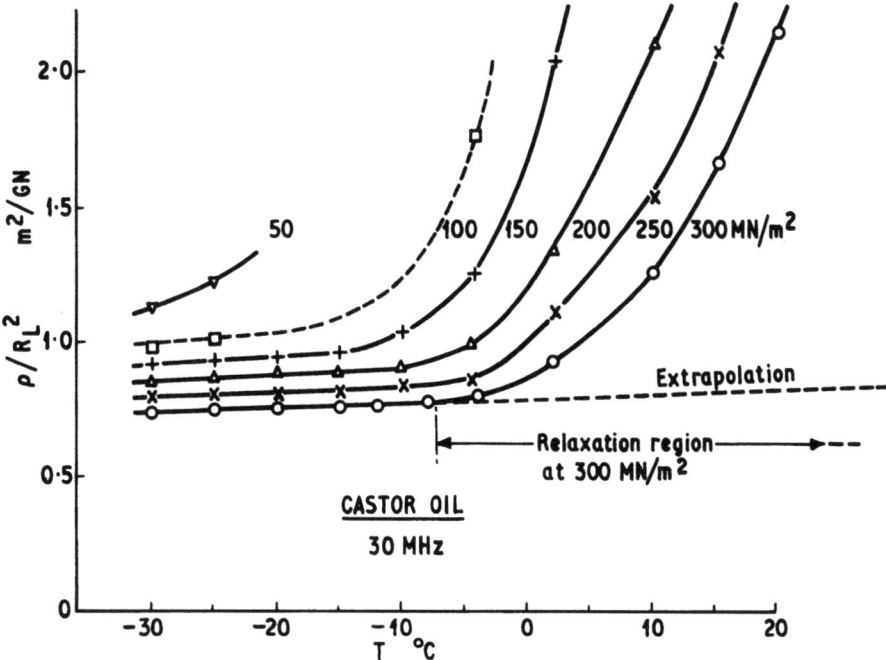

Fig. 22. The variation with temperature of ϱ/R_L^2 for castor oil at different pressures. In the elastic region J_∞ varies linearly with temperature at each pressure. Taken from Barlow et al. [13]. By permission of the Chemical Society.

9. Compressional Viscosity

The complex velocity v_L^* for the propagation of longitudinal plane waves in a liquid is related to the complex adiabatic compressional modulus K^* and the shear modulus G^* by:

$$\varrho(v_L^*)^2 = K^* + \tfrac{4}{3} G^*. \tag{17}$$

The phase velocity, v_L, and the absorption coefficient, α, of the excess pressure wave are quantities which can be measured and are related to v_L^* by

$$1/v_L^* = 1/v_L - j\alpha/\omega \tag{18}$$

K_0, the low-frequency, non-relaxing component of K^* is measured at effective frequencies below the relaxation region and is given by $K_0 = \varrho v_0^2$. Since the components of $G^*(j\omega)$ are known from viscoelastic experiments, it is therefore possible to determine the relaxational components of the compressional modulus:

$$K^*(j\omega) = K_0 + K' + jK''. \tag{19}$$

The limiting value of K' at high effective frequencies is generally denoted by K_2 ($= K_\infty - K_0$). For the condition $(\alpha v_L/\omega)^2 \ll 1$ which is normally valid, it follows that

$$\begin{aligned} K_0 + K' + \tfrac{4}{3}G' &= \varrho v_L^2 \\ K'' + \tfrac{4}{3}G'' &= 2\varrho v^3 \alpha/\omega. \end{aligned} \tag{20}$$

The shear viscosity, η, is equal to the limiting value at low frequencies of $G''(\omega)/\omega$ and likewise the volume or compressional viscosity, η_v, can be defined as the limiting

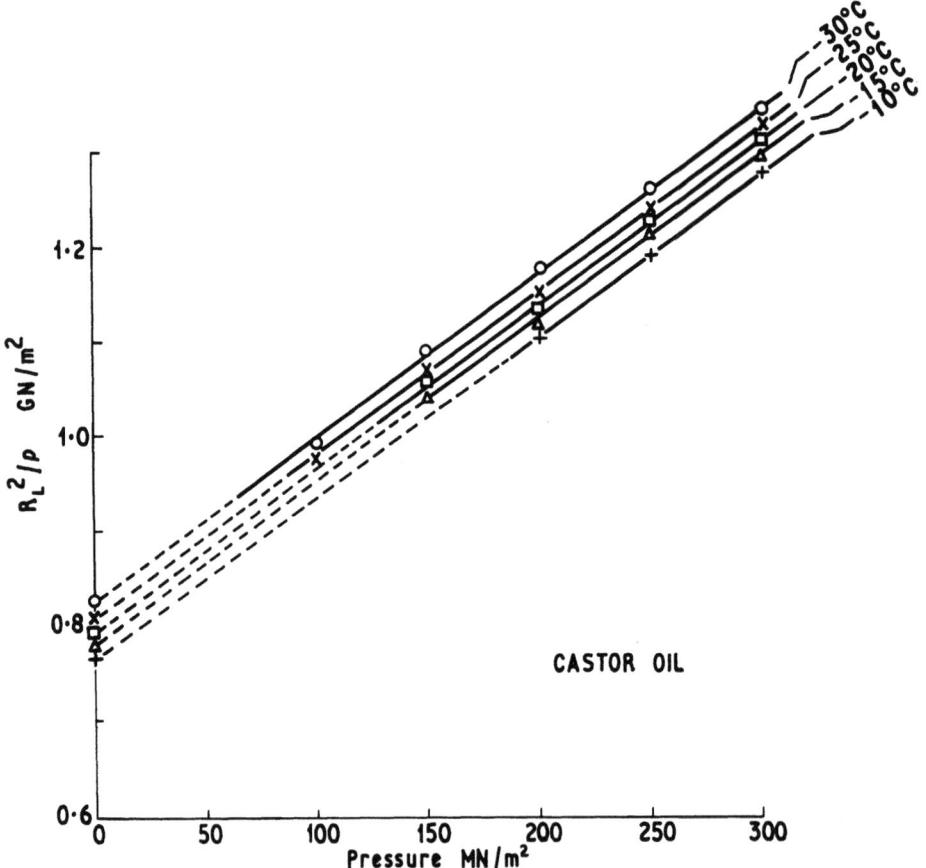

Fig. 23. Measured and extrapolated values of G_∞ for castor oil as a function of pressure at different temperatures. Taken from Barlow et al. [13]. By permission of the Chemical Society.

value at low frequencies of $K''(\omega)/\omega$ so that from Equation (20):

$$(\alpha/\omega^2)_{\omega \to 0} = \frac{1}{2\varrho v_0^3}(\eta_v + \tfrac{4}{3}\eta). \tag{21}$$

Since propagation of a longitudinal wave causes sinusoidal oscillation of temperature and pressure it is possible for such a wave to perturb both inter and intra molecular equilibria leading to extensive studies of thermal relaxation in liquids reviewed by Lamb [25]. However, in the supercooled highly viscous region, these relaxation

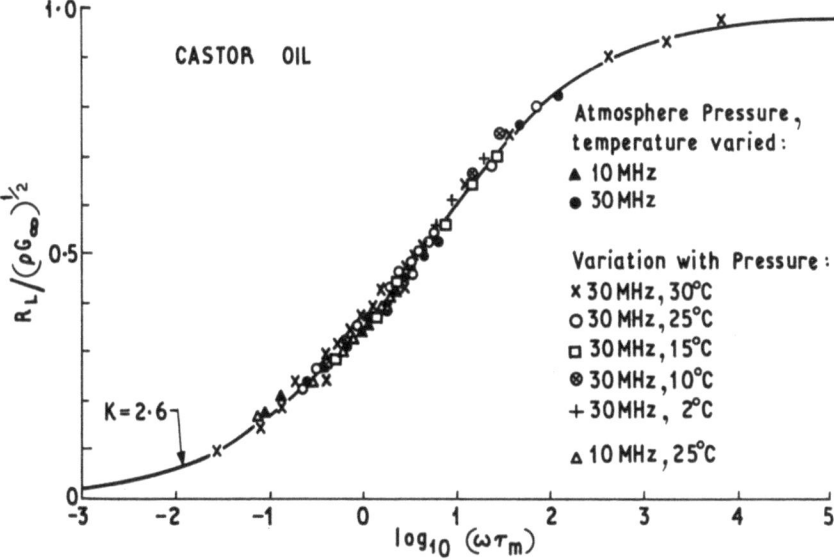

Fig. 24. Normalised shear resistance of castor oil plotted as a function of $\log_{10}(\omega\tau_m)$. The curve is calculated from the modified Barlow et al. [4] Equation (16) with $K = 2.6$. The results are obtained at different pressures and temperatures. Taken from Barlow et al. [13]. By permission of the Chemical Society.

processes are generally masked by the much greater effects due to the volume and shear viscosities. There are few liquids in which both the shear and compressional properties are known, but it would appear that the temperature dependence of the moduli is such that

$$\frac{1}{G_\infty}\frac{\partial G_\infty}{\partial T} = \frac{1}{K_2}\frac{\partial K_2}{\partial T},$$

with the values of G_∞ and K_2 being very similar. In addition η_v and η have the same temperature dependence with η_v being somewhat greater than η by a factor of 3 or 4. Thus, for two of the liquids mentioned previously, the results are as follows: bis(m-(m-phenoxy phenoxy) phenyl ether, [26], $K_2 = \tfrac{4}{3}G_\infty$, $\eta_r = 4 \cdot (\tfrac{4}{3} \cdot \eta)$, whilst for tri(o-tolyl) phosphate, [27] $K_2 = G_\infty$, $\eta_r = \tfrac{9}{4} \cdot (\tfrac{4}{3}\eta)$. These values are similar to those found for other supercooled liquids, a collection of results for which are given by Litovitz and Davies [28] and more recently by Piercy and Rao [29].

As demonstrated by the results for *tri*(*o*-tolyl) phosphate shown in Figures 25 and 26, satisfactory reduction of the data is obtained when these are plotted in normalised form vs $\log(\omega\tau_m)$. Moreover, these curves are the same as those for G'/G_∞ and G''/G_∞, respectively, but displaced along the axis of effective frequency by an amount which corresponds to placing $\eta_v = 3\eta$. The conclusion is drawn that the shear and compressional relaxations have a common origin in these supercooled liquids. Both relaxation

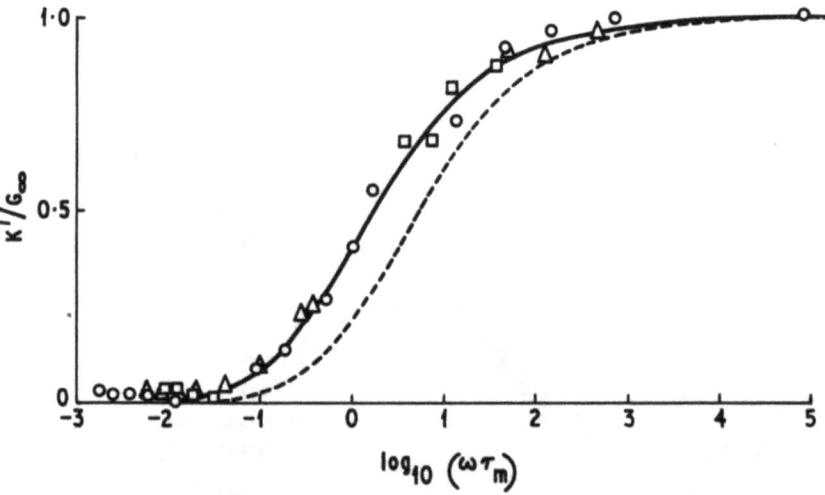

Fig. 25. Values of the normalised compressional modulus K'/G_∞ plotted vs $\log_{10}(\omega\tau_m)$ for tri(*o*-tolyl) phosphate: ○, 5 MHz; △, 15 MHz:, □, 25 mHz. The broken line represents the variation of the normalised shear modulus G'/G_∞ according to the Barlow *et al.* [4] equation. Taken from Barlow and Singh [27]. By permission of the Chemical Society.

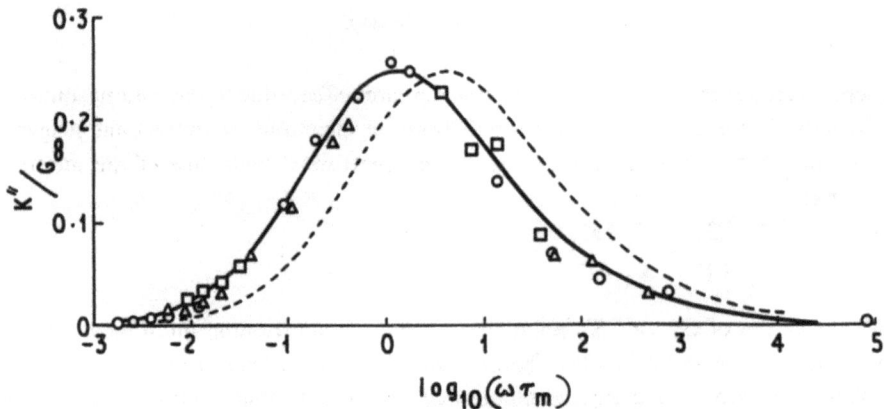

Fig. 26. Normalised values of the compressional loss modulus K''/G_∞ plotted vs $\log_{10}(\omega\tau_m)$ for tri(o-tolyl) phosphate: ○, 5 MHz; △, 15 MHz; □, 25 MHz. The broken line represents the variation of the normalised shear loss modulus G''/G_∞ according to the Barlow *et al.* [4] equation. Taken from Barlow and Singh [27]. By permission of the Chemical Society.

processes are therefore predominately controlled by the defect-diffusion mechanism as described by the BEL equation. Since the longitudinal modulus is dominated by the equilibrium compression modulus, K_0, it is unlikely that sufficient accuracy will be obtained from ultrasonic techniques to permit evaluation of the compressional retardation properties.

References

1. Longuet-Higgins, H. C. and Widom, B.: *Mol. Phys.* **8**, 549 (1964).
2. Cohen, M. H. and Turnbull, D.: *J. Chem. Phys.* **31**, 1164 (1959).
3. Vogel, H.: *Phys. Z.* **22**, 645 (1921).
4. Barlow, A. J., Erginsav, A., and Lamb, J.: *Proc. Roy. Soc. (London)* **A 298**, 481 (1967) (Paper II.)
5. Glarum, S.: *J. Chem. Phys.* **33**, 639 (1960).
6. Hunt, B. I. and Powles, J. G.: *Proc. Phys. Soc.* **88**, 513 (1966).
7. Phillips, M. C., Barlow, A. J., and Lamb, J.: *Proc. Roy. Soc. (London)* **A329**, 193 (1972).
8. Barlow, A. J., Lamb, J., Matheson, A. J., Padmini, P.R.K.L., and Richter, J.: *Proc. Roy. Soc. (London)* **A298**, 467 (1967). (Paper I.)
9. Lamb, J.: in S. Onagi (ed.), *Proc. Fifth Int. Congress on Rheology* **4**, Univ. Tokyo Press, 1970, p. 325.
10. Slie, W. M. and Madigosky, W. M.: *J. Chem. Phys.* **48**, 2810 (1968).
11. Barlow, A. J., Harrison, G., Irving, J. B., Kim, M. G., Lamb, J., and Pursley, W. C.: *Proc. Roy. Soc. (London)* **A327**, 403 (1972).
12. Hutton, J. F. and Phillips, M. C.: *Nature Phys. Sci.* **238**, 141 (1972).
13. Barlow, A. J., Harrison, G., Kim, M. G., and Lamb, J.: *J. Chem. Soc. Faraday Trans.* **69**, 1446 (1973).
14. Barlow, A. J., Erginsav, A., and Lamb, J.: *Proc. Roy. Soc. (London)* **A309**, 473 (1969).
15. Gross, B.: *Mathematical Structure of the Theories of Viscoelasticity*, Hermann, Paris, 1953.
16. Barlow, A. J. and Erginsav, A.: *Proc. Roy. Soc. (London)* **A327**, 175 (1972).
17. Davidson, D. W. and Cole, R. H.: *J. Chem. Phys.* **19**, 1484 (1951).
18. Plazek, D. J. and Magill, J. H.: *J. Chem. Phys.* **45**, 3038 (1966).
19. Shears, M. and Williams, G.: private communication (1972).
20. Litovitz, T. A. and McDuffie, G. E.: *J. Chem. Phys.* **39**, 279 (1963).
21. Kono, R., McDuffie, G. E., and Litovitz, T. A.: *J. Chem. Phys.* **47**, 5232 (1966).
22. Davies, D. B., Matheson, A. J., and Glover, G. M.: *J. Chem. Soc. Faraday Trans. II* **69**, 305 (1973).
23. Barlow, A. J., Lamb, J., and Matheson, A. J.: *Proc. Roy Soc. (London)* **A292**, 322 (1966).
24. Davies, D. B. and Matheson, A. J.: *J. Chem. Phys.* **45**, 1000 (1966).
25. Lamb, J.: in W. P. Mason (ed.), *Physical Acoustics*, Vol. IIa, Academic Press, New York, 1965, Chap. 4, p. 203.
26. Barlow, A. J., Lamb, J., and Taşköprülü, N. S.: *J. Acoust. Soc., America* **46**, 569 (1969).
27. Barlow, A. J. and Singh, R. P.: *J. Chem. Soc., Faraday Trans., II* **68**, 1404 (1972).
28. Litovitz, T. A. and Davies, C. M.: in W. P. Mason (ed.), *Physical Acoustics*, Vol. IIa, Academic Press, New York, 1965, Chap. 5, p. 281.
29. Piercy, J. E. and Rao, M. G. S.: *J. Acoustic Soc., America* **41**, 1063 (1967).

DISCUSSION

Litovitz: Is there real evidence for the existence of two distinct relaxation processes, one existing in each of the two separate non-Arrhenius regions? Can one argue that both exist in one region and only one in the other?

Lamb: In reply to the question Section posed by Prof. Litovitz I would refer him to the remarks made in Section 7 of the paper where this matter is discussed more fully than was possible in the lecture presentation. In summary I would emphasise the following points:

(1) Because of this close correspondence of dielectric relaxation results and those of viscoelastic

retardation, shown for example in Figure 15, it is argued that τ_r is associated with reorientation of the molecule in its local environment. Hence any significant changes which occur in the degree of rotational freedom should be reflected in the retardational processes.

(2) Certain liquids show a discontinuity in viscosity behaviour at a temperature T_K in the supercooled region which has been viewed by Davies and Matheson [24] as due to a restriction in two degrees of rotational motion.

(3) One of the liquids with a discontinuity at T_K in viscosity-temperature behaviour has been measured in our experiments. The liquid is di(isobutyl) phthalate for which the viscosity results are shown in Figure 17 and the viscoelastic retardation results in Figure 18.

(4) Because of the extreme accuracy required to evaluate *both* components of the retardational compliance, measurements have only been made at a frequency of 30 MHz. We do not as yet have a similar apparatus working at another frequency but it should be pointed out that no one has previously succeeded in evaluating both $J_2(\omega)$ and $J_1(\omega)$.

(5) The results of Figure 18 at 30 MHz show two distinct retardation processes, that at lower temperature corresponds to $\beta = 0.5$ in the Davidson-Cole equation while that at the higher temperatures corresponds to $\beta \simeq 1$.

(6) The change-over from one retardational process to the other occurs precisely at the temperature T_K. This may be coincidental but if so it would be a remarkable coincidence.

(7) This is the evidence from which Prof. Litovitz must draw his own conclusions. My own view is set forth in Section 7 of the paper but as emphasised there it would be unwise to place too much reliance upon the accuracy of the measurements in the higher temperature region.

Janik: You are introducing two parameters for interpreting viscoelastic relaxation (τ_r and τ_m). What is your opinion about a model approach similar to that used in quasielastic neutron scattering theories? one introduces there four parameters:

τ_0 – residence time for rotation,
τ_1 – observation of rotational jump,
τ'_0 – residence time for translation,
τ'_1 – observation of translational jump.

(See for instance R. E. Larsson, *Phys. Rev.* **71**, and also his new paper which is being published.)
You may then start speculations saying that in some cases $\tau_0 \gg \tau_1$, $\tau'_0 \gg \tau'_1$, and so on....

Lamb: I think that you have in mind the behaviour in the higher temperature Arrhenius region where molecular translational motion is considered to occur by a 'jump' over an energy barrier.

I have emphasised the restriction of our work to the supercooled region where I would focus attention on the motion of defects or holes. We then have two time-dependent procesess characterised by τ_m and τ_r and this is in line with the suggestion in the last sentence of Prof. Janik's remarks.

Friedman: Prof. Janik points out that one could also represent these phenomena by another class of physical models. I think may be that one can classify such theories in terms of the number of parameters one has to adjust. This seems clear from more stochastic theories: Mori's continued fraction expansion is one example. The way such a stochastic theory can accommodate many physical models is nicely illustrated in a recent paper by Kivelson (*J. Chem. Phys.*) concerning rotational correlations. It suggests that any theory with as many parameters as yours might represent the data equally well so it may not be possible to distinguish among physical models by studies of this kind.

Lamb: I would first draw your attention to the calculations of Phillips, Barlow and Lamb of reference [7] where the BEL. equation is derived from a one dimensional defect-diffusion model taking into account second as well as first nearest defects to a site. It is conceivable that this analysis could now be extended to a viscoelastic relaxation of a molecule in its local environment at a 'site' which is the 1-dimensional case is merely a point site.

My personal preference is to work with models which have include a physical basis but clearly as one progresses in understanding of the processes involved, the model is modified and even at times discarded in favour of a more acceptable analysis.

Perhaps it may not be possible to distinguish between different models but to date no one has yet produced an alternative so that the last sentence of Prof. Friedman's remarks is somewhat hypothetical.

I am sure the Prof. Friedman does not mean to imply, as has remarks suggest, that experiments of the type which I have described are a waste of time. The object of them is not to collect data but to advance understanding.

Finally I think that Prof. Friedman credits us with more parameters in our translative model than actually exist. I do not regard η and J_∞ as 'parameters' since these are physical quantities which are measured directly. We are then left only with two 'parameters' τ_m and τ_r and τ_m is determined uniquely by $\tau_m = \eta \cdot J_\infty$.

I admit that in certain cases β may not be 0.5 so that we must consider this as a parameter. τ_r is not arbitrary since it is measured directly without any ambiguity. In fact τ_r is also determined from experiments so we are only left with a single variable β, which in most cases is close to 0.5. I am naturally taking the view of an experimentalist aiming to interpret his results. These establish certain facts, such as the existence of a retardational compliance J_r at equilibrium, which are independent of any model which one may care to employ. In fact a model is useless if it does not first accept physically determined quantities which are beyond dispute.

ON CLUSTER EXPANSIONS AND AUTOCORRELATION FUNCTIONS: FURTHER REMARKS ON THE DIVERGENCE OF THE DENSITY EXPANSION OF TRANSPORT COEFFICIENTS

MANUEL G. VELARDE

Departamento de Física, Universidad Autónoma de Madrid, Canto Blanco (Madrid), Spain

Abstract. We discuss in simple physical terms still controversial aspects of the divergence in the density expansion of transport coefficients, in particular in the weak coupling case.

We clarify the distinction to be made between finite perturbation approximation and soft interaction. Contrary to a statement given in the literature we show that the presence of the logarithmic divergence does not crucially depend on the interaction potential. In particular, a divergent result is still present for soft repulsive potentials in classical as well as in quantum dynamics.

In finite perturbation approximation however, a clear distinction appears between classical and quantum dynamics. In this latter case the analysis leads to a logarithmic divergence whereas no such divergence is present for classical dynamics.

Résumé. Nous discutons, en termes physiques simples, le problème, sujet encore à controverse, de la divergence logarithmique du développement des coefficients de transport en puissances de la densité.

Nous mettons en évidence le différence qui existe entre l'approximation du *couplage faible* en calcul de perturbation et le calcul (exact) pour un *potentiel faible*. Nous montrons que l'existence la divergence logarithmique ne dépend pas de façon cruciale de la forme du potentiel (répulsif) d'interaction.

En particulier, la divergence existe pour des potentiels (répulsifs) faibles aussi bien en dynamique classique qu'en dynamique quantique.

Cependant, dans l'approximation du *couplage faible*, en calcul de perturbation, une nette différente apparaît entre les dynamiques classique et quantique. Dans ce dernier cas, l'analyse conduit à la divergence logarithmique alors que ce terme n'apparaît pas en dynamique classique.

In building a kinetic theory, one of the aims for a long while was to give a development of the transport coefficients as a power series in the density, ϱ. This kind of approach had already been quite successful in dealing with equilibrium properties. Since it became clear that, on the basis on Onsager's regression hypothesis, the transport coefficients could be expressed in terms of equilibrium averages of correlation functions [1], a formal expansion of transport coefficients in a power series in the density came within reach of the theory. Amongst others, Zwanzig [2] gave a simple scheme by which the coefficients of the power series could be calculated. When it was also shown that these expressions, starting from the Kubo formulae [1], formally agreed [3, 4] with those obtained from different kinetic theories (through a suitably generalized Chapman-Enskog procedure), the main issue of every formalism seemed to be the evaluation of the complicated expressions for particular models.

However, starting with the work of Weinstock [5], Kawasaki and Oppenheim [6], a number of authors [7] began to question the existence of such a power series. It is nowadays established, on fairly firm ground, that the formal power density expansion of transport coefficients diverges logarithmically for classical gases; these divergences occur for the three-body collision term in two dimensions and for the four-body collision in three dimensions.

For quantum gases, a similar singularity was indicated [8, 9], although this conclusion has not been generally agreed to as of yet. It has, in fact, been argued in various papers that this singular behavior is lost, both in classical [10] and in quantum systems [11, 12, 14] when the dynamics of the n-body collisions ($n=3, 4$) intervening in the calculation of transport coefficients is described by finite-order perturbation calculus in the coupling constant between the particles of the system: a weak-coupling model would thus be useless in the evaluation of the non-analytic behavior of transport coefficients in dense gases.

In this lecture we put forward a few remarks that will hopefully help to clarify some aspects of the problem. The main difficulty in discussing the present matter is that it generally involves a complicated mathematical formalism which hides the underlying physical picture. However, without going into any deep analysis of formalism, we argue that the belief that the presence of a divergence depends crucially on the interaction potential (in particular, a finite result should prevail for soft repulsive potentials) lacks relevant foundation.

To fix ideas, let us just recall the formal Kubo expression of a self-diffusion coefficient in a Lorentz gas. We take a system, enclosed in a volume Ω of a d-dimensional space ($d=2, 3$), consisting of one light particle (with mass m) and N fixed centers (with mass $M \gg m$). For simplicity, we neglect the static interactions between these centers and we consider their coordinates $r_1, ..., r_N$ as classical variables randomly distributed over the whole volume. These assumptions somewhat simplify the formulation of the problem but lead to no loss of generality in the final conclusion.

The Hamiltonian of the Lorentz gas model considered here is:

$$H = H_0 + \lambda V \tag{1}$$

$$H_0 = p^2/2m \tag{2}$$

$$\lambda V = \lambda \sum_{j=i}^{N} V(|\mathbf{r} - \mathbf{r}_j|), \tag{3}$$

where $V(|\mathbf{r}-\mathbf{r}_j|)$ denotes the interaction between the light particle located at point $\mathbf{r}$ and the fixed center j located at $\mathbf{r}_j$; in (3), this latter variable appears as a parameter and not as a canonical variable.

The autocorrelation function formula for the self-diffusion coefficient reads in the quantum case [1]

$$D = \lim_{T_1 \to \infty} \lim_{\substack{\Omega \to 0 \ N \to \infty \\ (N\Omega) = \varrho}} \frac{1}{kTd} \int_0^{T_1} dt \int d^{dN}r \times \text{Tr} \tfrac{1}{2}\hat{\mathbf{p}}(t)[\hat{\mathbf{p}}(0)\varrho^{eq} + \varrho^{eq}\hat{\mathbf{p}}(0)]. \tag{4}$$

Here ϱ^{eq} is the equilibrium distribution* of the system for a given configuration of the

---

* Not to be confused with the earlier introduced ϱ which refers to the mass density.

heavy centers:

$$\varrho^{eq} = \frac{\exp{-\beta[H_0 + \lambda V]}}{\int d^{dN}r \, \text{Tr} \exp{-\beta[H_0 + \lambda V]}} \tag{5}$$

and the symbol Tr indicates the trace over a complete set of states for the light particle (to be precise this was done in Reference 8 in the free-particle representation). Finally $\hat{\mathbf{p}}(t)$ is the velocity operator for the light particle in the Heisenberg picture:

$$\hat{\mathbf{p}}(t) = \exp\left(\frac{itH}{\hbar}\right) \hat{\mathbf{p}} \exp\left(\frac{-itH}{\hbar}\right) \tag{6}$$

One can rewrite (4) in a form which stresses the analogy with the classical case; we have

$$D = \lim_{T_1 \to \infty} \lim_{\substack{\Omega \to 0 \, N \to \infty \\ N/\Omega = \varrho}} \frac{(2\pi\hbar)^d}{(dT)\,d\Omega} \int_0^1 dt \int dN \, dr \, \text{Tr} \, \hat{\mathbf{p}} \delta\varrho(t) \tag{7}$$

with

$$\delta\varrho(t) = \exp(-i\mathcal{H}t)\,\delta\varrho(0) \tag{8}$$

In this latter formula, $\mathcal{H}$ denotes the von Neumann operator:

$$\mathcal{H} = \frac{1}{\hbar}[H, \ ]_- \tag{9}$$

and reduces to the Liouville operator in the classical limit, while we have defined $\delta\varrho(0)$ by

$$\delta\varrho(0) = \frac{\Omega}{2(2\pi\hbar)^d} [\hat{\mathbf{p}}\varrho^{eq} + \varrho^{eq}\hat{\mathbf{p}}]. \tag{10}$$

The formal density expansion of Equation (7) is carried out in Reference 8, and the reader is there referred for details. We shall not need them here as, because of the limitation of space, we should concentrate our interest on the conflicting three-body recollision event, leading classically to divergence. Indeed we shall reduce formalism to a minimum (invoking only the binary collision expansion of the triple collision, which is common to all existing works), and what we say seems valid to us no matter what formalism actually available is used. We also limit ourselves to the simple model of a 2d Lorentz gas; finally, we only discuss the existence of the divergence, not its removal [13]

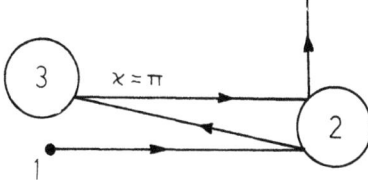

Fig. 1. The divergent triple collision.

Our notation will be the same as in References 8 and 14. The simplest and most convincing argument establishing the divergence is, without doubt, the geometrical analysis presented, for example, in the introduction of Reference 8: one simply evaluates on a dimensional basis the cross-section in which the scattered particle successively collides with the scattering centers 2, 3 and 2 again (in Figure 1, the circles represent the range of the interactions, not necessarily hard discs). This analysis is clearly independent of the details of the potential law and immediately leads to an expression for the first density correction involving a factor (see Reference 8, Equation 3.30)*

$$A(p) \propto \lim_{\delta \to 0} (lg K_0/\delta) \, \sigma(\pi, \varepsilon_p) \tag{11}$$

to be averaged, together with other p-dependent factors, over the Maxwellian distribution. In Reference 8, while K_0 represents an uninteresting high wave-number cut-off, the factor δ can be interpreted arbitrarily as a lower limiting wave-number K_{min} or as a limiting reduced frequency ε/p (=frequency). As transport coefficients have to be computed for $K_{min}=0$ and $\varepsilon=0$, the choice is irrelevant: the *divergence is present at the crucial condition that head-on collisions* ($\chi \approx \pi$) *have a finite cross-section*. This result, valid both in quantum and classical mechanics, is more precise than simply saying that the existence of the divergence depends on the potential.

In considering the case of weak potentials [$\sigma(\pi, p)$ depends, of course, on the coupling parameter], we see from (11) that a correct description of large angle scattering is needed, whatever λ. In the *classical* case, this is not feasible by a finite order Born approximation [15]. Consider, for example, a repulsive monotonous potential, like $\lambda V(r) = \lambda \exp -kr$; it is readily shown from the standard formula [15] that for $p^2/2m < \lambda$, a finite $\chi \approx \pi$ cross-section obtains [$V(0) \equiv 1$]:

$$\sigma(\pi, p) = \left[\frac{db}{d\chi}\right]_{\chi = \pi} = \left[2 \int_{r_{min}}^{\infty} dr/r^2 \sqrt{1 - 2(\lambda m/p^2) V(r)}\right]^{-1} \tag{12}$$

which is generally non-analytic in $m\lambda/p^2$.

It is quite true that for $p^2/2m > \lambda$, the particles will pass 'above' the repulsive potential with only weak deflections, leading to $\sigma(\pi, p) \approx 0$. Thus, for λ small, the values of p from which one gets an infinite A are limited to a small region in momentum space of order $\int_{p^2/2m < \lambda} d^2p \propto 2m\lambda$. Yet, we have to face the following alternative:

(a) Either we go *strictly* to the $\lambda=0$ limit, with all other parameters fixed, in which case we get (trivially) infinite transport coefficients.

(b) Or we consider λ small but *finite*. Then it is of no help to know, as shown in (14), that the first few terms in a formal Born expansion are finite, if we find, in the complete result, a λ-dependent *infinite* contribution (whatever this λ-dependence can be).

In the binary collision expansion formalism, the difficulty in the Born approximation

---

* Let us point out a very unfortunate misprint in Reference 8: the log factor is missing in Equation (3.30).

arises in the following way. Consider the quantity (see Reference 14 – Equation (5.1) and Figure 1 of this lecture):

$$\langle 0| G_0 T_{12} \{G_0 T_{13} G_0\} T_{12} G_0 |0\rangle, \tag{13}$$

where the bracket essentially represents the quantity $A(p)$. When evaluating T_{13} in the Born approximation, one gets integrals of the type (see Reference 14 – Equation (5.3)):

$$\int_0^\infty K\, dK \int_{-\pi}^{+\pi} d\phi\, [\varepsilon + iKp \cos \phi]^{-2} \tag{14}$$

and finite derivatives thereof, which are all *finite* (as already shown by Stecki [10]). On the contrary, the (Boltzmann-like) completely resumed T_{13} gives an integral of the type

$$\int_0^\infty K\, dK \int_{-\pi}^{+\pi} d\chi \sigma(\chi, p) \int_{-\pi}^{+\pi} d\phi\, [(\varepsilon + ikp \cos \phi)(\varepsilon + ikp \cos[\chi + \phi])]^{-1} \tag{15}$$

this leads to the divergent result (11), as shown, for example, by Oppenheim and Kawasaki [6]. Expanding, however, the bracket in (15) around $\chi=0$ is obviously meaningless when dealing with the region $\chi \approx \pi$.

The weak coupling theory does not fail in the *quantum* case; it is known that there, the Born approximation also applies to large scattering angles (provided other conditions are met [9, 15]). Yet, one still needs some care in evaluating (13). Indeed, if one takes each T_{12} to first order in λ only, one obtains a supplementary k^2 factor in (15), which renders the integral convergent: the λ^4 contribution to (13) is thus finite (as it is in the classical case); as soon, however, as T_{12} is treated to second order in λ, one gets a λ^6 diverging contribution to (13). This again is in complete agreement with our geometrical picture: taking T_{12} to first order in λ amounts to describing the interaction between the scattered particles and the scattering center 2 as a *virtual* transition which disappears when 1 goes out of the range of interaction of 2 (i.e., for $K=0$). On the contrary, our geometrical picture (see Figure 1) describes a *real* completed collision where 1 is infinitely apart from the range of 2: this is described in first approximation by the λ^2 Born approximation to T_{12}. This point was overlooked in the earlier work of Fujita [11] and Williams and Weinstock [12]. Here again, the smallness of λ is of no help, of course, for removing the infinite logarithmic coefficient [16].

The geometric argument also readily indicates that higher order terms in the binary collision expansion cannot remove the divergence (as confirmed in Reference 17).

Acknowledgements

This work was done in collaboration with Prof. P. Résibois of the Université Libre de Bruxelles (Belgium).

The travel expenses for attending this Conference were supported by G.I.F.T. (Spain).

References

1. Mori, H., Oppenheim, I., and Ross, J.: in J. de Boer and G. E. Uhlenbeck (eds.), *Studies in Statistical Mechanics* **1**, North Holland, Amsterdam, 1961.
2. Zwanzig, R.: *Phys. Rev.* **129**, 486 (1963).
3. Ernst, M. H., Dorfman, J. R., and Cohen, E. G. D.: *Physica* **31**, 493 (1965); Kawasaki, K. and Oppenheim, I.: *Phys. Rev.* **136A**, 1519 (1964).
4. Résibois, P.: *J. Chem. Phys.* **47**, 2979 (1964).
5. Weinstock, J.: *Phys. Rev.* **132**, 454 (1963); **140**, A460 (1965); *Phys. Rev. Letters* **17**, 130 (1966).
6. In fact, an improper reference is their paper in *Phys. Rev.* **139A**, 1763 (1965), as this paper contains an apparent trivial (conceptual) error. The error lays in the exchange of the $\varepsilon \to 0$ and the integration over k-momentum. As starting from the standard definitions in all 'non rigorous' available kinetic theories (up until 1965), one of the *rules* is to take the $\varepsilon \to 0$ limit last, as the K-integration is, in fact, the formal expansion of the identity in (in principle) the suitable functional space and the ε variable appears through a Laplace transform of the operators involved. To point out this mistake was indeed one of the purposes of Stecki's letter [10]. The proper reference should be Kawasaki, K. and Oppenheim, I.: in A. Bak (ed.), *Statistical Mechanics*, Benjamin, New York, 1967, p. 313. The proper calculation was also done independently by P. Resibois (unpublished 1965 results). Concerning this point see, however, the discussion for the Maxwell model of Pomeau, Y.: *Phys. Rev.* **A3**, 1174 (1971).
7. See for a review, Ernst, M., Haines, L., and Dorfman, J. R.: *Rev. Mod. Phys.* **41**, 296 (1969).
8. Résibois, P. and Velarde, M. G.: *Physica* **51**, 541 (1971). See also, M. G. Velarde and P. Résibois, *J. Chem. Phys.* **59**, 2166 (1973) where a shorter account of this lecture is given. A preliminary announcement was given by Velarde, M. G. and Résibois, P.: *Bull. Am. Phys. Soc.* **15**, 34 (1970).
9. Van Zuylen, H. J.: 'Electrical Conductivity of Imperfect Crystals (A Generalized Boltzmann Equation)', Ph. D. Dissertation, University of Amsterdam, 1973, Chap. II.
10. Stecki, J.: *Phys. Letters* **19**, 123 (1965).
11. S. Fujita, *Phys. Letters* **22**, 42 (1966); *Proc. Nat. Acad. Sci. (U.S.A.)* **56**, 794 (1966).
12. Williams, R. H. and Weinstock, J.: *Phys. Rev.* **169**, 196 (1968).
13. However, for a constant two body cross-section (say a formal s-wave approximation), a formal resumation of ring diagrams (the most divergent ones in the classical case) leads in the quantum case to a convergent k integral in the $\varepsilon \to 0$ limit and provides a $\log \varrho$ contribution to D. See also Albers, J. and Oppenheim, I.: *Physica* **59**, 187 (1972).
14. Eu, B. C.: *J. Chem. Phys.* **56**, 4613 (1971); **58**, 1352 (1973).
15. E. G., Bohm, D.: *Quantum Theory*, Prentice Hall, Englewood Cliffs, New Jersey, 1951.
16. It is perhaps worthwhile to still emphasize, and probably is not a trivial remark as it has indeed been overlooked by several authors, that with strong-coupling (Born series summed) 3 'collision-T-matrix factors' ($T_{12} T_{13} T_{12}$) appear in the cross-section of the recollision event. In finite perturbation calculation, this formally should merely mean replacing each T_{ij} factor by its $|V_{ij}|^2$ approximation (i.e., the square of the Fourier transform of the interaction potential). Although worded differently, this is stated in the main text. Now the 3-body recollision event computed by both Fujita and Eu does not involve 3 collision factors, but rather two. In the case of Williams and Weinstock for the 4-body problem in 3 dimensions, the argument refers to 4-collision factors instead of 3. Without the argument of the main text, this indeed is not a proof of anything, but you might easily agree with us that it is something that catches the eye of an attentive reader (no need to be a highly-trained physicist!).
17. Van Leeuwen, J. M. J. and Weijland, A.: *Physica* **36**, 457 (1967); Weijland, A. and Van Leeuwen, J. M. J.: *Physica* **38**, 35 (1968).

EVIDENCE OF COOPERATIVE PHENOMENA IN THE ASYMPTOTIC TIME BEHAVIOR OF CORRELATION FUNCTIONS: SELF-DIFFUSION COEFFICIENT OF METHANE*

NARINDER K. AILAWADI**

University of Rhode Island, Kingston, R. I., U.S.A.

Abstract. A long time tail of the velocity auto-correlation function (vcf) as proposed by Ernst et al. is assumed to exist and extend to infinite times. This long time tail is assumed to evolve smoothly from the vcf for shorter times as generated by molecular dynamics. Based on this behavior of the vcf, the self-diffusion coefficient of a hard sphere model of methane is calculated for several densities for the isotherm $T = 298.2$ K and compared with actual experimental data. Three different models are considered for density dependence of the time t_0 where the tail begins to appear.

Résumé. On suppose, après Ernst et al., l'existence dans la fonction d'auto-correlation des vitesses, d'une 'long time tail'; on admet aussi que cette queve est liée de façon continue aux les valuers fournies par la dynamique moléculaire.
Utilisant cette façon de construire la fonction d'autocorrelation des vitesses, on a calculé la constante de self-diffusion pour le méthane considéré comme un système de sphères dures. Le calcul est présenté ici pour plusieurs densités et $T = 298.2$ K, et il est comparés aux valeurs expérimentales. Trois modèles différents sont considéré pour l'influence de la densité sur le temps t_0 quand la 'long time tail' commence à apparaître.

1. Introduction

Recently, a great deal of interest in the long-time behavior of correlation functions has been generated by computer calculations of Alder and Wainwright [1]. They calculated the velocity auto-correlation function for a two- and three-dimensional system of hard spheres. Their main results are as follows:

(1) The asymptotic time-behavior of the velocity auto-correlation function is of the form $t^{-d/2}$ (where d is the dimensionality of the system).

(2) The velocity distribution around a particle is not irrotational, but shows a vortex pattern.

(3) On the basis of a numerical solution of Navier-Stokes equation, they proposed a hydrodynamical explanation of their observations.

This computer calculation led to a number of theoretical investigations on the asymptotic time dependence of various correlation functions by Ernst et al. [2], Dorfman and Cohen [3], Pomeau [4], Ailawadi and Berne [5], Kawasaki [6], Wainwright et al. [7], Zwanzig [8], Widom [9], and Berne [10].

The question now arises whether such long-time tails exist in real systems and how

---

* Initial stage of this work was supported in part by National Science Foundation under grant No. NSF-GP-22881.
** Present address: Institut für Theoretische Physik, Freie Universität Berlin, Berlin, F.R.G.

they will affect the transport coefficients. In particular, the self-diffusion coefficient (which is the time integral of the velocity auto-correlation function) could be enhanced due to the extra contribution from the long time tail of the velocity auto-correlation function. This should be particularly observable at intermediate densities where the velocity auto-correlation function does not go negative and kinematic viscosity shows a minimum as a function of density. Thus, as a function of density, the self-diffusion coefficient is expected to show a maximum.

This has been observed experimentally in the case of methane by Oosting [11] and Oosting and Trappeniers [12]. Computer calculation of Dymond and Alder [13] for a hard sphere model of methane also shows similar behavior. Thus, there is qualitative evidence of the existence of persistence of the vcf.

More recently, this behavior has been analyzed quantitatively by Alder, et al. [14] and by Dymond and Alder [15]. We present yet another attempt to analyze this behavior quantitatively for the following reasons:

(i) It is not clear how molecular dynamics hard sphere results should be extrapolated to compare with actual experimental results.

(ii) Since the magnitude of the hydrodynamic tail depends on density and temperature, by restricting ourselves to one isotherm we concentrate only on the density dependence of the self-diffusion coefficient. This is the quantity of interest in trying to identify the contribution of the tail in real systems.

We take a simple approach in relating molecular dynamics results to the actual system of gaseous methane. From the equation of state data on methane, a hard sphere diameter $\sigma = 3.48$ Å is obtained for the temperature $T = 298.2$ K. So we calculate the contribution of the tail using these parameters for the hard sphere model of methane and compare the self-diffusion coefficient with the experimental data on methane [11, 12].

2. Calculations

We assume that the velocity auto-correlation function is given by (in three dimensions)

$$\langle \mathbf{v}(0)\cdot\mathbf{v}(t)\rangle = \begin{cases} \langle \mathbf{v}(0)\cdot\mathbf{v}(t)\rangle_{\text{M.D.}} & t < t_0 \\ 2\frac{kT}{mn}\frac{1}{[4\pi(v+D)t]^{3/2}} & t > t_0 \end{cases} \quad (1)$$

where $\langle \mathbf{v}(0)\cdot\mathbf{v}(t)\rangle_{\text{M.D}}$ is that obtained from molecular dynamics calculation of Alder et al. [14] for 108 hard-sphere system and the second term is the asymptotic time limit of the vcf obtained from the hydrodynamic theory of Ernst et al. [2]. Here v is the kinematic shear viscosity ($=\eta/mn$), D is the self-diffusion coefficient, m is the mass of a single molecule, n is the number density, and k is the Boltzmann constant.

It is assumed that the long time tail of the vcf tags on continuously to the vcf for shorter times generated by molecular dynamics [16] and appears at a time t_0 measured in mean collision time t_c (Enskog value) where [17]

$$1/t_c = 4n\sigma^2 \left(\frac{\pi kT}{m}\right)^{1/2} g(\sigma) \tag{2}$$

and extends to infinite time.

To calculate the second term, the long time tail of the vcf, knowledge of both η and D is needed. We estimate the contribution of the hydrodynamic tail by substituting η and D for 108 particles computed from computer calculation of Alder et al.[14]. The radial distribution function $g(\sigma)$ at contact required for calculating the Enskog values of shear viscosity η_E and self-diffusion coefficient D_E is obtained from the data of Dymond and Alder [18].

The self-diffusion coefficient D

$$D = \tfrac{1}{3} \int_0^\infty dt \langle \mathbf{v}(0)\cdot\mathbf{v}(t)\rangle \tag{3}$$

is calculated from Equation (1)

$$D = D_{\text{M.D.}} + \frac{4kT}{3mn} \frac{1}{[4\pi(\nu_{\text{M.D.}} + D_{\text{M.D.}})]^{3/2}} \frac{1}{t_0^{1/2}}, \tag{4}$$

where

$$D_{\text{M.D.}} = \tfrac{1}{3} \int_0^{t_0} dt \langle \mathbf{v}(0)\cdot\mathbf{v}(t)\rangle; \tag{5}$$

t_0 is the time up to which molecular dynamics calculation has been done. It is also the time at which the long time tail first appears [19]. Measured in units of mean collision time t_c, Equation (4) can be written as

$$\frac{D}{D_E} = \left(\frac{D}{D_E}\right)_{\text{M.D.}} + \frac{4kT}{3mn} \frac{1}{[4\pi(\nu_{\text{M.D.}} + D_{\text{M.D.}})]^{3/2}} \frac{1}{t_c^{1/2}} \frac{1}{D_E} \frac{1}{(t_0/t_c)^{1/2}}. \tag{6}$$

Equation (6) is used to compute D/D_E which is compared with the experimental results on methane [11]. In order to compute the magnitude of the contribution from the long time tail (second term in Equation (6)), we need t_0/t_c.

3. Assumptions for t_0/t_c

Since the actual magnitude of t_0/t_c and its dependence on number density n are not known, we adopt an experimentalist's point of view. A density of $V_0/V = 0.20$ is chosen, V_0 being the close-packed volume, and the value of t_0/t_c is fixed such that D/D_E computed from Equation (6) matches the experimental value. This procedure yields $t_0/t_c = 9.26$ for $V_0/V = 0.20$. Next, we choose three different models for the density dependence of t_0/t_c.

(i) Assume t_0/t_c is independent of density. This is the simplest assumption we can make. Accordingly, $t_0/t_c = 9.26$ is used for computation of the tail for all densities.

(ii) The second model for t_0/t_c is based on the physical idea that the long time tail in the vcf is really due to the double vortex flow generated when the particle moves in the medium. The characteristic time for the vortex to decay is $t \sim 1/v^*$. If this characteristic time is related to the time t_0 when the long time tail in vcf first appears, then $t_0 \sim 1/v$. Furthermore, from the molecular dynamics calculations [14], it is known that $\eta D =$ constant, independent of density for $V_0/V \geqslant 0.2$. This is precisely the density region where, as a function of density, D/D_E is expected to show a maximum. Since the Enskog value of self-diffusion coefficient $D_E \sim 1/ng(\sigma)$, it is reasonable to assume that $v = \eta/mn \sim g(\sigma)$.

$$t_0/t_c \sim ng(\sigma)/g(\sigma) \sim n. \tag{7}$$

Thus, the second model for t_0/t_c is: knowing $t_0/t_c = 9.26$ for $V_0/V = 0.20$, t_0/t_c is scaled according to Equation (7) for all other densities. (iii) The third model is based on the assumption that t_0 is constant independent of density (e.g., measured in sec). Since $t_c \sim 1/ng(\sigma)$,

$$t_0/t_c \sim ng(\sigma). \tag{8}$$

4. Conclusion

We use these models for the density dependence of t_0/t_c to calculate the contribution of the long time tail to the self-diffusion coefficient as given by Equation (6). This contribution is added to the $(D/D_E)_{M.D.}$ for 108 particles calculated by molecular dynamics [14]. We list in Table I the values of t_0/t_c for different densities V_0/V for all the three models. In Figure 1, the computed D/D_E obtained by using Equation (6) for all the three models is plotted as a function of reduced density V_0/V and compared with the experimental results on methane [12] as well as $(D/D_E)_{M.D.}$ for a hard sphere system. For the purpose of comparison, the experimental data has been reduced to D_{exp}/D_E for several reduced densities V_0/V by choosing $\sigma = 3.48$ Å; $g(\sigma)$ at contact is obtained from the data of Dymond and Alder [18].

Alder et al. [14] have also done a similar calculation where they try to extend $(D/D_E)_{M.D.}$ to infinite systems on the basis of hydrodynamic theory. Their results listed in the last column of Table I in their paper are also shown in Figure 1. In their calculation, Alder et al. [14] have used the correction $(1 - 2/N)$ to account for number dependence of $(D/D_E)_{M.D.}$ as well as the hydrodynamic tail. In this way, the calculated D/D_E is larger than the experimental D/D_E for most reduced densities V_0/V. In a more recent paper, Dymond and Alder [15] have tried to extrapolate the infinite temperature value of D/D_E from the experimental data. Our procedure, on the other hand, is rather straightforward; once the hard sphere diameter of methane molecule is determined from the PVT data (critical density $V_0/V \approx 0.20$) we assume that methane can be approximated as a system of hard spheres at a fixed temperature and this diameter

---

* Dimensionally, we should have $t \sim l^2/v$ where l is some characteristic length. We simply neglect the l dependence in the present calculation.

TABLE I

Values of the parameter t_0/t_c as a function of density for different models: (i) t_0/t_c = constant, (ii) $t_0/t_c \sim n$, the number density, and (iii) $t_0/t_c \sim ng(\sigma)$; this is the case when t_0 is constant measured in ordinary units (s). One value of $t_0/t_c = 9.26$ is adjusted such that for $V_0/V = 0.20$, both the computed and experimental values of D/D_E agree

V_0/V	t_0/t_c (i)	(ii)	(iii)
0.01	9.26	0.46	0.31
0.05	9.26	2.31	1.69
0.10	9.26	4.63	3.74
0.20	9.26	9.26	9.26
0.25	9.26	11.57	12.83
0.33	9.26	15.43	21.05
0.40	9.26	18.42	29.94
0.50	9.26	23.14	50.15
0.55	9.26	25.72	66.08
0.62	9.26	28.93	95.90

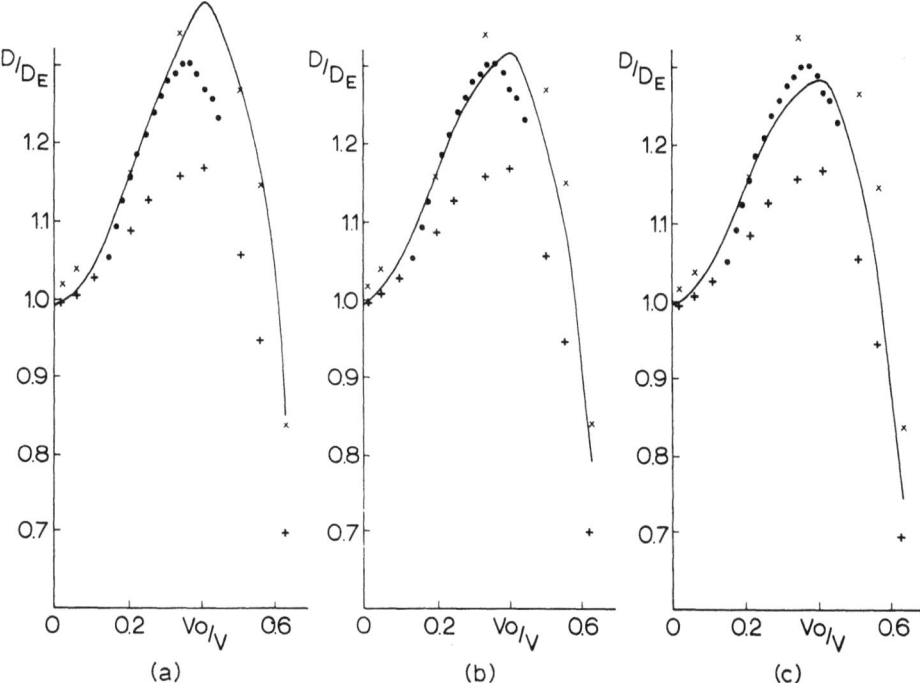

Fig. 1. The ratio of the self-diffusion coefficient D relative to the Enskog value D_E as a function of density, where the volume is measured relative to the close-packed volume V_0. ──── Computed D/D_E from Equation (6) using three different models for the density dependence of t_0/t_c. ● Experimental data on methane from Reference 11. × From Reference 14 (last column, Table I). This calculation is based on $(1 - (2/N))$ number dependent correction as well as hydrodynamic model of the asymptotic time behavior of the velocity correlation function. + $(D/D_E)_{M.D.}$ also from Reference 14 (Table I). (a) t_0/t_c = constant [model (i)]. (b) $t_0/t_c \sim n$ [model (ii)]. (c) $t_0/t_c \sim ng(\sigma)$ [model (iii)].

is used for our computation. Furthermore, the necessity of $(1-(2/N))$ correction for number dependence is not entirely clear.

From molecular dynamics calculations [14], t_0 the time where the tail first appears (measured in units of t_c) depends on density. This dependence on density is not known; accordingly, we use models for the density dependence of (t_0/t_c). From Figure 1 both models (ii) and (iii) agree reasonably with the experimental data.

It is worth pointing out that even without the addition of the tail, computer results $(D/D_E)_{M.D.}$ show a maximum at the same density as the computed D/D_E which includes the contribution of the tail. This points to the possibility [19] that some contribution of the tail has already been taken into account in calculating $(D/D_E)_{M.D.}$. This is particularly evident at low densities in models (ii) and (iii) for t_0/t_c. An inspection of Table I shows that for these models, t_0 at low densities ranges from a fraction of t_c to a few t_c (t_c being the average time between collisions). It is hard to imagine that at these short times, the hydrodynamic tail is fully developed. However, the contribution of the tail at these densities is so small as to make such errors insignificant.

Also, a series of contributions of the form $t^{-7/4}$, $t^{-15/8}$, ..., t^{-2} has been predicted theoretically by Pomeau [4]. It is reasonable to assume that the most important contribution of these intermediate terms (non-hydrodynamic part) of the vcf has already been taken into account in $(D/D_E)_{M.D.}$ since it does include deviations from the Enskog results. Thus, the essential contribution yet to be accounted for is the hydrodynamic tail of the form $t^{-3/2}$. Another way to look at our calculation is: since t_0/t_c in some cases (particularly in models (i) and (ii) for t_0/t_c) is quite low, this might lead one to estimate the contribution of these intermediate terms by comparison with the experimental results [11]. Considering the uncertainties involved in the present calculation, it is unfair, however, in our view, to comment on the existence or otherwise of these intermediate contributions. Finally, one might suggest that the 'non-hydrodynamic part' of the velocity autocorrelation function could be responsible for a maximum in (D/D_E) as a function of density.

Our paper should, in the end, be only regarded as an attempt to answer the question: do the hydrodynamical tails in correlation functions exist in real systems, and if so, do they affect the transport coefficients?

Acknowledgements

The author is indebted to Prof. Bruce J. Berne, who got him interested in this problem and for his valuable suggestions during the course of this work. He also wants to acknowledge several beneficial discussions and correspondence with Professors E. G. D. Cohen, B. J. Alder, and Dr T. E. Wainwright. Thanks are also due to Gerald Haines for his assistance in computation.

References

1. Alder, B. J. and Wainwright, T. E.: *Phys. Rev. Letters* **18**, 988 (1969); *J. Phys. Soc. Suppl.* **26**, 267 (1968); and *Phys. Rev.* **A1**, 18 (1970).

2. Ernst, M. H., Hauge, E. H., and van Leeuwen, J. M. J.: *Phys. Rev. Letters* **25**, 1254 (1970); *Phys. Rev.* **A4**, 2055 (1971).
3. Dorfman, J. R. and Cohen, E. G. D.: *Phys. Rev. Letters* **25**, 1257 (1970); *Phys. Rev.* **A6**, 776 (1972).
4. Pomeau, Y.: *Phys. Rev.* **A3**, 1174 (1971); *Phys. Rev.* **A5**, 2569 (1972); *Phys. Rev.* **A7**, 1134 (1973).
5. Ailawadi, N. K. and Berne, B. J.: *J. Chem. Phys.* **54**, 3569 (1971); IUPAP International Conference on Statistical Mechanics, University of Chicago, March 29–April 2, 1971; see also Ailawadi, N. K. and Harris, S.: *J. Chem. Phys.* **56**, 5783 (1972).
6. Kawasaki, K.: *Phys. Letters* **32A**, 379 (1970); **34A**, 12 (1971).
7. Wainwright, T. E., Alder, B. J., and Gass, D. M.: *Phys. Rev.* **A4**, 233 (1971).
8. Zwanzig, R. in S. A. Rice, K. F. Reed, and J. C. Light (eds.), *Statistical Mechanics – New Concepts, New Problems, New Applications* (Proceedings of the sixth IUPAP Conference on Statistical Mechanics), University of Chicago Press, 1972, p. 241.
9. Widom, A.: *Phys. Rev.* **A3**, 1394 (1971).
10. Berne, B. J.: *J. Chem. Phys.* **56**, 2164 (1972).
11. Oosting, P. H.: doctoral dissertation, Amsterdam, 1968.
12. Oosting, P. H. and Trappeniers, N. J.: *Physica* **51**, 418 (1971).
13. Dymond, J. H. and Alder, B. J.: *J. Chem. Phys.* **52**, 923 (1970).
14. Alder, B. J., Gass, D. M., and Wainwright, T. E.: *J. Chem. Phys.* **53**, 3813 (1971).
15. Dymond, J. H. and Alder, B. J.: *Ber. Bunsenges. physik. Chem.* **75**, 394 (1971).
16. That two parts of the velocity auto-correlation function given in Equation (1) join continuously has been shown by Alder and Wainwright in the study of the corresponding two-dimensional problem [Alder B. J. and Wainwright, T. E.: *Phys. Rev.* **A1**, 18 (1970)]. However, this has not been demonstrated in the three-dimensional problem.
17. Chapman, S. and Cowling, T. G.: *The Mathematical Theory of Non-Uniform Gases*, Cambridge University Press, New York, 1970, Ch. 16.
18. Dymond, J. H. and Alder, B. J.: *J. Chem. Phys.* **45**, 2061 (1966).
19. From molecular dynamics calculation (See Reference 14) the value of t_0 is not entirely clear. Hence, t_0 is to be considered as an adjustable parameter here. It is quite possible that for some (particularly lower) densities, a part of vcf might have been considered twice, once in $(D/D_E)_{M.D.}$ and once in the explicit contribution from hydrodynamic tail.

FULLY RENORMALIZED KINETIC THEORY OF THERMAL FLUCTUATIONS IN LIQUIDS

GENE F. MAZENKO and SIDNEY YIP

Dept. of Nuclear Engineering, Massachusetts Institute of Technology and Dept. of Physics, Harvard University, Cambridge, Mass. 02138, U.S.A.

Abstract. We describe a new method of analyzing dynamical properties of classical liquids. The method is developed in terms of phase-space time correlation functions which are calculated using a generalized kinetic equation. We develop exact and useful expressions for the kernel or memory function that appears in this kinetic equation. These exact expressions for the memory functions are written in terms of an effective two-body problem. We show how to develop an approximation scheme for evaluating this memory function and we explicitly make contact with the terms that give rise to the Boltzmann-Enskog collision integral, the ring terms associated with long-time tails and the mode-mode coupling terms associated with the divergence of transport coefficients near the critical point.

Résumé. Nous présentons dans ce rapport une méthode nouvelle pour analyser les propriétés des liquides classiques. La méthode est développée pour les fonctions de corrélation dans l'espace de phase. Pour calculer ces fonctions de corrélation nous utilisons une équation cinétique généralisée. Aussi, nous développons des expressions précises et practiques pour le noyau intégral ou fonction-mémoire (memory function) de cette équation cinétique. Ces expressions exactes pour les fonctions-mémoire ont été écrites en termes d'un problème effectif à deux corps. Nous montrons comment on peut construire un système d'approximations pour calculer cette fonction mémo. Nous entrons en contact explicitement avec les termes qui originent l'intégrale de collision de Boltzmann-Enskog, avec les termes annulaires (ring terms) associés aux variations seculaires et au couplage des modes (mode-mode couplings), et aussi avec les termes associés aux divergences des coefficients de transport à proximité du point critique.

1. Introduction

In this paper we describe a new method of analyzing dynamical properties of classical liquids. Our primary interest lies in the study of those time correlation functions which determine transport properties of liquids and which can be directly measured by neutron and light scattering or obtained in computer molecular dynamics experiments. While the discussion here will deal with simple liquids (such as argon), the techniques developed should be applicable also to more complicated systems (binary mixtures, quantum liquids, etc.)

Essentially all dynamical processes of interest can be discussed in terms of frequency and wavenumber dependent correlation functions. For definiteness one can consider the Fourier transform of the density-density correlation function $S_{nn}(\mathbf{k}, \omega)$. Any attempt to calculate the small wavenumber and frequency behavior of S by direct perturbation expansions will encounter serious difficulties. The main reason is that S shows a resonant structure in the hydrodynamical regime. A similar difficulty was encountered in recent efforts to obtain a density expansion for transport coefficients [1] using methods which involve the analysis of the dynamics of isolated clusters of interacting particles. The appearance of infinite coefficients in the density power series indicates that medium effects, or the influence of particles outside the cluster, cannot be ignored.

These 'medium' or self-energy type effects which are characteristic of resonant behavior are familiar from renormalization theory, and it is known that infinities can be removed by working with effective, not bare, interactions. This idea of an effective interaction is closely related to the idea of an effective few-body problem. Such a reduction is essential for progress. The best example is given by the Boltzmann equation which reduces the N-body problem to an effective 2-body problem. One would therefore like to reformulate the N-body problem in terms of a 'few' bodies that interact via effective or renormalized interactions. Our formalism consists essentially of two such renormalizations. The first renormalization involves development of an effective one-body problem.

One starts with the introduction of an equation of motion satisfied by a 'one-body' phase-space time correlation function [2]. This equation defines the memory function ϕ. If ϕ is set equal to zero, one finds a free-particle equation where particles move in straight-line trajectories at constant velocities. The memory function therefore represents the effects of the other $(N-1)$ particles on a given particle; in other words, it renormalizes the single-particle motions in the system. It is also useful to regard ϕ as a single-particle 'potential' which in general is non-local in space and time.

In our terminology this equation of motion constitutes a renormalized one-body problem. To develop a practical scheme for calculating ϕ we first obtain a relation between ϕ and a higher order phase-space density correlation function which satisfies an equation of motion in an effective two-body space. This equation contains an appropriate memory function or effective two-body potential and is therefore the result of a second renormalization.

Within this formalism one naturally obtains as the lowest order approximation the Boltzmann-Enskog theory which is known to give good results for hard sphere systems up to quite high densities [3]. In the limit of small density, frequency, and wavenumber, this part of the memory function reduces to the collision operator in the linearized Boltzmann equation [4]. The terms which appear in our full expression for the memory function, in addition to a Boltzmann-Enskog term, include mode-mode coupling terms [5], ring terms [6] which correspond to recollision processes, and terms which have not been studied previously.

An important feature in our theory is the treatment of static or thermodynamic properties of the system. The static effects enter into the calculation through the initial value of the time correlation functions, but more importantly they also appear in a proper calculation of ϕ. In our analysis the intrinsically static effects are separated from the dynamical effects in a natural way. This separation is particularly useful in view of recent progress in calculating static correlation functions both near [7] and away from [8] the critical point.

2. Formulation of the Memory Function Approach

We begin by considering a classical fluid system of volume Ω which has N particles of mass m. The time dependent correlation function of interest is

$$C(12; t) = \langle \delta f(2) e^{iLt} \delta f(1) \rangle, \tag{1}$$

where $\delta f(1)$ is the deviation of the phase-space density

$$f(1) = \sum_{j=1}^{N} \delta(1 - q_j) \tag{2}$$

from its equilibrium value $\langle f(1) \rangle = n f_0(p_1)$. We denote external phase coordinates $(\mathbf{r}_1, \mathbf{p}_1)$ by 1, the coordinates of molecule j $(\mathbf{R}_j, \mathbf{P}_j)$ by q_j, $n = N/\Omega$ is the fluid density, $f_0(p)$ the normalized Maxwellian distribution, L the Liouville operator, and $\langle \ \rangle$ indicates an average over an equilibrium ensemble. We will deal mostly with the Laplace transform of a time correlation function

$$C(12) = -i \int_0^{+\infty} dt\, e^{+izt} C(12; t) \tag{3}$$

and its Fourier transform

$$C(\mathbf{k}, \mathbf{p}_1 \mathbf{p}_2, z) = \frac{1}{\Omega} \int d^3 r_1\, d^3 r_2\, e^{+i\mathbf{k}\cdot(\mathbf{r}_1 - \mathbf{r}_2)} C(12). \tag{4}$$

We introduce the memory function ϕ as the kernel in the kinetic equation for C,

$$\left(z - \frac{\mathbf{k} \cdot \mathbf{p}}{m}\right) C(\mathbf{k}, \mathbf{p}\mathbf{p}', z) - \int d^3 \bar{p}\, \phi(\mathbf{k}, \mathbf{p}\bar{\mathbf{p}}, z) C(\mathbf{k}, \bar{\mathbf{p}}\mathbf{p}', z) = \tilde{C}(\mathbf{k}, \mathbf{p}\mathbf{p}'), \tag{5}$$

where $\tilde{C}(\mathbf{k}, \mathbf{p}\mathbf{p}')$ is the Fourier transform of Equation (1) at $t = 0$,

$$\tilde{C}(\mathbf{k}, \mathbf{p}\mathbf{p}') = n f_0(p) \delta(\mathbf{p} - \mathbf{p}') + n^2 f_0(p) f_0(p') h(k) \tag{6}$$

and $h(k)$ is the Fourier transform of $[g(r) - 1]$ with $g(r)$ being the static pair distribution function.

There are many formal properties of ϕ related to its symmetries, conservation laws, and asymptotic behavior [9]. We do not discuss them here except to note that in general the memory function can be separated into a time-independent (static) part and a time-dependent (collisional) part,

$$\phi = \phi^{(s)} + \phi^{(c)}. \tag{7}$$

The static part can be evaluated without approximation. The result is

$$\phi^{(s)}(\mathbf{k}, \mathbf{p}\mathbf{p}') = -\frac{\mathbf{k} \cdot \mathbf{p}}{m} C(k) n f_0(p). \tag{8}$$

Where $C(k)$ is the direct correlation function.

The calculation of $\phi^{(c)}$ is the major problem in any attempt to develop a microscopic theory of thermal fluctuation in fluids. Recently one of us [10] have shown that a systematic analysis of $\phi^{(c)}$ leads to the expression

$$\phi^{(c)}(12) \, n f_0(p_2) = - \int d3 \, d\bar{3} \, d4 \, d\bar{4} \, V(1; 3\bar{3}) \, G(3\bar{3}; 4\bar{4}) \, V(4\bar{4}; 2), \quad (9)$$

where V is an effective or renormalized interaction which we will call the 'end-point vertex', and G is a correlation function that describes the dynamical evolution of a two-particle field. (In contrast, $C(12)$ describes the evolution of a one-particle field.) The end-point vertex is a static quantity, so the time-dependence of $\phi^{(c)}$ is entirely governed by G. Equation (9) is an exact result and one can write down explicit expressions for V and G. We refer the interested reader to Reference 10 for details.

In order to obtain a tractable and realistic description of ϕ, we need to analyze the properties of V and G. One finds that V quite naturally separates into two parts, the first, V_s, involves only two particle static correlations and is proportional to $\nabla_r \exp(-\beta v(r))$, where $v(r)$ is the intermolecular potential. For molecules interacting through only short-range forces $V_s(1; 3\bar{3})$ is a function which tends to localize the spatial coordinates $\mathbf{r}_3$ and $\mathbf{r}_{\bar{3}}$. The other part, V_L, involves two- and three-particle correlations and is longer ranged by comparison since it contains a term proportional to $\nabla_r \tilde{g}(r)$, where $g(r) = \tilde{g}(r) \exp(-\beta v(r))$. The function $\tilde{g}(r)$ has a much smoother behavior than $\exp(-\beta v(r))$; it is continuous even for a hard-sphere potential.

The fact that the end-point vertex V has short- and long-ranged parts means that one should look for an approximate expression for $G(3\bar{3}; 4\bar{4})$ valid when the two particles are localized and another approximation for the case when they can be quite far apart. One finds that an appropriate expression for G to be used with V_s is a correlation function G_s which describes the propagation of two particles mostly under the influence of their mutual interactions. The medium effects still enter through static correlation functions, but the explicit effects of interaction between the two particles and the remaining $(N-2)$-particle medium are neglected. Intuitively one can see this is a reasonable approximation because when the two particles are close to each other their mutual interactions should dominate over whatever other interactions that they have with the medium. Conversely the interactions with the medium should become important when the two particles are allowed to be spatially well separated. Using the method of cumulant analysis which is particularly well-suited for analyzing the localization or clustering properties of correlation functions, we find that an appropriate expression for G, when it is combined with V_L, is given by its disconnected part G_D. G_D is the correlation function that describes the independent propagation of the two particles in the medium. Here the mutual interactions between the two particles are ignored, and one emphasizes the effects of interaction between each of the two particles and the remaining $(N-2)$ particles. One can show that G_D is simply related to the product of two correlation functions defined by Equation (1).

$$\tilde{G}_D(1\bar{1}; 3\bar{3}) \, G_D(3\bar{3}; 4\bar{4}) \, \tilde{G}_D(4\bar{4}; 2\bar{2}) =$$
$$-i \int_0^{+\infty} dt \, e^{+izt} \{C(12; t) \, C(\bar{1}\bar{2}; t) + C(1\bar{2}; t) \, C(\bar{1}2; t)\}, \quad (10)$$

where
$$\tilde{G}_D(1\bar{1}; 2\bar{2}) = \tilde{C}(12)\,\tilde{C}(\bar{1}\bar{2}) + \tilde{C}(1\bar{2})\,\tilde{C}(\bar{1}2).$$

The approximations just described qualitatively are the essential ingredients of a systematic iteration procedure which is operationally well-defined. In this approach one knows precisely the terms that are being neglected, so that corrections can be analyzed later, if desired. Using the iteration procedure one obtains an expression for the memory function which schematically may be written as

$$-\phi = TG_0 V_s + T(G_D - G_0) T^+ + V_L G_D V_L + V_L G_D T^+ + TG_D V_L,$$
(Boltz-   (ring)   (mode   (cross terms)   (11)
mann)   coupling)

where $T = V_s G_s G_0^{-1}$ is closely related to the usual classical T-matrix [2], $T^+ = G_0^{-1} G_s V_s$, and G_0 is given by Equation (10) with each C replaced by its free-particle expression (with correct initial values). Aside from the cross terms, Equation (11) contains terms which have received considerable attention in the recent literature. These will be discussed in some detail in the next section where we will also describe a number of applications.

3. Discussion and Applications

To gain some feeling for the physical content of the memory function given by Equation (11) we first consider the general effects of the first three terms in this expression. One can show that the first term, $TG_0 V_s$, corresponds to a generalized Boltzmann collision operator. The dynamical events described by this term are dynamically uncorrelated binary collisions of the type treated by the Boltzmann equation. These are microscopic and quite local events. It has been shown [10] that this term can be written as a generalization of the Boltzmann-Enskog collision operator (including static corrections) to systems other than hard spheres.

By contrast, the second and third terms in Equation (11) are much less localized and represent processes on a semi-microscopic level. One type of process is where a pair of molecules collide, propagate independently through the medium, and eventually recollide. The correlations that lead to this recollision are 'communicated' by the medium. Thus this 'collision' process is something of a collective effect. The essential difference between what we call the ring terms [6] and the mode-mode coupling [5] terms lies in the endpoint vertex which is either the T-matrix or V_L. We make this distinction because, as we will see in the following, T and V_L have quite different behavior when the fluid is thermodynamically near the critical point.

It is only recently that the importance of the 'collective' effects represented by the correlated collision sequences have been emphasized. From extensive studies [11] of transport coefficients in dense gases ($n > n^c$ = critical density) and high temperatures ($T \gg T^c$ = critical temperature), one knows that the Enskog-like term $TG_0 V_s$ is a good approximation for $n \sim 1.5 n^c$ to an accuracy to within 10 to 20%. The correlated collisions are not important at very low densities, but they become progressively more significant as we increase in density. The deviations in the self-diffusion coefficient from the Enskog values observed in computer experiments are good indications of the magnitude of such collective effects. At densities $n \gtrsim 2n^c$ Alder, et al. [3] find a 25% enhancement over the Enskog values for the diffusion coefficient. For densities less than about two times the critical density, but away from the critical point, it is a good approximation to assume that $T \gg V_L$ due to a difference in density dependence. In the neighborhood of the critical point however the dominant terms in Equations (11) are the mode-coupling terms. The reason is that V_L becomes anomalously large due to the correlations that develop near the critical point.

Another way of discussing the relative importance of the various terms in Equation (11) is to consider their effects on the time-dependence of various correlation functions. At very short times (about a tenth of a collision time) the behavior of all correlation functions is determined by precise knowledge of the potential. Thus one should look at the short-time sum rules using a Lennard-Jones like potential. Hard-sphere potentials lead to very different short-time behavior. For times between about 0.5 and 3 collision times the decay of the correlation functions is exponential with a decay constant given rather accurately by the Enskog theory. For long times the correlated recollisions lead to a power law decay. The effects of the ring terms and mode-coupling terms in the intermediate time domain are not yet well understood, but computer results [3, 12] indicate that significant deviations from the Enskog predictions exist. These deviations from the Enskog predictions become more pronounced as the density is increased.

We now consider specific calculations which are designed to investigate the detailed effects of each of the three contributions to $\phi^{(c)}$ discussed above.

If one considers only the first term in $\phi^{(c)}$, then one finds in the case of hard-sphere interactions and moderate densities that

$$\phi^{(c)}(\mathbf{k}, \mathbf{pp}') f_0(p') = \tilde{g}(\mathbf{r}_0) \phi_0^{(c)}(\mathbf{k}, \mathbf{pp}') f_0(p'), \tag{12}$$

where r_0 is the hard-sphere diameter and $\phi_0^{(c)}$ is the low-density hard-sphere memory function and is given by Equation (2.14) in Reference 13. Equations (5), (7), (8) and (12) have been recently studied as an appropriate basis for calculating time correlation functions [14]. Although these approximate equations lead to the same transport coefficients as the conventional Enskog theory, they do not have the defects of the Enskog equation at short times. The difference lies in the static or mean field part of the memory function $\phi^{(s)}$. This term is given incorrectly in the Enskog theory. The presence of this term does not affect the transport coefficients, but does modify the speed of sound as derived from the kinetic equation.

For very dilute systems only the $TG_0 V_s$ term contributes to the memory function. We can then easily show that to lowest order in the density one obtains the low density memory function discussed previously by one of us [4]. This equation can be considered as the generalization of the linearized Boltzmann equation to arbitrary frequencies and wavelengths. If one restricts the analysis to hard-spheres then the memory function is given by Equation (12) where $\tilde{g}(r_0)=1$ and $h(k)$ and $C(k)$ should be replaced by their low density values, Recently [13] this equation has been used to calculate the spectrum of density fluctuations, and the results used to examine the validity of the conventional linearized Boltzmann equation in the region where $kr_0 \sim 1$. Similar analysis is now being extended to binary mixtures in anticipation of line-shape measurements in quasielastic neutron scattering. It would be of interest to extend the low density calculation to the case where one uses the dense gas results for $\tilde{g}(r_0)$ and $C(k)$ (as given for example by the Percus-Yevick equation). One can then calculate $S_{nn}(\mathbf{k}, \omega)$ using the method of kinetic models and it is hoped that reasonable agreement can be obtained in explaining neutron scattering data and computer molecular dynamics experiments.

The calculation of the velocity autocorrelation function is one of the simplest problems in which one can study the effects of the ring terms over the entire time domain. An attack on this problem using the results of renormalized kinetic theory has been initiated, and work is still continuing. Basically the problem involves the evaluation of momentum matrix elements of $\delta\phi_s$, the ring-collision contribution to the memory function for the self-diffusion of test-particle problem. Since $\delta\phi_s$ is in the form $T(G_D - G_0)T^+$ one needs to find appropriate approximations for the time correlation functions C and the test-particle analogue C_s. Thus far explicit calculations are being carried out for hard spheres, and the results show the same general behavior as observed in the computer results. The long-time behavior of the velocity autocorrelation function is $t^{-3/2}$ and is determined essentially by the coupling of two-hydrodynamical correlation functions, the transverse current correlation function and the van Hove self (test-particle) correlation functions. In the intermediate time region a number of other correlation functions also contribute and give rise to a quite complicated structure in the velocity autocorrelation function. We find that the ring terms lead to a positive contribution to the velocity autocorrelation function for densities up to $n \sim 2n^c$ which is in qualitative agreement with the results of molecular dynamics calculations for low densities. We have also found that there is a negative structure that enters into the calculation for higher densities, but it would be premature to identify these contributions with the 'backscattering' or negative structure first observed by Rahman [15]. At very short time $\delta\phi_s$ appears to behave like t^4. Preliminary results for the self-diffusion coefficient indicate an enhancement behavior in qualitative agreement with computer calculations [3, 16].

In many cases one is interested in the long wavelength, low frequency regime where one expects the form of the time correlation functions to be given by hydrodynamics. It was shown [17] some time ago that the density-density correlation function is given in the hydrodynamical regime by

$$S_{nn}(\mathbf{k}, z) = mnV_0^2 \left(\frac{\delta n}{\delta \mathbf{p}}\right)_T \times$$

$$\times \left\{ \frac{1 - C_v/C_p}{z + ik^2 D_T} + \frac{C_v}{C_p} \frac{z + ik^2[\Gamma + D_T(C_p - C_v)/C_v]}{z^2 - C_0^2 k^2 + izk^2\Gamma} \right\}, \quad (13)$$

where $mV_0^2 = \beta^{-1} C_V$ and C_p are the specific heats, C_0 is the adiabatic speed of sound, D_T is the thermal diffusion coefficient, Γ is the sound attenuation coefficient. D_T is related to the thermal conductivity by $D_T = \lambda mn C_p$. We easily see from Equation (13) that the thermal conductivity (for example) can be calculated from S_{nn} as the limit

$$\lambda^{-1} = \frac{i}{mV_0^2 \left(\frac{\delta n}{\delta \mathbf{p}}\right)_T (mn)(C_p - C_v)} \lim_{k \to 0} \lim_{z \to 0} [k^2 S_{nn}(\mathbf{k}, z)]. \tag{14}$$

Since one can express S_{nn} in terms of the memory function, one can eventually find expressions for the thermal conductivity strictly in terms of momentum matrix elements of the memory function and certain thermodynamic quantities. As an interesting application of these relations between the transport coefficients and the memory function one can look at the contribution to λ from the mode-mode coupling terms in $\phi^{(c)}$ when the system is near the critical point. One then finds that the thermal conductivity is proportional to the correlation length and is therefore divergent at the critical point [5].

Acknowledgements

We would like to thank Prof. Paul Martin for his hospitality during our stay at Harvard. This work was supported in part by the National Science Foundation.

One of us (S.Y.) is grateful to the John Simon Guggenheim Foundation for a fellowship.

References

1. Ernst, M. H., Haines, L. K., and Dorfman, J. R.: *Rev. Mod. Phys.* **41**, 296 (1969).
2. A rather complete set of references on the phase-space method are given in Mazenko, G. F.: *Phys. Rev.* **A7**, 209 (1973).
3. Alder, B. J., Gass, D. M., and Wainwright, T. E.: *J. Chem. Phys.* **53**, 3813 (1970).
4. Mazenko, G. F.: *Phys. Rev.* **A5**, 2545 (1972).
5. Kadanoff, L. P. and Swift, J.: *Phys. Rev.* **166**, 89 (1968); Kawasaki, K.: *Ann. Phys. (N.Y.)* **61**, 1 (1970).
6. Dorfman, J. R. and Cohen, E. G. D.: *Phys. Rev.* **A6**, 2247 (1972).
7. Wilson, K. G.: *Phys. Rev. Letters* **28**, 548 (1972).
8. Anderson, H. C., Weeks, J. D., and Chandler, D.: *Phys. Rev.* **A4**, 1597 (1971).
9. Forster, D. and Martin, P. C.: *Phys. Rev.* **A2**, 1575 (1970).
10. Mazenko, G. F.: *Phys. Rev.* **A9**, 360 (1974).
11. Hanley, H. J. M., McCarty, R. D., and Cohen, E. G. D.: *Physica* **60**, 322 (1972).
12. Levesque, D., Verlet, L., and Kurkijarvi, J. *Phys. Rev.* **A7**, 1690 (1973).
13. Mazenko, G. F., Wei, T. Y. C., and Yip, S.: *Phys. Rev.* **A6**, 1981 (1972).
14. Sykes, J.: *J. Stat. Phys.* **8**, 279 (1973).
15. Rahman, A.: *Phys. Rev.* **136**, A405 (1964).
16. Mazenko, G. F.: *Phys. Rev.* **A7**, 222 (1973).
17. Kadanoff, L. P. and Martin, P. C.: *Ann. Phys. (N.Y.)* **24**, 419 (1963).

ISOTOPE EFFECTS UPON MOTIONS IN LIQUIDS IN THE CLASSICAL LIMIT*

HAROLD L. FRIEDMAN

Dept. of Chemistry, State University of New York, Stone Brook, N.Y. 11790, U.S.A.

Abstract. The theory of changes in transport and relaxation coefficients in fluids and dilute solutions due to isotope substitution in the absence of quantal effects is described. In some cases the effects are completely predictable while in others they can be used to study intermolecular interactions in liquids.

Résumé. On décrit la théorie des changements des coefficients de transport et de relaxation dans des fluides et des solutions diluées provenant de substitution isotopique en l'absence d'effets quantiques. Dans certains cas on peut prévoir parfaitement ces effets tandis que dans d'autres cas ceux-ci peuvent être utilisés pour l'étude des interactions intermoléculaires des liquides.

1. Introduction

It is well known that in the absence of quantal effects the thermodynamic functions of a system are not changed by isotopic substitution. On the other hand, this is not true for transport and relaxation coefficients, as exemplified by Graham's Law $D(m)/D(m') = (m'/m)^{1/2}$ for the diffusion coefficient of a dilute gas of molecular weight m. The purpose of the study reported here is to learn what may be expected for isotope effects upon transport and relaxation coefficients of liquids when quantal effects are negligible. Also, it is assumed in each case that when some or all of the atoms in each molecule are substituted isotopically the only changes in the hamiltonian function of the system are in the explicit mass factors. Even in this simple case there is a variety of isotope effects.

2. Linear Response Theory Formulas

From the linear response theory of the statistical mechanics of rate processes [1–3] one knows that various measurable transport and relaxation coefficients k_A can each be expressed in terms of a correlation time τ_A which is defined as follows [4].

$$\tau_A = \int_0^\infty dt \, \langle A(0) \, A(t) \rangle / \langle A^2 \rangle, \tag{1}$$

where $A = A(\Gamma)$ is the dynamical variable corresponding to k_A, where $\langle f \rangle$ means the equilibrium ensemble average of the function $f(\Gamma)$

$$\langle f \rangle = \int d\Gamma \, e^{-\beta H} f \Big/ \int d\Gamma \, e^{-\beta H}, \tag{2}$$

where H is the hamiltonian function, Γ is the phase

---

* Grateful acknowledgement is made for the support of this work by the National Science Foundation.

$$\Gamma = \mathbf{q}_1, ..., \mathbf{q}_N, \mathbf{p}_1, ..., \mathbf{p}_N \tag{3}$$

of the N-particle system, the integration is over all momenta and locations $\mathbf{q}_1, ..., \mathbf{q}_N$ for a system confined to a vessel of volume V, and $\beta = 1/k_B T$. We shall use the following explicit formulation of the time correlation function

$$\langle A(0) A(t) \rangle = \langle A e^{Lt} A \rangle, \tag{4}$$

where L is the Liouville operator corresponding to H.

$$Lf = -(H, f)_{PB}, \tag{5}$$

where $(...)_{PB}$ signifies Poisson bracket and where f is any dynamical variable.

As a particular example of this theory, the particle diffusion coefficient for particle 1, which is defined by the equation

$$\begin{aligned} D_1 &= \lim_{t=\alpha} \langle (\mathbf{q}_1(t) - \mathbf{q}_1(0))^2 \rangle / 6t \\ &= \lim_{t=\alpha} \langle (x_1(t) - x_1(0))^2 \rangle / 2t \end{aligned} \tag{6}$$

is given by

$$D_1 = \tau_{\dot{x}1} / \beta m_1 \tag{7}$$

with $\tau_{\dot{x}1}$ given by Equation (1) with $A = \dot{x}_1$, where x_1 is the x-coordinate of molecule 1 and $\dot{x}_1 = dx_1/dt$. To continue, it is convenient first to consider the simplest application to the calculation of isotope effects.

3. Homomolecular Fluid, Monatomic Molecules

In this case $D_i = D_1$ for each of the molecules $i = 1, ..., N$ and we may suppress the subscripts. We also have $m_i = m$ for the atomic weights. We ask how D changes when all of the atoms are changed from one isotope to another.

To calculate the m-dependence of $\tau_{\dot{x}}$ we notice that for a given phase volume $d\Gamma$ at a given point Γ, the product of equilibrium factors

$$d\Phi \equiv d\Gamma\, e^{-\beta H} \bigg/ \int d\Gamma\, e^{-\beta H}, \tag{1}$$

depends upon m through the hamiltonian function

$$H = \sum_i p_i^2 / 2m + U_N \tag{2}$$

but this dependence can be removed by redefining the phase ('scaling' [5–8]) by changing the momentum variables from p_i to

$$\mathbf{p}_i^* = \mathbf{p}_i / \sqrt{m}. \quad 1 \leq i \leq N. \tag{3}$$

Substitution of all the p_i by p_i^* in the variables f, Γ, H, $d\Phi$, and $\langle f \rangle$ results in scaled variables denoted by f^*, Γ^*, H^*, $d\Phi^*$, and $\langle f \rangle^*$ respectively. Because $d\Phi^*$ does not depend on m the massdependence of

$$\langle f \rangle^* \equiv \int d\Phi^* f^* \tag{4}$$

is all in f^*.

Now we find

$$\begin{aligned}\langle \dot{x}(0)\,\dot{x}(t)\rangle / \langle \dot{x}^2 \rangle &= \langle p(0)\,p(t)\rangle / \langle p^2 \rangle \\ &= \langle p(0)\,p(t)\rangle^* / \langle p^2 \rangle^* \\ &= \int d\Phi^* p^*\, e^{L^* t} p^* / \langle p^2 \rangle^* \\ &= \int d\Phi^* p^* \exp(L^* \sqrt{m} \cdot t/\sqrt{m})\, p^* / \langle p^2 \rangle^*.\end{aligned} \tag{5}$$

The last step is trivial but it is convenient because we have

$$L = \sum_i ((\mathbf{p}_i/m_i)\cdot(\partial/\partial \mathbf{q}_i) + \mathbf{F}_i \cdot (\partial/\partial \mathbf{p}_i)), \tag{6}$$

where

$$\mathbf{F}_i = -\partial U_N/\partial \mathbf{q}_i \tag{7}$$

and it follows that

$$L^* \sqrt{m} = \sum_{i=1}^{N} (\mathbf{p}_i^* (\partial/\partial \mathbf{q}_i) + \mathbf{F}_i (\partial/\partial \mathbf{p}_i^*)) \tag{8}$$

is independent of m. Therefore it follows from Equation (5) that

$$\tau_{\dot{x}}/\sqrt{m} = \int_0^\infty d(t/\sqrt{m})\, \langle \dot{x}(0)\,\dot{x}(t)\rangle / \langle \dot{x}^2 \rangle \tag{9}$$

is independent of m. Finally, in view of Equation (2.7) we find that $D\sqrt{m}$ is independent of m.

The same analysis applies to other transport and relaxation coefficients which may be defined for these systems. For example, the correlation time $\tau_{A\eta}$ corresponding to the shear viscosity (for which the dynamical variable $A\eta$ is a little complicated [9]) also is proportional to $\sqrt{m}$; i.e. $\tau_{A\eta}/\sqrt{m}$ is independent of m.

These results were derived earlier by various methods by Brown and March and by Rowlinson [8].

4. Homomolecular Fluid, Homoatomic Molecules

It is easy to remove the restriction to monatomic molecules and extend the results of the previous section to fluids such as N_2, O_3, or P_4 in which the atoms are bound in molecules of a given formula and to treat the effect of complete isotopic substitutions, e.g. $^{17}O_2$ for $^{16}O_2$. Now we consider that Equation (3.2) specifies the hamiltonian of the system of *N atoms* and the potential function U_N includes the intramolecular as well as the intermolecular interactions. Since the potential function was not specified in Section 3, the results obtained there apply in this system as well!

For this system we can also consider coefficients related to the molecular reorientational motion such as τ_θ and τ_J, for which the dynamical variables are, respectively, the (second-order Legendre polynomial of the) angle specifying the orientation about a particular molecular axis and the angular momentum about a molecular axis. We easily find that $\tau_\theta/\sqrt{m}$ and $\tau_J/\sqrt{m}$ are independent of m, using the method of Section 3.

The classical-model results for time-correlation functions of translational and rotational dynamical variables may be expected to apply quite accurately to many real systems, while for intramolecular vibrational dynamical variables the quantal effects are probably always important.

5. Homomolecular Fluids

Now we consider what happens when the restriction to one atomic species is removed and we compare the properties of a pair of fluids such as C_6H_6 and C_6D_6. Now isotopic substitution cannot be described in terms of the changes in a single mass variable. The same complication arises if one considers homoatomic molecules with incomplete isotopic substitution, e.g. the pair $^{16}O_2$ and $^{16}O^{17}O$.

For these systems it is convenient to write the hamiltonian in the following form rather than Equation (3.2).

$$H = \sum_{i=1}^{N} \sum_{d=1}^{v} p_{id}^2/2m_d + U_N(q_{1,1}, \ldots, q_{i,d}, \ldots, q_{N,v}), \qquad (1)$$

where index i specifies a molecule, index d specifies a degree of freedom, and m_d is the mass (or moment of inertia) appropriate to the canonical p_{id}, q_{id} pair. We scale each momentum variable by the appropriate mass

$$p_{id}^* = p_{id}/\sqrt{m_d} \qquad (2)$$

and proceed as in Section 3 to obtain

$$\tau_A/\sqrt{m_A} = \int_0^t ds \int d\Phi^* A^* \exp(L^* \sqrt{m_A} s) A^*/\langle A^2 \rangle^*, \qquad (3)$$

where we define $s = t/\sqrt{m_A}$ and m_A is the mass which is associated with the same degrees of freedom as the dynamical variable A. (We consider only cases in which A is simple enough for this to be possible.) We may write

$$L = L_A + L_B, \qquad (4)$$

where L_A is the part of L such that $L_A^* \sqrt{m_A}$ is unchanged upon isotopic substitution, while L_B is the remainder. For example, if A is $\dot{x}$, as in the calculation of D, then L_B consists of rotational and vibrational terms. Thus, $\tau_A/\sqrt{m_A}$ is not independent of m_A precisely because $L_B^*\sqrt{m_A}$ is not independent of m_A.

The effect of isotopic substitution upon $\tau_A/\sqrt{m_A}$ can be discussed in terms of the commutator

$$L_C = L_A L_B - L_B L_A. \tag{5}$$

We notice that we have

$$e^{Lt} = e^{L_A t} e^{-L_c t^2/2} e^{L_B t} + \cdots, \tag{6}$$

where we have neglected more complicated commutators which vanish when L_C does.
When L_C is neglected and $L_B A = 0$, we find [10]

$$\langle A(0) A(t) \rangle / \langle A^2 \rangle = \int d\Phi^* A^* \exp(L_A^* t) A^* / \langle A^2 \rangle^*. \tag{7}$$

and it follows that $\tau_A/\sqrt{m_A}$ is dependent of m_A.

More generally, L_C is not zero and no simple result for the isotope effect can be expected. This fact was noticed some time ago by Pople [11]. On the other hand, in these cases, the isotope dependence reflects the intermolecular interactions, as we can see from examining the structure of L_C. It is a sum of terms like

$$(\partial^2 U_N/\partial q_{iA} \partial q_{jB}) [q_{iA} (\partial/\partial p_{jB}) - q_{jB} (\partial/\partial p_{iA})], \tag{8}$$

where A and B represent degrees of freedom from the sets designated in Equation (4). The coupling of vibrations to other modes is presumably always relatively small and the important effects are expected for set A being translational degrees of freedom and set B rotational, or vice-versa. Then the energy derivative in Equation (8) is proportional to the change in torque on molecule j due to a translational displacement of molecule i, or the change in force on molecule i due to a rotation of molecule j.

So we may conclude that isotope effects of the kind discussed in this Section provide a way to investigate the coupling of translational and rotational molecular motions in fluids [12].

6. Diffusion of a Solute Molecule

In this section we consider the diffusion of a solute molecule, index 0, in a medium made up of identical solvent molecules $1, 2, \ldots, N$. We ask how the motion of the solute molecule changes when its mass m_0 is changed while $m_1 = m_2, \ldots, m_N = m$ and the potential function $U_{N+1}(\mathbf{q}_0, \mathbf{q}_1, \ldots, \mathbf{q}_N)$ are all fixed. We consider only spherical particles and examine only the diffusion coefficient D_0 of the solute which is defined as in Equation (2.6) but with index 1 replaced by 0.

The problem may be discussed in terms of Langevin's equation

$$\dot{\mathbf{p}}_0(t) = -(\zeta_0/m_0) \mathbf{p}_0(t) + \mathbf{R}_0(t). \tag{1}$$

Here ζ_0 is the friction coefficient and the first term on the right is the frictional force due to a steady component of the motion of the solute particle. Thus it is expected that the random force $\mathbf{R}_0$ is very nearly the fluctuating force on particle 0 when it is held fixed, in which case $\mathbf{R}_0$ is very nearly independent of m_0 in a classical system. From the general relation of D_0 to $\mathbf{R}_0$ given below in Equation (4) it follows that D_0 is very nearly independent of m_0.

Equally well one may expect D_0 and ζ_0 to be independent of m_0 on the basis of Stokes Law

$$\zeta_0 = 6\pi\eta a_0, \tag{2}$$

where η is the viscosity of the medium and a_0 the effective radius of the solute particle.

For a more reliable analysis, we examine the consequences of the exact generalized form of Langevin's equation [2]

$$\dot{\mathbf{p}}_0(t) = -\langle \mathbf{p}_0^2\rangle^{-1} \int_0^t ds \langle \mathbf{R}_0(t)\mathbf{R}_0(s)\rangle \mathbf{p}_0(s) + \mathbf{R}_0(t) \tag{3}$$

which leads to a general expression for D_0 in terms of the random force [3, 14]

$$D_0 = 1/\beta^2 \int_0^L dt \langle \mathbf{R}_0(0)\mathbf{R}_0(t)\rangle \tag{4}$$

$$\langle \mathbf{R}_0(0)\mathbf{R}_0(t)\rangle = \int d\Phi \mathbf{F}_0 \exp((1-P)Lt)\mathbf{F}_0, \tag{5}$$

where P is a projection operator which is defined by the following equation, where f is any dynamical variable.

$$Pf = \mathbf{p}_0 \langle \mathbf{p}_0 f\rangle / \langle \mathbf{p}_0^2\rangle. \tag{6}$$

The analysis of the m_0-dependence of the correlation function of the random force, Equation (5), is facilitated by scaling each p_i by m_i.

$$\mathbf{p}_i^* = \mathbf{p}_i/\sqrt{m_i}. \tag{7}$$

The scaled Liouville operator L^* is now given by a solvent term L_w^*, which is the same as the entire function L^* specified in Equation (3.8), plus a solute term L_0^*

$$L^* = L_w^* + L_0^* \tag{8}$$
$$L_0^* = (\mathbf{p}_0^*(\partial/\partial \mathbf{q}_0) + \mathbf{F}_0(\partial/\partial \mathbf{p}_0^*))\gamma/\sqrt{m}, \tag{9}$$

where we define

$$\gamma = (m/m_0)^{1/2}. \tag{10}$$

We now have

$$D_0 = 1/\beta^2 \sqrt{m} \int_0^\infty ds \int d\Phi^* \mathbf{F}_0 \exp((1-P^*)L^* \sqrt{ms}) \mathbf{F}_0. \tag{11}$$

Thus, if γ were independent of m we would recover the result obtained in Section 3. The m_0-dependence of the exponential in Equation (11) can be expressed as follows

$$(1-P^*)L^* = L_w^* + (\gamma/\sqrt{m}) K_0, \tag{12}$$

where the new operator K_0 is given by the equation, for arbitrary $f(\Gamma)$,

$$K_0 f = \left(\mathbf{p}_0^* \int d\Phi^* \mathbf{F}_0 f / \langle p_0^2 \rangle^*\right) + \mathbf{p}_0^* (\partial f/\partial \mathbf{q}_0) + \mathbf{F}_0 (\partial f/\partial \mathbf{p}_0^*). \tag{13}$$

Substituting (12) in (11), we find, after straightforward analysis,

$$D_0 = 1/\beta^2 \sqrt{m} [C_0 + \gamma C_1 + \gamma^2 C_2 + \cdots], \tag{14}$$

where we define

$$C_n = \int_0^\infty dt_0 \int_0^\infty dt_1 \cdots \int_0^\infty dt_n \int d\Phi^* \mathbf{F}_0 S(t_0) K_0 S(t_1) K_0 \cdots S(t_n) \mathbf{F}_0 \tag{15}$$

$$S(t) = \exp(t \sqrt{m} L_w^*). \tag{16}$$

Since all the C_n for odd n vanish, Equation (14) may be simplified by omitting the terms in odd powers of γ.

Results entirely equivalent to Equation (14) have been obtained before in another context [5-7, 14]. The remarkable conclusion is that D_0 becomes independent of γ as γ goes to zero, at least when the formal operations in passing from Equation (11) to Equation (14) are justified, i.e. when all of the integrals converge and the series in γ converges. Although the convergence problems are delicate [5-7] we note that in the $\gamma = 0$ limit of Equation (14) the diffusion coefficient D_0 is again independent of m_0. Finally we note that we find no basis for expecting that $D_0 \sqrt{m_0}$ is constant when m_0 is changed, contrary to what has often been assumed.

This sort of analysis has yet to be made for measurable coefficients, e.g. D_0, τ_θ, and τ_J, for a non-spherical solute. It seems likely that additional effects due to rotation-translation interaction will be found.

7. Selected Experimental Results

This section is given to a brief review of some results from the literature which are relevant to the preceding theory. (See also Reference 8.)

For both C_6H_6 and $c\text{-}C_6H_{12}$ liquids, the viscosity ratio for the completely deuterated compared to the normal form is $\eta_D/\eta_H = 1.061 \pm 0.004$ over a 40° temperature range [15].

The square root mass ratio $(m_D/m_H)^{1/2}$ is 1.038 for C_6H_6 and 1.070 for c-C_6H_{12} while the respective square root moment of inertia ratios are approximately $(I_D/I_H)^{1/2} = $ = 1.21 and 1.25. Quantum effects are unlikely to be important here and comparison with the results of Section 5 can be made, suggesting that rotation-translation coupling is much more important in C_6H_6 than in c-C_6H_{12}.

For liquid water, the ratio η_D/η_H (i.e. for D_2O/H_2O) varies from 1.305 at 5°C to 1.09 at 280°C [15] while $(m_D/m_H)^{1/2}$ is 1.054 and $(I_D/I_H)^{1/2} \simeq 1.42$. In liquid water the vibrations which may be described as hindered translations of the molecule are found near 60 and 170 cm^{-1} while the hindered rotations are found in the range 500–700 cm^{-1}. Thus, in the temperature range of the viscosity data, in which RT varies from 200 to 400 cm^{-1}, quantum effects are negligible in the translations, but perhaps not in the rotations. Thus with or without the quantum effects, the η_D/η_H data show that there is strong translation-rotation coupling in the cold liquid, but much less at higher temperatures. It is known from X-ray scattering that the structure of the liquid changes considerably over the same temperature range [16].

For liquid ammonia the ratio η_D/η_H is constant at 1.20±0.01 from −25 to 30°C (comparing ND_3, NH_3) [17]. The other parameters are $(m_D/m_H)^{1/2} = 1.085$ and $(I_D/I_H)^{1/2} \simeq 1.39$. In this case the interpretation is complicated by a large volume ratio, $V_H/V_D = 1.009$; it is assumed in the theory discussed above that the transport coefficients are compared at the same molar volume of the fluid. An estimate for the comparison at the same volumes tends to reduce η_D/η_H to 1.19 [17]. This implies that translation-rotation coupling contributes to the viscosity of NH_3. For the same system it has been found [18] that $(\tau_\theta)_D/(\tau_\theta)_H = 1.42$, just as would be expected from the theory if rotation-translation coupling were negligible and comparison were made at equal molar volumes. At first sight the η data and the τ_θ data seem to be inconsistent, for, in the terminology of Section 5, both involve the same L_C commutator. However, it seems quite possible in view of the structure of Equation (5.3) that the effect of the commutator is weighted quite differently in the calculation of different correlation times.

Data for the solute diffusion coefficient D_0 in aqueous solutions (with particle 0 either Li^+ or Na^+) show a very much weaker isotope dependence than $m^{1/2}$ [19]. While this effect is sometimes interpreted in terms of the concept that $D_0\sqrt{m_0}$ is constant, but there is a solvent contribution to m_0, Pikal presents evidence that this is not the case [19].

On the other hand, the data for D_0 of isotopically substituted benzenes as solutes in benzene solvent are fit very well by the equation $D_0\sqrt{m_0} = $ constant [20]. The theory of Section 6 is not complete for these systems, in which the relative change in moments of inertia is different than in the molecular weight when an isotope substitution is made. Thus, there might be a large contribution from translation-rotation coupling in benzene, for which we have already cited the evidence, but then it is only a coincidence that $D_0\sqrt{m_0}$ is so nearly constant.

Finally, some 'experimental data' of another kind are cited. Using the molecular dynamics method, Herman and Alder [21] calculated the diffusion coefficient of a solute hard sphere in a medium of solvent hard spheres of the same diameter, varying the mass of the solute. At the highest density of the 'fluid' their calculated D_0 varied by less than 25% while $\sqrt{m_0}$ was varied by a factor of 150 [22]. Rather similar results were obtained by Bishop and Berne [23] from a molecular dynamics calculation for a one-dimensional model of a fluid of particles interacting with a truncated Lennard-Jones potential. Of course these results fit in quite well with theory in Section 6.

Note added in proof. Additional studies of the mass dependence of D_0 for solutes in liquid N_2 [24] and hydrocarbons [25] have come to our attention. These data are consistent with the present theory.

Acknowledgements

I am grateful to S. Harris and H. C. Andersen for the benefit of discussions of the work reported here.

References

1. Zwanzig, R.: in H. Eyring (ed.), *Annual Reviews of Physical Chemistry* **16**, Annual Reviews, Inc., Palo Alto, California, 1965, p. 67.
2. Kubo, R.: *Reports on Progress in Physics (London)* **29**, 255 (1966). Berne, B. J.: in H. Eyring (ed.), *Physical Chemistry, An Advanced Treatise*, Academic Press, New York, 1971, Vol. 7.
3. Friedman, H. L.: *Zeits. physik. Chem.* (Leipzig) **228**, 318 (1965).
4. Equation (2.1) is a special case, but sufficient for our purpose. More generally, one also has cross correlations $\langle A(0)B(t)\rangle$.
5. Lebowitz, J. L. and Rubin, E.: *Phys. Rev.* **131**, 2381 (1963).
6. Resibois, P. and Davis, H. T.: *Physica* **30**, 1077 (1964).
7. Resibois, P. and Lebowitz, J. L.: *Phys. Rev.* **139**, A1101 (1965).
8. Brown, R. C. and March, H. N.: *Phys. Chem. Liquids* **1**, 141 (1969). Rowlinson, J. S.: *Physica* **19**, 303 (1953).
9. Alder, B. J., Gass, D. M., and Wainright, T. C.: *J. Chem. Phys.* **53**, 3813 (1970).
10. The discussion at this point is not complete. In simple cases L_BA vanishes. In other cases (e.g. calculation of the viscosity) $A = LA'$, where A' is a simpler dynamical variable such that L_BA' vanishes, and then L_BA is negligible when L_C is negligible. Even this discussion is not exhaustive but covers the cases specifically mentioned in this study.
11. Pople, J. A.: *Physica* **19**, 668 (1953).
12. If the center of mass moves relative to the atoms in an isotopic substitution (e.g. H_2 to HD) and if U_N is expressed in terms of center of mass coordinates, orientations and distortions of the molecules, then there appears to be a change in U_N even in the absence of quantum effects. It is assumed here that this complication can be avoided by an appropriate formulation of U_N.
13. Here as in the remainder of Section 6, dyads of cartesian vectors appear, for example $\mathbf{R}_0\mathbf{R}_0$ in Equation (4). In each dyadic correlation function we should specify one diagonal element, as $\langle \mathbf{1}_x \cdot \mathbf{R}_0\mathbf{R}_0 \cdot \mathbf{1}_x \rangle$ the auxiliary notation is suppressed. For further details on the vector aspects of these equations, reference should be made to the basic theory [2-4].
14. Harris, S.: *Molecular Physics* **23**, 861 (1972).
15. Rabinovitch, I.: '*Influence of Isotopy on the Physiochemical Properties of Liquids*', Consultants Bureau, New York, 1970.
16. Narten, A. H. and Levy, H. A.: in F. Franks (ed.), *Water, a Comprehensive Treatise*, Plenum Press, New York, 1972.
17. Alei, M. and Litchman, M. W.: *J. Chem. Phys.* **56**, 5818 (1972).
18. Atkins, P. W., Loewenstein, A., and Margalit, Y.: *Molecular Physics* **17**, 329 (1969).

19. Pikal, M. J.: *J. Phys. Chem.* **76**, 3038 (1972).
20. Eppstein, L. and Albright, J.: *J. Phys. Chem.* **75**, 1315 (1971).
21. Herman, P. T. and Alder, B. J.: *J. Chem. Phys.* **56**, 987 (1971).
22. Friedman, H. L.: *Chemistry in Britain* **9**, 300 (1973).
23. Bishop, M. and Berne, B. J.: *J. Chem. Phys.* **56**, 2850 (1972).
24. Ricci, F. P.: *Phys. Rev.* **156**, 184 (1967).
25. Thornton, S. J. and Dunlop, P. J.: *J. Phys. Chem.*, in press.

DISCUSSION

Rothschild: Supposing that you observe the orientational motion in a plastic crystal, where there is no translational motion, would you then expect that there is coupling with the lattice modes, particularly with those which represent predominantly translatory displacements of the isotopically substituted molecules?

Friedman: That seems reasonable, but the work described here doesn't give me any basis for a more definite answer.

Jonas: (a) In one of your slides you have shown the results for ammonia/ammonia- d_3 and benzene/benzene-d_6. I would like to point out that in the case of reorientational motion of a symmetric-top molecule of ammonia one should consider that the rotational correlation times are not necessarily equal for the motion about the main symmetry axis and about the axis perpendicular to the main symmetry axis.

(b) In the case of benzene, when one discusses the rotational/translational coupling; for the relatively free rotation about the 6-fold symmetry axis there will be very little, if any, rotational/translational coupling (T. E. Bull and J. Jonas, 1971). On the other hand, for rotation about axes lying in the plane of the benzene ring one can expect a strong rotational/translational coupling. In this connection I would also like to point out that we have recently studied the coupling between the rotational and translational motions in a series of monosubstituted benzenes (Assink, R. A., De Zwaan, J., and Jonas, J.: *J. Chem. Phys.*, 1972) and found that molecular shape plays a decisive role in this coupling.

Friedman: I believe that the three moments of inertia of NH_3 change in nearly the same ratio in going to ND_3. Then for the correlation time of the reorientation about any given axis the isotope effect is given by the square root of the moment of inertia ratio in the absence of rotation-translation coupling of the kind I described.

Harris: With reference to the data of Albright and Epstein for the tracer diffusion coefficient of benzene I would like to point out that the data of Harris, Pua and Dunlop (*J. Phys. Chem.*, 1970) do not show any dependence on the molecular weight of the tracer species. This point has been reinvestigated and confirmed by Dunlop and Allen (*Phys. Rev. Letters*, 1973) using the same samples of benzene used by Epstein and Albright. It would be experimentally interesting to determine the tracer diffusion coefficient of benzene in other solvents. Dr. Mills will be discussing isotope effects in a paper to be presented later during this conference. It should be also pointed out that the datum point of Birkett and Lyons is subject to an error of ± 1 to 2%.

Hertz: (1) You showed that there is no isotope effect in a Na^+ solution even if 6m LiBr is added. It would be interesting to see whether the isotope effect appears when the Na^+ concentration is increased, i.e. if we approach towards the pure 'Na^+ liquid'.

(2) I would have expected a certain residual inertial effect because a liquid is a system somewhere "between" the gas and the solid, and in the gas we have $(m)^{-1/2}$ dependence.

(3) Can there be an 'accidental' isotope effect due to appropriate balancing of mass and potential influences causing similar circumstances as if the particle were in the pure liquid?

Friedman: (1) I agree that further variations on that system would be interesting. (2) I presume that you are referring to the isotope effect in the diffusion of a solute. The theory I have described seems to be applicable to a fluid at any density except that the series in m/m_0 might have a radius of convergence depending on the density. But I don't know of any data on gases to compare it with. (3) I don't see any basis for this in the theory.

Mills: Later in this session, I shall be presenting experimental data for the self-diffusion of water and these show that a square-root mass dependence does not seem to be operative. Graham's law is, of course, only applicable to certain restricted cases of diffusion but equations from the kinetic theory of gases which incorporate the square root of the reduced mass do fit our data. I prefer to leave further discussion of this matter until our experimental material is given and explained.

COMPLEX PERMITTIVITY AND DIPOLE CORRELATIONS

ROBERT H. COLE

Chemistry Dept., Brown University, Providence, R.I. 02912, U.S.A.

Abstract. The relation of Glarum between macroscopic and microscopic relaxation functions has been derived by molecular response function theory, with effects of induced polarization included. A modified form of Zwanzig's theory of rotational diffusion of interacting dipoles, based on Kirkwood's theory of transport processes, is shown to give correlation functions in agreement with the relation, and with no new relaxation times from long range dipole forces.

Résumé. La theorie de réponse moléculaire est utilisée pour démontrer la relation de Glarum entre les fonctions de correlation dipolaire macroscopique et microscopique, y compris l'effet de polarisation induite. La théorie de Zwanzig pour la diffusion rotationnelle des dipoles à forces mutuelles est modifiée selon la théorie de transport de Kirkwood. Pour ce modèle, on obtien de fonctions de correlation en accord avec la relation de Glarum, et on ne retrouve plus des temps de relaxation nouveaux.

Time dependent electric polarization is related by linear response theory to a normalized macroscopic dipole correlation function $\Phi(t)$ defined by the relation

$$\left\langle f_N^0 \mu_{iz} \sum_j^N m_{jz}(t) \right\rangle = \left\langle f_N^0 \mu_{iz} \sum_j^N m_{jz} \right\rangle \Phi(t). \tag{1}$$

In this expression, μ_{iz} is the component in an arbitrary direction z of moment $\boldsymbol{\mu}_i$ of a representative dipole at time $t=0$, $\sum_j^N m_{jz}(t)$ is the sum over N molecules of a macroscopic sample of permanent and induced moments $\boldsymbol{\mu}_j + \boldsymbol{\eta}_j$ at a time t later and the ensemble average is with the equilibrium distribution function f_N^0.

The function $\Phi(t)$ involves both short range molecular correlations and effects of long range dipole forces, as is clear from the fact that its value depends on the sample shape. For a sphere in vacuum, which is simplest, the basic formalism of Kubo [1] as extended by Glarum [2] and the writer [3] relates $\Phi(t)$ to the complex permittivity $\varepsilon^*(\omega)$ by the equation

$$\frac{\varepsilon^* - \varepsilon_\infty}{\varepsilon_0 - \varepsilon_\infty} = \left\{ 1 + \frac{\varepsilon_0 + 2}{\varepsilon_\infty + 2} \left[\frac{1}{\mathscr{L}(-\dot\Phi)} - 1 \right] \right\}^{-1}, \tag{2}$$

where ε denotes the static permittivity and ε_∞ the permittivity of induced polarization. For simplicity, polarization resulting from moments induced by translational fluctuations of dipole and multipole fields has been neglected, but could be included. ε_∞ is related to molecular polarizability α by the Clausius-Massotti expression $(\varepsilon_\infty - 1)/(\varepsilon_\infty + 2) = 4\pi N \alpha / 3V$, where V is sample volume, and the equilibrium dipole contribution is given by

$$\varepsilon_0 - \varepsilon_\infty = \frac{4\pi N}{kTV} \left(\frac{\varepsilon_0 + 2}{3} \right) \left(\frac{\varepsilon_\infty + 2}{3} \right) \left\langle f_N^0 \mu_{iz} \sum_j^N (\mu_{iz} + \eta_{iz}) \right\rangle. \tag{3}$$

A long standing problem in dielectric theory with conflicting viewpoints and conclusions has been that of relating $\Phi(t)$ to a suitable molecular or microscopic relaxation function $\phi(t)$ not containing sample shape effects from long range interactions. Glarum [4] by macroscopic response arguments derived a generalization to the time dependent problem of Kirkwood's equilibrium relation [5] between $\langle f_N^0 \mu_{iz} \sum_i^N (\mu_{jz} + \eta_{jz}) \rangle$ and the function $\langle f_N^0 \mu_{iz} \sum_i^n (\mu_{jz} + \eta_{jz}) \rangle$ for the n dipoles of a smaller spherical region in the macroscopic sample. This relation as derived by the writer [6] by a dynamical treatment of the induced moments η_j is

$$\left\langle f_N^0 \mu_{iz} \sum_j^N (\mu_{jz} + \eta_{jz}) \right\rangle = (1 - A) \left\langle f_N^0 \mu_{iz} \sum_j^n (\mu_{jz} + \eta_{jz}) \right\rangle$$

$$A = \frac{2(\varepsilon_0 - 1)^2}{(2\varepsilon_0 + 1)(\varepsilon_0 + 2)}. \tag{4}$$

Glarum's relation derived without distinction between permanent and induced dipoles is

$$\phi(t) = \Phi(t) - A \int_0^t ds \dot{\phi}(s) \Phi(t - s), \tag{5}$$

where $\phi(t)$ is defined by the relation

$$\left\langle f_N^0 \mu_{iz} \sum_j^n m_{jz}(t) \right\rangle = \left\langle f_n^0 \mu_{iz} \sum_j^n m_{jz} \right\rangle \phi(t). \tag{6}$$

Glarum's approach was criticized by Fatuzzo and Mason [7] on the ground that he used the macroscopic sphere function $\Phi(t)$ to describe correlation of the shell surrounding the smaller spherical region. They and several other writers have proposed generalizations of Onsager's molecular cavity and reaction field equilibrium theory to the time dependent case, which lead to different and more complicated relations of $\Phi(t)$ to a single molecular response function. Glarum has recently answered these objections by macroscopic free energy and fluctuation theory arguments, [4] and has correctly in our opinion pointed out that use of cavity models introduces relaxation of artificial, in reality non-existent, surface polarizations.

The writer has succeeded in deriving Glarum's expression by molecular response theory arguments. This derivation, described in detail elsewhere, [8] considered permanent dipoles only, but can with minor changes be generalized to include induced moments. In the following, we outline this treatment, discuss the results, and compare it with results of rotational diffusion models with effects of long range dipole forces included.

1. Derivation of Relations Between Macroscopic and Microscopic Correlations

To relate $\phi(t)$ to $\Phi(t)$, we consider first the equation of motion of a dipole m_j in the small sphere. This is

$$\frac{d}{dt} m_{jz} = (L_N + L') m_{jz}, \tag{7}$$

where L_N is the Liouville operator for forces and momenta of dipoles in the small sphere and L' the interaction operator for forces from the surrounding shell on these dipoles. Integration over the phase space T_X of the surroundings with the distribution function f_N gives L' in the form

$$L' = \frac{\int dT_X f_N \sum_j^n \mathbf{F}_{j,x} \frac{\partial}{\partial \mathbf{p}_j}}{\int dT_X f_N} = \sum_j^n \overline{\mathbf{F}}_{j,x} \frac{\partial}{\partial \mathbf{p}_j}, \tag{8}$$

where $\overline{\mathbf{F}}_{j,x}$ is the average force on dipole j due to molecules in the surrounding shell. Strictly, the averaging is with the time-dependent distribution function, but we shall take the equilibrium average force as given by the Kirkwood reaction field expression

$$\overline{\mathbf{F}}_{j,x} = e_j \lambda \sum_j^n \mathbf{m}_j, \quad \lambda = \frac{2(\varepsilon_0 - 1)}{2\varepsilon_0 + 1} \frac{1}{R_n^3}, \tag{9}$$

where e_j is a formal charge of dipole m_j and R_n is the sphere radius.

A formal integration of Equation (7) and use of an operator identity gives:

$$m_{jz}(t) = e^{tL_n} m_{jz} + \int_0^t ds\, e^{(t-s)(L_n+L')} L' e^{sL_n} m_{jz}. \tag{10}$$

The correlation function in Equation (6) can then be written:

$$\left\langle f_N^0 \mu_{iz} \sum_j^n m_{jz}(t) \right\rangle = \int dT_n f_{n,x}^0 \mu_{iz} e^{tL_n} \sum_j^n m_{jz} + $$
$$+ \int_0^t ds \int dT_n \mu_{iz} e^{(t-s)(L_n+L')} L' e^{sL_n} \sum_j^n m_{jz}. \tag{11}$$

A partial integration over T_X has introduced the 'externally averaged' distribution function $f_{n,x}^0$ given by

$$f_{n,x}^0 = \frac{\exp\left[-\beta\left(H_n + \frac{1}{2}\lambda\left(\sum_j^n \mathbf{m}_j\right)^2\right)\right]}{\int dT_X \exp\left[-\beta\left(H_n + \frac{1}{2}\lambda\left(\sum_j^n \mathbf{m}_j\right)^2\right)\right]}, \tag{12}$$

where $\frac{1}{2}\lambda(\sum_j^n \mathbf{m}_j)^2$ is the free energy of interaction of $\sum_j^n \mathbf{m}_j$ with the surrounding shell.

In the first integral on the right hand side of Equation (11), we observe that the time evolution is the fluctuation of the total moment of the spherical region with forces and operator L_n for this region only, corresponding to the correlation function

$\Phi(t)$, and we write

$$\int dT_n f^0_{n,x} \mu_{iz} e^{tL_n} \sum_j^n m_{jz} = \int dT_n f^0_{n,x} \mu_{iz} \sum_j^n m_{jz} \Phi(t). \tag{13}$$

In the second integral of Equation (11), which we now denote by I_2 for brevity, the operator $e^{(t-s)(L_n+L')}$ operating on all the functions it precedes can be shifted to operate on μ_{iz} instead. Using Equation (8) for L' and Equation [9] for $F_{j,x}$ gives

$$I_2 = \lambda \int_0^t ds \int dT_n f^0_{n,x} \left[\sum_j^n e_j \frac{\partial}{\partial \mathbf{p}_j} e^{sL_n} m_{jz} \right] \cdot \left[\sum_j^n \mathbf{m}_j e^{-(t-s)(L_n+L)'} \mu_{iz} \right],$$

The averages over Γ_n of the two bracketed factors are independent except for terms of order $1/n$. The average of the second factor is seen on comparison with Equation (6) to give

$$\int dT_n f^0_{n,x} \sum m_{jz} e^{-(t-s)(L_n+L')} \mu_{iz} = \left\langle f^0_{n,x} \mu_{iz} \sum_j^n m_{jz} \right\rangle \phi(t-s). \tag{14}$$

The first factor can be integrated by parts with use of

$$\frac{\partial}{\partial p_j} f^0_{n,x} = -\beta p_j/M_j, \qquad e_j p_j/M_j = \dot m_j$$

to give

$$\frac{\beta \int dT_n \exp\left[-\beta\left(H_n + \tfrac12 \lambda \left(\sum_j^n m_j\right)^2\right)\right] \sum_j^n \dot m_{jz} e^{sL_n} \sum m_{jz}}{\int dT_n \exp\left[-\beta\left(H_n + \tfrac12 \lambda \left(\sum_j^n m_j\right)^2\right)\right]} = -\beta \left\langle f^0_n \left(\sum_j^n m_{jz}\right)^2 \right\rangle \dot\Phi(s), \tag{15}$$

where $f^0_n = \exp(-\beta H_n)/\int d\Gamma_n \exp(-\beta H_n)$ and we have recognized that averaging of the interaction free energy is essentially uncorrelated with that of $p_{jz} \exp sL_n \sum_j^n m_{jz}$. Use of Equations (13), (14), and (15) in Equation (11) gives

$$\left\langle f^0_N \mu_{iz} \sum_j^n m_{jz}(t) \right\rangle = \left\langle f^0_{n,x} \mu_{iz} \sum_j^n m_{jz} \right\rangle \phi(t) =$$
$$= \left\langle f^0_{n,x} \mu_{iz} \sum_j^n m_{jz} \right\rangle \Phi(t) +$$
$$- \beta \lambda \left\langle f^0_n \left(\sum_j^n m_{jz}\right)^2 \right\rangle \left\langle f^0_{n,x} \mu_{iz} \sum_j^n m_{jz} \right\rangle \int_0^t ds\, \dot\Phi(s)\, \phi(t-s).$$

Use of the Kirkwood relation $\langle f^0_n (\sum_j^n m_{jz})^2 \rangle = R_n^3 (\varepsilon_0-1)/(\varepsilon_0+2)$, gives the Glarum formula

$$\phi(t) = \Phi(t) - \frac{2(\varepsilon_0-1)^2}{(2\varepsilon_0+1)(\varepsilon_0+z)} \int_0^t ds\, \dot\Phi(s)\, \phi(t-s). \tag{16}$$

Our derivation by molecular response theory confirms Glarum's result and the validity of his derivation by macroscopic arguments. It also shows that a different correlation function for moments of the surrounding shell obtained by Fatuzzo and Mason [7] is not involved, as only the two time evolution operators L_n and $L_n + L'$ appear in Equation (11). The derivation moreover takes explicit account of permanent *and* induced moments in the total moments $\mathbf{m}_j = \boldsymbol{\mu}_j + \boldsymbol{\eta}_j$ which appear in the definitions of $\Phi(t)$ and $\phi(t)$.

Because the correlation function $\phi(t)$ does include induced moments, it is not quite the same as the time dependent correlation of dipole molecular axis orientations. Such a function $\gamma(t)$ is defined by the relation

$$\left\langle f^0_{n,x} \mu_{iz} \sum_j^n \mu_{jz}(t) \right\rangle = \left\langle f^0_{n,x} \mu_{iz} \sum_j^n \mu_{jz} \right\rangle \gamma(t). \tag{17}$$

A superposition response argument may be used to obtain a relation between $\phi(t)$ and $\gamma(t)$. At equilibrium, the n dipoles $\mathbf{m}_j$ of the spherical region have induced moments $\boldsymbol{\eta}_j$ which we take to be the product of polarizability α and the reaction field produced by $\sum_j \mathbf{m}_j$ polarizing the surroundings:

$$\sum_j^n \mathbf{m}_j = \sum_j^n \boldsymbol{\mu}_j + n\alpha\lambda \sum_j^n \mathbf{m}_j = \sum_j^n \boldsymbol{\mu}_j + \frac{n\alpha\lambda}{1 - n\alpha\lambda} \sum_j^n \boldsymbol{\mu}_j. \tag{18}$$

We write the moment $\sum_j^n m_{jz}(t)$ at a later time t as the permanent dipole term $\sum_j^n \mu_{jz}(t)$ plus the sum of delayed responses to the changing $\sum_j^n \mu_{jz}(t')$ for $t' < t$ to obtain

$$\left\langle f^0_{n,x} \mu_{iz} \sum_j^n m_{jz}(t) \right\rangle = \left\langle f^0_{n,x} \mu_{iz} \sum_j^n \mu_{jz}(t) \right\rangle +$$

$$+ \frac{n\alpha\lambda}{1 - n\alpha\lambda} \int_{-\infty}^t ds [1 - \phi(t-s)] \left\langle f^0_{n,x} \mu_{iz} \sum_j^n \dot{\mu}_{jz}(s) \right\rangle.$$

Evaluating the contribution of the integral from $-\infty$ to zero by the equilibrium condition (18) at $t = 0$ and introducing the definitions of $\phi(t)$ from Equation (6) and $\gamma(t)$ from (17) gives

$$\phi(t) = \gamma(t) - n\alpha\lambda \int_0^t ds\, \phi(t-s)\, \dot{\gamma}(s), \tag{20}$$

where

$$n\alpha\lambda = \frac{2(\varepsilon-1)}{2\varepsilon+1} \frac{n\alpha}{R_n^3} = \frac{2(\varepsilon-1)}{2\varepsilon+1} \frac{\varepsilon_\infty - 1}{\varepsilon_\infty + 2}. \tag{21}$$

The integral Equation (16) and (20) can because of the convolution integral be solved explicitly for Laplace transform relations of $\Phi(t)$ to $\phi(t)$ and $\gamma(t)$. These are conveniently written as

$$\frac{1}{\mathscr{L}(-\dot{\Phi})} - 1 = (1 - A)\left[\frac{1}{\mathscr{L}(-\dot{\varphi})} - 1\right] = \frac{1-A}{1-n\alpha\lambda}\left[\frac{1}{\mathscr{L}(-\dot{\gamma})} - 1\right]. \quad (22)$$

In terms of the permanent dipole relaxation function $\gamma(t)$, Equation (2) for the complex permittivity ε^* becomes

$$\frac{\varepsilon^* - \varepsilon_\infty}{\varepsilon_0 - \varepsilon_\infty} = \left\{1 + \frac{3\varepsilon_0}{2\varepsilon_0 + \varepsilon_\infty}\left[\frac{1}{\mathscr{L}(-\dot{\gamma})} - 1\right]\right\}^{-1}. \quad (23)$$

This equation shows that the macroscopic relaxation is simply related to $\gamma(t)$ by the local field factor $3\varepsilon_0/(2\varepsilon_0 + \varepsilon_\infty)$ proposed semi-empirically by Powles [9]. If $\gamma(t)$ is a simple exponential decay, $\gamma(t) = \exp(-t/\tau)$, then $1/\mathscr{L}(-\dot{\gamma}) = 1 + i\omega\tau$ and we have

$$\frac{\varepsilon^* - \varepsilon_\infty}{\varepsilon_0 - \varepsilon_\infty} = \frac{1}{1 + i\omega\left(\dfrac{3\varepsilon_0}{2\varepsilon_0 + \varepsilon_\infty}\right)\tau}. \quad (24)$$

Thus for a simple exponential decay of molecular dipole correlations, the relaxation is a simple Debye function with macroscopic relaxation time $T = 3\varepsilon_0\tau/(2\varepsilon_0 + \varepsilon_\infty)$ which differs from a molecular time τ by at most a factor $\tfrac{3}{2}$. This simplicity, and the simple form of the relations (23) more generally, is in contrast to results of cavity models.

2. Relaxation of Rigid Lattice Dipoles

The relations between $\Phi(t)$, $\phi(t)$, and $\gamma(t)$ which we have developed are quite general in that no particular models of molecular dynamics have been assumed. One way in which these and other relations can be tested is by calculating these functions independently for some model and determining whether the relations are satisfied.

Zwanzig [10] has evaluated the equivalent of $\Phi(t)$ for a model of rigid dipoles on a cubic lattice undergoing rotational diffusion and subject to the long range dipole-dipole fields and interaction. The model is thus an extension to the time dependent case of the dipole coupling problem discussed by Van Vleck [11], Rosenberg and Lax [12], and the writer [13]. Zwanzig postulated a diffusion equation for the N dipole rotational distribution function g_N with external force terms included to take account of the dipole forces. This equation has the form

$$\frac{\partial g_N}{\partial t} = \frac{1}{2\tau}\sum_{i}^{N} (\mathbf{r}_i \times \mathbf{\nabla}_{r_i}) \cdot \left[(\mathbf{r}_i \times \mathbf{\nabla}_{r_i}) - \frac{1}{kT}(\mathbf{r}_i \times \mathbf{F}_i)\right] 8N, \quad (25)$$

where $\mathbf{r}_i \times \mathbf{J}_{r_i}$ is the rotational configuration operator, τ the diffusion relaxation time, and $\mathbf{r}_i \times \mathbf{F}_i = \boldsymbol{\mu}_i \times \mathbf{E}_i$ is the torque on dipole i of dipole fields from all other dipoles:

$$\mathbf{E}_2 = \sum_{j \neq 2}^{N} R_{ij}^{-3}\left[\frac{3(\mathbf{R}_{ij}\boldsymbol{\mu}_j)\mathbf{R}_{ij}}{R_{ij}^2} - \boldsymbol{\mu}_j\right], \quad (26)$$

where R_{ij} is the distance between sites i and j.

In Zwanzig's calculation, the Laplace transform of $\Phi(t)$ is evaluated for the time evolution of $\mathbf{\mu}_i \cdot \sum_j^n \mathbf{\mu}_j(t)$ by expansion of powers of the dipole field terms treated as a perturbation. The Boltzmann factor $\exp(-\beta W_{\mu\mu})$, where $W_{\mu\mu} = \frac{1}{2}\sum_j^n \mathbf{\mu}_i \cdot \mathbf{E}_i$, is also expanded in powers of the interaction $W_{\mu\mu}$; mathematical complexity limits evaluation of the expansion to second order dipole interaction effects. The writer has extended the calculation [14] to obtain $\mathscr{L}[\phi(t)]$, i.e. the correlation function for the n dipoles of a spherical region in the large sphere of N dipoles. The result is:

$$\mathscr{L}[\phi(t)] = \frac{1}{i\omega + \frac{1}{\tau}} + 2a^2\lambda^2 \left[\frac{1}{\tau} \frac{1}{\left(i\omega + \frac{1}{\tau}\right)^2} + \frac{1}{\tau^2} \frac{1}{\left(i\omega + \frac{1}{\tau}\right)^3} \right] - \qquad (27)$$

$$- 2a^2 S \left[\frac{1}{\tau} \frac{1}{\left(i\omega + \frac{1}{\tau}\right)\left(i\omega_t + \frac{4}{\tau}\right)} + \frac{5}{2\tau^2} \frac{1}{\left(i\omega + \frac{1}{\tau}\right)^2 \left(i\omega + \frac{4}{\tau}\right)} \right],$$

where $a = \mu^2/3kT$, $\lambda = (4\pi N/3V)$, and $S = \sum_j^n R_{ij}^{-6}$ is the rapidly converging lattice sum appearing also in equilibrium theory. Zwanzig's result for $\mathscr{L}[\Phi(t)]$ is obtained by setting $\lambda = 0$ to eliminate the terms arising from interactions with dipoles in the surrounding shell.

Equation (27) and the result for $[\Phi(t)]$ are found *not* to satisfy Glarum's relation and, as Scaife [14] has pointed out, Zwanzig's result also disagrees with generalized Onsager models. The result is also puzzling in that it predicts a second faster relaxation with relaxation time $\tau/4$ in addition to a modified relaxation of the form $1/(i\omega + (1+b)/\tau)$, generated as an expansion in powers of $(i\omega + 1/\tau)^{-1}$. This discouraging situation prompted the writer [15] to examine the validity of the assumed Equation (25) for the configuration distribution function.

An equation similar to (25) is obtained from phenomenological Brownian motion theory (cf. Chandresekhar [16]) but with the difference that $\mathbf{F}_i$ is an external force on the particle, whereas in the present problem $\mathbf{F}_i$ is an intermolecular force from other particles in the system. The most ambitious attempt to *derive* distribution functions for diffusionlike motion from the fundamental molecular Liouville equation is Kirkwood's theory [17] of transport processes. For the present problem, we need the two particle distribution function $g_{ij}(t)$ to calculate $\mu_{iz}\mu_{jz}(t)$ and Kirkwood's result specialized to configuration distributions gives

$$\frac{\partial g_{ij}}{\partial t} = \frac{1}{2\tau}\left[(\mathbf{r}_i \times \mathbf{\nabla}_{r_i})\cdot(\mathbf{r}_i \times \mathbf{\nabla}_{r_i}) + (\mathbf{r}_j \times \mathbf{\nabla}_{r_j})\cdot(\mathbf{r}_j \times \mathbf{\nabla}_{r_j}) - \right.$$

$$\left. - \frac{1}{kT}(\mathbf{r}_i \times \langle \mathbf{F}_{i,j} \rangle) - \frac{1}{kT}(\mathbf{r}_j \times \langle \mathbf{F}_{j,i} \rangle)\right] g_{ij}, \qquad (28)$$

where $\mathbf{r}_i \times \langle \mathbf{F}_{i,j} \rangle$ is the torque of electric fields $\mathbf{E}_i$ on dipole i as in Equations (25) and (26), but now is averaged over coordinates of all dipoles other than i and j.

When Equation (28) and a similar one for the single particle distribution function $g_i(t)$ are used to obtain the time dependent correlations $\mu_{iz}\mu_{jz}(t)$ and $\mu_{iz}\mu_{iz}(t)$, the result for $\phi(t)$ is

$$\mathscr{L}[\phi(t)] = \frac{1}{i\omega + \dfrac{1}{\tau}} + 2a^2(\lambda^2 - S)\frac{1}{\left(i\omega + \dfrac{1}{\tau}\right)^2} + \cdots$$

$$= \frac{1}{i\omega + \dfrac{1 - 2a^2(\lambda^2 - S)}{\tau}} + O(a^3) + \cdots. \tag{29}$$

As before, the transform $\mathscr{L}[\Phi(t)]$ of the macroscopic correlation $\Phi(t)$ is obtained by setting $\lambda = 0$.

The present results for $\phi(t)$ and $\Phi(t)$ have three important properties: the first is that they *do* satisfy Glarum's relation [5], and the second is that no shorter time relaxation effects appear. The third feature is that for a continuum external to the central dipole in a cavity of radius given by $r_0^3 = 3V/4\pi N$ (as assumed in the Onsager dielectric model), $S = \lambda^2$ and the relaxation function $[\phi(t)] = (i\omega + (1/\tau)$, i.e. relaxation is that of the central dipole by assumed diffusion forces.

3. Conclusion

We believe that the recent work of Glarum [4] and the present results have resolved some long standing and basic questions in theory of dielectric relaxation and dipole correlation functions. Prior to them, there were three different approaches and contradictory conclusions as to the relation of macroscopic dipole correlations: Glarum's original relation [5] from macroscopic response function arguments, relations similar to those of Fatuzzo and Mason [7] derived by generalization of Onsager's local cavity field model, and Zwanzig's relaxation function for diffusional reorientations in long range dipole fields. Glarum's work in our opinion correctly identifies the error in use of molecular cavity arguments, our derivation of his original results is based on quite general molecular response theory, and our use of rotational correlation distribution functions from Kirkwood's theory gives results for Zwanzig's model in agreement with Glarum's relation. This consistent state of affairs is also more pleasant because of its comparative simplicity, as there are no more complex relaxation effects in $\phi(t)$ *vis-à-vis* $\Phi(t)$ or in both from long range dipole forces.

Remerciements

Une partie des théories présentées dans cet article s'est dévelopée au cours de mon séjour en France en 1969–70 en tant que Professeur Associé à la Faculté des Sciences d'Orsay. Je voudrais remercier M. Magat et particulièrement C. Brot pour tous les services qu'ils m'ont rendus.

References

1. Kubo, R.: *J. Phys. Soc. Japan* **12**, 570 (1957).
2. Glarum, S. H.: *J. Chem. Phys.* **33**, 1371 (1960).
3. Cole, R. H.: *J. Chem. Phys.* **42**, 637 (1965).
4. Glarum, S. H.: *Mol. Phys.* **24**, 1327 (1972).
5. Kirkwood, J. G.: *J. Chem. Phys.* **7**, 911 (1939).
6. Cole, R. H.: *J. Chem. Phys.* **27**, 33 (1957).
7. Fatuzzo, E. and Mason, P. R.: *Proc. Phys. Soc.* **90**, 729 (1967).
8. Cole, R. H.: *Mol. Phys.*, to be published.
9. Powles, J. G.: *J. Chem. Phys.* **21**, 633 (1953).
10. Zwanzig, R. W.: *J. Chem. Phys.* **38**, 2766 (1963).
11. Van Vleck, J. H.: *J. Chem. Phys.* **5**, 556 (1937).
12. Rosenberg, R. and Lax, M.: *J. Chem. Phys.* **21**, 424 (1953).
13. Cole, R. H.: *J. Chem. Phys.* **39**, 2602 (1963).
14. Scaife, B. K. P.: in J. G. Calderwood (ed.), *Complex Permittivity*, English Universities Press, Ltd. London, 1971.
15. Cole, R. H.: *Mol. Phys.*, to be published.
16. Chandresekhar, S. 1943 *Rev. Mod. Phys.* **15**, 1.
17. Kirkwood, J. G.: *J. Chem. Phys.* **14**, 180 (1946).

DISCUSSION

Kneubühl: In your paper, you conclude that the theory of Fatuzzo and Mason [1] is rather doubtful. This confirms the observations by Keller *et al.* ([2], Figure 1) on the equation found by Fatuzzo and Mason for the dielectric dispersion, that for high frequencies the absorption becomes negative. What is the theoretical background of this non-realistic phenomenon?

Cole: I believe the explanation of the negative absorption of Fatuzzo and Mason's theory lies in the fact that artificial surface states are introduced by their use of an Onsager model, and that energy emission results from decay of polarization of these unreal surface states.

Friedman: In the crucial step in which the perturbation term in the operator equation is factored to get a product of correlation functions, you need a second equilibrium distribution function factor and I wonder where it comes from?

Cole: One factor has the correlation of a single dipole with other dipoles, the other has the sum of momentum derivatives of the time evolution of all dipoles. Only for a very limited number, of order $1/n$, of the totality of terms when the operators are expanded in time series, for example, will the products be correlated, while for the others the averages of terms from the two factors can be performed separately.

Brot: I note that you changed your notation with respect to your previous papers: your $\gamma(t)$, which was the vector *ACF* (my $F_1(t)$) is now the collective *normalized CF* (corresponding to $g(t)/g$ in my notation). I prefer the *unnormalized CF* $g(t)$ because it contains the static information. Moreover the second moment of its spectral density is simple, being a one-molecule property.

Cole: These differences are of course a matter of taste. The preference for $g(t)$ has the advantages you mentioned, but one can also prefer to compare correlation functions normalized with respect to equilibrium in order to see differences in the form of time dependence more easily.

Rivail: Le raisonnement de Fatuzzo et Mason est basé sur des considérations de continuité entre la sphère et le milieu qui l'entoure. Cette continuité existe-t-elle pour les relations de Glarum?

Cole: The question of continuity is a subtle one. For a dipole in an Onsager cavity as used by Fatuzzo and Mason, Glarum showed in his 1972 paper that there is continuity only if the dipole is parallel to the reaction field, whereas across a Kirkwood boundary as used by Glarum the polarization is continuous and the reaction field is always parallel to the moment of the interior region. The surface states mentioned in the reply to Dr Kneubühl's question are related to these differences.

References

1. Fatuzzo, E. and Mason, P. R.: *Proc. Phys. Soc. London* **11**, 729 (1967).
2. Keller, B., Ebersold, P., and Kneubühl, F.: *J. Phys.* **B1**, 688 (1970).

MOUVEMENTS ORIENTATIONNEL: FONCTIONS DE CORRELATION ET ILLUSTRATIONS EXPERIMENTALES

CLAUDE BROT

Laboratoire de Physique de la Matière Condensée*, Faculté des Sciences, 06 Nice, France

Résumé. On rappelle brièvement les définitions des Fonctions de Corrélation, Monomoléculaires et Collectives, décrivant les mouvements orientationnels de molécules ainsi que les formules qui permettent de les atteindre à partir des différentes techniques expérimentales. On classe, un peu arbitrairement, les modèles *physiques* proposés en 4 types: 'Brownien', 'gazeux', 'quasicristallin' et 'microcristallin', et on donne les valeurs relatives des différents temps de corrélation correspondants.

On décrit des exemples de recherches très fouillées: acétonitrile, fluorobenzène, 'azote d'ordinateur': les précisions expérimentales devraient être améliorées; de plus la nature – ou sa simulation numérique – est plus riche que les modèles.

On termine par des conjonctures sur des couplages hydrodynamiques aux temps très longs.

Abstract. The definitions of the time – or space-time correlations functions (CF) useful for the description of the orientational motions of molecules are recalled. The formulae relating them to the various experimental data are briefly discussed. The stress is put on the distinction between the techniques which give an information about the individual motion, for example $F_1(t) = \langle \mathbf{u}(0) \cdot \mathbf{u}(t) \rangle$, and the ones reflecting collective CFs such as $g(t) = \langle \mathbf{u}_i(0) \cdot \Sigma_i \mathbf{u}_i(t) \rangle$. Short time expansions of the CFs (moments of their spectral densities) are recalled. A re-derivation of the Gordon's Sum Rule (integrated dipolar absorption) via the CF for the 'rotational velocities' $\langle \dot{\mathbf{u}}_i(0) \Sigma_i \dot{\mathbf{u}}_i(t) \rangle$ is sketched.

With some unavoidable arbitrariness a classification of the physical models in the litterature into four types, called 'Brownian' 'Gaseous', 'Quasicrystalline' and 'Microcrystalline' is proposed; the second type refers to the 'J diffusion model'; the last two ones imply perturbed oscillations ('librations') about a diffusive or life-time limited equilibrium orientation. The relative magnitude of the various correlation times (expressed in the natural thermal unit $(I/kT)^{1/2}$ are given for the 4 types (see Table I in English). A word of caution is given against hasty conclusion drawn from exponential behavior of the CFs, and against systematic use of the Hubbard relation between the CF for the angular momentum and the ones of the spherical harmonics.

The experimental illustrations are relative to thoroughly studied examples: these are acetonitrile, fluorobenzene and 'Computer nitrogen'. It is remarked that, for conclusions to be completely reliable, experimental accuracies still need improvement. Also, in the last example, which is a numerical simulation, the motion cannot be classified into any of the models.

Finally, recent conjectures about the influence of hydrodynamic couplings on the very-long-time tails of the CFs are briefly discussed.

Comme toutes les quantités dont on cherche à décrire la dynamique moyenne dans un ensemble statistique, le mouvement angulaire des molécules au sein d'un liquide se décrit à l'aide de fonctions de correlation (FC) spatiotemporelles. On peut imaginer un grand nombre de telles fonctions, qui refléteraient les mouvements angulaires (et leurs couplages) sous leurs différents aspects. Nous nous bornerons à rappeler la définition des plus utilisées, et brièvement la manière dont on peut les atteindre à l'aide des différentes techniques expérimentales qui sont aujourd'hui disponibles. (D'autres

* Associé au C.N.R.S.

exposés apporteront les justifications et les raffinements de détail nécessaires dans chaque cas.) Nous décrirons ensuite les modèles les plus couramment proposés pour décrire qualitativement le mouvement orientationnel et nous indiquerons quel type de comportement chaque modèle fait prévoir pour les FC qui nous intéressent. Des illustrations expérimentales, ou tirées d'études en Dynamique Moléculaire Numérique, suivront. Nous terminerons en évoquant certaines conjectures qui ont été faites sur le comportement des FC orientationnelles aux temps très longs.

Nous définirons les FC en mécanique classique, qui est (sauf pour H_2, HCl, etc.) en général suffisante pour décrire les mouvements orientationnels (une correction quantique du premier ordre étant toujours possible en remplaçant dans l'argument de la FC le temps t par la quantité complexe $t - i\hbar/2kT$).

Les FC les plus utilisées sont des FC monomoléculaires ou fonctions d'autocorrelation; la dépendance spatiale de la FC disparaît alors, le seul argument de la fonction étant le temps.

Quand on considère des fonctions multimoléculaires, il s'agit le plus souvent de FC construites à l'aide de sommes de quantités (vecteurs par exemple) monomoléculaires. Sauf exceptionnellement on n'a pas besoin de faire l'analyse de Fourier spatiale de ces FC collectives, car on manque de méthodes expérimentales en dehors de Q (vecteur d'onde)≈ 0 (l'échelle des distances étant ici fournie par la portée des corrélations orientationnelles qui n'est pas considérable dans les liquides macroscopiquement isotropes).

Nous noterons τ_x le temps de corrélation (intégrale prise de 0 à $+\infty$) de la fonction de corrélation F_x.

1. Fonctions d'autocorrelations

1.1. FAC du moment angulaire total J d'une molecule

$$F_J(t) = \langle \mathbf{J}(0) \cdot \mathbf{J}(t) \rangle / \langle J^2 \rangle. \tag{1}$$

Elle peut s'atteindre par la relaxation magnétique nucléaire d'origine interaction spin rotation: pour les liquides, où le maximum de densité spectrale de cette fonction se trouve à des fréquences très élevées par rapport à la fréquence de Larmor, on a [1], pour une molécule sphérique et un couplage isotrope C_0

$$T_1^{-1} = \frac{2IkT}{\hbar^2} C_0^2 \tau_J.$$

1.2. FAC du 1er harmonique spherique ou FAC vectorielle

$F_v(t) \equiv F_1(t) = \langle \mathbf{u}(0) \cdot \mathbf{u}(t) \rangle$ où $\mathbf{u}$ est un vecteur fixe dans la molécule, le plus souvent le long d'un axe de symétrie.

Elle s'obtient par transformée de Fourier de la forme de raie infrarouge $I(\omega)$ d'une vibration dont le moment de transition est parallèle à $\mathbf{u}$. On a en effet * [2, 3]

---

* Nous supposons ici l'absence de couplage des oscillateurs moléculaires à l'état liquide [4], supposition valable au moins en solution.

$$I(\omega) \propto \int_{-\infty}^{+\infty} dt\, e^{-i\omega t} G(t) F_1(t). \tag{2}$$

$G(t)$ est une fonction de relaxation vibrationnelle qui peut s'obtenir, pour la *même raie* à partir du spectre Raman isotrope $I_{\text{iso}}(\omega) \simeq TF\, G(t)$. L'équation (2) implique une déconvolution des $I(\omega)$, ou, si les raies sont à peu près lorentziennes, une soustraction des *demi* largeurs à mi hauteur que nous noterons $\delta v(\text{cm}^{-1})$:

$$\frac{1}{2\pi c\tau_1} = \delta v_{\text{IR}} - \delta v_{\text{iso}}.$$

Des corrections quantiques, de champ interne, de soustraction de 'bandes chaudes' etc. sont souvent nécessaires pour utiliser (2).

1.3. FAC DU 2E HARMONIQUE SPHERIQUE OU FAC TENSORIELLE

$$F_T(t) \equiv F_2(t) = \tfrac{1}{2}\langle 3[\mathbf{u}(0)\cdot\mathbf{u}(t)]^2 - 1\rangle \equiv P_2[\mathbf{u}(0)\cdot\mathbf{u}(t)].$$

(P_l polynome de Legendre).

On a, pour une vibration symétrique A_1 d'une toupie symétrique ($\mathbf{u}$ le long de l'axe), pour la forme de raie Raman dépolarisée [2, 5]

$$I(\omega)_{VH} \simeq \int_{-\infty}^{+\infty} dt\, e^{-i\omega t} G(t) F_2(t): \tag{3}$$

Mêmes remarques que pour l'obtention, par l'infrarouge, de $F_1(t)$.

Par RMN, pour certains types d'interaction [6], on peut obtenir aussi cette fonction, ou plutôt, dans les liquides non visqueux, seulement son intégrale τ_2 (souvent notée τ_c ou τ_θ):

(a) Si l'interaction est quadrupolaire électrique ($\mathbf{u}$ dans la direction du gradient de champ $\partial^2 V/\partial z'^2$), on a, pour un spin $S=1$:

$$T_1^{-1} = \frac{3}{8}\left(\frac{eQ}{\hbar}\frac{\partial^2 V}{\partial z'^2}\right)^2 \tau_2.$$

(b) Si l'interaction est dipolaire magnétique entre deux spins S ($\mathbf{u}/\!/\mathbf{r}$, vecteur joignant les 2 spins):

$$T_1^{-1} = \frac{2\gamma^4\hbar^2 S(S+1)}{r^6}\tau_2.$$

Des précautions sont nécessaires en RMN (par dilution isotopique par exemple), pour isoler l'effet intramoléculaire ci-dessus des contributions intermoléculaires.

1.4. LA DIFFUSION INCOHERENTE DES NEUTRONS

Signalons que la diffusion incohérente des neutrons permet, dans l'hypothèse où le mouvement translationnel est découplé du mouvement orientationnel, d'obtenir une somme pondérée des FAC des harmoniques sphériques successifs [7].

2. Les fonctions de corrélation collectives utiles

2.1. ORIENTATIONS VECTORIELLES

Pour préserver un degré de généralité qui pourrait être utile dans l'avenir, nous proposons de définir une fonction de corrélation spatiotemporelle décrivant, sous son aspect vectoriel, le caractère collectif de des réorientations moléculaires:

$$C_1(\mathbf{r}, t) = \frac{1}{n_0} \left\langle \sum_{i,j} \delta(\mathbf{r}_i(0)) \mathbf{u}_i(0) \cdot \mathbf{u}_j(t) \delta(\mathbf{r} - \mathbf{r}_j(t)) \right\rangle, \qquad (4)$$

où n_0 est la densité moyenne en nombre des molécules et où les $\mathbf{r}$ sont des vecteurs positions. La somme porte sur un domaine intérieur au liquide de préférence sphérique, plus grand que la portée des corrélations, et contenant $N = V n_0$ molécules. On a:

$$C_1(\mathbf{r}, t) = \frac{1}{N} \left\langle \sum_{i,j} \mathbf{u}_i(0) \cdot \mathbf{u}_j(t) \delta(\mathbf{r} + \mathbf{r}_i(0) - \mathbf{r}_j(t)) \right\rangle.$$

Considérons la transformée de Fourier spatiale $\hat{C}_1(\mathbf{Q}, t)$ de cette FC, et prenons la à $\mathbf{Q} = 0$:

$$\hat{C}(0, t) = \frac{1}{N} \left\langle \sum_{i,j} \mathbf{u}_i(0) \cdot \mathbf{u}_j(t) \right\rangle$$

$$= \left\langle \mathbf{u}_i(0) \sum_i \mathbf{u}_i(t) \right\rangle \equiv g(t). \qquad (5)$$

L'avant dernière égalité résulte de l'équivalence statistique des molécules i. La notation employée dans le membre de droite rappelle qu'au temps $t = 0$ notre fonction est identique au facteur de correlation statique de Kirkwood-Frohlich [8] habituellement noté g. La partie 'self' de $g(t)$ n'est autre que $F_1(t)$.

Notre fonction $g(t)$ peut s'obtenir par la mesure de l'absorption-dispersion due aux dipôles permanents moléculaires que nous noterons $\boldsymbol{\mu}_v = \mu_v \mathbf{u}$, l'indice v rappelant que nous utiliserons le dipôle permanent de la molécule isolée tel que mesuré in vacuo. Cette absorption étant située à des longueurs d'onde (micro-ondes et infra rouge lontain) beaucoup plus grande que la portée des corrélations, il est naturel que seule la FC à $Q = 0$, c'est-à-dire $g(t)$ soit d'ordinaire utilisée.

A la suite de Glarum [9] et Cole [10] on peut noter que l'on a, pour cette absorption - dispersion dipolaire, avec les notations habituelles pour les constantes diélectriques:

$$\mathscr{L}[-\dot{g}(t)] = (\varepsilon^*(\omega) - n^2) \frac{(2\varepsilon_0 + n^2)}{(2\varepsilon_0 + \varepsilon^*(\omega))} \left(\frac{3}{n^2 + 2}\right)^2 \frac{3kTV}{4\pi N_A \mu_v^2}, \qquad (6)$$

où $\mathscr{L}$ dénote la transformée de Laplace imaginaire $\int_0^\infty e^{-i\omega t} \ldots dt$ et $\dot{g}(t)$ est la dérivée de $g(t)$. Dans le membre de droite N_A et V sont respectivement le nombre d'Avogadro et le volume molaire.

D'autres expressions théoriques ont été proposées [11, 12] pour relier la constante diélectrique complexe $\varepsilon^*(\omega)$ aux fonctions de corrélation $g(t)$ ou $F_1(t)$, mais des travaux théoriques récents [13, 14] semblent montrer que l'expression ci-dessus est la seule correcte. Cette formule prise à $\omega=0$ redonne la formule de Onsager-Frohlich-Kirkwood pour la constante diélectrique statique ε_0:

$$\frac{(\varepsilon_0 - n^2)(2\varepsilon_0 + n^2)}{(n^2 + 2)^2} = \frac{4\pi N_A g \mu_v^2}{9kTV}.$$

On remarque que l'expression (6) donne pour un corps peu polaire une simple proportionnalité entre $\varepsilon^*(\omega) - n^2$ et $\mathscr{L}[-\dot{g}(t)]$; la partie réelle de cette dernière donne alors une proportionnalité entre ε''/ω et la transformée de Fourier (bilatérale) de $g(t)$. Ceci est la limite classique d'un résultat qui avait été obtenu directement par R. G. Gordon [2].

D'autre part, si $g(t)$ a une décroissance essentiellement exponentielle: $g(t) = ge^{-t/\tau_m}$, on a

$$\varepsilon^*(\omega) - n^2 = \frac{\varepsilon_0 - n^2}{(1 + i\omega\tau_D)} \text{ (loi de Debye)},$$

où le 'temps de relaxation diélectrique' τ_D vaut:

$$\tau_D = [\varepsilon_0/(2\varepsilon_0 + n^2)]\tau_m.$$

Ceci est la correction semi empirique antérieurement proposée par Powles [15]. *Elle est souvent indispensable.*

2.2. Orientations tensorielles

Les FC correspondantes sont reliées par transformées de Fourier à la forme de la raie Rayleigh dépolarisée diffusée par les molécules anisotropes. Les formulations qu'on en peut faire ne sont pas encore fixées par l'usage. La formulation de Pecora et Steele [16] est très analogue à celle que nous proposons pour $\hat{C}_1(Q, t)$. Une formulation un peu différente a été proposée par Ben Reuven et Gershon [17]. Pour des molécules toupies symétriques et si les corrélations sont négligeables toutes deux se réduisent à $F_2(t)$

2.3. Troisieme harmonique spherique

La FC collective correspondante peut s'obtenir, quoique difficilement, par l'effet 'hyper Rayleigh' [53].

2.3.1. *Comportement des FC aux temps courts*

Rappelons qu'on a pour une molécule linéaire de moment d'inertie I, les développements limités suivants [2]:

$$F_1(t) = 1 - (kT/I)\,t^2 + \left[\frac{1}{3}\left(\frac{kT}{I}\right)^2 + (24I^2)^{-1}\langle OV^2\rangle\right]t^4 + \cdots$$

$$F_2(t) = 1 - (3kT/I)\,t^2 + \left[4\left(\frac{kT}{I}\right)^2 + (8I^2)^{-1}\langle OV^2\rangle\right]t^4 + \cdots$$

$$g(t) = g - (kT/I)\,t^2 + \cdots .$$

Les coefficients sont les moments successifs des densités spectrales correspondantes. Le fait que le second moment de $g(t)$ soit le même que celui de $F_1(t)$ résulte du fait que l'énergie cinétique rotationnelle est diagonale par rapport aux molécules [18]. Cette remarque permet de retrouver immédiatement la règle de somme de Gordon [19] sur l'intégrale de l'absorption dipolaire α. En effet on peut définir une 'FC collective des vitesses rotationnelles' [18]:

$$C_{rv}(t) = \left\langle \dot{\mathbf{u}}_i(0)\cdot\sum_j \dot{\mathbf{u}}_j(t)\right\rangle \equiv -\ddot{g}(t) = (2kT/I) - \cdots .$$

Cette fonction (dont l'intégrale est nulle) a pour transformée de Fourier une quantité proportionnelle à $\omega\varepsilon''$, donc à peu près proportionnelle à l'absorption par unité de longueur α. L'intégrale de α est donc constante et proportionnelle au terme constant du développement ci dessus. Comme ε'' contient (μ_v^2/kT) en facteur, on trouve pour l'absorption intégrée une valeur proportionnelle à (μ_v^2/I), indépendante de la température et des interactions.

Une autre fonction très sensible à la nature du mouvement orientationnel aux temps courts est évidemment $F_j(t)$. On peut montrer [18] que si les perturbations du mouvement du rotateur sont très fréquents, $F_{J\perp}(t)$ approxime très bien la partie self de $C_{rv}(t)$ (sauf pour la longue queue négative de cette dernière*). Ce comportement commun peut être, comme on le verra, soit amorti dans les cas des collisions aléatoires fréquentes, soit oscillant amorti s'il existe des librations perturbées au sein d'une structure locale.** Dans les phases peu denses où les interactions sont rares, F_J décroit lentement tandis que C_{rv} montre quelques oscillations dont l'origine est le renversement des vitesses rotationnelles quand les molécules ont en moyenne effectué un demi-tour.

3. Types de modèles physiques[†] les plus fréquemment utilisés

Notre classement des différents modèles proposés est très schématique et un peu arbitraire, des situations intermédiaires entre les différents types pouvant être obser-

---

* Qui résulte du fait que son intégrale est nulle.
** Comme dans les modèles ③ et ④ du paragraphe suivant, ce qui donne lieu, pour l'absorption dipolaire dans l'IR lointain, à une absorption 'excédentaire' [28].
[†] Nous ne parlerons pas ici des modèles mathématiques tels que le formalisme des fonctions mémoires, qui sont d'emploi commode pour satisfaire certaines règles de somme, mais qui seront évoquées dans d'autres exposés.

vées. Nous le présentons sous forme de quatre types généraux et ne le donnons que pour fixer les idées (et peut être la terminologie). Nous nous intéresserons surtout dans ce paragraphe au comportement des FC aux temps assez longs. Pour une toupie symétrique ayant un moment d'inertie I autour d'un axe perpendiculaire à l'axe de symétrie, la vitesse angulaire thermique moyenne de l'axe de symétrie est $(2kT/I)^{1/2}$ rad s^{-1}. La quantité $(I/kT)^{1/2}$ peut donc servir d'échelle des temps. Nous noterons τ^* les temps τ exprimés dans cette unité.

Les modèles proposés, essentiellement monomoléculaires, ont en commun l'hypothèse que le mouvement angulaire de chaque molécule a lieu par une série de périodes de rotations libres, plus ou moins longues, interrompues soit par collisions, soit par des stades de 'piégeage orientationnel'. On peut dans chaque cas définir une durée moyenne des périodes de rotation libre, τ_{FL} et une durée moyenne des stades de piégeage (résidence autour d'une direction fixe) τ_{RES}. S'il n'y a que des collisions instantanées, $\tau_{RES}=0$. Le nombre moyen de sauts par unité de temps est évidemment $(\tau_{RES}+\tau_{FL})^{-1}$.

Les caractères essentiels des 4 types de modèles que nous considérons sont résumés dans la Table I. Quelques commentaires s'imposent:

Les dénominations proposées s'expliquent d'elles-mêmes dans le cas des modèles ① [20] et ② [21-23]; pour les modèles ③ [24] et ④ [25-28], elles veulent refléter le fait que, dans le modèle ③ le mouvement du rotateur est celui qu'il aurait dans une matrice quasicristalline qui se déformerait progressivement, tandis que dans le modèle ④ c'est celui qu'il aurait au sein d'une microstructure de durée de vie τ_{RES} [25, 28].

En considérant les 4 modèles on voit que le fait qu'une fonction de correlation orientationnelle se révèle expérimentalement exponentielle n'est pas un critère suffisant pour le choix du modèle ① par exemple. On aurait donc tort d'utiliser aveuglément la relation de Hubbard [29], qui s'écrit pour la FC d'un harmonique sphérique d'ordre l quelconque:

$$\tau_J^* \tau_l^* = [l(l+1)]^{-1}.$$

Cette relation n'est valable que pour le modèle ①. Dans le modèle ④ par exemple, au cas où τ_{RES} serait très long, elle ferait prévoir des τ_J absurdement courts.

Le problème de la marche au hasard, par sauts discrets, d'un rotateur sur la sphère unité a été traité par Ivanov [30] puis Montrose et Litovitz [31] dans le cas où on a une distribution des amplitudes de saut, et par Anderson [32] dans le cas où l'amplitude des sauts est constante. Seul est introduit dans le problème un temps moyen entre sauts, τ, qui pour nous vaut $\tau_{RES}+\tau_{FL}$. Le problème est donc applicable à notre modèle ④ ($\tau \approx \tau_{RES}$). Il l'est aussi au modèle ① en prenant $\tau^* \approx \tau_{FL}^* \ll 1$. La courbe τ_1/τ_2 en fonction de l'amplitude α des sauts, longtemps voisines de 3, tend évidemment vers zéro pour $\alpha = \pi$ dans l'hypothèse d'Anderson. Elle tend vers l'unité pour $\langle \alpha \rangle$ grand dans l'hypothèse de Montrose et Litovitz, probablement plus réaliste dans les liquides. La connaissance simultanée de τ_1^* et τ_2^* permet donc, en principe, de déterminer l'amplitude moyenne des sauts. La comparaison de ces quantités réduites avec l'unité permet éventuellement de choisir entre le modèle ② et le modèle ④.

TABLE I

Type	Description	τ^*_{RES}	τ^*_{FL}	Behavior of the CF (short time excepted)	Relations
(1) 'Brownian'	Rotational Diffusion by small uncorrelated steps (Langevin – Debye).	0	$\ll 1$	F_J exponential decaying fastly, F_1, F_2 exponential	$\tau_J = \tau_{FL} \ll 1$ $\tau^*_J \tau^*_2 = \frac{1}{6}$ (Hubbard) $\tau_1/\tau_2 = 3$
(2) 'gaseous'	Free rotation interrupted by instantaneous collisions randonizing J (or only its direction).	0	$\gtrsim 0.2$	F_J exponential F_1 can become negative if T^*_{FL} large	$\tau_J = \tau_{FL}$ (if $\tau^*_{FL} \approx 1$, $\tau_1/\tau_2 \approx 1$)
(3) 'quasicrystalline'	Perturbed Librations about diffusive equilibrium orientation.	elusive	$\ll 1$	F_J: (over)-damped oscillations F_1, F_2: exponential	$\tau^*_J \ll 1$ $\tau_1/\tau_2 = 3$
(4) 'microcrystalline'	Perturbed librations about fixed equilibrium orientation interrupted by large angle jumps.	$\geqslant 1$	$0.2 \gtrsim \tau_{FL} \ll \tau_2$	F_J: (over)-damped oscillations F_1, F_2 exponential if jumping times are governed by a Poisson's distribution.	$\tau_J \ll \tau_1 \approx \tau_2 \approx \tau_{RES}$

Disons quelques mots pour terminer des transitions possibles entre les différents modèles:

(a) Quand les sauts deviennent petits, le modèle ② redonne le modèle ①.

(b) Si, au lieu d'être non corrélés, les petits sauts du modèle ① présentent une anticorrélation (i.e. le renversement de la vitesse angulaire est privilégié), et si les collisions deviennent non-instantanées ('molles') on aboutit à la situation du modèle ③.

(c) Une combinaison des situations ③ et ④ est concevable, etc.

4. Quelques exemples

4.1. Etudes experimentales

4.1.1. *L'acétonitrile*

L'acétonitrile, CH_3–CN, toupie symétrique, a été étudiée en phase liquide pure, à l'aide d'un grand nombre de techniques, si bien qu'on devrait se faire une image claire, qualitativement et quantitativement, du mouvement orientationnel de ses molécules. Ce qui suit est relatif à la température de 25 °C. Nous nous efforcerons dans la mesure du possible, d'effectuer toutes les corrections particulières nécessitées par chaque technique.

Nous considérons d'abord le mouvement de l'axe moléculaire C_3 (nous envisagerons après le mouvement autour de cet axe).

Commençant par les techniques qui reflètent le mouvement individuel des molécules, nous sommes en présence de plusieurs études de *RMN*. Woessner *et al.* [33], Bopp [34], à partir de la relaxation quadrupolaire de l'azote, donnent respectivement $\tau_2 = 1,104$ ps et 1,23 ps pour CD_3CN. Les forces intermoléculaires étant les mêmes que pour la molécule non deutérée, et le moment d'inertie perpendiculaire à l'axe moléculaire étant peu différent, la valeur de τ_2 devrait être très voisine pour CH_3CN. C'est en effet ce que trouve Lyerla *et al.* [35] 1,2 ps.

Les largeurs de raie Raman, quand on constate qu'elles dépendent peu de la dilution dans un solvant inerte, peuvent alors être considérées comme d'origine purement monomoléculaire. C'est le cas de la fréquence v_4 (918 cm^{-1}) de l'acétonitrile [36]. Encore faut-il faire la correction de relaxation vibrationnelle. Pour des FC à peu près exponentielle, les raies sont sensiblement lorentziennes sauf éventuellement dans les ailes, et il suffit pour ce faire de mesurer la largeur vibrationnelle (raie diffusée isotropiquement). La demi largeur isotrope de v_4 est, après correction pour la fonction d'appareil, $\delta v_{iso} = 1,2$ cm^{-1} selon Fried et Perchard [37] (contre 2,3 cm^{-1} selon Bucaro, mais nous choisissons la valeur la plus faible car c'est probablement la moins déformée par la fonction d'appareil).

La demi largeur de la partie dépolarisée de cette même raie vaut 6,3 cm^{-1} selon [37], ce qui donne $6,3 - 1,2 = 5,1$ cm^{-1} pour l'élargissement orientationnel, soit $\tau_2 = 1,1$ ps, en accord avec les mesures RMN (Le τ_2 Raman de Bartoli et Litovitz [39], 1,4 ps est un peu plus élevé.)

Passant au premier harmonique sphérique, cette même vibration v_4 examinée en infra rouge doit, pour la raison mentionnée plus haut, donner un temps de corrélation

τ_1, sensiblement monomoléculaire. La demi largeur 3,8 cm^{-1}, mesurée par Vincent et al. [36] et par Rakov [40] dans le liquide pur, donne après soustraction de δv_{iso}, 2,6 cm^{-1} pour l'élargissement orientationnel, soit $\tau_1 = 2$ ps

$$\text{Nous trouvons donc } \frac{\tau_1}{\tau_2} = \frac{2}{1,1} = 1,8.$$

Avec le modèle à sauts distribués en amplitude, de Montrose et Litovitz [3], ceci correspond à une amplitude moyenne de sauts voisine de 40°.

Bartoli et Litovitz [39] proposent au contraire des sauts de 3° à partir de $\tau_2 = 1,3$ (Moyenne de la valeur RMN et de leur valeur Raman) et d'un $\tau_1 = 3,8$ emprunté à des mesures diélectriques [41], sur l'interprétation desquelles quelques commentaires nous semblent s'imposer: D'abord le temps diélectrique τ_D tiré par exemple d'un diagramme de Cole et Cole, est un temps de relaxation macroscopique. Comme on l'a vu la formule de Cole qui, dans le cas où les FC sont exponentielles, se réduit à la correction de Powles [15], permet de passer au temps de correlation microscopique τ_m. Avec $\tau_D = 4$ ps (valeur vers 25°C) et $\varepsilon_0 = 38$ et $n^2 = 2,1$, nous trouvons $\tau_m = 3,1$ ps. Le fait que ce chiffre soit différent de $\tau_1 = 2$ ps s'explique par la caractère collectif de l'interaction de l'onde électromagnétique hyperfréquence avec les dipoles moléculaires, si l'on admet que les mouvements de ceux-ci sont intercorrélés. Nous remarquons d'ailleurs que le même phénomène existe pour les tenseurs d'anisotropie permanente: le mouvement individuel de ces tenseurs donne $\tau_2 = 1,1$ ps comme on l'a vu, alors que collectivement (i.e. somme des anisotropies localement corrélées) on a par mesure de la largeur Rayleigh dépolarisée [42], $\tau_{2\,\text{collec}} = 1,8$ ps. Une présomption en faveur de l'existence de *corrélations dynamiques* d'orientation des molécules d'acétonitrile entre elles est aussi l'existence d'une *corrélation statique* non négligeable: En utilisant la valeur $\mu = 2,95$ D (mesurée dans le gaz) par le moment dipolaire, on trouve pour l'acétonitrile à température ambiante un facteur de correlation diélectrique de Kirkwood égal à 0,72. Il est probable que ces corrélations assez importantes sont dues à la fois à la valeur élevée du dipole et à l'anisotropie stérique de la molécule [43].

Revenant au mouvement orientationnel dans CH$_3$CN sous son aspect monomoléculaire, on peut essayer de choisir entre les différents modèles envisageables (Table I). Selon Bartoli et Litovitz [39], le modèle adéquat serait, puisque $\tau_1/\tau_2 \approx 3$, le modèle ① (Brownien). Puisque nous trouvons, au contraire, des sauts angulaires de grande amplitude, nous avons le choix entre le modèle ② et le modèle ④. Pour un saut de 40°, la vitesse angulaire moyenne étant $\sqrt{2}\sqrt{kT/I} = 3 \times 10^{12}$ rad s^{-1}, on obtient $\tau_{FL} = 0,23$ ps. D'autre part, en utilisant toujours le modèle à sauts distribués en amplitude de Ivanof-Montrose-Litovitz, au rapport $\tau_1/\tau_2 = 1,8$ correspond le rapport $\tau_1/\tau \approx 4$, où τ est le temps moyen entre sauts, c'est-à-dire, $\tau_{RES} + \tau_{FL}$. Avec $\tau_1 = 2$ ps, nous en tirons $\tau = 0,5$ ps d'où $\tau_{RES} = 0,27$ ps. Vu le caractère arbitraire de la forme de distribution adoptée par les auteurs cités, nous ne donnons ces chiffres qu'à titre indicatif.

Le temps de résidence dans un site orientationnel est ainsi loin d'être nul, le modèle ② est donc exclu. D'ailleurs, s'il était valable, on aurait $\tau_J \approx \tau_{FL}$. Une limite supérieure de τ_J peut être évaluée à partir des mesures de la contribution de l'interaction spin rotation à la relaxation nucléaire du carbone-13 effectuées par Lyerla et al. [44] : elle est nettement inférieure à $0,23 \text{ ps} = \tau_{FL}$; ceci implique aussi l'existence d'un temps de résidence non nul pendant lequel le moment angulaire perd son autocorrelation par des mouvements oscillatoires plus ou moins chaotiques (librations perturbées). Remarquons aussi que le modèle ④, contrairement au modèle ②, n'implique pas que $F_1(t)$ garde longtemps un comportement de rotation libre. Il est donc compatible avec certains résultats d'infra rouge de Rotschild [45].

En résumé nous *concluons que le modèle* ④ *semble valable* pour le mouvement de l'axe moléculaire de l'acétonitrile à 25°C.

Pour terminer disons quelques mots du mouvement *autour* de l'axe moléculaire (la séparation de ce mouvement et du précédent est licite tant que la rotation n'est pas très libre). Des études RMN utilisant surtout la relaxation des deuterons dans CD_3CN montrent que le temps $\tau_{2\parallel}$ est dix fois plus court que pour le mouvement perpendiculaire à l'axe. La vitesse angulaire thermique est 4 fois plus grande à cause du faible moment d'inertie. Le modèle ② semble s'appliquer pour ce mouvement, bien qu'on soit proche de la limite où un traitement quantique deviendrait nécessaire.

4.1.2. *Le fluorobenzene-d_5*

On dispose d'une étude RMN très complète, de 30 à 309°C et de 1 à 2000 bars, par Jonas et al. [46]. Ces auteurs supposent la rotation isotrope ce qui leur permet de déterminer τ_2 et τ_J à partir des T_1 des deutérium (interaction quadrupolaire) et du fluor (surtout interaction de spin rotation). A basse température et haute pression $\tau_J^* \approx 0,1$, ce qui donne des angles de sauts de 8°. Le produit $\tau_2^* \tau_J^*$ est voisin de 0,24, ce qui confirme que la situation est proche de celle du modèle ①. La petite différence avec la valeur théorique $\frac{1}{6}$ pourrait s'interpréter par une tendance à des sauts anticorrelés.

A haute température le produit $\tau_2^* \tau_J^*$ augmente et reste proche de la valeur que fait prévoir le modèle ② (J diffusion). Des mesures d'absorption dipolaire vers la région critique de C_6H_5F nous ont d'ailleurs permis une conclusion analogue [18].

Le caractère encore relativement conjectural de nos conclusions en particulier quant au mouvement de l'axe moléculaire de l'acétonitrile pour lequel nous divergeons d'avec Litovitz montre à quel point des efforts sont encore nécessaires pour affiner les théories et améliorer les précisions expérimentales. En effet :

(i) On est gêné par l'incertitude qui règne encore sur la valeur théorique de certaines corrections en principe nécessaires sur les données expérimentales.

(ii) On est trop souvent tenté de substituer l'équivalent multimoléculaire d'une quantité à sa valeur monomoléculaire (ou réciproquement).

(iii) Pour des angles de sauts pas trop grands, le rapport τ_1/τ_2 est assez insensible à l'angle moyen de saut exact [31, 32].

(iv) Les précisions expérimentales (souvent 10 à 20%) sont encore insuffisantes, à cause en particulier de (iii)) ci-dessus.

4.2. Dynamique moléculaire numérique

Nous citerons des résultats récents relatifs à une simulation de l'azote liquide [47]. Le potentiel adopté est un potentiel de Lennard-Jones entre les atomes des différentes molécules. Ce potentiel a permis d'obtenir des valeurs assez réalistes pour les quantités thermodynamiques du liquide, ainsi que pour certains coefficients de transport, en particulier la constante de diffusion translationnelle et le temps de correlation orientationnel τ_2.

La figure 1 représente certaines fonctions de correlation orientationnelles dans le liquide au voisinage du point triple. Une caractéristique frappante de ces courbes est que la FC vectorielle reste longtemps confondue avec la FC de rotation libre, du

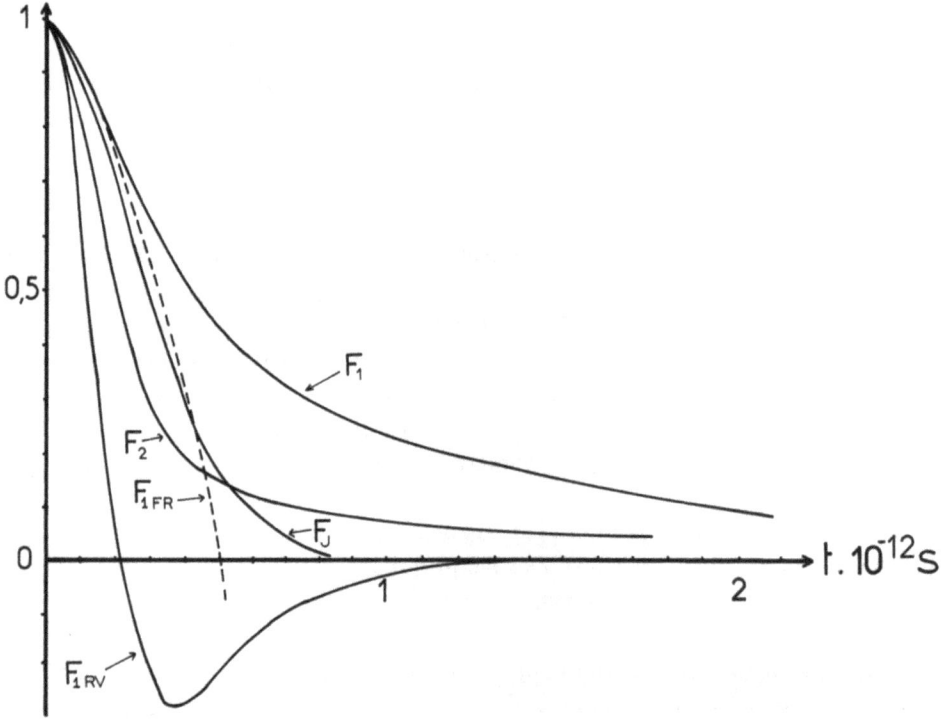

Fig. 1.

moins si on la compare à celles obtenues expérimentalement sur des molécules pluriatomiques. Le comportement aux temps longs est exponentiel, mais ne commence qu'assez tard, après que la FC a déjà décru de près de 60%. Aux temps longs F_2 et F_1 décroissent avec le même temps caractéristique. (cependant, définis comme les aires sous les courbes, τ_1 et τ_2 ne sont pas égaux).

La comparaison de F_J et de $F_{IRV} = \langle \dot{\mathbf{u}}(0) \dot{\mathbf{u}}(t) \rangle / (2kT/I)$ confirme que les interactions avec les molécules voisines sont moins fréquentes et moins fortes que pour des molécules plus grosses et plus allongées.

Des résultats complémentaires obtenus tout récemment [48] complètent cette image: sur la figure 2 la courbe a représente la fonction $\langle J^2(0)\,\mathbf{u}(0)\cdot\dot{\mathbf{u}}(t)\rangle/\langle J^2\rangle$. La fait qu'elle se trouve nettement au dessous de $F_1(t)$ confirme qu'une température rotationnelle initiale élevée facilite la réorientation. Par ailleurs les fluctuations *durables* de densité locale ont aussi une influence importante: la courbe b représente $F_1(t)$ échantillonné seulement sur les molécules pour lesquelles la densité locale* demeure supérieure de 10% au moins à la moyenne: la décroissance de $F_1(t)$ pour ces molécules est sensiblement ralentie.

Le mouvement orientationnel de molécules prises au hasard a été suivi individuel-

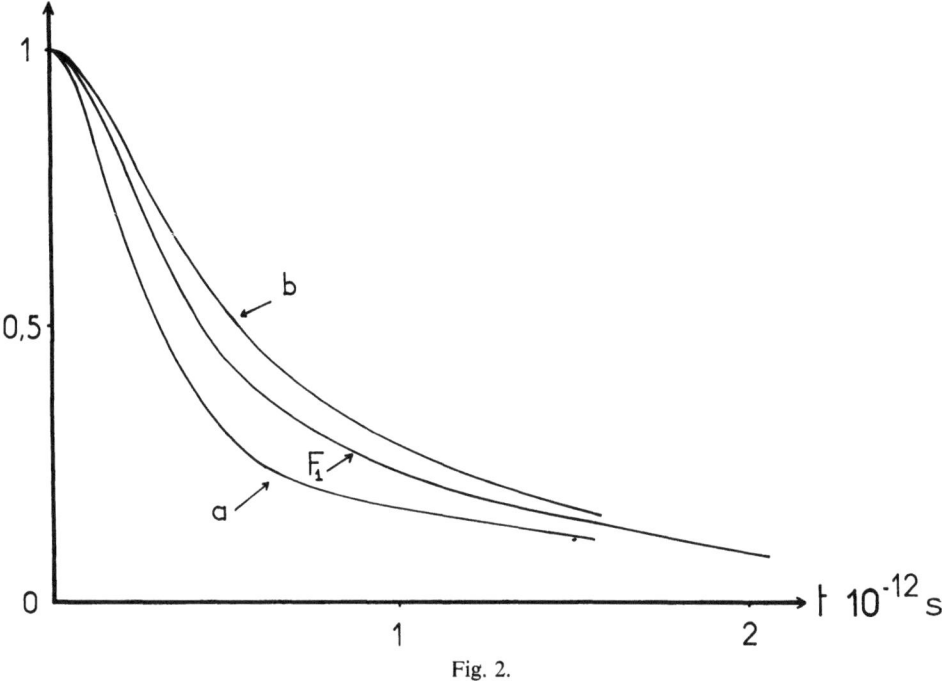

Fig. 2.

lement; on a constaté que les molécules restent assez longtemps autour d'une orientation moyenne, en effectuant des oscillations désordonnées autour de cette orientation, puis entrent dans des stades de rotation non uniforme mais continue pour des périodes assez longues, avant d'être de nouveau 'piégées'.

En résumé aucun des modèles précédemment cités ne semblent s'appliquer vraiment. Les deux modèles les moins éloignés des résultats obtenus sont soit le modèle ② (mais il n'y a pas vraiment de 'collisions') soit le modèle ④ (mais il faudrait pendre τ_{FL} du même ordre de grandeur que τ_{RES}).

Il est évident qu'il serait dangereux d'extrapoler ces résultats à des molécules d'une élongation supérieure ou d'une structure beaucoup plus complexe.

---

* Evaluée pour une sphère contenant en moyenne 9 molécules voisines.

5. Conjonctures sur le comportement 'hydrodynamique' des FC orientationnelles aux temps très longs

Des travaux théoriques récents font penser qu'aux temps très longs les fonctions de corrélation orientationnelles prennent des valeurs finales petites mais supérieures à la valeur extrapolée des temps moins longs. Ce comportement ultimement non stochastique serait dû au couplage avec des mouvements hydrodynamiques du liquide environnant. Dans le cas de la FC du moment angulaire moléculaire $F_J(t)$, le phénomène serait le suivant: A cause de la conservation du moment angulaire total, le moment angulaire initial d'une molécule se dissiperait en créant autour de la molécule un mouvement de vortex. Ce premier stade serait assez rapide et correspondrait à la partie 'temps moyen' de $F_J(t)$. Tant que la densité locale du vortex formé n'aurait pas décru par diffusion, le vortex entraînerait la molécule dans son mouvement de rotation. Ceci créerait pour $F_J(t)$ une longue queue décroissant selon la loi $t^{-5/2}$. Ce résultat a été obtenu d'une part en symbolisant la molécule par l'élément de volume élémentaire d'un fluide visqueux [49, 50], d'autre part en la représentant par une petite sphère rugueuse [51].

Des travaux récents [51, 52] semblent indiquer que certaines des fonctions de corrélation $F_l(t)$ auraient aussi des longues queues non exponentielles.

Bibliographie

1. Rigny, P. et Virlet, J.: *J. Chem. Phys.* **47**, 4645 (1967).
2. Gordon, R. G.: *J. Chem. Phys.* **43**, 1307 (1965).
3. Bratoz, S., Rios, J., et Guissani, Y.: *J. Chem. Phys.* **52**, 439 (1970).
4. Vincent-Geisse, J.: *J. Spectrochim. Acta* **24A**, 1 (1968).
5. Bratos, S. et Marechal, E.: *Phys. Rev* **A4**, 1078 (1971).
6. Abragam, A.: *Les principes du magnétisme nucléaire,* PUF, Paris, 1961.
7. Sears, V. F.: *Can. J. Phys.* **44**, 1299 (1966).
8. Fröhlich, H.: *Theory of Dielectrics,* Clarendon Press, Oxford, 1958.
9. Glarum, S. H.: *J. Chem. Phys.* **33**, 1371 (1960).
10. Cole, R. H.: *J. Chem. Phys* **42**, 637 (1965).
11. Fatuzzo, E. et Mason, P. R.: *Proc. Phys. Soc.* **90**, 741 (1967).
12. Rivail, J. L.: *J. Chim. Phys.* **66**, 981 (1969).
13. Glarum, S. H.: *Mol. Phys.* **24**, 1327 (1972).
14. Cole, R. H.: ce volume, p. 97.
15. Powles, J. G.: *J. Chem. Phys.* **21**, 633 (1953).
16. Pecora, R. et Steele, W. A.: *J. Chem. Phys* **42**, 1872 (1965).
17. Ben-Reuven, A. et Gershon, N. D.: *J. Chem. Phys.* **51**, 893 (1969).
18. Gerschel, A., Darmon, I., et Brot, C.: *Mol. Phys.* **23**, 317 (1972).
19. Gordon, R. G.: *J. Chem. Phys.* **38**, 1724 (1963).
20. Debye, P.: *Polar Molecules,* Dover, 1929.
21. Gordon, R. G.: *J. Chem. Phys.* **44**, 1830 (1966).
22. McClung, R. E. D.: *J. Chem. Phys.* **51**, 3842 (1969).
23. Fixman, M. et Rider, K.: *J. Chem. Phys.* **51**, 2425 (1969).
24. Hill, N. E.: *Proc. Phys. Soc.* **82**, 723 (1963).
25. Brot, C., Magat, M., et Reinisch, L.: *Koll. Zeits.* **134**, 101 (1953).
26. Anderson, J. E. et Ullman, R.: *J. Chem. Phys.* **47**, 2178 (1967) (third case).
27. McDuffie, G. E. et Litovitz, T. A.: *J. Chem. Phys.* **37**, 1699 (1962).
28. Lassier, B. et Brot, C.: *Chem. Phys. Letters* **1**, 581 (1968).
29. Hubbard, P. S.: *Phys. Rev.* **131**, 1155 (1963).

30. Ivanov, E. N.: *Sov. Phys. JETP* **18**, 1041 (1964).
31. Montrose, C. J. et Litovitz, T. A.: *Neutron Inelastic Scattering* 1, IAEA, Vienne, 1968, p. 623.
32. Anderson, J. E.: *Faraday Symp.* no. 6 (1972).
33. Woessner, D. E., Snowden Jr, B. S., et Strom, E. T.: *Mol. Phys.* **14**, 265 (1968).
34. Bopp, T. T.: *J. Chem. Phys.* **47**, 3621 (1967).
35. Lyerla Jr., J. R., Grant, D. M., et Harris, R. K.: *J. Phys. Chem.* **75**, 585 (1971).
36. Alliot, C., Cameo, M., et Vincent-Geisse, J.: *Compt. Rend. Acad. Sci. Paris* **266B**, 391 (1968).
37. Fried, F. et Perchard, J. P.: communication privée.
38. Bucaro, J. A.: Thèse, Catholic. Univers. Am., Washington, 1971.
39. Bartoli, F. J. et Litovitz, T. A.: *J. Chem. Phys.* **56**, 413 (1972).
40. Rakov, A. V.: *Opt. Spec. USSR* **13**, 203 (1962).
41. Krishnaji et Mansingh, A.: *J. Chem. Phys.* **41**, 827 (1964).
42. Bartoli, F. J. et Litovitz, T. A.: *Magnetic Relaxation Phenomena and Internal Kinetics of Fluid Systems*, 8th Colloquium, Aachen 1971, p. 516.
43. Weisbecker, A.: *J. Chim. Phys.* **66**, 1442 (1969).
44. Lyerla Jr, J. R., Grant, D. M., et Wang, C. H.: *J. Chem. Phys.* **55**, 4676 (1971).
45. Rothschild, W. G.: *J. Chem. Phys.* **57**, 991 (1972).
46. Assink, R. A. et Jonas J.: *J. Chem. Phys.* **57**, 3329 (1972).
47. Barojas, J., Levesque, D., et Quentrec, B.: *Phys. Rev.* **A7**, 1092 (1973).
48. Quentrec, B. et Brot, C.: à paraître.
49. Ailawadi, N. K. et Berne, B. J.: *J. Chem. Phys.* **54**, 3569 (1971).
50. Ailawadi, N. K. et Harris, S.: *J. Chem. Phys.* **56**, 5783 (1972).
51. Berne, B. J.: *J. Chem. Phys.* **56**, 2164 (1972).
52. Pomeau, Y. et Weber, J.: *Phys. Rev.* A, sous presse.
53. Maker, P. D.: *Phys. Rev.* **1A**, 923 (1970).

DISCUSSION

Gershon: There is a delicate point in Ivanov's theory. He assumes a certain sharp distribution of jump angles. If a different distribution is taken a different ratio between τ_1 and τ_2 is obtained. An example for this was worked out by Zamir and Ben-Reuven (*J. Chem. Phys.*, 1971). Also, Ivanov's theory was formulated for systems without dynamical pair correlations. Therefore there is the question whether it can be applied accurately to the comparison of correlation times deduced from depolarized light scattering and dielectric relaxation.

Brot: Of course the adopted jump angle distribution plays a role. The Anderson's distribution $\delta(\alpha)$ does not seem to be reasonable for liquids. The Ben-Reuven and Zamir's distribution, which is rectangular between 0 and π, hence completely random, should be reasonable for molecules of very high symmetry provided τ_{FL} be long enough. However for molecules with a particular steric shape it is possible that an Ivanov or Montrose and Litovitz distribution about some privileged value of the jump angle be the best choice.

I agree that τ_1/τ_2 considerations can only semi-quantitatively determine the mean jump angle, but even so they are interesting, because all *reasonable* distributions yield not too much different τ_1/τ_2 curves.

Papoular: (1) Dans quelle mesure peut-on intégrer dans ce(s) modèle(s) des phases à forte coopérativité, par exemple paranématiques, et en particulier, passer à une forme non locale des équations?
2) Dans quelle mesure peut-on corréler le degré de coopérativité avec l'amplitude moyenne des sauts et le rapport des temps caractéristiques de rotation libre et de résidence?

Brot: (1) Il faut d'abord employer les fonctions de corrélation convenables, comme j'ai proposé pour $C_1(\mathbf{r}, t)$, ou comme les auteurs que j'ai cités utilisent pour la diffusion de la lumière. Quant aux modèles, peu de tentatives ont été faites, sauf pour les phases cristallines réorientationnelles ('cogwheel rotations').

(2) Qualitativement on peut dire que si le temps de résidence est long, la coopérativité sera grande, l'arrivée d'un 'défaut' ou la rupture de l'ordre local faisant probablement relaxer plusieurs molécules à la fois.

Janik: I want to ask whether the quality of experimental data on τ_1 and τ_2 is good enough in order to conclude from the τ_1/τ_2 ratio about the geometry of rotational jumps.

Brot: I tried to stress the fact that experimental accuracies should still, indeed, be improved. Moreover the adopted assumption on the type of distribution of the jump angle about its mean value can slightly change the conclusion (see intervention by Gershon). Despite this, I think that such considerations of the τ_1/τ_2 ratio, if accurately and *correctly* done, are useful.

Kneubühl: As an experimental example, you mention the angular correlation function of C_6D_5F. This molecule possesses a low symmetry and thus represents an asymmetric top. Hence, in general the rotation of C_6D_5F cannot be described by a single angular correlation function. If you do so, you assume C_6D_5F to be or to behave like a spherical top. What is the experimental proof of this statement?

Brot: I did not do so myself, I just quoted Jonas. In the case of acetonitrile, on the other hand, I mentioned the tensorial character of the rotational diffusion constant.

Jonas: In our study of fluorobenzene-d_5 we assumed that the molecule reorients isotropically. In view of some other results obtained in our laboratory (see Reference in the original paper) we are sure that this assumption is reasonable. In the original work we give a detailed discussion of the analysis of our data; I only mention that the diagonal elements of the spin-rotation interaction tensor were available from molecular beam measurements.

Bratos: You discussed in your lecture the possible existence of the long-time tails in the rotational correlation functions. Is there any experimental possibility to detect such long time effects?

Brot: Perhaps by a costly 'experiment' of numerical molecular dynamics on a cleverly chosen model system.

Lascombe: Est-ce que les études de dynamique moléculaire de l'azote sur ordinateur suggèrent des modèles pour interpréter l'effet de la pression sur la dynamique orientationnelle d'un liquide?

Quentrec: Le comportement individuel des molécules d'azote que nous avons pu observer se caractérise de la façon suivante: la molécule se bloque suivant une certaine direction en faisant de petites oscillations de 20° maximum pendant des périodes variant de 2 à 6×10^{-13} s, puis tourne plus ou moins rapidement pendant des périodes du même ordre de grandeur, la molécule étant plus souvent et restant plus longtemps suivant une direction fixe lorsque le nombre de voisins est élevé.

D'autre part, il est intéressant de noter que si le comportement individuel rotationnel de la molécule est sensible aux fluctuations de densité de ce voisinage, il n'y a, par contre, aucune corrélation entre ce comportement rotationnel et les fluctuations de température (d'énergie cinétique moyenne) de ce même voisinage.

Brot: Quentrec vient d'évoquer les fluctuations locales de densité. J'ajoute, pour compléter sa réponse, que si, à une température supérieure au point triple, on prend successivement des densités différentes, on observe (1) que les pressions calculées varient de façon assez réalistes, (2) que les FC orientationnelles décroissent plus vite quand la densité décroît, la variation de τ_J étant en sens inverse, comme on devait s'y attendre.

Friedman: I had in mind the same problem as in the comment of Rahman, but I was going to discuss a different aspect, i.e. that those 'physical models' in which there are large jump angles may be physically inconsistent if the average time τ_J between jumps is very short, say a few picoseconds or less. The reason is that after a jump there is a certain local excess energy which, itself, cannot be distributed over a large number of molecules in less than a few picoseconds. This leads to memory effects: in the "hot spot" the probability density of another jump is greater than $1/\tau_J$, the average value. The molecular dynamics results refered to by Dr Rahman show that such memory effects are quite important in monoatomic fluids. In the systems you consider, if τ_J is as large as 50 ps, then translational diffusion may dissipate the hot spots and thus erase the memory. But for $\tau_J < 10$ ps I don't see why the memory effects may be neglected.

Brot: I agree, but if, by the memory effect you invoke, a first jump is immediately followed by a second one, the whole process can be called a single jump at non-uniform angular velocity. For 'computer nitrogen' I, in fact, mentioned that during their stages of continuous rotation the molecules do not rotate at constant angular speed. Unavoidably, models are schematic and a classification of them is still more.

Gershon: In relation to Prof. Janik's question, in addition to uncertainties due to experimental errors in the values of τ_1 and τ_2, there is also the influence of what theory is used to explain the experimental data.

Brot: τ_J, τ_1, τ_2 must be measured independently and one should avoid to use a collective correlation time instead of an auto-correlation time. Afterwards, any theory may be tried, but no hidden assumption should be made, for example the use of the Hubbard's relation to determine τ_J amounts to assuming rotational diffusion by small uncorrelated steps.

THE INFLUENCE OF THE FIR ABSORPTION ON THE DIELECTRIC BEHAVIOUR OF RIGID POLAR MOLECULES IN VERY DILUTE SOLUTIONS

H. KILP, G. KLAGES, and W. NOERPEL

Institut für Physik, Johannes Gutenberg-Universität, 65 Mainz, F.R.G.

Abstract. In view of the FIR absorption the dielectric loss of ten polar molecules – with different volume between furan and 4-bromobiphenyl – has been measured in very dilute solutions at 11 fixed frequencies over an extended range from 0.3 to 300 GHz. Solvents are heptane, cyclohexane, mesitylene, and decalin. For these rigid polar molecules a second high frequency absorption region with time constants of 1–3 ps can be separated from the predominant Debye absorption. This additional absorption decreases with increasing Debye relaxation time. Various molecular motions contributing to this absorption are discussed.

Résumé. Pour étudier l'absorption FIR, les pertes diélectriques de 10 molécules polaires – dont les volumes se situent entre ceux du furanne et de brom-4-biphényle – sont mesurées en solutions très diluées à 11 fréqunces fixes dans une bande de 0.3 à 300 GHz, les solvants étant: heptane, cyclohexane, mésitylène et décahydronaphalthene. Pour ces molécules polaires rigides un second domaine d'absorption à hautes fréquences avec des temps charactéristiques de 1 à 3 ps est trouvé à côté de l'absorption Debye principale. Cette absorption supplémentaire décroît avec le temps de relaxation Debye croissant. Plusieurs mouvements moléculaires contribuant à cette absorption sont discutés.

The dielectric behaviour of polar molecules in very dilute solutions of unpolar solvents can give additional information on the orientational motion in liquids [1]. This investigation with microwaves is aimed at rigid polar molecules which in the pure liquid show an excess absorption in the far infrared (FIR) [2], besides the predominant absorption due to a Debye relaxation process. It seems to be of interest to observe this behaviour in very dilute solutions where dipol-dipol interaction is greatly reduced. In dilute solution studies one is interested in the absorption difference to the pure solvent, which at very high frequencies must be measured in the presence of considerable solvent losses. For this reason the accuracy of the loss measurement in the millimeter wave range is of particular importance for the reliability of the results.

1. Experimental

The investigation includes polar molecules of different volume and dipole moment (0.69 to 4.11 D): furan, tetrahydrofuran, chlorobenzene, chlorocyclohexane, p-chlorotoluene, benzophenone, 2-bromonaphthalene, 2-cyanonaphthalene, 9-bromophenanthrene, and 4-bromobiphenyl. The aliphatic and aromatic solvents were: heptane at 20 °C, cyclohexane at 20 °C, mesitylene at 20 and -30 °C, and decalin (decahydronaphthalene) at 20 and 0 °C. The respective viscosities vary from 0.41 to 3.82 cP.

The dielectric loss $\Delta\varepsilon''$ that is the difference between solution and solvent was measured in the microwave range at 11 fixed frequencies between 0.3 and 300 GHz.

Depending on frequency the low loss measurements are based on different methods. Coaxial resonators are used from 0.3 to 3 GHz, waveguide transmission from 10 to 70 GHz and free space transmission at higher frequencies. These methods have been previously described [3, 4] except for the apparatus at 300 GHz, a preliminary description of which will be given here.

Figure 1 shows a schematic diagram of this apparatus. The output of a 70 GHz Klystron is fed into a harmonic generator from which the 4th harmonic is selected by a waveguide filter and transmitted into free space by a small horn antenna. The sample

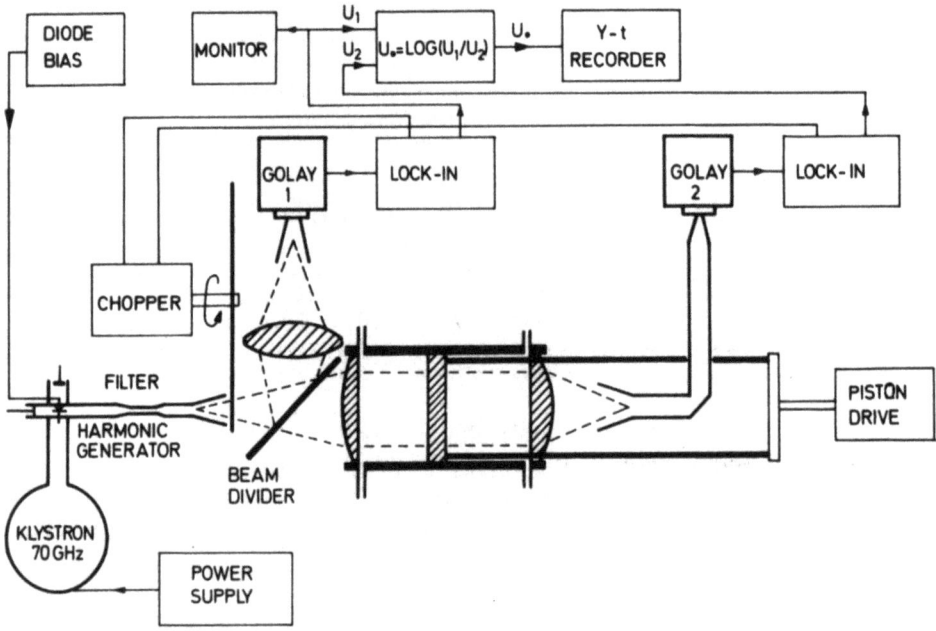

Fig. 1. Schematic diagram of the 300 GHz apparatus.

cell is a 60 mm long metal cylinder closed at both ends by planoconvex PTFE-lenses; it is transmitted by a nearly parallel beam which is fed via a light pipe (overmoded circular waveguide) into a Golay detector. A fraction of the signal is reflected by a beam divider to a second Golay detector. Thus, it is possible to eliminate small variations of signal power by recording the ratio of the detector outputs. A movable PTFE-piston divides the cell into two chambers containing solution and solvent, respectively. The power of the transmitted radiation then varies exponentially with the position of the piston, the exponent directly giving the difference between the absorption coefficients. Using a logarithmic amplifier and a synchronous motor to drive the piston, this difference is proportional to the slope of the recorded straight line. This method is advantageous at very high frequencies, where intrinsic losses of the unpolar solvents are larger than the loss contribution by polar molecules in very dilute solutions.

2. Results

At each frequency $\Delta\varepsilon''$ depends linearly on the concentration of the solute up to mole fractions of a few per cent. An example is given in Figure 2 for two frequencies in the millimeter wave range.

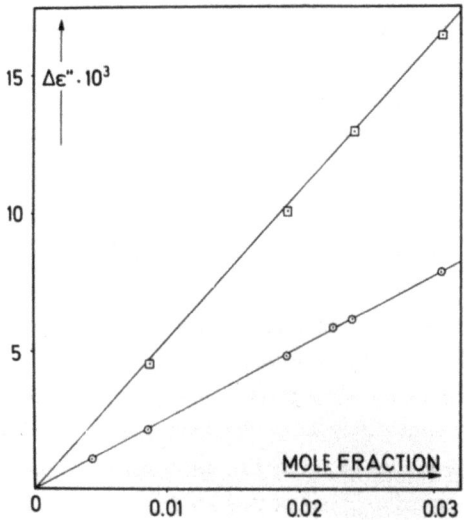

Fig. 2. Dielectric loss $\triangle\varepsilon''$ of 4-chlorotoluene in cyclohexane, 20°C, □ 3.96 mm, ○ 1.05 mm wave length.

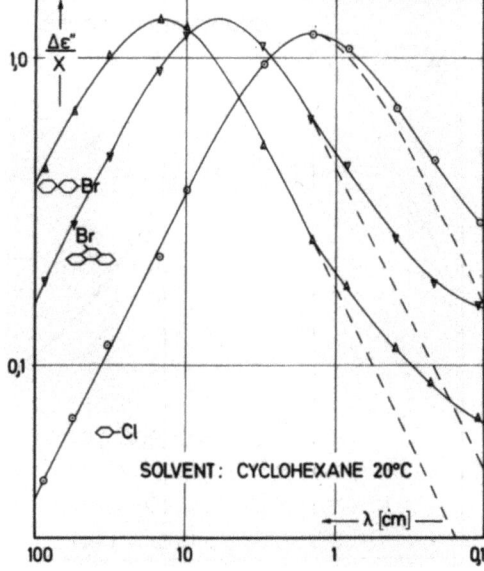

Fig. 3. Dielectric loss $\triangle\varepsilon''/x$ plotted against wavelength in cyclohexane, 20°C. △ 4-bromobiphenyl, ▽ 9-bromophenanthrene, ○ chlorobenzene. Broken lines: Debye absorption. x: mole fraction of polar molecules.

In Figure 3 the loss values $\Delta\varepsilon''/x$ (x is the mole fraction of the polar solute) have been plotted against frequency for three representative molecules in cyclohexane at 20°C. The deviation from a symmetric Debye or Fröhlich absorption curve at high frequencies is evident. It should be noticed, however, that using a logarithmic scale for $\Delta\varepsilon''/x$ the deviations have been exaggerated.

From the experimental data a distribution of relaxation times must be assumed. We have analysed all data superposing a symmetric absorption curve according to Fröhlich [5] for the predominant relaxation process and a Debye absorption curve for the fast orientation process:

$$\frac{\Delta\varepsilon''}{x} = \left(\frac{\Delta\varepsilon_0}{x} - \frac{\Delta\varepsilon_\infty}{x}\right) \times$$
$$\times \left[\frac{1-G_2}{p}\{\arctan(\omega\tau_b) - \arctan(\omega\tau_a)\} + G_2 \frac{\omega T_2}{1+\omega^2 T_2^2}\right]. \quad (1)$$

$\Delta\varepsilon_0$ and $\Delta\varepsilon_\infty$ are the differences between solution and solvent of the static and the high frequency permittivity, respectively. G_2 is the weight fraction and T_2 is the time constant of the high frequency absorption. $p = \ln(\tau_b/\tau_a)$ is a parameter describing the width of the relaxation time distribution between τ_a and τ_b. Here, this distribution is very narrow, p reaching values of up to 1.0, and only the relaxation time $\tau_1 = \sqrt{\tau_a \cdot \tau_b}$ will be considered. A typical analysis of this kind is shown in Figure 4, where the two symmetric curves represent the resolved absorption regions and the solid line their superposition giving a good fit to experimental data, except at the highest frequencies.

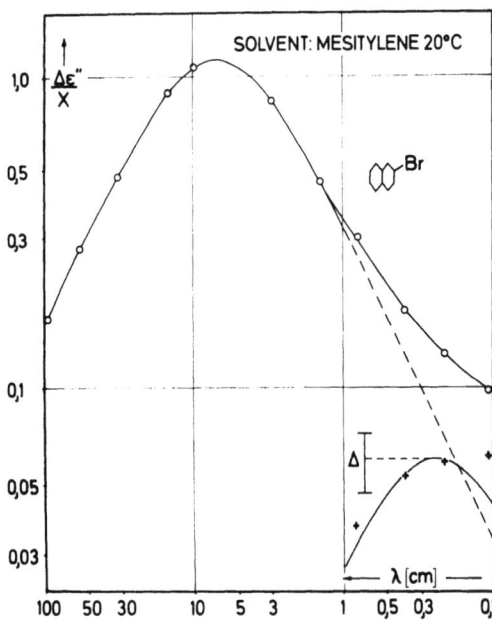

Fig. 4. Dielectric loss $\triangle\varepsilon''/x$ of 2-bromonaphthalene in mesitylene, 20°C. + difference between measured values and Debye absorption. ▽ variation of G_2 by ±0.01.

This can be seen by plotting the difference between experimental points and the values calculated from the first term of Equation (1) (crosses in Figure 4). Therefore, analysing the high frequency absorption by a Debye term is only an approximation. The measured curve has a larger half-width. In view of the experimental accuracy and the lack of data at frequencies higher than 300 GHz, however, it was not attempted to analyse this more exactly. The value of T_2 then must be regarded as the upper limit of a possible distribution of time constants. Despite this fact the weight fraction G_2 can be determined to better than ± 0.01, as indicated by Δ in Figure 4.

On the other hand one may ask if it is necessary to analyse the data by two separate absorption regions. Perhaps they can be fitted by a broad and continuous asymmetric relaxation time distribution. This has been tried using a distribution according to Cole-Davidson [6], but as Figure 5 shows it is not possible.

Figure 6 gives some additional dispersion data. Particularly the value at $\lambda_0 = 2.2$ mm is of interest. It shows that no significant contributions to the orientational polarization occur at frequencies higher than 300 GHz. This is evident from the small dielectric increment between 2.2 mm and the sodium-D-line.

For all investigated molecules the weight fraction G_2 of the fast orientation process shows a characteristic behaviour. It decreases in all solvents with increasing relaxation time τ_1 of the main absorption region. This is shown in Figure 7 for two different solvents.

Before we discuss the time constant T_2 of the high frequency absorption region it should be kept in mind that using a Debye term for its analysis not necessarily implies that it must be a relaxation process.

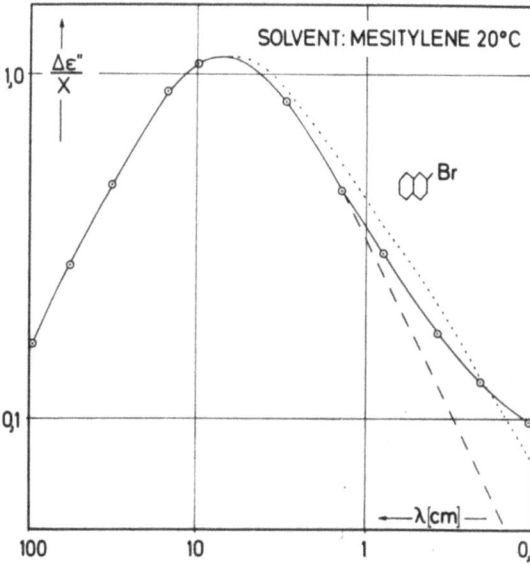

Fig. 5. Dielectric loss $\Delta\varepsilon''/x$ of 2-bromonaphthalene in mesitylene, 20°C. Broken line: Debye absorption; dotted line: Cole-Davidson curve with distribution parameter $\alpha = 0.8$.

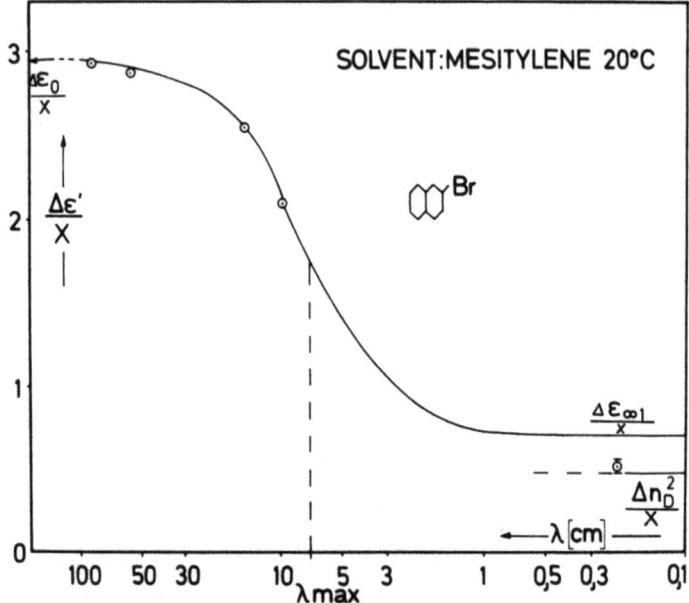

Fig. 6. Dispersion $\Delta\varepsilon'/x$ of 2-bromonaphthalene in mesitylene, 20°C. $\Delta\varepsilon_0/x$ static permittivity, $\Delta\varepsilon_\infty/x$ high frequency permittivity of the main Debye dispersion, $\Delta n_D^2/x$ permittivity at the sodium D-line.

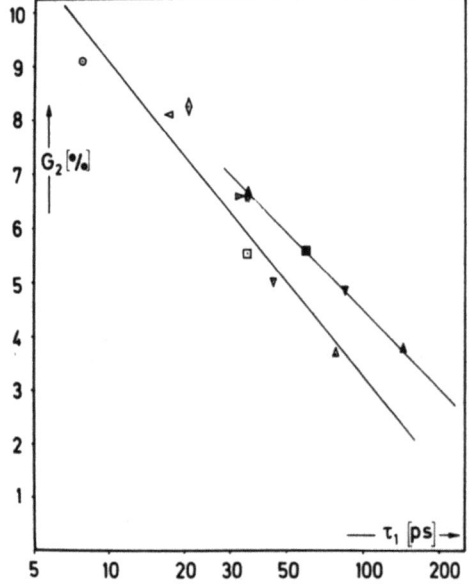

Fig. 7. Weight fraction G_2 of the additional high frequency process vs Debye relaxation time τ_1, 20°C. Filled marks decalin solution, open marks cyclohexane solution. Solute molecules: ○ chlorobenzene, △ 4-chlorotoluene, ◇ benzophenone, △ 2-cyanonaphthalene, □ 2-bromonaphthalene, ▽ 9-bromophenanthrene, △ 4-bromobiphenyl.

Then, within the experimental accuracy, these T_2 values are constant for all polar molecules in the same solvent. In Figure 8 they are plotted against solvent viscosity. The circles are average values of T_2 and the spread of individual values is indicated by error limits. It is apparent that T_2 increases with increasing viscosity of the solvent.

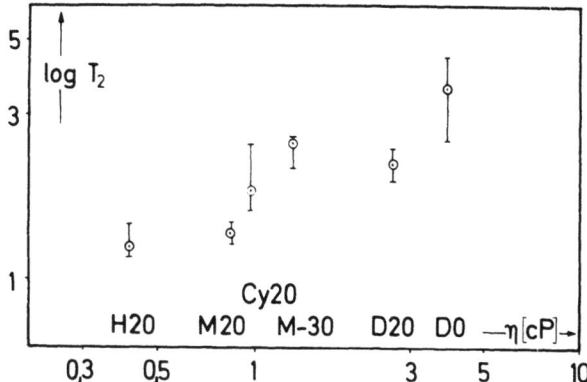

Fig. 8. Time constant T_2 vs viscosity of the solvent η. H heptane, M mesitylene, D decalin, Cy cyclohexane. Numbers indicate temperature in °C.

3. Discussion

For the interpretation of the additional millimeter wave absorption inertial effects, which become effective in the orientational motion of polar molecules [7] are not essential. They modify a Debye absorption curve in such a way that the high frequency slope is slightly smaller before it becomes much larger at very high frequencies. However, calculations show that a deviation of the observed magnitude, as shown in Figure 3, cannot be accounted for.

(1) The dipole orientation of rigid molecules can be described as a rate process, the relaxation time being determined by the activation energy. Broadening of the experimental absorption curve compared to the Debye shape indicates a distribution of activation energies [8]. In our case such a distribution would have to be unsymmetric decreasing slowly towards small energies. However, since a continuous Cole-Davidson distribution was shown not to be appropriate, we are led to assume a separate additional range of low activation energies. In thermal equilibrium a small number of molecules would then always be in excited states in which they can orient their complete dipol moment by overall rotation with very short time constants. In another context, Garton [9] has proposed the existence of such shallow temporary potential wells.

With this model the observed values of G_2 from Figure 7 would indicate that the number of such states decreases with increasing Debye relaxation time τ_1. In the same solvent, that is in the same environment, τ_1 increases with the maximum diameter of the molecule [10]. The data of Figure 7 then imply that such weakly bonded states becomes less probable with increasing size of the polar molecule.

More difficult to understand is the fact that the depth of these additional potential wells should not depend on molecular size. This follows from the experimental result which shows that the time constant T_2 does not vary significantly with the polar molecule but with solvent viscosity (Figure 8). Therefore, one could conclude that motions of the solvent molecules predominantly determine the time constant T_2.

(2) Alternatively to a complete orientation of a small number of polar molecules, rotations around small angles of all molecules could be possible. This corresponds to the model of Hill [11] who assumes the molecules to undergo damped librational motions in slowly fluctuating deeper potential wells. The corresponding orientational polarisation, which is proportional to G_2, should be inversely proportional to the force constant for rotational oscillation in a potential well. This will agree with the observation as shown in Figure 7, if the force constant increases with the size of the polar molecule.

The magnitude of G_2, however, is relatively large, e.g. a rotation of the molecule around more than 30° is required to explain a value of 0.07 for G_2. Also, the time constant T_2 is slightly too long for rotational oscillations and does not significantly increase with the moment of inertia. According to the experimental data at 1 mm wavelength T_2 is only the lower limit of a distribution. From this it may be expected that rotational oscillations will contribute to the observed additional absorption region, but they alone cannot explain its magnitude.

(3) Finally an absorption due to collision induced dipol moments as first observed by Whiffen [12] in unpolar liquids must be considered. In our case moments induced by the permanent molecular dipol in the adjacent solvent molecules will contribute to the absorption according to the local liquid structure. The time constant T_2 then describes cooperative motions of the solvent molecules and should not depend on volume and dipole moment of the polar molecule. This has been observed.

To estimate the magnitude of this contribution, we have investigated 1,4- and 1,3,5-derivatives of benzene in which partial moments compensate for the whole molecule. Cyclohexane was used as solvent because of its very low intrinsic loss. Figure 9 shows the absorption difference $\Delta\varepsilon''$ between solutions of two such substances and the pure solvent plotted against the mole fraction x of the solute. These measurements were made at 285 GHz, a frequency close to the maximum of the additional absorption for polar molecules. This maximum gives the square of the fast orienting moment, which will be compared with the same quantity calculated from the values of Figure 9 extrapolated to zero concentration. These squared moments only give a lower limit, if, as known for benzene and other unpolar liquids, the maximum loss is higher than the value at 285 GHz and occurs at higher frequencies.

We obtain for p-xylene $0.016\,D^2$ and for 1,4-dichlorobenzene $0.024\,D^2$, the increase being related to the larger partial moment of the Cl—C bond. The more symmetric 1,3,5-trichlorobenzene gives $0.022\,D^2$, benzene and carbontetrachloride only $0.015\,D^2$.

On the other hand, the square of this moment calculated for p-chlorotoluene from

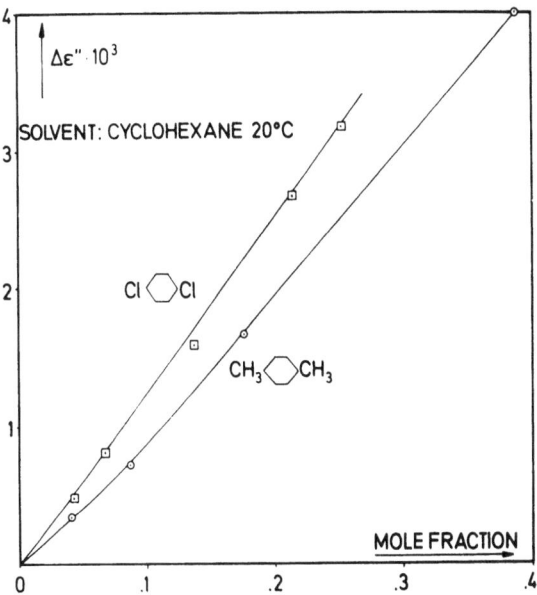

Fig. 9. Dielectric loss $\Delta\varepsilon''$ of nonpolar molecules in cyclohexane at 1.05 mm wavelength, 20°C. ▫ 1,4-dichlorobenzene, ○ p-xylene.

the permanent moment ($\mu = 1.97$ D) and from G_2 is 0.34 D^2. The corresponding values for some other polar molecules, all in cyclohexane solution are: 2-bromonaphthalene 0.19 D^2, 9-bromophenanthrene 0.13 D^2, and 4-bromobiphenyl 0.10 D^2. This decrease despite nearly equal permanent moments μ reflects the observed dependence on the Debye relaxation time τ_1 (Figure 7). Together with the absolute magnitude of these values it shows that induced moments alone cannot be the reason for the additional high frequency absorption. This is supported by comparing the much higher value for p-chlorotoluene with that for p-dichlorobenzene, both molecules having nearly the same size. That difference remains, even if one admits that for p-dichlorobenzene the induced moments can partially cancel.

Taken altogether, the additional high frequency absorption measured by us down to 1 mm wavelength sofar cannot be completely described in all observed details by only one of the molecular motions discussed. On the other hand, none can be excluded from experimental evidence. Therefore, it is very likely that all three types of motion do contribute to these dielectric losses of rigid polar molecules in very dilute solutions.

Acknowledgement

This investigation was supported by the Deutsche Forschungsgemeinschaft.

References

1. Klages, G. and Krauss, G.: *Z. Naturforsch.* **26a**, 1272 (1971); Eloranta, J.: *Z. Naturforsch.* **27a**, 1652 (1972).

2. Davies, Mansel: in *Dielectric Properties and Molecular Behaviour*, Van Nostrand, 1969.
3. Krauss, G.: Doctoral dissertation, Mainz, 1970. Kreuter, K.: *Z. Naturforsch.* **23a**, 1728 (1968).
4. Kilp, H.: *Z. angew. Physik* **30**, 288 (1970).
5. Fröhlich, H.: *Theory of Dielectrics*, Clarendon Press, Oxford, 1958.
6. Davidson, D. W. and Cole, R. H.: *J. Chem. Phys.* **19**, 1484 (1951).
7. Roccard, Y.: *J. Phys. Radium* **4**, 247 (1933).
   Powles, J. G.: *Trans. Faraday Soc.* **44**, 802 (1948).
8. Kauzman, W.: *Rev. Mod. Phys.* **14**, 12 (1942).
9. Garton, C. G.: *Trans. Faraday Soc.* **42a**, 56 (1946).
10. Hufnagel, F.: *Z. Naturforsch.* **25a**, 1143 (1970).
11. Hill, N. E.: *Proc. Phys. Soc.* **82**, 723 (1963).
12. Whiffen, D. H.: *Trans. Faraday Soc.* **46**, 124 (1950).

DISCUSSION

Gershon: What is the accuracy of the fitted parameters and how many are there compared with the number of data points?
To the previous remark of Prof. Davies: one never obtains information on the real molecules. It is too complicated. All what one gets is some statistical knowledge. The phenomenological models yield at most this information. They serve as computational tools although sometimes they might distort the physical picture if they are taken too seriously.

Kilp: By fitting the analytic expression (Equation (1)) to the dielectric loss $\Delta\varepsilon''/x$ measured at 11 frequencies between 0.3 and 300 GHz we obtain 5 parameters within the following error limits. The dielectric increment $(\Delta\varepsilon_0 - \Delta\varepsilon_\infty)/x$ is accurate to ± 0.02, the mean relaxation time τ_1 of the predominant absorption process to $\pm 3\%$, the relaxation time distribution parameter p to $\pm 0.5/(1+p)$, the weight fraction G_2 of the high frequency absorption to ± 0.01 and the time constant T_2 to about $\pm 20\%$.

Comment by M. Davies

It is difficult to see what justification there is for trying to characterise a second absorption beyond the Debye cooperative relaxation in the way described by Dr Kilp. Already in 1968 it was shown that all polar solutes have an absorption with α(max) near 60 cm^{-1} which can be converted to an ε''(max) at a much lower wavenumber. (1) This absorption mode has been followed from the free-rotational dipole absorption in the gas phase to its variously hindered forms in the liquid state. Dr Kilp is correct in emphasizing that collisionally induced dipoles can be contributing to this feature but that contribution is usually *relatively* small as can be seen by the much weaker absorptions in non-dipolar liquids even when large component moments (e.g. *p*-dicyanobenzene) are present. In one case, $N=N=O$, we have shown how the small permanent dipole moment provides conditions where the absorption induced by the quadrupole moment allows the latter to be measured (2).

In the usual case, e.g. chlorobenzene etc., as studied by Dr Kilp, the absorption he finds beyond the Debye process was predicted by Poley in 1955 and is molecularly represented by the Hill-Wyllie model (3) or, perhaps more satisfactorily by Brot's model (4) of hindered molecular angular jumps between the energy wells provided by the instantaneous molecular environment around each polar molecule in the liquid.

These absorptions have been shown to give quite acceptable parameters for the fluctuating energy wells in which the molecules librate – especially the depths of the wells are very similar in the Wyllie and Brot models for such significantly different molecular types as: (i) quasi-spherical polar molecules (5): (ii) rigid normally anisotropic molecules (chlorobenzene etc.) (6): (iii) very anisotropic, lathe-like molecules such as form a nematic liquid crystal phase (7). The different values found for the energy wells in these groups (i) < (ii) < (iii) conform to general expectations.

Many of the general features of this absorption were established by 1968, including the fact that the refractive index goes through a minimum near $\bar{\nu}$ for α(max) which establishes a quasi-resonant character in the absorption (1): often (e.g. chlorobenzene) n(minimum) $< n_D$.

UTILISATION DE LA FONCTION MEMOIRE DANS L'INTERPRETATION DES SPECTRES D'ABSORPTION DIPOLAIRE DES MOLECULES TOUPIES SYMETRIQUES EN PHASE LIQUIDE

PIERRE DESPLANQUES, EUGENE CONSTANT,
et RENAUD FAUQUEMBERGUE

Centre de Recherches sur les Propriétés Hyperfréquences des Milieux Condensés, Université des Sciences et Techniques de Lille I, B.P. No. 36, 59650 Villeneuve d'Ascq, France

Résumé. Nous proposons une interprétation et une exploitation globales des spectres d'absorption dipolaire des molécules toupies symétriques en phase liquide, fondées sur une théorie utilisant le formalisme de la fonction mémoire et ne dépendant que de deux paramètres physiques simples, le couple agissant sur l'orientation de la molécule, et sa dérivée temporelle.

Abstract. Molecular dynamics in liquid state is studied by us from hertzian and far-infrared dipolar absorption spectra. We particularly study non associated, rigid, symmetric top polar molecules diluted in different non polar solvents. The measured physical parameters are the absorption coefficient $\alpha(\omega)$ and the two components of complex permittivity $\varepsilon^*(\omega) = \varepsilon'(\omega) - i\varepsilon''(\omega)$; they can be coupled with the orientational molecular correlation function $\phi(t)$ by means of the Kubo method of fluctuation dissipation theorem. Under the condition to know $\alpha(\omega)$ in the whole frequency range, we can calculate $\phi(t)$ and so obtain informations on the rotational molecular dynamics of studied polar liquids.

In parts two and three, our theory bases are developed: the $\phi(t)$ function is defined by its associated memory function $K(t)$, and we choose for $K(t)$ an expression which takes account of the two $\phi(t)$ decreasing factors i.e. the statistical distribution of the rotational velocities and the perturbation of these by molecular interactions.

Thus, dipolar absorption spectra can be calculated as a function of only the temperature, the molecular moments of inertia, and two parameters: the mean square torque $\mathcal{N}^2$ acting on the molecular orientation, and its mean square temporal derivative $\dot{\mathcal{N}}^2$.

In part four, we show this theory describe qualitatively all the experimental results. For the most cases, the excess absorption observed in far infrared seems due to a more or less damped librational motion of molecule; in the other cases, the preponderant molecular motion is a rotational diffusion.

At least, in part five, the $\mathcal{N}$, $\dot{\mathcal{N}}$ parameters values are determined for many solutions which has been studied in our laboratory, and we give their variation as a function of the considered temperature, polar molecule and solvent. Then we compare quantitatively the theoretical and experimental spectra and correlation functions in two typical cases, and we observe they are in good agreement.

1. Introduction

L'étude de la dynamique moléculaire en phase liquide a connu ces dernières années une extension considérable que l'on peut apprécier à la fréquence des publications ainsi qu'au nombre des colloques, séminaires, écoles d'été, etc.... qui lui sont consacrés. Les causes de cet engouement sont multiples, et on peut citer les progrès réalisés dans l'étude thermodynamique des phénomènes irréversibles et dans la connaissance des potentiels intermoléculaires, le développement du calcul numérique sur ordinateur, l'avènement de nouvelles techniques expérimentales, l'amélioration et le raffinement progressifs apportés aux anciennes méthodes de mesure.

Parmi ces méthodes, la mesure de l'absorption dipolaire est la plus ancienne utilisée.

Si cependant ses enseignements restaient limités jusque naguère, c'est que le domaine de fréquence explorable ne permettait d'étudier que des phénomènes de durée de vie longue : c'était le cas par exemple des phénomènes d'association par liaison hydrogène dans les alcools et acides, et aussi des phénomènes de dynamique présentés par les très grosses molécules, beaucoup trop complexes pour autoriser des interprétations quantitatives.

Le développement des techniques en infrarouge lointain, jointes aux techniques hertziennes, permet maintenant d'explorer de façon pratiquement continue une gamme de fréquence comprise entre 0 et 400 cm^{-1}, et donc de décrire la dynamique moléculaire dans un domaine de temps compris entre 1 et 10^{-14} s ; on peut ainsi aborder l'étude de molécules suffisamment simples pour pouvoir être interprétées théoriquement.

C'est l'objet des recherches effectuées dans notre laboratoire depuis bientôt dix ans : notre étude porte sur des molécules polaires, toupies symétriques, rigides, non associées, en solutions diluées dans différents solvants non polaires. Les grandeurs physiques mesurées sont le coefficient d'absorption $\alpha(\omega)$ et les deux composantes de la permittivité complexe $\varepsilon^*(\omega) = \varepsilon'(\omega) - i\varepsilon''(\omega)$. Les travaux de Kubo [1, 2] et Cole [3] ont montré que, à partir des variations de ces grandeurs avec la fréquence, il était possible d'accéder à la microdynamique de la matière. Effectivement en utilisant la méthode du théorème de fluctuation-dissipation, on peut relier ε^* à la fonction de corrélation du moment dipolaire global du système à l'équilibre. Cette fonction dépend généralement des fluctuations de tous les degrés de liberté (translation et orientation) du système, et nous cherchons à déterminer plus particulièrement les variations liées au mouvement d'orientation de molécules possédant un moment dipolaire permanent, mouvement qu'on peut caractériser par la fonction de corrélation du vecteur unitaire **u** coaxial au moment dipolaire permanent :

$$\phi(t) = \langle \mathbf{u}(0) \cdot \mathbf{u}(t) \rangle. \tag{1}$$

Dans le cas de molécules toupies symétriques, ce vecteur **u** repère la direction de l'axe de symétrie de la molécule. Comme nous considérons le cas de solutions diluées de molécules polaires dans des solvants non polaires, nous pouvons supposer que le rapport entre champ local et champ de Maxwell est indépendant de la fréquence, et que le mouvement de chaque molécule polaire est indépendant de celui des autres molécules polaires. En utilisant ces hypothèses dans les résultats de Kubo et si on suppose négligeable la contribution des moments induits collisionnels, on peut relier simplement les grandeurs observables à la fonction $\phi(t)$:

$$\frac{\varepsilon^* - \varepsilon_\infty}{\varepsilon_0 - \varepsilon_\infty} = -\int_0^\infty \dot{\phi}(t) e^{-i\omega t} \, dt \tag{2}$$

et

$$\alpha(\omega) = \frac{\varepsilon'' \omega}{nc} = \frac{\varepsilon_0 - \varepsilon_\infty}{nc} \omega^2 \int_0^\infty \phi(t) \cos\omega t \, dt. \tag{3}$$

$\varepsilon_0 - \varepsilon_\infty$ étant la dispersion de la substance et n la partie réelle de $n^* = \sqrt{\varepsilon^*}$. A condition de connaître $\alpha(\omega)$ dans toute la gamme de fréquence jusqu'à l'infrarouge lointain, on peut calculer $\phi(t)$ et accéder ainsi à la dynamique monomoléculaire des liquides étudiés. Nous présentons dans ce travail les caractéristiques essentielles des résultats obtenus et leur interprétation à partir du formalisme de la fonction mémoire par une méthode originale que nous exposons en premier lieu.

2. Introduction de la fonction mémoire associée à $\phi(t)$

Soit $\boldsymbol{\omega}_\perp(t)$ le vecteur rotation du vecteur unitaire $\mathbf{u}(t)$ porté par le moment dipolaire; $\boldsymbol{\omega}_\perp(t)$ est défini vectoriellement par:

$$\boldsymbol{\omega}_\perp(t) = \mathbf{u} \wedge \dot{\mathbf{u}}.$$

Nous allons montrer que la fonction de corrélation normalisée de $\boldsymbol{\omega}_\perp(t)$:

$$\psi(t) = \frac{\langle \boldsymbol{\omega}_\perp(\tau) \cdot \boldsymbol{\omega}_\perp(\tau + t) \rangle}{\langle \omega_\perp^2(\tau) \rangle} \tag{4}$$

peut être assimilée, en modulation rapide *, à la fonction mémoire associée à $\phi(t)$ [4].

Par intégration entre 0 et t de l'équation $\dot{\mathbf{u}} = \boldsymbol{\omega}_\perp \wedge \mathbf{u}$, on a:

$$\mathbf{u}(t) = \mathbf{u}(0) + \int_0^t \boldsymbol{\omega}_\perp(\tau) \wedge \mathbf{u}(\tau) \, d\tau.$$

En reportant cette expression de $\mathbf{u}(t)$ dans l'équation suivante:

$$\dot{\phi}(t) = \langle \mathbf{u}(0) \cdot \dot{\mathbf{u}}(t) \rangle = \langle \boldsymbol{\omega}_\perp(t) \cdot (\mathbf{u}(t) \wedge \mathbf{u}(0)) \rangle$$

on obtient:

$$\dot{\phi}(t) = \left\langle \boldsymbol{\omega}_\perp(t) \cdot \int_0^t (\boldsymbol{\omega}_\perp(\tau) \wedge \mathbf{u}(\tau)) \wedge \mathbf{u}(0) \, d\tau \right\rangle =$$

$$= \int_0^t \langle (\mathbf{u}(0) \cdot \boldsymbol{\omega}_\perp(\tau))(\mathbf{u}(\tau) \cdot \boldsymbol{\omega}_\perp(t)) \rangle \, d\tau$$

$$- \int_0^t \langle (\mathbf{u}(0) \cdot \mathbf{u}(\tau))(\boldsymbol{\omega}_\perp(\tau) \cdot \boldsymbol{\omega}_\perp(t)) \rangle \, d\tau. \tag{5}$$

La première intégrale est nulle car $\mathbf{u}(\tau) \cdot \boldsymbol{\omega}_\perp(\tau) = 0$ pour τ quelconque, et pour

---

* On parle de modulation rapide lorsque les interactions moléculaires sont suffisamment importantes et fréquentes pour que la molécule n'ait pas le temps d'effectuer une rotation d'un angle appréciable avant que sa vitesse ne varie; dans ce cas le temps de corrélation τ_2 de $\psi(t)$ est beaucoup plus faible que celui τ_1 de $\phi(t)$. Lorsque $\phi(t)$ et $\psi(t)$ sont reliés par l'équation (6), on a la relation $\tau_1 \tau_2 = I/2kT = 1/\omega_m^2$.

$t > \tau$, $\boldsymbol{\omega}_\perp$ se déplace de façon isotrope en moyenne à partir de $\boldsymbol{\omega}_\perp(\tau)$. Dans l'hypothèse de modulation rapide, l'évolution de $\boldsymbol{\omega}_\perp(t)$ est beaucoup plus rapide que celle de $\mathbf{u}(t)$: on peut alors séparer les moyennes dans la seconde intégrale. Comme $\langle \omega_\perp^2(\tau) \rangle = 2kT/I = \omega_m^2$ quel que soit τ, la variable aléatoire $\boldsymbol{\omega}_\perp$ est stationnaire et donc $\psi(t)$ est indépendante de l'instant origine. Dans ces conditions, l'équation (5) s'écrit:

$$\dot{\phi}(t) = -\omega_m^2 \int_0^t \phi(\tau) \psi(t-\tau) d\tau$$

ou encore:

$$\dot{\phi}(t) = -\omega_m^2 \int_0^t \phi(t-\tau) \psi(\tau) d\tau. \tag{6}$$

Cette équation (6) est une équation intégro-différentielle dite de Volterra ou de Fredholm; elle est formellement identique à l'équation suivante:

$$\dot{\phi}(t) = -\int_0^t K(\tau) \phi(t-\tau) d\tau \tag{7}$$

qu'on peut déduire directement de l'équation de Langevin généralisée [5] appliquée à la variable $\mathbf{u}(t)$; dans cette équation $K(t)$ est la 'fonction mémoire' de $\mathbf{u}(t)$. Par suite, et dans la limite des hypothèses qui ont permis d'obtenir (6), $\omega_m^2 \psi(t)$ peut être considérée comme égale à $K(t)$.

Il est d'ailleurs possible de préciser davantage les conditions dans lesquelles on peut assimiler $\omega_m^2 \psi(t)$ à $K(t)$ en comparant les moments de ces deux fonctions:
- d'une part γ_{2n} de $\omega_m^2 \psi(t)$ tels que

$$\omega_m^2 \psi(t) = \langle \boldsymbol{\omega}_\perp(0) \cdot \boldsymbol{\omega}_\perp(t) \rangle = \sum_{n=0}^\infty (-1)^n \gamma_{2n} \frac{t^{2n}}{(2n)!}$$

$$\gamma_{2n} = \left\langle \left(\frac{d^n \boldsymbol{\omega}_\perp}{dt^n} \right)^2 \right\rangle.$$

- d'autre part μ_{2n} de $K(t)$, reliés aux moments m_{2n} de $\phi(t)$ par la relation issue de (6):

$$m_{2n+2} - \sum_{q=0}^n m_{2q} \mu_{2(n-q)} = 0 \qquad n \text{ entier} \geq 0.$$

En calculant μ_{2n}, on trouve que son expression est la somme de γ_{2n} et d'autres termes où figurent des dérivées de $\boldsymbol{\omega}_\perp$ d'ordre inférieur à n; si donc on veut que $\mu_{2n} \simeq \gamma_{2n}$, il est nécessaire et suffisant que ces autres termes soient négligeables devant γ_{2n}, ce qu'on peut exprimer par:

$$\left\langle \left(\frac{d^n \boldsymbol{\omega}_\perp}{dt^n} \right)^2 \right\rangle \gg \omega_m^2 \left\langle \left(\frac{d^{n-1} \boldsymbol{\omega}_\perp}{dt^{n-1}} \right)^2 \right\rangle \quad n \text{ entier} \geq 1. \tag{8}$$

Dans le cas $n=1$, cette condition peut s'écrire sous la forme $\omega_m \tau_2 \ll 1$ qui traduit bien l'hypothèse de modulation rapide; l'inégalité générale (8) est une extension de cette condition: c'est pourquoi nous l'appellerons condition de modulation rapide généralisée.

Lorsque cette condition est vérifiée, les deux fonctions $\omega_m^2 \psi(t)$ et $K(t)$ sont égales; mais il s'agit bien entendu d'un cas limite de comportement moléculaire. Dans la réalité, les phénomènes sont plus complexes, et il est nécessaire d'effectuer une étude plus approfondie de la fonction mémoire $K(t)$ pour pouvoir lui trouver une expression rendant mieux compte de la véritable dynamique moléculaire.

3. Présentation de la théorie générale proposée

3.1. Choix de $K(t)$

L'approximation $K(t) = \omega_m^2 \psi(t)$, justifiée dans le cas de la modulation rapide généralisée, est une approximation que nous qualifions 'd'ordre zéro': en effet, seul le moment μ_0 de $K(t)$ est exact dans cette expression. La voie qu'on emprunte naturellement pour obtenir une expression de $K(t)$ plus juste est de chercher à définir $K(t)$ par une approximation d'ordre deux, c'est-à-dire par une expression où le moment μ_2 serait également exact; ce moment peut s'exprimer sous la forme [6]:

$$\mu_2 = \langle \omega_\perp^4 \rangle - \omega_m^4 + \langle (\dot{\omega}_\perp)^2 \rangle. \tag{9}$$

Envisageons d'abord le cas le plus simple des molécules *linéaires*. Si nous étudions le mouvement d'*une* molécule en rotation libre à la vitesse moyenne $\omega_\perp = \omega_m$, la rotation s'effectue dans un plan fixe avec la pulsation ω_m, et $\boldsymbol{\omega}_\perp$ est un vecteur constant; le moment μ_2 est alors nul. Si maintenant on considère non plus une seule, mais un ensemble de molécules en rotation libre, chacune effectuant une rotation dans un plan fixe avec une pulsation constante $\omega_\perp$, cette fois $\omega_\perp$ est distribuée suivant la loi de probabilité:

$$f(\omega_\perp) = \frac{2\omega_\perp}{\omega_m^2} \exp\left(-\frac{\omega_\perp^2}{\omega_m^2}\right). \tag{10}$$

Dans ce cas μ_2 n'est pas nul puisque $\langle \omega_\perp^4 \rangle = 2\omega_m^4$. La comparaison de ces deux exemples montre que, en rotation libre, la décroissance de $K(t)$ (notée alors $K_{RL}(t)$) est déterminée par la distribution des vitesses de rotation et par elle seule.

Cette dernière affirmation n'est rigoureusement vraie que pour une molécule linéaire; pour une molécule toupie symétrique en rotation libre, le terme $\langle (\dot{\omega}_\perp)^2 \rangle$ n'est pas nul car le vecteur $\boldsymbol{\omega}_\perp$ 'précessionne' autour du moment cinétique constant, et on a [6]:

$$\langle (\dot{\omega}_\perp)^2 \rangle = \frac{I'}{2I} \omega_m^4 \tag{11}$$

ou I'/I est égal au rapport B/C des constantes de rotation.

Lorsque la rotation est perturbée, la distribution (10) est inchangée, mais un terme supplémentaire apparaît dans (11), et on obtient:

$$\mu_2 = \omega_m^4 \left(1 + \frac{I'}{2I}\right) + \frac{\langle N_\perp^2 \rangle}{I^2}. \tag{12}$$

$N_\perp$ étant la composante perpendiculaire à **u** du moment **N** des forces extérieures agissant sur la rotation de la molécule. On voit qu'un second mécanisme contribue à la décroissance de $K(t)$: c'est la variation au cours du temps des vitesses de rotation due aux interactions moléculaires. Ainsi, d'après (12), $K(t)$ sera toujours inférieure à $K_{RL}(t)$ aux temps courts (c'est le contraire pour $\phi(t)$).

Nous voyons donc que l'hypothèse de modulation rapide généralisée consiste à ne tenir compte que du second mécanisme, soit à négliger la distribution des vitesses de rotation. C'est en considérant également cette distribution que nous parviendrons à une théorie plus générale permettant de traiter l'ensemble des cas en modulation rapide, et éventuellement aussi les cas de modulation lente et intermédiaire.

Dans ce but, nous cherchons une expression de $K(t)$ qui satisfasse les conditions nécessaires suivantes:

(a) Donner la valeur exacte (12) de μ_2
(b) Etre compatible avec les cas limites
   – rotation libre $K(t) = K_{RL}(t)$
   – modulation rapide généralisée $K(t) = \omega_m^2 \psi(t)$.

Or la fonction $\psi(t)$ a pour développement à l'origine:

$$\begin{aligned}\psi(t) &= 1 - \frac{\langle(\dot{\omega}_\perp)^2\rangle}{\langle\omega_\perp^2\rangle}\frac{t^2}{2} + O(t^4) \\ &= 1 - \left(\frac{\langle N_\perp^2 \rangle}{I^2 \omega_m^2} + \omega_m^2 \frac{I'}{2I}\right)\frac{t^2}{2} + O(t^4).\end{aligned}$$

Dans un précédent article [6], nous avions proposé pour $K(t)$, dans le cas d'une molécule linéaire, l'expression $K(t) = K_{RL}(t) \cdot \psi(t)$. Cette relation peut aisément se généraliser au cas des molécules toupies symétriques en posant:

$$K(t) = \frac{K_{RL}(t)}{\psi_{RL}(t)} \cdot \psi(t), \tag{14}$$

où $\psi_{RL}(t)$ est l'expression de la fonction $\psi(t)$ en rotation libre.*

Cette expression est compatible avec les cas limites et elle redonne bien le développement à l'origine de $K(t)$. En modulation rapide, $\psi(t)$ décroît rapidement et il est alors plus simple de remplacer le quotient $K_{RL}(t)/\psi_{RL}(t)$ par sa forme exponentielle

---

* La fonction $\psi_{RL}(t)$, tout comme $\phi_{RL}(t)$, se calcule en appliquant la distribution de Maxwell aux énergies de rotation: nous avons établi en [7] leurs expressions qui ne dépendent que de ω_m et du rapport I'/I. Il en est de même pour $K_{RL}(t)$ qu'on calcule numériquement à partir de $\phi_{RL}(t)$.

aux temps courts, ce qui donne:

$$K(t) \simeq \omega_m^2 e\left(\frac{-\omega_m^2 t^2}{2}\right) \psi(t). \tag{15}$$

Les deux expressions (14) et (15) donnent également pour $\phi(t)$ un développement à l'origine exact jusqu'à l'ordre 5 inclusivement.

3.2. Choix de $\psi(t)$

Cette façon d'exprimer $K(t)$ laisse toute liberté quant à la détermination de $\psi(t)$ et on peut envisager celle-ci de différentes manières [7]. Nous avons choisi de déterminer $\psi(t)$ par sa propre fonction mémoire associée $k(t)$, en suivant une méthode proposée par Berne et Harp [5] qui revient en fait à exprimer $k(t)$ par une fonction gaussienne avec une approximation d'ordre 2. On a donc:

$$\dot\psi(t) = -\int_0^t k(\tau)\psi(t-\tau)\,d\tau \tag{16}$$

où

$$k(t) = \frac{\langle(\dot{\omega}_\perp)^2\rangle}{\omega_m^2} \exp\left[-\left(\frac{\langle(\ddot{\omega}_\perp)^2\rangle}{\langle(\dot{\omega}_\perp)^2\rangle} - \frac{\langle(\dot{\omega}_\perp)^2\rangle}{\omega_m^2}\right)\frac{t^2}{2}\right]. \tag{17}$$

Cette expression de $k(t)$ constitue le dernier maillon de la chaîne qui relie la dynamique moléculaire aux spectres d'absorption des substances étudiées. En effet, à partir de $k(t)$, on déduit $\psi(t)$, puis $K(t)$, $\phi(t)$ et enfin le coefficient d'absorption $\alpha(\omega)$ en utilisant les équations (16), (14) ou (15), (7), et (3).

Afin de simplifier l'écriture, on posera désormais

$$\mathcal{N} = I\sqrt{\langle(\dot{\omega}_\perp)^2\rangle} = I\sqrt{\frac{\langle N_\perp^2\rangle}{I^2} + \frac{I'}{2I}\omega_m^4}$$

$$\dot{\mathcal{N}} = I\sqrt{\langle(\ddot{\omega}_\perp)^2\rangle} = I\sqrt{\frac{\langle \dot N_\perp^2\rangle}{I^2} + \frac{I'}{I}\text{ (autres termes)}},$$

où $\mathcal{N}$ et $\dot{\mathcal{N}}$ sont deux grandeurs qui, dans le cas d'une molécule linéaire, représentent rigoureusement le couple moyen et sa dérivée temporelle moyenne agissant sur l'orientation du moment dipolaire, et qui, dans le cas d'une molécule toupie symétrique, sont reliés simplement au couple et à sa dérivée moyens par l'intermédiaire du rapport I'/I. Ainsi donc, il est clair que les spectres théoriques obtenus sont fonctions uniquement de cinq paramètres: $\mathcal{N}$, $\dot{\mathcal{N}}$, les moments d'inertie I et I', et la température T par l'intermédiaire de la pulsation moyenne de rotation ω_m.

4. Etude théorique du role de $\mathcal{N}$ et $\dot{\mathcal{N}}$

Nous montrons maintenant que la théorie proposée permet de décrire de façon satisfaisante les caractéristiques essentielles de la dynamique moléculaire en phase liquide et des spectres d'absorption dipolaire qui en résultent, et pour cela nous étudions les valeurs de $\mathcal{N}$ et $\dot{\mathcal{N}}$ pour les différents types de mouvements moléculaires, et les spectres d'absorption correspondants.

Au préalable il est nécessaire de faire une remarque importante. Quelle que soit la valeur du couple $\mathcal{N}$, la vitesse de rotation moyenne ω_m de la molécule est toujours la même. En conséquence, pour une valeur donnée de $\mathcal{N}$ il existe une durée de vie maximale pour le couple, donc une valeur minimale de $\dot{\mathcal{N}}$ qui peut se calculer facilement

$$\dot{\mathcal{N}}_{\min} = \frac{\mathcal{N}^2}{I\omega_m}.$$

A cette valeur minimale de $\dot{\mathcal{N}}$ correspond pour la molécule un mouvement de libration monochromatique de pulsation

$$\omega_0 = \frac{\mathcal{N}}{I\omega_m}. \tag{18}$$

Un tel mouvement n'existe bien entendu que si ω_0 est nettement supérieure à la pulsation de rotation moyenne ω_m, c'est à dire si $\mathcal{N} \gg kT$ ce qui suppose que la molécule considérée se trouve alors dans un milieu particulièrement dense (liquide ou solide) où les interactions sont très importantes.

En général cependant la libration n'est pas monochromatique et il est intéressant d'introduire le coefficient sans dimension K défini par:

$$K = \frac{1}{2}\left[I^2\omega_m^2 \frac{\dot{\mathcal{N}}^2}{\mathcal{N}^4} - 1\right] \tag{19}$$

tel que l'expression (16) de $k(t)$ s'écrive:

$$k(t) = \omega_0^2 \exp(-K\omega_0^2 t^2).$$

Ce coefficient K est nul lorsque le mouvement de libration est pur, c'est-à-dire non perturbé, non 'désordonné' ($\dot{\mathcal{N}}$ est alors minimum); lorsque le désordre du milieu augmente, $\dot{\mathcal{N}}$ va augmenter et K également. Dans le cas limite où le milieu est très fortement désordonné, la fonction $k(t)$ peut être assimilée à une fonction de Dirac: $\psi(t)$ est alors quasi-exponentielle avec un temps de corrélation $\tau_2 \ll \tau_m$ et l'on obtient un mouvement de diffusion rotationnelle pure. Nous pouvons ainsi donner au coefficient K la signification suivante: dans un milieu *dense* K représente le degré de perturbation des mouvements moléculaires; lorsque K est faible, la structure du milieu est relativement ordonnée et le mouvement de libration est prépondérant; lorsque K est élevé, le milieu est très désordonné et la cinétique rotationnelle est régie par un processus Gaussien et Markovien*.

* Pour tout processus à la fois Gaussien et Markovien la fonction de corrélation caractéristique du phénomène est exponentielle [8].

Si le couple moyen $\mathcal{N}$ agissant sur la molécule est inférieur ou peu supérieur à l'énergie moyenne de rotation kT, le mouvement prépondérant de la molécule est alors un mouvement de rotation plus ou moins perturbée: c'est le cas dans un milieu dilué tel un gaz. Dans ce cas, on observe [6] que la rotation est d'autant plus perturbée que $\dot{\mathcal{N}}$ diminue, c'est à dire que le couple perturbe d'autant plus la rotation que ses variations sont plus lentes; lorsque $\dot{\mathcal{N}} \to \infty$, le couple 'n'a pas le temps' d'agir et on retrouve le spectre de rotation libre (cf. figure 2).

Nous avons rassemblé ces résultats sur un graphique (figure 1) où les zones correspondant à chaque mouvement apparaissent clairement. Nous avons également représenté (figures 2 et 3) quatre spectres d'absorption théoriques illustrant la figure 1. Les spectres 2A et 2B montrent qu'on peut utiliser notre théorie pour interpréter les spectres de rotation perturbée des gaz comprimés; mais cette interprétation ne nécessite pas forcément un formalisme à deux paramètres, et Bliot et al. [9, 10] ont récem-

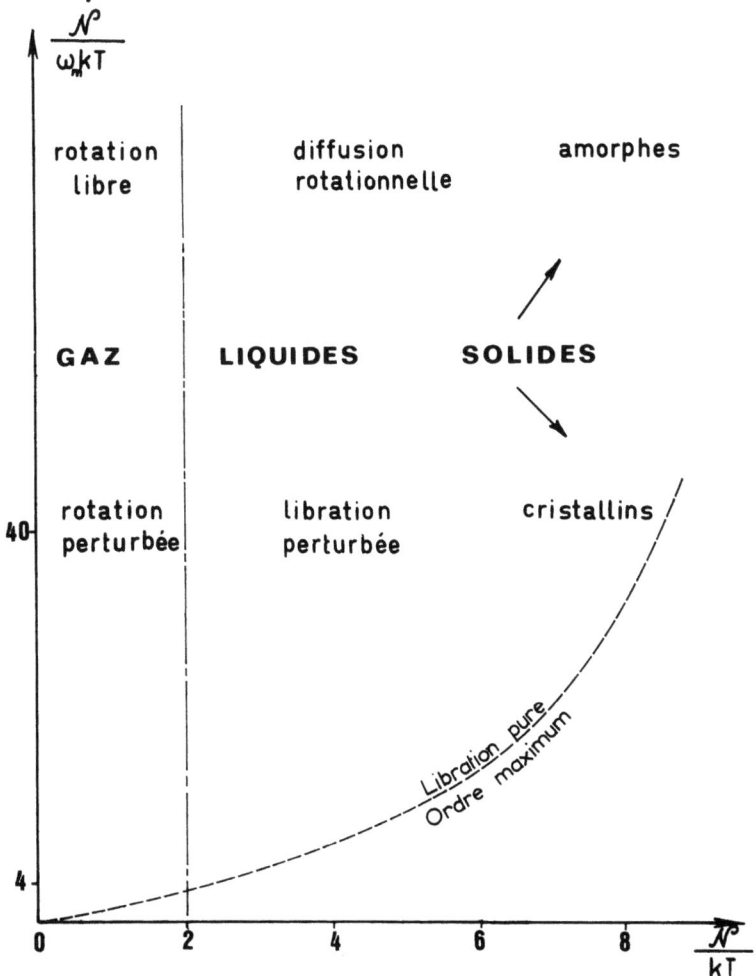

Fig. 1. Diagrammes des mouvements moléculaires selon les valeurs de $\mathcal{N}$ et $\dot{\mathcal{N}}$.

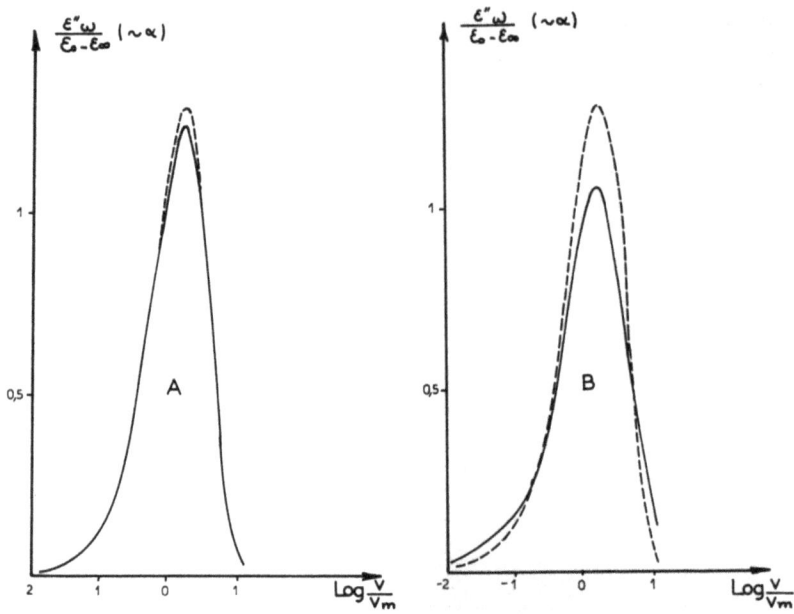

Fig. 2. Spectres d'absorption en modulation lente, calculés pour une molécule linéaire en utilisant l'équation (14). Cas A : $\mathcal{N} = 2\,kT$, $\dot{\mathcal{N}} = 72\,\omega_m kT$. Cas B : $\mathcal{N} = 2\,kT$, $\dot{\mathcal{N}} = 6\,\omega_m kT$. En trait pointillé, spectre de rotation libre

ment proposé une théorie très satisfaisante de ces phénomènes, n'utilisant qu'un seul paramètre physique.

Notre théorie permet également de décrire, pour des valeurs de $\mathcal{N}$ élevées, les spectres d'absorption dipolaire des liquides, et c'est là son principal objet. Nous illustrons ce fait en présentant les spectres $3C$ et $3D$: ces spectres reflètent qualitativement l'ensemble des phénomènes qu'on observe régulièrement dans les spectres expérimentaux. Nous pouvons donc tout aussi bien décrire les caractéristiques de l'absorption dipolaire à partir de ces spectres théoriques qu'à partir de cas concrets.

Dans le cas C où $\dot{\mathcal{N}}$ est élevé, le spectre obtenu est quasiment un spectre de diffusion rotationnelle dont le maximum d'absorption se situe autour de la pulsation moyenne de rotation ω_m. La décroissance en hautes fréquences de $\alpha(\omega)$, liée mathématiquement au fait que $\dot{\phi}(0) = 0$, correspond à un effet inertiel [11]. Le comportement spectral en basses fréquences suit la théorie de Debye [12] : on peut définir α_{MD} comme le maximum d'absorption prévu par cette théorie, et le temps de relaxation ou temps Debye $\tau_D = 1/\omega_D$ où ω_D est la pulsation du maximum de $\varepsilon''(\omega)$. Le diagramme de Cole et Cole $\varepsilon'' = f(\varepsilon')$ est un demi-cercle tant que la théorie de Debye est vérifiée ; mais en hautes fréquences, ε' tend vers ε_∞ par valeurs inférieures : c'est encore une conséquence de l'effet inertiel. Enfin la fonction de corrélation $\phi(t)$ est exponentielle (de constante de temps τ_D) sauf aux temps courts où elle présente un palier à l'origine ; si on prolonge la droite qui représente $\operatorname{Log}\phi(t)$ aux temps longs, son ordonnée à l'origine C est positive. Ce type de spectre et de fonction de corrélation s'observe expérimentalement

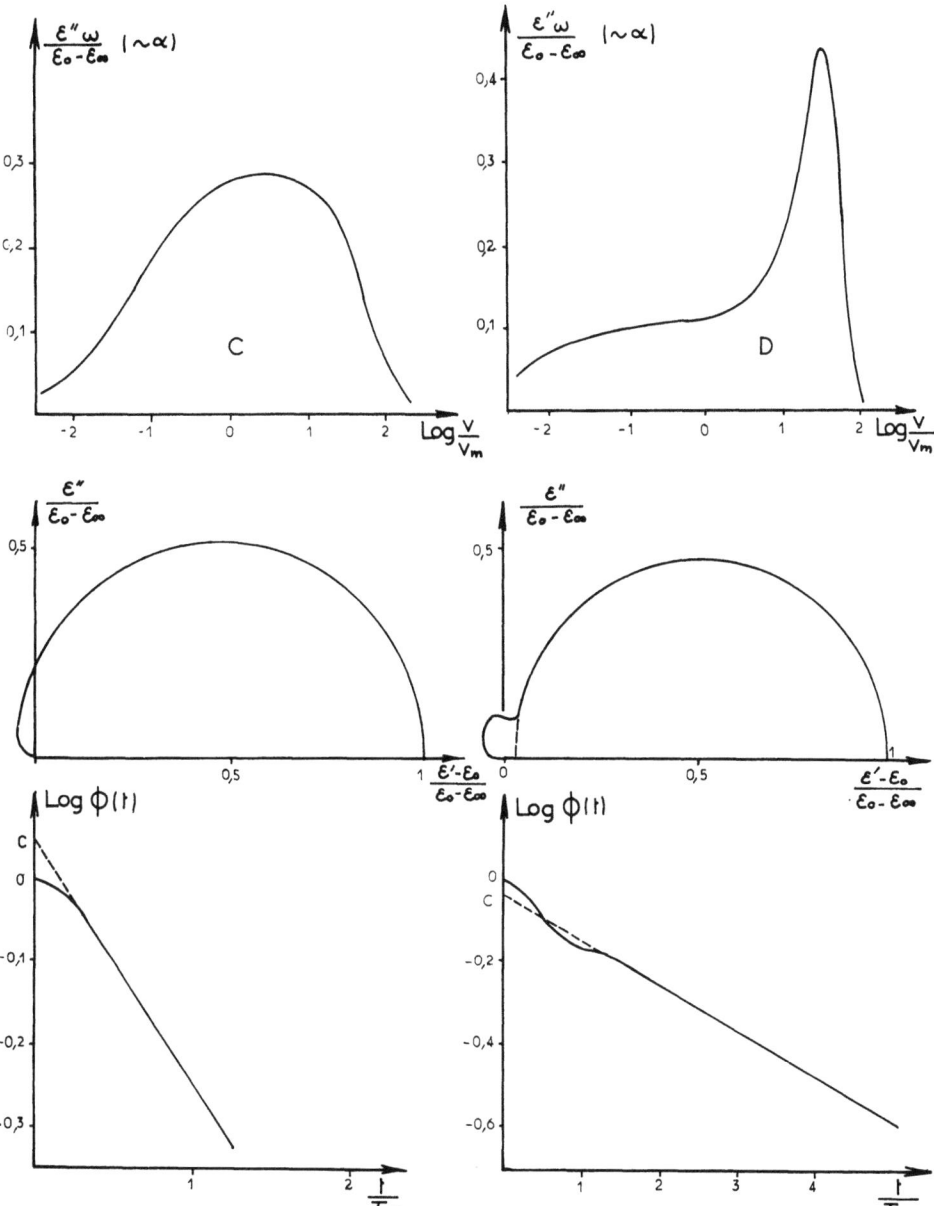

Fig. 3. Spectres d'absorption, diagrammes de Cole et Cole et fonctions de corrélation en modulation rapide, calculés en utilisant l'équation (15). Cas C: $\mathcal{N} = 8\,kT$, $\dot{\mathcal{N}} = 56\,\omega_m kT(K=0.15)$. Cas D: $\mathcal{N} = 8\,kT$, $\dot{\mathcal{N}} = 36\,\omega_m kT(K=1)$

dans des cas assez peu nombreux où la molécule polaire étudiée est de forme quasi-sphérique et ou le solvant est peu actif.

En fait, la plupart des spectres expérimentaux se rapprochent du cas D correspondant à des valeurs plus faibles de $\dot{\mathcal{N}}$. On observe une absorption excédentaire (par rapport au maximum α_{MD} prévu par la théorie de Debye) située dans la gamme infrarouge lointain, et on constate que la fréquence de son maximum est voisine de la fréquence de libration ω_0 établie en (18): il est donc naturel de penser que cette absorption excédentaire est un phénomène lié essentiellement à un mouvement de libration moléculaire plus ou moins amorti. Sur le diagramme de Cole et Cole, un second domaine de dispersion apparaît, correspondant à cette absorption excédentaire. Enfin la fonction de corrélation décroît toujours exponentiellement aux temps longs, mais cette fois le coefficient C est négatif.

Ainsi qualitativement la théorie proposée permet de décrire l'ensemble des phénomènes expérimentaux. Nous montrons dans la dernière partie qu'on peut étendre son application à la comparaison quantitative des spectres théoriques et expérimentaux.

5. Application de la théorie à l'interprétation et à l'exploitation des résultats expérimentaux

Nous avons montré dans l'exposé de la théorie que le calcul de $\alpha(\omega)$ ne dépendait que de deux paramètres $\mathcal{N}$ et $\dot{\mathcal{N}}$. Inversement, connaissant un spectre $\alpha(\omega)$ expérimental, il est possible de déterminer $\mathcal{N}$ et $\dot{\mathcal{N}}$ de façon à obtenir le meilleur accord possible entre spectres théorique et expérimental. Cette façon de procéder nous semble meilleure que celle basée sur la détermination des moments, bien souvent imprécise pour les moments d'ordre supérieur à 2.

En pratique, puisqu'il y a deux inconnues $\mathcal{N}$ et $\dot{\mathcal{N}}$ à déterminer, il suffit de s'imposer

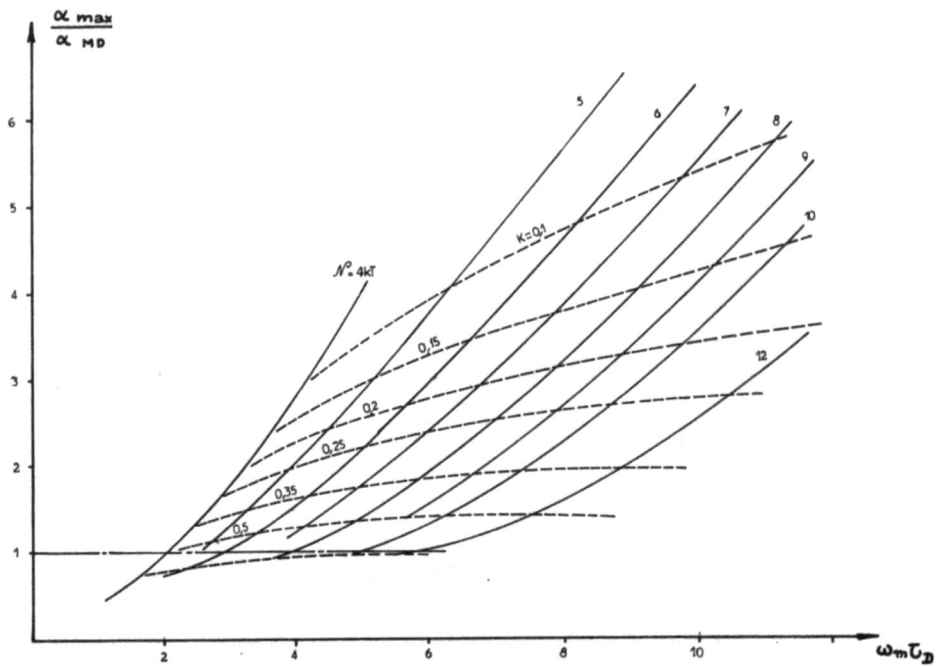

Fig. 4. Détermination de $\mathcal{N}$ et K (d'où $\dot{\mathcal{N}}$) à partir des valeurs de τ_D et α_{max}/α_{MD}.

deux contraintes à respecter, par exemple de retrouver par le calcul les valeurs exactes de deux grandeurs expérimentales caractéristiques du spectre. Nous avons choisi d'une part τ_D, temps de relaxation de Debye, qui caractérise les basses fréquences, d'autre part le rapport α_{max}/α_{MD}, qui caractérise les hautes fréquences. Nous avons tracé un abaque représentant ces deux grandeurs calculées en fonction de $\mathcal{N}$ et K (figure 4). Comme α_{max}/α_{MD} et τ_D s'obtiennent facilement d'après les spectres expérimentaux nous avons pu de cette façon déterminer $\mathcal{N}$ et K pour un grand nombre de solutions étudiées au laboratoire. Nous avons alors étudié l'évolution de ces deux paramètres en fonction des caractéristiques de la molécule et de son environnement, et nous avons regroupé les résultats les plus signifiants dans le tableau I.

On constate que, très normalement, le coefficient K augmente, c'est à dire l'ordre local diminue, lorsque T augmente. La forme géométrique de la molécule est également un facteur déterminant et de façon générale le couple est d'autant plus faible et sa dérivée d'autant plus élevée que la molécule se rapproche de la symétrie sphérique. Enfin naturellement l'activité du solvant favorise le mouvement de libration. Ces constatations suggèrent dans une certaine mesure que l'encombrement stérique, ou si l'on veut l'anisotropie des potentiels de repulsion, détermine de façon prépondérante le type de mouvement moléculaire que l'on rencontre en phase liquide.

Pour terminer, nous effectuons une comparaison quantitative entre théorie et expérience dans deux cas typiques. Dans le premier cas (figure 5a), la molécule polaire étudiée a une forme quasi-sphérique, le solvant est peu actif, et le spectre obtenu traduit un mouvement de diffusion rotationnelle; dans le second cas (figure 5b), la molécule polaire a une forme quasi linéaire, le solvant est plus actif, et la dynamique moléculaire est cette fois un mouvement de libration amortie. Dans chaque cas, les valeurs de $\mathcal{N}$ et K ont été choisies d'après l'abaque de la figure 4. On constate que dans l'ensemble l'accord entre théorie et expérience est bon, notamment dans le premier cas. Dans le second cas on observe qu'en infrarouge lointain le spectre expérimental est plus large que le spectre théorique; nous attribuons cette absorption 'supplémentaire' à l'influence des moments dipolaires induits qui se superposent aux moments dipolaires permanents et dont les fluctuations doivent contribuer à l'élargissement du spectre expérimental [7]; c'est pour cette même raison que le coefficient C expérimental est plus négatif que celui théorique.

6. Conclusion

En conclusion, la théorie proposée, basée sur l'utilisation du formalisme de la fonction mémoire, permet à partir de deux paramètres de signification physique simple, le couple moyen et sa dérivée moyenne, d'interpréter quantitativement l'essentiel des phénomènes de dynamique moléculaire présentés en phase liquide par des molécules toupies symétriques.

Nous avons utilisé cette théorie dans l'interprétation de l'absorption dipolaire, mais rien ne s'oppose naturellement à ce qu'elle puisse étendre son domaine d'application aux résultats expérimentaux obtenus par d'autres techniques.

	Lorsque la température augmente	Lorsqu'on passe d'un solvant peu actif à un solvant actif	Lorsque la forme de la molécule polaire se rapproche d'une forme sphérique
$\mathcal{N}$ EXEMPLES	CH$_3$I en solution dans CCl$_4$ (-20°C) (0°C) (20°C) 10,6 9,2 7,8	CHBr$_3$ en solution dans Hexane Cyclohexane CCl$_4$ 7,9 10 11	Solutions dans l'hexane de CHCl$_3$ CCl$_3$CH$_3$ CCl(CH$_3$)$_3$ 9,2 8,6 5,8
K EXEMPLES	CH$_3$I en solution dans CCl$_4$ (-20°C) (0°C) (20°C) 0,15 0,158 0,165	CHCl$_3$ en solution dans Hexane Cyclohexane CCl$_4$ 0,44 0,335 0,3	Solutions dans l'hexane de CHCl$_3$ CCl(CH$_3$)$_3$ CCl$_3$CH$_3$ 0,44 0,65 1,35

TABLEAU 1. EVOLUTION DU COUPLE MOYEN $\mathcal{N}$ (UNITE kT) AGISSANT SUR LA MOLECULE ET DU FACTEUR DE PERTURBATION K (LIE A $\dot{\mathcal{N}}$ PAR L'EQUATION (19))

(Température : 25°C sauf indications précises - Concentration molaire 20%)

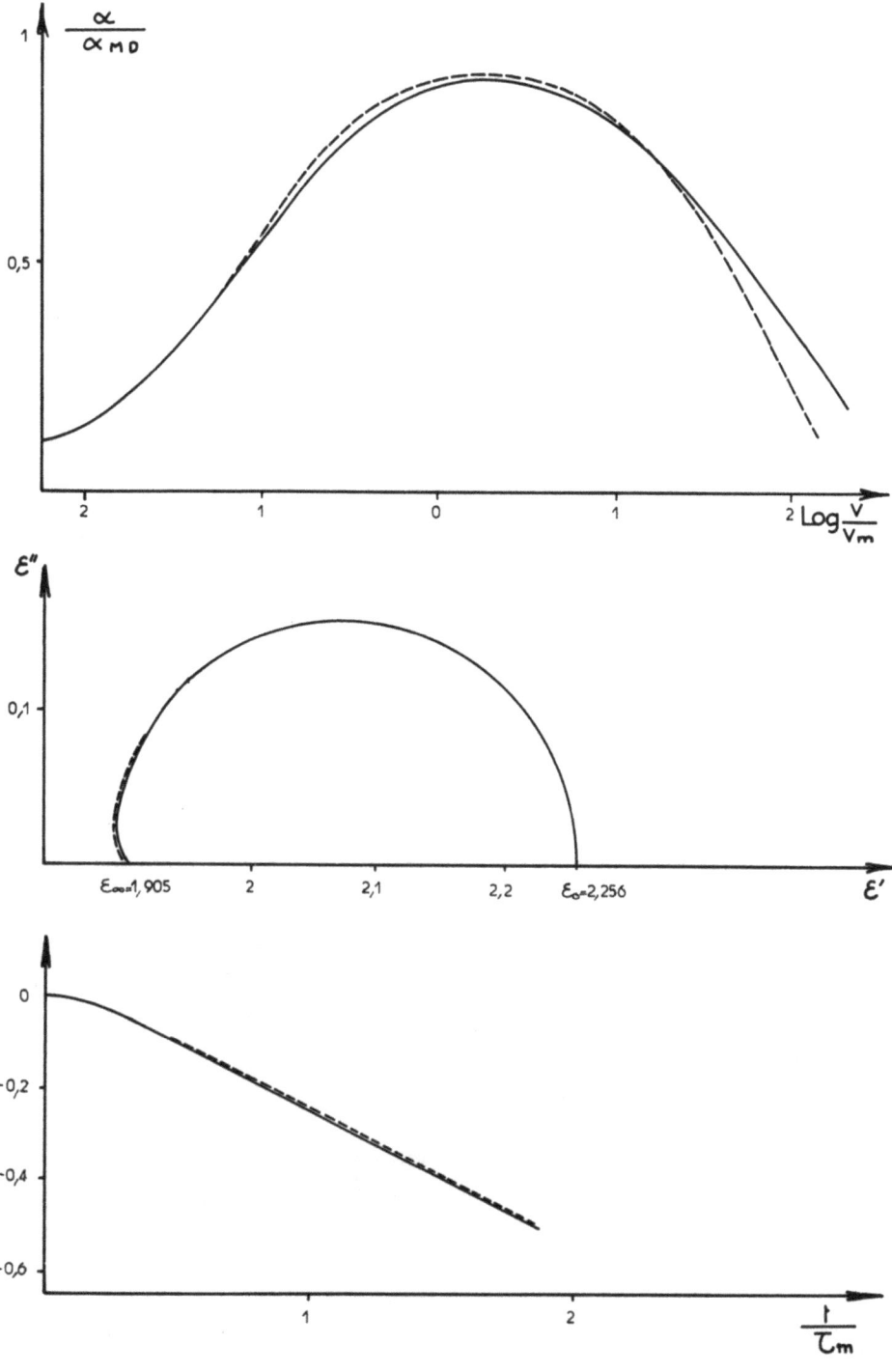

Fig. 5a.

Fig. 5a-b. Comparaison entre spectres et fonctions de corrélation expérimentaux (en trait plein) et théoriques (en trait pointillé, calculés pour les valeurs de $\mathcal{N}$ et K correspondantes dans le tableau I). (a) Trichloroethane Cl_3CCH_3 en solution dans l'hexane, concentration molaire 0.14, température 25°C ($\nu_m = 8.1$ cm^{-1}). (b) Iodure de Méthyle CH_3I en solution dans le tétrachlorure de carbone CCL_4, concentration molaire 0.2, température 20°C ($\nu_m = 14.3$ cm^{-1}).

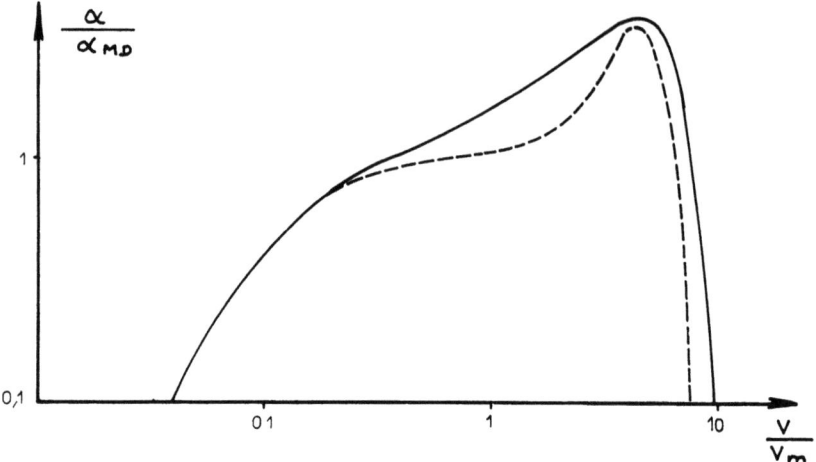
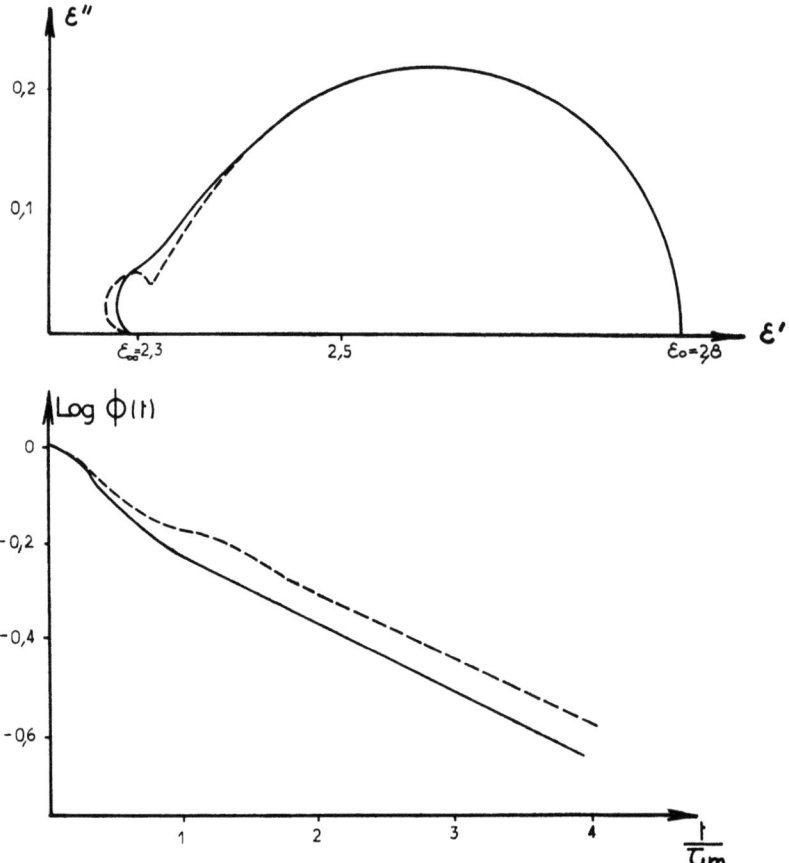

Fig. 5b.

Bibliographie

1. Kubo, R.: dans *Lectures in Theoretical Physics,* Interscience Publishers, 1959, p. 120.
2. Kubo, R.: *Reports on Progress in Physics* **29**, Part 1, 255 (1966).
3. Cole, R. H.: *J. Chem. Phys.* **42**, 637 (1965).
4. Desplanques P. et Constant, E.: *Compt. Rend. Acad. Sci., Série B* **272**, 1354 (1971).
5. Berne B. J. et Harp, G. D.: *Adv. Chem. Phys.* **17**, Interscience Publishers, 1970, p-63.
6. Desplanques P. et Constant, E.: *Compt. Rend. Acad. Sci., série B* **274**, 611 (1972).
7. Desplanques, P.: Thèse de Doctorat d'Etat, à paraître.
8. Doob, J. L.: *Ann. Math.* **43**, 351 (1942).
9. Bliot, F., Abbar, C., et Constant, E.: *Mol. Phys.* **24**, 241 (1972).
10. Bliot F. et Constant, E.: *Chem. Phys. Letters* **18**, 253 (1973).
11. Leroy, Y., Constant, E., et Desplanques, P.: *J. Chim. Phys.* **64**, 1499 (1967).
12. Debye, P.: *Polar Molecules,* Dover Publication, New York, 1929.

DISCUSSION

Bratos: Il serait intéressant de comparer votre théorie avec celle proposée par Nielsen et Keyes en vue d'étudier les rotations moléculaires dans les liquides.

Brot: I would like to make an historical remark: Twenty years ago the 'fashionable' language was the distribution of relaxation times. For ten years or so we use the time correlation functions. Now there seems to be a trend to adopt the formalism of a hierarchy of memory functions. The first and the third languages have, in my opinion, the drawback of biasing the intellectual approach: with the first one, people tended to think that the liquid was really made of a distribution of different species (weak associations, etc.); with the third one, one is tempted to implicitly assume that some quantity appearing in the nth order of the hierarchy has, 'naturally', a white spectrum (whereas all quantities are governed by the Liouville equation). I do not deny the convenience of being able, with this formalism, to introduce simulateneously parameters relative both to short and long times. Only the correlation function formalism is, in my feeling, objective, because it forces people to explicit their hypothesis. This general remark is not at all a criticism of Desplanques et al.'s excellent contribution. On the contrary, these authors have clearly stressed the physical meaning of the possible hypothesis concerning the value of their parameters $\mathcal{N}$ and $\dot{\mathcal{N}}$.

Friedman: I agree completely with Prof. Brot that your work is a beautiful illustration of the equivalence of various methods of representing the relaxation data. However I do not believe that all these representations are equivalent in a deeper sense, for we can hope that there is a representation whose parameters, adjusted to fit the relaxation data, change in a way that makes sense when we change from one system to another. For a kind of example I appeal to Prof. Davies remarks after the paper of Kilp. We 'believe' in molecules but we also believe in atoms. However it has been a long time since chemists realized that we cannot understand the properties of molecules simply in terms of atoms. We need a more abstract concept, the chemical bond, as well, at least for practical work in chemistry.

In the subject matter of this conference we are concerned with intermolecular interactions in liquids and I think we are searching for that concept, roughly analogous to chemical bonds, which will enable us to understand the changes in relaxation coefficients when we change from one system to another.

Desplanques: I agree with you: all these representations are not equivalent in a deeper sense. For us, the formalism or the language of memory function brings something more than the language of correlation function alone. It allowed us to obtain, in a great number of cases, the values of mean square torque and mean square derivative torque adjusted to fit the relaxation data or absorption spectra which change in a way that makes sense with temperautre, solvent and polar molecule used.

Rivail: Comment est définie l'activité du solvant?

Desplanques: Nous utilisons le langage employé par les spectroscopistes dans l'étude du déplacement de la fréquence des raies de vibration ou de leur élargissement, soit par ordre d'activité croissante: pentane, hexane, tétrachlorure de carbone, etc...

CHAMP DE REACTION DYNAMIQUE EN RELAXATION DIELECTRIQUE, APPLICATION AU CAS DU CHLOROFORME A 25°C

JEAN-LOUIS GREFFE, JOSE GOULON, JEAN BRONDEAU, et JEAN-LOUIS RIVAIL

Université de NANCY I, Laboratoire de Chimie Théorique, Equipe de recherche associée au C.N.R.S. no. 22, Case Officielle 140, 54037 Nancy-Cedex, France

Résumé. La formulation de la relaxation diélectrique à partir d'un modèle à un seul dipôle dans une cavité sphérique, est reprise en discutant l'approximation du continu. Une formule plus générale est établie. Applique au chloroforme, elle permet de sélectionner la meilleure fonction du continu à associer à une fonction d'autocorrélation donnée. Elle permet d'en déduire ε_∞ pour le chloroforme, la fonction d'autocorrélation et une valeur approchée du moment d'inertie perpendiculaire.

Abstract. The formulation of diélectric relaxation using a model of a dipole in a spherical cavity, is reexamined and the approximation of the continuum is discussed. A general formula is established. In the case of chloroform, this formula makes possible the choice of the best function for the continuum associated to a given auto-correlation function and an approximate value of the perpendicular moment of inertia are deduced for chloroform.

1. Résultats expérimentaux concernant le chloroforme à 25°C

J. Goulon a effectué 26 mesures à diverses fréquences sur le chloroforme à 25°C en relaxation diéléctrique. Ces mesures, effectuées par différentes techniques, s'étendent de 0,62 GHz à 120 cm^{-1}. Leurs résultats sont distribués de façon à permettre un tracé très précis du diagramme de Cole et Cole (figure 1). Ce diagramme présente un arc 'skewed' pour des fréquences comprises entre 50 et 800 GHz. Il met en évidence un maximum pour ε'' à 26 GHz et un minimum pour ε' à 40 cm^{-1}. Ce minimum rend impossible un raccordement normal avec l'axe des ε' qui serait compatible avec la valeur $\varepsilon_\infty = n_D^2 = 2{,}08$. La valeur extrapolée pour ε_∞, confirmée par des études de mélange à basse fréquence est de $2{,}24 \pm 0{,}1$. [1, 2, 3].

2. Les problèmes de l'interprétation

En théorie des phénomènes dissipatifs linéaires, la relaxation diélectrique est décrite à travers une formule du type:

$$F(\varepsilon_0, \varepsilon^*, \varepsilon_\infty) = L(-\dot{\Phi}), \qquad (2.1)$$

où F est une fonction du milieu, supposé continu, L symbolise la transformée de Laplace et Φ la fonction d'autocorrélation du moment dipolaire total d'un échantillon macroscopique. Le premier problème est posé par le traitement des interactions dipolaires lorsqu'on veut s'intéresser au comportement d'une seule molécule. Ce traitement peut se faire à travers le concept de champ de réaction qui permet alors de rap-

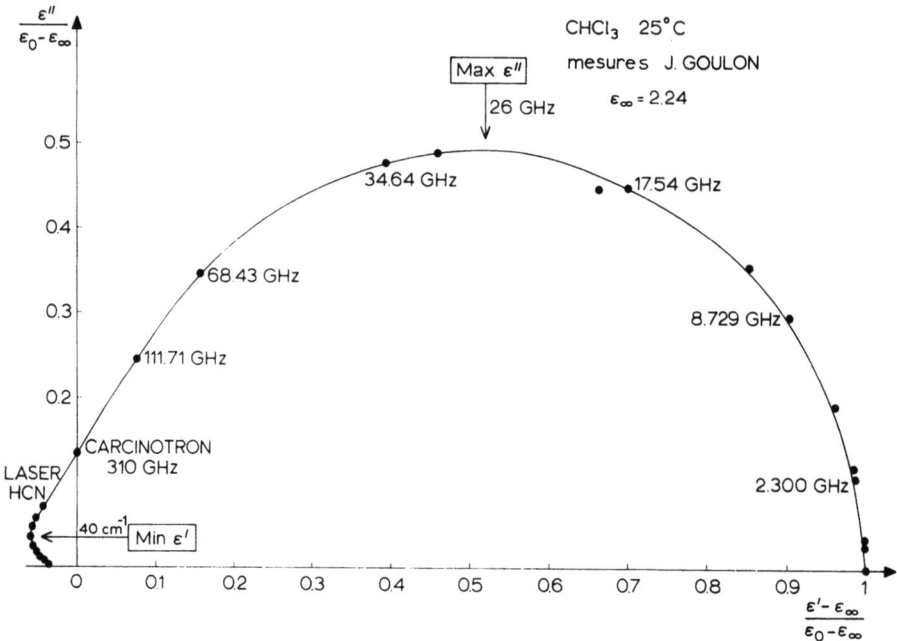

Fig. 1. Diagramme de Cole et Cole de la relaxation diélectrique du chloroforme à 25 °C.

porter l'équation précédente à une seule molécule:

$$f(\varepsilon_0, \varepsilon^*, \varepsilon_\infty) = L(-\dot\varphi), \tag{2.2}$$

où, cette fois, φ est la fonction d'autocorrélation d'un dipole unique supposé rigide.

On considère dans cette relation ε_∞ comme inconnue. On considère d'autre part qu'il existe parmi plusieurs fonctions f du continu données par la théorie du champ de réaction dynamique, une fonction optimale pour la représentation de la réalité discontinue. On considère enfin qu'il existe aussi une fonction φ optimale, dépendant d'assez peu de paramètres pour être aisément interprétée tout en restant soumise aux contraintes des conditions de Gordon aux temps courts et du comportement exponentiel aux temps longs.

Sur les données expérimentales on va donc optimiser ε_∞ et les formes mathématiques de f et φ compatibles avec les contraintes physiques. Les résultats pourraient être vraisemblablement améliorés si l'on recherchait plusieurs fonctions f et φ correspondant à des domaines de fréquence différents. Nous avons recherché ici une solution globale, unique pour toutes les fréquences.

3. Champ de réaction dynamique et formulation de la relaxation diélectrique dans le cadre d'un modèle à cavité monomoléculaire

Le champ interne **F** agissant sur la molécule est la résultante du champ de cavité **C**, et du champ de réaction **R**. Ce dernier est partagé en celui dû au moment permanent

R_0 et celui dû au moment induit R_1, tels que:

$$F = C + R_0 + R_1, \qquad (3.1)$$

où

$$R_0 = g_0 m_0 \quad \text{et} \quad R_1 = g_1 m_1 \qquad (3.2)$$

avec

$$m_0 = \mu + \alpha R_0 \quad \text{et} \quad m_1 = \alpha(C + R_1) \qquad (3.3)$$

donc

$$m_0 = \frac{\mu}{1 - \alpha g_0} \quad \text{et} \quad m_1 = \frac{\alpha C}{1 - \alpha g_1} \qquad (3.4)$$

α est la polarisabilité, g_0 et g_1 les coéfficients de réaction.

La polarisation acquise par N molécules par unité de volume dans le champ extérieur E devient en fonction de la pulsation:

$$P(\omega) = N\langle m(\omega)\rangle_E = N\langle m_0(\omega)\rangle_E + N m_1(\omega) \qquad (3.5)$$

$\langle m_0(\omega)\rangle_E$ est calculé par application de la théorie de Kubo qui donne l'expression de la susceptibilité:

$$\chi(\omega) = \frac{\langle \mu(\omega)\rangle_E}{C(\omega) + R_1(\omega)} = (1 - \alpha g_1^*)\frac{\langle \mu(\omega)\rangle_E}{C(\omega)} \qquad (3.6)$$

et

$$\langle m_0(\omega)\rangle_E = \frac{\langle \mu(\omega)\rangle_E}{1 - \alpha g_0^*} \qquad (3.7)$$

g_0^* et g_1^* désignant respectivement $g_0(\varepsilon^*)$ et $g_1(\varepsilon^*)$. D'autre part:

$$m_1(\omega) = \frac{\alpha C(\omega)}{1 - \alpha g_1^*}. \qquad (3.8)$$

On admet la validité de la formule de Clausius-Mossotti:

$$\frac{\alpha}{a^3} = \frac{\varepsilon_\infty - 1}{\varepsilon_\infty + 2}. \qquad (3.9)$$

La polarisation totale acquise par l'unité de volume dans le champ E s'écrit:

$$P(\omega) = N\langle m(\omega)\rangle_E + \frac{\varepsilon^* - 1}{4\pi} E(\omega) \qquad (3.10)$$

expression d'où l'on tire:

$$(1 - \alpha g_0^*)\left[(1 - \alpha g_1^*)\frac{\varepsilon^* - 1}{c^*} - 3\frac{\varepsilon_\infty - 1}{\varepsilon_\infty + 2}\right] = \frac{4\pi N\mu^2}{3kT} L(-\dot\varphi). \qquad (3.11)$$

La combinaison de cette relation avec celle que l'on établirait de façon similaire pour les diélectriques statiques, conduit à l'expression finale de la fonction du continu :

$$f(\varepsilon_0, \varepsilon^*, \varepsilon_\infty) = \frac{1 - \alpha g_0^*}{1 - \alpha g_0} \frac{(1 - \alpha g_1^*) \dfrac{\varepsilon^* - 1}{c^*} - 3 \dfrac{\varepsilon_\infty - 1}{\varepsilon_\infty + 2}}{(1 - \alpha g_1) \dfrac{\varepsilon_0 - 1}{c_0} - 3 \dfrac{\varepsilon_\infty - 1}{\varepsilon_\infty + 2}}. \tag{3.12}$$

A ce stade, aucune hypothèse sur le coefficient du champ de cavité ni sur ceux des champs de réaction n'a été faite.

4. Discussion des coefficients des champs de cavité et de réaction

Il faut d'abord noter que dans l'expression de la fonction du continu (3.12), il y a séparation de l'effet dû à la réaction du dipôle permanent décrit par le coefficient :

$$r = \frac{1 - \alpha g_0^*}{1 - \alpha g_0} = \frac{2\varepsilon_0 + 1}{2\varepsilon^* + 1} \frac{2\varepsilon^* + \varepsilon_\infty}{2\varepsilon_0 + \varepsilon_\infty}. \tag{4.1}$$

Dans la suite, nous considérons que le continu extérieur à la cavité possède une constante diélectrique, soit classiquement égale à ε^*, soit au contraire en permanence

c^*	g_1^*	g_0^*	f	
ε^*	ε^*	ε_0	$\dfrac{\varepsilon^* - \varepsilon_\infty}{\varepsilon_0 - \varepsilon_\infty} \dfrac{\varepsilon_0}{\varepsilon^*} \dfrac{2\varepsilon^* + 1}{2\varepsilon_0 + 1}$	B. C.
		ε^*	$\dfrac{\varepsilon^* - \varepsilon_\infty}{\varepsilon_0 - \varepsilon_\infty} \dfrac{\varepsilon_0}{\varepsilon^*} \dfrac{2\varepsilon^* + \varepsilon_\infty}{2\varepsilon_0 + \varepsilon_\infty}$	R. K.-K.-V.
ε^*	ε_∞	ε_0	$\dfrac{\varepsilon^* - \varepsilon_\infty}{\varepsilon_0 - \varepsilon_\infty} \dfrac{\varepsilon_0}{\varepsilon^*} \dfrac{2\varepsilon^* \varepsilon_\infty + 1}{2\varepsilon_0 \varepsilon_\infty + 1}$	G2
ε_∞	ε_∞	ε^*	$\dfrac{\varepsilon^* - \varepsilon_\infty}{\varepsilon_0 - \varepsilon_\infty} \dfrac{2\varepsilon_0 + 1}{2\varepsilon^* + 1} \dfrac{2\varepsilon^* + \varepsilon_\infty}{2\varepsilon_0 + \varepsilon_\infty}$	G1
		ε_∞	$\dfrac{\varepsilon^* - \varepsilon_\infty}{\varepsilon_0 - \varepsilon_\infty}$	D.

Fig. 2. Permittivités complexes choisies pour le calcul des coefficients des champs de cavité et de réaction. Fonctions du continu qui s'en déduisent. (B.=J. Barriol; C.=C. H. Colie et auteurs; R.= J.-L. Rivail; D.=P. Debye; K.−K.−V.=D. D. Klug, D. E. Kranbuehl, et W. E. Vaughan; G1 et G2 = les auteurs de cette note).

invariable et égale à ε_0 ou à ε_∞, et ceci pour le calcul des trois coéfficients: c^* pour rendre compte du champ de cavité, g_1^* pour rendre compte de la réaction du dipôle induit et g_0^* pour rendre compte de celle du dipôle permanent. Par le jeu des différentes combinaisons possibles, on retrouve les principales formules couramment employées [4, 5, 6, 7, 8, 9] et rappelées dans la figure 2. Ainsi, le fait de prendre pour g_0 une valeur réelle constante revient à ne considérer que les états où le dipôle est peu mobile.

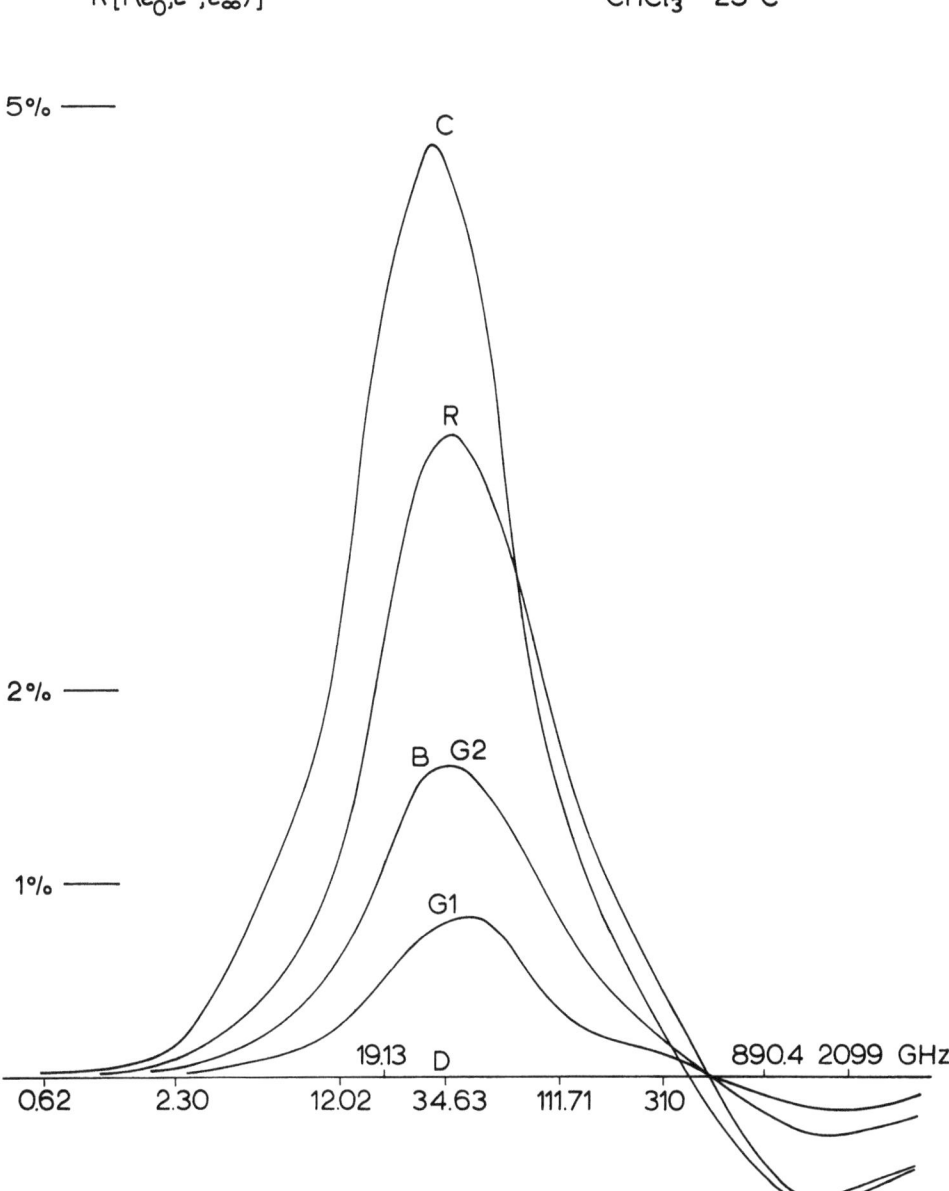

Fig. 3. Parties réelles des fonctions du continu extérieur à la cavité. Voir légende de la figure 2.

Il est aussi intéressant de noter que le premier membre de la formule de Debye:

$$\frac{\varepsilon^* - \varepsilon_\infty}{\varepsilon_0 - \varepsilon_\infty} \tag{4.2}$$

correspond, dans le cadre d'un modèle monomoléculaire à cavité, au cas où l'extérieur possède à toute fréquence la constante diélectrique ε_∞.

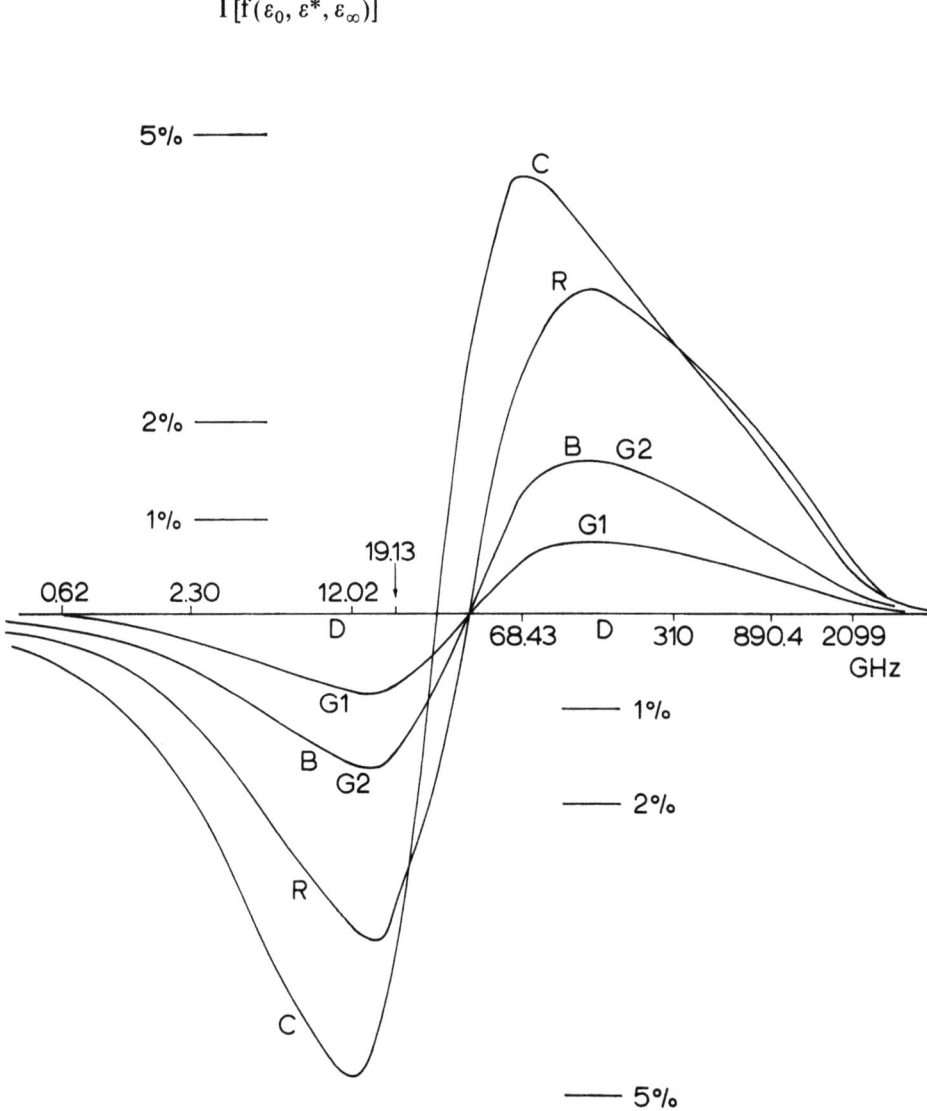

Fig. 4. Parties imaginaires des fonctions du continu extérieur à la cavité. Voir légende de la figure 2.

Nous avons enfin comparé nos résultats à ceux donnés par la formule de Cole [10]:

$$\frac{\varepsilon^* - \varepsilon_\infty}{\varepsilon_0 - \varepsilon_\infty} = \frac{1}{1 + \dfrac{3\varepsilon_0}{2\varepsilon_0 + \varepsilon_\infty}[L^{-1}(-\dot{\varphi}) - 1]} \qquad (4.3)$$

soit:

$$\frac{\varepsilon^* - \varepsilon_\infty}{\varepsilon_0 - \varepsilon_\infty} \cdot \frac{3\varepsilon_0}{2\varepsilon_0 + \varepsilon_\infty} = L(-\dot{\varphi}) \qquad (4.4)$$

que nous ne pouvons retrouver facilement à partir de la relation (3.12).

Les écarts entre les valeurs données pour $f(\varepsilon_0, \varepsilon^*, \varepsilon_\infty)$ par ces différentes formules sont de toute façon très faibles. Il nous a paru néanmoins intéressant de les matérialiser par les figures 3 et 4 qui indiquent, en pourcentage, les écarts à la formule de Debye pour la partie réelle et pour la partie imaginaire de f.

5. Fonctions d'autocorrélation dipolaires pour le chloroforme

La méthode employée ici ne consiste pas à déduire d'un mécanisme moléculaire une fonction d'autocorrélation mais au contraire à chercher en vue d'une interprétation ultérieure les fonctions d'autocorrélation les plus simples rendant compte des caractéristiques principales du comportement du chloroforme en fréquence.

Les propriétés que nous imposons aux fonctions d'autocorrélation sont les suivantes:
(1) $\varphi(0) = L(-\dot{\varphi})_{\omega=0} = 1$ et $\varphi(\infty) = 0$;
(2) Elles vérifient avec f les relations de Kramers-Kronig;
(3) Elles donnent un maximum et un seul pour ε'';
(4) Elles donnent un minimum et un seul pour ε' à fréquence finie;
(5) $\lim_{\omega \to \infty} IL(-\dot{\varphi})/RL(-\dot{\varphi}) = 0$;
(6) $\lim_{\omega \to \infty} |\omega \cdot IL(-\varphi)| \sim \lim_{\omega \to \infty} \alpha(\omega) = 0$;
(7) $\dot{\varphi}(0) = 0$.

Toutes les autres dérivées impaires devraient être théoriquement nulles mais nous voulons aussi une bonne représentation des faits aux temps longs pour lesquels l'exponentielle simple ne remplit pas cette condition.

Un certain nombre de fonctions d'autocorrélation dépendant d'un minimum de paramètres ont donc été sélectionnées. Elles sont présentées dans la figure 5 avec les valeurs de leurs premiers moments. Il s'agit des fonctions de Lassier et Brot [11], Birnbaum et Cohen [12], Kubo et Tomita [13], Sack [14], Greffe [9], Barriol [9], et Goulon [15].

6. Conclusions

L'optimisation de la formule de la relaxation (2.2) s'est donc faite sur l'ensemble des fonctions f et φ présentées. Elle montre d'abord que la valeur de ε_∞ à employer dans les études diélectriques du chloroforme est:

$$\varepsilon_\infty = 2{,}24 \pm 0{,}1 \text{ à } 25\,°\text{C}.$$

$\phi_\mu(t)$	m_1	m_2	m_3/j	m_4
$\dfrac{\tau_1}{\tau_1-\tau_2}e^{-t/\tau_1}-\dfrac{\tau_2}{\tau_1-\tau_2}e^{-t/\tau_2}$	0	$\dfrac{1}{\tau_1\tau_2}$	$\dfrac{1}{\tau_1^2\tau_2}+\dfrac{1}{\tau_1\tau_2^2}$	$-\dfrac{1}{\tau_1^3\tau_2}-\dfrac{1}{\tau_1^2\tau_2^2}-\dfrac{1}{\tau_1\tau_2^3}$
$\exp\left\{\dfrac{\tau_2}{\tau_1}-\sqrt{\left(\dfrac{\tau_2}{\tau_1}\right)^2+\left(\dfrac{t}{\tau_1}\right)^2}\right\}$	0	$\dfrac{1}{\tau_1\tau_2}$	0	$\dfrac{3}{\tau_1^2\tau_2^2}+\dfrac{3}{\tau_1\tau_2^3}$
$\exp\left[\dfrac{\tau_2}{\tau_1}\left(1-\dfrac{t}{\tau_2}-e^{-t/\tau_2}\right)\right]$	0	$\dfrac{1}{\tau_1\tau_2}$	$\dfrac{1}{\tau_1\tau_2}$	$\dfrac{6}{\tau_1^2\tau_2^2}-\dfrac{1}{\tau_1\tau_2^3}$
$e^{-t/\tau_1}+\dfrac{t}{\tau_1}e^{-t/\tau_2}$	0	$-\dfrac{1}{\tau_1^2}+\dfrac{2}{\tau_1\tau_2}$	$-\dfrac{1}{\tau_1^3}+\dfrac{3}{\tau_1\tau_2^2}$	$\dfrac{1}{\tau_1^4}-\dfrac{4}{\tau_1\tau_2^3}$
$e^{-t/\tau_1}+(1-e^{-t/\tau_1})e^{-t/\tau_2}$	0	$\dfrac{2}{\tau_1\tau_2}$	$\dfrac{3}{\tau_1^2\tau_2}+\dfrac{3}{\tau_1\tau_2^2}$	$-\dfrac{4}{\tau_1^3\tau_2}-\dfrac{6}{\tau_1^2\tau_2^2}-\dfrac{4}{\tau_1\tau_2^3}$
$e^{-t/\tau_1}+(1-e^{-t/\tau_1})e^{-t/\tau_2}\cos\dfrac{t}{\tau_3}$	0	$\dfrac{2}{\tau_1\tau_2}$	0	$\dfrac{2}{\tau_1\tau_2}\left(\dfrac{1}{\tau_1^2}+\dfrac{6}{\tau_1\tau_2}+\dfrac{1}{\tau_2^2}\right)$

Fig. 5. Fonctions d'autocorrelation dipolaire pour l'étude du chloroforme à 25 °C. Ce sont successivement les fonctions de Lassier-Brot, Birnbaum-Cohen, Kubo-Tomita ou Sack, Greffe, Barriol, et Goulon.

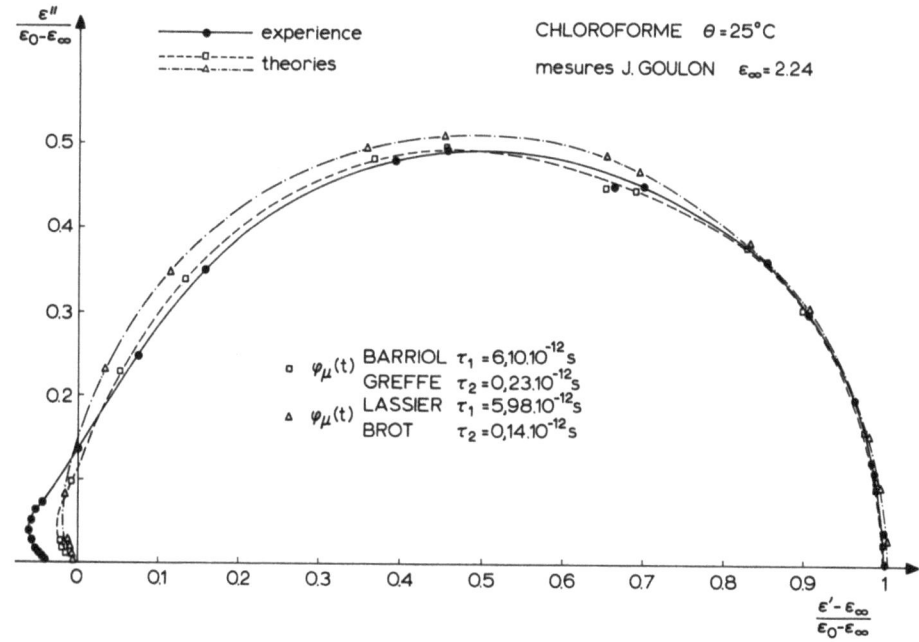

Fig. 6. Diagrammes de Cole et Cole du chloroforme.

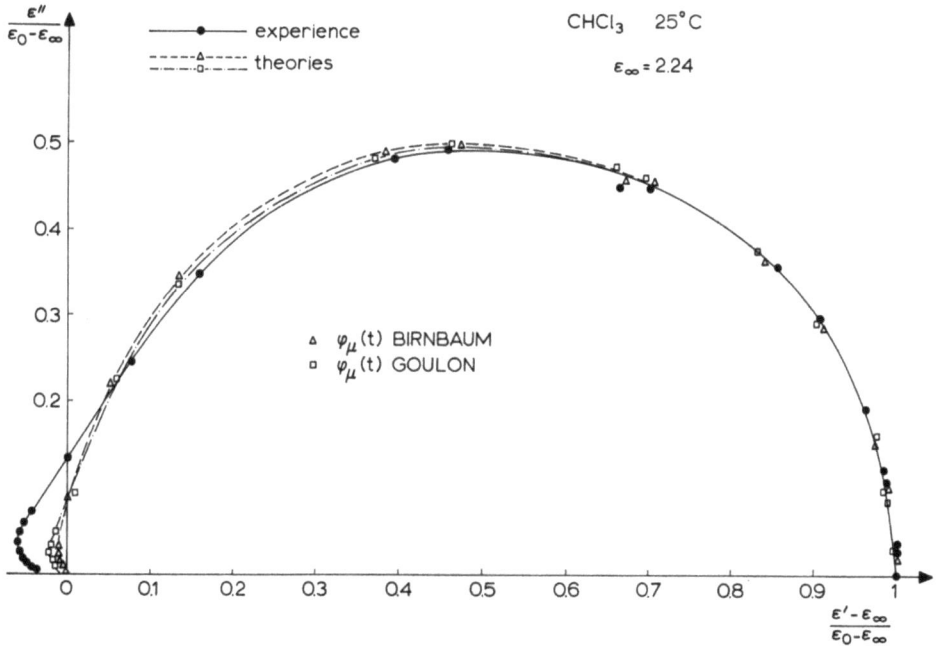

Fig. 7. Diagrammes de Cole et Cole du chloroforme interpretés par les fonctions de Birnbaum et de Goulon.

Elle montre que c'est la formule de Cole, de Klug, Kranbuehl, Vaughan ou de Rivail qu'il faut employer en basse et moyenne fréquence mais que ces formules surestiment le couplage du dipôle par rapport à la formule de Debye à très haute fréquence.

Pour les fonctions d'autocorrelation, des résultats ont déjà été présentés dans la référence 9.

Les figures 6 et 7 matérialisent l'ensemble des résultats sur les diagrammes de Cole et Cole. L'accord est très satisfaisant en basse et moyenne fréquence. Il l'est relativement moins en Infra-rouge lointain dans la mesure où l'on recherche une fonction unique pour tenter de représenter l'ensemble du comportement du chloroforme.

Il est néanmoins remarquable de noter que la valeur numérique du moment d'ordre 2 est sensiblement la même pour toutes les fonctions d'autocorrélation alors que leurs valeurs algébriques varient du simple au double. Par la théorie de Gordon [16], ce moment d'ordre 2 est lié au moment d'inertie perpendiculaire de la molécule. Par la formule de Lassier-Brot, qui postule un mouvement de libration perturbée par chocs:

$$I_\perp = 286{,}50 \times 10^{-40} \text{ g} \cdot \text{cm}^2$$

par celle de Birnbaum et Cohen:

$$I_\perp = 238{,}40 \times 10^{-40} \text{ g} \cdot \text{cm}^2$$

par celles de Barriol ou de Greffe:

$$I_\perp = 258{,}64 \times 10^{-40} \text{ g} \cdot \text{cm}^2$$

par celle de Goulon, dont se sert également J. C. Briquet:

$$I_\perp = 283{,}61 \times 10^{-40} \text{ g} \cdot \text{cm}^2$$

Les valeurs expérimentales sont:

$$254{,}15 \text{ et } 263{,}28 \times 10^{-40} \text{ g} \cdot \text{cm}^2.$$

L'ensemble de ces résultats montre que la corrélation des dipôles doit être faible dans le chloroforme. Il montre aussi que le moment d'ordre 4 est beaucoup plus sensible au choix de telle ou telle fonction mais que sa prise en considération conduirait à éliminer des fonctions qui pourtant globalement donnent des résultats satisfaisants. L'avenir nous paraît donc être l'examen de fonctions du type de celle de Birnbaum, mais à trois ou même à quatre paramètres.

Bibliographie

1. Goulon, J., Rivail, J. L., Feming, J. W., Chamberlain, J., et Chantry, G. W.: *Chem. Phys. Letters* **18**, (2) 211 (1973).
2. Boule, P.: *J. Chim. Phys.* **65** (5), 777 (1968).
3. Boyer-Donzelot, M.: *Bull. Soc. Chim.* **2**, 245 (1970).

4. Collie, C. H., Hasted, J. B., et Ritson, D. M.: *Proc. Phys. Soc.* **60**, 145 (1948).
5. Barriol, J.: *Compt. Rend. Acad. Sci.* **259**, 4010 (1964).
6. Rivail, J. L.: *J. Chem. Phys.* **66** (5) 981 (1969).
7. Klug, D. D., Kranbuehl, D. E., et Vaughan, W. E.: *J. Chem. Phys.* **50** (9), 3904 (1969).
8. Debye, P.: *Polar Molecules,* D. Publi, N.-Y. 1929.
9. Greffe, J. L., Goulon, J., Brondeau, J., et Rivail, J. L.: *J. Chim. Phys.* **70** (2), 282 (1973).
10. Cole, R. H.: *J. Chem. Phys.* **42**, 637 (1965).
11. Lassier, B. et Brot, C.: *Discuss. Faraday Soc.* **48**, 39 (1969).
12. Birnbaum, G. et Cohen, E. R.: *J. Chem. Phys.* **53**, 2885 (1970).
13. Kubo, R. et Tomita: *J. Phys. Soc. Japan* **7**, 888 (1954).
14. Sack, R. A.: *Proc. Phys. Soc. London* **70B**, 402 (1957).
15. Goulon, J.: à paraître.
16. Gordon, R. G.: *Adv. Chem. Phys.* **15**, 79 (1969).

ACTIVE INTRAMOLECULAR MOTION IN DIELECTRIC RELAXATION OF PURE LIQUID DIACETYL

JOSE GOULON and JEAN-LOUIS RIVAIL

Université de Nancy I, Laboratoire de Chimie Théorique, Case Officielle 140, 54037 Nancy-Cedex, France

and

JOHN CHAMBERLAIN and GEORGE W. CHANTRY

National Physical Laboratory, Teddington, Middlesen TW 11 OLW, England

Abstract. Interferometric measurements of the complex permittivity of pure liquid diacetyl in the range 17 GHz to 300 GHz are reported at four temperatures (5 °C; 25 °C; 50 °C; 75 °C), together with far infrared spectra on solutions of diacetyl in cyclopentane, at room temperature. Very short relaxation times have been observed, whereas a very strong and broad absorption appears in the far infrared region. The experimental results may be interpreted as being due to active intramolecular motions involving a considerable change of dipole moment: i.e. -cis/-trans isomerism or inversion of the cis forme. A simple analysis in terms of 'chemical relaxation processes' obeying the Curie principle allowed of the evaluation, from the microwave part of Cole-Cole plot, of the kinetic constants ($k_{cis \to trans}$ and $k_{trans \to cis}$). The nature of the far infrared absorption has been also discussed.

Résumé. Des mesures interférométriques de permittivité complexe sont présentées pour quatre températures (5 °C, 25 °C, 50 °C, 75 °C) sur le diacétyle pur, tandis que les spectres d'absorption en I.R. lointain de solutions diacétyle/cyclopentane sont également reproduits. Les éléments marquants de ces mesures sont, d'une part, les très faibles valeurs des temps de relaxation observés, et d'autre part, l'existence d'une bande d'absorption très intense et large en I.R. lointain. Ces résultats sont interprétés sur la base de mouvements intramoléculaires induisant une variation de la valeur scalaire du moment dipolaire: isomérisation -cis/-trans ou inversion de la forme cis. Pour ces deux types de relaxation d'origine chimique, une analyse simple, respectant le principe de symétrie de Curie, permet d'atteindre l'ordre de grandeur des constantes cinétique ($k_{cis \to trans}$ et $k_{trans \to cis}$). La nature de la bande d'absorption IR lointain est également discutée.

1. Introduction

Molecular motions are often considered as the main causes of relaxation processes in condensed media. However, some other mechanisms having a different nature ('internal processes') may also take place and compete with classical diffusion processes. *Intramolecular* motions, considered as the kinetic aspect of a chemical equilibrium between various species, appeared to us as the simplest example of such 'internal processes' giving rise to specific behaviour in dielectric relaxation.

Besides, far IR spectra of rigid molecules are often interpreted by librational motions but a quantitative analysis of experimental results is difficult because of the lack of knowledge on intermolecular potential wells. In contrast, one can assume that even in the liquid state, the movement of one part of the molecule with respect to the other is essentially governed by the intramolecular potential, on which one has more accurate data. Therefore, one would expect some interesting informations from studies on deformable molecules consisting of two symmetrical, identical polar groups.

The present study deals with diacetyl which is assumed to exist in two different configurations:

(i) *a non polar* one (trans form), in which all carbon and oxygen atoms lie in a plane.

(ii) *a polar* one which may be the corresponding cis form.

The rotation of the acetyl groups gives rise to a *cis/trans* isomerism and also to an inversion of the cis polar form. We therefore summarize the internal motion using the following scheme:

$$A_2^- \rightleftharpoons A_1^0 \rightleftharpoons A_2^+$$

2. Experimental

The diacetyl used was a Fluka product 'Puriss.' grade (>99.5%) freshly redistilled to remove small amounts of polymer.

The measurements below 112 GHz were made at the University of Nancy. In the millimetre and centimetre wave range, the measurements of the complex permittivity of the liquid were carried out at fixed frequencies produced by klystrons. These were coupled to a broad band Michelson interferometer featuring oversized waveguides [1]. The values of the real (ε') and the imaginary (ε'') parts of the complex permittivity were computed from the entire interferogram using a least squares optimisation program [2]. A complementary measurement at 308 GHz, using a carcinotron, was carried out at the University of Lille, by one of us (J.G.).

The far infrared instrumentation used at the National Physical Laboratory was of two kinds:

(1) A modular NPL-Grubb-Parsons-Michelson interferometer used in its conventional mode with digital Fourier transformation of the output [3]. This instrument was used to obtain the spectrum of the power absorption coefficient.

(2) A HCN laser refractometer, [4] which was used to give an accurate 'spot' frequency (890.76 GHz) value of the complex permittivity.

Because of the very strong absorption of liquid diacetyl in the far infrared region, the precision of the results calculated from far infrared measurements was considerably enhanced by the use of phase modulation [5].

Complex permittivity measurements in the microwave region were carried out at 4 temperatures (5°C, 25°C, 50°C, 75°C). Far infrared spectra have been obtained, at room temperature only, for solutions in cyclopentane.

3. Results

The results at centimeter and millimeter wave lengths can be fitted to semi-circular

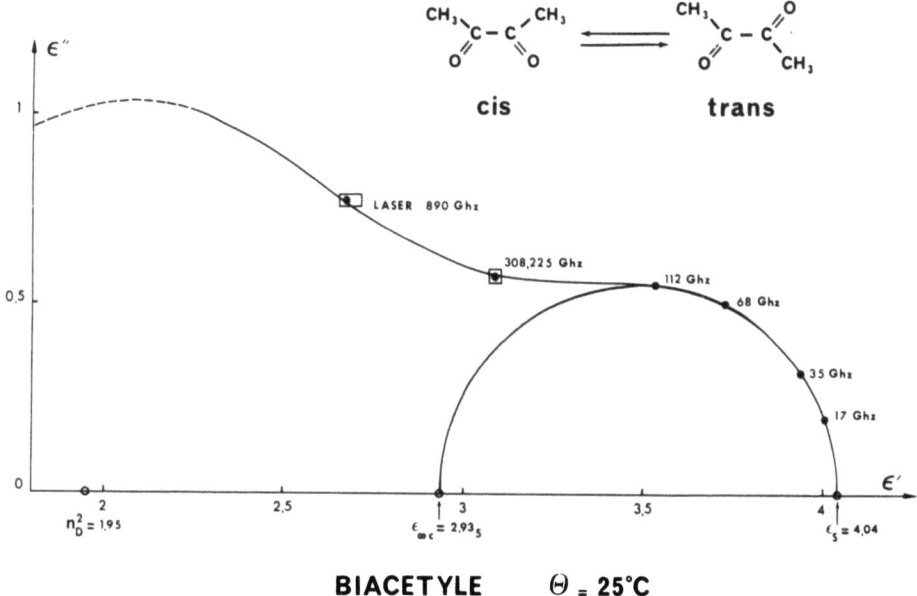

Fig. 1a. Cole-Cole plot of pure diacetyle liquid at 25°C.

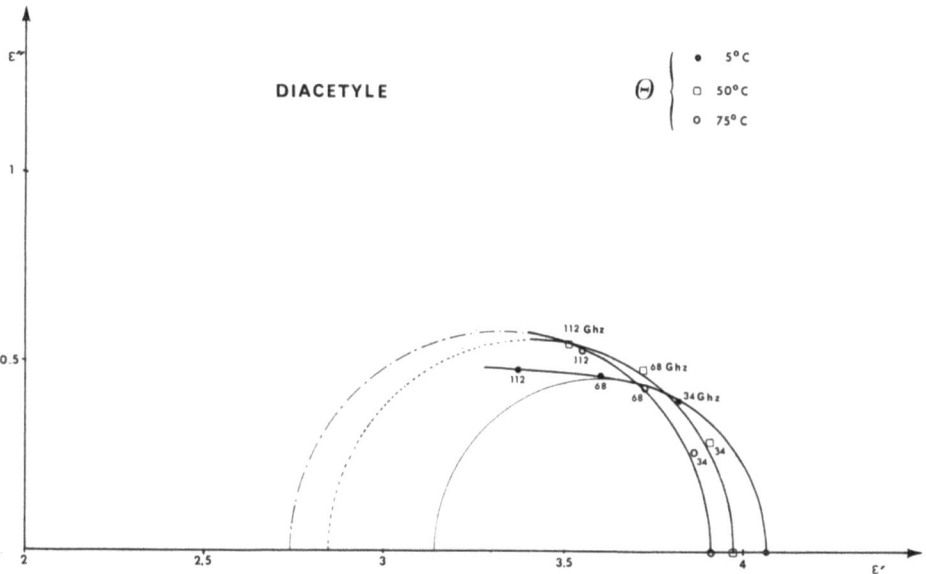

Fig. 1b. Temperature dependance of the relaxation processes.

Cole-Cole plots (Figures 1a, b). The characteristic parameters ε_0, ε_∞ and the observed relaxation time τ_0 (corresponding to the critical frequency), are reported in Table I. We have allowed for internal field effects using the corrections of Powles [6] and Klug-Vaughan [7] which in this case have opposite effects on the microscopic relaxation time τ_μ (due to the strong anisotropy of the cis diacetyl molecule).

TABLE I

T(°C)	ε_0	ε_∞	$10^{12} \cdot \tau_0(s)$	$10^{12} \cdot \tau_\mu(s)$
5	4.0_6	3.1_4	2.6_1	2.7_9
25	4.0_4	2.9_{35}	1.3_3	1.4_0
50	3.9_7	2.8_4	1.1_2	1.1_7
75	3.9_1	2.7_4	0.9_{25}	0.9_{65}

At higher frequencies, a strong deviation from the simple Debye process is observed as shown in Figure 2. This phenomenon corresponds to an extremely intense absorption ($\alpha_{max} > 300$ Np cm^{-1}) centred at about 90 cm^{-1}. This band, which is very broad, narrows when the temperature decreases.

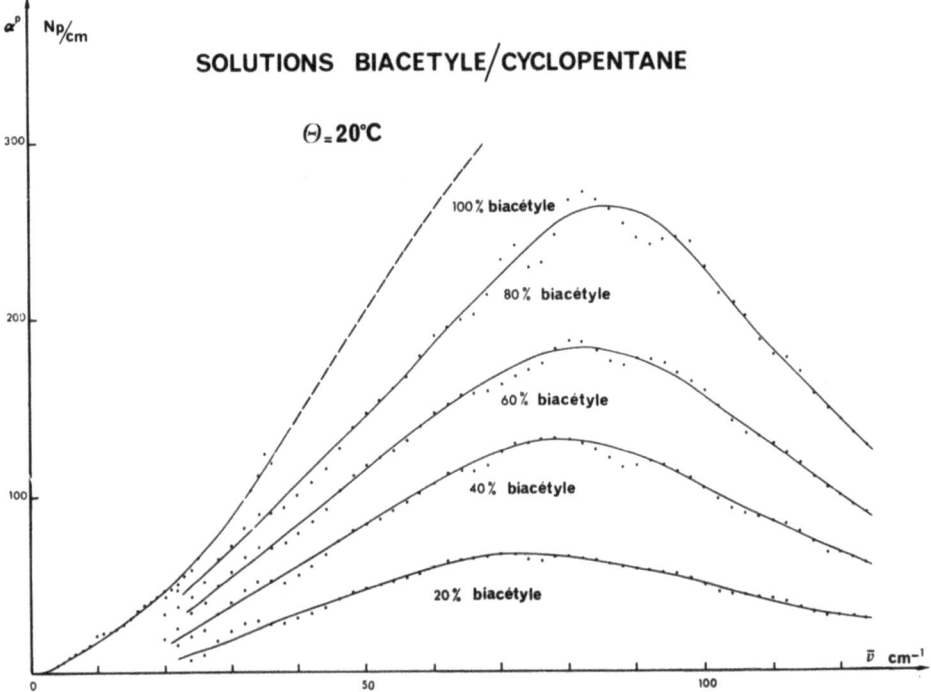

Fig. 2. Far infrared absorption spectra of diacetyle/cyclopentane solutions at room temperature.

4. Interpretation

Theories of 'Chemical Dielectric Relaxation' have been proposed especially by Schwarz [8], Williams [9] and Anderson [10]. They can be applied to the considered unimolecular reaction. However, we have been led to reexamine these theories in order to take into account the possibility of the molecular inversion, the various transformations being symbolized by the following symmetric scheme where the k_{ij}

refers to the corresponding kinetic constants:

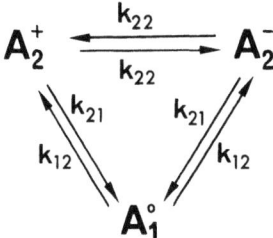

Furthermore, if one assumes that molecular reorientations (vector phenomenon) obey the classical Debye theory and compete with the chemical reactions (scalar phenomenon) which affect the modulus of the dipole moment, one is led to separate these two processes according to the 'Curie principle' stated by Prigogine and Mazur [11].

The dipolar autocorrelation function is then found to have the following time dependance [12]:

$$\langle \mu(0) \cdot \mu(t) \rangle \sim [e^{-t/\tau_1} + e^{-t/\tau_2}] \cdot e^{-t/\tau_{A_2}}$$

$$\frac{1}{\tau_1} = k_{21} + 2k_{12}; \quad \frac{1}{\tau_2} = k_{21} + 2k_{22}.$$

τ_{A_2} being the Debye relaxation time of the polar species A_2, considered as a rigid molecule.

The main difference between our results and the former ones is the absence of an extra term depending on τ_{A_2} only and therefore, one predicts two relaxation times which are both shorter than τ_{A_2}.

We see that τ_1 is equal to $\tau_{ch.}$ of References 8 and 9.

The ratio $2k_{12}/k_{21}$ is equal to the equilibrium constant for the cis/trans isomerism $K_{eq.}$ which has been determined by Thiebaut using non linear dielectric measurements [13].

Noting that $K_{eq.} \ll 1$ and assuming that $k_{22} \ll k_{12}$, we find one dominant internal relaxation time for which:

$$\frac{1}{\tau_{ch}} \simeq k_{21}.$$

The kinetic constant k_{21} can therefore be calculated if one knows τ_{A_2} or if τ_{A_2} is large enough compared with $\tau_{ch.}$. The relaxation time τ_{A_2} was evaluated using that of acetone, corrected for viscosity and molecular volume differences. At 25°C, one finds $\tau_{A_2} = 5.7 \times 10^{-12}$ s [12] which has to be compared with the τ_μ value of 1.33×10^{-12} s.

Using the relationship:

$$\frac{1}{\tau_{ch}} = \frac{1}{\tau_\mu} - \frac{1}{\tau_{A_2}}$$

one can calculate the kinetic constants at each temperature. The results are reported in Table II:

TABLE II

$T(°C)$	$K_{eq.}$ [13]	$10^{-11} \times k_{21}(s^{-1})$	$10^{-11} \times k_{12}(s^{-1})$
5	0.12_0	$2._2$	0.1_3
25	0.14_5	$5._4$	0.3_9
50	0.18_8	$6._4$	0.6_0
75	0.22_8	$7._7$	0.8_7

Neither the temperature dependance of τ_0^{-1} nor that of k_{21} and k_{12} can be fitted with a simple Arrhenius law. The measurements at 5°C and 25°C would lead to the following activation energies:

$$E_{12} \sim 6._1 \text{ kcal mole}^{-1} \qquad E_{21} \sim 4._2 \text{ kcal mole}^{-1}.$$

These values which appear in agreement with the picture of a torsional motion around a partly ethylenic C_1-C_2 bond have to be carefully considered because of the much lower ones calculated from higher temperature measurements ($E_{12} \sim 3._3$ kcal mole^{-1}, $E_{21} \sim 1._4$ kcal mole^{-1}). The discrepancy between low and high temperature results may have its origin in the overlapping of relaxation processes with far infrared phenomena. It should also be pointed out that the measurement at 5°C was carried out at only 3 deg above the freezing point of pure diacetyl.

5. Far Infrared Band

The very strong band observed in the far infrared region can be attributed to at least two kinds of phenomena:
(i) a distribution of torsional modes of the molecule, broadened by collisions in the liquid [14];
(ii) an absorption due to induced dipoles (collision induced, electric interactions between neighbours,...) [15].

The narrowing of the band when the temperature decreases seems to be in accordance with the first hypothesis (resonant phenomenon).

This absorption explains the great difference between n^2 and ε_∞ as shown on Figures 1a, b. The rather high values of ε_∞ listed in Table I are the only ones which give satisfactory results in non linear dielectric studies [13]. Therefore, it is not surprising that the calculations carried out on diacetyl, using a value of ε_∞ of the order of magnitude of 2, lead to unreasonable results. In particular, the interpretation of Kerr effect measurements by Le Fevre et al., giving the polar form a twisted configuration, appears to be a dubious exercise (Thiebaut's results agree with a fairly exact cis form, in which the oxygen and carbon atoms are in the same plane). Anyway, our interpretation of the chemical rate process is independent of any geometry of the polar form.

Acknowledgements

This work was partly supported by the C.N.R.S. (E.R.A. no. 22 Interactions Moléculaires).

The authors are indebted to Dr J. W. Fleming (N. P. L. Teddington), Pr. E. Constant, Dr R. Fauquembergue (University of Lille) and Miss M. Hollecker (University of Nancy) for their experimental help or valuable discussions.

J. G. and J. L. R. thank M. W. Evans for his linguistic assistance.

The measurements at N. P. L. and at the University of Lille were made whilst one of us (J.G.) was a guestworker.

References

1. Goulon, J., Brondeau, J., Sarteaux, J. P., and Roussy, G.: *Revue Phys. Appl.* **8**, 165 (1973).
2. Goulon, J., Roussy, G., Rivail, J. L.: *Revue Phys. Appl.* **4**, 413 (1969).
3. Chantry, G. W.: *Submillimeter Spectroscopy,* Academic Press, New York, 1971.
4. Chamberlain, J., Haigh, J., and Hine, M. J.,: *Infrared Phys.* **11**, 75 (1971).
5. Chamberlain, J.: *Infrared Phys.* **11**, 25 (1971).
6. Powles, J. G.: *J. Chem. Phys.* **21**, 633 (1953).
7. Klug, D. D. and Vaughan, W. E.: *J. Chem. Phys.* **56**, 5005 (1972).
8. Schwarz, G.: *J. Phys. Chem.* **71**, 4021 (1968).
9. Williams, G.: *Adv. Mol. Relaxation Processes* **1**, (4) 409 (1970).
10. Anderson, J. E.: *Ber. Bunsenges. Physik. Chem.* **75** (3/4), 294 (1971).
11. Prigogine, I. and Mazur, P.: *Physica* **19**, 241 (1953).
12. Goulon, J.: Thèse d'Etat, Université de Nancy, 1972.
13. Thiebaut, J. M.: unpublished results.
14. Durig, J. R., Hannum, S. E., and Brown, S. C.: *J. Phys. Chem.* **75** (13), 1946 (1971).
15. Davies, M., Pardoe, G. W. F., Chamberlain, J. E., and Gebbie, H. A.: *Trans. Faraday Soc.* **64**, 847 (1968).
16. Cureton, P. H., Le Fevre, C. G., and Le Fevre, R. J. W.: *J. Chem. Soc.*, 4447 (1961).

DIELECTRIC RELAXATION IN CRITICAL POLAR MIXTURES; DERIVATION OF GLARUM-COLE EQUATION BY PROJECTION OPERATOR METHOD*

ROBERT S. WILSON

*Chemistry Department, Michael Faraday Laboratories,
Northern Illinois University, DeKalb, Ill. 60115, U.S.A.*

and

Argonne National Laboratory, Argonne, Ill. 60439, U.S.A.

Abstract. We suggest a simple physical picture for the cause of the anomalous dielectric response of critical dipolar mixtures: The state of critical fluctuation of composition within the Ornstein-Zernike sphere of radius κ^{-1} is of a highly ordered state of orientation much like that found in liquid crystals. The insertion of an electric field will exert a torque on the *oriented* cluster thereby producing an intrinsic angular-momentum field. This field decays into a vortex field which subsequently back reacts on a molecule at its center, thereby producing an additional anomalous polarization. In the critical region, the assumption is that molecules can rotate cooperatively over the correlation length κ^{-1}, thereby setting up an enormous vortex field and a concomitant anomalous polarization. Comparison of this work with that of Ailawadi and Berne on the long-time persistence of intrinsic angular momentum in *non-critical* viscous fluids is made, along with suggestions for new experiments for the direct measurement of critical orientational correlations, including the rotational diffusion constant, by 'Pico-Second Kerr Fluctuation-Spectroscopy' *in equilibrium*.

Résumé. Nous suggérons une idée physique et simple comme cause d'une réponse diélectrique anormale de mélanges critiques dipolaires: L'état de fluctuation critiques de la composition compris dans la sphère Ornstein-Zernike de rayon κ^{-1} est dans un état d'orientation fortement ordonné très semblable à celui trouvé dans les cristaux liquides. L'insertion d'un champ électrique va exercer un couple de torsion sur l'agglomeration *orientée* Ornstein-Zernike produisant ainsi un champ angulaire de force vive intrinsèque.

Ce champ se perd dans un champ tourbillonnant qui ensuite réagit a son tour sur une molécule à son centre produisant ainsi une polarisation anormale additionnelle.

Nous suggérons egalement de nouvelles expériences pour mesurer directement des corrélations orientationnelles critiques par 'Pico-Second Kerr Fluctuation-Spectroscopy'.

1. Relaxation in Normal Dielectric Liquids: Derivation of Glarum-Cole Equation

Since Glarum [1] first applied Kubo's [2] statistical treatment of irreversible processes to the problem of dielectric relaxation, there has appeared a series of works on his approach. Cole [3] extended Glarum's work to polarizable dipolar molecules treated as harmonic oscillators in the relaxation counterpart of Kirkwood's [4] equilibrium theory. Cole's work is based on macroscopic arguments because of his use of a so-called 'superposition principle'. This superposition principle has a firm theoretical basis in the statistical theory of Kubo, only in the response of a system to an external field; however, in the above theories, use is made of the same superposition principle

---

* Acknowledgement is made to the Donors of the Petroleum Research Fund administered by the American Chemical Society, for partial support of this research and the U.S. Atomic Energy Commission through the Argonne Summer Faculty Research Program.

in the response of one part of a system to the motion of another part of the *same* system. The results of their work are rather interesting since a relatively simple integral equation follows in the relation between the macroscopic and microscopic relaxation functions which includes Kirkwood's electrostatic result as a boundary condition for equilibrium.

Since with Cole's result one would be able to calculate the dielectric relaxation of a macroscopic system from a knowledge only of short-range interactions, the microscopic relaxation function (in principle readily calculated from various models), and the static results it seems worthwhile to attempt to put his equation on a firm theoretical basis.

In an attempt to formally derive Cole's results we have found that such an equation is valid only under conditions which seem to be far removed from any physically reasonable basis. One is then led to suspect the superposition principle method of Glarum and Cole as applied to the electrical response of two different parts of the same system to each other, however very recent work of Cole (see this conference) and Glarum seem to have eliminated the need of such a surperposition principle and hence strengthened the foundation for it. We present below our more general formalism and compare it with Cole's results.

We consider a spherical sample of volume V containing N molecules with permanent moment μ and we ignore induced polarization of individual molecules in order to reduce the complexity of the problem. The Hamiltonian of the system can be written:

$$H = H^0 - \mathbf{M} \cdot \mathbf{E}^0(t), \tag{1}$$

where $\mathbf{M} \equiv \sum_{i=1}^{N} \boldsymbol{\mu}_i$, $\boldsymbol{\mu}_i$ is the dipole moment of particle i. The external field $\mathbf{E}^0(t)$ is assumed uniform where we neglect the wave character of the external field over the linear dimensions of the system. The mean moment of the system in the direction of the external field $\mathbf{e}$ can be shown to be:

$$\langle \mathbf{M}(t) \cdot \mathbf{e} \rangle = - \frac{N \langle \boldsymbol{\mu}(0) \cdot \mathbf{M}(0) \rangle}{3kT} \int_{-\infty}^{t} dt' E_0(t') \Phi(t - t'), \tag{2}$$

where the macroscopic decay function is defined as follows:

$$\Phi(t) \equiv \frac{\langle \boldsymbol{\mu}(0) \cdot \mathbf{M}(t) \rangle}{\langle \boldsymbol{\mu}(0) \cdot \mathbf{M}(0) \rangle}. \tag{3}$$

The brackets $\langle \ldots \rangle$ indicate an ensemble average over the phase space of the non-perturbed equilibrium system. As shown by Braun and Mazur [5], when the correct macroscopic susceptibility is calculated through the use of the true Maxwell field (consistently calculated from the external field through the use of the Kubo relation), one arrives at the following result for the spherical sample:

$$\frac{\varepsilon^*(\omega) - 1}{\varepsilon^*(\omega) + 2} = \frac{4\pi N \langle \boldsymbol{\mu}(0) \cdot \mathbf{M}(0) \rangle}{9kTV} L[-\Phi(t)], \tag{4}$$

where $\varepsilon^* - 1 \equiv 4\pi[\langle \mathbf{M} \cdot \mathbf{e}\rangle/V]E_M$. E_M is the Maxwell field and the relation between this field and the external field was shown to be:

$$E_M = \frac{3E^0}{\varepsilon^*(\omega) + 2}. \tag{5}$$

The operator L is defined as follows:

$$L[F(t)] \equiv \int_0^\infty dt \exp(-i\omega t) F(t). \tag{6}$$

We see then that knowledge of the macroscopic relaxation function allows one to calculate the dielectric susceptibility, or equivalently, the dielectric constant $\varepsilon^*(\omega)$.

Another important correlation function is the so-called microscopic correlation function which is given by:

$$\varrho(t) \equiv \langle \boldsymbol{\mu}(0) \cdot \mathbf{m}(t)\rangle = \varphi(t)\langle \boldsymbol{\mu}(0) \cdot \mathbf{m}(0)\rangle. \tag{7}$$

i.e., the correlation of a given dipole with a smaller sphere with moment $\mathbf{m}(t)$ contained in the large sphere. We derive now a formal result for the relation between the macroscopic and microscpic relaxation functions. Let us define for convenience the unnormalized macroscopic relaxation function:

$$R(t) \equiv \langle \boldsymbol{\mu}(0) \cdot \mathbf{M}(t)\rangle = \Phi(t)\langle \boldsymbol{\mu}(0) \cdot \mathbf{M}(0)\rangle. \tag{8}$$

Following the projection operator techniques of Zwanzig and Mori [6], we define the following so-called projection operator:

$$P_{\mu m} = \frac{\mathbf{m}\langle \mathbf{m} \cdots \rangle}{\mathbf{m} \cdot \boldsymbol{\mu}} \tag{9}$$

which is idempotent; i.e., $P_{\mu m}^2 = P_{\mu m} \cdot P_{\mu m}$ selects from $\mathbf{M}(t)$ that part which is relevant for $R(t)$ and projects it out onto $\mathbf{m}(0)$. The time dependence of $\boldsymbol{\mu}(t)$ is given by:

$$\frac{\partial \boldsymbol{\mu}(t)}{\partial t} = \frac{i}{\hbar}[H, \boldsymbol{\mu}(t)] = iL\boldsymbol{\mu}(t), \tag{10}$$

where L is the quantum mechanical Liouville operator. We have the following equation:

$$R(t) = \langle \boldsymbol{\mu} \cdot \mathbf{M}(t)\rangle = \langle \boldsymbol{\mu} \cdot \mathbf{m}(t)\rangle + \langle (\mathbf{M} - \mathbf{m}) \cdot \boldsymbol{\mu}(-t)\rangle = \varrho(t) + \\ + \langle X \cdot \mu(-t)\rangle \tag{11}$$

which defines $\mathbf{X}$ as:

$$\mathbf{X} = (\mathbf{M} - \mathbf{m}). \tag{12}$$

Taking derivatives of both sides we have:

$$\frac{\partial R}{\partial t} = \frac{\partial \varrho}{\partial t} + \frac{\langle \mathbf{X} \cdot \partial \boldsymbol{\mu}(-t)\rangle}{\partial t}. \tag{13}$$

Due to the fact that $P_\mu L P_\mu = 0$, we have the following integral equation for $\partial \mu(t)/\partial t$ which may be easily verified:

$$\frac{\partial \mu(-t)}{\partial t} = iL\mu_r(-t) - \frac{1}{\langle \mathbf{m} \cdot \boldsymbol{\mu} \rangle} \int_0^t L\, e^{i(1-P_\mu)L\tau} L\mu \varrho(t-\tau)\, d\tau, \tag{14}$$

where $\mu_r(t) \equiv P_\mu \mu(t)$.

Multiplying both sides of Equation (14) by $\mathbf{X} \cdot$ and averaging, we have:

$$\frac{\langle \mathbf{X} \cdot \partial \mu(-t) \rangle}{\partial t} = i \langle \mathbf{X} \cdot L P_{m\mu} \mu(-t) \rangle - \int_0^t \frac{\langle \mathbf{X} \cdot L\, e^{i(1-P_{m\mu})L\tau} L\mu \rangle \varrho(t-\tau)\, d\tau}{\langle \mathbf{m} \cdot \boldsymbol{\mu} \rangle}. \tag{15}$$

where the conditions: $\mu_{irr} = 0 = (1-P_{m\mu})\mu(0) = \mu_{irr}(0) = 0$ were used.

Substituting the value $\langle \mathbf{X} \cdot L P_{m\mu} \mu(t) \rangle = 0$ (which can easily be verified) we arrive at the following equation:

$$\frac{\partial R(t)}{\partial t} = \frac{\partial \varrho(t)}{\partial t} - \int_0^t \frac{\langle \mathbf{X} \cdot L\, e^{i(1-P_{m\mu})L\tau} L\mu \rangle \varrho(t-\tau)\, d\tau}{\langle \mathbf{m} \cdot \boldsymbol{\mu} \rangle}. \tag{16}$$

The Glarum-Cole result as found in Cole's article as Equation (10) can be put in the form:

$$\frac{\partial R(t)}{\partial t} = (1-A)\frac{\partial \varrho(t)}{\partial t} + \frac{A}{\langle \boldsymbol{\mu} \cdot \mathbf{m} \rangle} \int_0^t \ddot{R}(\tau) \varrho(t-\tau)\, d\tau. \tag{17}$$

If we express $\mathbf{X}$ in the following form:

$$\mathbf{X} = (-\mathbf{m} - \mathbf{F})A; \quad \text{or} \quad -\mathbf{F} \to \mathbf{m}, \quad \text{if} \quad A \to \infty;$$
$$-\mathbf{F} \to \mathbf{M} \quad \text{if} \quad A \to 1, \tag{18a}$$

where A is defined through Kirkwood's equilibrium result:

$$\langle \boldsymbol{\mu} \cdot \mathbf{X} \rangle = -A \langle \boldsymbol{\mu} \cdot \mathbf{m} \rangle \tag{18b}$$

we arrive at the following equation:

$$\frac{\partial R(t)}{\partial t} = (1-A)\frac{\partial \varrho(t)}{\partial t} + \frac{A}{\langle \mathbf{m} \cdot \boldsymbol{\mu} \rangle} \int_0^t \langle \mathbf{F} \cdot L\, e^{i(1-P_{m\mu})L\tau} L\mu \rangle \varrho(t-\tau)\, d\tau. \tag{19}$$

In order to obtain correspondence we would have to set the kernel of the exact solution equation equal to the following macroscopic function; i.e.,

$$\ddot{R}(\tau) = \langle \mathbf{F} \cdot (L\, e^{i(1-P_{m\mu})L\tau} L)\mu \rangle. \tag{20}$$

Notice when $A \sim 1$ and we neglect the projection operator in the exponent we do get $R(t)$, neither of which assumptions are we allowed to make on apriori grounds. There seems to be no direct and simple explanation which would allow such a correspondence, unless one were willing to introduce macroscopic reasoning to relate the average correlation of **X** (*t*) with **μ**(0) as Cole and Glarum [9] have done and recently justified on the basis of a lattice model.

In order to extend our knowledge of possible approximation schemes allowing one to calculate the dielectric relaxation function it appears worthwhile to study a model system.

We take for this model a rigid cubic lattice of permanent dipoles. Rosenberg and Lax [7] have considered the static susceptibility of a macroscopic spherical specimen in a high temperature limit as a function of the parameter:

$$\alpha = \mu^2/3kT,$$

where μ is the magnitude of the dipole moment and T is the absolute temperature. Zwanzig [8] has considered the effect of rotational Brownian motion on the frequency dependent dielectric susceptibility for such a model system. All dipolar interactions were taken into account by means of a high temperature perturbation expansion. His work is particularly interesting because he has been able to take into account the effects of dipolar interaction on not only the statistical correlation of dipoles, but also on the dynamical behavior of the dipoles. His results depend on the introduction of an additional parameter τ, his 'molecular' relaxation time. His work is based on the assumption that the system obeys a rotational diffusion equation in configuration space of N interacting dipoles.

In a preliminary study of the possibility of formally deriving such a diffusion equation in configuration space, we have found projection techniques similar in spirit to the ones discussed above which seem to give an exact statistical expression for Zwanzig's parameter τ. It will be interesting and worthwhile comparing our results with the recent work of Cole's perturbation treatment which uses Kirkwood theory of transport properties. Our results are exact and involve ultimately a projection operator in the tine evolution operator which is not easily disposed of except perhaps in a perturbative scheme.

2. Critical Response and Relaxation

In the remaining time I would like to give a very simple explanation for the anomalous response of critical polar mixtures based on an analogy with recent work of the author and Fixman on the critical mixture diffusion problem. In that work we showed that the Onsager diffusion transport coefficient diverged in the critical region due essentially to the fact that each species due to critical fluctuations in composition found itself in the fluid which itself was moving already in the direction of necessary diffusion flow. The basic assumption was that the critical fluctuations were so large that a generalized hydrodynamics could be used. The final result for this effect, a sort of inverse electrophoresiec effect, was that the diffusion flow should vanish as:

$$D_T = kT/6\pi\eta^0 \kappa^{-1},$$

where D_T is the translational diffusion constant, T is the temperature, η^0 is the 'non-critical' extrapolated viscosity, and κ^{-1} is the Ornstein-Zernike (O-Z) correlation length. Actually within the Debye theory $\kappa^{-1} \sim (T-T_c)^{-1/2}$ at the critical consolute point at critical composition in the one phase region.

If we generalize our previous result for the dielectric case along the same lines as before except we include 'dielectric friction' effects we find it reasonable to estimate the extent of orientation order of the size of the O-Z sphere κ^{-1}. With this in hand we find that the rotational diffusion constant should vary as:

$$D_R = \frac{kT}{8\pi\eta^0 (\kappa^{-1})^3} \sim |T - T_c|^{\phi = 3/2}$$

which is consistent with the dynamical scaling law result of Snider $\phi \geq \frac{1}{2}$.

Finally since we have no absolute guarantee that in fact the dipoles in the O-Z sphere are highly ordered and are able to move cooperatively over the correlation length it would be satisfying if not important to study the dipole ordering in the O-Z sphere.

Let us suggest the following two experiments. First it seems reasonable to assume the new Pico-Second Kerr Effect in polar liquids could be used to probe the critical fluctuations in orientational order. We have in mind measuring directly the decay of orientational fluctuations by Pico-Second Kerr spectroscopy along with the photon counting scheme as used by the Cornell group which recently studied concentration fluctuations in equilibrium. Any large fluctuations in the Kerr signal would signify large orientational fluctuations and the study of their time decay constant would via the Onsager assumption or rule lead to the rotational diffusion constant.

The other experiment would be to study the translational diffusion coeffiicent in a strong electric (static) field to see if one can observe any anisotropy as recently found in nematic liquid crystals in strong magnetic fields.

One also ought to point out that similar long-time persistance effects in orientation in normal fluids were studied by Ailawadi and Berne [11], among others, but there seems to be no evidence for divergences in transport properties in three dimensions, at least far away from the critical point.

Acknowledgements

The author would like to thank Prof. P. Mazur, Lorentz Institute, Leiden, Holland who originally suggested the possibility of deriving the Glarum-Cole equation by first principles when the author was an NSF Visiting Fellow (1966). The projection operator in the present work differs however from the one used by us previously (Mazur-Wilson, unpublished). Also he would like to thank Deans L. J. Sill and P. Burtness for a travel grant to attend this conference and Mr Donald Knauss for many stimulating discussion, on the critical polar mixture problem.

References

1. Glarum, S. H.: *J. Chem. Phys.* **33**, 1371 (1960).
2. Kubo, R. and Tomita, K.: *J. Phys. Soc. Japan* **9**, 888 (1954).
3. Cole, R. H.: *J. Chem. Phys.* **42**, 637 (1965); see paper of Cole in this volume, p. 97, for recent references.
4. Kirkwood, J. G.: *J. Chem. Phys.* **7**, 911 (1939).
5. Braun and Mazur, unpublished.
6. Zwanzig, R.: *Lect. in Theor. Phys.*, Interscience Publ. New York, 1961, Vol. III, p. 106; Mori, H : *Prog. Theor. Phys.* **33**, 423 (1965).
7. Rosenberg, R. and Lax, M.: *J. Chem. Phys.* **21**, 424 (1953).
8. Zwanzig, R.: *J. Chem. Phys.* **38**, 2766 (1963).
9. Glarum, S. H., *Mol. Phys.* **24**, 1327 (1922).
10. Magde, D., Elson, E., and Webb, W.: *Phys. Rev. Letters* **29**, 705 (1972).
11. Ailawadi, N. K. and Berne, B. J.: *J. Chem. Phys.* **54**, 3569 (1971).

DISCUSSION

Friedman: I wonder why dipoles add rather than cancel in a fluctuation to increasing density? To say it in another way, it would seem that if for some molecular geometry they add, there might be another kind of molecules for which they cancel.

Wilson: Actually no one really knows but one might use a generalized pico-second Kerr effect to study this. The fluctuations of light passed due to critical fluctuations in orientation would lead one to suspect lining-up. Actually one can also expect on thermodynamic grounds that the natural fluctuations should lead to low entropy states which implies ordering. Also one might check whether diffusion can be made anisotropic with the introduction of an electric field.

ETUDE, PAR RELAXATION DIELECTRIQUE, DU MOUVEMENT BROWNIEN DES IONS DANS LES LIQUIDES SIMPLES

JEAN-PIERRE BADIALI, HUBERT CACHET, ALAIN CYROT,

et

JEAN-CLAUDE LESTRADE

Groupe de Recherche no. 4 du C.N.R.S. 'Physique des liquides et Electrochimie',
associé à l'Université Paris VI, 4 place Jussieu, 75230 Paris Cedex 05, France

Résumé. L'étude de la relaxation diélectrique des solutions électrolytiques permet d'accéder au mouvement brownien des ions par l'intermédiaire des fonctions de corrélation temporelles. Toutefois, suivant que l'on considère le liquide comme étant ou non conducteur, ces fonctions portent sur des variables dynamiques différentes.

Nous comparons ici les deux approches actuellement proposées et nous illustrons les problèmes posés à partir de résultats expérimentaux récents.

Abstract. The motion of ions in an electrolyte solution can be related to the observed dielectric relaxation by means of correlation functions. When the dynamic variables are chosen as the velocities, the results of the Onsager and Debye-Falkenhagen (ODF) theories can be found. The ionic part of the complex permittivity shows then an asymmetrical distribution of relaxation time, when described phenomenologically in dielectrics language. Unfortunately, these theories are restricted to dilute solutions, too dilute for any experimental verification to be made. If the dynamic variables of the correlation function are chosen as the positions of the ions an *a priori* separation has to be made between a static conductivity $\sigma(0)$ and an ionic complex permittivity $\Delta\varepsilon(\omega)$. Only the latter can be deduced from the corresponding correlation function, and it represents the permittivity of a nonconductive medium. Within this frame, a dielectric model is built. The ions are first assumed to be gathered in complex aggregates which undergo deformations when an electric field is applied. These deformations are then assumed to come from a brownian linear motion of the ions, interrupted by collisions, and can be described by stochastic laws involving a diffusion coefficient and a mean collision time. The magnitude of both quantities can be estimated *a priori*, as well as the molecular part of the complex permittivity. This model fits fairly well the experimental data obtained between 137 MHz and 34 GHz with solutions of lithium perchlorate (0.6 M l^{-1}) in tetrahydrofuran-benzene mixtures.

It is not free from criticism since it precludes any information on $\sigma(0)$, which must be absent to be zero. However it leads to an asymmetrical distribution of relaxation time as the ODF theories do, and this is verified by experiment. In this respect, the dynamic features of the model can be used as an important element to extend the ODF theories to concentrated solutions.

L'étude du mouvement des ions en solution peut être abordée sous un jour nouveau grâce aux méthodes de la mécanique statistique qui se sont révélées les plus fructueuses dans le cas des liquides simples. Le but de cet article est de montrer comment la relaxation diélectrique des solutions électrolytiques peut s'interpréter en reliant la permittivité complexe mesurée à des fonctions de corrélation temporelles (FC). Nous comparerons en particulier les deux types d'approche actuellement proposées, qui font appel à des FC portant sur des variables dynamiques différentes.

1. Fonctions de corrélation temporelles ioniques

A partir de la formule de conductivité électrique donnée par Kubo, on a pu déduire les théories classiques dues à Onsager et Debye-Falkenhagen (théories ODF). Dans ces travaux (voir par exemple [1], [2]) la fonction de corrélation a pour expression:

$$\Phi_V(t) = \left\langle \sum_i e_i \mathbf{V}_i(0) \sum_j e_j \mathbf{V}_j(t) \right\rangle, \tag{1}$$

où les sommations sont étendues à tous les ions d'un volume V et où e_i est la charge de l'ion i animé à l'instant t de la vitesse $\mathbf{V}_i(t)$. Le calcul de $\Phi_V(t)$ a été effectué soit directement par une technique de diagrammes [1], soit par l'intermédiaire d'une fonction mémoire et d'un modèle pour la fonction de Van Hove [2]. La fonction $\Phi_V(t)$ n'est pas exponentielle et conduit à une conductivité ionique complexe $\sigma_i(\omega)$ qui ne dépend pas rationnellement de la pulsation ω. Cette forme de FC n'a jamais pu être confrontée à l'expérience car le domaine de validité des théories ODF est celui de la théorie de Debye-Hückel, c'est-a-dire limité à des concentrations d'au plus 10^{-3} M l^{-1} en solution aqueuse, et encore moins dans des solvants de plus faible permittivité statique: dans ces conditions, les effets de relaxation prévus sont trop faibles pour être observables avec les techniques de mesure actuelles, et aucun effet n'a effectivement été observé. Par ailleurs, même si cet obstacle pouvait être levé, la confrontation avec l'expérience ne pourrait être directe car $\Phi_V(t)$ ne représente qu'un effet dans la permittivité complexe mesurée $\varepsilon(\omega)$. On peut, en effet, montrer qu'à la permittivité ionique $\varepsilon_i(\omega)$, liée à $\sigma_i(\omega)$ par

$$\sigma_i(\omega) = j\omega\varepsilon_i(\omega) \quad (j^2 = -1)$$

s'ajoutent deux autres termes liés explicitement au mouvement moléculaire [3]. Le premier, $\varepsilon_m(\omega)$, représente la contribution du solvant en présence des ions, le second traduit les effets de couplage entre mouvements moléculaire et ionique.

Parallèlement à ces recherches fondamentales sur la théorie des électrolytes, on a observé des relaxations d'origine ionique avec des solutions concentrées s'écartant notablement des conditions où s'appliquent les théories ODF (par exemple, 0,01 à 1 M l^{-1} dans des solvants de permittivité statique 2 à 7 [4], [5], [6]). La permittivité mesurée, si l'on néglige les couplages entre mouvements moléculaires et ioniques, peut alors s'écrire:

$$\varepsilon(\omega) = \sigma(0)/j\omega + \Delta\varepsilon(\omega) + \varepsilon_m(\omega). \tag{2}$$

Dans (2) la conductivité statique $\sigma(0)$ et le terme de relaxation ionique $\Delta\varepsilon(\omega)$ ne sont pas indépendants puisqu'ils se déduisent de la même FC:

$$\sigma_i(\omega) = \sigma(0) + j\omega\Delta\varepsilon(\omega) = \frac{1}{3kTV} \int_0^\infty e^{-j\omega t} \Phi_V(t)\, dt. \tag{3}$$

Si $\Phi_V(t)$ est effectivement calculée, comme c'est le cas avec les théories ODF aux faibles concentrations, cette expression fournit simultanément $\sigma(0)$ et $\Delta\varepsilon(\omega)$. Pour les solutions concentrées sur lesquelles peuvent s'obtenir des résultats expérimentaux, aucun modèle n'a jusqu'ici été proposé pour calculer $\Phi_V(t)$. La présence des ions se traduisant par une polarisation supplémentaire ($\Delta\varepsilon(0)>0$), on a élaboré des modèles de type diélectrique où $\Delta\varepsilon(\omega)$ est relié à la FC du moment des ions dans un volume V:

$$\Phi_R(t) = \langle \mathbf{M}(0)\mathbf{M}(t) \rangle.$$

Il s'agit donc d'une FC sur les positions des ions. On écrit alors, comme pour un liquide diélectrique non conducteur

$$\Delta\varepsilon(\omega) = \frac{1}{3kTV} \int_0^\infty e^{-j\omega t} \dot\Phi_R(t)\,dt \qquad (4)$$

la conductivité $\sigma(0)$ étant ajoutée *a posteriori* pour rendre compte de l'expérience.

2. Modèles diélectriques de relaxation ionique

Si les ions sont dépourvus de moment dipolaire permanent, l'interprétation de $\Delta\varepsilon(\omega)$ à partir de (4) suppose l'existence d'effets collectifs. Le plus simple, et aussi le premier proposé historiquement, consiste à admettre que les ions de signe opposé s'associent par paires pour former des dipôles. La relaxation observée est alors reliée à la rotation de ces dipôles, et, dans la mesure où la durée de vie de ces édifices est suffisante, les modèles diélectriques utilisés pour les liquides moléculaires sont directement applicables. Cette approche a été largement utilisée lorsque $\Phi_R(t)$ est une fonction exponentielle (voir par exemple [7], [8]). Toutefois, certains résultats expérimentaux, en particulier lorsqu'ils sont obtenus dans une large gamme de fréquences et avec des solvants assez peu polaires [4], [5], [6], font apparaître une dispersion de temps de relaxation correspondant à une fonction $\Phi_R(t)$ non exponentielle. Ceci rappelle les théories ODF et suggère que le mouvement brownien de translation joue un rôle déterminant. Par ailleurs, la croissance simultanée de $\sigma(0)$ et $\Delta\varepsilon(0)$ avec la concentration en ions [5] ou la permittivité du solvant [4] laisse supposer l'existence d'agrégats plus complexes que des paires. A partir de ces éléments, on a pu proposer un nouveau modèle diélectrique pour calculer $\Phi_R(t)$ [4].

On admet que $\Delta\varepsilon(\omega)$ est lié à l'existence d'agrégats non rigides et formés de plusieurs ions. Ces derniers sont animés d'un mouvement de translation. La valeur du temps caractéristique associé à $\Delta\varepsilon(\omega)$, mais aussi l'existence de molécules de solvant qui empêchent de connaître $\Phi_R(t)$ aux temps courts, nous permettent de prendre une forme asymptotique pour le mouvement des ions. A partir des résultats obtenus par la dynamique moléculaire, aussi bien en ce qui concerne l'argon [9] que les sels fondus [10], nous supposerons que pour les temps qui nous concernent le carré moyen de la distance parcourue pendant Δt est de la forme $l^2 \sim D\Delta t$, où D est le coefficient de diffusion de la particule chargée (mouvement brownien linéaire). L'existence d'un mouvement de translation suggère d'autre part une durée limitée pour les effets collectifs. Nous supposerons que la durée de vie d'un agrégat est distribuée suivant une loi de Poisson. Si on admet que le mouvement de diffusion de l'ion j s'effectue le long d'une droite d'orientation $\mathbf{U}_j$ durant toute la vie de l'agrégat, la fonction $\Phi_R(t)$ s'écrit:

$$\Phi_R(t) = e^{-t/\tau} \left[\langle \mathbf{M}^2(0) \rangle + \langle \mathbf{M}(0) \frac{2}{\sqrt{\pi}} \sum_j e_j \mathbf{U}_j \sqrt{D_j} \rangle \sqrt{t} \right] \quad (6)$$

où τ est la durée de vie moyenne des effets collectifs.

Dans [4] et [5] un modèle plus général a été proposé dans lequel le mouvement ionique est approximé par une ligne brisée pendant la durée des interactions collectives. Dans ce cas, il existe dans $\Phi_R(t)$ deux temps caractéristiques. Le premier a la même signification que dans (6) et le second $\tau'(\tau_i < \tau)$ représente le temps moyen durant lequel la trajectoire d'un ion conserve son orientation et que, pour simplifier, on appellera le temps moyen entre collisions. La FC a alors pour expression:

$$\Phi_R(t) = e^{-t/\tau}$$
$$\left[\langle \mathbf{M}^2(0) \rangle + \left\langle \mathbf{M}(0) \sum_j e_j \mathbf{U}_j \sqrt{D_j \left(\frac{\tau' \tau}{\tau - \tau'} \right)} \right\rangle \operatorname{erf} \sqrt{t \frac{(\tau - \tau')}{\tau \tau'}} \right]. \quad (7)$$

Nous voyons que (6) et (7) se composent chacune de deux termes. La partie purement exponentielle de $\Phi_R(t)$ correspond à une relaxation déterminée uniquement par la durée de vie des entités. Le second terme qui accélère la thermalisation de la FC est lié aux mouvements à l'intérieur de l'agrégat.

3. Résultats expérimentaux et discussion

A titre d'exemple, ce modèle peut être confronté à des résultats récents concernant des solutions de perchlorate de lithium (0.6 M l^{-1}) dans des mélanges de tétrahydrofuranne (THF) et de benzène. Les mesures sont effectuées entre 0.1 et 34 GHz en ce qui concerne $\varepsilon(\omega)$, et $\sigma(0)$ est déterminé aux fréquences acoustiques [4]. Le comportement diélectrique des solvants eux-mêmes est simple; il peut s'expliquer par la rotation de dipôles rigides (molécules de THF) ne présentant pas de corrélation d'orientation suivant le critère du facteur de Kirkwood [4].

Nous avons comparé (6) et (7) aux résultats expérimentaux en supposant que la relaxation des molécules de THF s'effectuait avec le même temps caractéristique τ_m que dans le solvant binaire correspondant. L'amplitude de cette relaxation a été estimée à partir d'un modèle de solvatation développé par ailleurs [3], [5]. Il reste trois paramètres ajustables dans le cas de (6): l'amplitude de la relaxation ionique, la durée de vie des agrégats et le terme γ qui multiplie $\sqrt{t}$ et caractérise le caractère non exponentiel de (6). L'ajustement de ces paramètres par des méthodes statistiques [4] ne permet pas de retrouver les résultats expérimentaux. Dans le cas de (7), nous avons estimé a priori τ', temps moyen entre collisions. Pour cela, nous avons admis que les collisions avaient lieu avec les ions, les molécules de benzène et les molécules de THF non solvatantes; ceci permet de calculer la distance quadratique moyenne entre collisions. Le coefficient de diffusion étant calculé à partir de la viscosité [4] et de la rela-

TABLEAU I

	$\varepsilon_m(0)$	$\varepsilon_m(\infty)$	$10^{12}\tau_m$ (s)	$\Delta\varepsilon(0)$	γ	$10^{12}\tau$ (s)	$10^{12}\tau'$ (s)	$10^3\sigma(0)$ (ohms^{-1} cm^{-1})
THF pur + 0,6 M LiClO$_4$	6,24	2,296	2,52	10,38	0,47	200	150	1,61
THF (7,2 M l^{-1}) + Benzène + 0,6 LiClO$_4$	4	2,288	2,58	9,55	0,60	310	175	0,903
THF (4,8 M l^{-1}) + Benzène + 0,6 M l^{-1} LiClO$_4$	3,15	2,283	2,62	8,34	0,61	380	190	0,562

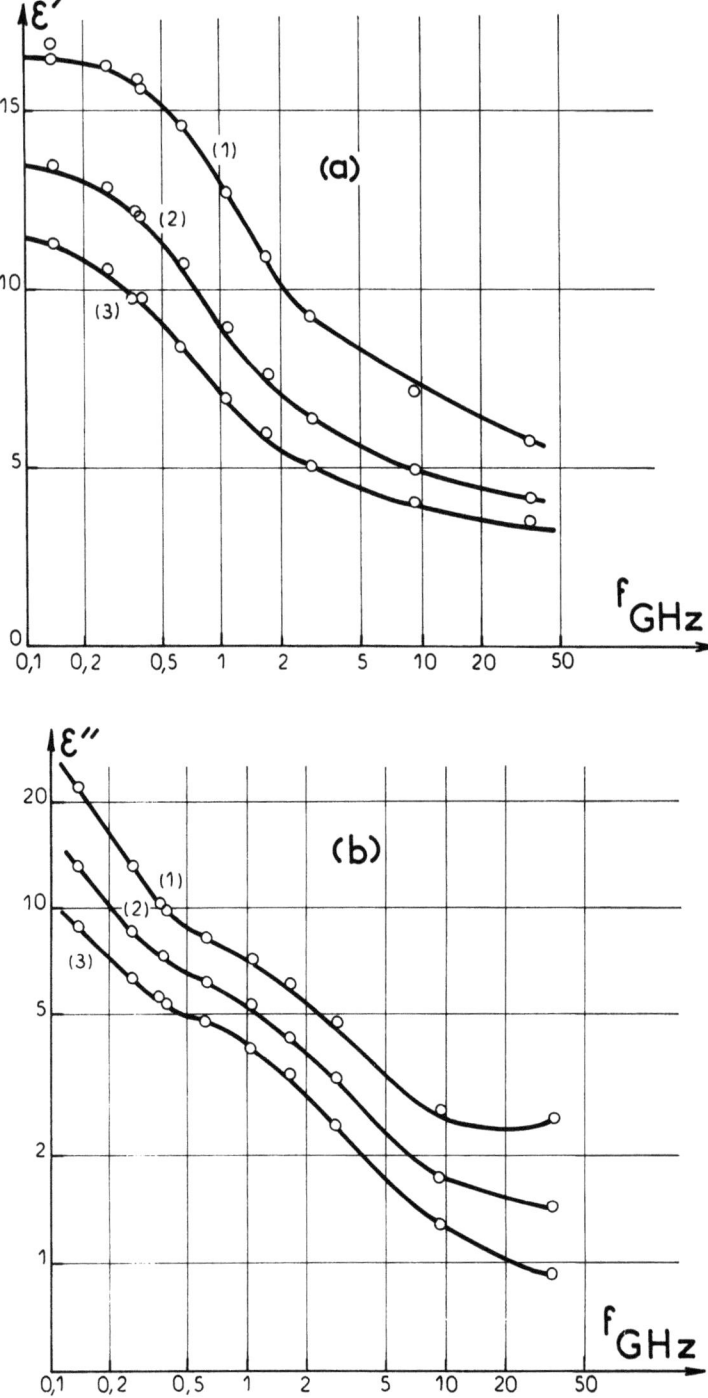

Fig. 1. Solutions de perchlorate de lithium (0,6 M l^{-1}) dans les solvants suivants: (1), tetrahydrofuranne (THF); (2), THF (7,2 M l^{-1}) + benzène; (3), THF (4,8 M l^{-1}) + benzène. Parties réelle (a) et imaginaire (b) de la permittivité complexe. Les courbes correspondent aux valeurs théoriques déduites de (7).

tion de Stokes, on a ainsi une estimation de τ' et (7) contient seulement les trois mêmes paramètres ajustables que (6). Les valeurs trouvées sont données dans le tableau I, et on constate sur les figures 1a et 1b le bon accord entre valeurs calculées et valeurs expérimentales de la permittivité complexe.

Le modèle proposé, dans la version correspondant à (7), permet donc de retrouver les résultats expérimentaux avec un minimum de paramètres ajustables. Parmi ceux-ci $\Delta\varepsilon(0)$, selon (4), est proportionnel aux fluctuations $\langle M^2(0)\rangle$ du moment des ions, fluctuations qui sont reliées ici à une polarisation de déformation des agrégats ioniques. Toutefois, et ceci tient au caractère diélectrique du modèle, les mouvements des ions décrivant cette déformation ne contribuent pas à la conductivité. Cette difficulté apparaît d'une autre façon si l'on cherche à déduire $\Delta\varepsilon(0)$ de $\langle M^2(0)\rangle$ à l'aide de (4), en calculant $\langle M^2(0)\rangle$ pour un liquide conducteur. Cette dernière grandeur peut, en effet, se déduire de la fonction de distribution radiale de Debye-Hückel; nous avons trouvé à partir de (4):

$$\Delta\varepsilon(0) = \varepsilon_m(0) \tag{8}$$

résultat qui peut aussi s'obtenir en calculant $\langle M^2(0)\rangle$ à partir de la restriction générale sur les fonctions de distribution radiale introduite par Stillinger et Lovett [11], reprise par Waisman et Lebowitz [12]. La relation (8) est en complet désaccord avec l'expérience, confirmant ainsi qu'un modèle diélectrique utilisant (4) pour relier $\Phi_R(t)$ à $\Delta\varepsilon(\omega)$ ne s'applique qu'à un milieu isolant.

Le modèle proposé soulève donc des objections: il ne fournit aucun moyen de calculer $\sigma(0)$, et, même si l'on néglige la conductivité, il pose la question de savoir pourquoi il existe des agrégats et comment la structure reflétée par $\langle M^2(0)\rangle$ se déduit d'un système de forces interioniques autorisant les déformations décrites par le mouvement brownien. Il faut toutefois remarquer que la variation de permittivité prédite par le modèle diélectrique est une fonction non rationnelle de ω, comme dans les théories ODF, et que de plus elle décrit parfaitement les résultats expérimentaux. On peut donc penser que certains éléments de ce modèle, plus particulièrement ceux qui décrivent le mouvement des ions par un processus stochastique, sont susceptibles d'être utilisés pour élaborer un modèle permettant le calcul de $\Phi_V(t)$ aux concentrations élevées, c'est-à-dire pour généraliser les théories ODF.

4. Conclusion

L'utilisation d'une fonction de corrélation des vitesses permet seule de prédire à la fois la conductivité des solutions électrolytiques et la variation de cette conductivité avec la fréquence. Toutefois, les résultats obtenus jusqu'ici restent limités aux solutions diluées et ne sont pas applicables aux solutions concentrées, seules susceptibles d'être étudiées expérimentalement. L'utilisation d'une fonction de corrélation des positions, bien qu'imparfaitement adaptée au problème, conduit à penser que les relaxations observées se rattachent au mouvement brownien des ions en solution, et peuvent donc constituer une méthode intéressante pour étudier ce mouvement.

Bibliographie

1. Friedman, M.: *Physica* **30**, 509, 530 (1964).
2. Badiali, J. P. et Rosinberg, M. L.: *Compt. Rend. Acad. Sci.* **276B**, 825 (1973).
3. Badiali, J. P. et Lestrade, J. C.: *J. Chim. Phys.*, no. Spécial, octobre, 107 (1969).
4. Badiali, J. P., Cachet, H., Cyrot, A., et Lestrade, J.C.: *J. Chem. Soc. Faraday II* **69**, 1339 (1973).
5. Badiali, J. P., Cachet, H., et Lestrade, J. C.: *Ber. Bunsenges. physik. Chem.* **75**, 297 (1971).
6. Davies, M. et Williams, G.: *Trans. Faraday Soc.* **56**, 1619 (1960).
7. Pottel, R.: *Ber. Bunsenges. physik. Chem.* **69**, 363 (1965).
8. Cavell, E. A. S.: *Trans. Faraday Soc.* **512**, 1578 (1965).
9. Rahman, A.: *Phys. Rev.* **136**, 405 (1964).
10. Woodcock, L. V.: *Chem. Phys. Letters* **10**, 257 (1971).
11. Stillinger, F. M. et Lovett, R.: *J. Chem. Phys.* **49**, 1991 (1968).
12. Waisman, E. et Lebowitz, S.: *J. Chem. Phys.* **56**, 3086 (1972).

DISCUSSION

Rivail: Il y aurait un troisième mécanisme qui pourrait intervenir dans la relaxation de solutions de sels dans des milieux organiques: ce serait un mécanisme chimique entrant dans le cadre de l'exposé de J. Goulon. En effet, dans la mesure où il existe des espèces moléculaires dans les solutions étudiées, on peut imaginer toute une série d'équilibres d'associations entre molécules de soluté et entre soluté et solvant. Les théories de la relaxation chimique laissant prévoir plusieurs temps de relaxation ce qui pourrait expliquer la forme du type Cole-Davidson observée. Un bon système pour mettre à l'épreuve de telles considérations serait par exemple les solutions de bromure de lithium dans l'éther diéthylique pour lesquelles Chabanel* a montré qu'il existait des tétramères de soluté plus ou moins associés avec l'éther.

Badiali: Pour les solutions que nous avons étudiée la conductibilité équivalente Λ est faible, mais beaucoup plus importante que celle correspondant aux solutions de LiBr dans l'éther étudiées par Chabanel. On constate que Λ et $\Delta\varepsilon_1(0)$ croissent simultanément avec la concentration en sel ou la teneur en tétrahydrofuranne. L'accroissement de Λ avec la concentration en sel est attribué habituellement à des déplacements d'équilibre entre différentes sortes d'agrégats (paires d'ions, triplets) mais l'existence et la nature de ces agrégats constituent des hypothèses. Le spectre observé ne fait pas apparaître plusieurs relaxations et sa décomposition, si elle est possible, n'est pas unique. Ainsi il nous a paru plus intéressant de relier cette dispersion de type Cole-Davidson à un nouveau mécanisme de diffusion translationnelle comme Glarum l'a proposé pour les diélectriques. Cette approche nous semble d'autant plus justifiée que le mouvement brownien de translation joue un rôle déterminant dans la théorie de la conductivité.

Friedman: If I am not mistaken your theory includes a particular model for the motions of the ions. I wonder whether you have found that this model is consistent with the experimentally known relation of the ohmic conductance and the Hall conductance in such solutions, which shows that Langevin's equation is an adequate representation of the ionic motions in very dilute solutions.

Badiali: Pour interpréter la relaxation d'origine ionique dans les solutions électrolytiques, on ne possède de théorie réellement cohérente que pour les solutions diluées où la théorie de Debye-Hückel est applicable. Dans ce cas, le calcul de la fonction de corrélation du courant fournit à la fois $\sigma(0)$ et $\Delta\varepsilon_i(\omega)$. Pour les solutions dont on peut étudier expérimentalement la relaxation, la concentration est trop élevée pour que l'on puisse appliquer la théorie précédente. Les interprétations proposées séparent alors $\sigma(0)$ et $\Delta\varepsilon_1(\omega)$ en deux grandeurs relativement indépendantes, ce qui est évidemment une approximation. Dans le travail présenté ici la conductivité est de 50 à 100 fois inférieure à celle qu'elle serait dans les solutions aqueuses de même concentration. Ainsi on s'intéresse à la permittivité complexe $\Delta\varepsilon_1(\omega)$ d'un milieu non conducteur qui peut être considéré comme constitué d'agrégats ioniques de charge totale nulle. La polarisation résulte alors d'une déformation de ces agrégats résultant d'un mouvement de translation diffusionnelle des ions qui les constituent. La nature du modèle proposé ne permet donc pas de déduire d'information sur la conductivité.

---

* Chabanel, M.: Thèse, Nancy 1966; *J. Chim. Phys.* **63**, 1143 (1966).

MOLECULAR VIBRATIONS IN LIQUIDS

SAVO BRATOS, YVES GUISSANI, and JEAN-CLAUDE LEICKNAM

Laboratoire de Physique Théorique des Liquides, Université Paris VI, Paris, France

Abstract. Experimental methods and basic theories permitting the study of molecular vibrations in liquids are briefly reviewed. Techniques such as IR absorption, spontaneous Raman scattering, ultrasonic absorption, Rayleigh-Brillouin scattering and stimulated Raman scattering are discussed in this context. The physical nature of different vibrational relaxation times measured by these experiments is discussed in detail.

Résumé. Ce travail représente une mise au point sur les méthodes expérimentales et les théories utilisées en vue d'étudier les vibrations moléculaires dans les liquides. Les techniques telles que l'absorption IR, la diffusion Raman spontanée, l'absorption ultrasonore, la diffusion Rayleigh-Brillouin et la diffusion Raman stimulée sont discutées dans ce contexte. La nature physique des différents temps de relaxation vibrationnelle détectés par ces méthodes est discutée en détail.

1. Introduction

Molecular vibrations represent a particular form of molecular motions in liquids. In the past they received only a limited attention, much less than other characteristic modes such as translational diffusion, rotational diffusion or sound waves. The situation improved in the last few years when, in addition to classical ultrasonic absorption techniques, IR, Raman, Rayleigh-Brillouin and stimulated Raman spectroscopies have been successfully applied to the study of vibrational dynamics in liquids. The purpose of this paper is to present a comparative analysis of results supplied by these methods. In particular, the physical nature of different vibrational relaxation times is discussed in detail. Only a limited number of references are quoted; the very unequal development of techniques under discussion makes any other choice difficult.

2. IR and Raman Spectroscopies

(A) The basic theory relating the observed profiles to molecular dynamics has been established for inert solutions of diatomic [1, 2] and, in part, polyatomic molecules [3]. The moment analysis of the IR and Raman bands has also been worked out [4, 5]. The limitation to dilute solutions is presently a necessity; the problem of the diffusion, in pure liquids, of vibrational excitons has not yet been mastered theoretically.

(B) The IR absorption and Raman scattering are both due to vibrational and rotational motions of the active molecule. These motions generate a partially chaotic time variation of its dipole moment vector and of its polarisability tensor, therefore the IR and Raman bands of liquids have a finite width. As a rule, rotational broadening effects predominate for small molecules dissolved in very inert solvents and vibrational broadening effects are preponderant in the case of heavy polyatomics and associated liquids. Thus, if an analysis of vibrational motions in liquids is desired, vibrational and rotational broadening effects must somehow be separated; two procedures have been proposed.

An isolated $0 \to \alpha$ IR band of a small polyatomic molecule is recorded and the corresponding Raman band is recorded for VV and VH scattering geometries [2, 3]. All these bands are Fourier inverted and the results are compared with theoretical expressions for IR or Raman correlation functions:

$$G_{0\alpha}(t) = M_{0\alpha}^2 \langle e^{-i\omega_{0\alpha}t} \rangle \langle \mathbf{u}_{0\alpha}(t) \mathbf{u}_{0\alpha}(0) \rangle \quad \text{(IR)} \quad (1)$$

$$G_{0\alpha}(t) = A\alpha_{0\alpha}^2 \langle e^{-i\omega_{0\alpha}t} \rangle + B \langle e^{-i\omega_{0\alpha}t} \rangle \langle \text{Tr}[\boldsymbol{\beta}_{0\alpha}(t) \boldsymbol{\beta}_{0\alpha}(0)] \rangle. \quad \text{(R)} \quad (2)$$

In these equations, $\mathbf{M}_{0\alpha} = M_{0\alpha}\mathbf{u}_{0\alpha}$ is the transition moment vector and $\boldsymbol{\alpha}_{0\alpha} = \alpha_{0\alpha}\mathbf{1} + \boldsymbol{\beta}_{0\alpha}$ the Raman diffusion tensor associated with the $0 \to \alpha$ vibrational transition; the notions such as mean polarisability α and magnitude of anisotropy β are frequently used in what follows. Finally, A and B are coefficients which depend on the geometry of the experimental arrangement; their values for VV and VH geometries are given in Table I. It results from Equations (1) and (2) that, if $\alpha_{0\alpha}$ is non-vanishing, the vibrational correlation function $\langle e^{-i\omega_{0\alpha}t} \rangle$ as well as the rotational correlation functions $\langle \mathbf{u}_{0\alpha}(t) \times \mathbf{u}_{0\alpha}(0) \rangle$, $\langle \text{Tr}[\boldsymbol{\beta}_{0\alpha}(t)\boldsymbol{\beta}_{0\alpha}(0)] \rangle$ are separately obtainable from the Fourier inverted band profiles, i.e. rotations and vibrations are separated. If $\alpha_{0\alpha}$ vanishes or if the band under consideration is not isolated, this procedure meets with some difficulties.

TABLE I

Coefficients A and B indicate the contribution of $G_i(t)$ and $G_a(t)$ to the Raman correlation function $G(t)$; these contributions are associated with the isotropic and with the anisotropic component of the scattered radiation, respectively

	A	B
Polarized light, VH geometry	0	$1/15$
Polarized light, VV geometry	1	$4/45$
Natural light, VH geometry	0	$1/15$
Natural light, VV geometry	$1/2$	$7/90$

(ii) The half-width $\Delta\omega_{1/2}$ of an isolated IR or Raman band is studied as a function of temperature T [6, 7]. Assuming vibrational broadening mechanisms to be much less sensitive to the temperature variations than rotational broadening mechanisms and

---

* A number of hypotheses underly the above procedures (i, ii). Separate determination of vibrational and rotational correlation functions following the method (i) is only legitimate if the rotation-vibration coupling is absent and if the rotational and vibrational motions can be considered to be statistically independent. Moreover, the theories on which the separation (i) is based do not consider the solvent-induced components in $\mathbf{M}$, α; the effect of these components on the separation (i) has not yet been studied. On the other hand the splitting of $\Delta\omega_{1/2}$ into $\Delta\omega_V$ and $\Delta\omega_R$ following the method (ii) stands and falls with the hypothesis that $\Delta\omega_V$ does not vary with T. Unfortunately, this assumption can not always be considered valid and the procedure (ii) has a semi-quantitative character.

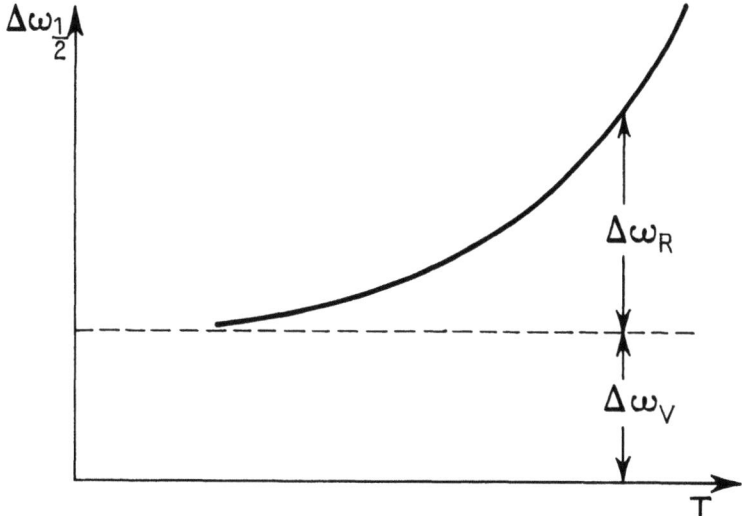

Fig. 1. The half-width $\Delta\omega_{1/2}$ of an IR or Raman band is plotted as a function of temperature T. The separation of $\Delta\omega_{1/2}$ in $\Delta\omega_V$, $\Delta\omega_R$ is obtainable by (i) assuming $\Delta\omega_{1/2}$ to be additive in $\Delta\omega_V$, $\Delta\omega_R$, (ii) $\Delta\omega_V$ to be independent of temperature.

using the construction of the Figure 1 gives the separation of the total half-width $\Delta\omega_{1/2}$ into its vibrational and rotational contributions $\Delta\omega_V$, $\Delta\omega_R$*.

(C) Once the vibrational relaxation functions $\langle e^{-i\omega_{0\alpha}t} \rangle$ have been determined, the study of vibrational motions in liquids becomes accessible. (i) The physical nature of IR and Raman vibrational relaxation times is examined first; the following elements can be used to guide this analysis. (i) Physical characteristics of IR and Raman vibrational band profiles are determined by the relaxation of quantities such as $M_{0\alpha}(t)$, $\alpha_{0\alpha}(t)$, $\beta_{0\alpha}(t)$. (ii) Except for very low frequency transitions, the solvent-solute interaction energy $V_S(t)$ acts in an adiabatic way. This is due to the fact that, for times of the order of $1/\Delta\omega$ where $\Delta\omega$ is the total band width and for ω's characterizing vibrational transitions, the Fourier components $\langle 0| V_s[\omega] |\alpha\rangle$ are too small to induce a non-negligible number of transitions between vibrational levels. (iii) In most cases, with some rare exceptions in isotropic Raman spectra, the transition frequencies $\omega_{0\alpha}(t)$ are either very slowly or slowly varying random functions of time. The correlation times τ_ω of $\omega_{0\alpha}(t)$ most often are long compared to the reference time $1/\Delta\omega$; it is generally legitimate to write $\Delta\omega_{0\alpha}(t) \to \Delta\omega_{0\alpha}$. Thus, in these experiments, the liquid appears as an amorphous solid rather than a liquid. Combining the points (i-iii) one concludes that, for dilute solutions at least, the decay of vibrational relaxation functions $\langle e^{-i\omega_{0\alpha}t} \rangle$ is due to a progressive loss of phase coherence of excited vibrators; this loss is, in turn, originating in the spread of transition frequencies $\omega_{0\alpha}$. Vibrational relaxation times τ in IR and Raman are thus dephasing times and have no bearing to the time vibrational excitation spends on a given molecule. They are rather short and range in the interval 10^{-11}–10^{-14}s; compare with Table II.

(D) In times measurable in IR and Raman, molecular vibrations in liquids appear

TABLE II

Times τ measured by IR and Raman spectroscopies; these times vary in wide limits

System	Normal mode	τ(in 10^{-12} s, $T \sim 293$ K)	Reference
SO_2/CS_2	$\nu(SO)$	10	9
CH_3I/hexane	$\nu(CI)$	5.3	8
CH_3I/CCl_4	$\nu(CI)$	3.0	8
CH_3CN	$\nu(CC)$	2.2	6
	$\nu(C\equiv N)$	1.9	6
$CH Cl_3$	$\nu(CCl_3)$	1.9	6
	$\nu(CH)$	1.0	6
Phenol	$\nu(OH)$	0.02	

like to vibrations of amorphous solids. The spectral density function $I(\Delta\omega)$ of an isolated band is deducible from the distribution function $P(\omega_{0\alpha})$ of transition frequencies $\omega_{0\alpha}$; this can be inferred from Equations (1, 2). In these circumstances an explicit calculation of $I(\Delta\omega)$ and τ is practically possible; the following cases are worth attention. (i) In inert solutions the solvent molecule interacts with a large number of solvent molecules i and $\omega_{0\alpha} - \omega_{0\alpha}^0 = \Delta\omega_{0\alpha} = \sum_i \Delta\omega_{0\alpha}^i$. According to the central limit theorem $P(\omega_{0\alpha})$ is expected to be close to a Gaussian and $I(\Delta\omega) \sim$
$\sim \exp[-\Delta\omega^2/2\beta_v]$; here $\Delta\omega = \omega - \langle\omega_{0\alpha}\rangle$, $\beta_v = \langle(\omega_{0\alpha} - \langle\omega_{0\alpha}\rangle)^2\rangle$. For transitions $(00 \cdot 0_\alpha \cdot 0) \to (00 \cdot v_\alpha \cdot 0)$ and $(00 \cdot 0_\alpha 0_\beta \cdot 0) \to (00 \cdot v_\alpha v_\beta \cdot 0)$ one finds:

$$\omega_{0\alpha} - \omega_{0\alpha}^0 = v_\alpha \left[\sum_{u=i}^{3N-6} \frac{\lambda_{u\alpha\alpha}}{2\omega_u^2 \omega_\alpha} \left(\frac{\partial V_s}{\partial n_u}\right)_0 - \frac{1}{2\omega_\alpha} \left(\frac{\partial^2 V_s}{\partial n_\alpha^2}\right)_0 + \sum_{u=i}^{3N-6} 0 \left(\frac{\partial^3 V_s}{\partial n_u^3}\right)_0 \right] \tag{3}$$

$$\omega_{0\alpha} - \omega_{0\alpha}^0 = v_\alpha \left[\sum_{u=i}^{3N-6} \frac{\lambda_{u\alpha\alpha}}{2\omega_0^2 \omega_\alpha} \left(\frac{\partial V_s}{\partial n_u}\right)_0 - \frac{1}{2\omega_\alpha} \left(\frac{\partial^2 V_s}{\partial n_\alpha^2}\right)_0 + \sum_{u=i}^{3N-6} 0 \left(\frac{\partial^3 V_s}{\partial n_u^3}\right)_0 \right] +$$
$$+ v_\beta \left[\sum_{u=i}^{3N-6} \frac{\lambda_{u\beta\beta}}{2\omega_u^2 \omega_\beta} \left(\frac{\partial V_s}{\partial n_u}\right)_0 - \frac{1}{2\omega_\beta} \left(\frac{\partial^2 V_s}{\partial n_\beta^2}\right)_0 + \sum_{u=i}^{3N-6} 0 \left(\frac{\partial^3 V_s}{\partial n_u^3}\right)_0 \right]. \tag{4}$$

In these equations n_u, n_v, n_w are the normal coordinates of the active molecule and λ_u, λ_{uvw} the harmonic and cubic force constants associated with them; moreover, $\lambda_u = \omega_u^2$. Equations (3, 4) depend on $\partial V_s/\partial n_u$, $\partial^2 V_s/\partial n_u^2$ etc. One concludes that the experimental observation of vibrational band profiles and of corresponding relaxation times $\tau \sim 1/\sqrt{\beta_v}$ gives a valuable information on V_s; similarly, the analysis of their variation when going from one band to another gives a qualitative idea about the site in the molecule which is most exposed to the action of intermolecular forces. The increase of

T may either produce an increase in τ or leave it unchanged [8–10]. (ii) If the solvent-solute interaction is not inert, the frequency distribution function $P(\omega_{0\alpha})$ and the vibrational band profile are broad and asymmetrical. The effects are particularly pronounced for associated liquids and may be spectacular if hydrogen bonds are present. Applying Equation (1) to the study of an isolated complex due to a weak hydrogen bond leads to the following expression for the intensity of a bonded ν_{OH} band:

$$I(\Delta\omega) \sim \frac{\omega_0 + \Delta\omega}{1 + \left[\dfrac{\omega''}{\omega'}\right]\left(\dfrac{\Delta\omega}{\omega'}\right)} \times$$

$$\times \exp\left\{-\left[\frac{\Lambda}{2KT}\right]\left(\frac{\Delta\omega}{\omega'}\right)^2 - \left[\frac{\Lambda\omega''}{2KT\omega'} - \frac{F}{6KT}\right]\left(\frac{\Delta\omega}{\omega'}\right)^3\right\}. \quad (5)$$

Λ, F, ω', ω'' are parameters defining the hydrogen bond under study: Λ and F are harmonic and cubic force constants associated with $r_u = r(OH\cdots O')$, ω' and ω'' are derivatives around the equilibrium configuration of the $\Delta\omega(\nu OH)$ vs $R(OH\cdots O')$ curve, ω_0 determines the position of the band center and $\Delta\omega = \omega - \omega_0$. The profile (5) is a asymmetrically distorted Gaussian. It describes the distribution of the frequency shifts brought about by the fluctuation in the hydrogen bond length $R(OH\cdots O')$; its half-width increases, approximately, with $\sqrt{T}$. Spectral manifestation of hydrogen bonded liquids represent the most extreme case of vibrational broadening.

3. Ultrasonic Absorption and Rayleigh-Brillouin Scattering

(A) The theory of ultrasonic absorption in liquids is described in detail in several excellent text books; see, e.g. [11–13]. The three main mechanisms responsible for the sound absorption are (i) viscosity and heat conduction, (ii) thermal relaxation and (iii) structural relaxation. Only thermal relaxation is related to the present subject and is discussed in what follows.

In polyatomic liquids the total energy is distributed between the translational-rotational (external) and vibrational (internal) degrees of freedom of molecules. A sound wave, compressing and decompressing periodically elementary volumes of the liquid, disturbs the equilibrium distribution of energy. There results an instantaneous deviation from equilibrium and an irreversible flow of energy between these energy modes; one observes the absorption of sound. This is what is called thermal relaxation. The theory of sound absorption is then based on the three following essential assumptions: (i) There is a rapid energy redistribution amongst the external degrees of freedom themselves and between the internal degrees of freedom themselves. Thus these two sets of degrees of freedom can be regarded as two distinct thermodynamic systems characterized, between other, by the temperatures T and T_i. (ii) The internal energy $E(T_i)$ depends only on T_i and not, for example, on the pressure of the liquid. (iii) The exchange of energy between the internal and the external systems are governed by a simple rate equation $dE(T_i)/dt = -(E(T_i) - E(T))/\tau'$ where τ' is the relaxation time. The constant volume specific heat is then frequency dependent, $C_v = C_v^0 + C^1/ \times (1 + i\omega\tau')$. Using the assumptions (i–iii) and neglecting the classical absorption due to viscosity and heat conduction leads to the following formulas for the velocity of sound and absorption per wave-length α:

$$\frac{v_0^2}{v^2} = 1 - \frac{r\omega^2\tilde{\tau}^2}{1+\omega^2\tilde{\tau}^2}; \quad \frac{\alpha v_0^2}{v^2} = \pi r \frac{\omega\tilde{\tau}}{1+\omega^2\tilde{\tau}^2}. \tag{6a, b}$$

v, v_0 are sound velocities at $\omega = \omega$ and $\omega = 0$, $\tilde{\tau} = \tau' \, C_p^\infty/C_p^0$, $r = C^1(C_p^0 - C_v^0)/C_p^\infty C_v^0$, where C_p^0, C_p^∞, C_v^0, C_v^∞ are constant pressure and constant volume specific heats at $\omega = 0$ and $\omega = \infty$; the quantity r is known as the relaxation strength. The theory just described can easily be generalized to include several relaxation processes and classical mechanisms of sound absorption. Moreover, the validity of the energy rate equation has been investigated in some special situations [14, 15]. All these theories predict, like Equation (6a, b), that the absorption per wave-length will attain its peak value at $\omega \sim 1/\tau'$.

(B) Thermal relaxation in liquids can also be measured by setting up a very different experiment, the Rayleigh-Brillouin scattering. A detailed theory has been elaborated for the case of a single relaxation time [16, 17]; see, however [18]. The description of the phenomenon is as follows.

When a beam of monochromatic light is passed through a liquid, some of the light is scattered since the density is not uniform*. There would be no frequency shift of the scattered light if the density non-uniformities were static. As frozen-in non uniformities are not possible in a liquid, density fluctuations are time dependent and shifted frequencies occur in the scattered light. The spectrum most often consists of the well known Rayleigh-Brillouin triplet; however, if there is some coupling between vibrational and translational degrees of freedom, an additional band is found superimposed over the central band of this triplet. The theory of this phenomenon contains the following elements. (i) The spectral density $I(\omega)$ of the scattered light is expressed by using the space and time Fourier transform of the density autocorrelation function. (ii) The density fluctuations are described by linearized hydrodynamic equations comprising the continuity equation, the Navier-Stokes equation an the equation of the heat transfer. (iii) Thermal relaxation is described either by introducing a frequency dependent bulk viscosity, $\eta_v = \eta_v^0 + \eta^1/1 + i\omega\tau'$ [16], or by adding a hydrodynamic equation for an additional state variable for the internal degree of freedom [17]. The theory gives a simple expression for the spectral density function of the scattered light; the comparison with the experiment then permits the determination of the relaxation time τ'.

(C) The times τ' detected by ultrasonic absorption and Rayleigh-Brillouin scattering experiments are due to the relaxation of the vibrational specific heat and bulk viscosity, respectively. Both of them are, essentially, energy relaxation times. The

---

* Orientation fluctuations have no connection with the present problem and are not considered here. From the experimental point of view, their spectral effects are most often easily separable from those brought about by density fluctuations.

exact identification of the normal modes implied on a given particular problem is difficult, except for diatomics. As a rule, many modes are active with different relaxation times and strengths. Their activity depends on whether they exchange energy directly with the sound wave (parallel excitation) or one mode gets its energy through the disturbance of another one (series excitation). Most frequently relaxation processes with comparable relaxation times merge to a unique, average relaxation process and a very restricted number of such processes are needed to reproduce experimental data. The identification is further facilitated by calculating C^1, and r, by means of the well known Einstein formula for the vibrational specific heat:

$$C^1(T) = R(\hbar\omega_\alpha/KT) \frac{\exp(-\hbar\omega_\alpha/KT)}{[1 - \exp(-\hbar\omega_\alpha/KT)]^2}. \quad (7)$$

Here, ω_α is the frequency of the α'th normal mode. Empirical estimations of relaxation times τ are often based on the formula $1/\tau' \sim NP$ where N is the number of collisions per second and P, essentially, the gas phase vibrational transition probability. Recent theoretical work shows the practical and conceptual limits of these theories and proposes a statistical approach to this problem [19]. Typically, τ's range in an interval of 10^{-9}–10^{-11}s; compare with Table III.

TABLE III

Vibrational specific heat and bulk viscosity relaxation times τ' for some simple liquids: these times have been measured by applying the ultrasonic absorption and the Rayleigh-Brillouin scattering techniques, respectively

System	Method	τ' (in 10^{-10}s, $T \sim 293$ K)	Reference
CS_2	ultrasons	28	Cited in Reference 12, p. 196
CH_3Cl	ultrasons	7.3	ibid
CH_2Cl_2	Rayleigh-Brillouin	2.8	
		1.7	18
$CH Cl_3$	ultrasons	1.3	Cited in Reference 12, p. 196
Benzene	ultrasons	2.7	ibid
	Rayleigh-Brillouin	1.4	18

The results of this discussion is that vibrational energy relaxation times τ' detected by ultrasonic absorption and Rayleigh-Brillouin scattering experiments have no simple relationship with the vibrational relaxation times τ furnished by IR and Raman spectroscopies. No theory relating these quantities has yet been proposed.

4. Stimulated Raman Scattering

(A) This technique, applied only recently to the study of vibrational relaxation times in liquids [20, 21], rests on the following basic principles.

The liquid is subjected to action of the light produced by a laser emitting powerful, picosecond pulses. The light beam induces Stokes-type vibrational transitions with a probability which increases nonlinearly with the input intensity; this is the stimulated Raman effect. If the laser pulses are short in duration compared to vibrational relaxation times in the liquid, there results an impulse-type, or shock, coherent vibrational wave. Two useful experiments can be set up in these circumstances. (i) An increase in the population of the excited vibrational state produced by the pump pulse gives rise to enhanced incoherent anti-Stokes Raman scattering of a second short light pulse of different frequency. Varying the delay time between pump pulse and probe pulse, the rise and decay of the excess density of excited molecules is observed. The decay of the scattering signal is a direct measure of the energy relaxation time τ'. (ii) The rise and decay of the vibrational amplitude is measured with delayed pulses of the same time dependence. The observed anti-Stokes Raman signal is scattered under phase matching conditions close to forward direction. If the pulses are short enough, molecular dephasing times τ are obtained. The theory of these phenomena involves four basic quantities, (i) the laser field amplitude $E_L(\mathbf{r}t)$ supposed known, (ii) the Stokes field amplitude $E_S(\mathbf{r}t)$, (iii) the amplitude of the coherent vibrational field $Q(\mathbf{r}t)$ where $Q(\mathbf{r}t)$ is the average amplitude in a small volume centred on $\mathbf{r}$, of the normal mode responsible for transition, and (iv) the number density $n(\mathbf{r}t)$ of excited molecules. E_L, E_S, Q and n are related by means of the following, partially phenomenological, differential equations [22–24]:

$$\frac{\partial E_S}{\partial z} + \frac{1}{v_S}\frac{\partial E_S}{\partial t} = \frac{i\pi\omega_S^2 N}{c^2 k_S}\frac{\partial \alpha}{\partial Q}Q^* E_L; \quad \frac{\partial Q}{\partial t} + \frac{Q}{2\tau} = \frac{i}{4\omega_0}\frac{\partial \alpha}{\partial Q}E_L E_S^*(1-2n);$$

$$\frac{\partial n}{\partial t} + \frac{n}{\tau'} = \frac{i}{8\hbar}\frac{\partial \alpha}{\partial Q}(E_L E_S^* Q^* - E_L^* E_S Q). \qquad (8\text{–}10)$$

In Equations (8)–(10), v_S, k_S, ω_S are the group velocity, the wave vector and the frequency of the Stokes light, ω_0 is the frequency of the normal mode involved in the transition, N is the number of molecules per unit volume and $\partial \alpha/\partial Q$ represents the change of polarisability with vibrational amplitude. These equations can be solved, for a given form of the imput pulse and a given set of boundary conditions using numerical methods. Comparing measured signals with those determined theoretically gives τ and τ'. Some results are listed in Table IV.

TABLE IV

Times τ, τ' brought about by the stimulated Raman experiments: τ is the dephasing time and τ' is the life-time of the excited molecule

System	Normal mode	τ	τ' (in 10^{-12} s $T \sim 293$ K)	Reference
CH_3CCl_3	$\nu(CH)$	1.3	5	21
C_2H_5OH	$\nu(CH)$		20	21

(B) The following comparison can be made between the times τ, τ' detected by the stimulated Raman scattering and by other experimental methods discussed in this paper. For dilute solutions at least, the dephasing times are essentially similar in stimulated Raman, spontaneous Raman and IR spectroscopies. This can be seen by (i) assimilating the stimulated Raman dephasing time τ with the correlation time of the normal coordinate n_u responsible for the transition, (ii) calculating the IR and spontaneous Raman vibrational relaxation time for the same transition by the help of Equations (1, 2). There results:

$$\tau \sim \int_0^\infty \langle n_u(t) n_u(0) \rangle \, dt / \langle n_u(0)^2 \rangle \sim n_{01}^2 \int_0^\infty \langle e^{-i\omega_u t} \rangle \, dt / n_{01}^2 =$$

$$= \int_0^\infty \langle e^{-i\omega_u t} \rangle \, dt \quad \text{(stimulated Raman)} \quad (11)$$

$$\tau = M_{01}^2 \int_0^\infty \langle e^{-i\omega_u t} \rangle \, dt / M_{01}^2 = \int_0^\infty \langle e^{-i\omega_u t} \rangle \, dt \quad \text{(IR)}. \quad (12)$$

In these equations ω_u indicates the frequency of the u'th normal mode involved in the transition. The comparison of energy relaxation times is more ambiguous. While there is no doubt of the intrinsic similarity between the stimulated Raman τ' and relaxation times of vibrational specific heat and bulk viscosity, the detailed identification is delicate. Ultrasonic and Rayleigh-Brillouin experiments rarely give a unique relaxation time ascribable to a single well defined vibrational transition and stimulated Raman data generally refer to various intermolecular decay processes. More data are needed before a definitive conclusion can be reached.

References

1. Bratos, S., Rios, J., and Guissani, Y.: *J. Chem. Phys.* **52**, 439 (1970).
2. Bratos, S. and Marechal, E.: *Phys. Rev.* **A4**, 1078 (1971).
3. Nafie, L. A. and Peticolas, W. L.: *J. Chem. Phys.* **57**, 3145 (1972).
4. Gordon, R. G.: *J. Chem. Phys.* **38**, 1724 (1963).
5. Gordon, R. G.: *J. Chem. Phys.* **40**, 1973 (1964).
6. Rakov, A.: *Proc. Lebedev Phys. Inst.* **27**, 111 (1965).
7. Bartoli, F. J. and Litowitz, T. A.: *J. Chem. Phys.* **56**, 404 (1972).
8. Constant, M., Delhaye, M., and Fauquembergue, R.: *Compt. Rend.* **271**, 1177.
9. Ouillon, R.: Thèse 3ème cycle, Université Paris VI, 1972.
10. Levant, R. and Bratos, S.: *Compt. Rend.* **276**, 603 (1973).
11. Hertzfeld, K. F. and Litowitz, T. A.: *Absorption and Dispersion of Ultrasonic Waves*, Academic Press, New York, 1959.
12. Bhatia, A. B.: *Ultrasonic Absorption*, Clarendon Press, Oxford, 1967.
13. Sette, D.: in H. N. V. Temperley, J. S. Rowlinson, and G. S. Rusbrooke (eds.), *Physics of Simple Liquids*, North Holland Publishing Press, Amsterdam, 1968, p. 325.
14. Kneser, H. O.: *Ann. Phys.* **16**, 360 (1933).
15. Landau, L. D. and Teller, E.: *Phys. Zeit. Soviet Union* **10**, 34 (1936).
16. Mountain, R. D.: *J. Research Natl. Bur. Std* **70A**, 207 (1966).

17. Mountain, R. D.: *J. Research Natl. Bur. Std* **72A**, 95 (1968).
18. Caloin, M. and Candau, S.: *J. Phys.*, to be published.
19. Davies, P. K. and Oppenheim, I.: *J. Chem. Phys.* **57**, 505 (1972).
20. Von der Linde, D., Laubereau, A., and Kaiser, W.: *Phys. Rev. Letters* **26**, 954 (1971).
21. Laubereau, A., Von der Linde, D., and Kaiser, W.: *Phys. Rev. Letters* **28**, 1162 (1972).
22. Maier, M., Kaiser, W., and Giordmaine, J. A.: *Phys. Rev.* **177**, 580 (1969).
23. Carman, R. L., Shimizu, F., Wang, C. S., and Bloembergen, N.: *Phys. Rev.* **A2**, 60 (1970).
24. Lallemand, P.: in A. Anderson (ed.), *The Raman Effect*, Marcel Dekker, Inc., New York, 1971, p. 287.

DISCUSSION

Sillescu: I should like to remark that there is a formal similarity between the time constants $1/\omega_{\alpha 0}$, τ and τ' discussed by Prof. Bratos and relaxation times defined in spin relaxation theory. Thus, τ' is similar to the spin lattice relaxation time T_1, τ is similar to the spin-spin relaxation time T_2 for homogeneous broadening, and $\omega_{\alpha 0}$ for inhomogeneous broadening $1/T'_2$. With respect to $\omega_{\alpha 0}$, do you think there is the possibility of additional contributions from the coupling of normal vibrations that may occur, e.g. in hydrogen bonded networks?

Bratos: I agree with the remark by Prof. Sillescu relative to the comparison of times $1/\omega_{\alpha 0}$, τ and τ' from one side and of times T'_2, T_2 and T_1 from the other side. It is generally useful to compare different phenomena in this way.

As for the hydrogen bonded liquids here concerned, the coupling between different ν_{OH} vibrations are not believed to contribute an important contribution to the ν_{OH} band width. This can be shown by isolating a H-bonded complex in an inert solvent and comparing its ν_{OH} band width with that of the pure liquid. One never observes order of magnitude changes.

Janik: (1) If the low temperature plateau of the $\Delta \nu$ vs T dependence appears to be in the solid phase, is it still justified to conclude anything about orientational broadening in the liquid phase, by the Rakov method?

(2) How important in all these methods is the knowledge of instrumental function?

Bratos: (1) The Rakov method is a semi-quantitative method. If this limitation is considered, one can use the plateau in the $\Delta \omega_{1/2}$ vs T curve observed on amorphous solid and estimate the vibrational width $\Delta \omega_R$ in this way. (2) In principle all observed band profiles have to be deconvoluted by the instrumental function. This is practically important for narrow bands and only has a small effect in broad bands. The correction for finite slit-width is always practically possible.

Gershon: It is worthwhile to mention that in addition to Mountain's theory there is a more accurate theory by Weinberg and Oppenheim (*Physica* **62**, 1 (1972)), for the polarized spectrum of systems with internal degrees of freedom. Among the important points of this theory are that the simple relaxation equation of the vibrational energy is not always complete. There are additional terms which depend on the wave vector **k** and that $1/\tau$ might depend on k or k^2 in some experimental situations.

Bratos: What I tried in my lecture is to describe briefly the most essential features of the theory of the effect of the thermal relaxation on the absorption of ultrasonic waves and on the Rayleigh-Brillouin scattering. For simplicity, the questions such as the existence of several relaxation times, the effect of **k**, etc. have not been included.

Perchard: Vous avez précisé que la largeur vibrationnelle ne dépend pas de la durée de vie des états excités. Comment expliquez-vous que, dans le cas des dilutions isotopiques, cette largeur vibrationnelle varie parfois de façon notable (HCl dans DCl, C_6H_6 dans C_6D_6...)?

Bratos: Les systèmes auxquels vous faites allusion ne sont pas les solutions diluées et la théorie présentée ici ne s'y applique pas. Il est certain que, dans le cas des liquides purs, le transfert d'exciton vibrationnel limite la durée de vie de l'état vibrationnel excité.

Vincent: (à propos de la remarque de Perchard). La largeur vibrationnelle dépend probablement de l'amplitude de la vibration et, en particulier, il semble que les bandes correspondant à des vibrations de valence C–H présentent une largeur vibrationnelle plus grande que les autres. Si cette interprétation est juste, la largeur vibrationnelle d'une bande faisant intervenir H devrait être plus grande que la bande homologue du composé deutérié.

DETERMINATION PAR SPECTROMETRIE DE VIBRATION DES FONCTIONS DE CORRELATION ORIENTATIONNELLES MONOMOLECULAIRES DANS LES LIQUIDES

J. LASCOMBE, M. BESNARD, P. B. CALOINE, J. DEVAURE, et M. PERROT

Laboratoire de Spectroscopie Infrarouge, associé au C.N.R.S.
Université de Bordeaux I, 33405 – Talence, France

Résumé. Le domaine temporel dans lequel peuvent être déterminées, avec une précision suffisante, les fonctions de corrélation à partir des spectres infrarouges ou Raman est discuté. Une analyse des conditions dans lesquelles on peut extraire, des données des spectrométries vibrationnelles, l'information concernant les fluctuations orientationnelles des molécules dans les liquides est également présentée. Enfin, quelques résultats expérimentaux ayant trait à des solutions d'oxyde de carbone, ou de chlorure d'hydrogène et à la triméthylamine liquide sont présentés.

Abstract. The times in which the correlations functions are determined from infrared and Raman spectra with a suffisant accuracy are discussed. It is also discussed about conditions needed to obtain informations on molecular orientational motions in liquids from vibrational spectroscopy data. Some experimental results of carbon oxyde or hydrogen chloride solutions and liquid trimethylamine are presented.

Les travaux de Kubo [1] puis de Gordon [2] ont montré que les bandes d'absorption infrarouge ou de diffusion Raman d'un liquide sont les transformées de Fourier des fonctions de corrélation du moment ou de la polarisabilité de transition vibrationnelle.

$$I(\omega)_{IR} \alpha TF \langle (\varepsilon \mu_{(0)})(\varepsilon \mu_{(t)}) \rangle$$

$$I(\omega)_{Raman} \alpha TF \langle (\varepsilon_I \alpha_{(0)} \varepsilon_D)(\varepsilon_I \alpha_{(t)} \varepsilon_D) \rangle .$$

ε représente le vecteur unitaire définissant la direction du champ électrique du rayonnement electromagnétique et les moyennes sont prises sur toutes les orientations possibles du système au temps zéro et sur son évolution du temps zéro au temps t. Ces fonctions de corrélation caractérisent les fluctuations temporelles du système et leur détermination permet donc en principe l'étude de certains aspects de la dynamique moléculaire à l'état liquide.

Deux problèmes méritent toutefois d'être discutés préalablement à une étude de la dynamique moléculaire par cette méthode: en premier lieu, le domaine temporel durant lequel les fonctions de corrélation peuvent être déterminées avec précision à partir des mesures infrarouge et Raman; en second lieu, la séparation des effets des différents types de fluctuation qui affectent polarisabilité ou moment de transition afin d'accèder, en particulier, à la dynamique orientationnelle des molécules. Ces deux questions constituent les deux premières parties de cette communication qui sera complétée ensuite par la présentation de quelques résultats expérimentaux ayant trait aux solutions d'oxyde de carbone et de chlorure d'hydrogène dans divers solvants et à la triméthylamine liquide.

1. Domaine temporel des fonctions de corrélation des liquides accessibles par spectrométrie infrarouge et Raman

Représentons par $C(t)$ la fonction de corrélation correspondant au profil $I(\omega)$, et pour simplifier, prenons pour origine des pulsations le centre de la bande. La fonction $I(\omega)$ n'est pas parfaitement connue. Elle n'est en effet déterminée que dans un intervalle spectral limité $(-a, +a)$; elle est déformée par la fonction d'appareil; il se superpose un bruit de fond au signal; enfin, il y a une erreur sur la ligne de base.

Ces différentes erreurs limiteront la connaissance de $C(t)$ en particulier aux temps courts et aux temps longs [3].

C'est surtout la limitation du domaine spectral d'observation de $I(\omega)$ qui entraîne une mauvaise connaissance de la fonction de corrélation aux temps courts. Tout se passe, en effet, comme si on déterminait seulement:

$$I'(\omega) = \hat{I}(\omega) \cdot R(\omega),$$

où $R(\omega)$ est une fonction valant 1 dans l'intervalle $(-a; +a)$ et 0 en dehors de cet intervalle. La transformée de Fourier de $I'(\omega)$ s'écrit:

$$C'(t) = C(t) * \varrho(t).$$

ou $\varrho(t) = 2(\sin at)/t$. On détermine donc la fonction de corrélation convoluée par une fonction du type de celle qui apparaît quand on étudie les phénomènes de diffraction. En terme qualitatif, on peut définir une résolution de la fonction de corrélation ou un intervalle temporel de l'ordre de π/a à l'intérieur duquel les informations ne peuvent plus être analysées. Pratiquement, les bandes sont observées dans des intervalles dont l'ordre de grandeur peut varier entre $(-30, +30 \text{ cm}^{-1})$ et $(-300 \text{ à} +300 \text{ cm}^{-1})$; cela correspond à des résolutions de la fonction de corrélation allant de 0,5 à 0,05 ps.

Ce temps définit la borne supérieure du domaine temporel dans lequel les fonctions de corrélation peuvent être connues. Ajoutons que l'erreur sur la ligne de base se traduit également par une erreur sur la fonction de corrélation aux temps courts. En effet, celle-ci, de la forme $k(\sin at)/at$, est maximum pour $t=0$. Lorsque le domaine spectral d'étude de $I(\omega)$ est relativement important, cette erreur peut faire apparaître une quasi singularité à l'origine de la fonction de corrélation.

Dans la plupart des cas, il est donc difficile de déterminer les coefficients du développement des fonctions de corrélation au temps courts, qui pour la partie réelle après normalisation s'écrit:

$$C(t) = 1 - \frac{1}{2} M(2) t^2 + \frac{1}{4} M(4) t^4.$$

On peut la vérifier facilement en étudiant:

$$F_n(\Omega) = \int_{-\Omega}^{+\Omega} I(\omega) \omega^n \, d\omega$$

en fonction de Ω. Il est fréquent de ne pas observer de convergence de $F(\Omega)$ dans l'intervalle $(-a, +a)$. Si la fonction $F_0(\Omega)$ correspondant à l'intensité intégrée de la bande ou moment d'ordre zéro converge presque toujours, il n'en est plus nécessairement ainsi de la fonction correspondant au moment d'ordre deux, et presque jamais, de la fonction correspondant au moment d'ordre 4.

Ce qui limite la connaissance des fonctions de corrélation aux temps longs peut avoir plusieurs origines. Dans l'hypothèse ou l'on a choisi une résolution convenable pour étudier les bandes, il reste néanmoins l'effet du bruit de fond, de l'erreur sur la ligne de base et de la convolution par la fonction $2(\sin at)/t$. Il en résulte des oscillations qui brouillent complètement la fonction de corrélation aux temps longs. Néanmoins, en choisissant de bonnes conditions d'enregistrement, tout en utilisant un appareillage classique, on peut limiter l'amplitude de ces oscillations, sur une fonction de corrélation normalisée, à 0.01 environ. Dans ces conditions, la limitation aux temps longs dépend de la largeur de la bande. Pour une bande dont la largeur est quelques cm^{-1} on peut déterminer la transformée de Fourier jusqu'à des temps qui peuvent atteindre 20 ps. Si par contre, la bande a une largeur à la demi intensité de 20 à 30 cm^{-1}, ces temps ne sont plus que de 3 à 4 ps.

On voit ainsi l'ordre de grandeur des temps de fluctuations qui pourront être étudiées par spectrométrie infrarouge ou Raman. Seuls des phénomènes relativement rapides de l'ordre de la picoseconde seront accessibles, comme par exemple, les fluctuations orientationnelles de molécules relativement légères.

2. Séparation des effets des fluctuations vibrationnelles et orientationnelles

Les fluctuations des moments ou des polarisabilités de transition à l'état liquide ont deux causes principales: les fluctuations des états rotationnels et celles des états vibrationnels [2, 4, 5]. On peut représenter aisément les phénomènes en utilisant l'image classique d'un moment de transition.

$$\boldsymbol{\mu} = \mathbf{u}\left(\frac{d\mu}{dq}\right) q \cos(\omega_0 t + \varphi)$$

le terme de fluctuation vibrationnelle recouvre à la fois les fluctuations temporelles de la charge de l'oscillateur $d\mu/dq$ s'il y a une charge induite par le solvant, de son amplitude q et de sa phase φ. Les fluctuations orientationnelles sont celles du vecteur unitaire $\mathbf{u}$ qui porte le dipole. Des considérations analogues sont possibles pour le tenseur de polarisabilité.

Deux hypothèses simples peuvent alors être faites. Dans la première on suppose négligeable les fluctuations vibrationnelles. Il vient alors pour les bandes infrarouge $I(\omega)$ $0 \to n$ correspondant aux transitions vibrationnelles $0 \to n$ et pour les bandes Raman isotropes $I(\omega)_{RI}$ et anisotropes $I(\omega)_{RA}$ correspondant à la transition vibrationnelle $0 \to 1$ les relations suivantes:

$$I(\omega) \propto TF\, G_R(t)$$

$$I(\omega)_{RI} \propto \delta(\omega, 0)$$
$$I(\omega)_{RA} \propto TF\, G_{2R}(t)$$

où $G_R(t)$ et $G_{2R}(t)$ sont respectivement les fonctions de corrélation orientationnelles associées à un vecteur ou à un tenseur symétrique de rang 2, et ou le centre de la bande a été choisi comme origine des pulsations.

Dans la seconde, particulièrement utilisée par Bratos et ses collaborateurs [4, 5] on suppose qu'il n'y a pas interaction vibration-rotation et que les fluctuations vibrationnelles ont essentiellement pour origine les fluctuations de phase. On néglige donc la relaxation de l'énergie vibrationnelle et tout effet de moment ou de polarisabilité induite. Il vient alors:

$$I(\omega)_{0 \to n} \propto TF\, G_{v0 \to 1}(nt) G_R(t)$$
$$I(\omega)_{RI} \propto TF\, G_{v0 \to 1}(t)$$
$$I(\omega)_{RA} \propto TF\, G_{v0 \to 1}(t) G_{2R}(t).$$

Dans ces expressions $G_{v\,0 \to 1}(t)$ est la fonction de corrélation représentant les fluctuations vibrationnelles; elle détermine essentiellement le spectre Raman isotrope. Cette même fonction apparaît dans les expressions donnant le spectre infrarouge de la transition $0 \to 1$. Pour les transitions $0 \to n$ la fonction de corrélation vibrationnelle est obtenue à partir de $G_{v\,0 \to 1}(t)$ en faisant le changement de variable t en nt.

Ces relations montrent l'intérêt d'une comparaison des spectres infrarouge et Raman. Quand l'une ou l'autre de ces hypothèses est vérifiée il est aisé d'extraire des données de la spectrométrie vibrationnelle les fonctions de corrélation $G_R(t)$ et $G_{2R}(t)$.

3. Résultats expérimentaux

3.1. Solutions d'oxyde de carbone dans C_2Cl_4 et SO_2 liquide [3]

Les résultats expérimentaux obtenus avec ces solutions sont portés sur les figures 1 et 2. On remarque que la bande Raman isotrope est très étroite si on la compare à la bande infrarouge et peut-être assimilée à une raie δ. Quant aux profils infrarouge correspondant aux transitions $0 \to 1$ et $0 \to 2$, ils sont presque identiques. On peut donc admettre comme l'avait fait Gordon [6] que les profils infrarouge ont essentiellement pour origine les fluctuations orientationnelles.

Les fonctions de corrélation $G_R(t)$ sont données sur la figure 3. Elles sont déterminées avec une résolution de 0,1 ps car les bandes ont pu être étudiées dans un intervalle de $(-150\ \text{à}\ +150\ \text{cm}^{-1})$. Cette résolution est suffisante pour observer une rotation quasi libre aux temps courts. Toutefois, elle ne semble pas suffisante pour déterminer à partir du 4ème moment de la bande la moyenne quadratique du couple s'exerçant sur la molécule [6]. En effet, la figure 4 montre que si la fonction $F_2(\Omega)$ correspondant au moment d'ordre 2 converge approximativement dans l'intervalle $(-150;\ +150\ \text{cm}^{-1})$ il n'en est plus de même de la fonction $F_4(\Omega)$. A l'aide d'une fonction de la forme $|\omega| \exp(A\omega^2 + B\omega)$ une tentative de paramétrage des ailes de la bande a été faite et grace à une extrapolation l'intégration a pu être poursuivie au delà

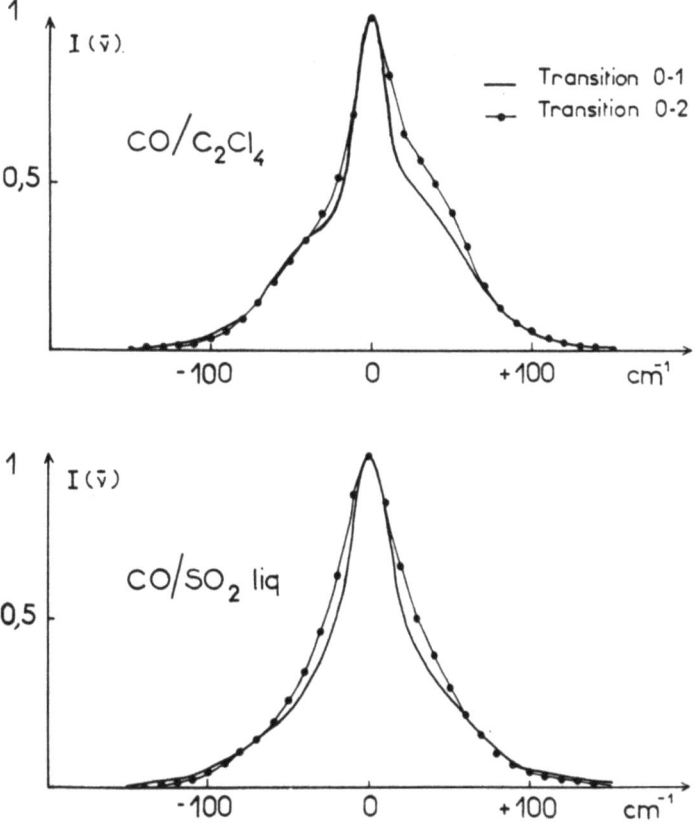

Fig. 1. Spectre infrarouge de l'oxyde de carbone dans C_2Cl_4 et SO_2 liquide. Concentration voisine de 0,1 mole par mole. L'intensité des maximums a été prise comme unité et leurs fréquences comme origine des abscisses.

de l'intervalle d'observation; les pointillés sur les graphes représentent $F_2(\Omega)$ et $F_4(\Omega)$ correspondent à ce calcul. De cette manière une valeur du couple quadratique moyen a été déterminée, elle est évidemment tributaire de la fonction choisie pour représenter les ailes de la bande (tableau I).

Vers les temps longs, les fonctions de corrélation ont pu être déterminées jusqu'à 2 ps; un comportement exponentiel apparaît très nettement.

Il est remarquable, que les temps de corrélation orientationnels $\tau_R = \int_0^\infty \mathrm{Re}\, G_R(t)\, dt$ dans C_2Cl_4 et SO_2 liquide soient très voisins bien que les profils spectraux et l'allure des fonctions de corrélation soient très différents (tableau I) (figures 1 et 3).

Dans le premier solvant ou la bande présente des ailes rotationnelles marquées et la fonction de corrélation un comportement aux temps courts très caractéristique d'une rotation libre, il est probable que, par suite de la masse des molécules du solvant, la durée de vie moyenne des cages qui emprisonnent les molécules d'oxyde de carbone est relativement longue. Dans le dioxyde de soufre, au contraire, les fluctuations des potentiels d'interaction soluté-solvant sont beaucoup plus rapides, de sorte que la dynamique orientationnelle est plus proche du type diffusionnel.

TABLEAU I

Moments de bande, couple moyen et temps de corrélation de CO dans C_2Cl_4 et dans SO_2

	$M(2)^a$ 10^2 cm^{-2}	$M(4)$ 10^6 cm^{-4}	$\langle C^2 \rangle^{1/2}$ 10^2 cm^{-1}	τ_R 10^{-12} s
CO/C_2Cl_4 $0 \to 1$	19 ± 2	14 ± 3	6,6	0,24
CO/C_2Cl_4 $0 \to 2$	$16,5 \pm 2$	9 ± 3	5,2	0,22
CO/SO_2 $0 \to 1$	18 ± 2	$12,5 \pm 3$	6,3	0,24
CO/SO_2 $0 \to 2$	$16,5 \pm 2$	10 ± 3	5,6	0,22

[a] Moment d'ordre 2 théorique 16×10^2 cm^{-2}.

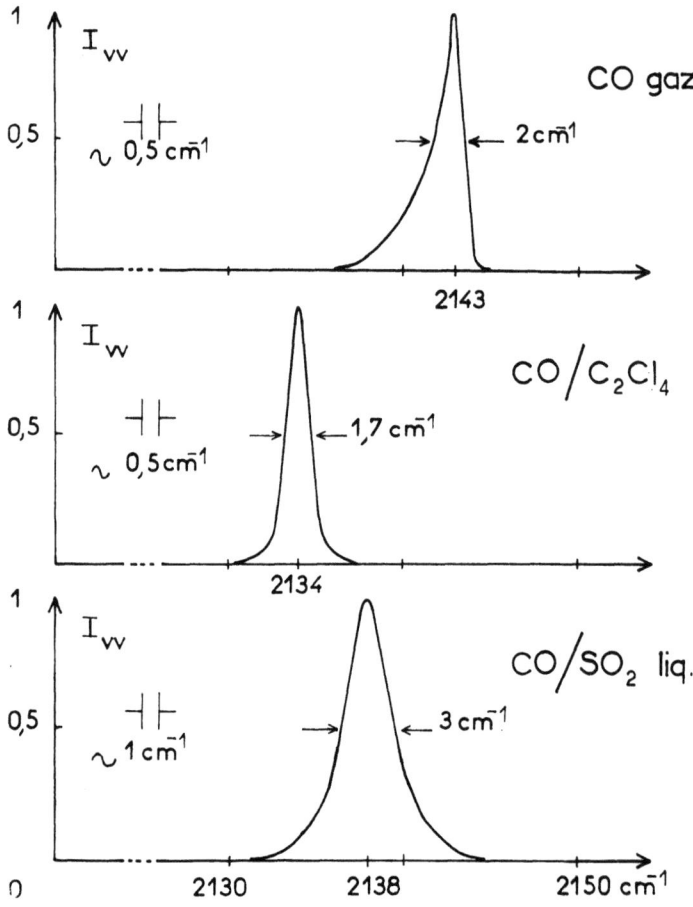

Fig. 2. Spectre Raman I_{vv} de l'oxyde de carbone à l'état gazeux et en solution. *Solution*: Concentration voisine de 0,1 mole par mole. *Gaz*: pression = 1 atm. L'intensité des maximums a été prise comme unité.

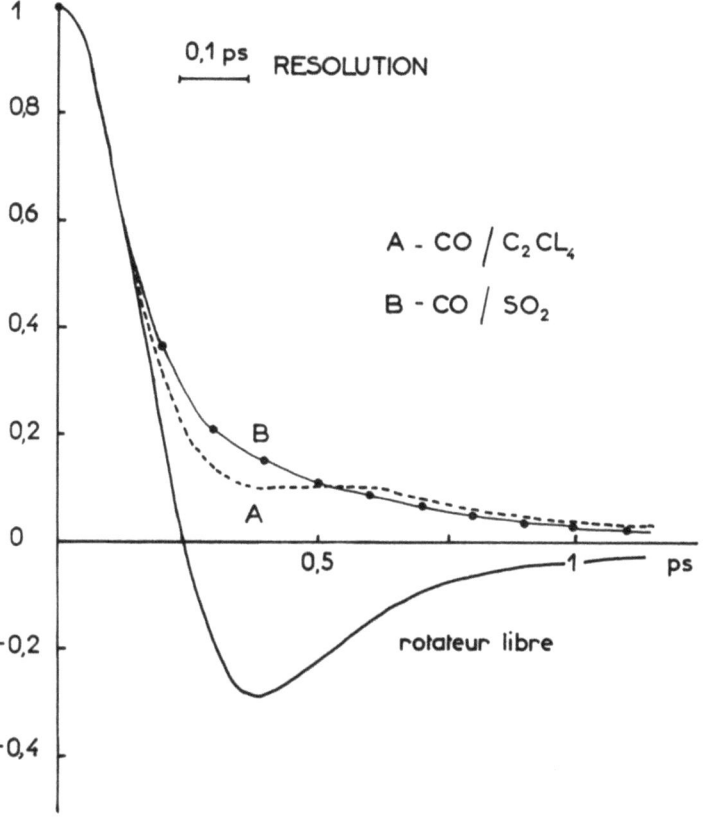

Fig. 3. Parties réelles des fonctions de corrélation orientationnelles de l'oxyde de carbone en solution déduites des spectres infrarouge correspondant à la transition vibrationnelle $0 \to 1$.

Notons que les modèles J et M de Gordon [7] ne rendent pas compte de façon satisfaisante des résultats obtenus dans les deux solvants.

3.2. Solutions de HCl [9]

Dans ces solutions, la largeur de la bande Raman isotrope n'est plus négligeable; elle augmente d'ailleurs fortement quand on passe d'un solvant comme SF_6 liquide à des solvants plus perturbateurs comme SO_2 ou DCl (figure 5). Il faut donc tenir compte de l'influence des fluctuations vibrationnelles. A l'appui de cette conclusion on observe bien que les profils infrarouge des transitions vibrationnelles $0 \to 1$ et $0 \to 2$ ne sont plus identiques (figure 6 et 7). Le cas des solutions dans SF_6 liquide est intéressant à cet égard (figure 6). Dans le lointain infrarouge Birnbaum [8] a observé une structure fine de rotation nettement marquée. Cette structure fine est moins visible sur la bande infrarouge correspondant à la transition $0 \to 1$ par suite des fluctuations vibrationnelles. Elle disparaît presque totalement sur la bande $0 \to 2$ car, comme on l'attend, les fluctuations vibrationnelles sont encore plus importantes.

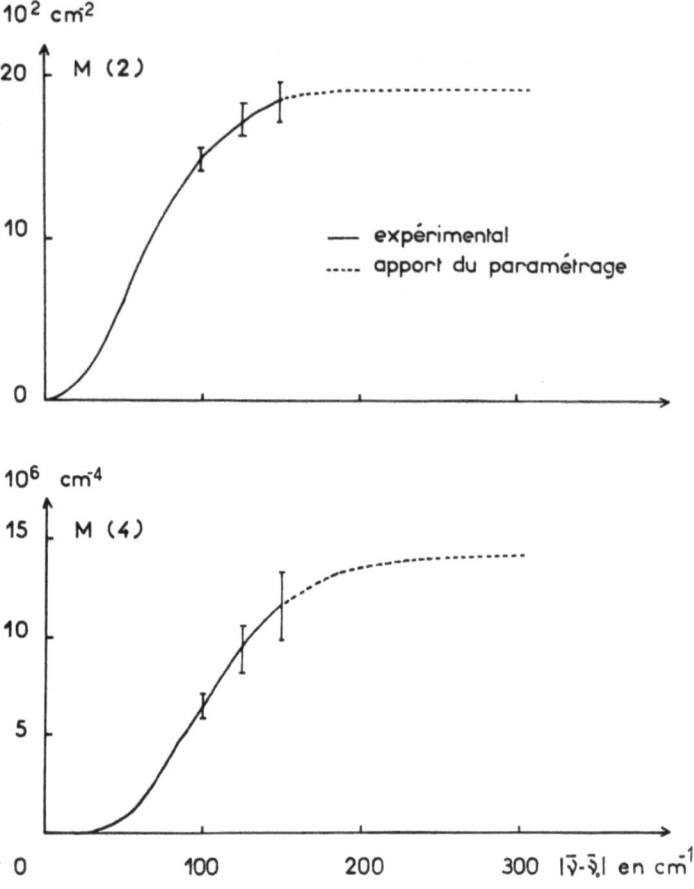

Fig. 4. Détermination des moments d'ordre 2 et 4 du spectre infrarouge de l'oxyde de carbone dans C_2Cl_4 (transition $0 \to 1$) par étude des fonctions $F_2(\Omega)$ et $F_4(\Omega)$.

Malheureusement, les solutions de HCl constituent un cas où les conditions dans lesquelles on peut aisément séparer les effets vibrationnels et orientationnels ne sont pas remplies.

Dans les solvants peu perturbateurs comme SF_6 ou le fréon FC75 (mélange d'isomères du perfluoro-n-propyl pyranne) on met en évidence une interaction vibration-rotation du type de celle de l'état gazeux. En effet, le rapport $I(\omega)/I(-\omega)$ en prenant pour origine des fréquences le centre de la bande, n'est pas égal à $\hbar\omega/kT$ (figure 8). On peut rendre compte des courbes expérimentales en faisant l'hypothèse d'une modulation lente des niveaux de rotation et en prenant un coefficient d'interaction vibration-rotation voisin de celui du gaz [9, 10].

Notons que le spectre Raman isotrope de ces solutions ne peut s'interpréter au contraire que dans l'hypothèse d'une modulation rapide des niveaux de vibration-rota-

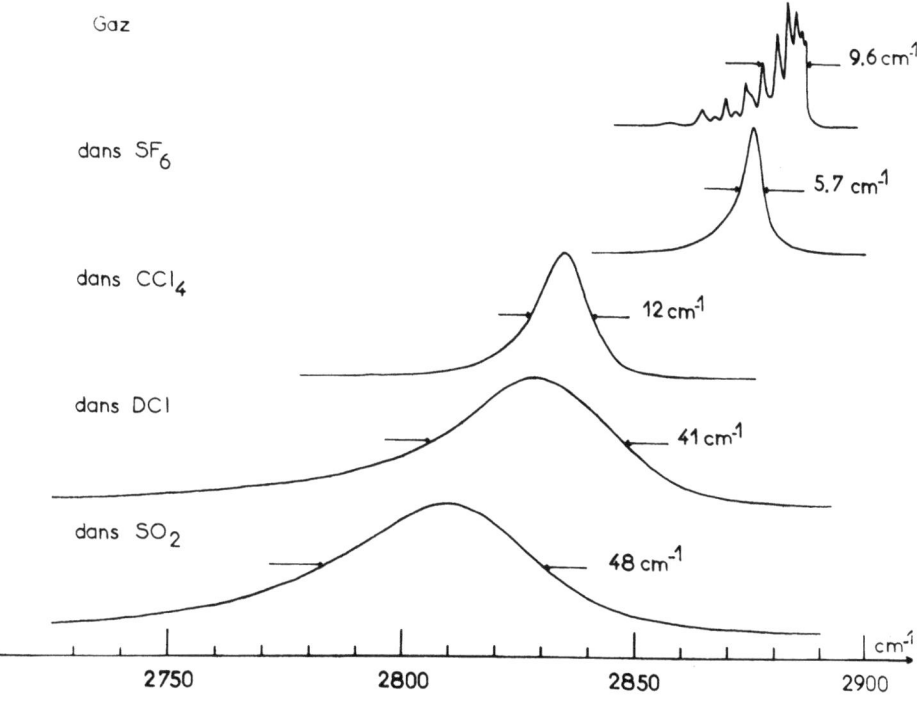

Fig. 5. Spectre Raman $I_{vv} + I_{vh}$ de HCl à l'état gazeux et dans différents solvants: Largeur spectrale de fente: 0,5 à 1 cm^{-1}. Concentration des solutions: 2×10^{-2} mole/mole.

tion; par suite d'un retrécissement dynamique, l'effet d'interaction vibration-rotation ne s'y manifeste pas de façon importante et la largeur de la raie provient surtout des fluctuations de phase vibrationnelles [9, 11].

Néanmoins, l'effet d'interaction vibration-rotation provenant de l'énergie cinétique de la molécule de soluté se traduit surtout sur les bandes infrarouge par l'apparition d'une dissymétrie et sur les fonctions de corrélation par une partie imaginaire importante; la partie réelle de la fonction de corrélation est moins affectée. De la sorte, on peut évaluer des temps de corrélation orientationnels voisins de 0,54 et de $0{,}73 \times 10^{-13}$s dans SF$_6$ et dans FC75; les résultats obtenus sont les mêmes que l'on étudie la bande fondamentale ou la première harmonique [9].

Dans les solvants plus perturbateurs par contre, il est très difficile d'accéder à la partie réelle de la fonction de corrélation orientationnelle. En effet, la déconvolution des profils infrarouge $0 \to 1$ par $I(\omega)_{RI}$ et $0 \to 2$ par la fonction $I(\omega/2)_{RI}$ ne conduit pas au même résultat. Deux hypothèses peuvent être proposées: ou bien il existe une interaction vibration-rotation ayant pour origine le potentiel d'interaction soluté-solvant; ou bien les fluctuation des moments induits sont corrélées à celles des états rotationnels. Nous avons montré, en effet, que les moments induits sont importants dans ces solutions et sont différents pour les transitions $0 \to 1$ et $0 \to 2$ [9, 12].

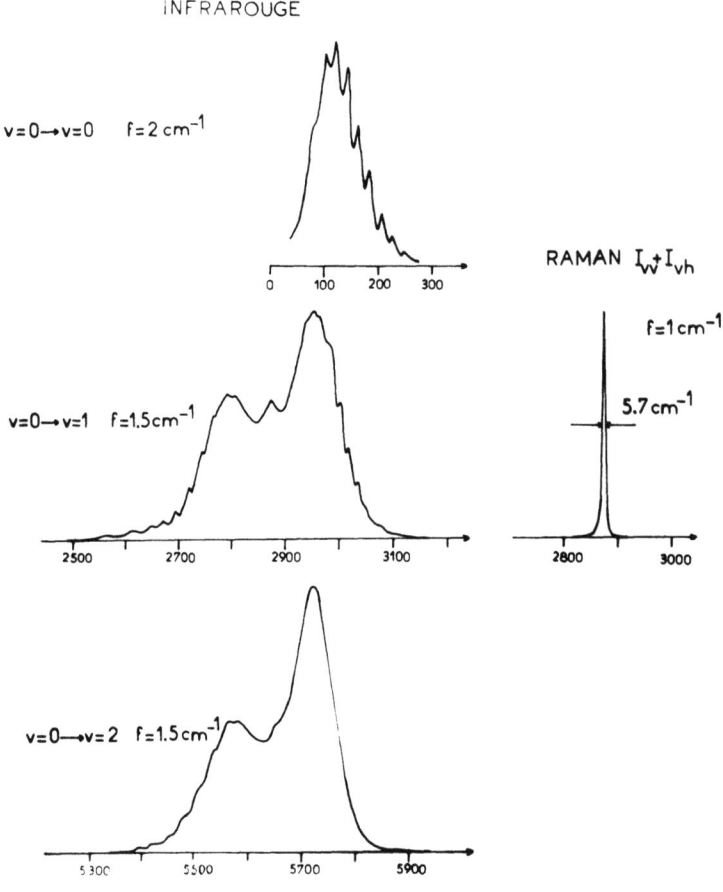

Fig. 6. Spectres infrarouge et Raman de HCl dans SF_6. f = largeur spectrale de fente, concentration = 2×10^{-2} mole/mole.

3.3. Triméthylamine liquide

La triméthylamine, molécule relativement légère et de faible moment dipolaire, qui est liquide à température ordinaire sous 2 atm environ et qui cristallise à $-110°C$ nous a paru constituer un liquide se prêtant bien à des études de fluctuations orientationnelles. Toutefois, on ne peut appliquer sans précautions aux liquides purs des théories élaborées pour des solutions. C'est particulièrement vrai en absorption infrarouge où la variation d'indice de réfraction affecte le profil des bandes [13]. En Raman, par contre, ce dernier effet n'existe pas et nous avons donc choisi cette technique. En outre, pour analyser les résultats on est obligé de supposer l'absence de tout effet d'interaction vibration-rotation et de moment induit. Afin de nous assurer de la validité de ces hypothèses, il nous a paru bon de faire porter notre étude sur plusieurs bandes: deux bandes de symétrie A_1, à 830 et 1180 cm^{-1} correspondant respectivement à une vibration d'élongation des liaisons NC et de balancement des groupements méthyl; 3 bandes de symétrie E à 450, 1050 et 1270 cm^{-1} correspondant à des vibrations de déformation des angles CNC, de balancements des groupements méthyl et d'élongation des liaisons NC.

Notre étude a été faite à différentes températures de $-110\,°C$ à $25\,°C$ et à différentes pressions de 1 à 3000 atm. La fonction d'appareil très proche d'une fonction trian-

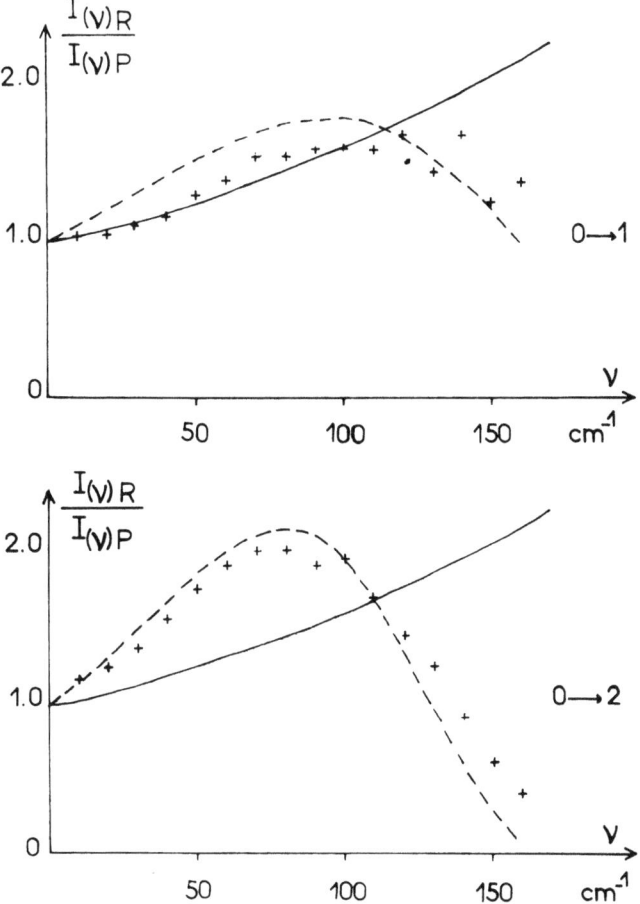

Fig. 7. Etude des rapports $I(v)/I(-v)$ en prenant pour origine de fréquence le maximum de la bande pour les spectres infrarouge de HCl dans SF$_6$. *Traits pleins*: $\exp(hv/kT)$. *Traits pointillés*: rapport calculé en utilisant le coefficient d'interaction vibration-rotation du gaz $\alpha = 0{,}30$ cm^{-1}
Croix: rapport mesuré.

gulaire avait dans la plupart des cas une largeur de 2 cm^{-1}; afin d'améliorer la précision, les techniques d'acquisition de données numériques ont été utilisées, ces dernières étant traitées ensuite par ordinateur. Quelques-uns de nos résultats sont présentés sur les figures 9 et 10. Les bandes ont pu être étudiées dans un intervalle spectral de 150 cm^{-1} de part et d'autre du maximum central, la résolution des fonctions de corrélation est, dans ces conditions, voisine de 0,1 ps. Les profils sont très sensiblement lorentziens et il est difficile d'évaluer les moments d'ordre 2.

On obtient la fonction de corrélation $G_{2R(t)}$ associée aux réorientations de l'axe de la molécule en divisant la transformée de Fourier de la partie anisotrope des bandes de types A_1 par la transformée de Fourier de la partie isotrope; les résultats obtenus avec les bandes à 830 et 1180 cm^{-1} sont les mêmes: la fonction de corrélation est une exponentielle, le temps de corrélation $\tau_{2R}\perp$ à la température ordinaire et sous la pression ordinaire est de $0,75\pm0,1$ ps; enfin, l'énergie et le volume d'activation sont respectivement de $1,7\pm0,5$ kcal mole^{-1} et de 11 ± 4 cm^3. Ces résultats sont en bon accord avec ceux obtenus par résonance magnétique nucléaire [14].

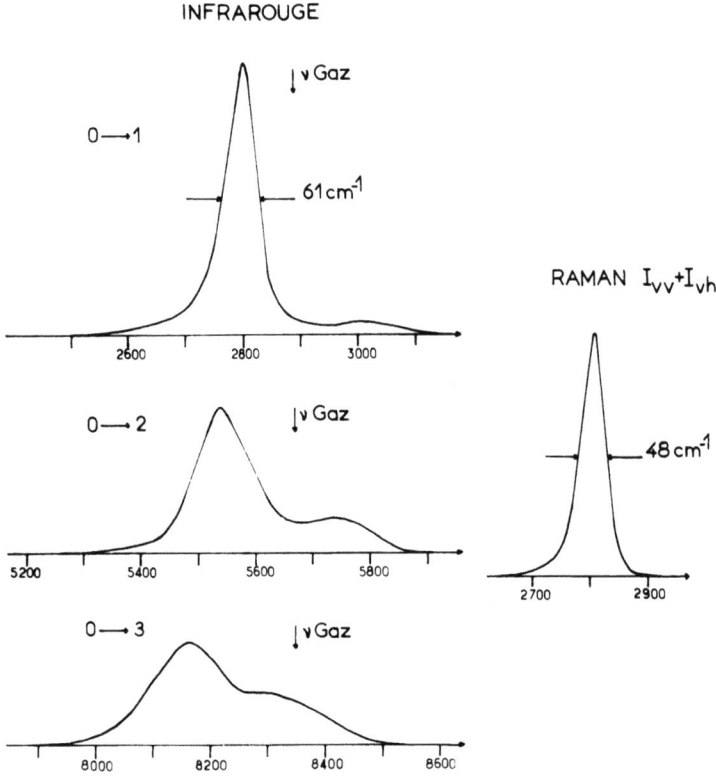

Fig. 8. Spectre infrarouge et Raman de HCl dans SO$_2$. Concentration des solutions: IR $0 \to 1$, IR $0 \to 2$, Raman $c = 2 \times 10^{-2}$ mole/mole. IR $0 \to 3 C = 0,3$ mole mole. Largeur spectrale de fente: 1,5 à 2 cm^{-1}.

L'analyse des bandes E est plus difficile. En effet, il n'existe pas de méthode directe pour déterminer les profils vibrationnels. En outre, l'effet des fluctuations orientationnelles dépend de la valeur relative des différents coéfficients du tenseur de polarisabilité associé à ces transitions [15, 16]. Nous avons toutefois tenté une analyse en faisant trois hypothèses:

(a) à −110°C les fluctuations orientationnelles sont négligeables;

(b) l'effet des fluctuations vibrationnelles dépend peu de la température et de la pression;

(c) le tenseur de diffusion est isotrope.

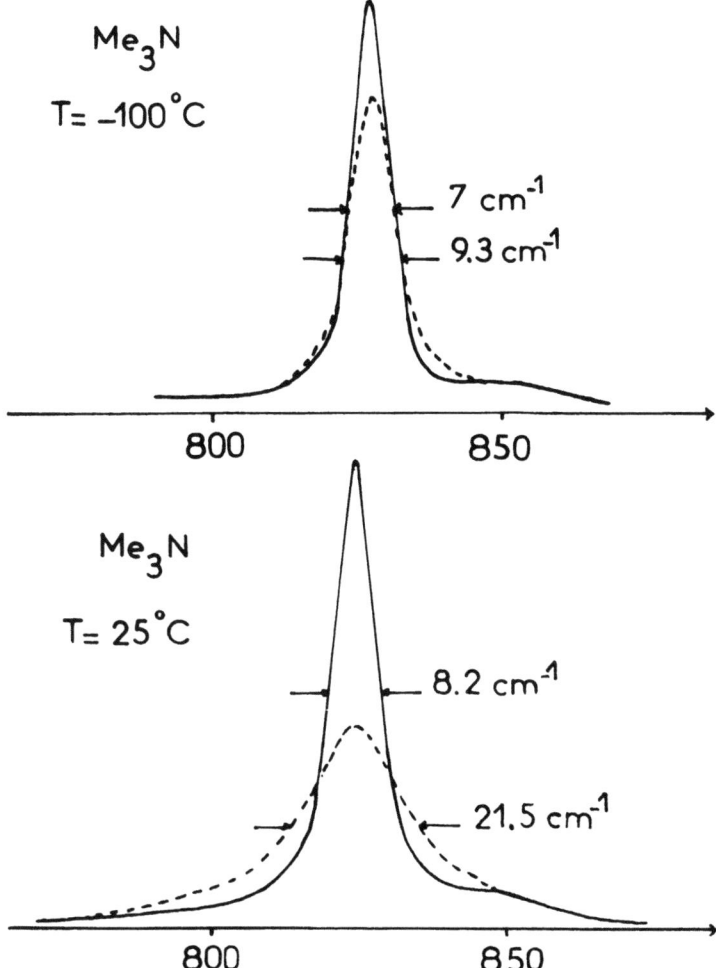

Fig. 9. Partie polarisée (trait plein) et partie dépolarisée (pointillés) du spectre Raman de la triméthylamine à l'état liquide (vibration de type A_1 centrée à 830 cm^{-1}). Les bandes sont normalisées à la surface, la résolution est de 2 cm^{-1}.

Les deux premières hypothèses ont été suggérées par les observations que nous avons faites sur les bandes de symétrie A. Notons cependant que la première entraîne une surévaluation du temps de corrélation et de l'énergie d'activation. Quant à la troisième, nos résultats semblent la justifier. En effet, à partir des bandes de type E on obtient à nouveau des fonctions de corrélation orientationnelles exponentielles et un temps de

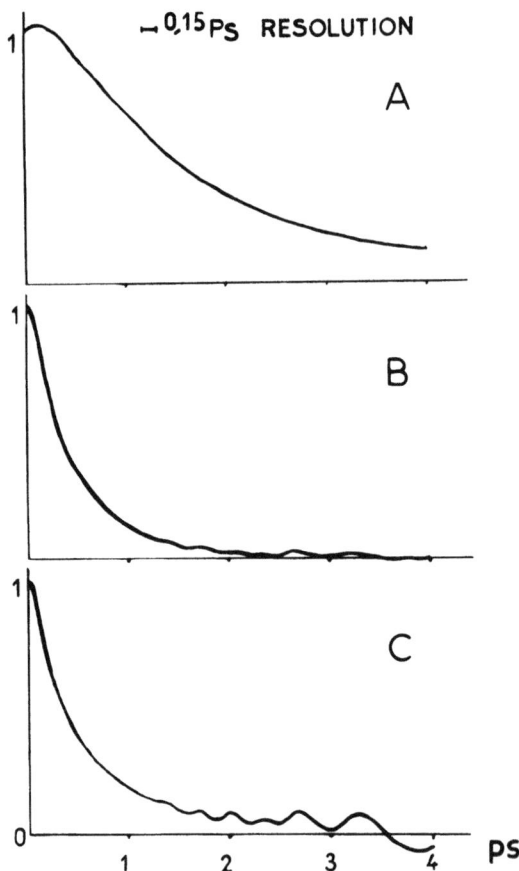

Fig. 10. Partie réelle de la transformée de Fourier de la bande isotrope (A) et anisotrope (B) à 830 cm^{-1} de la triméthylamine liquide à $-20\,°C$ et fonction de corrélation orientationnelle (C). Le comportement anormal de la transformée de Fourier du spectre isotrope à l'origine provient d'une erreur sur la ligne de base; des oscillations assez fortes apparaissent au-delà de 2 ps sur la fonction de corrélation orientationnelle qui empêchent une détermination très précise du temps de corrélation.

corrélation à température et à pression ordinaire $\tau_{2R} = 0,85 \pm 0,15$ ps; les énergies et les volumes d'activation trouvés sont respectivement de $2,1 \pm 0,5$ kcal mole^{-1} et de 11 ± 4 cm^3. Ces résultats sont très proches de ceux obtenus à partir des bandes de type A.

Bibliographie

1. Kubo, R.: *J. Phys. Soc. Japan* **12**, 570 (1957).
2. Gordon, R. G.: *Adv. Magn. Resonance* **3**, 1 (1968).
3. Caloine, P. B.: Thèse Docteur Ingénieur, Bordeaux, 1972.
4. Bratos, S., Rios, J., et Guissani, Y.: *J. Chem. Phys.* **52**, 439 (1970).
5. Bratos, S. et Marechal, E.: *Phys. Rev.* **A4**, 1078 (1971).
6. Gordon, R. G.: *J. Chem. Phys.* **43**, 1307 (1965).
7. Gordon, R. G.: *J. Chem. Phys.* **44**, 1830 (1966).
8. Birnbaum, G. et Ho, W.: *Chem. Phys. Letters* **5**, 334 (1970).
9. Perrot, M.: Thèse Doctorat es Sciences Physiques, Bordeaux, 1973.

10. Perrot, M. et Lascombe, J.: *J. Chim. Phys.* **70**, 5 (1973).
11. Perrot, M. et Lascombe, J.: *Compt. Rend. Acad. Sci.* **276**, 25 (1973).
12. Perrot, M. et Lascombe, J.: *J. Chim. Phys.*, sous presse.
13. Fulton, R. L.: *J. Chem. Phys.* **55**, 1390 (1971).
14. Loewenstein, A. et Waiman, R.: *Mol. Phys.* **25**, 49 (1973).
15. Nafie, L. A. et Peticolas, W. L.: *J. Chem. Phys.* **57**, 3145 (1972).
16. Leicknam, J. C. et Guissani, Y.: ce volume, p. 257.

DISCUSSION

Quentrec: J'aimerais que vous précisiez ce que vous appelez effet de pression sur le comportement rotationnel des molécules. Ne s'agit-il pas plutôt d'un effet de densité?

Lascombe: Il s'agit en effet de densité. Les variations de pression ont été faites à température constante (25 °C).

RAMAN SPECTROSCOPY AND MOLECULAR REORIENTATION IN LIQUIDS: $CDCl_3$ AND CH_3Br

I. LAULICHT and S. MEIRMAN

Dept. of Physics, Bar-Ilan University, Ramat-Gan, Israel

Abstract. Raman scattering experiments on symmetrical stretching lines of liquid $CDCl_3$ (Figure 1) and CH_3Br (Figure 2) have been carried out at room temperature. Angular autocorrelation functions (Figure 3 and Figure 4) were calculated by Fourier transforming the spectra, and by eliminating the vibrational function [1]. It is found that vibrational relaxation contribution to band shapes is important for these molecules. Correlation times were found to be $\tau_2 = 1.59 \pm 0.03$ ps for $CDCl_3$ and 1.33 ± 0.07 ps for CH_3Br in agreement with previous results obtained in different ways [2, 3]. The times between collisions were found to be much shorter (~ 0.01–0.03 ps) than those obtained from infrared absorption experiments [4] but in fair agreement

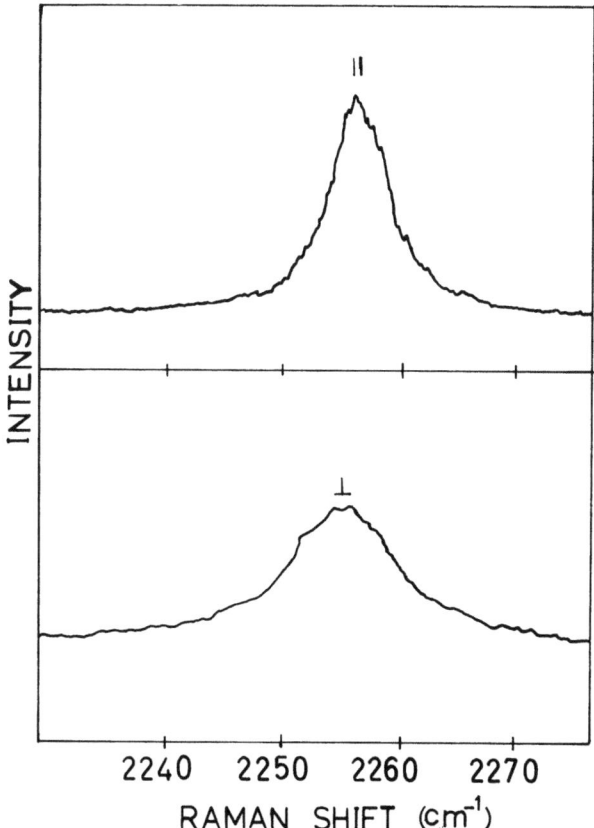

Fig. 1. The two components of the 2256 cm^{-1} line of $CDCl_3$: ∥ parallel geometry ⊥ perpendicular geometry.

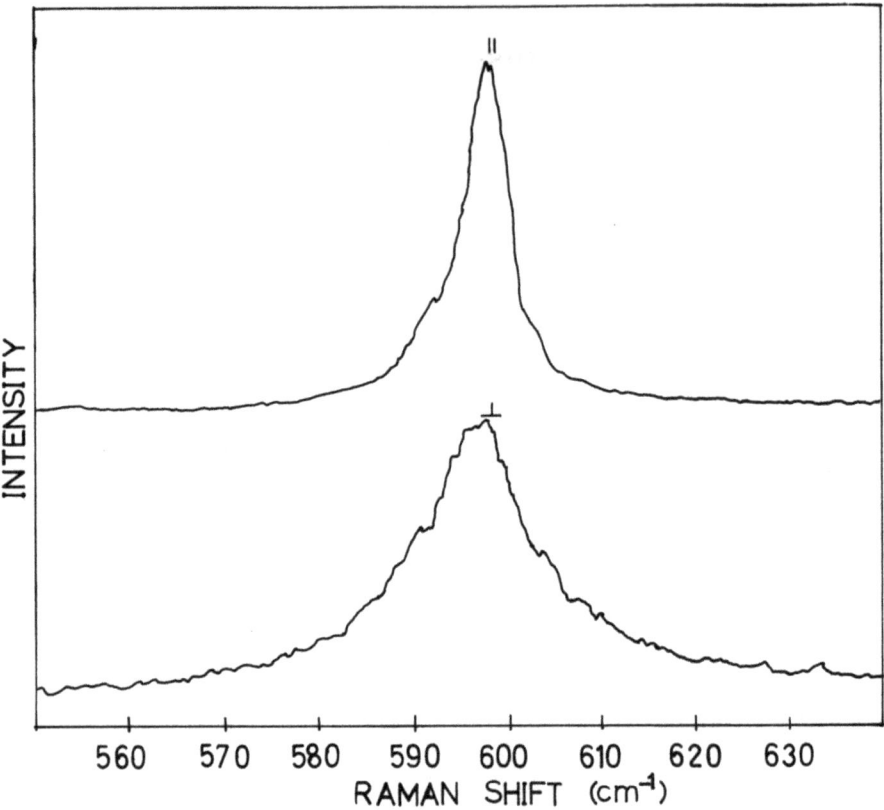

Fig. 2. The two components of the C–Br stretching band of CH_3Br: ∥ parallel geometry; ⊥ perpendicular geometry.

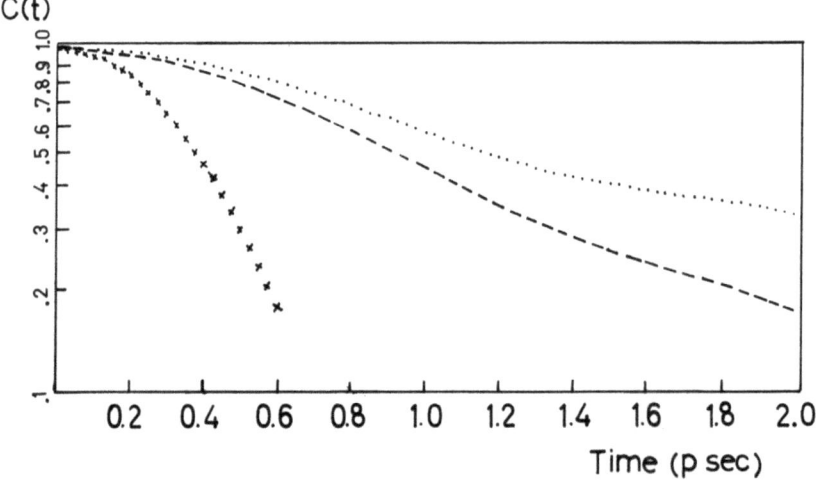

Fig. 3. $CDCl_3$ correlation functions. ---- $G_\perp(t)$; the Fourier transform of the perpendicular component of the spectra; .... $G_R(t)$ The angular autocorrelation function of liquid $CDCl_3$; the angular autocorrelation function for free rotating $CDCl_3$.

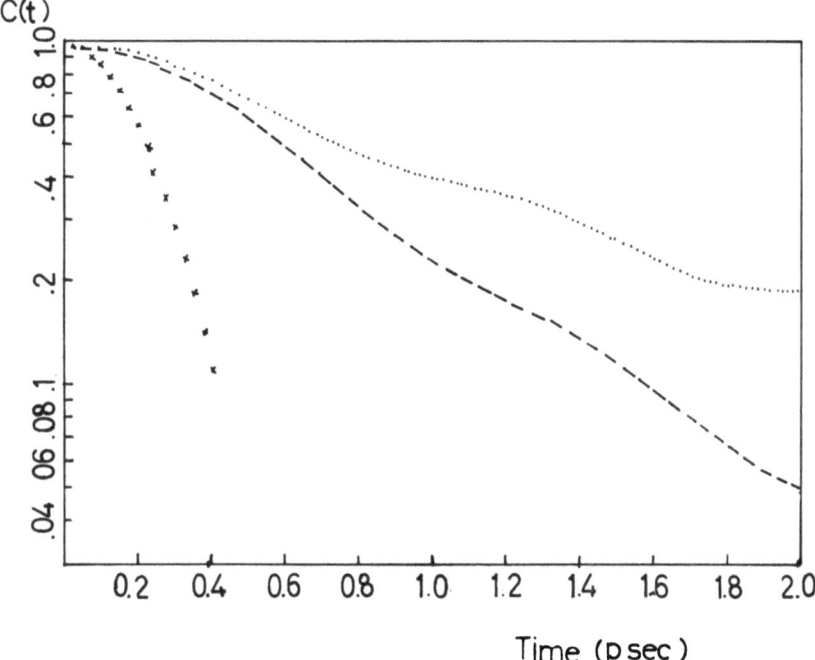

Fig. 4. CH$_3$Br correlation functions. ---- $G_\perp(t)$ the Fourier transform of the perpendicular component of the spectra; ..... $G(t)$ The angular autocorrelation function of liquid CH$_3$Br; The angular autocorrelation function of free rotating CH$_3$Br.

with recent results obtained by Bartoli and Litovitz [3]. Second moments were found to be smaller than those predicted for pure rotational Raman bands in the classical limit. The last result may partly be explained by the important contribution of the vibrational relaxation to the band shape.

References

1. Bratos, S. and Marechal, E.: *Phys. Rev.* **A4**, 1078 (1971).
2. Hogenboom, D. L., Oriely, D. E., and Peterson, E. M.: *J. Chem. Phys.* **52**, 2793 (1970).
3. Bartoli, F. J. and Litovitz, T. A.: *J. Chem. Phys.* **56**, 404 and 413 (1972).
4. Rothschild, W. G.: *J. Chem. Phys.* **53**, 3265 (1970).

DISCUSSION

De Graaf: In méthylbromide are there any affects of the isotope effect in the bromine atoms, as there are participating in the deformation mode?

Laulicht: Isotope shifts exist of course in the C–Br stretching mode we measured. However, since the mass of the carbon is 12 and that of bromine is 79, the effect of the bromine isotopes on the reduced mass of this normal coordinate is very small and had been neglected.

Rothschild: You mentioned in your talk that you find an experimental value of the second moment which is too small. In such a case, if one wishes to play the game of 'jump angles', your data underestimate the jump angles. Possibly, your measurements did not go far enough into the wings of the band, the most reason for too small a second moment in the absence of strong intermolecular interactions. The same holds for intermolecular torques.

Laulicht: You may be right. But we believe that going farther into the wings will not change the situation, and the second moment will yet be smaller than the theoretical one for pure rotation. Anyhow we are not speaking about 'jump angles' but on 'free rotation angles'. We still believe that they are much smaller than those one gets by infrared absorption measurements, where, by neglecting the decay of the vibrational correlation function, one certainly overestimates the time between collision and the free rotation angle.

Rahman: In dense systems where a particle is in strong interaction with many neighbours at every moment, how far is it justified to use the 'collision' terminology?

Laulicht: I think that since Gordon's models for M diffusions and J diffusion have been found to give good agreement with experiments, it means that the collision terminology is at least practical. I agree however that the concepts like 'collision times', 'free rotations between collisions', etc. are simplifications at least for polar liquids.

Frenkel: Referring to Dr Rahman's remark on the applicability of the collision concept in liquids, I would like to stress the following; collision models can be used to describe a wide variety of experimental data on liquids. This should however not be considered as an indication that short collisions do indeed occur at these densities. Rather the situation resembles the description of monoatomic liquids by hard spheres in the following respect: in both cases some of the essential characteristics of the molecule motion are accounted for, be it in a very crude way.

Laulicht: I agree with that. Whatever is the physical meaning of collision in liquids, I think that we all agree that in some sense one may look on simple liquids as dense gases, and in the case of 'inert' liquids where no indication exists for a large amount of polymerization, one may speak with a considerable amount of confidence about single molecular motion, which implies automatically concepts as 'the mean time between collisions', 'the mean free rotation angle', etc.

RAMAN SPECTROSCOPIC STUDIES ON TEMPERATURE DEPENDENCE OF REORIENTATIONAL MOTIONS OF CH_3J AND CD_3J MOLECULES

REINER ARNDT, R. MOORMANN, and A. SCHÄFFER

Institut für Physikalische Chemie der Technischen Universität Braunschweig, Braunschweig F.R.G.

Abstract. We describe the experimental investigations of the molecular motions of CH_3J and CD_3J molecules about the molecular x, y and z axis in the liquid phase. By comparing Raman and Rayleigh correlation times one finds a kind of collective reorientation about the x, y-axis. As the x, y correlation time is quite long the correlation function of the v_4 vibration is simplified to a sum of two terms describing the reorientation about the z-axis. Following the Abragam formalism we separate the two terms and explain the temperaturedependent short and long time behaviour of the correlation functions and find an inertial effect on this quasi free rotation.

Résumé. Nous décrirons la recherche expérimentale du mouvement moléculaire de CH_3J et de CD_3J autour d'essieu moléculaire x, y et z dans la phase liquide. En comparant les temps de corrélation Raman avec Rayleigh on trouve une sorte de réorientation collective autour d'essieu x, y. Comme le temps de corrélation x, y est assez long la fonction de corrélation de la vibration v_4 est simplifiée en une somme de deux termes qui décrivent la réorientation autour d'essiau z. En suivant le formalisme d'Abragam nous séparons les deux termes et expliquons la conduite dépendant de la température pour les temps brefs et longs de la fonction de corrélation et nous trouvons un effet inertiel sur cette rotation quasi libre.

1. Introduction

Following the theoretical predictions of Peticolas and Nafie [1] it is very difficult to measure and calculate the different correlation times of a rotating molecule except of those of very high symmetry. For our researches we chose the small molecules CH_3J and CD_3J of the symmetry point group C_{3v} as caused by the symmetry one only finds two correlation times τ_z and $\tau_{x,y}$ which are quite different because the ratio of the two inertial moments is 1:20. $\tau_{x,y}$ is found by measuring the depolarized part of Rayleigh scattering or of a total symmetric vibration. The reorientation about the z-axis one can calculate after some simplifications of the complicated correlation function of the v_4 vibration.

2. Theory

For the depolarized Rayleigh scattering Bartoli and Litovitz [2] found using the polarisation theory the following classical correlation function:

$$C_\gamma^{Ray}(t) = \langle \text{Tr} \, \boldsymbol{\beta}_0^A(t) \cdot \boldsymbol{\beta}_0^A(0) \rangle + \text{interference term}, \tag{1}$$

where the second term the interference term of Rayleigh scattering is non-zero if there exists an angular correlation between the molecules.

For the depolarized Raman scattering results

$$C_\gamma^{Ram}(t) = \langle \text{Tr} \, \boldsymbol{\beta}^{vA}(t) \boldsymbol{\beta}^{vA}(0) \rangle \langle q^{vA}(t) q^{vA}(0) \rangle + \text{interference term} \tag{2}$$

and the second term called the interference term of Raman scattering vanishes if there is no angular correlation or if the phases of the normal vibrations are independent, i.e.:

$$\langle q^{vA}(t) q^{vA}(0) \rangle = 0. \tag{3}$$

The intensity of scattered light is:

$$I_\gamma(\omega) = \int_0^\infty \exp(-i\omega t) C_\gamma(t) \, dt. \tag{4}$$

The broadenings of the depolarized Rayleigh wing and of the total symmetric $A 1$ vibration of C_{3v} are only caused by the reorientation of the molecular z-axis and one can compare the results of those measurements.

For the v_4 vibration of the symmetry race E one can write [1]:

$$C(t) = [1/(a^2 + b^2)] \langle a^2 [u^x(0) u^x(t)]^2 + b^2 \{[u^x(0) u^x(t) u^z(0) u^z(t)] + [u^x(0) u^z(t)]^2\} - 2ab [u^x(0) u^y(t)][u^z(0) u^y(t)] \rangle. \tag{5}$$

$u^x(0) u^x(t)$ is the scalar product of the unit vectors which are fixed to the x-axis of the molecule.

The reorientation correlation function for a classical free rotation can be written [3]:

$$C(t) = \langle \exp(i\omega t) \rangle \tag{6}$$

and for even functions:

$$C(t) = \langle \cos \omega t \rangle. \tag{7}$$

With ω as the Gaussian distributed angular velocity one finds:

$$C(t) = \exp - \frac{\langle \omega^2 \rangle}{2} t^2 \tag{8}$$

$$\langle \omega^2 \rangle = \frac{k \cdot T}{I}. \tag{9}$$

If the ω_i are stochastically changing but fullfilling the stationarity condition:

$$C(t) = \exp - \left[\left\langle \left(\int \frac{\omega(t)}{2} dt \right)^2 \right\rangle \right]. \tag{10}$$

For a Gaussian distribution results:

$$C(t) = \exp - \left[\langle \omega^2 \rangle \int_0^t (t - t') g_\omega(t') \, dt' \right] \tag{11}$$

with $g_\omega(t)$ as an exponential angular velocity correlation function with the correlation time $t_{c\omega}$:

$$g_\omega(t) = \exp - t/t_{c\omega}. \tag{12}$$

Herefrom follows:

$$C(t) = \exp - \{\langle \omega^2 \rangle [t_{c\omega} t + t_{c\omega}^2 (\exp(-t/t_{c\omega}) - 1)]\}. \tag{13}$$

On the assumption that:

$$t_{c\omega} \gg (\langle \omega^2 \rangle)^{-1/2} \qquad t_{c\omega} \gg t \tag{14}$$

results Equation (8) which predicts a Gaussian correlation function for free rotation.
If:

$$t_{c\omega} \ll (\langle \omega^2 \rangle)^{-1/2} \qquad t_{c\omega} \ll t \tag{15}$$

one gets for (13) an exponential function:

$$C(t) = \exp - [\langle \omega^2 \rangle t t_{c\omega}] = \exp - t/\tau. \tag{16}$$

In the case of non free rotation one should find a $C(t)$ with gaussian beginning which decays exponentially at long times.

3. Experiments and Interpretation

We measured the temperature dependent band profiles of the depolarized Rayleigh wing (90°-view) of the v_4 Raman band of CH_3J (90°-view) and CD_3J (0°-view) and the band width of the v_3 band of CH_3J (90°-view) from -60°C up to 30°C. As the profiles are symmetric we only measured one half of the bands. A Cary 81 with an 53 A Argon laser at 514.5 nm was used. The temperature was constant of ± 0.3°C and the slit width were for the Rayleigh wing 0.5 cm^{-1} the v_3 band 1.5 cm^{-1} and the v_4 bands 4 cm^{-1}. Avoiding the truncation effect we recorded the bands up to seven times of the half width and stored them on a punch tape. By numerical Fourier transform of the band with a Fortran-IV-Program and deconvolution from the apperatus profile we got the normalized time correlation functions $\bar{C}(t)$ which were plotted semilogarithmically. On the assumption that the band has Lorentz profile we calculated the correlation times for the v_3 band although there were some difficulties caused by the intrinsic line width and the existence of a hot band. For the destination of the intrinsic line width of the v_4 vibration the Rakov method [4] cannot be used [5]. As one finds in the crystal only a weak site group splitting of 2 cm^{-1} this leads to an intrinsic line width which can be neglected.

4. Results and Discussions

Interpreting the measurements of the depolarized part of the v_3 Raman band of CH_3J we used the formula:

$$\tau = 1/(2\pi c \Delta \omega_{1,2}) \tag{17}$$

and calculated the correlation times [6]:

T	$\Delta \omega_{1,2}$	τ
$-30\,°C$	$2.20\ cm^{-1}$	2.42 ps
$0\,°C$	$3.50\ cm^{-1}$	1.51 ps
$30\,°C$	$4.09\ cm^{-1}$	1.296 ps

In Figure 1 one sees the logarithmic time correlation functions $\ln \bar{C}(t)$ of the depolarized Rayleigh wing at $-54.8\,°C$ and $30.4\,°C$.

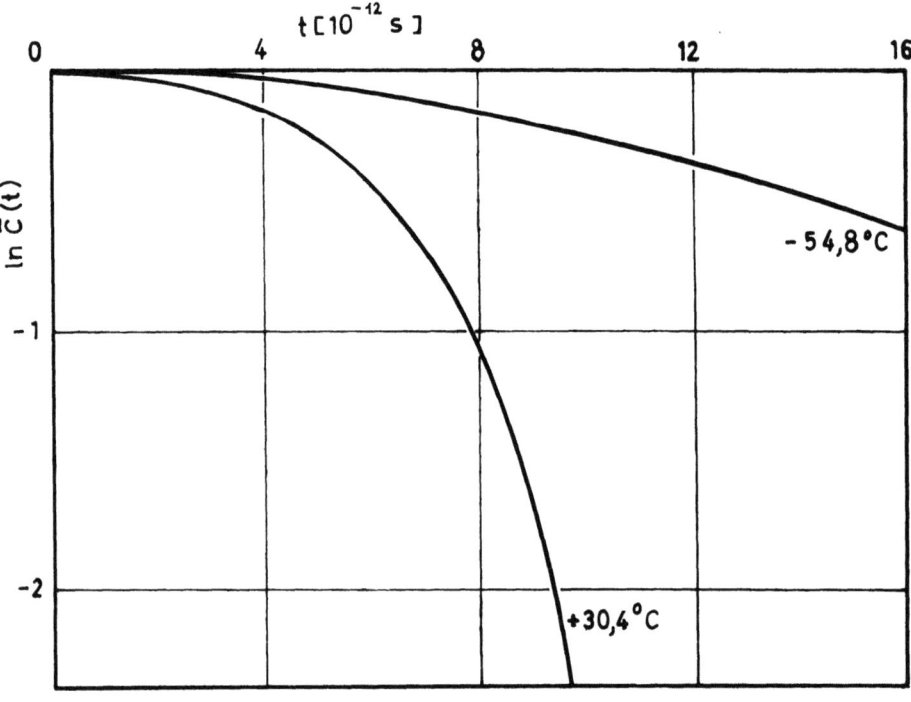

Fig. 1.

For the higher temperature on finds τ on the point where $\ln \bar{C}(t)$ has decayed to e^{-1} as $\tau = 7.9$ ps while the time at low temperature is not very reliable as the resolution of the apperature doesn't allow to measure such long correlation times.

By comparing τ_{Ram} with τ_{Ray} at $30\,°C$ one finds for the reorientation of the z-axis:

$$\tau_{Ray} > \tau_{Ram}$$

while $\tau_{Ram} = 1.29$ ps is rather long. This is caused by the high inertial moment of this rotation. We interpret the longer τ_{Ray} on the assumption that here exists a kind of collective reorientation by orientational exchange [7] because the volume needed for the rotation is larger than the free volume. Therefore exists a non zero interference term of Rayleigh scattering which means that the molecule rotates some times before it 'forgets the remembering' of the averaged orientation of its neighbours [7]. As the band is not broadened by resonant vibration transition [6] the interference term of Raman scattering is zero as $\langle q^{vA}(t) q^{vB}(0)\rangle = 0$ and one measures the self correlation time. If the total symmetric vibration has a non zero interference term one gets the self correlation time from neutron diffraction measurements.

In Figures 2 and 3 one sees the correlation functions dependent on temperature of the v_4 vibration of CH_3J and CD_3J.

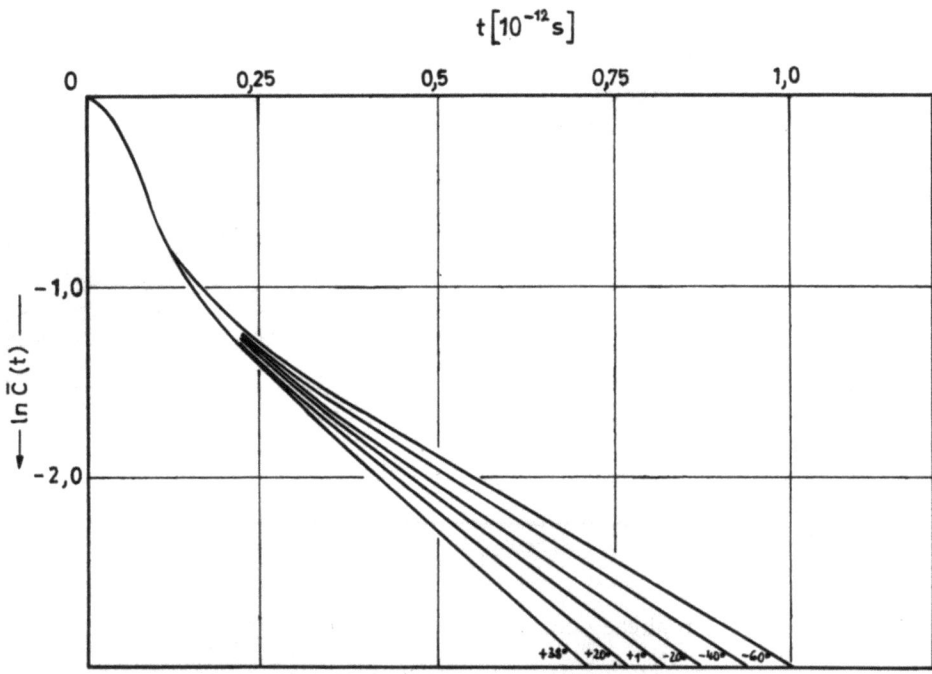

Fig. 2.

All curves show the same behaviour starting like parables and nearly independent on temperature and they have a linear shape at long times being $\sim 1/T$.

As the reorientation about the x, y-axis is rather slow we calculate in the first approximation from 0–1 ps with $\langle u^z(t) u^z(0)\rangle \approx 1$ which simplifies formula (5):

$$C(t) = A \langle u^x(t)^2 u^x(0)^2\rangle + B \langle u^x(t) u^x(0)\rangle \qquad (18)$$

$$A = \frac{a^2}{a^2 + b^2} \qquad B = \frac{b^2}{a^2 + b^2}.$$

The first term is a tensor correlation function $C_2(t)$ and the second term is a vector correlation function $C_1(t)$. Each of these two functions one can divide in a short and a long time part. As one gets for the reorientation velocities:

$$\langle \omega_2^2 \rangle = \langle (2\omega_1)^2 \rangle \qquad (19)$$

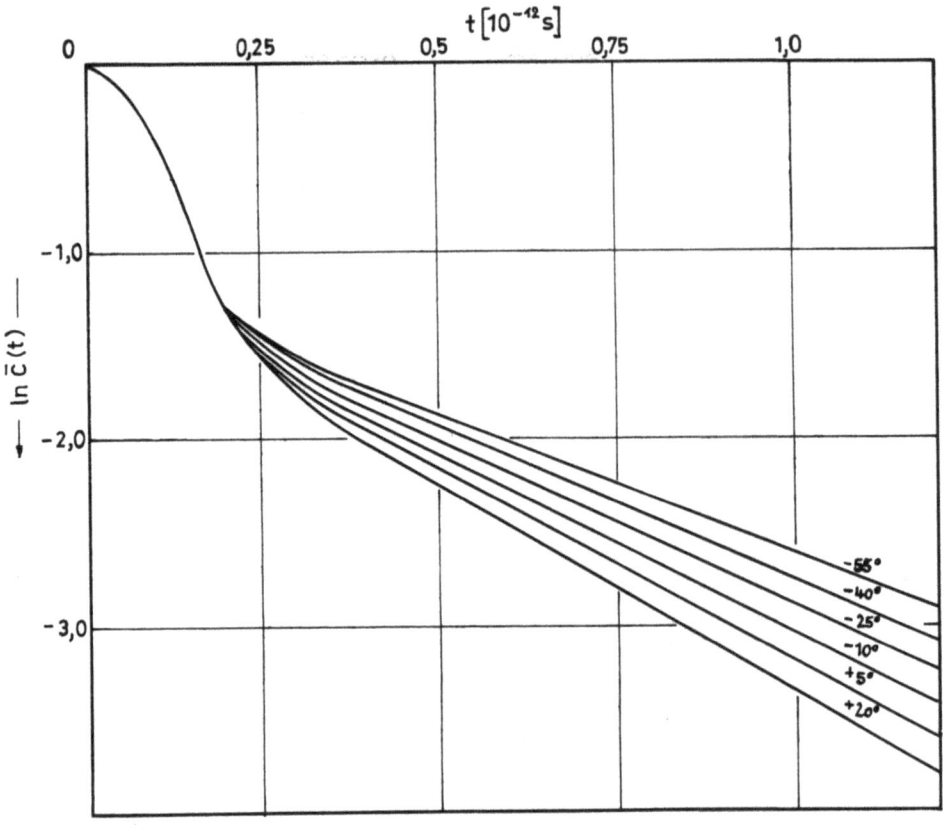

Fig. 3.

at long times $C_2(t)$ has already decayed and only $C_1(t)$ exists.

$$C_1(t) = B \exp[-\langle\omega_1^2\rangle t_{c\omega} t] = B \exp - t/\tau_1. \tag{20}$$

By interpolation to $t=0$ results $B=0.420$ for CH_3J and by normalizing $\ln \bar{C}_1(0)=0$ one can determine $\tau_1(T)$ measuring the slopes of the straight lines. For τ_1 at 20°C of CD_3I we calculate 0.46 ps which is in good agreement with the IR results of Crawford [8] $\tau_1 = 0.45$ ps. For both substances one gets $\tau_1 \sim 1/T$ which means that $t_{c\omega}$ is nearly independent on T as $\langle\omega_1^2\rangle \sim 1/T$. Comparing the results of CH_3J and CD_3J one can examine the influence of the inertial moment on the rotation and the theory predicts [3].

$$\tau_1^D/\tau_1^H = 1.97 \, \tau_{c\omega}^D(\Theta)/\tau_{c\omega}^H(\Theta) \tag{21}$$

and one finds quite a good agreement with 1.8.

After a transition area at short times the $\bar{C}(t)$ consist of a sum of two gaussian functions nearly independent on T declining as $\ln\bar{C}(t)$ like parables.

$$C(t) = A \exp[-\tfrac{1}{2}\langle\omega_2^2\rangle t^2] + B \exp[-\tfrac{1}{2}\langle\omega_1^2\rangle t^2]. \tag{22}$$

In the case of CH_3J $\ln \bar{C}(t)$ reaches e^{-1} after 0.13 ps and for CD_3J after 0.17 ps which is signifying quasi free rotation as the greatest decay of correlation takes place in the Gaussian region of $\bar{C}(t)$.

Near $t=0$ one can write for (22):

$$C(t) \underset{t\to 0}{=} \exp -\left[\frac{(A\langle\omega_2^2\rangle + B\langle\omega_1^2\rangle) t^2}{2}\right] = \exp\left[-\tfrac{1}{2}\langle\omega_{1/2}^2\rangle t^2\right]. \qquad (22a)$$

By calculating one gets for CH_3J at $-40\,°C$:

$$\langle\omega_{1/2}^2\rangle = 162 \times 10^{24} \text{ rad}^2 \text{ s}^{-2}.$$

With formula (22a) one can calculate A and B using the starting slopes of the functions and in good agreement with the long time result one finds $B=0.41$. The influence of the inertial moment on the rotation is predicted [3] as:

$$\Delta\omega_{1/2}^H/\Delta\omega_{1/2}^D = \sqrt{\Theta_H^z/\Theta_D^z} = 0.71 \qquad (23)$$

and from the experiment results 0.65. The rather good agreement between theory and experiment and the consistence of the results for B seem to justify the simplifications on the correlation function of the v_4 vibration.

Acknowledgements

We wish to thank Dr G. Döge for the helpful discussions. The work was supported by the DFG.

References

1. Nafie, L. A. and Peticolas, W. L.: *J. Chem. Phys.* **57**, 3145 (1972).
2. Bartoli, F. J. and Litovitz, T. A.: *J. Chem. Phys.* **56**, 413 (1972).
3. Abragam, A.: *The Principles of Nuclear Magnetism*, Oxford, 1961.
4. Rakov, A. V.: *Optika i Spektrosk.* **7**, 202 (1959). *Tr. Fiz. Inst. Akad. Nauk SSSR* **27**, 111 (1964).
5. Döge, G.: *Z. Naturforsch.* **28a**, 919 (1973).
6. Döge, G.: *Lecture on the Chemiedozententagung*, Heidelberg, 1972.
7. Egelstaff, P.: *J. Chem. Phys.* **53**, 2590 (1970).
8. Crawford, B.: *J. Phys. Chem.* **73**, 4040 (1969).

DISCUSSION

Yarwood: Have you fitted your correlation times (for reorientation) to an Arrhenius type relationship in order to calculate the so-called 'Activation energy' or 'Barrier' for rotational diffusion? The reason I ask this is because we have found that the rotational diffusion coefficients for acetonitrile are much smaller (pure liquid) than those obtained for methyl iodide by Crawford *et al.** Our data, obtained from the infrared spectrum, are:

$$\left. \begin{array}{l} D_x = 1.20 \pm 0.1 \text{ (ps)}^{-1} \\ D_z = 0.37 \pm 0.02 \text{ (ps)}^{-1} \end{array} \right\} \text{ for acetonitrile at } 25\,°C$$

compared with $\left. \begin{array}{l} D_x = 1.97 \text{ (ps)}^{-1} \\ D_z = 0.53 \text{ (ps)}^{-1} \end{array} \right\}$ for methyl iodide at room temperature

* Crawford, B. L., Favelukes, E. E., and Clifford, A. A.: *J. Phys. Chem.* **72**, 962 (1968).

We believe the difference between these results may be due to larger intermolecular forces between the acetonitrile molecules. It would therefore be interesting to compare the Arrhenius energy parameter inherent in your data with the values of 0.2 and 1.0 kcal mole^{-1} obtained for acetonitrile from the temperature dependence of D_x and D_z respectively (x is the direction of the C_3 axis).

Arndt: No, because for this quasi free rotation there doesn't exist a potential well and therefore it is useless to calculate an activation energy. The observed temperature dependence of $1/\tau_1$ is nearly described by the preexponential factor of $T \cdot e^{-\Delta H/RT}$ It is possible formally to calculate an activation energy neglecting the preexponential factor but that has no physical meaning.

Janik: Was your statement about the quasi-free rotation in CH$_3$I based upon the small value of τ or upon a fit to a free rotator correlation function?

Arndt: It is well known that the free rotation correlation function is Gaussian. As most of the decay of our measured correlation functions to e^{-1} takes place in the Gaussian area we conclude that the reorientation is quasi free.

RAMAN SPECTROSCOPIC OBSERVATION OF COLLECTIVE EXCITATION OF MOLECULAR VIBRATIONS IN LIQUID METHYL IODIDE

G. DÖGE

Institut für Physikalische Chemie der Technischen Universität, Braunschweig,
3300 Braunschweig Hans Sommerstr. 10, F.R.G.

Abstract. The ν_2 Raman band of CH_3J is chosen as an example to show how coupling of intramolecular vibrations by transition dipole-transition dipole interaction can influence the profile of the band in the liquid phase. The resulting contribution to the intrinsic line width is temperature dependent. In some cases it aggravates the determination of the reorientational part of the band width. On the other hand it may give some information about the structure of pure liquids and liquid mixtures.

Résumé. Par l'exemple de la bande Raman ν_2 de CH_3J est démontré de quelle façon l'accouplement des vibrations intramoléculaires par la 'transition dipole – transition dipole interaction peut influencer le profil des bandes de vibration en phase liquide. La contribution résultante pour la largeur de la ligne intérieure dépend de la température. Elle rend ainsi plus difficile la recherche des fonctions de corrélation orientationnelles. En cas contraire sa recherche permet des rapports certains pour la structure du liquide pur et surtout pour des mélanges liquides.

1. General Considerations

It is the aim of this paper to show that collective excitation of intramolecular vibrations, caused by transition dipole – transition dipole interaction, in liquids may give remarkable contributions to the intrinsic line width and that this has some consequences even on the problem of reorientational broadening.

The intrinsic part of the shape of a vibration band is caused by molecular interactions. In the pair interaction approximation the Hamiltonian for the whole system is

$$H = H_0 + V = \sum_i H_{0i} + \sum_{j>i} V_{ij}. \tag{1}$$

It is convenient to use the following model. One considers the excitation of a single molecule i, which is exposed to the interactions with its neighbours. The corresponding perturbation term is ${}^\alpha V_i = \sum_j {}^\alpha V_{ij}$. It determines the frequency shift ${}^\alpha \omega_i$ relative to the unperturbed αth vibrating frequency ${}^\alpha \omega_0$. The pair potentials V_{ij} are to expand in terms of the normal coordinates ${}^\alpha q_i$ and ${}^\alpha q_j$. If the first order perturbation treatment is applicable, the frequency corrections are:

$${}^\alpha \omega_i = \sum_j {}^\alpha \omega_{ij} \tag{2a}$$

$${}^\alpha \omega_{ij} = \Delta E_{ij}/\hbar = (\Delta \mathcal{D}_{ij} \pm \mathcal{E}_{ij})/\hbar = V_{ij}'' \frac{1}{2\gamma} \tag{2b}$$

$$\mathscr{E}_{ij} = \langle {}^{\alpha}\psi_i^+ \, {}^{\alpha}\psi_j^0 | \, V_{ij} \, | {}^{\alpha}\psi_i^0 \, {}^{\alpha}\psi_j^+ \rangle = \frac{\partial^2 V_{ij}}{\partial^{\alpha} q_i \partial^{\alpha} q_j} \cdot \frac{\hbar}{4\pi\omega_0} = V_{ij}'' \frac{1}{2\gamma} \tag{2c}$$

$$\Delta \mathscr{D}_{ij} = \langle {}^{\alpha}\psi_i^+ \, {}^{\alpha}\psi_j^0 | \, V_{ij} \, | {}^{\alpha}\psi_i^+ \, {}^{\alpha}\psi_j^0 \rangle - \langle {}^{\alpha}\psi_i^0 \, {}^{\alpha}\psi_j^0 | \, V_{ij} \, | {}^{\alpha}\psi_i^0 \, {}^{\alpha}\psi_j^0 \rangle = V_{ii}'' \frac{1}{2\gamma}. \tag{2d}$$

The term, which is of interest here, is $\mathscr{E}_{ij}$. It describes the coupling of the vibrations α_i and α_j. Therefore a perceptible contribution to ${}^{\alpha}\omega_i$ is to expect only if i and j are identical molecules.

In a disordered liquid phase each reference molecule i sees another perturbation ${}^{\alpha}V_i$. This leads to a frequency distribution $I(\omega)$ about ${}^{\alpha}\omega_0$. The Fourier transform is the relaxation function

$$\Phi_v(t) = \langle \exp(i\omega t) \rangle \quad \text{or} \quad \left\langle \exp\left(i \int \omega(t)\, dt\right) \right\rangle. \tag{3a+b}$$

Now it is necessary to make some considerations about the nature of V_{ij}. For solid CH_3J Hexter [1] found that the coupling mechanism is mainly due to transition dipole-transition dipole interaction. According to the comparable intermolecular distances this type of coupling should be present in the liquid phase, too.

(2c) then becomes:

$$\mathscr{E}_{ij} = \frac{M_i M_j}{R_{ij}^3} K_{ij}; \qquad M_i = M_j = \frac{\partial \mu}{\partial^{\alpha} q} \cdot \frac{1}{2\gamma} \tag{4}$$

$$K_{ij} = 2 \cos {}_i^z\theta \cos {}_j^z\theta - \cos {}_i^x\theta \cos {}_j^x\theta - \cos {}_i^y\theta \cos {}_j^y\theta.$$

if point dipoles are assumed. R_{ij} = intermolecular distance, ${}_i^z\theta$ = angle between the direction of the transition moment of i and the z-axis of a coordinate system in which R_{ij} lies in the z-direction.

The different signs of $\mathscr{E}_{ij}$ in (2b) predict a splitting of the molecular vibration frequency if we consider the excitation of a pair of molecules. The two splitting components correspond in-phase and anti-phase vibrations of the two pair molecules. Other phase relations are not allowed if one excites the vibrations with light, the wavelength of which is large compared with the intermolecular distance. It was shown recently [4] that only the in-phase component appears in isotropic Raman scattering. In this case K_{ij} not only describes the mutual orientation of the transition moments but also of the corresponding molecular axes.

For pure liquids it is allowed to make the assumption that all molecules have the same coordination number z. In fact this is not true, but it does not lead to remarkable mistakes. So we can get $I(\omega_i)$ by a z-fold convolution of $I(\omega_{ij})$. $I(\omega_{ij})$ is the frequency distribution if all molecules have only one neighbour in the first coordination shell. It depends on the distributions of R_{ij} and K_{ij}. If one assumes that the relative orientation is completely random, the distribution function $n(K_{ij})$ is to obtain. This function forms the main part of the width of the distribution $n(\mathscr{E}_{ij})$ or $I(\omega_{ij})$.

Its moments are very close to those of a Gaussian function. $I(\omega_j)$ becomes still more similar to a Gaussian function if one takes into consideration the influence of the distribution of R_{ij} within the first coordination shell. A somewhat detailed estimation [4] indicates that we can write:

$$I(\omega_{ij}) = \exp\left(-\frac{\omega_{ij}^2}{2\langle\omega_{ij}^2\rangle}\right) \quad \text{with} \quad \langle\omega_{ij}^2\rangle = (0.885\, M_i^2/R_{ij}^3, \hbar)^2 \qquad (5a)$$

and

$$I(\omega_i) = \exp\left(-\frac{\omega_i^2}{2z\langle\omega_{ij}^2\rangle}\right) \qquad (5b)$$

or with consideration of the more distant coordination shells:

$$I(\omega) = \exp\left(-\frac{3\omega^2}{\pi z\langle\omega_{ij}^2\rangle}\right) = \exp\left(-\frac{\omega^2}{2\langle\omega^2\rangle}\right). \qquad (5c)$$

These equations are only valid if molecular motions can be neglected. Introducing them, one should go back to the relaxation function.

The Gaussian type of (5) which is mainly caused by $n(K)$ and not introduced by central limit theorem, gives $\Phi_v(t)$ in the following form:

$$\Phi_v(t) = \exp(-\tfrac{1}{2}\langle\omega^2\rangle t^2) \quad \text{or} \quad \exp\left\{-\tfrac{1}{2}\left\langle\left(\int_0^t \omega(t)\,dt\right)^2\right\rangle\right\}. \qquad (6a, b)$$

The use of 6a or 6b depends on the validity of

$$t_c \gg (\langle\omega^2\rangle)^{-1/2} \qquad (7)$$

if t_c is the correlation time of $\langle\omega(t)\omega(0)\rangle$. One can generally equate t_c with $_k t_c$, which is the correlation time of $\langle K(t)K(0)\rangle$. If the reorientations occur independently, we can write:

$$_k t_c = 0.5\, _\rightarrow t_c. \qquad (8)$$

$_\rightarrow t_c$ is the correlation time for a vector lying in the direction of $\mathbf{M}_i$.

It can be obtained from IR-measurements. If the corresponding correlation functions are exponentials, one can write (6b) [2]:

$$\Phi_v(t) = \exp\{-\langle\omega^2\rangle[tt_c + t_c^2(\exp\{-t/t_c\} - 1)]\}. \qquad (9)$$

If $t \ll t_c$ (6c) passes over in (6a). Thus one obtains $\langle\omega^2\rangle$ by the second derivative of $\ln \Phi_v(t)$ at short times. On the other hand, if t is large enough, (9) simplifies to

$$\Phi_v(t) = \exp(-\langle\omega^2\rangle t_c t) \qquad (10)$$

and so t_c is determinable, too.

2. Experimental Results

Investigations were carried through by analysing the band shape of v_2, because experiments with diluted solutions in CD_3J indicate by a nearly complete narrowing that only vibration coupling broadens this band. Figure 1 and Figure 2 show the $\Phi_v(t)$-curves at a high and a low temperature. They are separated from the effect of the apparatus. The form of the experimental curves ——— is nearly the same as that given by (9) ------. The dotted lines show the $\Phi_v(t)$-curves for a 1:1 CH$_3$J–CD$_3$J

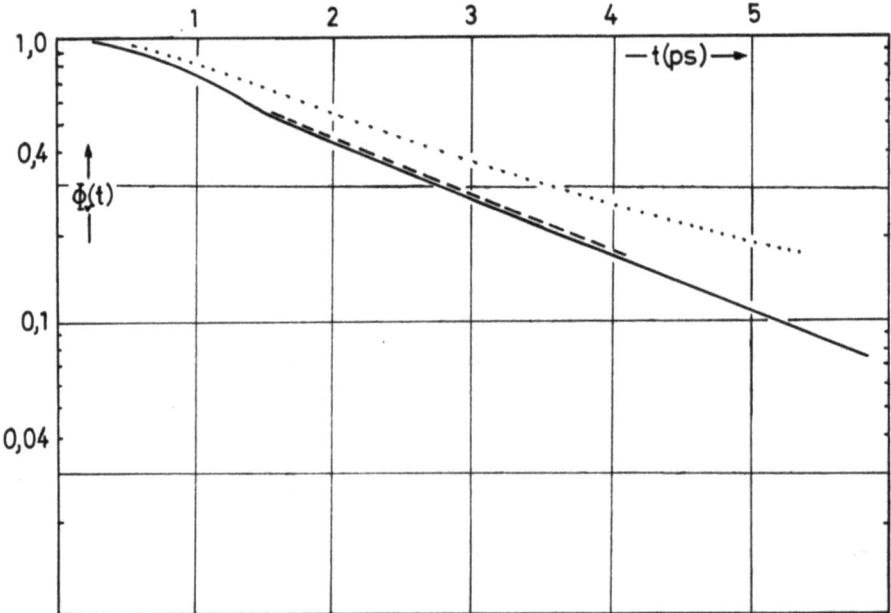

Fig. 1. $\Phi_v(t)$; CH₃J; + 30 °C.

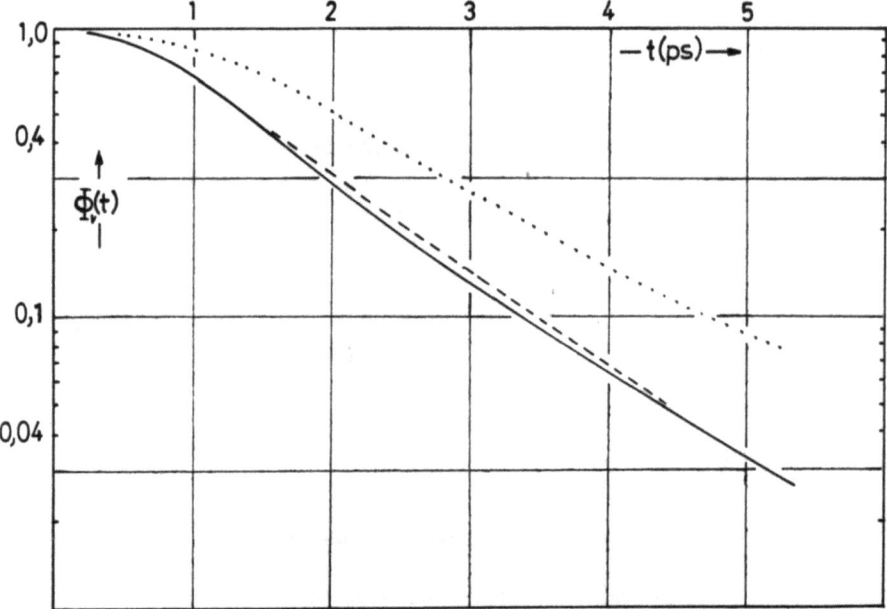

Fig. 2. $\Phi_v(t)$; CH₃J; − 70 °C.

mixture. At the higher temperature $\Phi_v(t)$ decays more slowly, which is a result of interaction perturbation by molecular reorientations. This means that $I(\omega)$ suffers a distinct motional narrowing. So the band width becomes smaller with increasing temperature. If one determines $\langle \omega^2 \rangle$ and t_c in the way described previously, one obtains:

Temperature	$\langle \omega^2 \rangle$	t_c
$-70\,°C$	0.77×10^{23} rad^2 s^{-2}	1.16×10^{-12} s
$+30\,°C$	0.67	0.63

Van Konynenburg and Steele [3] estimated t_c from orientational broadening of the IR band shape of v_3 at room temperature. Their result, 1.2×10^{-12} s, agrees with the given value for t_c at $30\,°C$ if one notes (8).

The relaxation functions for the CH_3J–CD_3J mixtures decrease more slowly because here the possibility for an excitation exchange is restrained. In the 1:1 mixtures one should expect that the slope in the Gaussian part of the curves (short times) reduces to $1/\sqrt{2}$ and in the exponential part to $1/2$ of the value of pure CH_3J. The experiments confirm this expectation. At the lower temperature the exponential part is not attained within the represented time range as a result of the larger t_c-value. Experiments were also carried through at $-40\,°C$, $-20\,°C$, $0\,°C$ and $+20\,°C$. The band width decreases monotonously with increasing temperature according to the circumstance that the decreasing t_c-value causes a smaller slope of $\Phi_v(t)$.

Summarizing one can state, that this part of the intrinsic line width depends strongly on temperature and concentration. This is also valid for contributions by transition dipole – permanent dipole interaction. Such a dipole contribution to ΔD_i is to expect if electrical or mechanical anharmonicity is present. The v_3-vibration of CH_3J is an example for such a behaviour [4, 5], which complicates the determination of orientation correlation functions. For if the intrinsic line width is temperature dependent, it is not allowed to use Rakov's method [6] to eliminate this part from the reorientational part. Moreover, as above-mentioned and previously shown [4] there are different selection rules for the two splitting components in a pair of molecules for isotropic and anisotropic Raman scattering. If the mutual orientations of the molecules are not random, the intrinsic part of the line shape then does not become equal for the two scattering components. So the usual method for separation of the reorientational part of the shape from the intrinsic part is not allowed, too. Finally one has to take into consideration that, if vibration coupling broadens the profile of the band, the phase relations between the interacting molecules may cause the appearance of interference terms in the orientational correlation function. For, if $\langle q_i(t) q_j(0) \rangle \neq 0$, the term $\langle \alpha_i(t) \alpha_j(0) \rangle$ may contribute to preserve correlation. α_i is the transition polarizability tensor of the molecule i.

On the other hand the dipole interaction part of the intrinsic line width can give some informations about the structure of the liquid. The band shape depends on the distribution of K_{ij}, and so it can indicate if there are favoured mutual orientations. In the case of liquid CH_3J one finds that such favoured orientations are not present. This result is not unexpected, if one compares the interaction of the permanent dipoles with kT.

Better statements are possible if one considers liquid mixtures. Equation (5b) indicates that the band width depends on the coordination number of molecules of the same species. The concentration dependence of the width gives information if the molecular distribution is statistical or if selfassociation or the opposite case is present in the mixture. Figure 3 shows an example, which does not refer to CH_3J. It shows the variation of $\Delta_{1/2}$ (=half width at half height) of the band of v_6 of thiophene in liquid mixtures with benzene and p-xylole. Nearly the whole width of the band of this

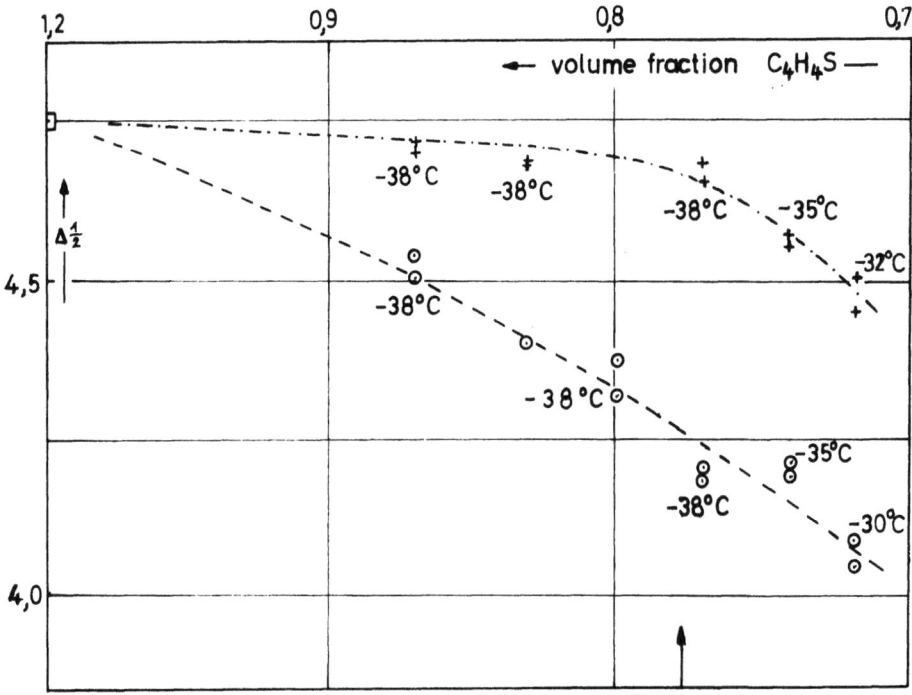

Fig. 3. Half band width of thiophene-v_6 in mixtures with benzene and with p-xylole-·-·-·-.

vibration is due to vibration coupling. One can assume that reorientations do not perturb the interactions, because the temperature is very low. One sees that the band width remains constant in a certain concentration range in the mixture with p-xylole indicating self association of the thiophene molecules. In the mixture with benzene we find a statistical distribution of the thiophene molecules. It seems possible that this behaviour corresponds with the fact that the thiophene-p-xylole mixture is a simple eutectic system (immiscibility in the solid phase), whilst thiophene-benzene shows full miscibility in the solid phase, too. The arrow in Figure 3 indicates the eutectic concentration of the thiophene-p-xylole system.

Acknowledgement

Thanks are due to the Deutsche Forschungsgemeinschaft for financial support.

References

1. Hexter, R. M.: *J. Chem. Phys.* **33**, 1833, (1960).
2. Abragam, A.: *The principles of Nuclear Magnitism*, Oxford, 1961, p. 432ff.
3. Van Konynenburg, P. and Steele, W. A.: *J. Chem. Phys.* **56**, 4776 (1972).
4. Döge, G.: *Z. Naturforsch.* **28a**, Nr. 7 (1973).
5. Döge, G.: *Z. Naturforsch.* **23a**, 1405 (1968).
6. Rakov, A. V.: *Optika i Spektrosk.* **7**, 202 (1959); *Tr. Fiz. Inst. Akad. Nauk SSSR* **27**, 111 (1964).

INTERMOLECULAR VIBRATIONAL RELAXATION IN LIQUIDS, MEASURED BY THE ISOTOPIC DILUTION METHOD

P. C. M. VAN WOERKOM, J. DE BLEYSER, and J. C. LEYTE

Gorlaeus Laboratoria, Afdeling Fysische Chemie III, Rijksuniversiteit Leiden, The Netherlands

As pointed out before, the isotopic dilution method is suitable for getting information on the contribution of intermolecular vibrational relaxation to IR-bandshapes [1].

Especially it is possible in this way to study the coupling between identical oscillators in the liquid.

Measurements on various liquids have shown that the magnitude of the isotopic dilution effect is strongly dependent on the integrated intensity of the investigated absorption bands.

Furthermore the experimental data show a large temperature dependence for the isotopic dilution effect. The measured effect increases with decreasing temperature.

A detailed report of the experimental results will be published shortly.

The measured isotopic dilution effect can be interpreted in terms of the transition dipole-transition dipole interaction between identical oscillators [1].

This is shown by expanding the quantummechanical autocorrelation function of the total electric moment of the system until second order in the transition dipole-transition dipole interaction [1].

The following expression for the autocorrelation function, obtained from an IR-bandshape, can then be derived [2],

$$G(t) = G^\circ(t) \exp\left[-\frac{2\pi N \left(\frac{\partial \mu}{\partial Q}\right)^4}{27 \omega^2 d^3} \{\tfrac{1}{2}t^2 + 2\tau_2(t + \tau_2(e^{-t/\tau_2} - 1))\}\right]. \quad (1)$$

Deriving this equation, it is assumed that the molecular reorientation may be described by the rotational diffusion model.

In Equation (1) $G(t)$ is the correlation function, obtained from an investigated absorption band in the pure liquid. $G^\circ(t)$ is the correlation function of the same absorption band, measured in infinite isotopic dilution.

The exponential function in Equation (1) describes the contribution of intermolecular vibrational relaxation, due to the transition dipole-transition dipole interaction, to the autocorrelation function of an IR-absorption band in the pure liquid.

In the expression for $G(t)$ N is the number of identical oscillators per cm^3, μ is the permanent dipole moment, Q the normal coordinate of the investigated vibration, ω the frequency (in rad s^{-1}) and d is the closest distance of approach between the oscillators in the liquid.

The most important result of the calculation is the prediction that the isotopic dilution effect strongly depends on the magnitude of the integrated intensity of an absorption band (by the factor $(\partial \mu/\partial Q)^4$ in Equation (1)).

Furthermore the derived equation shows a modulation of the intermolecular vibrational relaxation by the rotational motions of the liquid molecules. (by τ_2). In fact, this result means an indication of motional narrowing in the IR-spectrum.

In a forthcoming paper [2] it will be shown by a numerical calculation of Equation (1), that the derived expression predicts an isotopic dilution effect of the same order of magnitude as experimentally determined.

Finally it can be concluded that the experimental results as well as the theoretical calculations show that intermolecular vibrational relaxation is dependent on the reorientational motions of the molecules and therefore is strongly temperature dependent.

From this it follows, that determination of the intrinsic vibrational lineshape according to the Rakov method [3, 4] is impossible for intense IR-absorption bands.

For these absorption bands, the assumption, that vibrational relaxation is temperature independent, is incorrect.

References

1. Van Woerkom, P. C. M., De Bleyser, J., and Leyte, J. C.: *Chem. Phys. Letters* **20**, 592 (1973).
2. A theoretical treatment will be published shortly.
3. Rakov, A. V.: *Optika i Spektrosk.* **7**, 202 (1959).
4. Bartoli, F. J. and Litovitz, T. A.: *J. Chem. Phys.* **56**, 404 and 413 (1972).

MOUVEMENTS MOLECULAIRES DU SULFURE DE CARBONYLE A L'ETAT DISSOUS PAR SPECTROMETRIES RAMAN ET INFRAROUGE

J. P. PERCHARD, C. PERCHARD, et D. LEGAY

Laboratoire de Spectrochimie Moléculaire, Université de Paris VI, 8 rue Cuvier, Paris Ve, France

Dans l'étude des mouvements moléculaires à partir du profil des bandes Raman ou infrarouge, il convient dans un premier temps d'identifier la partie du profil d'origine réorientationnelle. Dans un seul cas, celui des bandes Raman polarisées, le problème peut être résolu de façon satisfaisante grâce à la connaissance simultanée des intensités isotropes et anisotropes. Pour les bandes infrarouges ou Raman dépolarisées, la méthode habituellement employée, à la suite de Rakov [1], consiste à distinguer dans le profil une partie dépendante de la température, associée aux mouvements moléculaires, et une partie indépendante de ce facteur, liée à la distribution des fréquences de vibration résultant des fluctuations du potentiel intermoléculaire.

Nous nous proposons dans cette communication de montrer, sur l'exemple d'une raie polarisée, que la méthode de Rakov est dans certains cas peu correcte et qu'en réalité les deux parties du profil sont fonction de la température. Nous considérons ici le cas de la vibration $v_1(\Sigma^+)$ du sulfure de carbonyle à l'état dissous (fraction molaire: 5 à 10%) dans quelques solvants à bas point de fusion. La bande Raman associée à cette vibration est fortement polarisée ($\varrho = 0,06$) de sorte que nous avons identifié la composante isotrope au spectre non analysé dont le profil varie de façon importante selon le solvant; comme la largeur à mi-hauteur de la bande varie de 0,5 cm^{-1} pour les solvants peu perturbateurs (carbures saturés, dérivés perfluorés) à 3–3,5 cm^{-1} pour les solvants actifs (SO_2, CH_2Cl_2), nous présentons les résultats relatifs à chacun des deux cas extrêmes: solution de sulfure de carbonyle dans l'isopentane et solution dans le chlorure de méthylène.

1. Influence de la température sur le profil des bandes Raman et infrarouge

La figure 1 montre l'évolution du profil Raman isotrope en fonction de la température dans le cas de la solution dans CH_2Cl_2. A température ordinaire on note la présence d'une bande chaude centrée vers 850 cm^{-1} ainsi qu'une bande de l'espèce ^{34}SCO vers 845 cm^{-1}; par suite, nous avons admis que la bande était symétrique et étudié la distribution d'intensité de la partie haute fréquence non perturbée. Par refroidissement la bande s'élargit de façon très notable et tend vers une distribution gaussienne presque parfaite sauf dans la partie lointaine des ailes. La figure 2 montre l'évolution quasi linéaire de la largeur isotrope symétrisée en fonction de la température ainsi que l'évolution opposée de la largeur anisotrope. En raison du point de fusion trop élevé du solvant, le point de convergence du profil des deux bandes n'a pu être atteint.

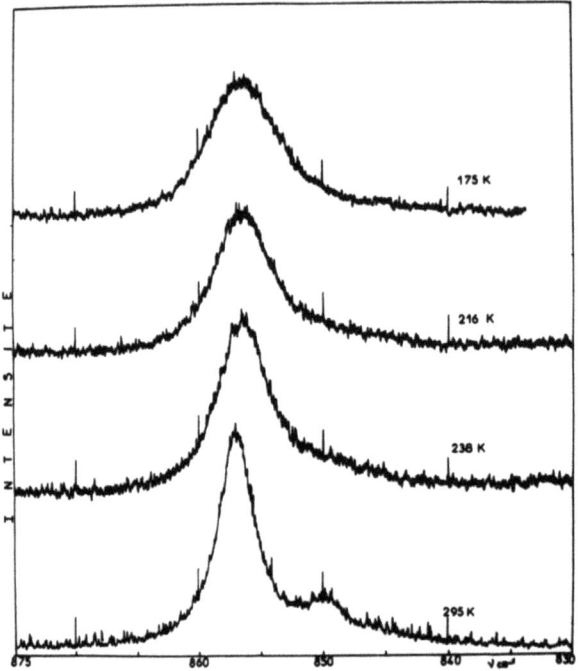

Fig. 1. SCO dans CH_2Cl_2 – Bande non analysée de la vibration ν_1 à 858 cm^{-1} en fonction de la température. Largeur de fente spectrale 0,5 cm^{-1}.

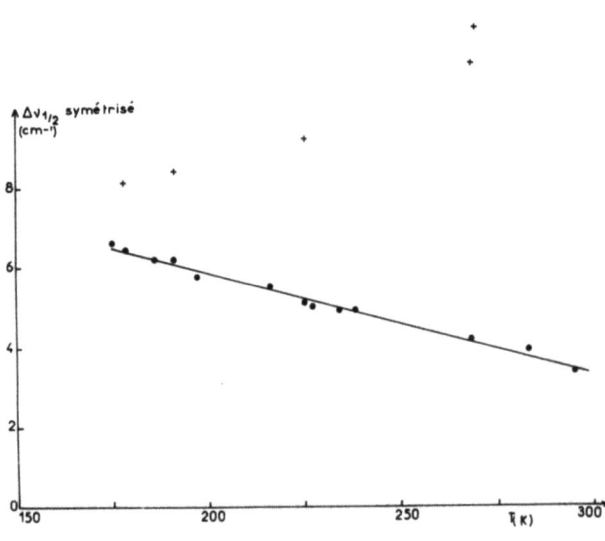

Fig. 2. SCO dans CH_2Cl_2 – largeurs symétrisées des parties isotropes (●) et anisotropes (+) de la bande à 858 cm^{-1}.

La figure 3 rassemble les résultats obtenus dans le cas de la solution dans l'isopentane. Ici encore le profil isotrope varie de façon notable avec la température; mais, bien qu'ayant atteint des températures proches du point de fusion de l'isopentane (140 K environ), nous n'avons pu obtenir une larguer résiduelle commune pour les raies Raman et infrarouge. En particulier cette largeur résiduelle qui semble être égale

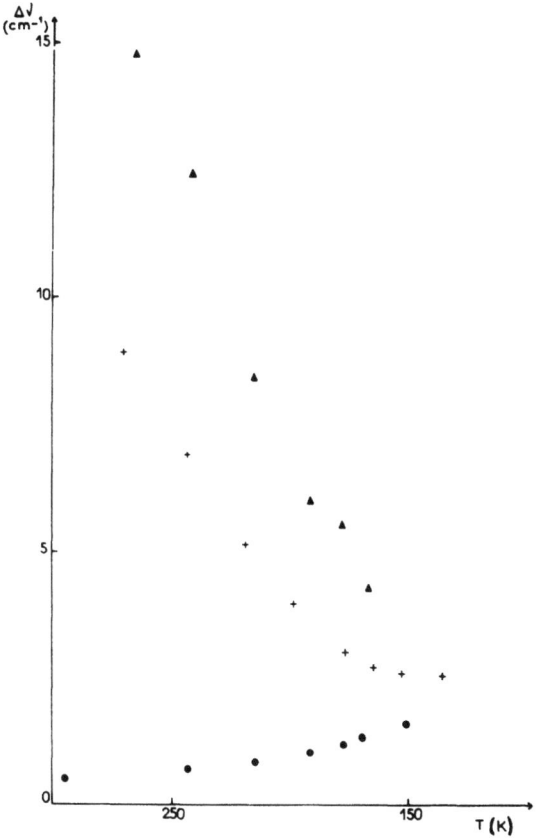

Fig. 3. SCO dans l'isopentane – largeurs symétrisées des bandes v_1 Raman (● isotrope, ▲ anisotrope) et infrarouge (+).

à 2,5 cm^{-1} en infrarouge ne correspond à la largeur Raman isotrope, contrairement à ce que suggère la théorie [2].

2. Etude des fonctions de corrélation et des temps de corrélation

La connaissance simultanée des profils Raman et infrarouge nous a permis de déterminer les fonctions de corrélation rotationnelle G_{1R} et G_{2R} ainsi que les temps de corrélation correspondants. Le rapport τ_1/τ_2 est de l'ordre de 1,5 pour les solutions dans l'isopentane et 2 pour les solutions dans le chlorure de méthylène. Il est possible d'en conclure que les réorientations moléculaires s'effectuent par des mouvements angulaires d'amplitude importante et que, pour cette molécule, on se trouve assez loin du modèle de diffusion rotationnelle très fréquemment utilisé.

Par ailleurs, une étude de ces temps de corrélation en fonction de la température a permis de déterminer l'énergie d'activation associée à la réorientation; dans le cas de l'isopentane, la valeur obtenue, tant à partir τ_{1R} que de τ_{2R} (figure 4), est de l'ordre de 1,4

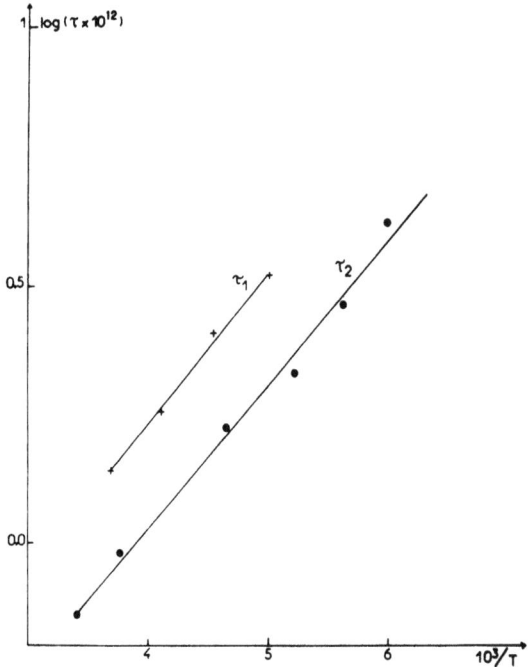

Fig. 4. SCO dans l'isopentane – évolution des temps de corrélation τ_{1R} et τ_{2R} en fonction de la température.

kcal mole^{-1}. Il est bon de rapprocher cette valeur de celle calculée selon la méthode de Rakov, à partir des résultats infrarouges, en prenant pour largeur vibrationnelle la valeur 2,5 cm^{-1}. La valeur alors obtenue est 2,1 kcal mole^{-1}, soit de moitié supérieure à la mesure correcte.

Pour terminer, nous nous proposons de comparer les fonctions de corrélation G_{1R} et G_{2R} en utilisant une théorie développée par Berne *et al.* [3] et appliquée avec succès dans le cas de la molécule d'iodure de méthyle par Constant [4]. Selon cette théorie, basée sur la notion de désordre maximum dans la structure du liquide, il est possible d'obtenir une représentation paramétrique de G_{1R} et G_{2R}, le paramètre étant fonction implicite du temps. Nous avons ainsi pu calculer G_{2R} à partir de G_{1R}; mais cette fonction calculée est assez différente de celle tirée des résultats Raman (figure 5). Cette divergence provient peut être du fait que la théorie ne s'applique correctement que dans le cas de la diffusion rotationnelle, cas précisément envisagé par Constant.

En conclusion, cette étude de la molécule de sulfure de carbonyle à l'état dissous a

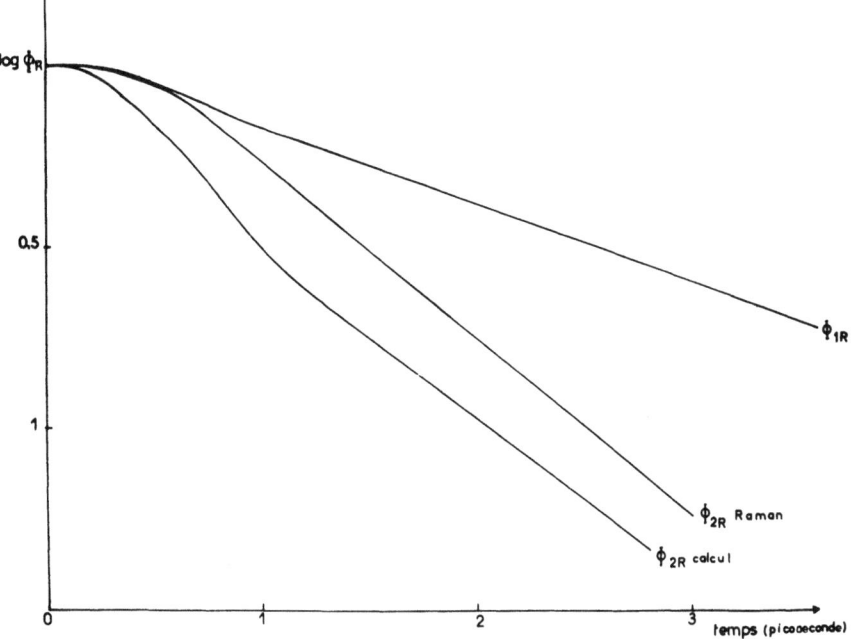

Fig. 5. SCO dans CH$_2$Cl$_2$ à 297 K – Comparaison entre G_{2R} expérimental et G_{2R} calculé à partir de G_{1R}.

d'une part permis de montrer le peu de précision des informations obtenues par la méthode de Rakov sur la partie réorientationnelle du profil d'une bande de vibration, et d'autre part constitue une des premières tentatives pour relier de façon quantitative les résultats Raman et infrarouge. Nos efforts ultérieurs tendront à développer ce second point.

Bibliographie

1. Rakov, A. V.: *Proc. P.N. Lebedev Phys. Inst.* **27**, 109 (1965).
2. Bratos, S. et Marechal, E.: *Phys. Rev.* **4**, 1078 (1971).
3. Berne, B. J., Pechukas, P., et Harp, G. D.: *J. Chem. Phys.* **49**, 3125 (1968).
4. Constant, M.: Thèse de troisième cycle, Lille, 1972.

DISCUSSION

Laulicht: I was impressed by the anomalous temperature dependence that you have found for the line width of the isotropic component of the v_1 line of SCO. Have you or someone else found such a decrease of the linewidth with increasing temperature in other compounds? It reminds me in some sense of the pressure narrowing of the Q branch of N$_2$ at high pressures found by Welsh's group at Canada. Do you have a theoretical explanation of this phenomenon?

Perchard: We observe a broadening of the isotropic component when the temperature decreases; the same observation has been done for SO$_2$ (Ouillon) and CH$_3$I (Constant, Wang). A theory, developed by Bratos and Levant, explains this effect in terms of increasing intermolecular potential when the density of the liquid increases (shorter intermolecular distances).

VIBRATIONAL AND REORIENTATIONAL RELAXATION OF HYDROGEN-BONDED SYSTEMS

B. BORŠTNIK, D. PUMPERNIK, and A. AŽMAN

Chemical Institute Boris Kidrič and University of Ljubljana, Ljubljana, Yugoslavia

Abstract. The molecular dynamics method was used to evaluate the reorientational and vibrational relaxation functions. The calculations were done at extreme dilution of the donor phase and at various temperatures, densities, and hydrogen bond strengths. The results show that in the case of strong hydrogen bonds the rotational relaxation is negligible while at weak bonds the rotational diffusion or even jump diffusion may influence the spectral bandshapes.

Résumé. Pour obtenir les fonctions de corrélation rotationnelles et vibrationnelles nous avons utilisé la méthode de dynamique moléculaire. Le calcul était fait sur une solution du donneur extrêmement dilué à differentes températures, densités et forces de la liaison hydrogène. Les résultats indiquent que dans les cas de liaison hydrogène forte la relaxation rotationnelle peut être négligée, alors que dans les cas de liaison faible les mouvements orientationnels influent sur les profils des bandes de vibrations.

The hydrogen-bonded systems are well known to the spectroscopists because of complicated vibrational band shapes which are difficult to analyse. As there exist the hydrogen bonds with different strengths there are also different theoretical approaches to the study of the line shapes. One can expect that in the case of diluted solutions of $A-H\cdots B$ complex with weak hydrogen bond and small moment of inertia I_A and I_B, the reorientational relaxation may take part in the relaxation process of $A-H$ stretching vibration. To study this effect the molecular dynamics method [1] which gives us the direct picture of the motion of $A-H\cdots B$ complex can be accepted. To realize this kind of calculation one must make several simplifications. Since the motion of the proton can not be treated classically and its amplitude of vibration in the ground state is rather small, one can accept the supposition of rigid $A-H$ bond as feasible when dealing with reorientational motion of $A-H\cdots B$ complex.

The next problem concerns the $A\cdots B$ stretching which may also behave as quantum oscillator rather than the classical one. In the case of weak hydrogen bond when the bond energy does not exceed the thermal energy for a factor which is much greater than unity, $A\cdots B$ stretching vibration behave more or less classically, and the molecular dynamics method is justified. There is one further fact which prohibite the treating of strong hydrogen bonds. With growing strength of the bond the frequency of $A\cdots B$ stretching exceeds appreciably the characteristic frequencies of the thermal motion. These two time scales become too different shich means that in the process of numerical integration of the Newton equation of motion the time increment must be shortened which increases the computing time.

The studied system was representing an extremely diluted donor phase $(A-H)$ in the acceptor (B) phase, and consisted of 44 acceptor molecules and one $A-H$ molecule. The acceptor molecules were arranged to simulate the first and the second coordination sphere. The geometrical neighbours of third and higher order were replaced by the rigid wall of the spherical cavity whose diameter was related to the density of the sample. The pairwise interaction between acceptor molecules was taken to be of Lennard-Jones type. Since the equations of the motion were written in dimensionless form, the Lennard-Jones parameters do not need to be specified and are taken as an arbitrary scale parameters defining time and space scaling. The acceptor-donor interaction was described by the Lippincot-Schroeder potential [2]:

$$V = V_{AH}[1 - \exp(-\alpha_{AH}(r_{AH} - a_{AH}))]^2 + \\ + V_{HB}[1 - \exp(-\alpha_{HB}(r_{HB} - a_{HB}))]^2 + \\ + V_{AB}\exp(-2\alpha_{AB}(r_{AB} - R_{AB})) - V_{HB}. \quad (1)$$

The parameters appearing in the upper expression were determined by the requirement that $A-H$ and $A-B$ stretching frequencies, AB bond lengths and bond strengths attained the values corresponding to various types of hydrogen bonds. Since the supposition of rigid $A\cdots H$ bond was accepted the proton coordinate was not an independent variable. The potential energy for the linear $A-H\cdots B$ system was obtained by placing the proton on the equilibrium position: $r_{AH} = a_{AH} + \Delta r(r_{AB})$ where $\Delta r(r_{AB})$ was determined by vanishing the first derivative of the potential with respect to r_{AH}. In the case of bent hydrogen bond (see Figure 1.) the potential energy was expressed again as the sum of three terms (1) with $r_{AH} = a_{AH} + \Delta r(r_{AB}) \cdot \cos\alpha$.

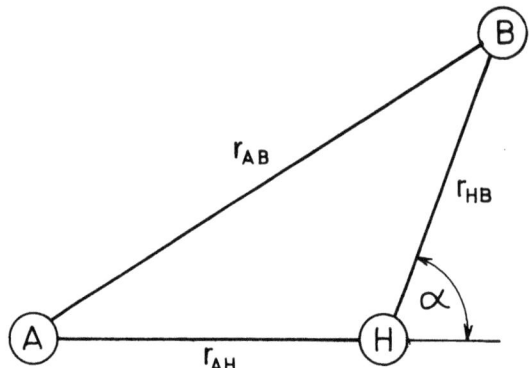

Fig. 1. The geometry of the $A-H \ldots B$ complex. The mass of A and B are equal 1 (unit of mass).

The Newton equations of motion were solved using simple algorithm described elsewhere [3]. The equations of motion corresponding to the reorientational motion of $A-H$ molecule were solved using the Lagrange multipliers. The time increment for the numerical integration was taken to be about 10^{-2} in dimensionless units. The system was followed for about 1000 steps and the coordinates of $A-H$ molecule and the nearest neighbours were analysed to obtain different kinetic properties of the $A-H\cdots B$ complex. The following quantities were evaluated: dipole $P_1(\cos\vartheta)$ and $P_2(\cos\vartheta)$ correlation functions, angular velocity correlation function and the correlation function of the force acting on the donor molecule in paralell direction with respect to $A-H$ bond. Because the averages were done with one $A-H$ molecule only, the statistics is rather poor [4], and only short time parts of the correlation functions can be accepted with confidence. The dipole correlation functions $P_1(\cos\vartheta)$ are shown on Figure 2. It is evident that the correlation functions depend upon the bond strength, density and temperature in the expected manner. Except in the case of

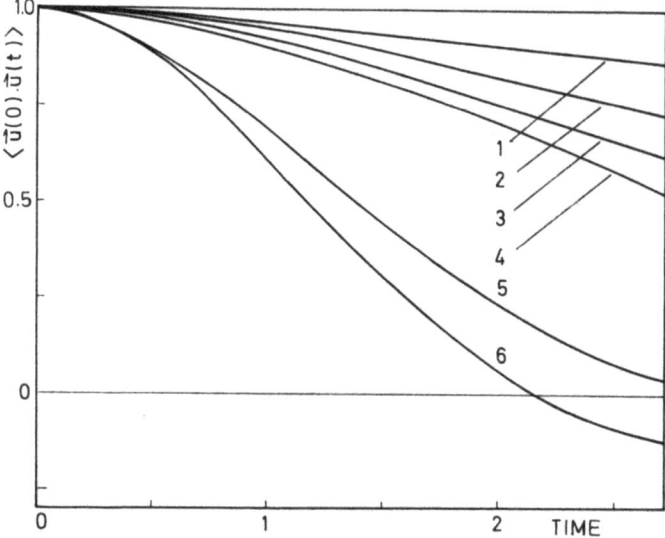

Fig. 2. The dipole correlation function. 1, $\varrho = 0.75$, $T = 2.5$, $V_0 = 48$; 2, $\varrho = 0.5$, $T = 2.5$, $V_0 = 10$; 3, $\varrho = 0.5$, $T = 2.5$, $V_0 = 48$; 4, $\varrho = 0.5$, $T = 2.5$, $V_0 = 21$; 5, $\varrho = 0.75$, $T = 1.2$, $V_0 = 5$; 6, $\varrho = 0.5$, $T = 2.5$, $V_0 = 5$, where ϱ denotes numerical density, T is the temperature and V_0 is the strength of the hydrogen bond. Time unit is $\sqrt{m\sigma^2/48\varepsilon}$.

the weakest hydrogen bond the reorientational correlation times are rather long and thus the reorientational relaxation is negligible in comparison with the vibrational relaxation. On the other hand in the case of weak hydrogen bonds (curves 5 and 6, Figure 2), where the thermal energy is comparable with the hydrogen bond strength, the bond is continually broken and re-formed. This means that the dipole moment is more mobile and the contribution of reorientational relaxation is appreciable. The ratio of the correlation times τ_{P_1}/τ_{P_2} provides [5] a check on models for the molecular reorientation. In the case of weak hydrogen bonds this ratio is about 1.4 and indicates the presence of jump diffusion mechanism. At strong hydrogen bonds τ_{P_1}/τ_{P_2} is about 2.0, supporting the rotational diffusion motion. The comparison between τ_{P_1} and τ_J leads us to similar conclusion. τ_J being about 0.5 at strong hydrogen bonds is much smaller than τ_{P_1} (about 2.7) while these two values are comparable at weak hydrogen bonds.

According to Klier [6] one expects that the time dependence of the dipole correlation function would be modulated with higher frequency due to vibrationally perturbed rotatory motion. Since at strong hydrogen bonds the dipole correlation function decays slowly, and librational frequencies are rather high, the modulation of $\langle \mathbf{u}(0)\cdot\mathbf{u}(t)\rangle$ appears at short times. This modulation is very weak (not depicted on Figure 2) and can be evaluated from the computer output data. The frequency of this modulating factor corresponds to the time variation of the angular velocity correlation functions (Figure 3).

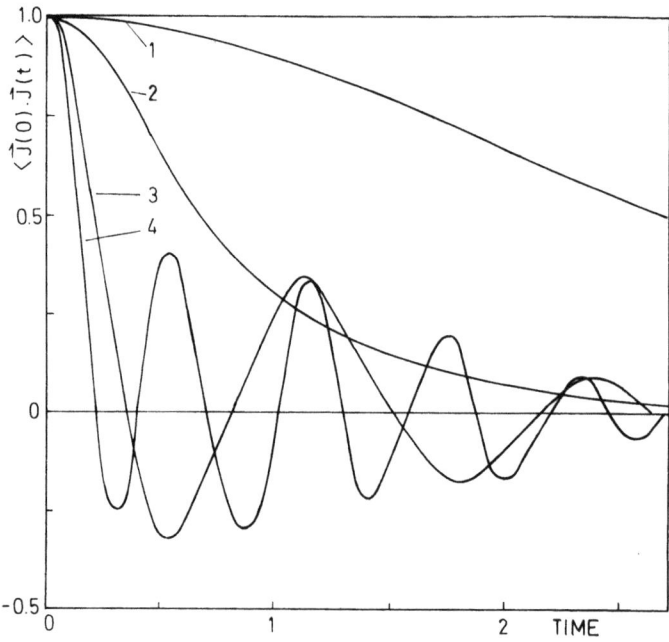

Fig. 3. Angular velocity correlation function. 1, $\varrho = 0.5$, $T = 2.5$, $V_0 = 5$; 2, $\varrho = 0.75$, $T = 1.2$, $V_0 = 5$; 3, $\varrho = 0.75$, $T = 1.2$, $V_0 = 10$; 4, $\varrho = 0.5$, $T = 2.5$, $V_0 = 48$.

Since the vibrational motion of $A-H\cdots B$ complex behave as a quantum mechanical phenomenon, it is not possible to interpret the vibrational relaxation in terms of molecular dynamics results. It must stressed be that the supposition of slow modulation [8] is not satisfied in this case. Because of complexity of the problem the semiclassical approach [7] is often used. We tried to estimate the influence of the time dependence of $\omega_{0\alpha}$ [9](the solvent induced part of transition frequency of $A-H$ stretching vibration) on the vibrational relaxation function. Using the cumulant expansion theorem, the vibrational relaxation function can be written as follows:

$$G_v(t) = \exp\left(\sum_{k=1}^{\alpha} \omega_{ck}(it)^k\right).$$

Calculating the cumulant averages [8] in the limit of slow modulation, dependence of the solvent induced stochastic part of the potential is usually suppressed. From the computer experiment we obtained the correlation function of the force acting on

$A-H$ molecule parallel to the $A-H$ bond, which may be treated as a quantity proportional to the solvent induced potential influencing the transition frequency. The correlation function of this parallel force was then expanded in a power series of time, retaining the first three terms.

$$\langle \mathbf{F}(0) \cdot \mathbf{F}(t) \rangle = \langle F^2 \rangle \cdot (1 + a_1 t + \tfrac{1}{2} a_2 t^2).$$

Adopting some further approximations, the cumulants were expressed as follows:

$$\omega_{c1} = \Omega_1$$
$$\omega_{c2} = \Omega_2 - \Omega_1^2$$
$$\omega_{c3} = \Omega_3 - 3\Omega_1 \Omega_2 + 2\Omega_1^3 + i\Omega_2 a_1$$
$$\omega_{c4} = \Omega_4 - 3\Omega_2^2 - 4\Omega_1 \Omega_3 + 12\Omega_1^2 \Omega_2 -$$
$$- 6\Omega_1^4 - \Omega_2 a_2 + i(8/3 \Omega_3 a_1 - 4\Omega_1 \Omega_2 a_1),$$

where Ω_i is equal to the ensemble average of ith power of solvent induced contribution to the transition frequency. According to Figure 4, the first derivative of the parallel force correlation function vanish ($a_1 = 0$) and thus the only contribution due to the time dependence of the solvent induced potential is the term $-\Omega_2 a_2$ in

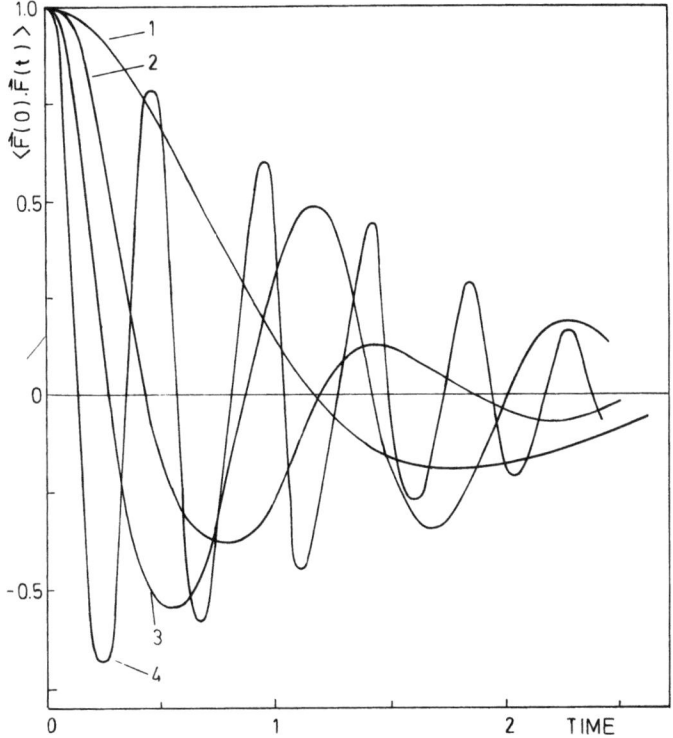

Fig. 4. Paralell force correlation function. 1, $\varrho = 0.5$, $T = 2.5$, $V_0 = 5$; 2, $\varrho = 0.75$, $T = 1.2$, $V_0 = 5$; 3, $\varrho = 0.5$, $T = 2.5$, $V_0 = 21$; 4, $\varrho = 0.5$, $T = 2.5$, $V_0 = 48$.

the expression for ω_{c4}. The approximative values of a_2 range from -2 (in units $48\varepsilon/(m\sigma^2)$) at weak hydrogen bonds, to -100 at strong ones. Molecular dynamics method is not capable to provide the data necessary to evaluate the averages Ω_i and we can only conclude that the term $-\Omega_2 a_2$, which is a positive quantity, diminishes the convergence of cumulant expansion series at longer times.

Acknowledgement

The financial support of Boris Kidrič Fund is gratefully acknowledged.

References

1. Rahman, A. and Stillinger, F. H.: *J. Chem. Phys.* **55**, 3336 (1971).
2. Schroeder, R. and Lippincott, E. R.: *J. Phys. Chem.* **61**, 921 (1957).
3. Verlet, L.: *Phys. Rev.* **159**, 98 (1967).
4. Zwanzig, R. and Ailawadi, N. K.: *Phys. Rev.* **182**, 280 (1969).
5. Gordon, R. G.: *Adv. Magn. Resonance* **3**, 1 (1967).
6. Klier, K.: *J. Chem. Phys.* **58**, 737 (1973).
7. Robertson, G.: Presented at International Summer School on *Molecular Dynamics in Simple Liquids*, Menton, 1972.
8. Bratož, S., Rios, J., and Guissani, Y.: *J. Chem. Phys.* **52**, 439 (1970).
9. See Reference 8, Equation (8).

VIBRATIONAL RELAXATION IN LIQUID AND SOLID QUINOLINE

WALTER G. ROTHSCHILD

Scientific Research Staff Ford Motor Company, Dearborn, Mich. 48121, U.S.A.

Abstract. The infrared and Raman contours of a fundamental of quinoline (1033 cm^{-1}) in its liquid and solid (crystalline and glassy) phases have been Fourier-inverted. The correlation functions show essentially only vibrational relaxation which is greatly influenced by the state of the surrounding molecules. A large distribution of frozen-in molecular configurations (glassy state) leads to relatively rapid vibrational relaxation; a more isotropic environment (crystal and dilute solution) lead to much slower vibrational decay. The quantitative aspects can be described by a 'true' vibrational correlation function and by a modulation factor arising from the medium-induced perturbations of the vibrational level.

Résumé. Nous avons transformé par la méthode de Fourier les contours d'une bande fondamentale de la quinoléine en Raman et infrarouge à l'état liquide (258–368 K) et solide (cristallin et vitreux, 115–183 K). Les fonctions de corrélation donnent une description de la relaxation vibrationnelle qui dépend fortement du couplage de la molécule active avec son environnement. A l'état vitreux la distribution des multiples configurations 'congelées' aboutit à un amortissement rapide de la fonction de corrélation; à l'état liquide, solution diluée (en CS$_2$) et corps cristallin, on constate une relaxation vibrationnelle beaucoup moins efficace. On obtient un aspect quantitatif en calculant une simple fonction modulatoire à l'origine des perturbation du niveau de la molécule active.

1. Introduction

Fourier transformation of infrared and Raman band contours of molecules in their condensed as well as vapor phases has been shown to be a very useful approach for obtaining certain information on molecular dynamics. For instance, we now realize that the rotational mobility in the liquid phase of many molecules is appreciably greater than had been believed in the past [1], that vibrational relaxation processes in most liquids are much faster than hitherto expected [2], and that there is almost always a short but significant initial time interval during which the molecular relaxation process is essentially determined by inertial effects [1, 3].

As a general rule, it seems that in smaller molecules the rotational relaxation is more rapid than the vibrational relaxation [4]. Of course, the reversed case or, at least, near equal contributions of both processes can be expected to prevail in cases of particularly strong or efficient intermolecular coupling, such as often occurs in hydrogen bonding [5]. Furthermore, it is sometimes possible to select in the same molecular system those very transitions for Fourier analysis which either do not or do reflect strong intermolecular coupling. Thereby we have a means to stress either the rotational or the vibrational relaxation phenomena of the particular system [4]. Directional solvent effects have also been observed [6].

There are good reasons to suspect that the corresponding relaxation behavior of large rigid molecules is greatly different. Firstly, on the longest time scales of the observations (say 10–20 ps), the molecules have probably not reoriented through appreciable angles (say about 1 rad) because of their large size and rigidity. Secondly, the great number of vibrational degrees of freedom (relative to that in smaller molecules) favors vibrational decay as the most efficient pathway of molecular relaxation during the significant time intervals of observation.

Since in condensed phases the vibrational relaxation phenomena are lesser understood than the rotational relaxation phenomena, and since the number and the diversity of molecules consisting of relatively few atoms is restricted, we considered it worthwhile to study the relaxation behavior of a large rigid molecule in its condensed phases. We have chosen quinoline,

a conjugated ring molecule, for several reasons. From the point of view of accomplishing the analyses, (i) dielectric and NMR relaxation data clearly indicate that rotational reorientation in liquid quinoline is rather insignificant during the first 5–10 ps [7, 8], (ii) the infrared and Raman spectra are well known [9], (iii) all infrared-active modes are also Raman-active (no center of inversion), and (iv) several Raman shifts are highly polarized and thus lend themselves particularly well to an analysis of the pure vibrational correlation functions [2, 10] without undue experimental difficulties.

From the point of view of molecular dynamics, quinoline (i) is an interesting planar molecule of, effectively, high structural symmetry (considering the N atom equivalent to the C–H group) but which possesses appreciable electric anisotropy ($\mu = 2.2$ D), (ii) it can be easily cooled into a glassy phase to very low temperatures, (iii) and its liquid-phase range stretches over 250°.

We were also tempted to undertake this study by the curiosity to see to what extent the concept of the simple correlation functions can be stretched in considering systems for which the exact theory [2, 11] no longer seems to be valid.

2. Experimental

2.1. Infrared Spectra

The infrared spectra were taken with a Perkin-Elmer spectrophotometer, model 180. The quinoline, distilled at atmospheric pressure prior to use, was contained in a variable-temperature cell of about 0.02 mm pathlength, equipped with CsI windows into which an iron-constantan thermocouple was inserted. The temperature of the assembly (Beckman RIIC VLT2/Tem-L) was set and controlled within 0.5° between 115 and 368 K. Scanning speeds and spectral slit widths were kept sufficiently small to avoid distortion of the band contours.

The vibrational fundamental chosen was a symmetric in-plane deformation mode, $v_{23}(A')$, at 1033 cm^{-1} [9]. Because in the liquid state the band overlaps to lower frequencies with the adjacent v_{24} mode, its high-frequency wing was folded into the lowfrequency half for the Fourier inversion. Since the band contour at the temperatures of the liquid phase appeared to be fairly symmetric, this procedure – although approximate – is considered a valid expedience in the context of our work. At temperatures below the freezing point of quinoline, the overlap region between v_{23} and v_{24} can be more readily corrected without significantly influencing the resulting correlation function.

2.2. Raman Spectrum

The Raman spectrum of the intense 1033-cm^{-1} mode, at ambient temperature, was determined by 90° scattering using a Spex 1401 double monochromator (slit width 1 cm^{-1}). The 514.5 nm line from an argon laser at powers of approximately 90 mW served as excitation source. The contour of the Raman shift of the v_{23} mode is very symmetric and, since there is also some overlap with the v_{24} shift, the same folding procedure as described for the infrared case (Section 2.1) was adopted. A measurement of the degree of depolarization yielded 0.004, which is lower than a literature value [9] – we believe on account of our much narrower slit widths [12]. Since the contribution of the anisotropic component of the scattered radiation to the correlation function in the parallel scattering geometry is therefore small (0.64 × 4/45), we make only a minor error by assuming that the Fourier transform of the contour of the 1033-cm^{-1} Raman shift, measured under parallel scattering geometry, yields

$$\hat{C}(t) = \langle u(0)u(t) \rangle, \quad (1)$$
(u = amplitude of the vibrational transition moment)

the 'pure vibrational' correlation function [2, 10]; that is – to be more specific – a correlation function which directly describes all effective relaxation processes other than reorientational motion of the individual molecule.

3. Results and Discussion

3.1. Reorientational Motion

We consider first the correlation functions (of the 1033-cm^{-1} fundamental) at room temperature from the Raman and the infrared spectrum of the pure liquid. They are shown in Figure 1 by the triangular (Raman) and open circular (infrared) points. It is evident that the correlation functions are essentially identical during the first 5 ps; the corresponding correlation times [13],

$$\tau = \int \hat{C}(t)\,dt, \tag{2}$$

amount to 2.3 and 2.1 ps, respectively.

These results fully support the expectation that rotational relaxation is not observed to be significant during the quoted time intervals: (i) The infrared and Raman correlation functions are near identical – which should not occur if a significant degree of rotational relaxation were present since this would induce the infrared correlation function [10],

$$\hat{C}(t) = \langle u(0)u(t)\rangle\langle \hat{u}(0)\cdot\hat{u}(t)\rangle, \tag{3}$$
($\hat{u}$ = orientation of the vibrational transition moment)

to decay faster than the (isotropic) Raman correlation function (see Equation (1)). (ii) The correlation times given above are, at least, one order of magnitude shorter than some reported orientational correlation times of liquid quinoline, for instance

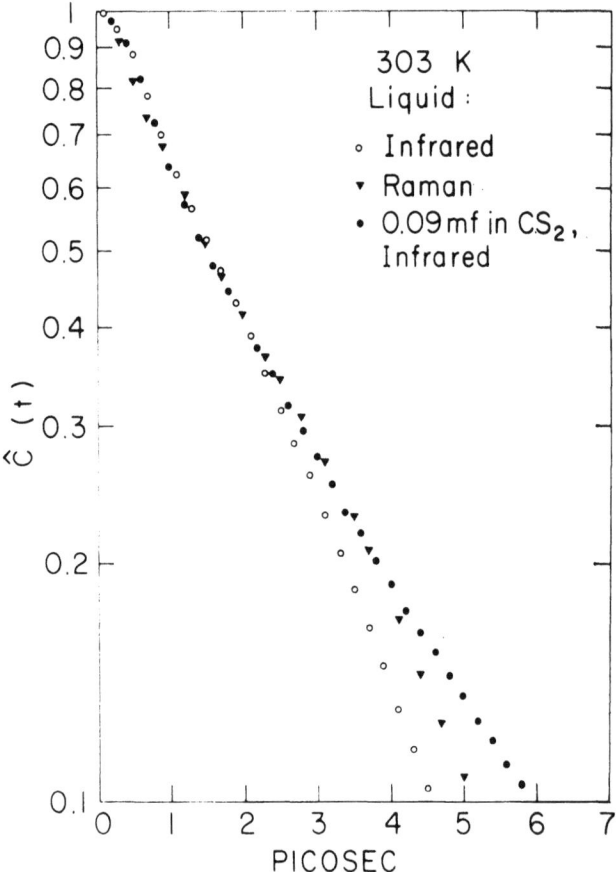

Fig. 1. Vibrational correlation functions of the 1033-cm^{-1} in-plane bending fundamental (A') of quinoline at ambient temperature. Infrared: pure liquid and 0.09 mole fraction solution in CS$_2$, respectively. Raman: pure liquid.

$\tau(\text{NMR}) = 1.2 \times 10^{-10}$ s (from ^{14}N quadrupole relaxation) [8] and $\tau(\text{dielec}) = 0.45 \times \times 10^{-10}$ s [7]. Similar values were obtained from light-scattering measurements [14]. Although these measurements do not observe the same transition moment vector in the quinoline molecule, we are convinced that we make a negligible error by setting the pure rotational correlation function

$$\langle \hat{\mathbf{u}}(0) \cdot \hat{\mathbf{u}}(t) \rangle = 1 \qquad (4)$$

for t up to about 5 ps.

3.2. VIBRATIONAL RELAXATION AND MOLECULAR ENVIRONMENT

The appearance of the infrared spectra of the v_{23} mode as a function of temperature (spectra not shown here) and the corresponding correlation functions are now described.

(i) On increasing the temperature of the liquid up to 368 K (the highest temperature tolerated by the cell), a slight increase of the band width (about 5%) and a small red-shift of the peak frequency (0.5 cm^{-1}) were observed. Furthermore, the band contour became more symmetric.

(ii) An interesting effect was noticed upon dissolution of quinoline in CS$_2$ (which has flat, very weak absorption in the frequency range of the v_{23} mode of quinoline): Relative to the pure liquid, the peak frequency of v_{23} in a 0.09 mole fraction solution has now red-shifted by 2 cm^{-1} and its width has narrowed by 30% (ambient temperature). The correlation function of quinoline computed from these CS$_2$ solution data is displayed in Figure 1 by the solid circles. For times up to 2 ps, it is seen to coincide with that of the pure quinoline liquid at this temperature. Beyond this time interval, correlation in neat liquid quinoline is lost faster than in its dilute CS$_2$ solution, the correlation times being 2.1 and 2.5 ps, respectively. We have here apparently an indication that if the quinoline molecule finds itself in a more 'isotropic' environment, its vibrational relaxation is slowed down.

(iii) More striking evidence on the influence of the environment on the relaxation rate of the v_{23} mode of quinoline is afforded by cooling the compound below its freezing point (258 K). If the cooling process into the solid state is done rapidly, the molecules apparently persist in some kind of glassy or amorphous state, as is manifest by an increase in band width and by a considerable asymmetry in the band contour towards higher frequencies (spectra not shown here). A slight blue-shift of the peak frequency is also observed. On the other hand, gradual cooling below 258 K leads at once to the narrow and symmetric band contour usually encountered in the crystalline state. The infrared correlation functions of solid quinoline computed from the band contour of these 'glassy' (open circles) and crystalline states (solid circles) are shown in Figure 2 for a temperature of 115 K. We see that the vibrational relaxation function decays appreciably faster for the glassy than for the crystalline phase; this phenomenon is already observable at short times (below 1 ps). The respective correlation times are 3.5 (crystalline) and 1.9 (glassy) ps.

The explanation we wish to offer for this phenomenon is as follows: the glassy

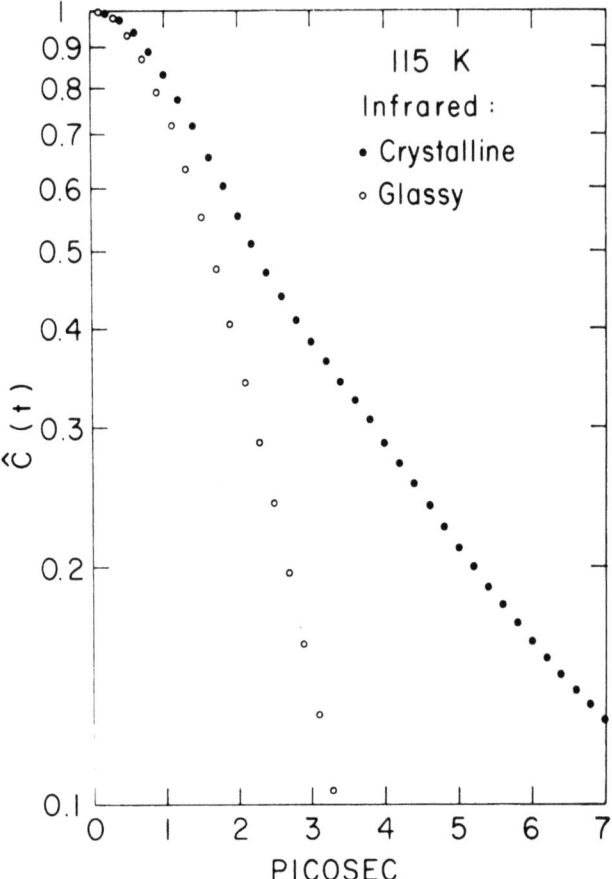

Fig. 2. Infrared vibrational correlation function of the 1033-cm^{-1} fundamental of solid quinoline at 115 K in its crystalline and glassy state.

state yields the greatest number of different environments (=distributions of configurations and distances between adjacent molecules of quinoline). The solid crystalline as well as the liquid and the dilute solution phases offer a smaller number of such different environments, be it through the regular arrangement in the crystal or through the more effective configuration-averaging motions in the liquid and solution phases, respectively [15]. To account for this effect in a more quantitative manner, we consider the vibrational Hamiltonian to consist of two terms,

$$H^v = H_0^v + H_1^v, \tag{5}$$

where H_0^v is the Hamiltonian of the vibrational state v of an individual molecule,

$$H_0^v |v\rangle |e\rangle = E_0^v |v\rangle |e\rangle \quad (|e\rangle = \text{external state}), \tag{6}$$

and where H_1^v is some type of perturbation of the vibrational level brought about by the multiplicity of environments. The Heisenberg operator of the vibrational transition moment is then

$$\exp[i(H_0^v + H_1^v)t/\hbar] u \exp[-i(H_0^v + H_1^v)t/\hbar]. \tag{7}$$

Writing [16]

$$\begin{aligned}(H_1^v) = \langle v| H_1^v |v\rangle, (H_1^{v'}) = \langle v'| H_1^v |v'\rangle \\ (\Delta H_1^v) = \langle e| (H_1^v) - (H_1^{v'}) |e\rangle \ (' = \text{upper level})\end{aligned} \tag{8}$$

the corresponding correlation function (real part) is

$$\hat{C}(t) = \cos(\Delta\omega_1 t) \langle u(0)u(t)\rangle. \tag{9}$$

Here, $\langle u(0)u(t)\rangle$ is the usual vibrational correlation function of the individual molecule taken with respect to the band center, and $\cos(\Delta\omega_1 t)$ is the real part of the modulation factor $\exp[i(\Delta H_1^v)t/\hbar]$ arising from the perturbation term.

For $\langle u(0)u(t)\rangle$, the 'true' vibrational correlation function of the individual quinoline molecule, we take the crystalline-phase correlation function at the given

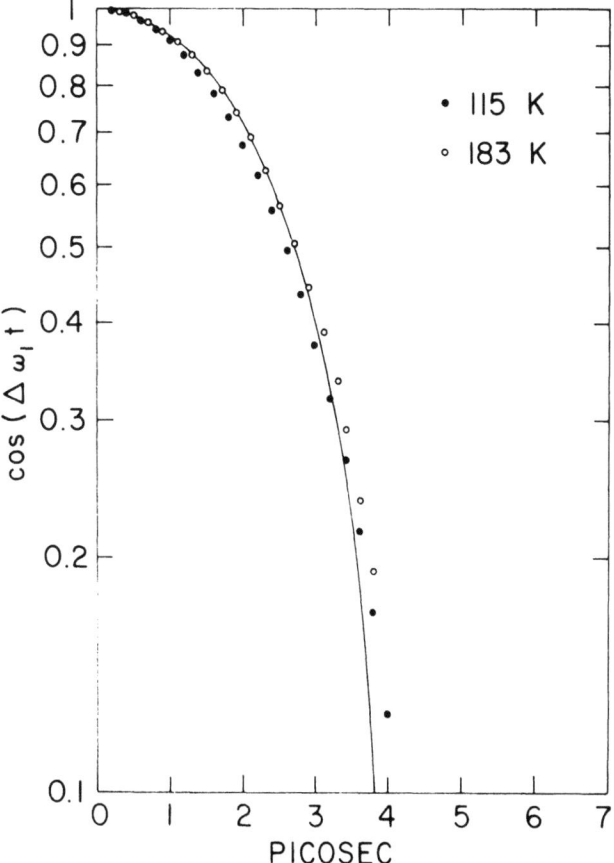

Fig. 3. Modulation factor of the infrared vibrational correlation function of the 1033-cm^{-1} fundamental of glassy quinoline. The curve represents the least-square fit to the data at 115 K ($\Delta\omega_1 = 2.04$ cm^{-1}).

temperature. Evaluation of the data displayed in Figure 2 with the help of Equation (9) then leads to the modulation functions shown in Figure 3 for the temperatures 115 (solid points) and 183 K (open circles). The solid curve in the figure shows the least-square cosine fitted to the experimental points at 115 K: The resulting value of $\Delta\omega_1$ amounts to 2.04 cm^{-1}. The same procedure for quinoline at 183 K (open circles) yields $\Delta\omega_1 = 1.96$ cm^{-1}. We see that the modulation function is slightly weaker at the higher temperature – according to expectation – although the experimental errors accumulating at the longer times would not permit to attach undue significance to the value of this difference.

4. Summary and Conclusions

We have shown that the relaxation processes observable from the infrared and Raman spectra of the condensed phases of quinoline are essentially vibrational in character. If, for instance by rapid cooling, the molecules are brought into a state where the active molecule is surrounded by many different environments, the overall vibrational relaxation rate is greatly increased and can approximately be described in terms of the 'true' vibrational relaxation function (crystalline state) and a slow modulation function which accounts for the environmental effects perturbing the vibrational transition at this temperature. The cause of this behavior appears to be mainly steric: measurements under increased pressures are predicted to lead to a similar phenomenon.

Acknowledgment

I would like to express my deep gratitude to Dr Robert C. Livingston (Washington, D.C.) for furnishing me with the Raman data.

References

1. Cabana, A., Bardoux, R., and Chamberland, A.: *Can. J. Chem.* **47**, 2915 (1969); Rothschild, W. G.: *J. Chem. Phys.* **53**, 3265 (1970); *ibid.* **55**, 1402 (1971); Rossi-Sonnichsen, I., Bouanich, J.-P. and Thanh, N. V.: *Compt. Rend.* **273C**, 19 (1971); Vincent-Geisse, J., Soussen-Jacob, J., Tai, N. T., and Descout, D.: *Can. J. Chem.* **48**, 3918 (1970); Yarwood, J.: *Spectry. Letters* **5**, 193 (1972); Lévi, G., Chalaye, M., Marsault-Herail, F., and Marsault, J. P.: *Mol. Phys.* **24**, 1217 (1972).
2. Valiev, K. A.: *Opt. Spectry.* **11**, 253 (1961); Döge, G.: *Z. Naturforsch.* **23a**, 1405 (1968); Bratos, S. and Marechal, E.: *Phys. Rev.* **A4**, 1078 (1971); Bartoli, F. J. and Litovitz, T. A.: *J. Chem. Phys.* **56**, 413 (1972).
3. Favelukes, C. E., Clifford, A. A., and Crawford, B., Jr.: *J. Phys. Chem.* **72**, 962 (1968); Rothschild, W. G.: *J. Chem. Phys.* **49**, 2250 (1968); Marabella, L. and Ewing, G. E.: *J. Chem. Phys.* **56**, 5445 (1972); Van Aalst, R. M., Van der Elsken, J., Frenkel, D., and Wegdam, G. H.: *Faraday Symp. Chem. Soc.*, No. 6, 94 (1972); Perchard, J. P., Murphy, W. F., and Bernstein, H. J.: *Mol. Phys.* **23**, 499 (1972).
4. Perrot, M., Van Huong, P., and Lascombe, J.: *J. Chim. Phys.* **68**, 614 (1971); Rothschild, W. G.: *J. Chem. Phys.* **57**, 991 (1972).
5. Terpilovskii, D. N.: *Opt. Spectry.* **28**, 380 (1970); Wall, T. T.: *J. Chem. Phys.* **52**, 2792 (1970).
6. Kondilenko, I. I., Pogorelov, V. E., and Khue, K.: *Opt. Spectry.* **27**, 518 (1969); Constant, M., Delhaye, M., and Fauquembergue, R.: *Compt. Rend.* **271B**, 1177 (1970).

7. Holland, R. S. and Smyth, C. P.: *J. Phys. Chem.* **59**, 1088 (1955).
8. Dinesh and Rogers, M. T.: *J. Magn. Resonance* **7**, 30 (1972).
9. Wait, S. C., Jr. and McNerney, J. C.: *J. Mol. Spectry.* **34**, 56 (1970).
10. Nafie, L. A. and Peticolas, W. L.: *J. Chem. Phys.* **57**, 3145 (1972).
11. Gordon, R. G.: *Adv. Magn. Resonance* **3**, 1 (1968).
12. Hess, W. R., Hacker, H., Schrötter, W. H., and Brandmüller, J.: *Z. angew. Phys.* **27**, 233 (1969).
13. The upper limit of this integral is equivalent to a time where the correlation has dropped to 0.01–0.05.
14. Ben-Reuven, A. and Zamir, E.: *J. Chem. Phys.* **55**, 475 (1971).
15. Shin, H. K.: *Chem. Phys. Letters* **1**, 443 (1967).
16. See, for instance, Klier, K.: *J. Chem. Phys.* **58**, 737 (1973).

DISCUSSION

Carneiro: Since you attribute the difference in memory function between the crystalline and the glassy states to the difference in structural environment, will you then be able to distinguish between a 'microcrystalline glass' and a 'random network glass'?

Rothschild: The data are too preliminary to give a satisfactory answer to your question. Furthermore, the correlation function is an average over all environments, that is to say, of the amorphous and any existing polycrystalline regions which may exist side by side. At this moment I just do not know anything about the nature of this glassy state of quinoline.

Döge: You found that the line shape in the solution is narrower than in the pure liquid. Can it be possible that this behaviour is due to a strong self-association?

Rothschild: Yes, this is quite possible and certainly entirely reasonable. In solution, the decrease of close coupling between adjacent molecules would be lessened and, as you suggest, would lead to a narrower band. In terms of the proposed model, this would again mean that the active molecule 'sees' a more isotropic environment and therefore experiences a slower rate of vibrational relaxation.

FONCTIONS DE CORRELATION TENSORIELLES : FORMALISME GENERAL POUR LES MOLECULES POLYATOMIQUES EN SOLUTION ET APPLICATION AU CAS DE LA DIFFUSION DU ROTATEUR ASYMETRIQUE

JEAN-CLAUDE LEICKNAM et YVES GUISSANI

Laboratoire de Physique Théorique des Liquides, Tour 16, Université Paris-VI, 75230 Paris Cedex 05, France

Résumé. Les fonctions de corrélation monomoléculaires de tenseurs d'ordre un, deux, trois,..., qui sont liées aux formes des bandes de processus à plusieurs photons (IR, Raman, hyper-Raman,..., par exemple) sont mises sous une forme adaptée à l'étude des mouvements moléculaires dans les liquides. L'influence des géométries expérimentales est traitée analytiquement; on peut ainsi déterminer le nombre et la nature des conditions expérimentales nécessaires pour obtenir séparément toutes les fonctions d'autocorrélation qui relèvent d'un même phénomène. On calcule ensuite ces fonctions de corrélation dans le cas limite de la diffusion rotationnelle en utilisant une méthode directe mettant en valeur les hypothèses stochastiques de ce cas limite.

Abstract. Monomolecular correlation functions of first, second and third rank tensors are investigated in the case of liquid systems. The property of liquids to be isotropic is used to write these functions in a form well adapted to the study of molecular motions. There is no restriction concerning the form of tensors which may be complex, symmetrical or asymmetrical. The effect of experimental geometry is studied in an analytical way. As a result, autocorrelation functions of scalars, vectors, ..., are written in a way suitable for experimental determination of these functions. This method represents a generalisation of well known theories valid for usual Raman spectroscopy.

These correlation functions are calculated in the rotational diffusion limit. Explicit use is made of the Kubo cumulant expansion of ordered exponentials of operators. Stochastic hypotheses introduced in this connection are shown to be identical with the usual assumptions. As an example of application, the hyper-Raman correlation function is analysed.

1. Introduction

Les profils des spectres vibrationnels sont très dépendants des conditions expérimentales (géométrie du montage et états de polarisation de la lumière incidente ou diffusée). Pour l'effet Raman ordinaire, où le tenseur de polarisabilité est symétrique, on analyse généralement les spectres à l'aide des géométries VV et VH [1–3]. Dans le cas général, les tenseurs en cause ne sont pas nécessairement symétriques. L'analyse des intensités totales diffusées a été faite récemment par McClain *et al.* [4–7] en termes de processus à un, deux ou trois photons, dans le cas plus spécifique des transitions électroniques.

Nous voulons ici, dans une première partie, enrichir cette analyse en la reformulant complètement en termes de théorie des groupes (ainsi que Placzek lui-même formulait en 1934 [8] les fondements théoriques de l'effet Raman) et en l'adaptant à la description des profils des bandes vibrationnelles qui seront interprétés en termes de fonctions de corrélation simples. Dans une deuxième partie on cherchera à appliquer ces résultats dans le cas d'un modèle bien précis décrivant les mouvements moléculaires dans le liquide. On calculera toutes les fonctions de corrélation nécessaires à l'analyse complète des profils des spectres vibrationnels à un, deux ou trois photons.

2. Fonctions de corrélation

2.1. Généralités sur les profils vibrationnels

L'analyse des profils des bandes vibrationnelles conduit à introduire directement par l'intermédiaire du théorème de fluctuation-dissipation [9] ou par un raisonnement du type de celui employé par Gordon [10] pour l'effet Raman, une fonction de corrélation $G(t)$ dont l'intensité mesurée est la transformée de Fourier. Cette fonction dans un certain nombre de phénomènes physiques peut s'écrire si on peut négliger les corrélations croisées (solutions à faible concentration):

$$G(t) = \langle \lambda_i \mu_j \nu_k \cdots \xi_{i,j,k\ldots}(t) \, \xi^*_{l,m,n\ldots}(0) \, \lambda^*_l \mu^*_m \nu^*_n \cdots \rangle.$$

$\lambda, \mu, \nu \ldots$ étant les vecteurs polarisation des photons en interaction avec la matière, le tenseur ξ étant défini sur un référentiel moléculaire. Ce tenseur peut être, suivant le phénomène étudié, le moment dipolaire, le tenseur de polarisabilité, le tenseur d'hyperpolarisabilité... Les indices répétés indiquent une sommation muette. La moyenne est prise à l'équilibre thermique en l'absence des champs.

2.2. Isotropie du milieu

2.2.1. Méthode

Pour des milieux isotropes la formule générale ci-dessus se simplifie en faisant apparaître des fonctions d'autocorrélation de certaines grandeurs physiques déduites du tenseur ξ. (Dans le cas le plus simple de l'absorption à un photon par exemple, il est bien connu qu'on obtient la fonction d'autocorrélation du vecteur moment dipolaire $G(t) = = \frac{1}{3}\langle \mathbf{M}(t) \cdot \mathbf{M}(0) \rangle$).

En effet, l'isotropie du milieu se traduit au niveau mathématique par une analyse en termes d'invariants isotropes d'un tenseur d'un ordre donné. Dans la référence 6, cette analyse est effectuée sur une base standard d'invariants isotropes d'un tenseur d'ordre 2, 4 et 6. Nous pensons préférable de reprendre le problème par quelques considérations issues de la théorie des groupes. En substance, si l'on sait décomposer le tenseur ξ sur les représentations irréductibles du groupe R_3 [11] (c'est-à-dire, quelles sont les parties de ξ qui se transforment comme un scalaire ($l=0$), comme un vecteur ($l=1$), comme $l=2$, $l=3$ etc. ...) on peut alors écrire $G(t)$ comme une somme de fonctions d'autocorrélation $G_i(t)$ de scalaires, de vecteurs, etc.... On est en effet, assuré qu'il n'y a pas de corrélations croisées:

$$G(t) = \sum_i A_i G_i(t),$$

où les A_i sont des fonctions simples de la géométrie expérimentale (λ, μ, ν, etc....).

Outre le fait que cette méthode ne nécessite que peu de calculs, elle est très bien adaptée à l'analyse des mouvements moléculaires en solution.

2.2.2. Effet Raman

Dans ce cas la séparation est bien connue pour les intensités totales [12] dans le cas d'un tenseur quelconque et pour le problème du profil des bandes vibrationnelles dans le cas d'un tenseur symétrique et réel [1-3].

On trouve :

$$G(t) = A_0 \langle \alpha^H(t)\alpha^H(0)\rangle + A_1 \langle \pmb{\eta}^H(t)\cdot\pmb{\eta}^H(0)\rangle + A_2 \operatorname{Tr}\langle \pmb{\beta}^H(t)\pmb{\beta}^H(0)\rangle$$

avec

$$A_0 = |\pmb{\lambda}\cdot\pmb{\mu}|^2$$
$$A_1 = \tfrac{1}{3}|\pmb{\lambda}\times\pmb{\mu}|^2$$
$$A_2 = \tfrac{1}{30}\{3(1 + |\pmb{\lambda}\cdot\pmb{\mu}^*|^2) - 2|\pmb{\lambda}\cdot\pmb{\mu}|^2\}$$

α, $\pmb{\eta}$ et $\pmb{\beta}$ sont les grandeurs déduites de ξ et se transformant comme $l=0$, $l=1$, $l=2$.

Pour un tenseur symétrique et réel $\pmb{\eta} \equiv 0$. On retrouve ainsi les résultats bien connus pour les géométries parallèles et perpendiculaires.

2.2.3. Effet hyper-Raman

Par la méthode décrite ci-dessus on obtient facilement les formules pour un tenseur quelconque du 3ème ordre. W. L. Peticolas et al. [3] ont examiné cet effet dans le cas d'un tenseur hyper-Raman symétrique. Il a, à notre avis, donné des formules erronnées pour les fonctions de corrélation. Pour un tenseur du 3ème ordre symétrique la décomposition suivant les représentations irréductibles de R_3 ne fait intervenir qu'un vecteur $\pmb{\eta}(l=1)$ et un tenseur $\gamma(l=3)$ on a

$$G(t) = A_1 \langle \pmb{\eta}^H(t)\cdot\pmb{\eta}^H(0)\rangle + A_3 \langle \operatorname{Tr}\gamma^H(t)\gamma^H(0)\rangle$$

avec

$$A_1 = \tfrac{1}{3}|\pmb{\lambda}(\pmb{\mu}\cdot\pmb{v}) + \pmb{\mu}(\pmb{v}\cdot\pmb{\lambda}) + \pmb{v}(\pmb{\lambda}\cdot\pmb{\mu})|^2$$
$$A_3 = \tfrac{1}{7}\{\tfrac{1}{6}[1 + |\pmb{\lambda}\cdot\pmb{\mu}^*|^2 + |\pmb{\lambda}\cdot\pmb{v}^*|^2 + |\pmb{\mu}\cdot\pmb{v}^*|^2 + (\pmb{\lambda}\cdot\pmb{v}^*)(\pmb{v}\cdot\pmb{\mu}^*)(\pmb{\mu}\cdot\pmb{\lambda}^*) +$$
$$+ (\pmb{\lambda}\cdot\pmb{\mu}^*)(\pmb{\mu}\cdot\pmb{v}^*)(\pmb{v}\cdot\pmb{\lambda}^*)] - \tfrac{1}{15}|\pmb{\lambda}(\pmb{\mu}\cdot\pmb{v}) + \pmb{\mu}(\pmb{v}\cdot\pmb{\lambda}) + \pmb{v}(\pmb{\lambda}\cdot\pmb{\mu})|^2\}.$$

Dans les géométries particulières examinées dans la référence (13) on retrouve les bonnes valeurs numériques pour A_1 et A_3 (compte tenu évidemment des définitions de $\pmb{\eta}$ et γ)*.

2.2.4. Conclusion

L'intérêt principal de cette méthode réside dans la possibilité d'extraire à partir des spectres, par une série de géométries expérimentales comprenant l'utilisation de polarisations circulaires ($\pmb{\lambda}$, $\pmb{\mu}$, $\pmb{v}$ complexes), les fonctions de corrélation entre scalaires, vecteurs, etc....

On peut ainsi comparer les données obtenues par les différentes spectroscopies vibrationnelles et ainsi mieux tester les modèles décrivant les mouvements moléculaires en solution.**

---
* Pour l'étude détaillée du cas général voir J. C. Leicknam, et Y. Guissani – article à paraître.
** Pershan et Callendar [20] ont utilisé une séparation du même genre mais dans un cristal de symétrie O_h pour l'effet Raman.

3. Rotation diffusionnelle

3.1. Generalites

Les fonctions générales $G_i(t)$ ci-dessus définies seront calculées dans le cas particulier ou la diffusion rotationnelle est le processus dominant. Nous ne tiendrons pas compte ici des effets vibrationnels qui peuvent être séparés moyennant certaines hypothèses[*] et proposons une méthode d'approche pour l'étude de la limite de diffusion pour la rotation des molécules polyatomiques.

3.2. Methode

Généralement [14–17] cette étude se fait par le biais d'une équation de diffusion sur les probabilités conditionnelles, qu'il faut ensuite résoudre par la méthode standard des fonctions de Green. L'emploi d'exponentielles ordonnées d'opérateurs et la technique du développement en cumulants [18] permettent d'éviter ce détour et donnent aux hypothèses de nature stochastique leur sens physique plus directement assimilable.

On montre que si

$$G_{ijk}(t) = \langle \xi_{ijk}(t) \xi_{ijk}(0) \rangle$$

alors

$$G_{ijk}(t) = \xi_{ijk}(0) \langle exp\left[i \int_0^t \Omega(t') \cdot \mathbf{L}\, dt'\right]\rangle \xi_{ijk}(0),$$

ou $\mathbf{L}$ est l'opérateur moment cinétique agissant sur les composantes du tenseur ξ et $\Omega(t)$ le vecteur vitesse instantanée de rotation à l'instant t.

Les deux hypothèses de modulation gaussienne et des temps longs ($t \gg \tau_c$) de ce formalisme sont respectivement identiques aux hypothèses dites des petits angles et d'un processus de Markov, dans l'analyse des équations de diffusion [19].

On obtient alors directement

$$G_{ijk}(t) = \xi_{ijk}(0)\, e^{-\mathbf{LDL}t}\, \xi_{ijk}(0)$$

avec

$$\mathbf{D} = \int_0^\infty \langle \Omega(0) \otimes \Omega(t) \rangle\, dt.$$

La partie proprement dynamique du problème est à ce niveau résolue. Un des intérêts de l'écriture de $G(t)$ comme somme de fonctions de corrélation de grandeurs scalaires, vectorielles, etc.... devient ici apparent. Il ne reste alors qu'à diagonaliser $\mathbf{LDL}$ sur une base de fonctions se transformant comme $l=0$, puis $l=1$, etc.... ($\mathbf{LDL}$ commute avec $\mathbf{L}^2$)

3.3. Resultats

3.3.1. *Scalaires* ($l=0$)

$$G_0(t) = |\alpha|^2.$$

---

[*] Bratos dans sa conférence a montré les limites de telles hypothèses et les résultats qu'elles permettent d'obtenir.

3.3.2. Vecteurs ($l=1$)

x, y, z sont vecteurs propres de LDL avec les valeurs propres respectives D_y+D_z, D_x+D_z, D_x+D_y d'où:

$$G_1(t) = |\eta_x|^2 e^{-(D_y+D_z)t} + |\eta_y|^2 e^{-(D_x+D_z)t} + |\eta_z|^2 e^{-(D_x+D_y)t}.$$

3.3.3. Tenseurs du 2ème ordre ($l=2$)

xy, xz et yz sont fonctions propres de LDL avec respectivement les valeurs propres $D_x+D_y+4D_z$, $D_x+4D_y+D_z$, $4D_x+D_y+D_z$.

Il reste à diagonaliser LDL sur x^2 et y^2 (compte tenu de la relation $x^2+y^2+z^2=0$, c'est-à-dire trace nulle).

Les vecteurs propres sont:

$$\alpha_1 = (D_z - D_x)x^2 + [(D_y - D_x) - \sqrt{\Sigma_1^2 - 3\Sigma_2}]y^2$$
$$\alpha_2 = (D_z - D_x)x^2 + [(D_y - D_x) + \sqrt{\Sigma_1^2 - 3\Sigma_2}]y^2$$

où: $\quad \Sigma_1 = D_x + D_y + D_z, \quad \Sigma_2 = D_x D_y + D_x D_z + D_y D_z.$

avec les valeurs propres

$$\lambda_{1,2} = 2\Sigma_1 \pm 2\sqrt{\Sigma_1^2 - 3\Sigma_2}$$

On obtient ainsi:

$$G_2(t) = \sum_{i=1}^{5} |\alpha_i|^2 e^{-\lambda_i t}.$$

3.3.4. Tenseurs du 3ème ordre ($l=3$)

xyz est fonction propre de LDL ayant pour valeur propre $4(D_x+D_y+D_z)$.

La diagonalisation de LDL s'effectue sur 3 blocs 2×2 indépendants. On diagonalise donc LDL sur x^3 et xz^2 (compte tenu de la relation $x^3+xz^2+xy^2=0$). Par permutation circulaire sur x, y, z on obtient les autres vecteurs propres et valeurs propres.

On obtient ainsi:

$$G_3(t) = \sum_{i=1}^{7} |\alpha_i|^2 e^{-\lambda_i t}.$$

Nous donnons ici[*] le tableau des valeurs propres λ_i

$$\lambda_1 = 4(D_x+D_y+D_z)$$
$$\lambda_{2,3} = (2D_x + 5D_y + 5D_z) \pm \sqrt{(D_x - D_y)(D_x - D_z) + 4(D_y - D_z)^2}$$
$$\lambda_{4,5} = (2D_y + 5D_z + 5D_x) \pm \sqrt{(D_y - D_z)(D_y - D_x) + 4(D_z - D_x)^2}$$
$$\lambda_{6,7} = (2D_z + 5D_x + 5D_y) \pm \sqrt{(D_z - D_x)(D_z - D_y) + 4(D_x - D_y)^2}.$$

---

[*] Pour une discussion complète des résultats en fonction de la symétrie de la molécule envisagée voir Y. Guissani, J. C. Leicknam – article à paraître.

On aurait pu se référer au problème bien connu du rotateur asymétrique quantique libre [15-17]. Cependant, l'introduction d'un axe z préférentiel cache la symétrie entre x, y et z, alors que celle-ci est préservée tout au long d'un calcul direct qui ne nécessite jamais plus que la diagonalisation d'une matrice 2×2 même dans le cas d'un tenseur du 3ème ordre.

3.4. Discussion

Les résultats obtenus ici sont connus [17] sauf pour $l=3$. La méthode semble plus directe que celle utilisée généralement. D'autre part, le formalisme permet en mettant en lumière les hypothèses de nature stochastique d'espérer généraliser ce traitement à d'autres modèles de rotation.

4. Conclusion

Dans cet article, nous avons cherché, dans les milieux isotropes, à définir les fonctions de corrélation nécessaires à l'analyse des profils des spectres vibrationnels dans le cas général où le tenseur en jeu n'est pas nécessairement réel et symétrique. Dans un second temps, nous avons donné en exemple d'application ces fonctions de corrélation dans le cas limite de rotation diffusionnelle.

Bibliographie

1. Bratos, S. et Marechal, E.: *Phys. Rev.* **A4**, 1078 (1971).
2. Bartoli, F. J. et Litovitz, T. A.: *J. Chem. Phys.* **56**, 404 et 413 (1972).
3. Nafie, L. A. et Peticolas, W. L.: *J. Chem. Phys.* **57**, 3145 (1972).
4. Monson, P. R. et McClain, W. M.: *J. Chem. Phys.* **53**, 29 (1970).
5. McClain, W. M.: *J. Chem. Phys.* **55**, 2789 (1971).
6. McClain, W. M.: *J. Chem. Phys.* **57**, 2264 (1972).
7. McClain, W. M.: *J. Chem. Phys.* **58**, 324 (1973).
8. Placzek, G.: 'The Rayleigh and Raman Scattering', in *Handbuch der Radiologie* 2, Leipzig Akademische Verlagsgesellschaft, 1934, VI, p. 209.
9. Kubo, R.: *Report. Progress. Phys.* **24**, Part, 255 (1966).
10. Gordon, R. G.: *J. Chem. Phys.* **42**, 3658 (1965).
11. Gel'fand, I. M., Minlos, R. A., et Shapiro, Z. Ya.: *Representations of the Rotation and Lorentz Groups and Their Applications*, Pergamon Press, 1963, Chap. 1, Section 5.
12. Landau, L. D. et Lifshitz, E. M.: *Relativistic Quantum Theory*, Pergamon Press, 1971, Vol. 4, Part 1, Chap. VI.
13. Cyvin, S. J., Rauch, J. E., et Decius, J. C.: *J. Chem. Phys.* **43**, 4083 (1965).
14. Perrin, F.: *J. Phys. Radium* **5**, 497 (1934).
15. Favro, L. D.: *Phys. Rev.* **119**, 53 (1960).
16. Huntress, W. T., Jr.: *Adv. Magn. Resonance* **4** (1970).
17. Valiev, K. A.: *Opt. Spectry.* **13**, 282 (1962).
18. Kubo, R.: *J. Phys. Soc. Japan* **17**, 1100 (1962).
19. Chandrasekhar: *Rev. Mod. Phys.* **15**, 1 (1943).
20. Callendar, R. et Pershan, P. S.: *Phys. Rev.* **A2**, 672 (1970).

DISCUSSION

Kneubühl: Quelle est la différence entre votre théorie des groupes de la diffusion rotationnelle et la théorie de Keller *et al.*? Est-ce qu'il y a des erreurs dans la théorie de Keller?

Guissani: Nous pensons que Keller ne s'est absolument pas posé le même problème que nous. Il a cherché à tenir compte, à l'aide de la théorie des groupes, de la symétrie de la molécule en question et en a déduit des relations entre fonctions de corrélation pour l'infrarouge et en appendice pour le Raman dans le cas d'un tenseur symétrique. Alors que, tenant compte, à l'aide de la théorie des groupes, uniquement de l'isotropie du milieu pour un tenseur quelconque de rang 2 et 3, notre problème est en amont. Nos sujets étant disjoints, nous nous garderons ici d'apprecier s'il y a d'éventuelles erreurs dans la théorie de Keller et Kneubühl.

THE INFLUENCE OF VIBRATION-ROTATION COUPLING ON CORRELATION FUNCTIONS OF MOLECULES

K. MÜLLER, P. ETIQUE, and F. KNEUBÜHL
Solid State Physics Laboratory ETH, Zürich, Switzerland

Abstract. The influence of the Coriolis coupling of the triply degenerate vibrations of spherical top molecules on the angular autocorrelation functions is studied theoretically. The results are applied to experimental correlation functions of carbonyls [Cr(CO)$_6$, Mo(CO)$_6$] dissolved in gaseous N_2 and liquid CCl_4. In many cases the vibrational relaxation can be separated from the rotational correlation functions and compared with relevant theories.

Résumé. L'influence du couplage de Coriolis sur les fonctions de corrélation angulaire est étudiée théoriquement dans le cas des vibrations trois-fois dégénérées des molécules sphériques.

Les résultats sont appliqués aux fonctions de corrélation expérimentales de carbonyls [Cr(CO)$_6$, Mo(CO)$_6$] dissous dans du N_2 gazeux et dans du CCl_4 liquide.

Dans quelques cas la fonction de relaxation vibrationnelle est séparable de la fonction de relaxation rotationnelle et on peut la comparer à des théories relevantes.

The interpretation of the absorption line shapes of rotating molecules with the aid of time-dependent correlation functions [1, 2] was initiated by Kubo's linear response theory [3, 4]. These correlation functions illustrate the average of the molecular rotation and the vibrational relaxation. They can be determined from the Fourier transform of the corresponding line profiles.

Many authors [5–13, 18] constructed specific theoretical models for the rotational relaxation processes of molecules in liquids. The gamut of their models extends from the free rotation studied by Steele [7, 15] to the pure rotational diffusion according to Debye [14]. Most of the theoretical and experimental work on rotational relaxation is restricted to diatomic or spherical top molecules.

Only recently Keller and Kneubühl [12] developed a group theoretical method to define angular correlation functions for symmetric and asymmetric top molecules and the experimental procedure to determine them from vibration-rotation or Raman spectra. However, their considerations are strictly valid for rigid-top molecules only.

To our knowledge, none of the papers on correlation functions of rotating molecules deals with the coupling of the rotational and the vibrational angular momenta, which may occur in spherical or symmetric top molecules. This coupling is also called Coriolis coupling or vibration-rotation coupling. It is the purpose of this paper to discuss the influence of the Coriolis coupling on correlation functions. For simplicity we restrict our considerations to spherical top molecules.

In the rigid-rotor approximation, the Hamiltonian of a spherical top molecule can be represented by

$$H = H_{\text{vib}} + \frac{(\mathbf{J} - \boldsymbol{\kappa})^2}{2I} \tag{1}$$

with the corresponding eigenvalues

$$E = \hbar\omega \cdot (\bar{n} + \tfrac{1}{2} - \kappa) + \frac{\hbar^2}{2I} \cdot J(J+1) + \frac{\hbar^2}{2I} \cdot \kappa(\kappa+1) - \frac{\zeta}{I}(\mathbf{J}\cdot\mathbf{\kappa}), \qquad (2)$$

where

- $\mathbf{J}$ = total angular momentum
- $\mathbf{\kappa}$ = vibrational angular momentum
- I = moment of inertia
- ζ = constant of the Coriolis coupling
- $\bar{n}$ = vibrational quantum number.

The first term gives the vibrational energy of a three-dimensional harmonic oscillator. The last term describes the Coriolis coupling [17].

The vibrational angular momentum $\mathbf{\kappa}(t)$ is defined in the molecule-fixed coordinate system. The corresponding equation of motion is

$$\frac{\partial}{\partial t}\mathbf{\kappa}(t) = \mathbf{M}_{\text{internal}}, \qquad (3)$$

where the internal torque, $\mathbf{M}_{\text{internal}}$, is caused by the angular momentum $\mathbf{J}(t)$. It can be evaluated by analogy to the torque a magnetic dipole $\mathbf{\mu}$ suffers in a magnetic field $\mathbf{B}$.

The potential energy of $\mathbf{\mu}$ is

$$E_P = -\mathbf{\mu}\cdot\mathbf{B}$$

and its torque

$$\mathbf{M} = \mathbf{\mu} \times \mathbf{B}.$$

Thus the torque corresponding to the Coriolis term

$$E_P = -\frac{\zeta}{I}\cdot\mathbf{\kappa}\cdot\mathbf{J} \qquad (4)$$

is

$$\mathbf{M}_{\text{internal}} = -\frac{\zeta}{I}\cdot[\mathbf{J} \times \mathbf{\kappa}]. \qquad (5)$$

If we change our frame of reference to a space-fixed system, our equation of motion is transformed to:

$$\frac{d}{dt}\mathbf{\kappa}(t) = \frac{\partial}{\partial t}\mathbf{\kappa}(t) + [\mathbf{\omega} \times \mathbf{\kappa}(t)], \qquad (6)$$

where the vector $\mathbf{\omega}$ is the angular frequency of the rotating molecule. Because the molecule is spherical, $\mathbf{\omega}$ corresponds to $\mathbf{J}/I$. Thus we obtain:

$$\frac{d}{dt}\mathbf{\kappa}(t) = \frac{1-\zeta}{I}\cdot[\mathbf{J} \times \mathbf{\kappa}(t)]. \qquad (7)$$

Before we discuss the solutions of this equation, we pay attention to the autocorrelation function (ACF) of the transition dipole moment **m** which can be calculated by taking the Fourier transform of the measured absorption band shape. This ACF is defined by:

$$\left\langle \frac{\mathbf{m}(t) \cdot \mathbf{m}(0)}{m(0)^2} \right\rangle = \left\langle \frac{m(0) \cdot m(t)}{m(0) \cdot m(0)} \cdot \mathbf{u}(t) \cdot \mathbf{u}(0) \right\rangle, \tag{8}$$

where $\mathbf{u}(t)$ indicates the direction of the transition dipole and $m(t)$ its magnitude. $\mathbf{u}(t)$ describes the rotation of the transition dipole and $m(t)$ denotes its vibrational relaxation.

Assuming weak coupling between the vibrational and the rotational-translational modes of the molecules, Bratož [18] and Morawitz [10] find a separation of the above ACF into a vibrational and a rotational relaxation function:

$$\left\langle \frac{\mathbf{m}(0) \cdot \mathbf{m}(t)}{m(0) \cdot m(0)} \right\rangle = \left\langle \frac{m(0) \cdot m(t)}{m(0) \cdot m(0)} \right\rangle \cdot \langle \mathbf{u}(0) \cdot \mathbf{u}(t) \rangle = G^{\text{vib}}(t) \cdot G^{\text{rot}}(t). \tag{9}$$

We now assume the rotational relaxation of the transition dipole $\mathbf{m}(t)$ to correspond to the relaxation of the vibrational angular momentum $\kappa(t)$:

$$G^{\text{rot}}(t) = \left\langle \frac{\mathbf{\kappa}(0) \cdot \mathbf{\kappa}(t)}{\kappa(0) \cdot \kappa(0)} \right\rangle. \tag{10}$$

We now return to the problem of the solution of the equation of motion of $\kappa(t)$ in a space-fixed frame of reference. It can be solved exactly for:

$$\mathbf{J}(t) = \mathbf{J} = \text{constant}. \tag{11}$$

By taking the average over all possible initial directions of κ and the Boltzmann distribution of the **J**'s, we find the following ACF, also shown in Figure 1:

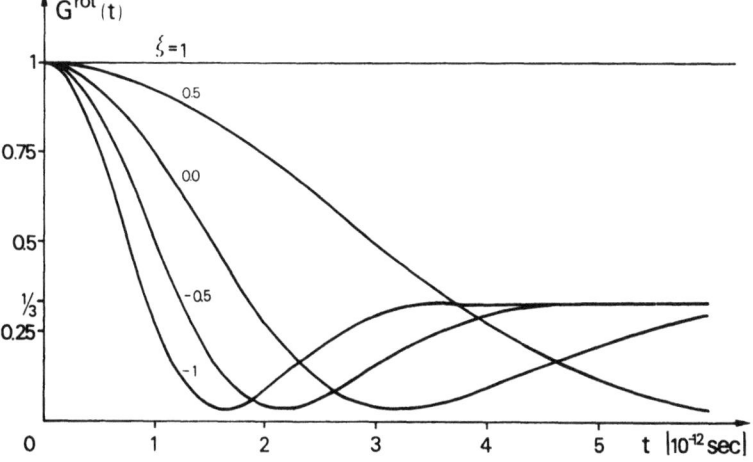

Fig. 1. The influence of the Coriolis coupling constant ζ on $G^{\text{rot}}(t)$.

$$G_0^{rot}(t) = \tfrac{1}{3} + \tfrac{2}{3} \cdot \left[1 - \frac{kT}{I} \cdot (1-\zeta)^2 \, t^2\right] \exp\left[-\frac{kT}{2I} \cdot (1-\zeta)^2 \cdot t^2\right]. \qquad (12)$$

For $\zeta=0$, this rotational relaxation equals the ACF obtained by Steele [7] for the proper rotator.

In liquids the angular momentum **J** of a molecule is statistically varying. For missing Coriolis coupling ($\zeta=0$), solutions of the Equation (7) have been found by various authors [6, 7, 8, 9, 11, 13].

Fortunately, the correlation function calculated for $\zeta \neq 0$ can be obtained from those with $\zeta=0$ by replacing

J by **J**$(1-\zeta)$

and

T by $T(1-\zeta)^2$.

As an example, we present the ACF $G^{rot}(t)$ for spherical molecules with small rotational friction ζ originally evaluated by Steele [7]

$$G^{rot}(t) = G_0^{rot}(t) + t^3 \cdot \frac{\zeta}{3I^2} \cdot kT \cdot (1-\zeta)^2 \cdot \left[3 - t^2 \cdot \frac{kT}{I}(1-\zeta)^2\right] \times$$
$$\times \exp\left[-\frac{t^2}{2} \cdot \frac{kT}{I}(1-\zeta)^2\right]. \qquad (13)$$

The vibrational relaxation $G^{vib}(t)$ can be understood with aid of a model developed by Anderson [19] and Kubo [20]: The time-dependent magnitude **m**(t) of the transition dipole moment obeys the equation

$$\frac{d}{dt} m(t) = -i \cdot \Delta\omega(t) \cdot m(t) \qquad (14)$$

with the solution

$$G^{vib}(t) = \left\langle \frac{m(t) \cdot m(0)}{m(0) \cdot m(0)} \right\rangle = \left\langle \exp\left[-i \int_0^t \Delta\omega(t) \cdot dt\right] \right\rangle. \qquad (15)$$

If $\Delta\omega(t)$ describes a Gaussian process, the relaxation function $G^{vib}(t)$ can be represented by

$$G^{vib}(t) = \exp\left[-\langle(\Delta\omega)^2\rangle \cdot \int_0^t (t-\tau) \cdot \psi(\tau) \, d\tau\right] \qquad (16)$$

with

$$\psi(\tau) = \left\langle \frac{\Delta\omega(t) \cdot \Delta\omega(t+\tau)}{\Delta\omega(t) \cdot \Delta\omega(t)} \right\rangle. \qquad (17)$$

Two cases are of interest:

(a) Gaussian-Gaussian modulation of $\Delta\omega(t)$:

$$\psi(t) = \exp\left[-(t/\tau_c)^2\right]. \tag{18}$$

(b) Gaussian-Markoffian modulation of $\Delta\omega(t)$:

$$\psi(t) = \exp\left[-t/\tau_c\right]. \tag{19}$$

The relaxation time τ_c is defined by

$$\tau_c = \int_0^\infty \psi(t)\,dt.$$

For an experimental study of the influence of the Coriolis coupling on angular and vibrational relaxation functions we selected hexa- and tetracarbonyls. They represent spherical top molecules with large Coriolis coupling of some triply degenerate ir-active vibrational modes [21, 22]. Table I shows the ir-active vibrations and the corresponding Coriolis coupling constants of the carbonyls investigated.

TABLE I [21, 22]

σ_i [cm^{-1}] (ζ_i) [1]	σ_5	σ_6	σ_7	σ_8	σ_9
Ni(CO)$_4$	2057.8	458.9	423.1	80	–
	(−0.01)	(0.42)	(0.38)	(0.21)	–
Cr(CO)$_6$	–	2000.4	668.1	440.5	97.8
	–	(0.00)	(0.77)	(0.30)	(−0.07)
Mo(CO)$_6$	–	2003.0	595.6	367.2	81.6
	–	(0.00)	(0.67)	(0.28)	(0.05)
W(CO)$_6$	–	1997.6	586.6	374.4	82.0
	–	(0.00)	(0.62)	(0.18)	(0.20)

We measured in detail the band shapes of σ_5 of Ni(CO)$_4$ and σ_6, σ_7, σ_8 of the three other carbonyls dissolved in gaseous N$_2$ and in liquid CCl$_4$ at room temperature. Subsequently we determined the ACFs of the transition dipole moments [1, 2]. For Cr(CO)$_6$ in gaseous N$_2$ the result is presented in Figure 2, whereas Figure 3 shows the correlation functions for Mo(CO)$_6$ in liquid CCl$_4$.

For the carbonyls in the gas phase the angular momenta J of the molecules can be assumed constant within the short times considered. Thus $G_0^{\mathrm{rot}}(t)$ is given by Equation (12). Because the experimental correlation functions correspond to the product $G_0^{\mathrm{rot}}(t) \cdot G^{\mathrm{vib}}(t)$, the vibrational correlation function $G^{\mathrm{vib}}(t)$ can be deduced. The $G^{\mathrm{vib}}(t)$ corresponding to σ_6, σ_7, σ_8 of Cr(CO)$_6$ in N$_2$ are plotted on Figure 4. The shapes of the $G^{\mathrm{vib}}(t)$ can be approximated by one or two Gaussians. They correspond to the model of Anderson and Kubo for a Gaussian-Gaussian modulation of the vibrational frequency.

The relatively fast vibrational relaxations of the carbonyls are probably due to the

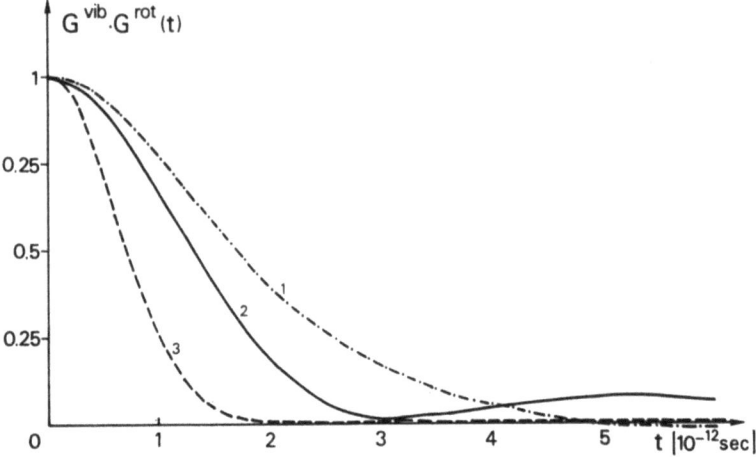

Fig. 2. Experimental autocorrelation functions of the transition dipoles of $\sigma_7(1)$, $\sigma_6(2)$ and $\sigma_8(3)$ of $Cr(CO)_6$ in gaseous N_2.

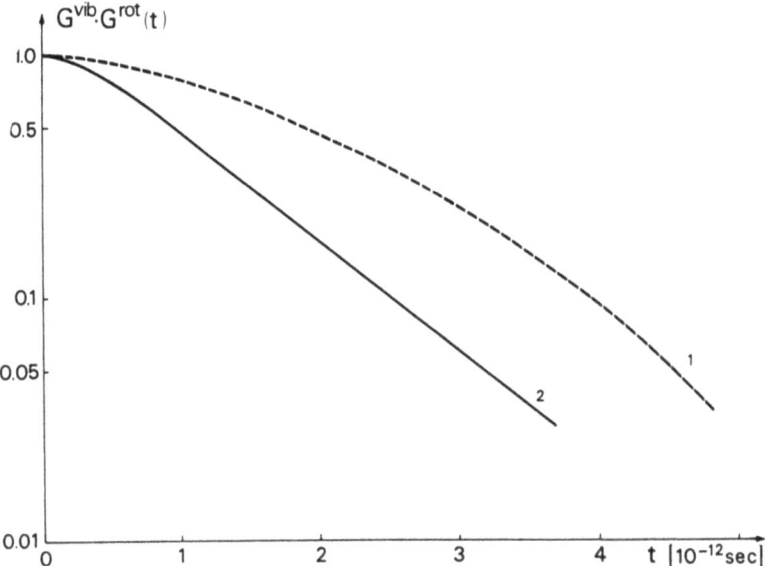

Fig. 3. Experimental autocorrelation functions of the transition dipoles of $\sigma_7(1)$ and $\sigma_{6,8}(2)$ of $Mo(CO)_6$ dissolved in liquid CCl_4.

anharmonic terms of the intramolecular potentials of these large molecules. Some of the decay channels within the insulated molecules allowed by the conservation of energy are:

$$\sigma_6 \to 2\sigma_7 + 2\sigma_8 + \sigma_9$$
$$\sigma_7 \to \sigma_9 + \sigma_{12}$$
$$\sigma_8 \to 3\sigma_9 + 2\sigma_{13}.$$

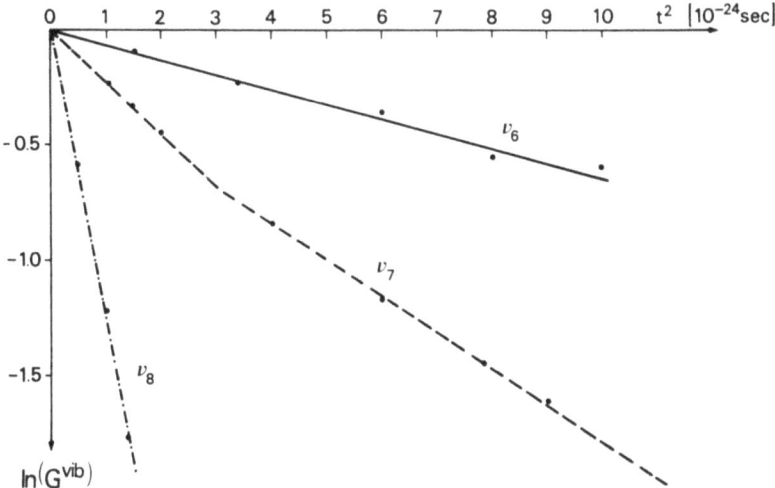

Fig. 4. Vibrational relaxation functions $G^{vib}(t)$ of σ_6, σ_7, σ_8 of $Cr(CO)_6$ in gaseous N_2.

The ACF of the carbonyls in liquids Figure 3 differ considerably from those in the gas phase. The ACFs of σ_6 and σ_8 of $MO(CO)_6$ become exponentials at times greater than 0.7×10^{-12} s. This is in agreement with the theories of Bratož [18] for diatomic molecules and Morawitz [10] for polyatomic molecules, which predict an exponential decay for both functions, $G^{vib}(t)$ and $G^{rot}(t)$, for times corresponding to a large number of collisions.

The ACF of σ_7 of $Mo(CO)_6$ remains Gaussian for times up to 5×10^{-12} s. This observation can be understood qualitatively:

(i) The vibration σ_7 is concentrated on the central Mo-atom and the six neighbouring C-atoms. In contrast, the six O-atoms perform only minor oscillations [21]. Thus the O-atoms protect the σ_7-vibration against the collisions with the solvent molecules.

(ii) The Coriolis coupling constant ζ_7 has a large positive value. Hence, the decay of the rotational relaxation function of the vibrational angular momentum **κ** and the direction **u** of the transition dipole moment is slow. It differs noticeably from the rotational relaxation of the molecule itself.

These two phenomena cause a delay of up to 5×10^{-12} s in the change from the Gaussian part of the ACF to the exponential section. For all triply degenerate vibrations of carbonyls dissolved in liquids, the vibrational relaxation can be described adequately by the model of Anderson and Kubo [19, 20], assuming Gaussian-Markoffian modulation of the vibrational frequency. This is a consequence of the relatively large size of the carbonyl molecules. From the above theoretical and experimental investigations we conclude that the Coriolis coupling has a considerable influence on the angular correlation functions of spherical top molecules in gases as well as in liquids. At present these studies are being extended to symmetric top molecules.

Acknowledgement

We wish to thank Dr B. Keller for valuable discussions.

References

1. Gordon, R.: *J. Chem. Phys.* **43**, 1307 (1965).
2. Shimizu, H.: *J. Chem. Phys.* **43**, 2453 (1965); **48**, 2494 (1968).
3. Kubo, R.: *J. Phys. Soc. Japan* **12**, 570 (1957).
4. Kubo, R.: in W. E. Brittin and L. G. Dunham (eds.), *Lectures in Theoretical Physics*, Vol. I, Interscience, New York, 1961, p. 20.
5. Gordon, R.: *J. Chem. Phys.* **44**, 1830 (1966).
6. Furry, W. H.: *Phys. Rev.* **107**, 7 (1957).
7. Steele, W. A.: *J. Chem. Phys.* **38**, 2404 and 2411 (1963).
8. Fixman, M. and Rider, K.: *J. Chem. Phys.* **51**, 2425 (1969).
9. Mc Clung, R. E. D.: *J. Chem. Phys.* **51**, 3842 (1969).
10. Morawitz, H. and Eisenthal, K. B.: *J. Chem. Phys.* **55**, 887 (1971).
11. Hubbard, P. S.: *Phys. Rev.* A**6**, 2421 (1972).
12. Keller, B. and Kneubühl, F.: *Helv. Phys. Acta* **45**, 1127 (1972).
13. Van Konynenburg, P. and Steele, W. A.: *J. Chem. Phys.* **56**, 4776 (1972).
14. Debye, P.: *Polar Molecules*, Dover Publ., New York, 1954.
15. St. Pierre, A. G. and Steele, W. A.: *Phys. Rev.* **184**, 172 (1969).
16. Teller, E. and Tisza, L.: *Z. Physik* **73**, 791 (1932).
17. Allen, H. C. and Cross, P. C.: *Molecular Vib. Rotors*, J. Wiley, New York, 1963.
18. Bratoz, S., Rios, J., and Guissani, Y.: *J. Chem. Phys.* **52**, 439 (1970).
19. Anderson, P. W.: *J. Phys. Soc. Japan* **9**, 3 and 316 (1954).
20. Kubo, R.: in K. E. Shuler (ed.), *Stochastic Process in Chemical Phyiscs*, Interscience Publishers, 1969.
21. Jones, L. H., McDowell, R. S., and Goldblatt, M.: *Inorganic Chemistry* **8**, 2349 (1969).
22. Jones, L. H., McDowell, R. S., and Goldblatt, M.: *J. Chem. Phys.* **48**, 2663 (1968).

DISCUSSION

Vincent: En ce qui concerne le couplage de Coriolis, si j'ai bien compris, vos formules sont seulement applicables aux bandes F des rotateurs sphériques. Avez-vous étudié le cas des bandes E des rotateurs symétriques?

Kneubühl: Nous sommes en train d'étudier l'effet Coriolis des rotateurs symétriques. En outre, nous mesurons les spectres rotationnel-vibrationnel de $CHCl_3$.

RAMAN STUDIES OF MOLECULAR REORIENTATION IN LIQUID SULFUR HEXAFLUORIDE*

S. SUNDER and R. E. D. McCLUNG

Dept. of Chemistry, University of Alberta, Edmonton, Alberta, Canada

Abstract. The contour of the v_2 band of SF_6 has been studied over the temperature range 230–300 K. Reorientational and angular momentum correlation times were obtained by comparing the Fourier transforms of the Raman bands with reorientational correlation functions calculated from the extended J-diffusion model. A reanalysis of Hackleman and Hubbard's nuclear resonance data for SF_6 yields the spin-rotation asymmetry $C_\parallel - C_\perp = \pm\, 6.0 \times 10^4\ \text{s}^{-1}$ which is comparable to those of other hexafluorides.

Résumé. Le contour de la bande v_2 de SF_6 a été étudié dans l'intervalle de température 230–300 K. Les temps de correlation réorientationnel et moment angulaires ont été obtenus en comparant le transformé Fourier des bandes de Raman avec les fonctions de correlation réorientationnelles calculées à partir du modèle de diffusion-J. Un nouvel analyse des données de résonance nucléaire de Hackleman et Hubbard a résulté dans l'asymétrie spin-rotationnelle $C_\parallel - C_\perp = \pm\, 6.0 \times 10^4\ \text{s}^{-1}$ qui est comparable avec des autres hexafluorures.

1. Introduction

Fourier analysis of Raman band contours for molecules in liquids gives reorientation-vibration correlation functions [1–5]. The reorientational correlation functions contained in these Fourier transforms characterize the time correlation of the spherical harmonics of second order [1, 3, 5]. The interpretation of the reorientational correlation functions in terms of the details of the motion of the molecules in the liquid is usually based on theoretical models for this motion [3]. The reorientational correlation functions and correlation times obtained from the analysis of Raman bands are also important in the interpretation of nuclear relaxation data [6, 7, 8].

In this paper, a study of the v_2 Raman band of liquid sulfur hexafluoride over the temperature range 230–300 K is reported. Nuclear relaxation measurements on liquid SF_6 in this temperature range have been reported by Hackleman and Hubbard [9] and interpreted in terms of the rotational diffusion equation model [6] for molecular rotation in liquids. Our results are compared with the correlation functions and spectral densities predicted by the extended rotational diffusion model [10], and the results of this comparison are used to reinterpret the nuclear relaxation data.

2. Experimental

Samples for Raman studies were prepared by distilling sulfur hexafluoride (Matheson research grade) *in vacuo* into thick-wall pyrex capillaries (T. H. Garner Co., Claremont, Calif., U.S.A.). The capillary was immersed in liquid nitrogen and sealed off under vacuum.

---

* Research supported in part by the National Research Council of Canada under Operating Grant A 5887.

Raman scattering excited by the 4880 Å excitation line from a Coherent Radiation *52GA* argon ion laser was recorded with a 90° scattering geometry using a Spex model 1401 dual monochromator, a photomultiplier tube (RCA C31034) cooled to $-30\,°C$ and photon-counting electronics. The laser output was filtered to remove extraneous emission lines and the power at the sample was about 200 mV. The slit width of the monochromator was adjusted for 2 cm^{-1} resolution for all experiments.

Temperature control was achieved using a cold nitrogen gas stream flowing over a heater-sensor system (Thermoelectric model 401) in an unsilvered vacuum-jacketed transfer tube. The sample temperature was measured with an iron-constantan thermocouple placed near the sample capillary.

The spectra were recorded in analog form with a Hewlett-Packard model 7100B strip chart recorder, and in digital form with a Hewlett-Packard model 34701A digital voltmeter connected to a Hewlett-Packard model 5055A digital recorder. The signal-to-noise ratio of each spectrum was enhanced by averaging four or more digitized spectra run under identical experimental conditions.

3. Results

The contours of the v_2 band of liquid SF$_6$ at 232 and 290 K, and the v_1 band at 290 K are shown in Figure 1. Since the background scattering increases slightly across the v_2 band, we have applied a linear interpolation procedure for baseline correction. The v_2 band contours for parallel and perpendicular polarizations were found to be es-

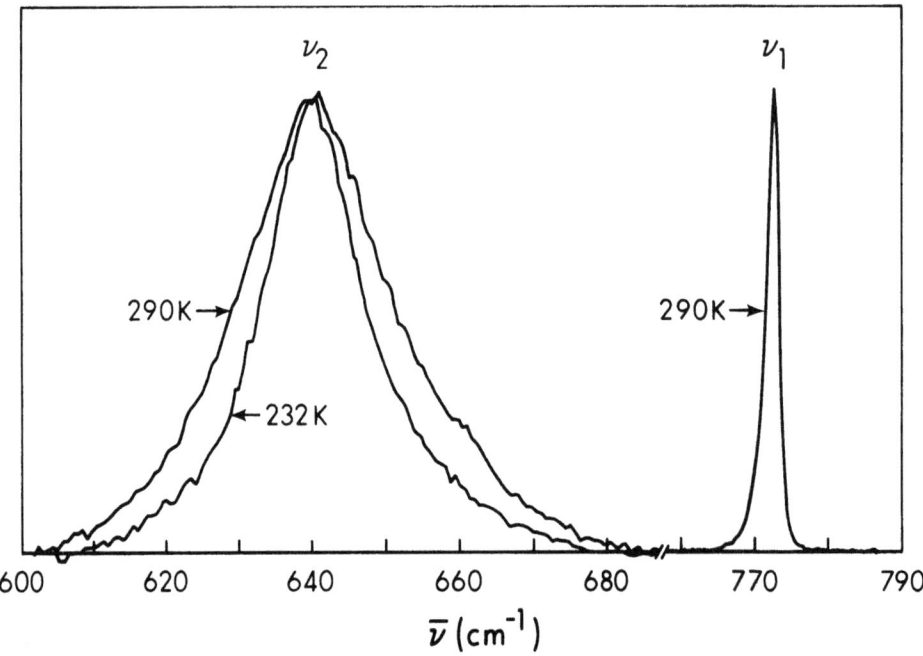

Fig. 1. Raman bandshapes of the v_1 and v_2 bands of SF$_6$.

sentially identical. The band contours presented in Figure 1 and those for Fourier analysis were recorded without a polarizer.

The Fourier transform of the v_2 band is a composite of vibrational and rotational correlation functions [5, 11] and instrumental effects, but the transform of the v_1 band is determined only by vibrational correlations and instrumental effects. Studies of the v_1 band with different slit widths indicated that most of the v_1 band shape could be attributed to instrumental effects for the slit widths used here in recording the v_2 band. The separation of the Fourier transform of the v_2 band into vibrational, reorientational and instrumental contributions cannot be made without knowledge of the vibrational correlation function, $G_{\text{vib}}(t)$, for the v_2 mode. If we assume that the vibrational correlation functions for the different modes exhibit similar time-dependence, the Fourier transform of the v_1 band will contain the same vibrational and instrumental functions as the transform of the v_2 band. We can therefore approximate the reorientational correlation $G^{(2)}(t)$ by the ratio of the Fourier transforms of the v_2 and v_1 bands. The corrected Fourier transforms of the v_2 band at 232 and 290K, and the transform of the v_1 band at 290K are shown in Figure 2. It is apparent from this figure that the vibrational and instrumental corrections are of the same order of magnitude as the uncertainties in the correlation functions for the v_2 band.

In order to extract meaningful parameters from the reorientational correlation func-

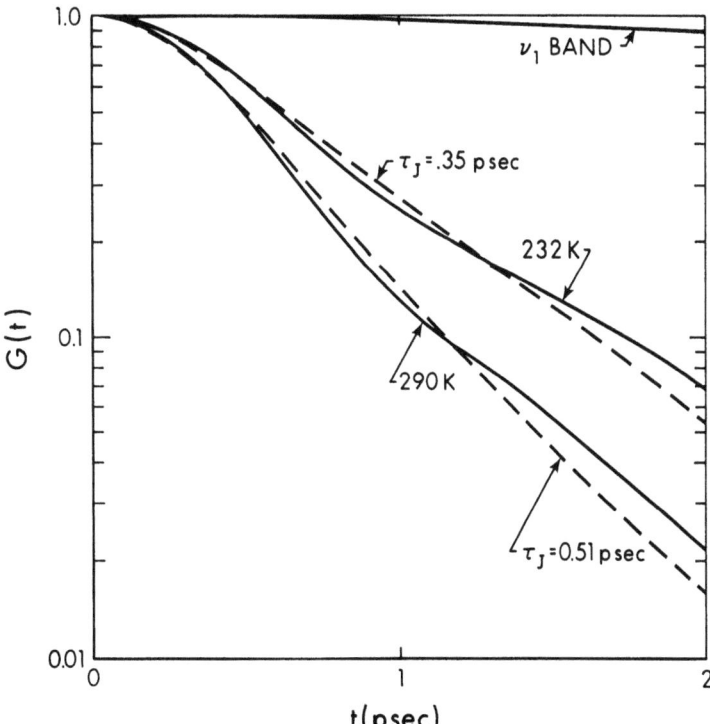

Fig. 2. Reorientational correlation functions for SF₆. —— corrected Fourier transforms of v_2 band, ---- calculated with extended J-diffusion model. The Fourier transform of the v_1 band at 290K is also included.

tions, the Fourier transforms were compared with correlation functions predicted by the M-and J-diffusion limits of the extended rotational diffusion model [10]. The spectral densities and correlation functions predicted with the M-diffusion model were incompatible with the observed data for SF_6. The calculated J-diffusion correlation functions which fit the observed ones more closely are shown in Figure 2. The comparison of calculated and observed correlation functions gives τ_J, the angular momentum correlation time, as well as $\tau_\theta^{(2)}$, the reorientational correlation time which characterizes $G^{(2)}(t)$. The values of these correlation times for SF_6 liquid are given in Table I below.

TABLE I

Reorientational and angular momentum correlation times[a] for liquid SF_6

$T(K)$	$\tau_\theta^{(2)}$ (ps)	τ_J (ps)
290	0.61	0.51
282	0.62	0.53
273	0.66	0.45
259	0.69	0.43
245	0.76	0.37
232	0.81	0.35

[a] Estimated errors in correlation times are $\pm 5\%$.

4. Discussion

The values for the correlation times for SF_6 obtained from our Raman studies indicate that the Hubbard relation [7]

$$\tau_J \tau_\theta^{(2)} = \tfrac{1}{6} kT \tag{1}$$

is not valid over the temperature range investigated. In Equation (1), I is the moment of inertia, k the Boltzmann constant and T the absolute temperature. Equation (1) is expected to be valid when $\tau_J \ll \tau_\theta^{(2)}$ where small angle diffusion is a good approximation. In liquid SF_6, the molecules rotate through angles of 20–35° in an average diffusive step, and $\tau_\theta^{(2)} \cong \tau_J$. This relatively free rotation of the molecules in liquid SF_6 is not too surprising in the light of recent observations of resolved rotational structure in the rotational [12] and vibration-rotation [13] spectra of small molecules in liquid SF_6.

Hackleman and Hubbard [9] have measured the ^{19}F nuclear relaxation times in liquid SF_6 and have interpreted them in terms of a rotational diffusion model [7, 9] which implicitly assumes that $\tau_J \ll \tau_\theta^{(2)}$ so that Equation (1) applies, and that τ_J is related to the translational diffusion coefficient D by

$$\tau_J = 3ID/4a^2 kT, \tag{2}$$

where a is the hydrodynamic radius (2.58 Å) of the SF_6 molecule. Equation (2) was derived from the Stokes relations for translational and rotational diffusion coefficients and friction constants. The extension of this relation by Kivelson et al. [14, 15]

$$\tau_J = \kappa(3ID/4a^2kT) \tag{3}$$

introduces the parameter κ which is related to the ratio of the autocorrelation functions for the torque and the force on the molecules in the liquid. The values of τ_J obtained in this work and the translational diffusion coefficients of Hackleman and Hubbard [9] follow Equation (3) approximately with $\kappa=0.1$. This value is comparable to that obtained in the analysis [15] of the infrared spectrum of liquid $CH_4 (\kappa=0.024)$

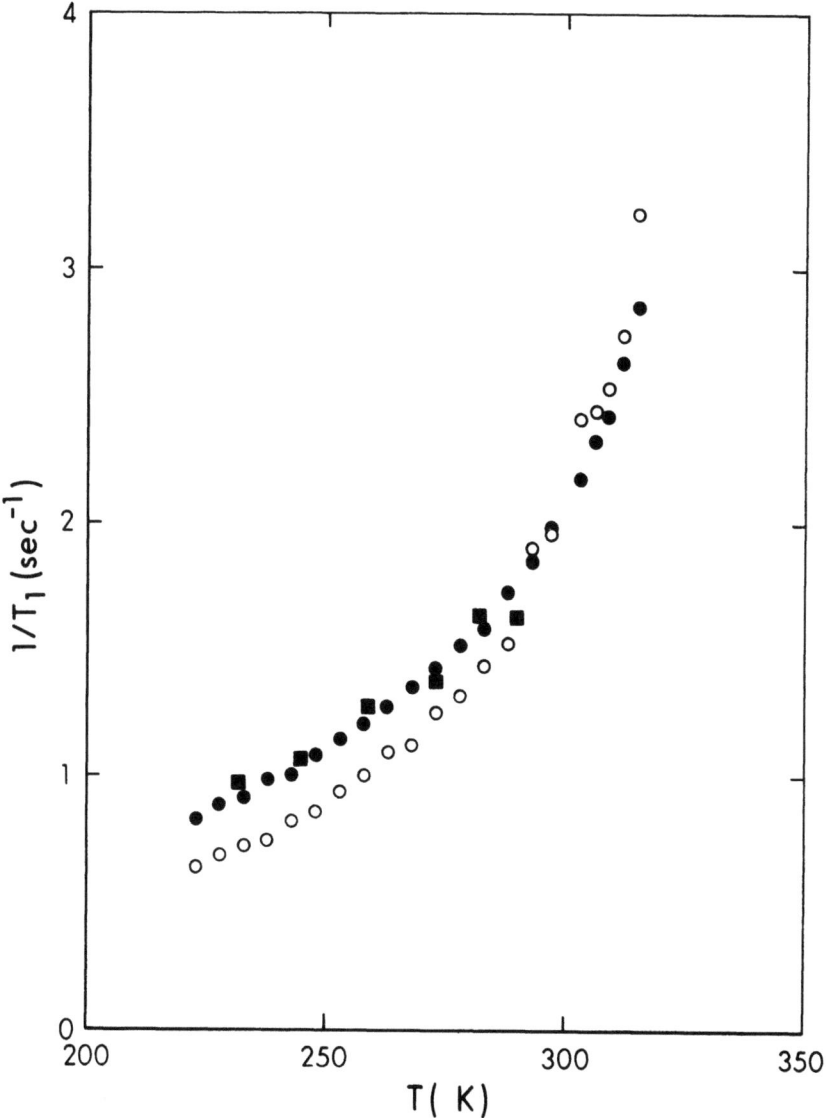

Fig. 3. Comparison of calculated and observed nuclear relaxation rates. ● – observed values (Hackleman and Hubbard [9]). ■ – values calculated with correlation times obtained from Raman data, ○ – values calculated with rotational diffusion equation theory [9].

and indicates that the anisotropies of the intermolecular potentials in CH_4 and SF_6 are comparable.

A reanalysis of the nuclear relaxation data [9] in terms of intermolecular and intramolecular dipole-dipole, and spin-rotational relaxation mechanisms using the values of τ_J and $\tau_\theta^{(2)}$ from the Raman data and the densities and diffusion coefficients of Hackleman and Hubbard, is shown in Figure 3 together with the experimental relaxation times and the values calculated using Equations (1) and (2). The asymmetry, $\Delta C = C_\parallel - C_\perp$, of the spin-rotation interaction tensor, which is required in order for the observed nuclear relaxation times to be compatible with the correlation times obtained, is $\Delta C = \pm 6.0 \times 10^4$ s^{-1}. This value differs significantly from that used by Hackleman and Hubbard [9], but is much more in keeping with the asymmetries of the spin-rotation tensors of other hexafluorides which have been determined from ^{19}F chemical shift studies of the solids [16].

References

1. Gordon, R. G.: *J. Chem. Phys.* **40**, 1973 (1964); **42**, 3658 (1965); **43**, 1307 (1965).
2. Bartoli, F. J. and Litovitz, T. A.: *J. Chem. Phys.* **56**, 404 (1972).
3. Bartoli, F. J. and Litovitz, T. A.: *J. Chem. Phys.* **56**, 413 (1972).
4. Peichard, J. P., Murphy, W. F., and Bernstein, H. J.: *Mol. Phys.* **23**, 499 (1972).
5. Nafie, L. A. and Peticolas, W. L.: *J. Chem. Phys.* **57**, 3145 (1972).
6. Abragam, A.: *Principles of Nuclear Magnetism*, Oxford University Press, London, 1961, Chapter VIII.
7. Hubbard, P. S.: *Phys. Rev.* **131**, 1155 (1963).
8. Gillen, K. T. and Griffiths, J. E.: *Chem. Phys. Letters* **17**, 359 (1972).
9. Hackleman, W. R. and Hubbard, P. S.: *J. Chem. Phys.* **39**, 2688 (1963).
10. McClung, R. E. D.: *J. Chem. Phys.* **51**, 3842 (1969); **54**, 3248 (1971).
11. Bratos, S. and Marechal, E.: *Phys. Rev.* **A4**, 1078 (1971).
12. Birnbaum, G. and Ho, W.: *Chem. Phys. Letters* **5**, 334 (1970); Birnbaum, G.: *Mol. Phys.* **25**, 241 (1973).
13. Perchard, J. P., Murphy, W. F., and Bernstein, H. J.: *Mol. Phys.* **23**, 519 (1972).
14. Kivelson, D., Kivelson, M. G., and Oppenheim, I.: *J. Chem. Phys.* **52**, 1810 (1970); McClung, R. E. D. and Kivelson, D.: *J. Chem. Phys.* **49**, 3380 (1968).
15. McClung, R. E. D.: *J. Chem. Phys.* **55**, 3459 (1971).
16. Rigny, P. and Virlet, J.: *J. Chem. Phys.* **47**, 4645 (1967).

DISCUSSION

Bratos: It is generally not justified to suppose that vibrational widths $\Delta\omega_n$ are the same for different normal modes. In fact $\Delta\omega_n$'s depend on the derivatives of the solvent-solute interaction energy V_S ($n_1, n_2 \ldots n_{3n+6}, t$) over the normal coordinates $n_1, n_2, \ldots, n_{3n+6}$, and vary according to the transition under consideration. This effect is not necessarily small. The bands which are most strongly broadened are those which imply motions of the part of the molecule which is most exposed to intermolecular forces.

McClung: I am in agreement with your views, but do not feel that the vibrational correction is very important in this system.

Janik: I want to give one example from our experiments, in connection with Prof. Bratos comment. We made an IR-width study of solid Ni(NH$_3$)$_6$Cl$_2$ in order to get rotational correlation time for NH$_3$ groups. The peak at 680 cm^{-1} showed a behaviour useful for the Rakov method. Its width varied from ca. 70 cm^{-1} (room temperature) to ca. 30 cm^{-1} (liquid nitrogen temperature). The 1200 cm^{-1} peak on the other hand had a constant width of ca. 15 cm^{-1}, significantly different from the low temperature value of the 680 cm^{-1} peak width.

ETUDE DYNAMIQUE PAR DIFFUSION RAMAN-LASER DES HEXAFLUORURES DE METAUX ET METALLOIDES

M. GILBERT et M. DRIFFORD

Service de Chimie-Physique C.E.N., B.P. no 2 91190 Gif sur Yvette, Saclay, France

Résumé. L'étude par diffusion Raman des hexafluorures de métalloides (SF_6, SeF_6, TeF_6) et de métaux (MoF_6, WF_6) en phase liquide est présentée. L'analyse des différentes contributions à l'élargissement des raies Raman a permis de préciser l'influence des mouvements de rotation sur les modes dépolarisés. Les valeurs des discontinuités des largeurs des modes dépolarisés à la transition liquide → → plastique sont discutées et justifiées par les valeurs des entropies de fusion.

L'analyse des formes de raies par transformée de Fourier a permis de préciser les différents types de rotation et de déduire des temps de corrélation orientationnels $\tau_\theta^{(2)}$. La relaxation nucléaire du ^{19}F fournit le temps de corrélation du moment cinétique τ_J. A partir des deux temps de corrélation déduits expérimentalement, le modèle J de MacClung semble le plus convenable pour décrire la dynamique de ces composés.

Abstract. The study by Raman spectroscopy of metalloid hexafluorides: SF_6, SeF_6, TeF_6, and metal hexafluorides: MoF_6, WF_6 in the liquid state, is reported. The different contributions to the broadening of Raman lines other than molecular rotations are examined and found to be small so that depolarised lines $\nu_2(e_g)$ et $\nu_5(f_{2g})$ are considered as pure rotational contours. The discontinuities in the widths of these lines at melting are in agreement with melting entropy values. Depolarized lines have been Fourier transformed. With the experimental orientational correlation times $\tau_\theta^{(2)}$ deduced from the ν_2 Raman line and the angular momentum correlation times τ_J calculated from ^{19}F NMR results, the extended J diffusion model seems the most adequate to describe molecular reorientations in liquid hexafluorides.

1. Introduction

L'étude par spectroscopie Raman des mouvements de rotation des hexafluorures de métalloïdes et de métaux est présentée respectivement pour SF_6, SeF_6, TeF_6 et MoF_6, WF_6.

Les hexafluorures dont une phase solide 'plastique' caractérisée par des mouvements de réorientation rapide. Les premiers résultats expérimentaux sur les rotations moléculaires de ces phases ont été exposés [1] ainsi que la méthode expérimentale.

Cette étude nous a permis de déduire des formes de raies des temps de corrélation orientationnels $\tau_\theta^{(2)}$ qui ont été reliés aux temps de corrélation du moment cinétique τ_J obtenus à partir de la RMN et ainsi de préciser un modèle dynamique convenable pour ces molécules globulaires.

2. Spectre de diffusion Raman des hexafluorures

La molécule XF_6 ($X = S$, Se, Te, Mo, W) octaédrique de symétrie ponctuelle O_h possèdent trois modes de vibration actifs en Raman (figure 1). Le spectre se compose ainsi:

$v_2(e_g)$, $v_5(f_{2g})$ modes dépolarisés de largeur à mi-hauteur importante:
$(2\Gamma(v_2) \simeq 20$ cm – figure 1)

$v_1(a_{1g})$ mode totalement polarisé de largeur a mi-hauteur très faible $(2\Gamma(v_1) \simeq 2$ cm^{-1} – tableau I).

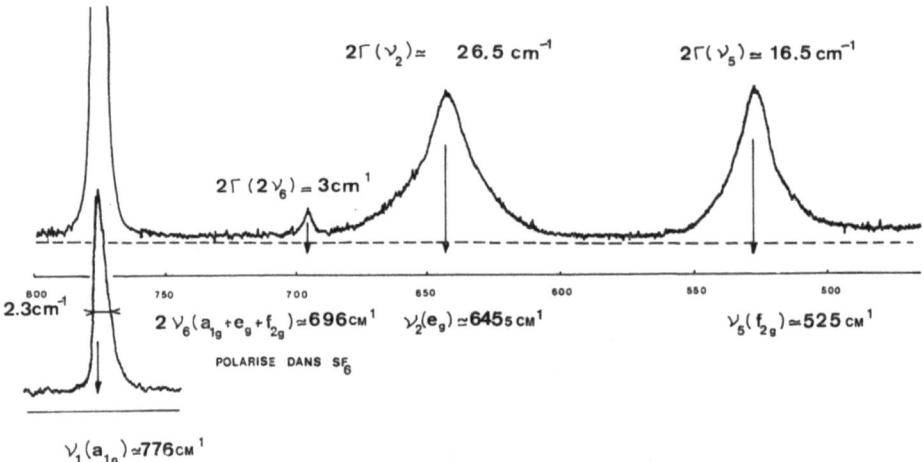

Fig. 1. Le spectre Raman de SF$_6$ liquide à la température ambiante.

TABLEAU I

Largeurs de v_1 dans SF$_6$, TeF$_6$, MoF$_6$, WF$_6$ à différentes températures

Largeur de fente: 0.8 cm^{-1}	SF$_6$	T(°C)	TeF$_6$	T(°C)	MoF$_6$	T(°C)	WF$_6$	T(°C)
Liquide	3.3	45						
	2.3	25	2.35	25				
	1.8	47	2.15	– 32	3.1	25	2.7	25
Cristal plastique	1.6	65	1.8	– 70	3.1	0	2.5	0

Les causes d'élargissement des raies Raman peuvent être séparées en diverses contributions.

Tout d'abord, la raie v_1 correspondant au mode de respiration de la molécule, doit sa largeur en phase liquide:

(a) A la dispersion des fréquences de vibration due aux interactions moléculaires (largeur intrinsèque) [2]

(b) A l'interaction vibration – rotation provenant de la variation du moment d'inertie moléculaire avec l'état de vibration.

(c) A l'anharmonicité du potentiel de vibration et à la présence de bandes chaudes.

Les raies dépolarisées $v_2(e_g)$ et $v_5(f_{2g})$ sont en plus, sensibles aux mouvements de réorientation dont la contribution est la même pour les deux modes.

La comparaison des largeurs de v_1 (tableau I) et des raies dépolarisées v_2 et v_5 (figures 1 et 2) semble montrer que les profils de v_2 et v_5 sont essentiellement des enveloppes de rotation légèrement perturbées par les interactions a, b et c.

Une difficulté est de préciser l'importance relative des contributions a, b et c à la largeur des raies.

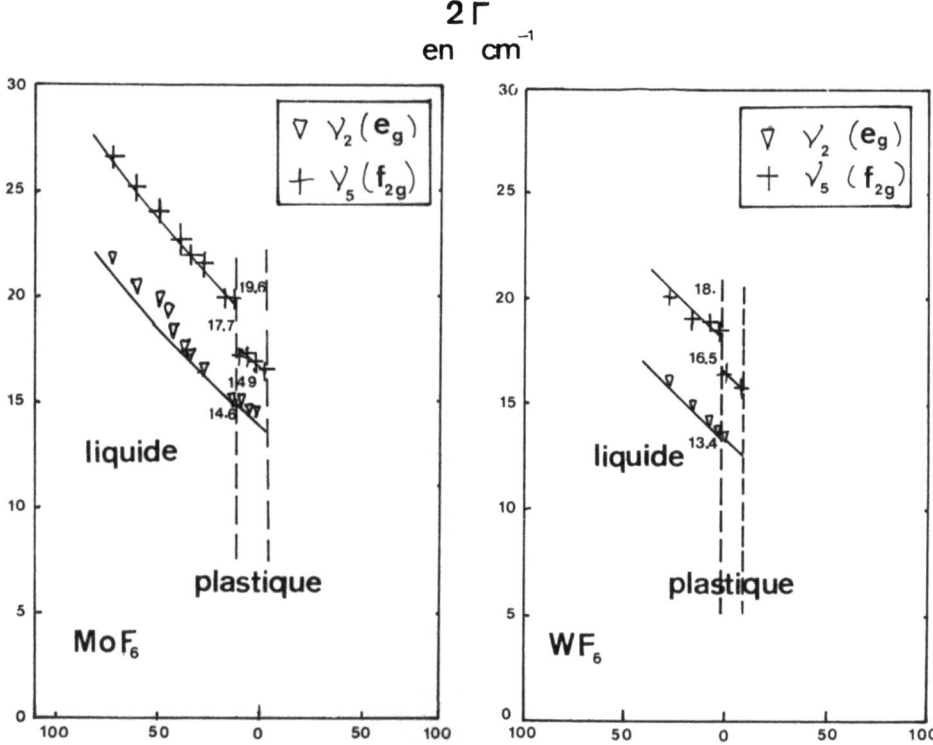

Fig. 2. Evolution des largeurs 2Γ des raies dépolarisées v_2 et v_5 dans les phases liquide et plastique des hexafluorures de métaux MoF$_6$ et WF$_6$.

L'interaction vibration – rotation b, qui peut se manifester sur les branches Q des spectres Raman de liquides légers comme HCl [3] est faible pour les hexafluorures. Son effet sur les spectres des gaz est connu [15]. Pour SF$_6$ gazeux cette interaction provoque un élargissement de 0,2 cm^{-1} par le mode $v_1(a_{1g})$ la contribution doit être encore plus réduite pour le liquide.

Les interactions moléculaires a interviennent faiblement sur la largeur du mode v_1. A la transition liquide-plastique $2\Gamma(v_1)$ varie de 0,2 cm^{-1}. En phase liquide, quand la température croît, $2\Gamma(v_1)$ augmente contrairement à ce que prévoit la théorie lorsque les interactions moléculaires sont responsables de la largeur des raies [4].

Les fréquences de raies Raman en phase liquide et plastique ont été mesurées avec précision en enregistrant simultanément les spectres Raman des hexafluorures et les raies d'une lampe au néon. D'après les glissements de fréquence aux changements de phase, groupés au tableau II, les interactions moléculaires ont un effet du même ordre

TABLEAU II

Glissements de fréquence des raies Raman aux changements de phase; (a) les fréquences dans le gaz proviennent de la référence 17; (b) référence 18.

a) Glissements de fréquence liquide-gaz:

$\Delta v = v_{\text{gaz}} - v_{\text{liquide}}$	$\Delta v_1 (a_{1g})$	$\Delta v_2 (e_g)$	$\Delta v_5 (f_{2g})$
SF_6	− 2	2.1	− 1.2
SeF_6	0.8	3.2	3.2
TeF_6	0.1	0.6	− 1.4
MoF_6	1.3	7.8	− 3.5
WF_6	− 0.4	4.8	− 3.8
NpF_6	3	11	− 10

b) Glissements de fréquence liquide-plastique: incertude sur Δv estimée à 0.8 cm^{-1}

$\Delta v = v_{\text{liquide}} - v_{\text{plastique}}$	$\Delta v_1 (a_{1g})$	$\Delta v_2 (e_g)$	$\Delta v_5 (f_{2g})$
SF_6	− 0.5	− 2.4	1.6
SeF_6	− 0.7	− 1.5	1.2
TeF_6	− 0.8	− 2.	0.5
MoF_6	− 1.9	− 1.7	1.4
WF_6	− 0.8	− 3.7	1.6

de grandeur sur les trois raies Raman, quoique légèrement plus grand pour les modes dégénérés $v_2(e_g)$ et $v_5(f_{2g})$ que pour $v_1(a_{1g})$. Les interactions moléculaires devraient avoir un effet relativement négligeable sur les largeurs des raies dépolarisées v_2 et v_5.

A partir de ces considérations, il semble qu'en phase liquide, les contributions a, b et c sont faibles et pratiquement négligeables devant les mouvements de rotation qui provoquent ainsi l'élargissement important des modes dépolarisés.

3. Mouvements de réorientation en phase liquide

3.1. Evolution des largeurs des modes depolarises γ_2 et γ_5 en fonction de la temperature

Cette évolution a été décrite [1] pour les hexafluorures SF_6, SeF_6 et TeF_6. Au passage liquide → plastique, une discontinuité importante des largeurs des raies v_2 et v_5 est observée.

Dans MoF_6 et WF_6, la discontinuité est faible (figure 2) inférieure à 2 cm^{-1}. La variation faible des mouvements de rotation à la transition, ainsi mise en évidence, est à rapprocher des faibles valeurs des entropies de fusion [7].

$$\Delta S_{\text{fusion}}(MoF_6) = 3{,}55 \text{ u.e}$$
$$\Delta S_{\text{fusion}}(WF_6) = 3{,}56 \text{ u.e.}$$

Dans les hexafluorures de métalloïdes, la variation d'entropie [8]: importante $\Delta S_{\text{fusion}}(\text{SF}_6) \simeq 5,4$ u.e correspond à un changement dans les mouvements de rotation entre les phases liquide et plastique.

3.2. ETUDE DES FORMES DE RAIES

Les mouvements de rotation seront étudiés essentiellement à partir de la raie $v_2(e_g)$ puisque le mode $v_5(f_{2g})$ est perturbé par l'interaction de Coriolis avec une constante de Coriolis théorique $\zeta_5 = -0,5$: ce qui correspond à une réduction de la largeur de γ_5 par rapport à celle de γ_2, en phase gazeuse [6].

Si l'on tient compte seulement des mouvements de rotation, le spectre $I^{(2)}(\omega)$ de la raie de vibration v_2 de symétrie e_g est la transformée de Fourier [5]:

$$I^{(2)}(\omega) \alpha (\omega_{\text{diff}})^4 \int_{-\infty}^{+\infty} \exp(-i\omega t) \hat{C}^{(2)}(t) \, dt \tag{1}$$

où $\omega_{\text{diff}} = \omega_{\text{laser}} - \omega^{(2)}\omega$ est la fréquence de la lumière diffusée et $\hat{C}^{(2)}(t)$ la fonction d'autocorrélation normalisée:

$$\hat{C}^{(2)}(t) = \tfrac{1}{2}\{3[\mathbf{u}(0)\cdot\mathbf{u}(t)]^2 - 1\}.$$

$\mathbf{u}$ désigne un vecteur unitaire lié à la molécule. On définit le temps de corrélation correspondant:

$$\tau_\theta^{(2)} = \int_0^\infty \hat{C}^{(2)}(t) \, dt = \pi \frac{I^{(2)}(0)}{\int_{-\infty}^{+\infty} I^{(2)}(\omega) \, d\omega}.$$

Les transformées de Fourier de v_2 et v_5: $\hat{C}^{(2)}(t)$ et $\hat{C}^{(5)}(t)$ pour SF_6 sont représentées figure 3, à trois températures et on distingue deux intervalles de temps où les fonctions d'autocorrélation ont un comportement différent:

aux temps courts ($t < 10^{-12}$ s) la perte de corrélation est essentiellement due à la dispersion des vitesses de rotation libre et $\hat{C}(t)$ a une forme quasi parabolique;

aux temps longs ($t < 10^{-12}$ s) la décroissance est exponentielle, caractéristique d'une diffusion rotationnelle.

3.3. MODELE DE DIFFUSION DE ROTATION

Le temps de corrélation du moment cinétique:

$$\tau_J = \frac{1}{\langle J_x^2 \rangle} \int_0^\infty \langle J_x(0) J_x(t) \rangle \, dt,$$

où J_x est une composante du moment cinétique a été estimé à partir des mesures *de RMN du* ^{19}F [9-11] en utilisant les hypothèses de Rigny et Virlet sur les tenseurs d'interaction spin – rotation [9, 12].

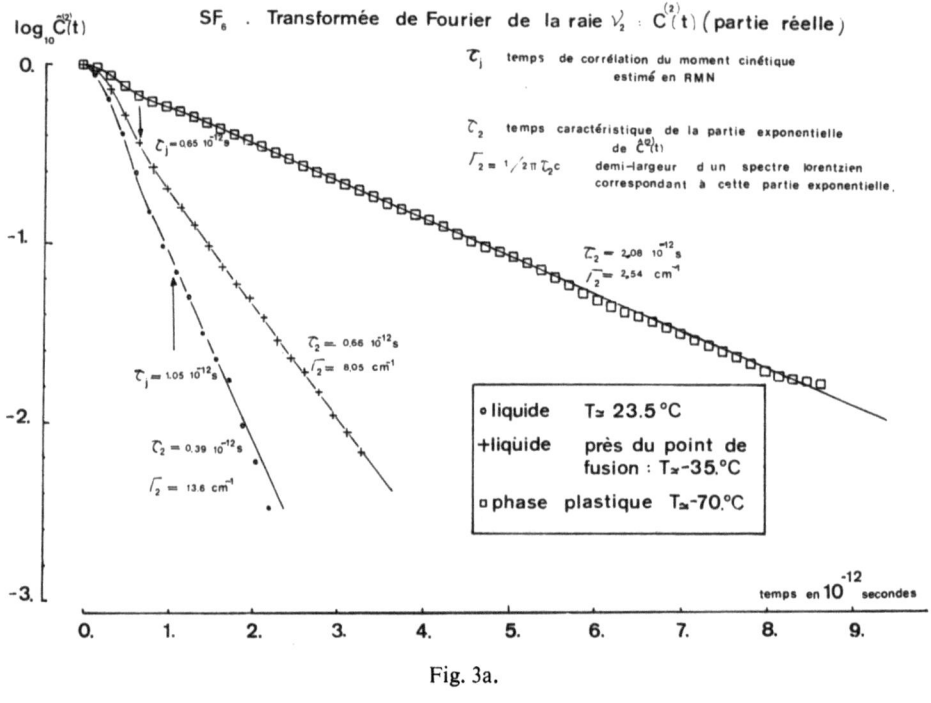

Fig. 3a.

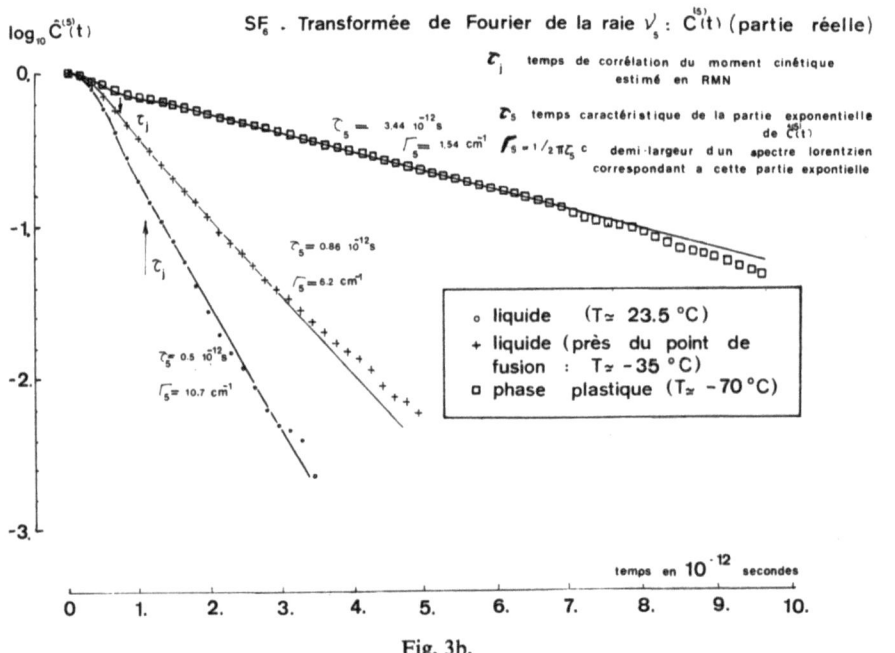

Fig. 3b.

Fig. 3a–b. Transformée de Fourier des raies dépolarisées de SF$_6$ en phase liquide et plastique: (a) ν_2, (b) ν_5.

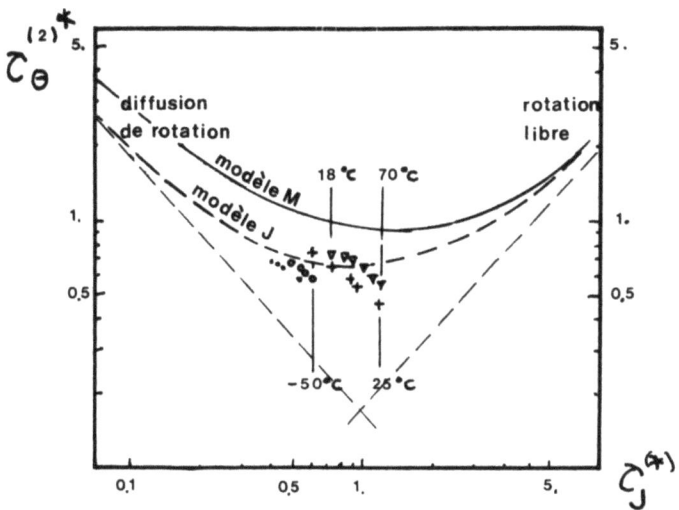

a) $\tau_\theta^{(2)}$ calculé à partir de la formule approchée : $\tau_\theta^{(2)} \simeq 1/2\pi \, \Gamma(\nu_2) c$

+ SF_6 ; ● SeF_6 ; ▽ MoF_6 ; ○ WF_6 ;

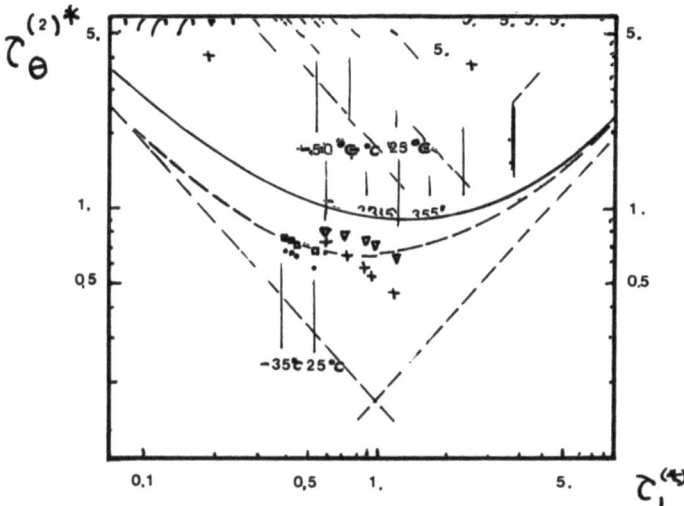

b) comparaison des deux méthodes de calcul de $\tau_\theta^{(2)}$: pour SF_6 et SeF_6 :

$$\tau_\theta^{(2)} = \frac{\pi \, I^{(a)}(0)}{\int_{-\infty}^{+\infty} I^{(a)}(\omega) \, d\omega} \quad \begin{cases} \triangledown \ SF_6 \\ \square \ SeF_6 \end{cases} \qquad \tau_\theta^{(2)} \simeq \frac{1}{2\pi \Gamma(\nu_2) c} \begin{cases} + \ SF_6 \\ \bullet \ SeF_6 \end{cases}$$

Fig. 4. Comparaison des résultats expérimentaux aux prévisions des modèles J et M de MacClung.

Dans MoF$_6$ liquide à 70°C [9]: $\tau_J = 0.97 \times 10^{-12}$ s. L'angle de rotation libre correspondant $\delta\varphi$ est grand:

$$\delta\varphi = \tau_J^* = \tau_J \sqrt{\frac{kT}{I}} \simeq 0.97 \text{ rad} = 55°.$$

Le modèle de diffusion de rotation de Debye [13], qui suppose de petits angles de réorientation, ne peut donc convenir à l'interprétation des spectres Raman des hexafluorures.

Par contre, les modèles J et M développés par Gordon pour des molécules diatomiques, et étendus aux polyatomiques par MacClung [14], autorisant de grands angles de rotation sont plus adaptés pour décrire les rotations des hexafluorures.

Les deux temps de corrélation $\tau_\theta^{(2)}$ et τ_J déduits expérimentalement, l'un des données de RMN et l'autre de la diffusion Raman à différentes températures sont représentés figure 4.

Dans les quatre hexafluorures étudiés, la rotation peut être considérée comme intermédiaire entre la diffusion de rotation et la rotation libre existant dans les gaz sous faible pression. Les angles de rotation varient entre 0,4 et 1,2 rad, et la liberté de rotation est semblable à celle existant dans CD$_4$ et CH$_4$ liquides [14] et paraît plus grande que dans CCl$_4$ [20] ou CCl$_3$ [19].

Le modèle J apparaît donc comme un modèle privilégié dans l'étude dynamique des hexafluorures de métaux ou métalloïdes.

Bibliographie

1. Gilbert, M. et Drifford, M.: voir [16], p. 204.
2. Bratos, S. et Marechal, E.: *Phys. Rev.* **A4**, 1078 (1971).
3. Perrot, M. et Lascombe, J.: *J. Chim. Phys.* **70**, 5 (1973).
4. Levant, R.: Thèse 3è cycle U.E.R. Paris VI, 1971.
5. Gordon, R. G.: *J. Chem. Phys.* **42**, 3658 (1965).
6. Masri, F. N. et Fletcher, W. H.: *J. Chem. Phys.* **52**, 5759 (1970).
7. Westrum, E. F.: *J. Chim. Phys.* **62**, 47 (1966).
8. Staveley, L. A.: *Ann. Rev. Phys. Chem.* **13**, 351 (1962).
9. Rigny, P. et Virlet, J.: *J. Chem. Phys.* **47**, 4645 (1967).
10. Blinč, R. et Lahajnar, G.: *Phys. Rev. Letters* **19**, 685 (1967).
11. Hackleman, W. R. et Hubbard, P. S.: *J. Chem. Phys.* **39**, 2688 (1963).
12. Virlet, J.: Thèse U.E.R. Paris IX – Orsay, 1971.
13. Favro, L. D.: *Phys. Rev.* **119**, 53 (1960).
14. MacClung, R. E. D.: *J. Chem. Phys.* **37**, 5478 (1972); et références citées.
15. Gaufres, R. et Sportouch, S.: voir [16], p. 478.
16. *Adv. Raman Spectry* **1**, *Proc. of 3rd Int. Conf. on Raman Spectroscopy, Reims, 1972*, Heyden et Sons Ltd., London.
17. Claassen, H. H., Goodman, G. L., Holloway, J. H., and Selig, H.: *J. Chem. Phys.* **53**, 341 (1970).
18. Gasner, E. L. et Frlec, B.: *J. Chem. Phys.* **49**, 5135 (1968).
19. Gillen, K. T., Douglass, D. C., Malmberg, M. S., and Maryott, A. A.: *J. Chem. Phys.* **57**, 5170 (1972).
20. Gillen, K. T., Noggle, J. H., and Leipert, T. K.: *Chem. Phys. Letters* **17**, 505 (1972).

REORIENTATION DE MOLECULES TOUPIE SYMETRIQUE CH₃D ET CD₃H DISSOUTES DANS DES FLUIDES SIMPLES DU POINT TRIPLE AU POINT CRITIQUE

J. P. MARSAULT, F. MARSAULT-HERAIL, J. L. SAULNIER, et G. LEVI

*Physique Expérimentale Moléculaire, T. 22 – 1er étage, Université de Paris VI,
4 place Jussieu, 75230 Paris, Cedex 05, France*

Résumé. Nous avons étudié par spectroscopie infrarouge les profils des bandes parallèles (A_1) et perpendiculaires (E) de CD₃H et CH₃D dissous dans quelques fluides simples (Ar, N₂, CH₄) du point triple au point critique et le long d'une courbe isobare au dessus de la pression critique. Nous avons calculé les fonctions de corrélation, les moments spectraux et les couples quadratiques moyens. Nous interprêtons nos résultats à l'aide des modèles de diffusion 'm' et 'j' de Gordon et en utilisant le formalisme de la fonction mémoire.

Abstract. Extensive measurements of parallel (A_1) and perpendicular (E) fundamental infrared bands of methanes CH₃D and CD₃H dissolved in various simple fluids (Ar, N₂, CH₄) from the triple point to critical point and along an isobaric line above the critical pressure, have been carried out.

Rotational correlation functions, band moments and intermolecular mean squared torques have been determined from the band shapes. We used the 'm' or 'j' Gordon extended diffusion models to interprete the rotational correlation functions. In argon and CH₄ or CF₄ the 'j' model gives a very good fit with experiments except in the critical region and at low temperature in argon solution. If $\tau_J^{(1)}$ denotes the angular momentum correlation time, the ratio $1/\tau_J^{(1)}$, which represents the frequency of the collisions varies almost linearly with density, whatever the state may be. We interpreted this feature by assuming that the frequency of the collision depends only on the density of the solvent.

We also computed the correlation functions by the memory function method which leads to a rather good agreement with experimental results, but the mean squared torque values deduced from the calculated curves do not agree with the experimental ones.

As the density, the mean squared torques change continuously from a very low value to the liquid value along an isobaric line, while they break off passing through the liquidus curve.

Lastly we also investigated the liquid-solid transition.

1. Introduction

Une étude précédente [1] a montré que les réorientations moléculaires de CH₃D comprimé par He, Ar, O₂, N₂ et CH₄ à température ordinaire (pression maximum $P = 150$ bars) sont bien décrites par le modèle de diffusion 'j' [2]. La valeur du couple quadratique moyen $\langle(O_\perp V)^2\rangle$ (s'exerçant autour d'un axe perpendiculaire à l'axe C_3) n'est pas mesurable car inférieure aux erreurs expérimentales.

Dans un récent travail [3], nous avons également étudié les profils des bandes d'absorption infrarouge de CH₃D et CD₃H en solution liquide, au voisinage du point triple dans des gaz monoatomiques (Ar, Kr, Xe), diatomiques (N₂, O₂) et dans CH₄ ou CF₄. Les résultats obtenus sont analogues pour les deux molécules CH₃D et CD₃H.

L'ordre de grandeur du couple quadratique moyen est sensiblement le même pour tous les solvants étudiés: $\langle(O_\perp V)^2\rangle \sim 2$ à 3×10^4 cm⁻².

Le modèle de diffusion 'j' est encore satisfaisant pour les solutions de CH₃D dans CH₄ et de CD₃H dans CF₄. L'étude en fonction de la température de CD₃H dissous dans CF_4 montre cependant que ce modèle est de moins en moins satisfaisant au fur et à mesure que l'on se rapproche du point triple. Dans le cas des solutions de CH_3D et CD_3H dans l'oxygène, et l'azote liquide, les modèles de diffusion généralisée ne rendent plus compte des fonctions de corrélation expérimentales. Dans les liquides monoatomiques (Ar, Kr, Xe) enfin, la fonction de corrélation théorique calculée dans le modèle de diffusion 'm' déjà proposé par McClung pour le méthane [4] traduit les réorientations moléculaires aux temps courts mais aux temps longs elle s'éloigne de la courbe expérimentale alors exponentielle.

Dans les solvants étudiés, il apparaît donc que les modèles de diffusion généralisée ne décrivent qu'imparfaitement les mouvements de réorientations de CH_3D et CD_3H près du point de fusion. Ces modèles supposent en effet des chocs de durée infinitésimale et on peut considérer qu'au voisinage de la transition liquide-solide, les molécules sont piégées dans des puits de potentiel pendant un temps non négligeable par rapport à l'intervalle séparant deux collisions.

Ces résultats nous ont suggéré de poursuivre l'étude de ces solutions:

– d'une part jusqu'au point critique et dans le fluide hypercritique. A cet effet, nous avons choisi trois solvants dont le comportement est particulièrement caractéristique et donnons les premiers résultats obtenus pour les solutions de CD_3H dans l'argon et l'azote et de CH_3D dans le méthane.

– d'autre part la transition liquide-solide est envisagée dans le cas de CD_3H dans l'argon, l'azote (phase β) et CF_4 (phase plastique).

2. Résultats – Discussion

Nous avons supposé que la relaxation vibrationnelle était suffisamment lente pour négliger sa contribution aux profils des bandes. La transformée de Fourier des bandes de symétrie A_1, auxquelles nous nous sommes plus particulièrement intéressés, est alors assimilable à la fonction de corrélation $\hat{C}(t)$ d'un vecteur unitaire porté par la direction du moment dipolaire de transition [5] (suivant l'axe C_3).

Les deuxième et quatrième moments expérimentaux permettent d'évaluer à 25% près la valeur du couple quadratique moyen $\langle (O_\perp V)^2 \rangle$.

Nous avons calculé les fonctions de corrélation du modèle 'j' par une évaluation numérique des intégrales multiples [1, 6, 7] pour les bandes de symétrie A_1 et pour les bandes de symétrie E à l'aide d'un programme proposé par McClung [8]. Ce programme nous a également permis de calculer les fonctions de corrélation de diffusion 'm' pour les bandes de symétrie A_1 et E.

2.1. Etude des solutions jusqu'au point critique et dans le fluide hypercritique

2.1.1. *Fonctions de corrélation*

2.1.1.1. Solutions dans l'argon

2.1.1.1.1. *Passage continu du gaz au liquide à une pression constante $P = 80$ bars supérieure à la pression critique $P_c = 48$ bars. Le modèle de diffusion 'J' décrit bien les* réorientations de CD_3H à la fois dans le gaz et dans le liquide jusqu'à une température (119 K) nettement inférieure à la température critique $T_c = 150,6$ K. L'inverse du paramètre $\tau_J^{(j)}$, assimilable à la fréquence des chocs, est directement proportionnel à la densité comme le montre la figure 1A. Sur la figure 2, nous avons comparé les fonctions de corrélation expérimentales et calculées par le modèle 'j' dans le cas d'une bande A_1 et d'une bande E enregistrées dans les mêmes conditions ($p = 70$ bars, $T = 150$ K)). La bande E que nous avons choisie ($v_4(E) = 2245$ cm^{-1}) est celle qui a le plus faible couplage de Coriolis à température ambiante ($\zeta = 0,14$). [3].

L'accord entre les courbes expérimentales et calculées est excellent avec la même

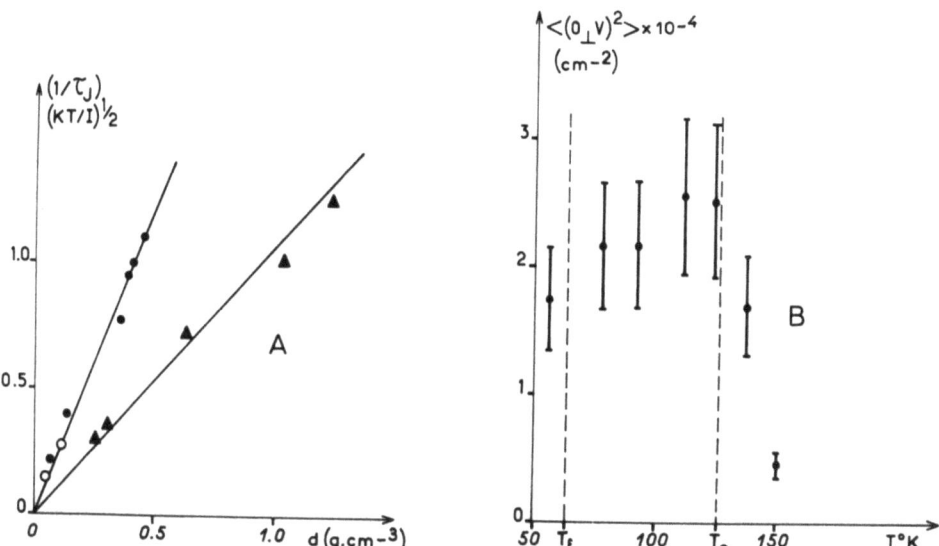

Fig. 1. (A) $1/\tau_J^{(j)} = f$(densité): ▲ ▲ ▲ CD_3H dans l'argon; ● ● ● CH_3D dans le méthane; ○ ○ ○ données de la référence 1. (B) Courbe $\langle(O_\perp V)^2\rangle = f(T)$ CD_3H dans l'azote; $P = 60$ bars.

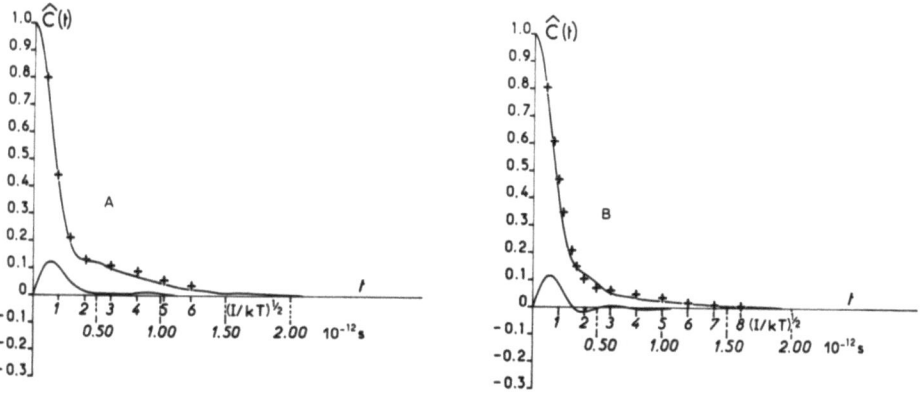

Fig. 2. Fonctions de corrélation de CD_3H dans l'argon, —— courbe expérimentale (parties réelle et imaginaire) $T = 150$ K, $P = 70$ bars. +++ Modèle de diffusion 'j' $\tau_J^{(j)} = 1,4$ $(I/kT)^{1/2}$. (A) Bande de symétrie A_1; (B) Bande de symétrie E.

valeur du paramètre $\tau_J^{(j)}$ pour les bandes A_1 et E. En effet, comme le prévoient ces modèles de diffusion généralisée [8], il faut admettre que les temps de corrélation des moments angulaires sont identiques pour les deux rotations autour de l'axe C_3 et autour d'un axe perpendiculaire à C_3.

2.1.1.1.2. Comportement du liquide le long de la courbe de coexistence liquide-gaz. Tout près du point critique, l'accord entre les fonctions de corrélation expérimentales et calculées, qui est excellent en phase gaz, devient brusquement médiocre (figure 5) au passage à l'état liquide. Cet accord redevient tout à fait satisfaisant quand on s'éloigne de la région critique. Par contre, à basse température, nous constatons que le modèle de diffusion 'm' décrit mieux la fonction de corrélation expérimentale, du moins aux temps courts ($t \leqslant 3(I/kT)^{1/2}$) ainsi que nous l'avons déjà observé [3].

2.1.1.2. Solutions dans le méthane

Nous avons également étudié les solutions de CH_3D dans le méthane à pression constante ($P = 60$ bars $> P_c = 45{,}9$ bars) et le long de la courbe de coexistence liquide-gaz.

Le modèle de diffusion 'j' est satisfaisant dans le gaz [1] et dans le liquide jusqu'au voisinage du point triple [3] (figure 3). Cependant dans la région critique l'accord entre les fonctions de corrélation expériméndales et calculées est nettement moins bon, comme en temoigne le tableau I où le temps de corrélation du moment dipolaire de transition $\tau_\theta = \int_0^\infty \hat{C}(t)\,dt$ mesuré sur la courbe expérimentale est comparé à sa valeur calculée à l'aide du modèle de diffusion 'j'. Nous suggérons donc qu'il y a une discontinuité de modèle à la transition gaz-liquide au niveau du point critique.

La courbe $1/\tau_J^{(j)} = f$ (densité) (figure 1A) qui réunit des données expérimentales obtenues en liquide et en gaz, suggère que, comme dans le cas de l'argon, la fréquence des chocs ne dépend pas de l'état, mais seulement de la densité du solvant.

Dans le cas de certains liquides purs, Gerschel [9] a montré que, du point de vue des mouvements rotationnels, la transition gaz-liquide s'effectuait dans le liquide.

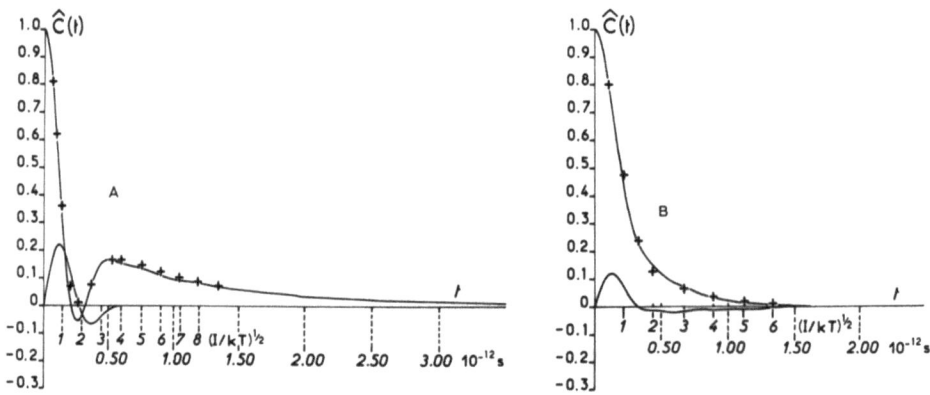

Fig. 3. Fonctions de corrélation de CH_3D dans le méthane, —— courbe expérimentale (parties réelle et imaginaire). (A) $T = 235$ K; $P = 60$ bars; (B) $T = 106$ K; $P = 1$ bars. +++ Modèle de diffusion 'j'.

(A) $\tau_J^{(j)} = 4{,}5\ (I/kT)^{1/2}$ (B) $\tau_J^{(j)} = 0{,}9\ (I/kT)^{1/2}$.

TABLEAU I

(a) CH$_3$D dans la méthane liquide le long de la courbe de coexistence liquide-gaz;
(b) CH$_3$D dans le méthane le long d'une isobare ($P = 60$ bars)

Pression (bar)	Température (K)	(I/kT) (10^{-13} s)	$\tau_J^{(j)}$ (u.r)	$\tau_\theta^{(j)}$ (u.r)	$\tau_{\rm cor}^{\rm (expe)}$ (u.r.)
(a)					
–	91	2,41	0,7	1,3	1,3
–	106	2,23	0,9	1,2	1,2
4	119,5	2,09	1,0	1,2	1,2
7	133	1,99	1,1	1,2	1,2
10	145	1,89	(1,3)	1,2	1,0$_5$
15	157,5	1,82	(1,5)	(1,2$_5$)	1,0
35	180	1,70	(1,5)	(1,2$_5$)	1,0
46	189,5	1,66	(1,5)	(1,2$_5$)	1,0
46 (gaz)	191,5	1,65	2,5	1,3$_5$	1,3
(b)					
60	91	2,41	0,7	1,3	1,3
60	180	1,70	(1,2)	(1,2)	1,0
60	193	1,64	(1,3)	(1,2)	1,0
60	201	1,61	(1,8)	(1,2$_5$)	1,1
60	235	1,49	4,6	2,1$_5$	2,1

2.1.1.3. Solutions dans l'azote

Nous observons un écart important entre les fonctions de corrélation expérimentales et calculées dans l'hypothèse d'une diffusion 'm' ou 'j', pour CD$_3$H dissous dans l'azote ($P = 60$ bars $> P_c = 33,5$ bars, $T \leqslant 150$ K) et ceci quelque soit l'état, liquide ou gaz. Nous envisageons de compléter cette étude jusqu'à la température ordinaire où le modèle de diffusion 'j' est alors satisfaisant.

2.1.2. Couples quadratiques moyens

Pour toutes les solutions étudiées, la traversée de la courbe de coexistence liquide-gaz ($P = f(T)$) entraîne une brusque variation du couple carré moyen. Par contre, à pression constante, supérieure à la pression critique, la densité du solvant augmentant moins brusquement du gaz au liquide quand la température diminue, nous constatons (figure 1B) un accroissement continu du couple d'une valeur non mesurable à la valeur 2 à 3×10^4 cm^{-2} déjà obtenue en liquide à la pression ordinaire [3]. La figure 4 montre l'évolution du couple en fonction de la densité le long de la courbe de coexistence liquide-gaz et à pression constante supérieure à P_c. Sur ce dernier exemple, CD$_3$H dans l'argon, nous remarquons que le couple croit tout d'abord rapidement avec la densité, puis reste sensiblement constant jusqu'au point de fusion.

2.1.3. Calcul de la fonction de corrélation à partir de la fonction mémoire

Dans les cas où les modèles de diffusion généralisée ne décrivent pas correctement les réorientations moléculaires, on peut penser que le temps pendant lequel le couple agit sur la molécule n'est plus négligeable, contrairement à ce que ces modèles supposent. D'autre part nous pouvons évaluer un ordre de grandeur du couple quadratique moyen. Ceci nous a conduit à utiliser le formalisme de la fonction mémoire $K(t)$ pour calculer les fonctions de corrélation du moment dipolaire de transition.

Desplanques et Constant [10] ont, en effet, étendu l'approximation proposée par Berne et Harp [11] aux cas où le couple s'exerçant sur la molécule est faible, et ont proposé d'écrire:

$$K(t) = K_{RL}(t) \cdot \phi_J(t),$$

où $K_{RL}(t)$ est la fonction mémoire du rotateur libre et $\phi_J(t)$ la fonction de corrélation de la composante perpendiculaire du moment angulaire, elle-même obtenue à partir de sa fonction mémoire, qui, dans la théorie de l'information, s'écrit:

$$\frac{\langle (O_\perp V)^2 \rangle}{2IKT} \exp\left\{ -\left[\frac{\langle (\dot{O}_\perp V)^2 \rangle}{\langle (O_\perp V)^2 \rangle} - \frac{\langle (O_\perp V)^2 \rangle}{2IKT} \right] \frac{t^2}{2} \right\},$$

où $\langle (\dot{O}_\perp V)^2 \rangle$ désigne la carré moyen de a dérivée du couple par rapport au temps.

Nous avons d'abord été tenté d'introduire le couple carré moyen expérimental le seul paramètre restant alors étant sa dérivée par rapport au temps. Dans ce cas, l'accord entre les fonctions de corrélation expérimentales et calculées n'est pas réalisable. Pour l'améliorer, il faut augmenter considérablement la valeur du couple quadratique moyen bien au delà de l'erreur expérimentale. Nous montrons ainsi sur la figure 5, un exemple de simulation de la fonction de corrélation expérimentale de CD_3H dissous dans l'argon où:

$$\langle (O_\perp V)^2 \rangle_{\text{expé}} = 2 \times 10^4 \text{ cm}^{-2}, \qquad \langle (O_\perp V)^2 \rangle_{\text{théo}} = 9 \times 10^4 \text{ cm}^{-2}.$$

2.2. Transition liquide-solide

Lorsque l'on refroidit la solution de CD_3H dans l'argon au delà de la transition liquide-solide, on observe une brusque diminution du temps de corrélation (figure 6A) et le couple quadratique moyen tend à augmenter légèrement (figure 4A).

Par contre, dans le cas de solutions dans CF_4 où la phase solide I est plastique et dans l'azote où le solide β présente également un désordre réorientationnel, la transition liquide-solide ne semble pas modifier sensiblement les mouvements de réorientation. Ainsi la variation du temps de corrélation $\tau_\theta(A_1)$ en fonction de l'inverse de la température (figure 6B) ne présente pas de discontinuité à la transition. Nous avons déjà remarqué pour la solution liquide dans CF_4, qu'à l'intérieur des limites d'erreur, le couple quadratique moyen tendait à diminuer avec la température au voisinage du point triple [3]. Il semble que pour les deux solutions (dans CF_4 et N_2) le rapport $\langle (O_\perp V)^2 \rangle / T$ devienne constant dans le liquide dense et ne soit pas modifié à la cristallisation. On peut penser alors que seule la dérivée du couple, $\langle (\dot{O}_\perp V)^2 \rangle$, diminue uniformément du liquide au solide. Ceci pourrait expliquer l'écart observé entre le modèle de diffusion 'j' et les résultats expérimentaux obtenus pour CF_4 à basse température [3].

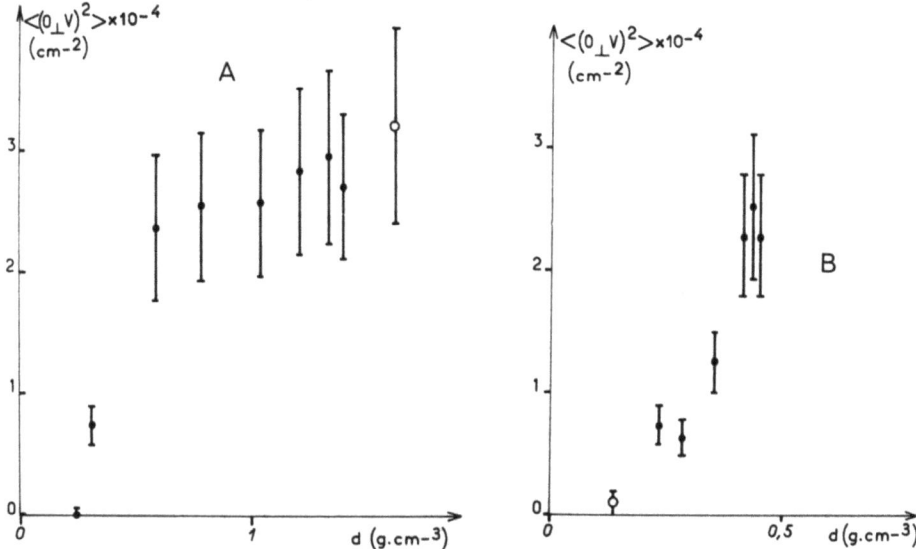

Fig. 4. Courbes $\langle (O_\perp V)^2 \rangle = f$(densité). (A) CD$_3$H dans l'argon; $P = 80$ bars: ● ● gaz et liquide, ○ solide; (B) CH$_3$D dans le méthane: ● ● ● liquide le long de la courbe de coexistence liquide-gaz, ○ fluide critique.

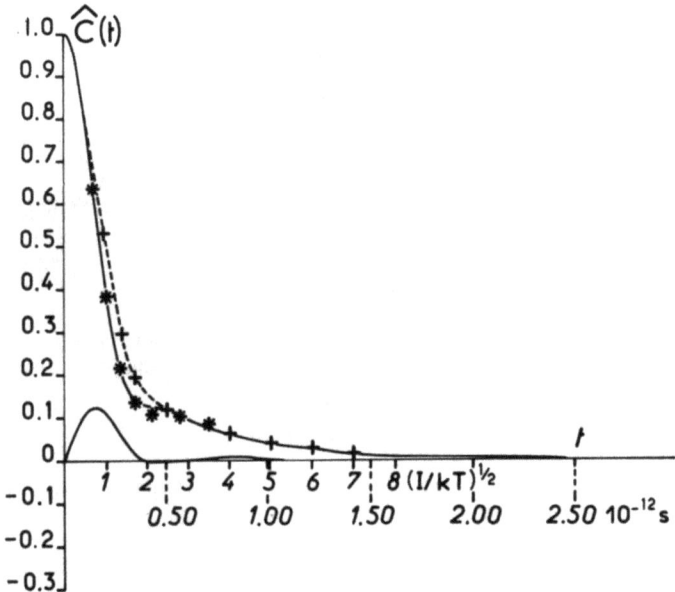

Fig. 5. Fonctions de corrélation de CD$_3$H dans l'argon; $P = 48$ bars, $T = 149$ K . ——— courbe expérimentale (parties réelle et imaginaire); +++ Modèle de diffusion 'j' $\tau_J{}^{(j)} = 1,1 \ (I/kT)^{1/2}$; ✱ ✱ courbe calculée par le formalisme de la fonction mémoire:

$$\langle (O_\perp V)^2 \rangle = 9 \times 10^4 \, \text{cm}^{-2} \qquad \frac{\langle (\dot{O}_\perp V)^2 \rangle}{\langle (O_\perp V)^2 \rangle} \frac{I}{2kT} = 3.$$

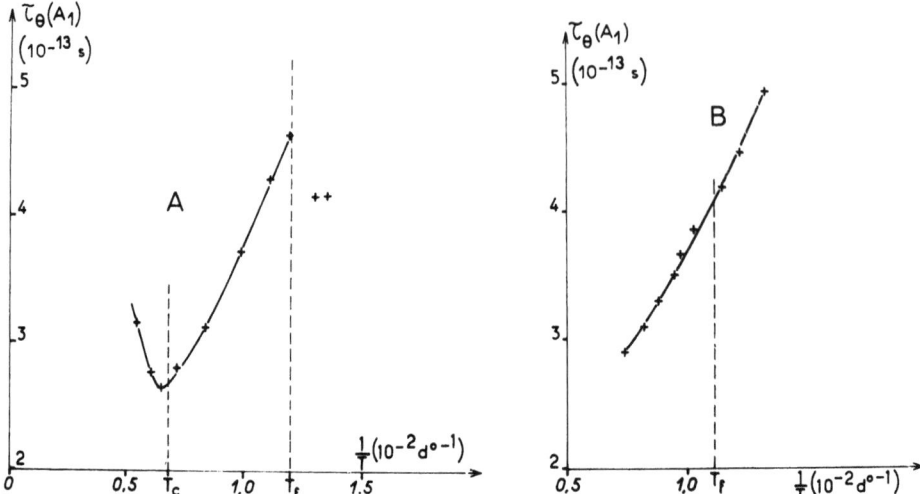

Fig. 6. Temps de corrélation du moment dipolaire τ_θ en fonction de $1/T$. (A) CD$_3$H dans l'argon ($P = 80$ bars); (B) CD$_3$H dans CF$_4$.

Remerciements

Nous remercions le Professeur R. E. D. McClung qui nous a aimablement communiqué ses programmes de calculs et Monsieur A. Bullier qui a contribué à la réalisation des expériences.

Bibliographie

1. Levi, G. et Chalaye, M.: *Chem. Phys. Letters* **19**, 263 (1973).
2. Gordon, R. G.: *J. Chem. Phys.* **44**, 1830 (1966).
3. Marsault, J. P., Marsault-Herail, F., et Levi, G.: *Mol. Phys.* **26**, 997 (1973).
4. McClung, R. E. D.: *J. Chem. Phys.* **51**, 3842 (1969); *ibid.* **55**, 3459 (1971).
5. Gordon, R. G.: *J. Chem. Phys.* **42**, 3658 (1965); *J. Chem. Phys.* **43**, 1307 (1965).
6. Bliot, F., Abbar, C., et Constant, E.: *Mol. Phys.* **24**, 241 (1972).
7. Bliot, F. et Constant, E.: *Chem. Phys. Letters* **18**, 253 (1973).
8. McClung, R. E. D.: *J. Chem. Phys.* **57**, 5478 (1972).
9. Gerschel, A.: Thèse, Paris, 1971.
10. Desplanques, P. et Constant, E.: *Compt. Rend. Acad. Sci.* **B274**, 611 (1972).
11. Berne, B. J. et Harp, G. D.: *Adv. Chem. Phys.* **XVII**, John Wiley and Sons, 1970, p. 63.

DISCUSSION

Lascombe: Comment concilier le modèle 'j' qui semble permettre d'interpréter vos résultats et l'existence d'un couple? Avez-vous calculé la valeur du quatrième moment qu'impliquerait un modèle 'j' ?

Marsault: Effectivement il y aurait une certaine contradiction entre le fait d'utiliser le modèle 'j' et ensuite de calculer les couples quadratiques moyens. Il est cependant bien évident qu'il existe un couple s'opposant à la rotation de la molécule. Nous avons donc essayé d'en déterminer l'ordre de grandeur, d'ailleurs faible, à l'aide des deuxième et quatrième moments expérimentaux. Le fait que le quatrième moment dans le modèle 'j' soit inférieur au quatrième moment du rotateur libre peut être dû à un défaut de ce modèle qui n'est pas rigoureusement conforme aux lois de la mécanique statistique. Notons enfin que le modèle 'j' décrit une fonction de corrélation impaire alors que nous le comparons à la fonction de corrélation expérimentale paire (transformée de Fourier en cosinus).

MOMENTS DE BANDES ET FONCTIONS DE CORRELATION ROTATIONNELLES EN INFRAROUGE POUR UNE VIBRATION PERPENDICULAIRE D'UNE MOLECULE LINEAIRE

JOSETTE VINCENT-GEISSE et CATHERINE DREYFUS

Groupe Infrarouge du Laboratoire de Recherches Physiques, associé au C.N.R.S., Université Paris VI, 4 place Jussieu, 75230 Paris Cedex 05, France

Résumé. L'objet du présent travail est la comparaison des fonctions de corrélation rotationnelles et des moments des vibrations perpendiculaires et parallèles d'une molécule linéaire. L'application de la théorie de Gordon permet tout d'abord de calculer les moments d'ordres 2 et 4 relatifs à une vibration perpendiculaire. D'autre part on peut établir, en mécanique classique, une relation entre les fonctions de corrélation $\Phi_\parallel(t)$ et $\Phi_\perp(t)$ du rotateur libre. Cette relation permet de calculer $\Phi_\perp(t)$ dans l'hypothèse des modèles M et J de Gordon. Il est également possible, dans le cas du modèle de Steele, d'obtenir une relation entre les deux fonctions de corrélation. En dernier lieu on procède à une comparaison entre résultats théoriques et expérimentaux.

Abstract. In this work the rotational correlation functions and band moments of the perpendicular and parallel vibrations of linear molecules are compared. The application of existing theories to the parallel bands involves no difficulty, whereas the degenerate perpendicular vibrations are to be treated from the beginning in the particular case of the linear molecule.

The second and fourth infrared rotational moments of the bands have been calculated by Gordon in the most general case of the asymmetric top. The relations thus obtained may be applied to all particular cases, except to the perpendicular vibrations of linear molecules. Calculations similar to Gordon's yield $M_\perp = \frac{1}{2} M_\parallel$.

Correlation functions involve the consideration of several cases. First, in the case of the free rotator, one obtains: $\Phi_\perp(t) = \frac{1}{2}[1 + \Phi_\parallel(t)]$. The others imply models. In the case of Gordon's M and J models we have evaluated $\Phi_\perp(t)$, apart from $\Phi_\parallel(t)$. Besides, the application of Steele's method to this model leads to: $[\Phi_\perp(t)]^2 = \Phi_\parallel(t)$, a relation which exhibits a satisfactory behaviour both to short and long times.

Finally all the above results are compared to experimental data.

De nombreuses études expérimentales de profils de bandes et de fonctions de corrélation ont été effectuées sur des molécules linéaires. L'application des théories existantes ne présente aucune difficulté dans le cas des bandes parallèles des molécules linéaires mais il n'en est plus de même en ce qui concerne les vibrations perpendiculaires dégénérées qui exigent généralement de reprendre les calculs à la base en se plaçant, dès le départ, dans le cas particulier de la molécule linéaire.

1. Moments de bandes

Dans le cadre des théories thermodynamiques, Gordon [1] a calculé les moments rotationnels d'ordres 2 et 4 des bandes de vibration rotation infrarouges dans le cas le plus général du rotateur asymétrique. Les relations obtenues sont applicables à tous les cas particuliers sauf à celui des vibrations perpendiculaires des molécules linéaires. En effet le moment d'inertie autour de l'axe moléculaire est alors nul et les expressions des moments deviennent infinies.

Il suffit, en reprenant les calculs, d'envisager le cas de la molécule linéaire lorsqu'on applique le théorème d'équipartition de l'énergie cinétique. On obtient alors une valeur finie pour les moments spectraux et l'on a [2]

$$M_\perp(2) = \tfrac{1}{2} M_\parallel(2) = \frac{kT}{I} \tag{1}$$

$$M_\perp(4) = \tfrac{1}{2} M_\parallel(4) = \frac{4k^2 T^2}{I^2} + \frac{\langle (OV)^2 \rangle}{2I^2}. \tag{2}$$

2. Fonctions de corrélation $\Phi_\parallel(t)$ et $\Phi_\perp(t)$ du rotateur libre

La vibration perpendiculaire est dégénérée et le vecteur unitaire $\mathbf{m}_\perp(t)$ de l'une de ses composantes est perpendiculaire au vecteur unitaire $\mathbf{m}_\parallel(t)$ porté par l'axe nucléaire et fait un angle α avec le moment cinétique L, fixe dans l'espace. On a :

$$\Phi_\perp(t) = \langle \mathbf{m}_\perp(0) \cdot \mathbf{m}_\perp(t) \rangle = \langle \mathbf{m}_\parallel(0) \cdot \mathbf{m}_\parallel(t) \cos^2\alpha + \sin^2\alpha \rangle.$$

En faisant la moyenne sur l'angle α, compris entre 0 et 2π on obtient :

$$\Phi_\perp(t) = \tfrac{1}{2}[1 + \Phi_\parallel(t)]. \tag{3}$$

Si l'on fait un développement limité aux temps courts, de $\Phi_\perp$ et $\Phi_\parallel$ on retrouve les rapports des moments $M_\parallel$ et $M_\perp$ trouvés plus haut.

En posant $\varphi = [\mathbf{m}_\parallel(0), \mathbf{m}_\parallel(t)]$ et $\psi = [\mathbf{m}_\perp(0), \mathbf{m}_\perp(t)]$ la relation (3) peut encore s'écrire :

$$\langle \cos\psi \rangle = \langle \cos^2 \varphi/2 \rangle \tag{4}$$

et l'on retrouve, d'une manière plus rigoureuse, une expression donnée antérieurement [3].

3. Fonctions de corrélation $\Phi_\parallel(t)$ et $\Phi_\perp(t)$ dans le cas des modèles M et J de Gordon [4]

Il est possible d'appliquer les modèles M et J de Gordon à la vibration perpendiculaire de la molécule linéaire en prenant l'expression (3) de $\Phi_\perp(t)$ dans les calculs puisque la rotation est supposée libre entre deux chocs. Dans le cas des modèles M et M généralisé $\Phi_\parallel$ et $\Phi_\perp$ se mettent sous la forme d'une intégrale calculable numériquement. Dans le cas du modèle J, l'intensité $I(\omega)$ de la bande s'évalue en utilisant les propriétés des transformations de Fourier et de Laplace, et, par inversion de Fourier, on en déduit $\Phi_\perp(t)$ et $\Phi_\parallel(t)$.

4. Fonctions de corrélation $\Phi_\parallel(t)$ et $\Phi_\perp(t)$ dans le cas de la diffusion rotationnelle (modèle de Steele [5])

Steele fait l'hypothèse de frottements visqueux et calcule la fonction d'autocorrélation rotationnelle d'un rotateur sphérique. En appliquant sa méthode de calcul au cas des molécules linéaires nous obtenons

$$[\Phi_\perp(t)]^2 = \Phi_\parallel(t). \tag{5}$$

Cette relation présente un comportement satisfaisant aux temps courts car elle se ramène alors à (3) et il en est de même aux temps longs car on obtient pour $\Phi_\perp(t)$ une décroissance exponentielle dont la constante de temps est le double de celle correspondant à la vibration parallèle.

La relation (5) peut se mettre sous la forme

$$\langle\cos\psi\rangle^2 = \langle\cos\varphi\rangle \tag{6}$$

et l'on retrouve une expression obtenue précédemment [6] en faisant l'hypothèse que les probabilités de rotation autour de deux axes perpendiculaires à l'axe nucléaire et perpendiculaires entre eux sont égales.

5. Comparaison entre la théorie et l'expérience

Celle-ci porte sur les différents points étudiés ci-dessus.

Nous avons tout d'abord rassemblé dans le tableau I les moments d'ordres 2 et 4 calculés pour C_2H_2 et N_2O gazeux sous pression de gaz inertes [7] et pour N_2O, OCS et C_2D_2 dissous dans des solvants inertes [6, 8, 9, 10]. Les seuls exemples retenus sont ceux pour lesquels on connaît à la fois les moments pour les deux types de bandes et

TABLEAU I

(Les unités sont $10^{24} S^{-2}$ pour $M(2)$ et $10^{48} S^{-4}$ pour $M(4)$)

Solvant	Soluté	N_2O			OCS		C_2D_2		C_2H_2	
		ν_1	ν_2	ν_3	ν_2	ν_3	ν_3	ν_5	ν_3	ν_5
gaz[7]	$M(2)$		7,1	14,2					39	19,5
	$M(4)$		215	530					3660	1440
CS_2[9]	$M(2)$						27	12		
	$M(4)$						3800	1000		
C_6H_{14} [8, 10]	$M(2)$	13,1	6,1	15,0	3,5	6,6				
	$M(4)$	960	310	1330	120	260				
C_6H_{16} [6, 8]	$M(2)$				2,8	7,0				
	$M(4)$				70	310				

pour lesquels les moments d'ordre 2 sont voisins de leur valeur théorique. On remarque sur ce tableau que la relation entre les moments d'ordre 2 est partout vérifiée, compte tenu de l'incertitude expérimentale mais qu'il n'en est pas de même pour les moments d'ordre 4, $M_\perp(4)$ étant systématiquement inférieur à $\frac{1}{2}M_\parallel(4)$. Dans certains cas la différence est de l'ordre de grandeur de l'incertitude mais dans d'autres (N_2O dans C_6H_{14}, OCS dans C_6H_{16} et C_2D_2 dans CS_2) elle lui est largement supérieure.

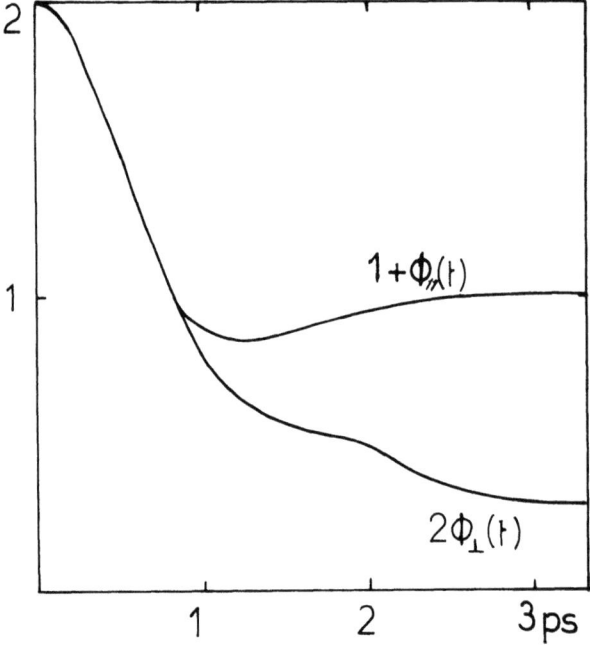

Fig. 1. Comparaison de $2\Phi_\perp(t)$ et de $1 + \Phi_\parallel(t)$ pour OCS dans l'argon à 80 bars.

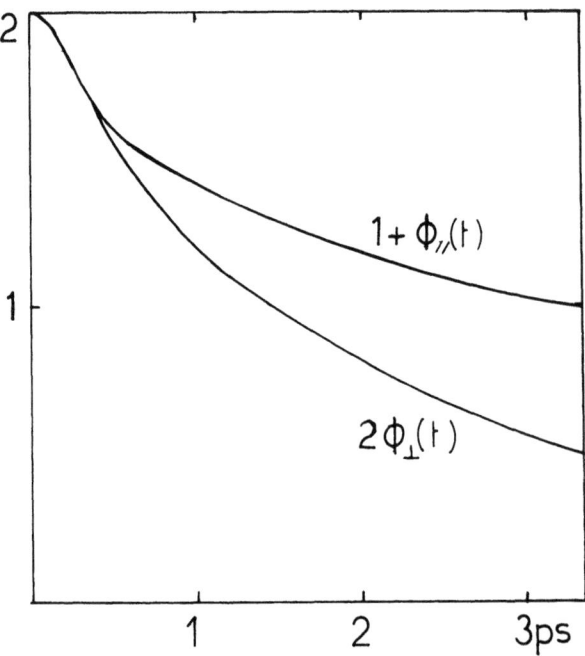

Fig. 2. Comparaison de $2\Phi_\perp(t)$ et de $1 + \Phi_\parallel(t)$ pour OCS dans l'hexane.

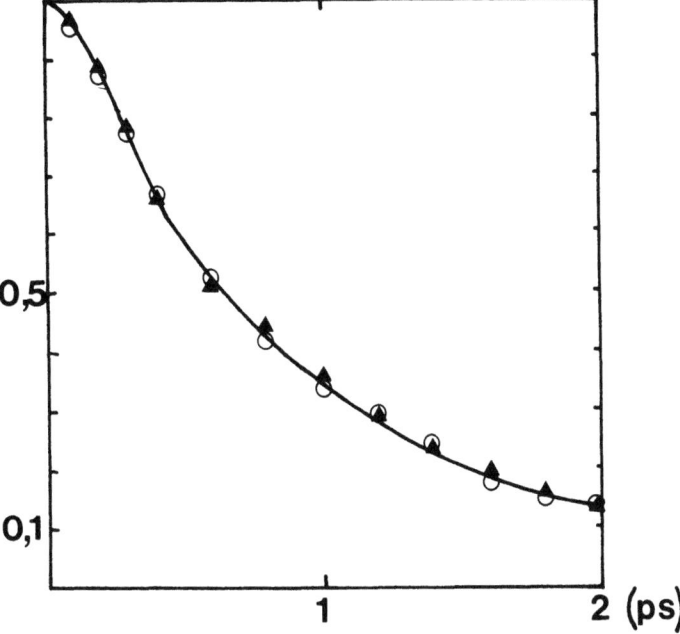

Fig. 3. Comparaison de $\Phi_\perp^2(t)$ et de $\Phi_\parallel(t)$. OCS dans C_6H_{14} [$\Phi_\perp^2(t)$: ○ et $\Phi_\parallel(t)$: ▲].

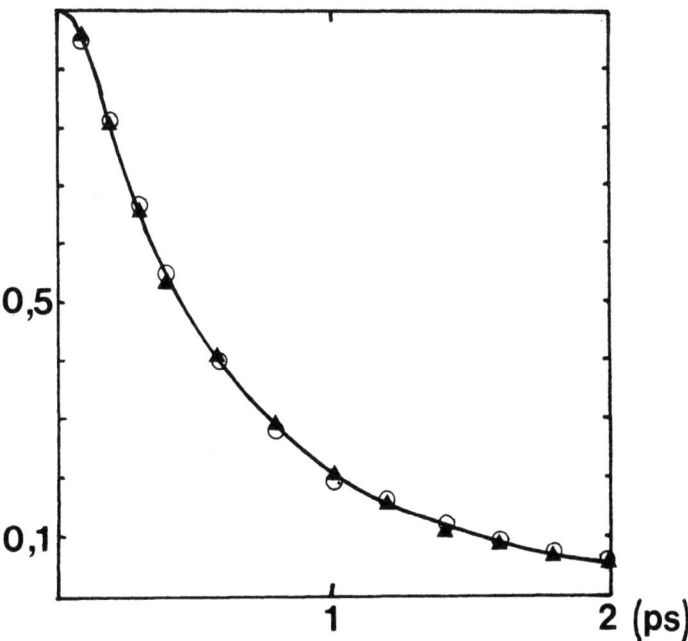

Fig. 4. Comparaison de $\Phi_\perp^2(t)$ et de $\Phi_\parallel(t)$. N_2O dans C_6H_{14} [$\Phi_\perp^2(t)$: ○ et $\Phi_\parallel(t)$: ▲].

Pour expliquer cet écart on peut mettre en cause l'existence d'un effet vibrationnel non négligeable, les relations (1) et (2) n'étant valables que pour les moments rotationnels. Dans ce cas il faudrait supposer que les vibrations parallèles, correspondant à des allongements de liaisons, seraient plus sensibles que les vibrations perpendiculaires aux fluctuations du potentiel intermoléculaire.

La vérification de la relation (3) a été faite sur des liquides et sur des gaz. La figure 2 montre les résultats obtenus dans le cas de OCS dissous dans l'hexane. Les fonctions $2\Phi_\perp$ et $1+\Phi_\parallel$ coïncident aux temps courts conformément à ce que l'on attendait, et pendant un temps de 0,45 ps très supérieur à celui de rotation libre. La figure 1 représente le cas de OCS gazeux, comprimé par l'argon sous une pression de 80 bars. La coïncidence des deux courbes s'étend plus loin, jusqu'à $t=0,7$ ps, correspondant à une rotation d'un angle de 100°.

La vérification des formules obtenues pour les modèles M et J consiste à calculer les valeurs du paramètre τ de Gordon, qui rendent le mieux compte de $\Phi_\parallel(t)$ et de $\Phi_\perp(t)$, ces deux valeurs doivent être identiques. Ce travail est en cours.

La vérification de la formule (5) a été effectuée sur les molécules N_2O et OCS en solution dans l'hexane, cas dans lesquels il a été montré que l'élargissement de bande d'origine rotationnelle était largement prépondérant; les figures 3 et 4 montrent que l'accord est excellent. Il existe également un autre moyen indirect. De la formule (5), en tenant compte du fait qu'aux temps longs les deux fonctions $\Phi(t)$ sont des exponentielles, on tire:

$$(\Delta v_{1/2})_\parallel = 2(\Delta v_{1/2})_\perp. \tag{8}$$

Cette relation entre les largeurs des bandes parallèles et perpendiculaires avait été remarquée depuis longtemps expérimentalement [11]; elle est très bien vérifiée dans les cas favorables où la largeur d'origine vibrationnelle reste faible.

En conclusion on peut dire que les relations obtenues ci-dessus se vérifient très bien en ce qui concerne les fonctions de corrélation et les moments d'ordre 2 et moins bien pour les moments d'ordre 4 qui devront faire l'objet d'études ultérieures.

Bibliographie

1. Gordon, R. G.: *J. Chem. Phys.* **39**, 2788 (1963); **40**, 1973 (1964); **41**, 1819 (1964).
2. Dreyfus, C. et Vincent-Geisse, J.: *Chem. Phys. Letters* **21**, 170 (1973).
3. Vincent-Geisse, J., Soussen-Jacob, J., et Nguyen Tan, T.: *Compt. Rend. Acad. Sci. Paris* **268**, 1020 (1969).
4. Gordon, R. G.: *J. Chem. Phys.* **44**, 1830 (1966).
5. Steele, W. A.: *J. Chem. Phys.* **38**, 2404 et 2411 (1963).
6. Bize, A. M.: Thèse 3ème cycle, 1971.
7. Levi, G., Chalaye, M., et Dayan, E.: *Chem. Phys. Letters* **12**, 462 (1972).
8. Bize, A. M., Soussen-Jacob, J., Vincent-Geisse, J., Legay, D., et Perchard, J. P.: *Can. J. Chem.* **50**, 217 (1972).
9. Elbez, G.: Thèse 3ème cycle, 1972.
10. Vincent-Geisse, J., Soussen-Jacob, J., Nguyen Tan, T., Descout, D.: *Can. J. Chem.* **48**, 3918 (1970).
11. Vincent-Geisse, J., Soussen-Jacob, J., Pelerin, D., et Bize, A. M.: *Ber. Bunsenges. physik. Chem.* **75**, 348 (1971).

ETUDE DES MOUVEMENTS MOLECULAIRES DANS LES LIQUIDES AU MOYEN DU PROFIL DES BANDES DE VIBRATION DANS L'INFRAROUGE

JANINE SOUSSEN-JACOB, JOSETTE VINCENT-GEISSE,
CATHERINE ALLIOT, ANNE-MARIE BIZE, JEAN-CLAUDE BRIQUET,
ELISE DERVIL, JACQUELINE LOISEL, et JEAN-PAUL PINAN-LUCARRE

Service infrarouge, Département de recherches physiques,
Tour 22, 4pl. Jussieu, 75005 Paris, France

Résumé. Au cours de ces dernières années, nous avons étudié une dizaine de molécules polyatomiques dissoutes dans différents solvants [1 à 10] et nous avons essayé de donner ici une vue d'ensemble des résultats fournis par l'examen des largeurs et des profils de bandes. Nous avons examiné d'abord la validité de l'approximation de Rakov dans la séparation des effets vibrationnel et rotationnel, par comparaison avec les résultats apportés par d'autres techniques. Nous avons ensuite abordé plus particulièrement l'étude des phénomènes rotationnels à partir du profil de bande et de la fonction de corrélation Φ_R. On interprète en général les résultats en terme de diffusion rotationnelle par sauts de plus ou moins grande amplitude, sauf dans quelques cas particuliers où l'on se rapproche quelque peu de la rotation libre.

Abstract. In recent years, we have investigated about ten polyatomic molecules in solution in various solvents [1–10]. We are now attempting to give a comprehensive view of the results provided by the study of the profiles and bandwidths. As a first step, we have examined the validity of Rakov's approximation, as applied to the separation between vibrational and rotational effects, by comparison with the results obtained through other techniques. Then, we have more particularly taken up the investigation of rotational phenomena, from the profiles of the bands and the correlation function Φ_R. The results have generally been interpreted in terms of rotational diffusion by jumps of more or less large amplitude, to the exception of a few cases where free rotation is somewhat approached.

Pour une molécule diatomique, en solution diluée dans un solvant inerte, la transformée de Fourier du profil de bande, qui représente la fonction de corrélation du moment dipolaire de transition, peut se mettre d'après Bratos [11], en l'absence d'interaction entre vibration et rotation, sous la forme

$$\Phi(t) = \Phi_V(t) \cdot \Phi_R(t)$$

($\Phi_V(t)$ et $\Phi_R(t)$, fonctions de corrélation vibrationnelle et rotationnelle). La largeur de la bande est alors donnée en première approximation par

$$\Delta v_{1/2} = (\Delta v_{1/2})_V + (\Delta v_{1/2})_R$$

relation vérifiée en toute rigueur si les profils des contributions vibrationnelle et rotationnelle sont tous deux lorentziens.

Généralisant ces expressions à une molécule de forme quelconque nous nous attacherons ici principalement aux deux points qui suivent:

1. Séparation des largeurs vibrationnelle et rotationnelle

C'est un problème fondamental pour interpréter correctement les renseignements fournis par les profils observés. En utilisant l'infrarouge seul, on peut obtenir la largeur vibrationnelle en étudiant la variation de la largeur totale avec la température et en lui appliquant la loi $\Delta v_{1/2} = \delta + A \exp(-U/RT)$ proposée par Rakov [12]; celle-ci suppose que (1) Φ_R est de la forme $\exp(-t/\tau)$, donc que le profil rotationnel est lorentzien et $(\Delta v_{1/2})_R = 1/c\Pi\tau$, (2) $\tau = \tau_0 \exp(U/RT)$ où U est la barrière de potentiel s'opposant à la réorientation, et (3) δ correspond à $(\Delta v_{1/2})_V$ et reste constant quand la température varie, ou sa variation reste faible devant celle de $(\Delta v_{1/2})_R$. Cette dernière condition est réalisée de toutes manières lorsque $(\Delta v_{1/2})_V$ est faible vis-à-vis de $(\Delta v_{1/2})_R$ et il semble alors vraisemblable que l'application de la formule de Rakov donne de bons résultats.

Lorsque $(\Delta v_{1/2})_V$ est plus important, il y a doute sur la validité de la méthode. Or, nous disposons parfois de résultats fournis par d'autres techniques, qui nous permettent une comparaison et une vérification. Il s'agit d'une part des largeurs de bandes obtenues par diffusion Raman, qui dans certains cas doivent être égales à la largeur vibrationnelle mesurée en infrarouge, d'autre part des largeurs rotationnelles mesurées à partir du spectre hertzien, ultrahertzien et infrarouge lointain.

Sur le tableau I, nous avons rassemblé les différentes valeurs des largeurs vibrationnelles dans tous les cas où l'on disposait, outre des résultats fournis en infrarouge par la méthode de Rakov, δ_{IR}, de ceux correspondant à l'une ou l'autre des deux autres méthodes, δ_{RAM} pour les largeurs vibrationnelles en Raman, δ_{CALC} pour la différence entre la largeur totale $\Delta v_{1/2}$ en infrarouge et la largeur rotationnelle $(\Delta v_{1/2})_R$ déduite du spectre 'basses fréquences'. L'examen de ce tableau nous indique que, comme l'on pouvait s'y attendre, l'application de la formule de Rakov donne de bons résultats, lorsque, comparativement à la largeur totale, l'élargissement vibrationnel est très faible, même si celui-ci varie avec la température comme cela a été mis en évidence dans quelques cas par diffusion Raman; il en est ainsi pour la bande v_1 de SO_2. Lorsque l'élargissement vibrationnel devient prépondérant, on remarque que la méthode de Rakov donne encore un ordre de grandeur, évidemment d'autant moins correct que la variation de $(\Delta v_{1/2})_V$ avec la température est plus importante, la valeur obtenue par cette méthode se rapprochant alors plutôt de la largeur vibrationnelle mesurée en Raman à basse température (CH_3CN, OCS dans CH_2Cl_2).

Les tableaux I et II font apparaître la plupart des résultats de nos études en fonction de la température. Ils suscitent quelques remarques: $(\Delta v_{1/2})_R$ est très important pour des bandes mettant en évidence des rotations autour d'axes par rapport auxquels les moments d'inertie sont faibles; pour une même bande, $(\Delta v_{1/2})_R$ diminue quand l'activité du solvant augmente. Quant à $(\Delta v_{1/2})_V$ sa variation pour une même bande en fonction du solvant est en sens inverse; de plus, sa valeur est plus grande pour les vibrations faisant intervenir des atomes H ou D que pour les vibrations d'atomes plus lourds.

En conclusion, malgré les hypothèses intervenant dans la formule de Rakov (additivité des largeurs, indépendance de $(\Delta v_{1/2})_V$ avec la température), l'infrarouge nous

TABLEAU I

Détermination des largeurs rotationnelle et vibrationnelle par différentes techniques expérimentales

Soluté	Solvant	Vibration	δ_{IR} (cm^{-1})	δ_{RAM} (cm^{-1})	δ_{CALC} (cm^{-1})	$(\Delta\nu_{1/2})_R$ (cm^{-1})
SO$_2$ [8]	CCl$_4$	ν_1	0 ± 0,5	0,60 ± 0,05 [14] (250 K)	0,50 ± 0,05 [14] (300 K)	17,9
	CS$_2$	ν_1	1 ± 1	0,53 ± 0,05 [14] (250 K)	0,45 ± 0,05 [14] (300 K)	21
CH$_3$CN [7]	liquide pur	ν_1	9		11	3* [18]
	liquide pur	ν_4 sym.	3,0		3,2	3* [18]
	CCl$_4$	ν_2 sym.	4,5	4,6 [15] (250 K)	3,5	2* [17]
	CCl$_4$	ν_4 sym.	4,0	2,9 [15] (250 K)	3,1	2* [17]
OCS [13]	CH$_2$Cl$_2$	ν_1	7	7 [16] (180 K)		1,9
C$_4$H$_4$O [10] (furanne)	liquide pur	ν_4	2	4 [16] (300 K)	1,2 (313 K) 2,2 (193 K)	9,0* [19]

ν_{sym} : bande pour laquelle on a éliminé la bande chaude par symétrisation.
$(\Delta\nu_{1/2})_R = \Delta\nu_{1/2} - \delta_{IR}$, sauf pour les valeurs marquées d'un astérisque obtenues à partir des spectres 'basses fréquences'.
$\delta_{CALC} = \Delta\nu_{1/2} - (\Delta\nu_{1/2})_R$. $\Delta\nu_{1/2}$: largeur à température ordinaire, mesurée avec une incertitude de 0,3 à 0,5 cm^{-1}.

TABLEAU II

Séparation entre largeurs rotationnelle et vibrationnelle par la méthode de Rakov

Soluté	Vibration	Solvant	δ_{IR} (cm^{-1})	$(\Delta v_{1/2})_R$ (cm^{-1})
SO_2 [8]	v_2	CS_2	2 ± 1	19,5
		CH_2Cl_2	$2,5 \pm 0,2$	8,7
	v_3	CCl_4	$0 \pm 0,5$	8,7
		CS_2	$3,0 \pm 0,5$	9
		CH_2Cl_2	$6,0 \pm 0,5$	4
CO_2 [13]	v_3	C_5H_{12}	0	17,4
C_2D_2 [6]	v_3	CS_2	5 ± 1	20,5
	v_5	CS_2	5 ± 1	9
OCS [5]	v_3	c-C_5H_{10}	$2,3 \pm 0,3$	7,9
		C_7H_{16}	$3,1 \pm 0,1$	7,4
		iso-C_5H_{12}	3,9	7,2
		CS_2	$6,3 \pm 0,5$	4,1
		C_2Cl_4	$6,7 \pm 0,5$	2,1
		CH_2Cl_2	$11,6 \pm 0,6$	3,6

$(\Delta v_{1/2})_R = \Delta v_{1/2} - \delta_{IR}$.
$\Delta v_{1/2}$: largeur à température ordinaire, mesurée avec une incertitude de 0,3 à 0,5 cm^{-1}.

conduit à une séparation relativement convenable entre contributions rotationnelle et vibrationnelle de la largeur.

2. Etude des phénomènes rotationnels à partir du profil de bande et de la fonction de corrélation Φ_R

Dans l'ensemble pour les molécules polyatomiques, on peut penser que les mouvements de rotation sont de la forme diffusion rotationnelle. En l'absence d'autres causes d'élargissement, le profil de la bande est alors lorentzien et la fonction de corrélation est une exponentielle de la forme $\Phi_R = e^{-t/\tau_{\theta 1}}$, où $\tau_{\theta 1}$ est le temps de corrélation angulaire.

En fait, pour la plupart des bandes que nous avons étudiées, le profil était effectivement lorentzien, au moins dans un domaine de fréquences égal à 1,5 fois la largeur et la fonction de corrélation exponentielle à partir d'un temps de l'ordre de 0,3 à 0,4 ps correspondant en gros aux ailes de la bande. Ces fonctions de corrélation ont été souvent exploitées, comme si elles représentaient effectivement Φ_R et on a déduit directement $\tau_{\theta 1}$ de la pente de la partie linéaire de $\text{Log}\,\Phi(t)$. Or, il est rare que ceci soit exact en toute rigueur. Le plus souvent, Φ_V intervient dans la fonction de corrélation calculée. Dans l'ignorance où nous nous trouvons en général de sa forme exacte (gaussienne ou complexe, exponentielle d'après la théorie, exponentielle dans les rares cas étudiés expérimentalement), nous évaluerons $\tau_{\theta 1}$ en utilisant la largeur rotationnelle, lorsqu'on la connaît, et en appliquant la relation

$$\tau_{\theta 1} = \frac{1}{c\Pi\,(\Delta v_{1/2})_R}.$$

A partir de $\tau_{\theta 1}$, on peut calculer le temps de corrélation des vitesses angulaires, τ_ω.

TABLEAU III

Paramètres décrivant les mouvements de rotation de différentes molécules

Soluté	Solvant	Vibration	τ_{θ_1} (ps)	ω_R (s^{-1})	τ_ω (ps)	$\omega_R \tau_\omega$
CH$_3$CN [7]	CCl$_4$	Spectre basses fréquences	5,7 [17]	$3,0 \times 10^{12}$	0,02	3°
CHBr$_3$ [9]	liquide pur	Spectre basses fréquences	19 ± 1,9 [21]	$1,1 \times 10^{12}$	0,04	3°
CHCl$_3$ [9]	liquide pur	Spectre basses fréquences	6,0 ± 0,6 [22]	$1,8 \times 10^{12}$	0,05	5°
SO$_2$ [8]	CH$_2$Cl$_2$	ν_3	2,68	$3,1 \times 10^{12}$	0,04	7°
	CCl$_4$		1,22	$3,1 \times 10^{12}$	0,08	15°
	C$_6$H$_{14}$		0,43	$3,1 \times 10^{12}$	0,23	41°
	C$_6$H$_{12}$	ν_1		$6,1 \times 10^{12}$	0,18	56°
OCS [5]	c-C$_5$H$_{10}$	ν_3	1,34	$2,5 \times 10^{12}$	0,12	17°
C$_2$D$_2$ [6]	CS$_2$	ν_3	0,52	$5,1 \times 10^{12}$	0,07	22°

Dans le cas d'une molécule diatomique, Bratos [11] a montré que si la modulation de la rotation est rapide, on a $\tau_{\theta 1} = 1/\omega_R^2 \tau_\omega$; cette relation peut aussi être déduite de la formule de Hubbard [20]. A partir de τ_ω, on calcule l'angle de rotation libre, $\omega_R \tau_\omega$.

Il n'est pas toujours possible de traiter les spectres de cette façon. En effet, le profil de la bande s'écarte parfois de manière importante d'une lorentzienne, avec une base nettement plus évasée (v_1 de SO_2 dans C_6H_{12} [8], v_5 de CH_3CN dans CCl_4 [7], par exemple). On se trouve alors dans des cas intermédiaires entre modulation lente et rapide. La courbe $\text{Log}\,\Phi(t)$ décroît très vite aux temps courts, avant d'atteindre une partie linéaire qui passe loin de l'origine et dont la pente n'est plus reliée à $\tau_{\theta 1}$. Cependant, τ_ω peut être lu directement sur la courbe $\text{Log}\,\Phi(t)$; c'est le temps pendant lequel cette courbe reste confondue avec celle du rotateur libre, temps malheureusement déterminé avec une mauvaise précision, cette partie de la courbe correspondant aux ailes de la bande. On calcule, comme précédemment, l'angle de rotation $\omega_R \tau_\omega$.

Sur le tableau III, nous donnons quelques exemples des résultats obtenus de ces différentes manières. Les valeurs de $\tau_{\theta 1}$ nous fournissent des renseignements intéressants si on les confronte à celles de $\tau_{\theta 2}$ (temps de corrélation du 2è harmonique sphérique) mesuré en Raman ou en RMN: un rapport $\tau_{\theta 1}/\tau_{\theta 2}$ égal à 3 correspond au cas type de diffusion rotationnelle par petits angles; il en est ainsi pour la rotation de l'axe ternaire de CH_3CN, $CHCl_3$ et $CHBr_3$. Pour ces molécules, les angles de rotation libre ont effectivement des valeurs faibles (tableau III); pour les autres molécules étudiées, ils sont souvent supérieurs. Si l'on compare les rotations de deux molécules différentes, pour lesquelles ω_R a des valeurs voisines, nous voyons que les angles de rotation libre dans un même solvant ne sont pas nécessairement proches. Ainsi, pour la rotation de l'axe correspondant à la vibration v_3 de SO_2, qui se produit autour de l'axe de symétrie portant le moment dipolaire, $\omega_R \tau_\omega$ est bien plus grand que pour la rotation de l'axe ternaire de CH_3CN qui porte un moment dipolaire important. On peut de plus constater sur l'exemple de v_3 de SO_2 que l'angle obtenu à partir d'une même bande dans des solvants différents est d'autant plus petit que le solvant est plus actif.

3. Conclusion

Après avoir montré sur l'ensemble des molécules étudiées comment séparer les contributions vibrationnelle et rotationnelle de la largeur, nous nous sommes attachés plus particulièrement à l'étude des phénomènes rotationnels. La diffusion rotationnelle rend compte en première approximation de nos résultats, sauf dans quelques cas; on ne trouve jamais de rotation libre, bien que peut-être la rotation de CH_3CN autour de son axe s'en approche [7].

Il faut noter que, si l'infrarouge est une bonne méthode et se montre suffisant, d'une part pour étudier les processus de rotation lorsque la contribution rotationnelle est prépondérante, d'autre part pour déterminer l'élargissement vibrationnel lorsqu'il n'est plus négligeable mais pas encore important, il est en général intéressant sinon nécessaire de faire appel à d'autres techniques pour parvenir à une étude plus approfondie. Ainsi, grâce à la connaissance précise d'une bande Raman presque totalement polarisée et du spectre 'basses fréquences', on peut interpréter correctement toutes les bandes infrarouges de même symétrie. En outre, la confrontation de ces trois différentes techniques peut faire apparaître, ce dont nous n'avons pas parlé ici (voir [9]), des effets autres que rotationnels et vibrationnels. Cependant, les spectres 'basses fréquences' ne peuvent être obtenus que pour les vibrations parallèles au moment permanent de la molécule; les spectres Raman, encore difficiles à mesurer, ne donnent en principe des résultats interprétables simplement que pour les vibrations complètement polarisées. La méthode infrarouge, elle, présente l'avantage d'être applicable à un plus grand nombre de bandes.

Bibliographie

1. Jacob, J., Leclerc, J., et Vincent-Geisse, J.: *J. Chim. Phys.* **66**, 970 (1969).
2. Soussen-Jacob, J., Vincent-Geisse, J., Beaulieu, D., et Tsakiris, J.: *J. Chim. Phys.* **67**, 1118 (1970).
3. Vincent-Geisse, J., Soussen-Jacob, J., Pélerin, D., et Bize, A. M.: *Ber. Bunsenges. physik. Chem.* **75**, 348 (1971).
4. Vincent-Geisse, J., Soussen-Jacob, J., Tan Tai, Nguyen, et Descout, D.: *Can. J. Chem.* **48**, 3918 (1970).
5. Bize, A. M., Soussen-Jacob, J., Vincent-Geisse, J., Legay, D., et Perchard, J. P.: *Can. J. Chem.* **50**, 217 (1972).
6. Elbez, G.: Thèse de 3è cycle, Paris, 1972.
7. Breuillard-Alliot, C. et Soussen-Jacob, J.: à paraître.
8. Briquet, J. C. et Soussen-Jacob, J.: à paraître.
9. Soussen-Jacob, J., Dervil, E., et Vincent-Geisse, J.: à paraître.
10. Pinan-Lucarré, J. P., Loisel, J., et Vincent-Geisse, J.: *Molecular Crystals and Liquid Crystals*, sous presse.
11. Bratos, S., Rios, J., et Guissani, Y.: *J. Chem. Phys.* **52**, 439 (1970).
12. Rakov, A. V.: *Opt. Spectr.* **13**, 203 (1962); *Proc. P. N. Lebedev, Phys. Inst.* **27** (1965).
13. Goulay-Bize, A. M.: communication privée.
14. Ouillon, R. et Le Duff, Y.: *Advances in Raman Spectrosc.*, vol. 1, Heyden and Son, London, 1973, p. 428.
15. Ouillon, R.: communication privée.
16. Perchard, J. P. et Legay, D.: communication privée.
17. Fauquembergue, R., Leroy, Y., et Desplanques, P.: *Compt. Rend. Acad. Sci. Paris* **269**, 701 (1969). Desplanques, P. et Constant, E.: Congrès Herrenalb, 1970.
18. Mansigh, K. et Mansigh, A.: *J. Chem. Phys.* **41**, 827 (1964).
19. Bezot, P.: Thèse de 3è cycle, Orsay, 1971.
20. Hubbard, P. S.: *Phys. Rev.* **131**, 1155 (1963).
21. Fauquembergue, R.: Thèse de 3è cycle, Lille, 1968.
22. Greffe, J. L., Goulon, J., Brondeau, J., et Rivail, J. L.: *J. Chim. Phys.* **70**, 282 (1973).

THE TEMPERATURE EFFECT ON THE WIDTH AND INTENSITY OF THE RAMAN SPECTRA

N. I. REZAEV

Dept. of Physics, Moscow State University, U.S.S.R.

Abstract. The contours of the Raman lines of *o*-xylene, *m*-xylene, *p*-xylene and cyclohexane have been investigated depending on temperature.
It was established that the temperature broadening of the lines and the anomalous temperature change of its integrated intensities were due to different causes. The temperature effect on depolarized lines was explained by a chaotic reorientation of molecules in a liquid. The time τ of the chaotic reorientation of the molecules and some other parameters were calculated. It was supposed that the anomalous temperature change of the line intensities was caused by the intermolecular interaction.

Résumé. Les contours des raies Raman des spectres, des *o*-xylène, *m*-xylène, *p*-xylène et cyclohexane ont été analysés en fonction de la température. On a établi que des causes différentes provoquent l'élargissement des raies et les variations anormales de leurs intensités intégrées avec la température. L'élargissement avec la température des raies dépolarisées est expliqué par l'influence de la réorientation chaotique des molécules. On a utilisé la largeur des raies pour le calcul du temps de réorientation chaotique des molécules et quelques autres caractéristiques de la substance. On suppose que les changements anormaux des intensités de raies avec la température sont provoqués par l'action moléculaire.

1. Introduction

According to the kinetic theory of liquids offered by Frenkel [1] the motion of small molecules can be considered as following process. A molecule performs rotating oscillations near an equilibrium position in a period of time τ. Then the molecule jumps in a new equilibrium position. A chaotic reorientation of the molecule takes place when it jumps from an equilibrium position to another one.

Sobelman [2] proved theoretically that the chaotic reorientation of the molecules in a liquid must cause the broadening of the depolarized line of the Raman spectrum because the polarizability derivative of a molecule corresponding to this line was anisotropic. The shape of the line must be of the Lorentz type and the line width δ must be inversely proportional to the time τ of the chaotic reorientation of the molecules, i.e. $\delta \sim 1/\tau$.

In order to examine this theory the dependence of the Raman line contours on temperature was investigated in this work. It is known that the time τ of the chaotic reorientation of the molecules depends on temperature. For calculating the time τ one can use Debye's equation

$$\tau = \frac{4\pi a^3 \eta}{3kT}, \tag{1}$$

where a is the effective radius of the molecule, η is the viscosity, k is Boltzmann's constant, T is the absolute temperature, or Frenkel's equation

$$\tau = \tau_0 e^{u/kT}, \qquad (2)$$

where τ_0 is a period of the oscillation of the molecule near the equilibrium position, U is the activation energy which is necessary for transition of the molecule from the equilibrium position to another one. According to Sobelman's theory, the width of the depolarized line should increase with temperature. The previous measurements confirmed it [3–6]. The present work is a continuation of these investigations.

2. Experimental

Cyclohexane, o-xylene, m-xylene, and p-xylene were chosen as objects of the investigation. The molecules of these substances are small, and one has reason to suppose that they, can jump chaotically and reorientate simultaneously. Besides it these substances, have many intense lines with various depolarization factors which is very convenient for the investigation. The liquid substance to be analyzed was placed into a special thermostatic cell.

The measurements of the Raman line contours were performed by a photoelectric grating spectrometer with a dispersion of 6.4 Å mm^{-1}. The Raman spectra were excited by a low-pressure mercury lamp. The exciting line was Hg 4358 Å. The widths of both the entrance and exit slits of the spectrometer were chosen narrow in comparison with the width of the Raman line.

The width δ of the Raman line was found by the following method. The observed contour was corrected for the distortion caused by the slits [7]. Then the observed width of the exciting line together with the instrumental function was substracted from the observed width of the Raman line [8].

3. Investigation of the Widths of the Raman Lines

The results of the measurements of the line widths at various temperatures are shown in Figures 1, 2, 3 and 4. It can be seen that the temperature effect depends on the depolarization factor ϱ of the line. The greater the depolarization factor ϱ the greater is the temperature effect on the line width. The temperature effect is maximum for the completely depolarized lines ($\varrho = \frac{6}{7}$).

The experimental results show (Figure 5) that the contour of the Raman lines is very near to the type given by the Lorentz equation

$$I(v) = I_0 \frac{(\delta/2)^2}{(v - v_0)^2 + (\delta/2)^2}, \qquad (3)$$

where I_0 is the intensity at the maximum of the line (the peak intensity), v(cm^{-1}) is the frequency, δ(cm^{-1}) is the width of the line which is measured at the half height of its contour.

These data confirm that the width of the depolarized line depends on a chaotic reorientation of molecules. For calculation of the time τ the following equations were used

$$\delta = \delta_0 + \delta(T), \quad \delta(T) = \frac{\gamma}{\pi c \tau}, \qquad (4)$$

where δ is the width of the Raman line, $\delta(T)$ is the part of the width which depends on temperature, γ is some function of the depolarization factor ϱ with following limits: $0 \leqslant \gamma \leqslant 1$ when $0 \leqslant \varrho \leqslant \frac{6}{7}$, c is the light velocity, $\pi = 3.14$.

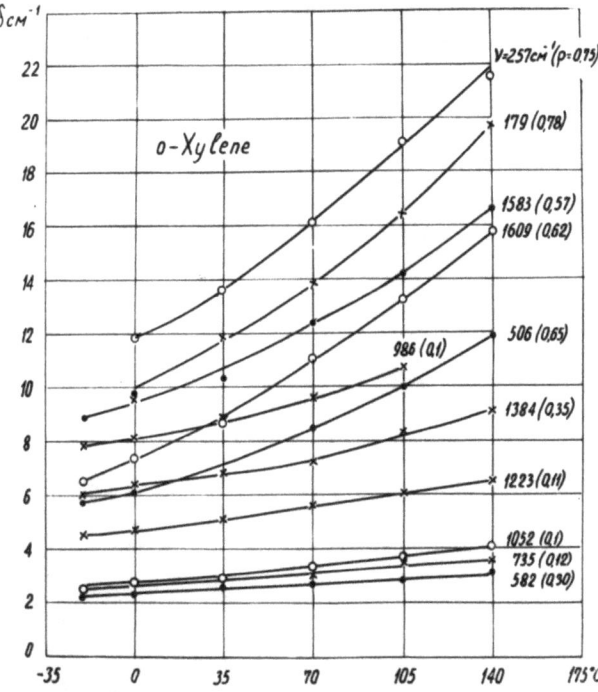

Fig. 1. The temperature effect on the line widths of o-xylene. v is the line frequency, ϱ is the depolarization factor.

δ_0 is own width of the Raman line which depends on the structure of the molecule and the type of its vibration [9]. According to the recent investigation [10] one can suppose that the value δ_0 is caused by vibrational lifetimes of molecules in liquids.

The time τ was calculated for the completely depolarized lines ($\varrho = \frac{6}{7}$). In this case $\gamma = 1$. Previously the width δ_0 was calculated in the following way. On the basis of the Equations (2) and (4) the width of the Raman line was expressed as

$$\delta = \delta_0 + \frac{\gamma}{\pi c \tau_0} e^{-u/kT}. \qquad (5)$$

Using the experimental data of the width δ the diagram $\delta = f(1/T)$ was plotted. On this diagram three points were selected which were quite a distance from each other and answered the condition

$$\frac{1}{T_3} = \frac{1}{2}\left(\frac{1}{T_1} + \frac{1}{T_2}\right). \tag{6}$$

In this case [11]

$$\delta_0 = \frac{\delta_1 \delta_2 - \delta_3^2}{\delta_1 + \delta_2 - 2\delta_3}. \tag{7}$$

The width δ_0 can be also found by an other method. Using the formulas (1) and (4) one can get the equation

$$\delta = \delta_0 + \frac{3kT\gamma}{4\pi^2 a^3 c\eta}, \tag{8}$$

The experimental data can be expressed in the form of the linear diagram, on which

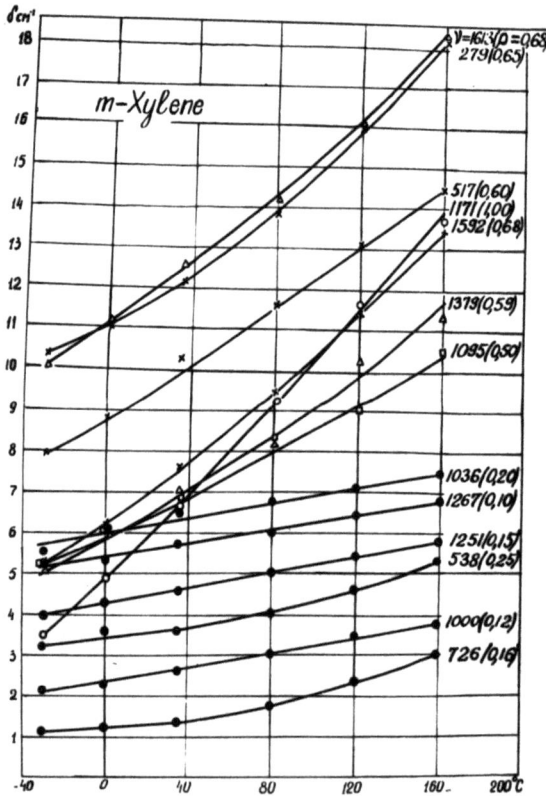

Fig. 2. The temperature effect on the line widths of *m*-xylene. ν is the line frequency. ϱ is the depolarization factor.

the width δ depends on T/η. Assuming $T/\eta \to 0$ one can find the width δ_0 by this diagram. The diagram also allows to calculate the effective radius a of the molecule. The viscosity η was taken from other work [12]. It is necessary to note that both methods give close results for the width δ_0.

The period τ_0 and the energy u were found by the linear diagram $\ln(\delta - \delta_0) = f(1/T)$ which have been plotted by using the equation

$$\ln(\delta - \delta_0) = \ln\left(\frac{\gamma}{\pi c \tau_0}\right) - \frac{u}{kT}. \tag{9}$$

For each of the investigated substances the values of τ, a, τ_0 and u were calculated by using two depolarized lines. As can be seen from Table I the data, which were calculated for a pair of lines, are close to each other. All calculated data of τ, a, τ_0 and u have reasonable values and are in agreement with data of other experimental methods. For example the times τ of the chaotic reorientation of the o-xylene, m-xylene and p-xylene molecules calculated on the basis of the measurement of the depolarized wings of the Rayleigh line [13] are 7.6×10^{-12}, 6.3×10^{-12} and 9.1×10^{-12} s respectively.

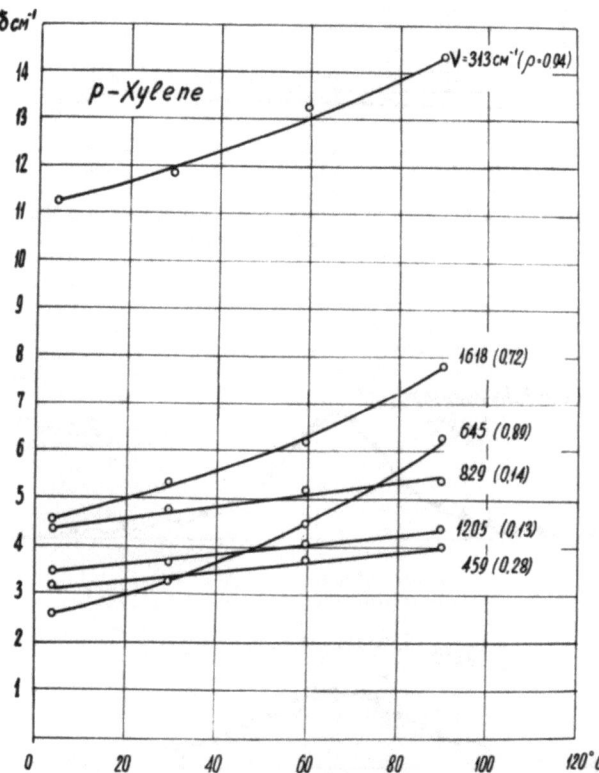

Fig. 3. The temperature effect on the line widths of p-xylene, ν is the line frequency, ϱ is the depolarization factor.

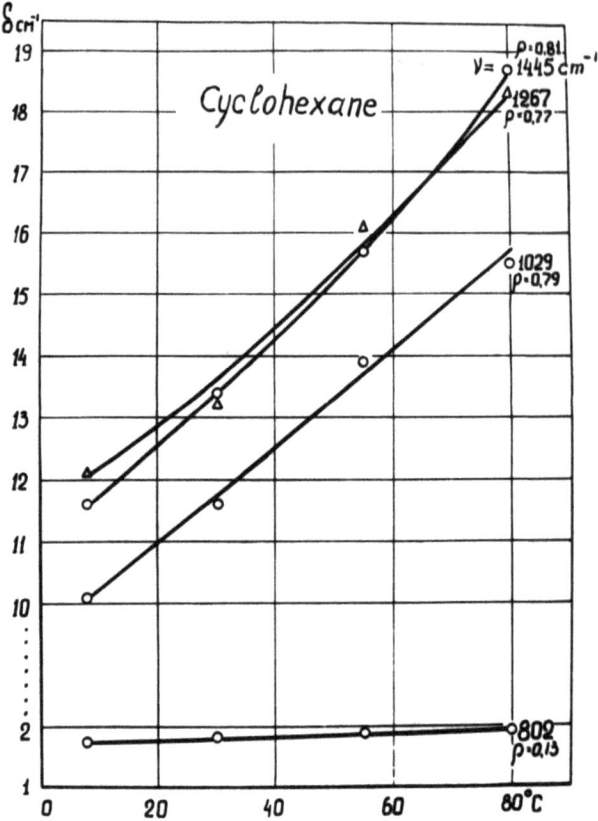

Fig. 4. The temperature effect on the line widths of cyclohexane. v is the line frequency, ϱ is the depolarization factor.

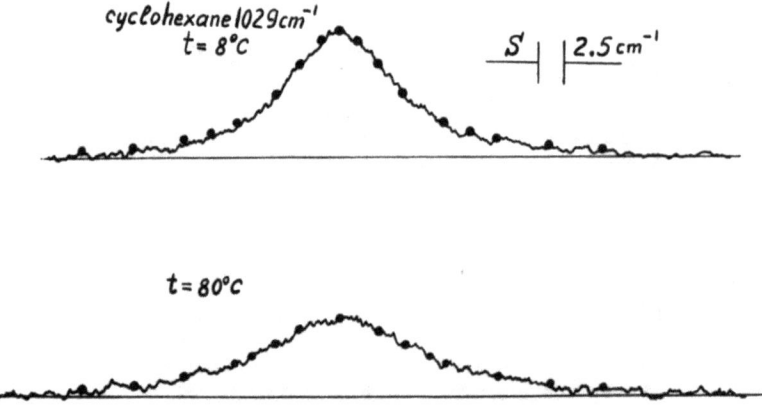

Fig. 5. The contour of the Raman line ($v = 1029$ cm^{-1}) of cyclohexane at temperatures 8 °C and 80 °C. S is the slit width. The Lorentz function is plotted by dots.

TABLE I
The calculated data

	o-xylen		m-xylen		p-xylen		Cyclohexane	
Frequency, cm^{-1}	179	257	1171	1592	313	645	1029	1445
δ_0, cm^{-1}	8.2	9.8	1.5	3.9	10	1.3	6.7	8.3
τ_0, 10^{14} s	2.5	2.6	9.6	10	4.8	2.9	2.2	1.4
τ, 10^{12} s ($t = 20\,°\mathrm{c}$)	4.0	3.5	2.4	3.7	6.4	6.2	2.5	2.5
a, Å	1.7	1.7	1.7	1.9	2.1	2.0	1.5	1.4
U, kcal/mole	2.9	2.9	1.9	2.0	2.9	3.1	2.8	1.3

4. Investigation of the Raman Intensities

In the present work the temperature effect on the Raman intensity was investigated too. It was established that the integrated intensity $I_\infty = \int_{-\infty}^{+\infty} I(\nu)\,d\nu$ of the Stokes lines decreased with temperature but didn't increase according to the known theory

$$I_\infty \sim \frac{1}{1 - e^{-h\nu/kT}}. \tag{10}$$

The results of this investigation are shown in Table II. The ratio of the intensities for two limit temperatures for experiment and theory and the vibration forms [14] of molecules are given in this Table.

As can be seen from Table II the anomolous temperature changes of the line intensities of xylenes are much larger than those of cyclohexane. These changes don't depend on the frequency ν and the depolarization factor ϱ of the lines.

However they depend on the vibration forms of the xylenes molecules. The Raman intensites of the valency vibrations of the carbon skeleton ($Q(CC)$ form) have a least declination from theory. The anomalous changes of the line intensities, which correspond to the deformational vibration of the $\beta(CCH)$, $\beta(CH_3)$ and $\varrho(C-H)$ forms, are the greatest. It is necessary to note that the peripheral atoms of molecules take part in these vibrations.

The obtained data give the possibility to suppose that the anomalous temperature changes of the Raman intensities of the investigated liquids are caused by the intermolecular interaction. The intermolecular interaction acts weakly on the line intensities of cyclohexane.

It is interesting to note that the Raman intensities of cyclohexane change also a little when this substance change its condition from liquid to solid.

The anomalous temperature changes of the Raman intensities of some liquids were observed by other authors [15–19]. The experimental results of the present work agree with their data. In this way the investigation of the temperature effect on the line intensities shows that the theoretical Equation (10) is inapplicable to the liquids.

In conclusion it should be noted that the peak intensities of the Raman lines decrease with temperature more greatly than the integrated intensities because of the

TABLE II

The temperature effect on the Raman intensities

o-xylen										
Frequency, cm⁻¹	257	506	582	735	986	1052		1223	1384	1583	1609
Vibration form	ϱ_{C-H}	γ_{CCC}	γ_{CCC}	Q_{CC}	ϱ_{C-H}	Q_{CC}		q_{C-C}	β_{CH_3}	Q_{CC}	Q_{CC}
$\overline{I_{+140°C} \over I_{-25°C}}$ experiment	0.49	0.66	0.46	0.71	0.36	0.62		0.42	0.43	0.59	0.72
$\overline{I_{+140°C} \over I_{-25°C}}$ theory	1.32	1.14	1.11	1.07	1.03	1.02		1.01	1.01	1.00	1.00
Declination from theory, %	63	42	59	34	65	39		58	57	41	28
m-xylen											
Frequency, cm⁻¹	279	517	538	726		1000	1036	1171	1251	1379	1592	1613
Vibration form	β_{CCH}	γ_{CCC}	γ_{CCC}	Q_{CC}		Q_{CC}	β_{CH_3}	β_{CCH}	q_{C-C}	β_{CH_3}	Q_{CC}	Q_{CC}
$\overline{I_{+160°C} \over I_{-30°C}}$ experiment	0.27	0.57	0.63	0.84		0.61	0.39	0.25	0.57	0.38	0.65	0.52
$\overline{I_{+160°C} \over I_{-30°C}}$ theory	1.33	1.66	1.14	1.08		1.03	1.03	1.02	1.01	1.00	1.00	1.00
Declination from theory, %	80	51	45	22		41	71	76	43	62	35	48
p-xylen											
Frequency, cm⁻¹	313	645	459	829					1205	1379	1618	
Vibration form	ϱ_{C-H}	γ_{CCC}	γ_{CCC}	Q_{CC}					q_{C-C}	β_{CH_3}	Q_{CC}	
$\overline{I_{+90°C} \over I_{+4°C}}$ experiment	0.77	0.85	0.86	0.86					0.78	0.69	0.81	
$\overline{I_{+90°C} \over I_{+4°C}}$ theory	1.14	1.05	1.09	1.02					1.01	1.00	1.00	
Declination from theory, %	33	19	21	16					23	31	19	
cyclohexane											
Frequency, cm⁻¹				802		1029			1267	1445		
Vibration form				Q_{CC}		Q_{CC}			χ_{CH_2}	α_{HCH}		
$\overline{I_{+80°C} \over I_{+8°C}}$ experiment				0.89		0.92			0.96	0.94		
$\overline{I_{+80°C} \over I_{+8°C}}$ theory				1.02		1.01			1.00	1.00		
Declination from theory, %				13		9			4	6		

temperature effect on the line widths. The temperature change of the peak intensity depends on the depolarization factor of the line.

It is need to take this into account in the molecular spectroscopic analysis.

References

1. Frenkel, Y. I.: *Kinetic Theory of Liquids* (in Russian), Moscow, 1945.
2. Sobelman, I. I.: *Izv. Akad. Nauk SSSR, Ser. Fis.* **17**, 554 (1953).
3. Rezaev, N. I.: *Mater. X Vsesouzn. Sovesch. po spectoscopii* **1**, 230 (1957).
4. Rezaev, N. I.: *Vestnik Moscow. Univer.*, No. 2, 145 (1957).
5. Rezaev, N. I. and Andreev, N. S.: *Optika i spectroscopia* **7**, 119 (1959).
6. Rakov, A. V.: *Optika i spectroscopia* **7**, 202 (1959).
7. Strutt, J.: *Phys. Mag.* **42**, 441 (1871).
8. Suschinskii, M. M.: *Zhurn. Eksperim i Teor. Fiz.* **25**, 87 (1953).
9. Rezaev, N. I.: *Optika i spectroscopia* **5**, 561 (1958).
10. Laubereau, A., Von der Linde, D., and Kaiser, W.: *Phys. Rev. Letters* **28**, 1162 (1972).
11. Worthing, A. G. and Geffner, J.: *Treatment of Experimental Data*, New York-London, 1943.
12. Vargaftik, N. B.: *Handbook of Gases and Liquids* (in Russian), Moscow, 1963.
13. Vuks, M. V. and Atahodjaev, A. K.: *Optika i spectroscopia* **5**, 51 (1958).
14. Sverdlov, L. M., Kovner, M. A., and Krainov, E. P.: *Vibrational Spectra of Polyatomic Molecules* (in Russian), Moscow, 1970.
15. Venkateswarlu, K.: *Curr. Sci.* **16**, 15 (1947).
16. Fishkova, L. M.: *Dokladi Akademii Nauk SSSR* **75**, 523 (1950).
17. Bolovich, J. S. and Arhipenko, D. K.: *Dokladi Akademii Nauk SSSR* **86**, 247 (1952).
28. Stehanov, A. I. and Chiler, E. V.: *Zhurn. Eksperim. i Teor. Fiz.* **25**, 2209 (1955).
19. Sokolovskaja, A. I. and Bazhulin, P. A.: *Mater. X Vsesouzn. Sovesch. po spectroscopii* **1**, 225 (1957).

ETUDE PAR SPECTROSCOPIE INFRAROUGE ET PAR SPECTROSCOPIE RAMAN DES FLUCTUATIONS ORIENTATIONNELLES DE LA MOLECULE DE FURANNE EN SOLUTION

PATRICK DORVAL et PIERRE SAUMAGNE

Laboratoire de Spectrochimie Moléculaire, Université de Bretagne Occidentale, 6, Avenue Le Gorgeu, 29283 Brest-Cedex, France

Résumé. Une étude par spectroscopie infrarouge des largeurs des bandes d'absorption de la molécule de furanne dissoute dans divers solvants a permis de conclure à une réorientation préférentielle autour d'un axe perpendiculaire au plan de la molécule. Ce résultat a été confirmé par un effet de température.

Fonction de corrélation et temps de corrélation ont été calculés par spectroscopie Raman à partir de la théorie établie par Bratos et Marechal. Une comparaison est faite avec les résultats obtenus à partir de la théorie de Rakov.

Abstract. A study of reorientational motions by infrared spectroscopy and Raman scattering is undertaken for the molecule of furan in solution. This molecule is a near oblate top molecule: the values of its moments of inertia are:

$$I_x = 54,6720 \quad I_y = 53,5126 \quad I_z = 108,2301 \; (\text{amu Å}^2).$$

According to Rakov, the examination of width of the infrared vibration bands of substances in a condensed medium gives informations concerning molecular motions. Measurements performed on the vibration bands ν_7 of A_1 type, ν_{16} of B_1 type and ν_{19} of B_2 type, are described here, for furan dissolved in different solvents. It is concluded that the molecule reorients around the z axis perpendicular to the plane of the molecule. This result is confirmed by a temperature effect: a variation of temperature from $-60\,°\text{C}$ to $+30\,°\text{C}$ does not affect practically the width of the B_2 band, whereas the width of the A_1 and B_1 bands increase of 50%.

Raman scattering has been used to determine the reorientational correlation time. The Rakov theory applied to the width of the polarized and depolarized parts of the ν_3 and ν_4 Raman lines of the furan molecule dissolved in carbon disulfide at $25\,°\text{C}$ gives the following correlations times:

$$\tau(1138 \; \text{cm}^{-1}) = 1,15 \times 10^{-12}\,\text{s}.$$
$$\tau(1384 \; \text{cm}^{-1}) = 1,65 \times 10^{-12}\,\text{s}.$$

According to a recent theory due to Bratos and Marechal, the orientational correlation function $G_{2R}(t)$ can be obtained only after elimination of the vibrational correlation function determined from the isotropic Raman scattering. This theory has been applied to the ν_3 and ν_4 bands of furan. The reorientation correlation time, the same for the two bands, is:

$$\tau = 1,50 \times 10^{-12}\,\text{s}.$$

These results show that the reorientational correlation times obtain from the $G_{2R}(t)$ function and from the Rakov theory are in good agreement.

Nous avons analysé par spectroscopie infrarouge et par spectroscopie Raman les fluctuations orientationnelles de la molécule de furanne en solution. Cette molécule est un rotateur asymétrique proche du type toupie. En effet, le furanne, C_4H_4O, est une molécule plane dont les valeurs des différents moments d'inertie sont les suivants:

$$I_x = 54,6720; \quad I_y = 53,5126; \quad I_z = 108,2301 \,(\text{UMA Å}^2) \; [1].$$

1. Détermination de l'axe de réorientation par spectroscopie infrarouge

Rakov a montré a montré que l'étude des largeurs de bandes fournissait des informations sur la réorientation moléculaire [2]. Nous avons étudié les largeurs des bandes de vibration $\delta(C-H)$, ν_7 de type A_1, ν_{16} de type B_1, centrées respectivement à 996 cm^{-1} et 1172 cm^{-1}, ainsi que celle de la vibration $\gamma(C-H)$, ν_{19} de type B_2 centrée à 745 cm^{-1}.

Les largeurs des bandes observées pour différents solvants sont rassemblées dans le tableau suivant:

TABLEAU I

Largeurs des bandes infrarouges $\Delta\nu_{\frac{1}{2}}$ (cm^{-1})

Solvant vibration	Tétrachlorure de carbone	Sulfure de carbone	Cyclohexane	Hexane	Pentane
ν_7 (A_1)[a]	8	9	9,6	11,6	12
ν_{16} (B_1)[b]	7,2	7,6	9,2	10	10
ν_{19} (B_2)[c]	–	7,6	6	8	–

[a] Concentration $\sim$ 0,5 mole l^{-1}.
[b] Concentration $\sim$ 1 mole l^{-1}.
[c] Concentration $\sim$ 0,1 mole l^{-1}; (mesures effectuées sur spectromètre Beckman IR9).

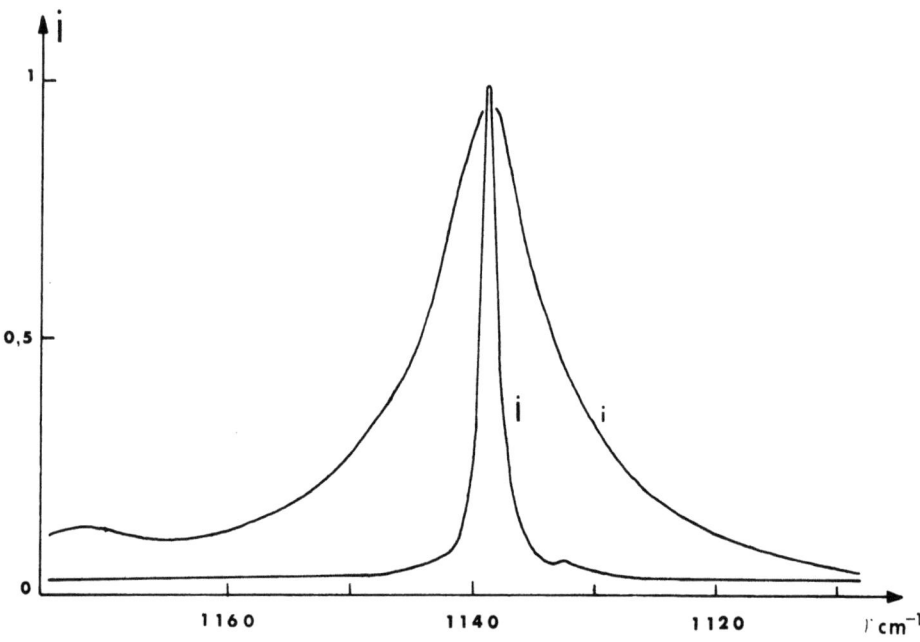

Fig. 1. Spectres des parties polarisée (I) et dépolarisée (i) de la bande Raman ν_3 du furanne dissous dans le sulfure de carbone. $t = 25\,°C$. Spectromètre Raman Coderg PHl, laser Argon ionisé, 200 mW pour I et 2 W pour i. Fente spectrale: 0,5 cm^{-1}.

Lorsque l'on passe des solvants polarisables comme le tétrachlorure de carbone et le sulfure de carbone aux solvants véritablement inertes comme les alcanes, plus favorables aux réorientations, on constate que la largeur des bandes de type A_1 et B_1 croît, cependant que celle de la bande de type B_2 reste sensiblement constante. On conclut donc à une réorientation préférentielle autour de l'axe z perpendiculaire au plan de la molécule.

La théorie de Rakov prévoit en outre que pour toute vibration qui correspond à une variation du moment dipolaire perpendiculaire à l'axe de réorientation, la largeur de la bande augmente avec la température.

Nous avons enregistré les spectres des trois absorptions précédentes à différentes températures dans le sulfure de carbone. On observe alors que la largeur de la bande de type B_2 n'est quasiment pas affectée par une variation de température de $-60\,°C$ à $+30\,°C$, cependant que les largeurs des bandes de type B_1 et A_1 augmentent de 50% environ. Ces résultats confirment l'existence d'une réorientation préférentielle autour de l'axe z [3].

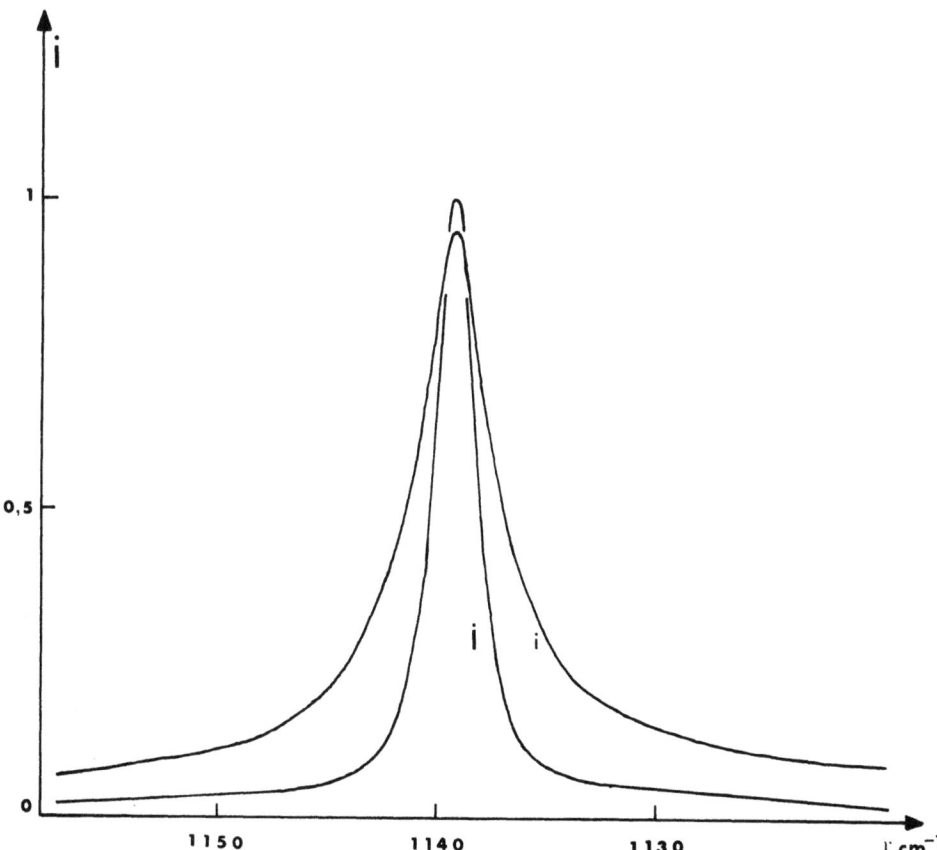

Fig. 2. Spectres des parties polarisée (I) et dépolarisée (i) de la bande Raman v_3 du furanne dissous dans le sulfure de carbone. $t = -70\,°C$. Mêmes conditions expérimentales que pour la figure 1.

2. Mesure du temps de corrélation par spectroscopie Raman

L'étude par spectroscopie Raman a porté sur les vibrations v_3 et v_4 de type A_1, polarisées et centrées respectivement à 1138 cm^{-1} et 1384 cm^{-1}. Sur la figure 1 sont représentés les spectres des parties polarisées I et dépolarisée i de la bande v_3 enregistrés dans le sulfure de carbone à 25 °C. Nous avons également enregistré les spectres de diffusion des bandes v_3 et v_4 toujours dans le sulfure de carbone, à -70 °C (figure 2) et de la bande v_3 dans le diméthylsulfoxyde à 25 °C (figure 3). Les largeurs

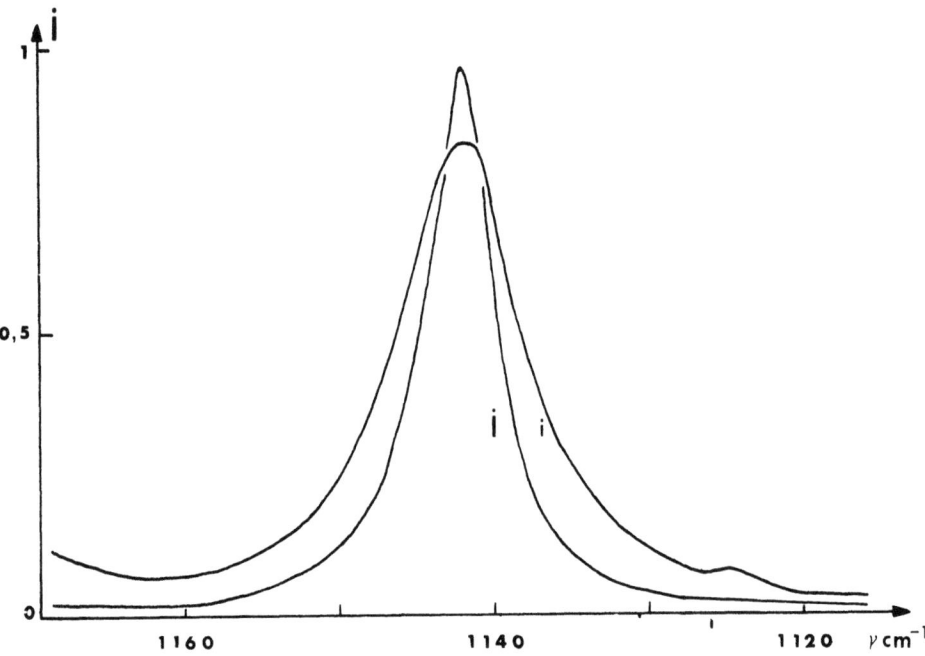

Fig. 3. Spectres des parties polarisée (I) et dépolarisée (i) de la bande Raman v_3 du furanne dissous dans le diméthylsulfoxyde $t = 25$ °C. Mêmes conditions expérimentales que pour la figure 1.

des parties polarisées $\Delta v_{\frac{1}{2}}(I)$ et dépolarisées $\Delta v_{\frac{1}{2}}(i)$ ainsi mesurées sont rassemblées dans le tableau II.

On constate tout d'abord que la différence $\Delta v_{\frac{1}{2}}(i) - \Delta v_{\frac{1}{2}}(I)$ observée à basse température ou dans le diméthylsulfoxyde est nettement moins grande que celle observée dans le sulfure de carbone à 25 °C. Ceci s'explique par le blocage de la réorientation aléatoire, dans le premier cas par effet de froid, dans le second cas par la formation de liaisons hydrogène faibles à caractère électrostatique prédominant entre le diméthylsulfoxyde et les groupements C—H de la molécule de furanne, ainsi que nous l'avons montré par ailleurs [3].

Si l'on applique maintenant la théorie de Rakov aux largeurs des parties dépolarisées et polarisées, mesurées dans le sulfure de carbone à 25 °C, le temps de corrélation

TABLEAU II

Largeurs des parties polarisées (I) et dépolarisées (i) des bandes ν_3 et ν_4 du furanne

(A)		(B) CS$_2$ ($T = 25\,°C$)	(C) CS$_2$ ($T = -70\,°C$)	(D) DMSO ($T = 25\,°C$)
ν	$\Delta\nu_{\frac{1}{2}}$			
ν_3	$\Delta\nu_{\frac{1}{2}}(i)$	10,8	4,2	9
1138 cm^{-1}	$\Delta\nu_{\frac{1}{2}}(I)$	1,6	2,2	5,2
	$\Delta\nu_{\frac{1}{2}}(i) - \Delta\nu_{\frac{1}{2}}(I)$	9,2	2,4	3,8
ν_4	$\Delta\nu_{\frac{1}{2}}(i)$	8	4,4	–
1384 cm^{-1}	$\Delta\nu_{\frac{1}{2}}(I)$	1,6	1,8	–
	$\Delta\nu_{\frac{1}{2}}(i) - \Delta\nu_{\frac{1}{2}}(I)$	6,4	2,6	–

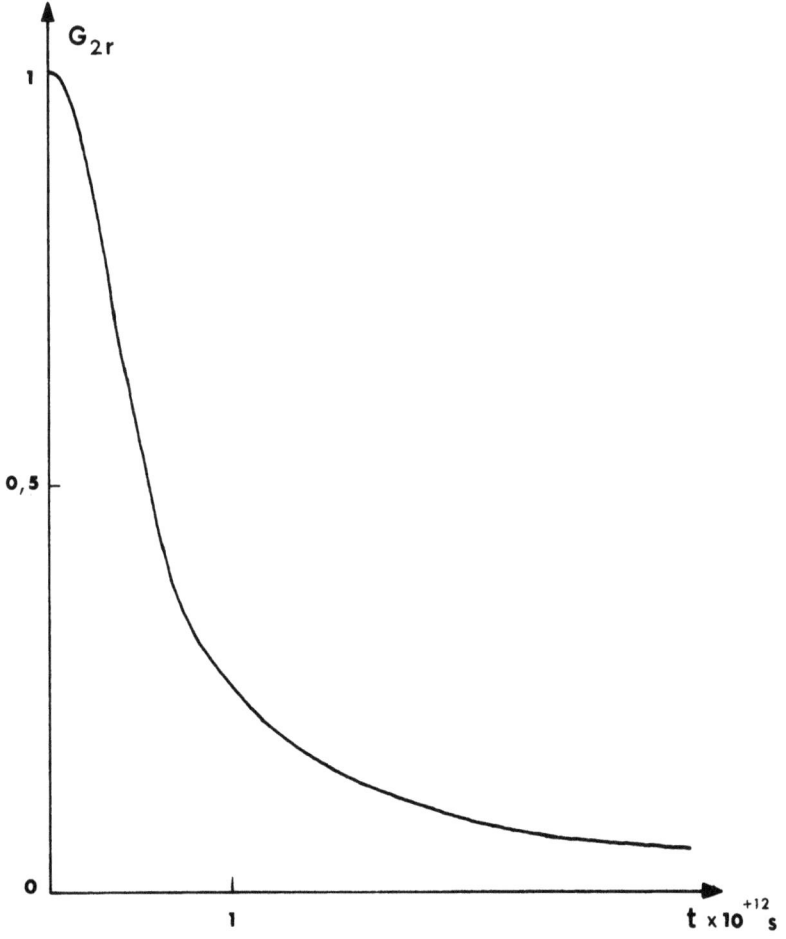

Fig. 4. Fonction de corrélation orientationnelle $G_{2R} = f(t)$ obtenue pour la bande ν_3 du furanne dans le sulfure de carbone à 25 °C.

est donné par la relation:

$$\tau = \frac{1}{\pi \cdot C \{\Delta v_{\frac{1}{2}}(i) - \Delta v_{\frac{1}{2}}(I)\}}$$

ce qui donne:

$$\tau(1138 \text{ cm}^{-1}) = 1{,}15 \times 10^{-12} \text{ s}$$
$$\tau(1384 \text{ cm}^{-1}) = 1{,}65 \times 10^{-12} \text{ s}.$$

Par ailleurs, les récents travaux de Bratos *et al.* permettent l'analyse des profils des bandes de diffusion Raman [4]. Ces auteurs ont montré que la diffusion de trace traduit uniquement les fluctuations vibrationnelles, tandis que la diffusion anisotrope est influencée par les fluctuations vibrationnelles et orientationnelles. La déconvolu-

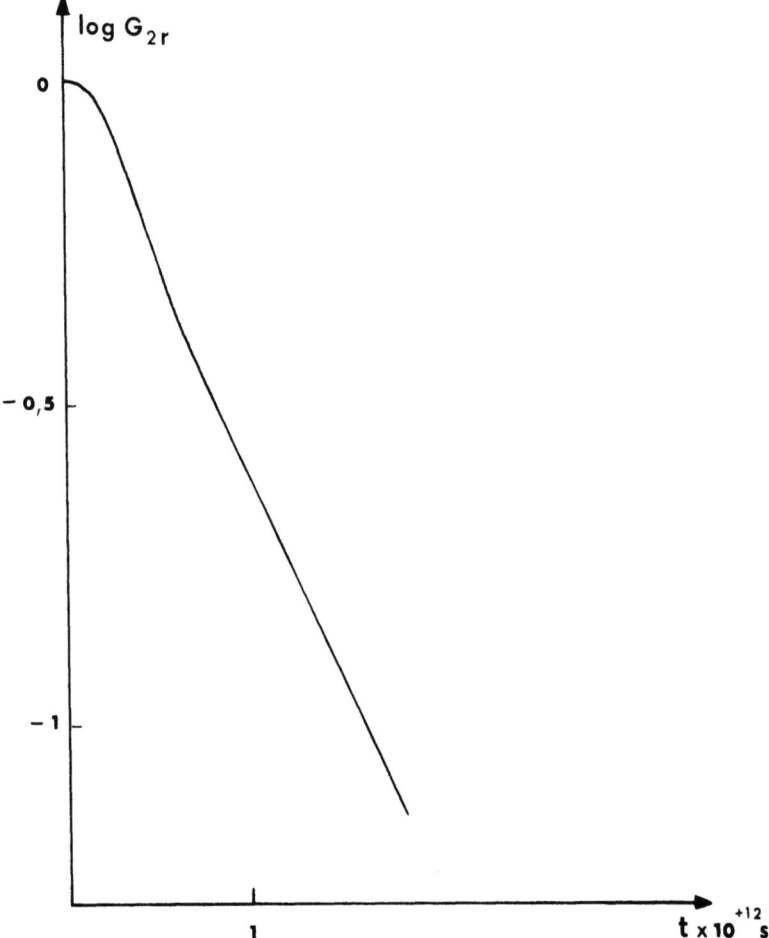

Fig. 5. Fonction $\log G_{2R} = f(t)$ obtenue pour la bande v_3 du furanne dissous dans le sulfure de carbone à 25°C.

tion de la bande anisotrope par la bande isotrope permet alors d'obtenir la fonction de corrélation orientationnelle $G_{2R}(t)=\langle(\frac{3}{2})\cos^2\theta-\frac{1}{2}\rangle$ expression où $\theta(t)$ représente la variation aléatoire caractérisant l'angle de rotation de l'axe moléculaire entre les instants 0 et t.

Sur les figures 4 et 5 nous avons représenté les graphes correspondant aux fonctions $G_{2R}=f(t)$ et $\log G_{2R}=f(t)$ pour la bande v_3 dissoute dans le sulfure de carbone. Le temps de corrélation obtenu à partir de ces fonctions est $\tau=1{,}50\times 10^{-12}$ s. Le calcul effectué sur la bande v_4 à titre de vérification a donné le même résultat.

On constate donc avec d'autres auteurs [5] que les temps de corrélation obtenus à partir de la fonction de corrélation $G_{2R}(t)$ et ceux calculés à partir de la théorie de Rakov sont voisins.

Remerciements

Nous remercions Monsieur le Professeur Lascombe d'avoir mis à notre disposition l'équipement Raman de son laboratoire de spectrochimie moléculaire de la Faculté des Sciences de Bordeaux, et de nous avoir fait bénéficier au cours de discussions fructueuses de ses judicieux conseils.

Bibliographie

1. Bak, B., Hansen, L., et Basrup-Andersen, J.: *Disc. Faraday Soc.* **19**, 30 (1955).
2. Rakov, A. V.: *Proc. P. N. Lebedev Physics Inst.* **27**, 109 (1964).
3. Dorval, P.: Thèse, Brest, 1971.
4. Bratos, S. et Marechal, E.: *Phys. Rev.* **A4**, 1078 (1971).
5. Devaure, J. et Lascombe, J.: résultats non publiés.

ANISOTROPY IN MOLECULAR REORIENTATIONAL MOTIONS IN SIMPLE LIQUIDS

JAMES E. GRIFFITHS
Bell Laboratories, Murray Hill, N.J. 07974, U.S.A.

Abstract. Anisotropy in the reorientational motions of simple symmetric top molecules in the liquid state have been studied at various temperatures. Raman and ^{2}D NMR spin relaxation techniques were used to evaluate the individual components $D_\perp$ and $D_\parallel$ of the total diffusion tensor D_t and the corresponding activation energies for reorientational motions perpendicular and parallel to the major symmetry axis of the molecules. The molecular systems studied include CH_3F, CD_3F, CH_3CN, CD_3CN, CH_3I, C_6H_6, C_6D_6, s-$C_6H_3F_3$, s-$C_6D_3F_3$, s-$C_3N_3F_3$ and C_6F_6.

Résumé. L'anisotropie dans les mouvements d'orientations des molécules toupies symétriques en phase liquide a été étudiée à plusieurs températures. Les techniques de Raman et ^{2}D NMR relaxation de spin sont employées pour évaluer les composantes individuelles $D_\perp$ et $D_\parallel$ du tenseur de diffusion complète D_t et les energies d'activation correspondantes perpendiculaires et parallèles aux axes majeurs de symétrie des molécules. Les systèmes moléculaires étudiés sont les suivants: CH_3F, CD_3F, CH_3CN, CD_3CN, CH_3I, CD_3I, C_6H_6, C_6D_6, s-$C_6H_3F_3$, s-$C_6D_3F_3$, s-$C_3N_3F_3$ et C_6F_6.

1. Introduction

A casual glance at the contents of this meeting shows that there are a wide variety of techniques available for the study of molecular motion in liquids. In this discussion, however, I shall restrict myself to only two of these, Raman spectroscopy and NMR spin relaxation spectroscopy with the specific intension of showing how the combined methods give valuable information about the anisotropy in the reorientational motions of the molecules in the liquid state.

Since this is the last paper in the series devoted to vibrational spectroscopy, I assume that the theory and experimental techniques for the Raman method have already been presented and do not need to be repeated. The next session is to be devoted to the NMR method so I will not dwell on the details since they will surely be presented later. It is necessary to introduce appropriate equations, the derivation and validity of which are assumed to be correct except when noted otherwise.

I plan to discuss specific parts of our work on a variety of molecules which I hope demonstrate most of the important principals and of equal importance the limitations of the techniques. The first diagram (Figure 1) shows the molecules studied. These were selected because of the interesting contrasts in dipole moment, geometrical shape and in the ratios of their principle moments of inertia.

Acetonitrile was selected because it is one of the few molecules for which the anisotropy in the reorientational motions has already been well established. This was accomplished from NMR spin relaxation studies of the nitrogen and deuterium nuclei in CD_3CN. The reorientational motion perpendicular to the three fold axis was found to be in the small-step diffusion limit while that around the figure axis (spinning mo-

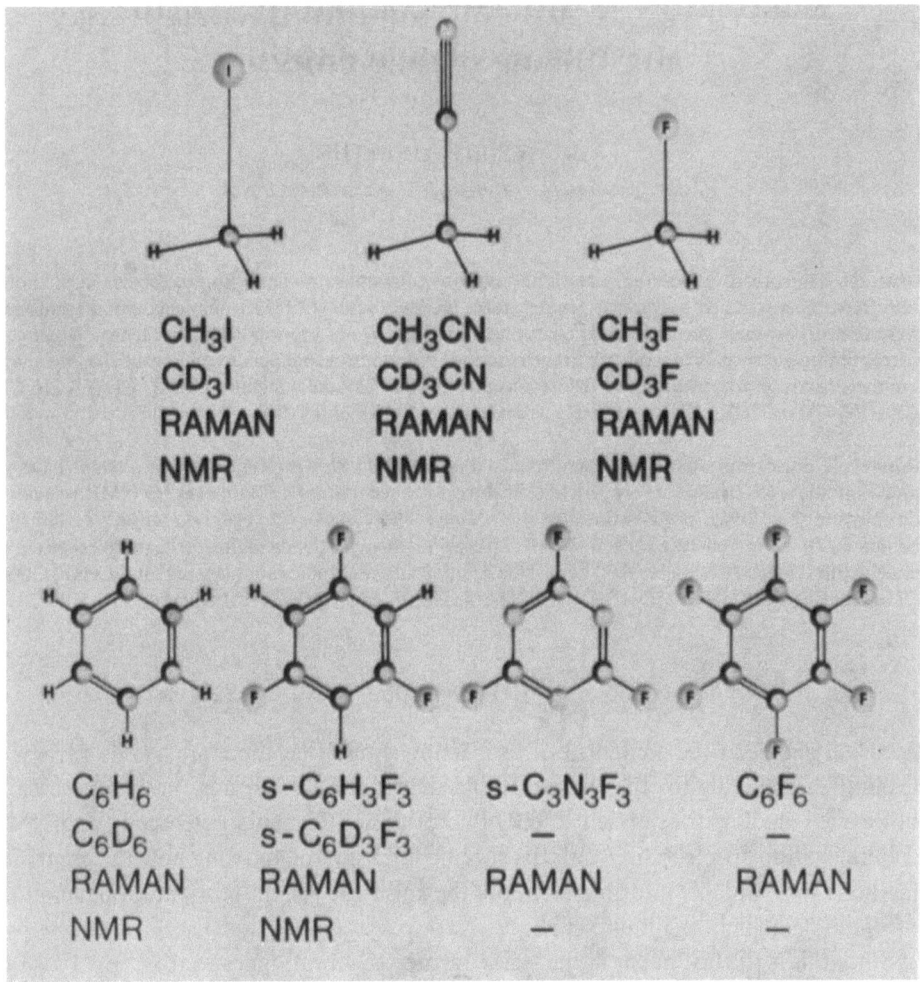

Fig. 1. Prolate and oblate symmetric top molecules.

tion) was essentially free. The ratio of the components of the diffusion constant, $D_\parallel/D_\perp \simeq 10$ at room temperature.

This can also be expressed in terms of the correlation times $\tau_\parallel/\tau_\perp \sim 0.1$. Our first point is to show that $D_\perp(\text{Raman}) = D_\perp(\text{NMR})$ for acetonitrile.

↔Measurements and data analysis are summarized in Figure 2. The analysis outlined in Figure 2 neglects the first fraction of a picosecond of the reorientational process but this is not too serious for our purposes and in any case the information is available if one wishes to carry the analysis further. The analysis is model independent but there is the fundamental assumption that vibrational and reorientational motions are not correlated that will lead to complications later.

At this point it is appropriate to examine some data on acetonitrile. Figure 3 shows the Raman spectrum of the $v_1(a_1)$ fundamental of CD_3CN. Figure 4 shows the Lorentzian test of this data where the solid curve is a Lorentzian calculated on the basis of the observed half width. The solid points are taken from the observed curves and the fit is satisfactory.

Before these results can be interpreted in terms of specific models, I wish to introduce the appropriate NMR equations. In molecular species containing quadrupolar nuclei

Measurements (a) $I_{\parallel} = I_{\alpha}(\omega) + \frac{4}{3} I_{\beta}(\omega)$

(b) $I_{\perp} = I_{\beta}(\omega)$

for parallel bands where $\rho \ll 0.75$

Data analysis

(a) if I_{α} and I_{β} are Lorentzian

$\tau_V = (2\pi c \omega_{\alpha, 1/2})^{-1}$

$\omega_{or} = (\omega_{\beta, 1/2} - \omega_{\alpha, 1/2})$

$\tau_{or} = (2\pi c \omega_{or, 1/2})^{-1}$

Model – Small step diffusion limit

$\tau_{or} = (6D_{\perp})^{-1}$

Fig. 2. Expressions used in data analysis.

with spin $I > \frac{1}{2}$, nuclear relaxation is usually dominated by quadrupolar relaxation and the correlation times for molecular reorientation in principle can be derived from measured spin lattice relaxation times according to the equation

$$(T_1)^{-1} = [3\Pi^2 (2I + 3)/10 I^2 (2I - 1)]\left(1 + \frac{\eta^2}{3}\right)\left(\frac{e^2 qQ}{h}\right)\tau_c. \qquad (1)$$

When the reorientational process proceeds in the small step diffusion limit, the correlation time τ_c can be expressed in terms of the various components of the rotational diffusion tensor D. For symmetric top molecules we have

$$\tau_c = \frac{(3\cos^2\theta - 1)^2}{24 D_{\perp}} + \frac{3 \sin^2\theta \cos^2\theta}{5 D_{\perp} + D_{\parallel}} + \frac{3 \sin^4\theta}{4(2 D_{\perp} + 4 D_{\parallel})}, \qquad (2)$$

where θ is the angle between the principal axis of the field gradient tensor eq and the symmetry axis of the top and $D_{\parallel}$ and $D_{\perp}$ are the diffusion constants for rotation about axes parallel and perpendicular to the symmetry axes. The reorientational correlation time on the one hand is a function of the nuclear spin which is known, the spin lattice relaxation time which is measured and the nuclear quadrupole coupling constant

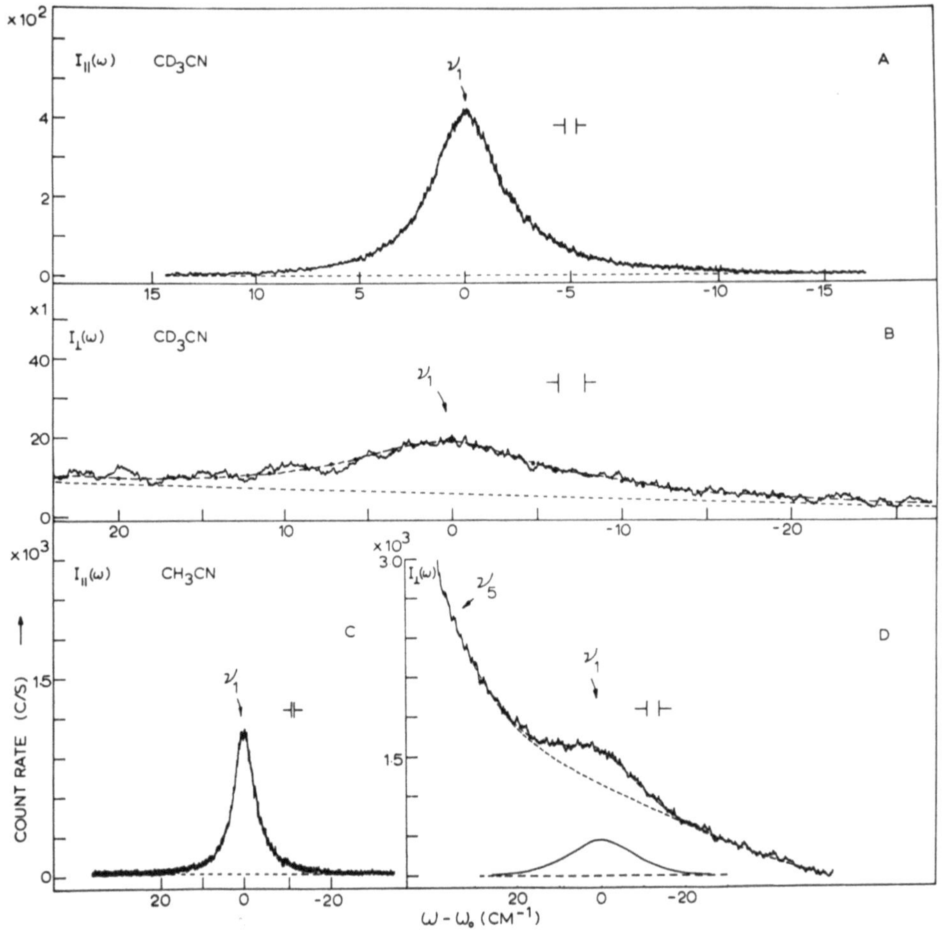

Fig. 3. $I_\parallel$ and $I_\perp$ spectra of ν_1 of CD$_3$CN (A and B) and CH$_3$CN (C and D).

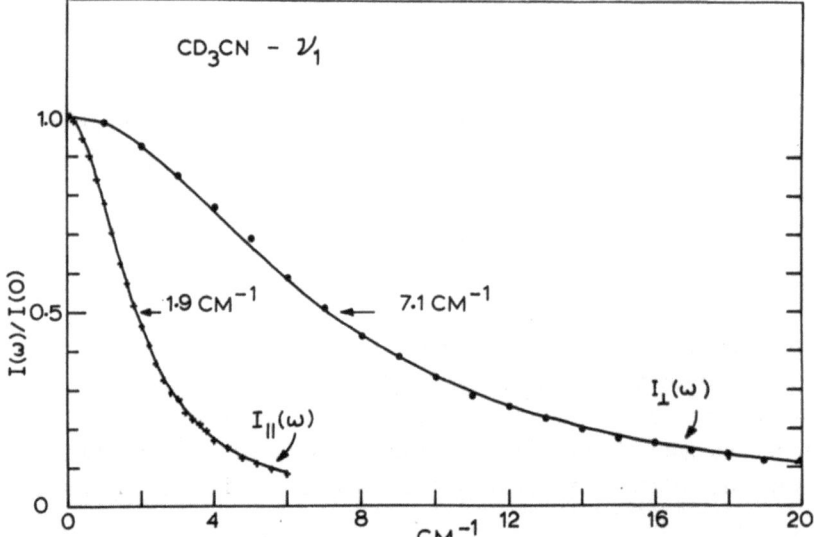

Fig. 4. Lorentzian line shape analysis of ν_1 of CD$_3$CN. Solid lines calculated, dots and crosses are data prints.

which is measured independently. The asymmetry parameter η, if known is usually very small and is commonly neglected without serious consequences. In order to solve uniquely for $D_\perp$ and $D_\parallel$ in order to evaluate the anisotropy in the reorientational motions from NMR results alone, these must be two quadrupolar nuclei with different θ values. This is a practical difficulty with this technique because very few molecular species satisfy this condition. Acetonitrile-d_3 is one such species and it has been extensively studied by others with the result that $D_\parallel \simeq 10 D_\perp$ or in other words, the reorientational motions are highly anisotropic.

On the other hand, for reorientational motions in the small step diffusion limit the Raman technique gives a unique value for $D_\perp$ through the analysis of the strongly polarized bands. Knowing $D_\perp$, and using this result in consort with NMR results for quadrupolar nuclei off the figure axis allows us to evaluate $D_\parallel$ under the more relaxed requirements of one quadrupolar nuclei in the molecular species.

A serious problem still exists however and involves the quantity $(e^2qQ/h)^2$ which is not always known and this is a difficult quantity to calculate theoretically with sufficient accuracy for our purposes. There are several useful empirical ways of dealing with this, however, and perhaps I will have time to dwell on it later.

Returning to the analysis of data for acetonitrile, Figure 5 shows a comparison of data for a model assuming the diffusion limit. Absolute magnitudes and temperature dependences are essentially in agreement for the NMR and Raman techniques and also for D_μ calculated on the basis of hydrodynamic theory. The activation energy for the reorientational process is 2 kcal mole.$^{-1}$

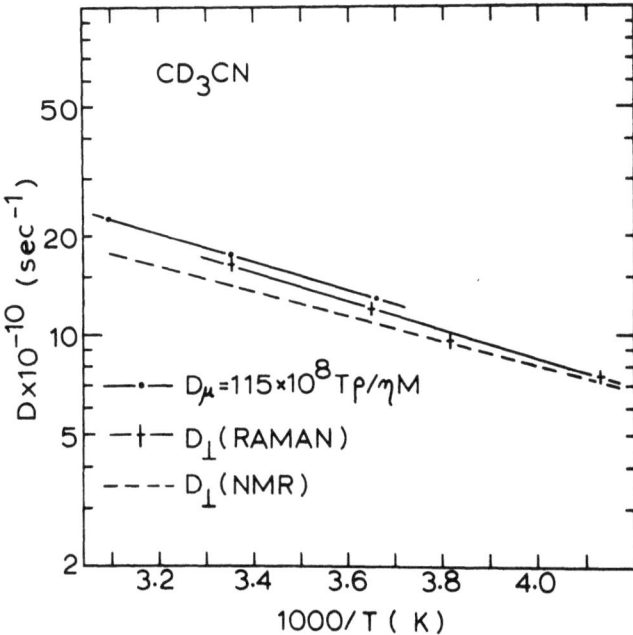

Fig. 5. Variation of the perpendicular diffusion constant for reorientation with temperature.

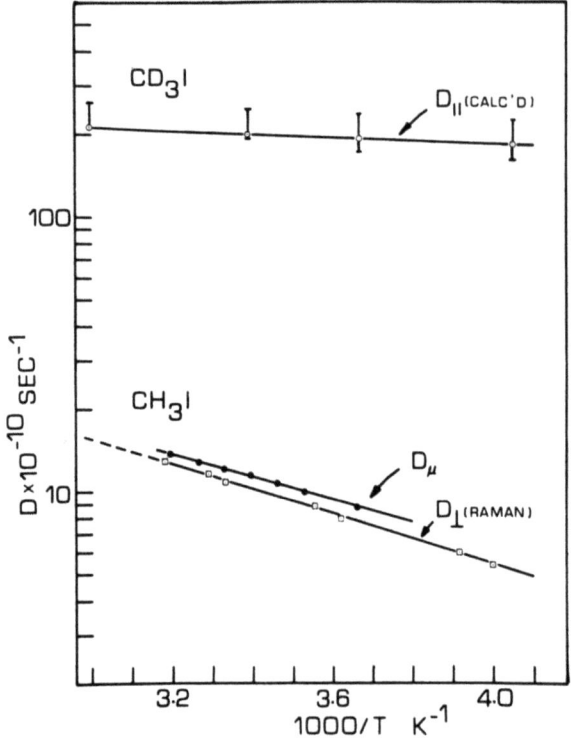

Fig. 6. Experimental and theoretical values of $D_\perp$ and $D_\parallel$ for CH$_3$I and CD$_3$I.

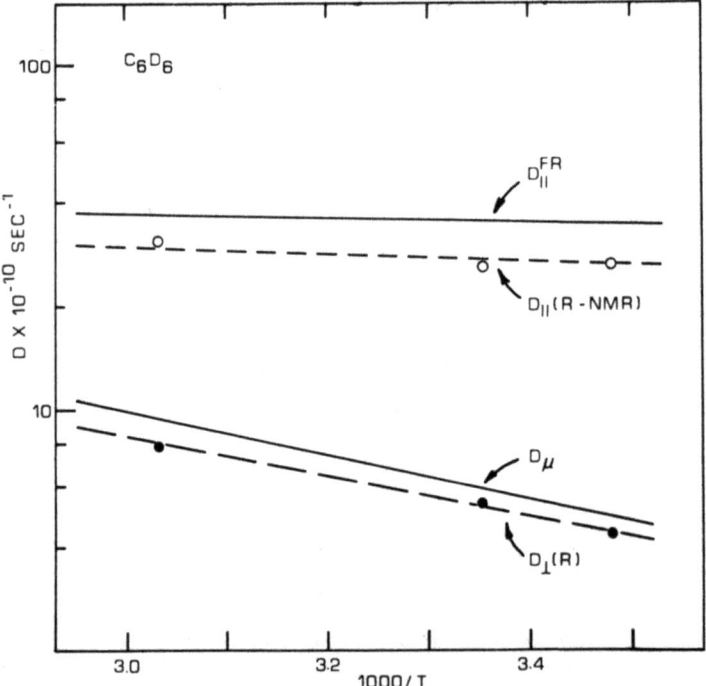

Fig. 7. Variation of $D_\perp$ and $D_\parallel$ with temperature

Figure 6 shows the comparable analysis for methyl iodide although in this case we use the two techniques to their maximum advantage in order to obtain temperature dependent data for $D_\perp$, $D_\parallel$, $E_\perp$ and $E_\parallel$. The results are self evident. Reorientational motions are highly anisotropic with the perpendicular motion being in the diffusion limit and the parallel motion being essentially free. Anisotropy in the reorientational motions are conveniently detected by the significant differences in the activation energies.

Results for benzene are shown in Figure 7. This molecule was the first one studied by the present method and the results were presented last year at the International Conference on Raman spectroscopy held in Reims. Previous to this work, it had been concluded by others on the basis of Raman results alone that the reorientational motions were isotropic. These results show that the earlier conclusion was in error and in fact these motions are highly anisotropic.

Figure 8 shows more recent results for s-$C_6D_3F_3$ for which two vibrational bands were analyzed. In this case we are at a clear disadvantage for we have no information on the nuclear quadrupole coupling constant. We will have this information in the

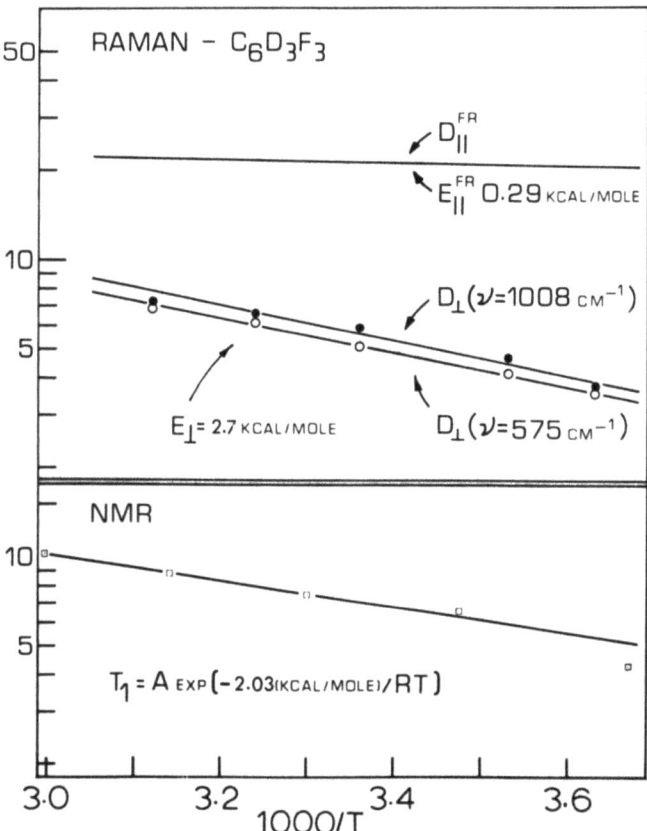

Fig. 8. Experimental (Raman) and theoretical values of $D_\perp$ and $D_\parallel$ as a function of temperature (top) and the temperature dependence of T_1 (bottom) for S–$C_6D_3F_3$.

next few weeks but in the meantime all is not lost. One can calculate, for example, the activation energy for free rotation about the three fold axis and this is 0.3 kcal mole^{-1}. The observed value for the perpendicular reorientation motion in the diffusion limit is 2.7 kcal mole^{-1} and the value obtained from the NMR spin relaxation results is 2.0 kcal mole^{-1}. Since the last two are unequal, we immediately conclude that the motions in question are quite anisotropic ($D_\perp \neq D_\parallel$) and from this it is a straight-forward exercise to calculate the upper and lower bounds expected for (e^2qQ/h). For free rotation the value is 242 kHz and for isotropic reorientation $e^2qQ/h = 172$ kHz. For those very few benzene compounds where e^2qQ/h is known the values are between 175 and 190 kHz. Accordingly, one might expect e^2qQ/h to be near 180–185 kHz.

I had hoped to be able to discuss other molecules such as $C_3N_3F_3$, C_6F_6 and the very important CH_3F species but time will not permit. I will be quite willing to discuss these privately with anyone interested in the results. Even without this information, I believe that the use of the Raman and NMR techniques will prove to be especially valuable in studying molecular motions in liquids. With the added insight obtained from these methods, it may be possible to get around the problem of gaining reliable and accurate information from the analysis of band shapes arising from degenerate fundamentals. This will circumvent the normal situation where accurate values of the nuclear quadrupole coupling constant are unknown.

DISCUSSION

Loewenstein: There seems to be no absolute evidence that molecular motion in symmetric top molecules such as CH_3CN is anisotropic. The reasons are: (1) experimental inaccuracies in the determination of the T_1's (cf. A. Allerhand), (2) the possibility of applying different procedures to interpret NMR results (cf. O'Reilly).

Griffiths: The experimental accuracy in the measurement of T_1 is no more than 5% and I do not know right at this moment the extent of the resulting error in the extracted value of $D_\parallel$. Bull and Jonas, however, in their NMR study of the effects of pressure in acetonitrile have estimated this error to be roughly 30% in $D_\parallel$. If this is correct, and I have no reason to doubt it, our conclusions about the anisotropy in the reorientational motions are not changed but the absolute values of $D_\parallel$ are not exact. This is because $D_\parallel/D_\perp$ is about 10.

Yarwood: Our infrared data for acetonitrile show that the so called 'activation energy' referred to by Griffiths is a factor of five greater for rotation about the perpendicular axis ($D_{\perp\,ar}$) than for rotation about the C_3 axis ($D_\parallel$). (These values are 1 kcal mole^{-1} and 0,2 kcal mole^{-1} respectively). This would seem to support the idea of anisotropy of the rotational diffusion for such a molecule even if the absolute values of these Arrhenius parameters are somewhat in error (due to lack of correction of the infrared data for vibrational effects). I should like to ask how one is to interpret these 'activation parameters' especially when they turn out to be less than, or of the same order as, kT. Is it valid to speak of a 'small step diffusion' model under such circumstances?

Griffiths: The error in $D_\perp$ (infrared) is 100% because of the inability to correct for vibrational relaxation and the error in $D_\parallel$ (infrared) is about 20 to 30%. This can account for the large differences between the Raman and infrared results. One can consider only the perpendicular component of the reorientational process to be in the diffusion limit and the close agreement between the magnitude and temperature dependence of $D_\perp$ (Raman), D_μ (microviscosity) and $D_\perp$ (NMR) supports this view completely. Since the 'activation energy' for the parallel motion is so small (kT) one cannot speak of this in terms of the diffusion limit and we do not do this. The so-called 'activation energy' is perhaps a misnomer because it conjures up images of a model which may not be consistent with all of the evidence. The slope of a plot of the form $A = Ae^{-E/kT}$ has historically been called an 'activation energy' and the units are in fact energy units but as far as the parallel motion is concerned the meaning of the expression in terms of a particular model is not valid.

Jonas: In recent years we have carried out measurements in liquids where we separated the effects of temperature and density on the rotational correlation times. If one carries out the experiment at constant density then the activation energy for a roerientational process is close to the value of kT. One wonders then what is the physical meaning of this quantity. Arrhenius-type plots and calculation of activation energies are useful ways to treat the experimental data but one has to be always aware of the limitations of this procedure.

Kneubühl (comment): The behaviour of a rotating molecule with a kinetic energy (kT) near the potential barrier E_B is illustrated by the behaviour of the diatomic ion OD$^-$ in the crystal $K\varepsilon$. For $kT \ll E_B$ one observes a libration, for $kT \gg E_B$ a rotational diffusion [1].

Friedman: I am curious as to whether the description of the methyl group motion which you find is similar to that found for methyl groups in other molecules which are not linear but in which there is a very small barrier to rotation about the bond to the methyl group. An example might be CH_3NO_2 or $(CH_3)_2CO$.

Griffiths: We have not studied any of the asymmetric molecules you mention although during the next year we expect to investigate a molecule with a barrier to internal rotation in excess of the 'activation energy' for perpendicular reorientational motion. At the conclusion of that study it may be possible to answer your question. At the moment, however, we share your curiosity and interest.

Janik: (1) In the *solid* CH_3I the torsional frequency (C_3 axis) is ca. 100 cm^{-1}. It implies a relatively low barrier to rotation (ca. 1000 cal mole^{-1}) already in the *solid* CH_3I.

(2) In the liquid CH_3I there exist indirect (I must admit) neutron data, indirect in a sense that these are total scattering experiments. From the slopes of neutron total cross sections vs neutron wave length obtained for liquid CH_3I in our laboratory and also by Dr Fischer in the Hahn-Meitner Institute in Berlin, one may conclude that the barrier to rotation in liquid CH_3I is smaller than 500 cal mole^{-1}.

Griffiths: The results in the second comment are in excellent agreement with the combined Raman and spin relaxation experiments.

Reference

1. Keller, B. and Kneubühl, F.: *Solid State Communications* **8**, 867–871 (1970).

Comment by M. R. Hoare

(Earlier) Profs. Rahman and Brot quite reasonably poured scorn on the idea of rotational relaxation occurring through a 'binary collision' mechanism. Perhaps, in defence of Prof. Litowitz and others who have spoken in terms of 'collisions', it should be said that a collision-like relaxation behaviour may well arise from mathematical models of somewhat less crudity.

Again I would suggest that the correct viewpoint is from some form of rotational transport equation giving the time-dependence of the populations of orientational and/or rotational states in an ensemble, possibly with position and translational velocity coupled in as well. My point is that, however fearsome this equation might look (and I will spare the details here), it will usually, within a whole range of optional assumptions, contain at least three distinguishable terms viz.: (a) the streaming in positional and angular space, (b) the scattering *into* a specified state of interest and (c) the scattering out. This is standard in ordinary transport theory and various severely contracted forms have been given for rotational transport e.g. by Kivelson, St. Pierre and Steele, Cupier and Lakatos-Lindenberg, Rider and Fixman, and others.

The crucial point is that, for a class of useful, linear theories, the scatering-out term will be a flux of probability which looks like a collision-number (possibly translational and rotational velocity-dependent) multiplying the distribution function of interest. What matters is not so much whether this term need arise from physical binary collisions as whether the hidden statistical assumptions built into the transport equation – linearity, Markovicity, decoupling of variables and so on – are justified in a more general way. There seems to be a case for believing that a statistical collision-like mechanism may have a much wider range of validity than binary collisions could possibly have. This point of view seems altogether in the spirit of autocorrelation function computations in which, in any case, information about the particulars of individual scattering events is renounced in a systematic way.

TRANSLATIONAL MOTIONS AS STUDIED BY NUCLEAR MAGNETIC RESONANCE

H. G. HERTZ

Institut für Physikalische Chemie und Elektrochemie der Universität, Karlsruhe, F.R.G.

Abstract. This article treats translational motion in the laboratory coordinate system, i.e. self-diffusion, and relative translational motion with respect to a given reference molecule. In the first topic emphasis lies on the possibility of studying dynamical details with the NMR techniques. Relative translational motion may be studied by aid of the intermolecular nuclear magnetic relaxation rate. Some techniques to obtain the intermolecular relaxation rate are described and the general background for the evaluation of $(1/T_1)_{\text{inter}}$ with respect to the desired dynamical information is given. Lastly, the problem to predict the intermolecular relaxation rate at frequency 0 is discussed.

Résumé. Cet article traite des mouvements de translation dans le système de coordonnées du laboratoire, c'est à dire auto-diffusion, ainsi que des mouvements de translation relatifs par rapport à une molécule de référence: dans le premier cas l'accent est mis sur la possibilité d'étudier par la RMN des détails de dynamique. Les mouvements de translation relatifs peuvent être étudiés d'après la vitesse de relaxation nucléaire intermoléculaire. Quelques techniques permettant de déterminer les vitesses de relaxation intermoléculaires sont décrites, et un arrière-fond général pour l'évaluation de $(1/T_1)_{\text{inter}}$ en vue d'obtenir les informations souhaitées sur la dynamique est présenté. Finalement est discuté le problème de la prévision de la vitesse de relaxation intermoléculaire à la fréquence zéro.

1. Introduction

If we wish to learn something about translational molecular motions from nuclear magnetic resonance phenomena we have to study the relaxation behaviour of the nuclear magnetization. The longitudinal and transverse component of the nuclear magnetization, $\mathfrak{M}_\parallel$ and $\mathfrak{M}_\perp$, respectively, relax according to the equations (see e.g. [1–3])

$$\frac{d\mathfrak{M}_\parallel}{dt} = -\frac{1}{T_1}(\mathfrak{M}_\parallel - \mathfrak{M}_\parallel^0) \tag{1}$$

$$\frac{d\mathfrak{M}_\perp}{dt} = -\frac{1}{T_2}\mathfrak{M}_\perp. \tag{2}$$

$1/T_1$ and $1/T_2$ are termed as longitudinal and transverse relaxation rates, respectively, T_1 and T_2 are the corresponding relaxation times. T_1 and T_2 do not differ very much for most of the systems of interest in connection with our subject. In certain cases the parameters $1/T_1$ and $1/T_2$ depend on the *relative* translational motion of the molecules in the liquid. The corresponding details will be presented in Section 4 of this article. The techniques which have been applied to separate the intermolecular relaxation rate from the intramolecular relaxation rate will be the subject of Section 3. In a large number of cases it is also possible to study the absolute translational diffusive motion (i.e. relative to the laboratory coordinate system). A brief outline of the fundamental background together with some particular aspects regarding deviations from the ordinary diffusion law will be given in Section 2.

J. Lascombe (ed.), Molecular Motions in Liquids, 337–357. All Rights Reserved
Copyright © 1974 by D. Reidel Publishing Company, Dordrecht-Holland

2. The Investigation of Absolute Translational Motion

The qualitative basis of this experimental method is as follows. $\mathfrak{M}_\perp$ decays $\to 0$ because the phase coherence between the spins in the sample is lost due to spin flipping and to spin-spin interaction (in the case of magnetic dipole-dipole interaction). Equation (2) implies that the static magnetic field is constant over the sample volume. If this is not the case, then the magnetic field at the nucleus in question varies slightly as the molecule diffuses in the liquid which causes an additional decay of the transverse magnetization. Assume that at $t=0$ the nucleus is at $\mathbf{r}=0$, then the probability to find the same nucleus in $d\mathbf{r}$ at $\mathbf{r}$ at time t is $P(\mathbf{r}, t)$. After sufficiently long time we may write

$$P(\mathbf{r}, t) = (4\pi Dt)^{-3/2} e^{-r^2/4Dt} \tag{3}$$

which is the solution of the translational diffusion equation

$$\frac{\partial P(\mathbf{r}, t)}{\partial t} = D\nabla^2 P(\mathbf{r}, t). \tag{4}$$

D is the self-diffusion coefficient of the molecule which carries the nuclei forming the nuclear magnetization. In general, we shall denote any quantity $P(\mathbf{r}, t)$ which gives the conditional probability to find a particle in $d\mathbf{r}$ at $\mathbf{r}$ at time t, if at time $t=0$ it was at $\mathbf{r}=0$, as the translational propagator. Thus Equation (3) represents a special form of the propagator. It is the long time limiting form into which all translational propagators must pass for sufficiently long times.

The variation with time of the transverse magnetization is usually observed via the decrease of the amplitude of the spin-echo [1, 4, 5]. In the presence of translational diffusion the irreversible change of $\mathfrak{M}_\perp$ with time is given by

$$\mathfrak{M}_\perp = \mathfrak{M}_\perp^0 e^{-t/T_2} e^{-\gamma^2 g^2 Dt^3/12}. \tag{5}$$

g is the linear magnetic field gradient, γ is the gyromagnetic ratio of the nucleus studied. Equation (5) permits the evaluation of D. The spin echo method has the advantage that it is less time consuming than the classical methods. However, the precision of the classical methods is usually greater. This is partly due to the fact that they are older and thus the technical details are studied and developed more thoroughly. It is seen from Equation (5) that for the spin echo method to be applicable T_2 must be sufficiently long and γ must be sufficiently large. Both these facts are indeed limitations to the applicability of the method in all those cases where the diffusing particle does not contain protons (I=spin=$\frac{1}{2}$ for protons). Very often T_2 is short for those nuclei which have $I>\frac{1}{2}$. In this event the mechanism of the nuclear magnetic relaxation is the nuclear quadrupole interaction [1, 3]. Thus, as typical examples, the self-diffusion coefficients of Cl^-, Br^-, I^- and Rb^+ [6] in electrolyte solutions cannot be measured by NMR techniques. Furthermore, weak nuclear magnetic signals due to small concentrations or small magnetic moments of the nuclei to be studied in many cases present appreciable difficulties whereas with the direct diffusion methods the concentration – in particular for radioactive tracers – may be very small.

Fortunately, the techniques of Fourier transform nmr spectroscopy [3], which has recently been developed, now opens new possibilites for the measurement of D even in those systems which contain small amounts of the resonating nuclei. Moreover, more than one diffusing species in a mixture may be studied at the same time.

From inspection of Equation (5) one might conclude that the measurement of D can always be accomplished if the gradient of the static magnetic field, g, can be made sufficiently large. But the amplitude of the rf magnetic field, which produces the transverse component of the magnetization, must still be larger than the variation of the static field over the sample volume. This requirement confines the magnitude of g. However, the static field gradient may only be applied during the time when the rf magnetic field is switched off. This characterizes the method of pulsed field gradients [7, 8] as depicted schematically in Figure 1. Now the amplitude of the spin echo, ψ,

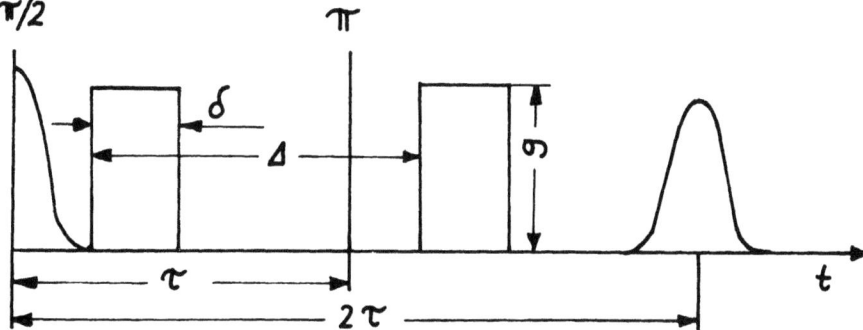

Fig. 1. Schematic representation of the method of pulsed field gradients. The ordinate represents the NMR signals (at $t = 0$ and $t = 2\tau$) and the magnetic field gradient g, but both quantities are of totally different orders of magnitude (and dimension). The signal at $t = 2\tau$ is the spin echo, at $t = 0$ and at $t = \tau$ a $\pi/2$ and π RF pulse, respectively, is applied. The gradient pulses have a duration δ and a separation Δ in time.

is given by the relation

$$\psi(t = 2\tau) = \exp\left\{-\frac{2\tau}{T_2} - \gamma^2\delta^2g^2D(\Delta - \tfrac{1}{3}\delta)\right\}. \tag{6}$$

g may be made as large as 1000 G cm^{-1}, thus the lower limit of measurable D's is appreciably shifted downwards ($D \gtrsim 10^{-9}$ cm^2 s^{-1} for protons). A number of modifications of the method of pulsed field gradients has been described in the literature [9–11]. Furthermore, it has been observed that D may depend on the diffusion time Δ. In this event we have restricted diffusion of a molecule in a rigid cage or similar situations [12, 13]. But those applications are not of importance for the study of simple homogeneous liquids.

There is, however, another aspect of these experiments which may be of interest for the subject of this conference [14, 15]. Assume we have $\delta \ll \Delta$ and let the diffusion time Δ approach towards 2τ, furthermore, assume that $\gamma^2\delta^2g^2D \gg 1/T_2$, then Equation

(6) becomes

$$\psi(\Delta) = \exp\{-\gamma^2\delta^2 g^2 D\Delta\}, \qquad (7)$$

i.e. the echo damping is an exponential function of Δ, the diffusion times, which we may rewrite as

$$\psi(\Delta) = e^{-\Delta/\tau_{obs}}, \qquad (8)$$

where we have defined $(\gamma^2\delta^2 g^2 D)^{-1} = \tau_{obs}$, τ_{obs} we call observation time. On the other hand, it may be shown that the relation holds [14, 15].

$$\psi(\Delta) = \int \cos(\gamma\delta \mathbf{gr}) P(\mathbf{r}, t = \Delta) \, d\mathbf{r}, \qquad (9)$$

i.e. the damping of the echo under these particular circumstances is the Fourier transform of the propagator $P(\mathbf{r}, t) = P(\mathbf{r}, \Delta)$. Indeed, if we introduce Equation (3) into Equation (9) we obtain Equation (7).

Furthermore, it is known from the theory of neutron scattering that the intermediate scattering law $I(\varkappa, t)$ is the space Fourier transform of the translational propagator of the scattering particle. The intermediate (incoherent) scattering law [16] describes the intensity of the scattered neutrons in the momentum transfer-time language – i.e. it is the frequency-time Fourier transform of the usual scattering law. $\varkappa$ is the momentum transfer and we see that $\varkappa$ and $\gamma\delta g$ correspond to one another.

$$I(\varkappa, t) \leftrightarrow \psi(\gamma\delta g, t) = \psi(\gamma\delta g, \Delta; \quad \varkappa \leftrightarrow \gamma\delta g. \qquad (10)$$

$\gamma\delta g$ is the phase displacement per cm of the precessing spin in this particular form of experiment. Whilst with neutron scattering we have $\varkappa \approx 10^8$ cm^{-1}, now $\gamma\delta g \lesssim 5 \times 10^4$ cm^{-1} may be achieved. The observation time for neutron scattering experiments is $\tau_{obs} = (D\varkappa^2)^{-1}$, i.e. with $D = 10^{-5}$ cm^2 s^{-1} $\tau_{obs} \approx 10^{-11}$ s, whereas from Equations (7) and (8) with the mnr experiments we have $\tau_{obs} = (\gamma^2 g^2 \delta^2 D)^{-1} \approx 10^{-4}$ s, i.e. now the observation time is at least by seven orders of magnitude longer. With the neutron scattering experiments τ_{obs} is the time after which the mean square displacement of the diffusing particle is of the order of $\varkappa^{-2}$. With our particular pulsed gradient experiment τ_{obs} is about the time the nucleus travels per unit phase displacement (relative to the mean precession phase).

Is it still possible to observe dynamical details of translational motion by aid of the observation of the echo damping? This can only be done if the observation time is smaller than or of the same order as the duration of the elementary process constituting the dynamical details of translational motion.

To give an example we consider the exchange of a diffusing particle between two environments during diffusion through the volume of the liquid. In state B the particle is bound to another diffusing particle, in state A the particle diffuses 'freely' in the liquid. Then we may write for the A and B contribution to the intermediate scattering law (or to the echo amplitude)

$$\frac{dI_A}{dt} = -D_A\varkappa^2 I_A - \frac{I_A}{\tau_A} + \frac{I_B}{\tau_B} \tag{11}$$

$$\frac{dI_B}{dt} = -D_B\varkappa^2 I_B + \frac{I_A}{\tau_A} - \frac{I_B}{\tau_B}, \tag{12}$$

where τ_A, τ_B are the residence times in the environments A and B, respectively, D_A and D_B are the corresponding self-diffusion coefficients. The mean value of the intermediate scattering law (= the mean damping of the echo amplitude) may be shown to be [13, 17]

$$\overline{I(\varkappa, t)} = \mathscr{P}'_1 e^{-D'_A \varkappa^2 t} + \mathscr{P}'_2 e^{-D'_B \varkappa^2 t}, \tag{13}$$

with

$$\mathscr{P}'_1 + \mathscr{P}'_2 = 1$$

$$D'_A \atop (B) = \frac{1}{2}\left\{D_A + D_B + \frac{1}{\varkappa^2}\left(\frac{1}{\tau_A} + \frac{1}{\tau_B}\right)\genfrac{}{}{0pt}{}{-}{+}\left[\left[D_A - D_B + \frac{1}{\varkappa^2}\left(\frac{1}{\tau_A} - \frac{1}{\tau_B}\right)\right]^2 + \frac{4}{\varkappa^2 \tau_A \tau_B}\right]^{1/2}\right\},$$

$$\mathscr{P}'_2 = \frac{1}{D'_B - D'_A}\{p_A D_A + p_B D_B - D'_A\},$$

$$p_A = \frac{\tau_A}{\tau_A + \tau_B}, \quad p_A + p_B = 1.$$

The limiting behaviour of $\overline{I(\varkappa, t)}$ is:

$$\overline{I(\varkappa, t)} = p_A e^{-D_A \varkappa^2 t} + p_B e^{-D_B \varkappa^2 t} \tag{14}$$

if

$$\tau_A, \tau_B \gg (D_A \varkappa^2)^{-1}, (D_B \varkappa^2)^{-1}$$

(slow exchange, residence times much longer than observation times);

$$\overline{I(\varkappa, t)} = e^{-(p_A D_A + p_B D_B)\varkappa^2 t} = e^{-\bar{D}\varkappa^2 t} \tag{15}$$

if

$$\tau_A, \tau_B \ll (D_A \varkappa^2)^{-1}, (D_B \varkappa^2)^{-1}$$

(fast exchange, i.e. residence times much shorter than observation times), where $\bar{D}$ is the mean self-diffusion coefficient.

Thus, for relatively slow exchange processes $\tau_A, \tau_B \gtrsim 10^{-3}$ s the dynamical detail of the translational motion: 'exchange between two environments' may be observed via the damping of the spin echo, whereas in the situation of fast exchange only the mean self-diffusion coefficient is available. However, as stated above, we must fulfill the condition $\gamma^2 \delta^2 g^2 D_i \gg 1/I_\perp$; $i = A, B$. If we are concerned with diamagnetic solutions where the proton magnetization is utilized, this requirement can always be satisfied by replacing most of the protons in the system by deuterons. To the author's knowledge this proposed method has not been applied yet for the study of homogeneous liquid systems.

Finally we may generalize Equations (11), (12) to more than two different microscopic environments:

$$\frac{dI_i(\varkappa, t)}{dt} = -\left(D_i\varkappa^2 + \sum_j^s \frac{1}{\tau_{ij}}\right) I_i + \sum_j^s \frac{I_j}{\tau_{ji}} \qquad i = 1, ..., s. \tag{16}$$

The formal solutions for the mean $I(\varkappa, t)$ is

$$I(\varkappa, t) = \sum_{i=1}^s \mathscr{P}'_i(\varkappa) e^{-D'_i\varkappa^2 t}, \tag{17}$$

where the $\mathscr{P}'_i$ and $D'_i\varkappa^2$ are functions of D_i, τ_{ij}, τ_{ji}, and of the occurrence probabilities p'_i of the ith configurational environment. Due to the secular problem implicit in Equation (16) closed expressions for D'_i, $\mathscr{P}'_i$ cannot be given for $s > 3$. However, the limiting behaviour corresponds exactly to that given in Equations (14), (15). Now, the propagator has the form:

$$P(\mathbf{r}, t) = (2\pi)^{-3} \int \sum_{i=1}^s \mathscr{P}'_i(\varkappa) e^{-D'_i\varkappa^2 t} e^{-i\mathbf{r}\varkappa} d\varkappa. \tag{18}$$

For long diffusion times, only terms with long observation times $(D\varkappa^2)^{-1}$ (see Equation (17)) contribute to the integral over $\varkappa$, thus we obtain

$$P(\mathbf{r}_i t) = (4\pi \bar{D} t)^{-3/2} \exp\left(\frac{r^2}{4\bar{D} t}\right). \tag{18a}$$

On the other hand, if we have a mixture of 'non-reacting' molecules, an equation of type (14) holds always, thus

$$P(\mathbf{r}_i t) = \sum p_i (4\pi D_i t)^{-3/2} \exp\left(\frac{r^2}{4D_i t}\right). \tag{18b}$$

These formulae, in particular Equation (17) corresponds to a 'distribution of correlation times'

$$\hat{p}_i = p(\tau_i) = p((D_i\varkappa^2)^{-1}).$$

However, one sees that, apart from D_i and p_i, in general also the rate of mixing of states, i.e. the residence times τ_{ij}, enters into the observable quantity. It seems to be important to point out this fact, in particular with regard to so many treatments of rotational motion which use the concept of a distribution of correlation times or distribution of relaxation times. The only difference is that the length of rotational observation times cannot be chosen at will, the longest observation times being the correlation times of spherical harmonics of first order (dielectric relaxation) (see also [18]).

3. The Intermolecular Relaxation Rate, Experimental Determination

Usually the total relaxation rate $1/T_1$ of a nucleus is separated into two contributions

$$\frac{1}{T_1} = \left(\frac{1}{T_1}\right)_{intra} + \left(\frac{1}{T_1}\right)_{inter}.$$

The intramolecular relaxation rate, $(1/T_1)_{intra}$ is caused by the interaction of the relaxing nucleus with the nuclei and the electrons of the same molecule in which the nucleus resides. The intermolecular relaxation rate, $(1/T_1)_{inter}$, is due to the interaction of the nucleus considered with all other nuclei and electrons of the system. The experimental separation of $(1/T_1)_{intra}$ and $(1/T_1)_{inter}$ is accomplished as follows [19, 20, 21]. If the proton relaxation rate is concerned, one has to measure $1/T_1$ in a mixture of the normal molecule with the same, however fully deuterated, molecule. Let x_H be the mole fraction of the proton containing species, then

$$\frac{1}{T_1} = \left(\frac{1}{T_1}\right)_{inter} [x_H + \varepsilon(1 - x_H)] + \left(\frac{1}{T_1}\right)_{intra} \tag{19}$$

which allows the determination of $(1/T_1)_{inter}$ and $(1/T_1)_{intra}$.

$$\varepsilon = \frac{2\gamma_D^2 I_D(I_D + 1)}{3\gamma_H^2 I_H(I_H + 1)} = 0.042,$$

where γ_D, γ_H = gyromagnetic ratio of the deuteron and proton, respectively, $I_H = \tfrac{1}{2} =$ = spin of the proton, $I_D = 1$ spin of the deuteron.

In a mixture of two (or more) chemically different molecules, A, B, Equation (19) may also be applied. Now the amounts of both components of the mixture have to be kept constant whilst the fraction of proton containing species A is varied. B has to be fully deuterated. In a further series of experiments the isotopic composition of B is to be varied leaving A fully deuterated. In this manner one obtains the intermolecular relaxation rates of the nucleus on A (A–A interaction) and B (B–B interaction), respectively, at a given composition A/B of the mixture. Finally, the measurement of the additional relaxation rate of the proton on molecule A in the presence of normal B yields the intermoleculer relaxation rate of A due to interaction with B. The reverse experiment must give the same result [22].

Evidently, the procedure of measuring the entire set of intermolecular relaxation rates in a binary mixture is very tedious. Therefore it is more convenient to measure the deuteron relaxation rate $(1/T_1)_d$ which is proportional to the intramolecular proton relaxation rate:

$$\left(\frac{1}{T_1}\right)_d = K\left(\frac{1}{T_1}\right)_{intra}.$$

Thus, it is sufficient to measure one $(1/T_1)_{intra}$, as described, in order to determine the factor K and then to continue the measurements at other compositions A/B as deuteron relaxation rates. Then the intermolecular relaxation rate may be obtained from the total relaxation rate $1/T_1$

$$\left(\frac{1}{T_1}\right)_{\text{inter}} = \frac{1}{T_1} - K^{-1}\left(\frac{1}{T_1}\right)_d.$$

This technique implies that the quadrupole coupling constant does not depend on the composition A/B of the mixture. It seems that this is not a critical condition.

If the intermolecular relaxation rate of ^{19}F in a pure liquid is desired one has to use mixtures of the liquid in question with an inert solvent not containing magnetic nuclei because the isotope ^{19}F has 100% natural abundance.

$$\frac{1}{T_1} = \left(\frac{1}{T_1}\right)_{\text{inter}} x_F + \left(\frac{1}{T_1}\right)_{\text{intra}}. \tag{20}$$

x_F is the mole fraction of the ^{19}F containing molecule. In Equation (20) we have assumed that the inert solvent has the same viscosity as the liquid to be investigated.

So far the methods quoted are based on the condition that the molecule considered is unchanged when diluted in the liquid not containing the magnetic nucleus in question. Thus, for the determination of $(1/T_1)_{\text{inter}}$ of water, D_2O cannot be used. H_2O has to be diluted in a deuterated aprotic solvent [21].

This gives $(1/T_1)_{\text{intra}}$ for $x_{H_2O} \to 0$. At the same time $(1/T_1)_d = k(1/T_1)_{\text{intra}}$ may be measured for $0 < x_{H_2O} \leq 1$. At $x_{H_2O} = 0$ the constant k can be determined which gives $(1/T_1)_{\text{intra}}$ for $0 < x_{H_2O} \leq 1$. Using different aprotic solvents it has been checked that the quadrupole coupling constant of D_2O does not vary between $x_{H_2O}=0$ and $x_{H_2O}=1$. Another method has been reported [23] which makes use of the fact that $(1/T_1)_{\text{inter}}$ is proportional to the number of spins per cm³. If one approaches towards the critical temperature the proton relaxation rate of H_2O becomes increasingly smaller than to be expected from the deuteron relaxation rate of D_2O. The difference must be due to the decreasing number density of protons as the temperature increases. The intermolecular relaxation rate has been estimated from this deviation [23].

4. Translational Motion from the Intermolecular Relaxation Rate

4.1. General Relations, Frequency Dependence for Diamagnetic Liquids

As in general the nuclear magnetic relaxation rate, so too, the intermolecular relaxation rate is a linear combination of a number of spectral intensities [1]

$$\left(\frac{1}{T_1}\right)_{\text{inter}} = \gamma_I^2 \hbar^2 \{\tilde{a}J(\omega_1) + \tilde{b}J(\omega_2) + \tilde{c}J(\omega_3)\} \tag{21}$$

$\tilde{a}$, $\tilde{b}$, and $\tilde{c}$ being constants given by the theory and

$$\omega_1 = \omega_I - \omega_S$$
$$\omega_2 = \omega_I$$
$$\omega_3 = \omega_I + \omega_S,$$

where ω_I and ω_S are the nuclear magnetic resonance frequencies of the relaxing nucleus and that of the interaction partner, respectively (in the case of like spins $\omega_I = \omega_S$, $\tilde{a} = 0$). $J(\omega)$ is the Fourier transform of the time correlation function

$$g(t) = N \frac{\overline{Y_2^m(0) Y_2^{m*}(t)}}{r^3(0) \; r^3(t)} \tag{22}$$

$$J(\omega) = \int_{-\infty}^{+\infty} g(t) e^{-i\omega t} \, dt. \tag{23}$$

For $g(t)$ we write

$$g(t) = N \iint \frac{Y_2^{m*}(\theta, \phi) \; Y_2^m(\theta_0, \phi_0)}{r^3} \frac{1}{r_0^3} p(\mathbf{r}_0) P(\mathbf{r}_0, \mathbf{r}, t) \, d\mathbf{r}_0 \, d\mathbf{r} \tag{24}$$

N = number of interaction partners in the system.

$Y_2^m(\theta, \phi)$ is the spherical harmonic of order 2. On the left hand side of Equations (22) and (24) the superscript m is dropped which indicates that for the isotropic system to be treated here $g(t)$ does not depend on m. θ and ϕ are the polar and azimuthal angles of the vector $\mathbf{r}$ relative to the laboratory system, the z direction being given by the magnetic field. $\mathbf{r} = \{\theta, \phi, r\}$. $\mathbf{r}$ connects the reference nucleus with another particle which is the interaction partner. Both particles undergo diffusion relative to one another, thus $\mathbf{r} = \{\theta(t), \phi(t), r(t)\}$. θ_0, ϕ_0 and r_0 stand for $\mathbf{r}_0 = \{\theta(0), \phi(0), r(0)\}$. $p(\mathbf{r}_0)$ is the probability density to find the interaction partner at $\mathbf{r}_0$. Here the specification that $t = 0$ is concerned is of no significance, thus $p(\mathbf{r}_0)$ is the pair distribution function. $P(\mathbf{r}_0, \mathbf{r}, t)$, the propagator, determines the probability that a particle is in $d\mathbf{r}$ at $\mathbf{r}$ *relative to the reference* nucleus at time t if we know that it was at $\mathbf{r}_0$ at time 0.

We see two important facts from Equation (24): (i) $P(\mathbf{r}_0, \mathbf{r}, t)$ is the quantity which contains the information regarding the translational motion of the particles, thus it is the quantity we wish to know. But now it is the relative translational motion of one particle as seen from another one which occurs in Equation (24), it is not the 'absolute' translational motion measured in the laboratory coordinate system. (ii) If we strive for the knowledge of $P(\mathbf{r}_0, \mathbf{r}, t)$ we have to know the pair distribution function regarding the nuclei which interact by magnetic dipole-dipole interaction.

One way out of this difficulty is to assume a simple model pair distribution function. Three such model pair distribution functions are depicted in Figure 2. After this the procedure must be as follows: We assume a propagator $P(\mathbf{r}_0, \mathbf{r}, t)$ which is described by a set of parameters $\beta_1, \beta_2, \ldots$ and Equations (21), (23), and (24) may be used to determine the parameters from the experimental $(1/T_1)_{\text{inter}}$ as a function of the magnetic field strength. The ω_i are all connected to the magnetic field strength through the respective gyromagnetic ratios.

Another way of treatment is to use at the same time a certain number of parameters for the determination of $p(\mathbf{r}_0)$.

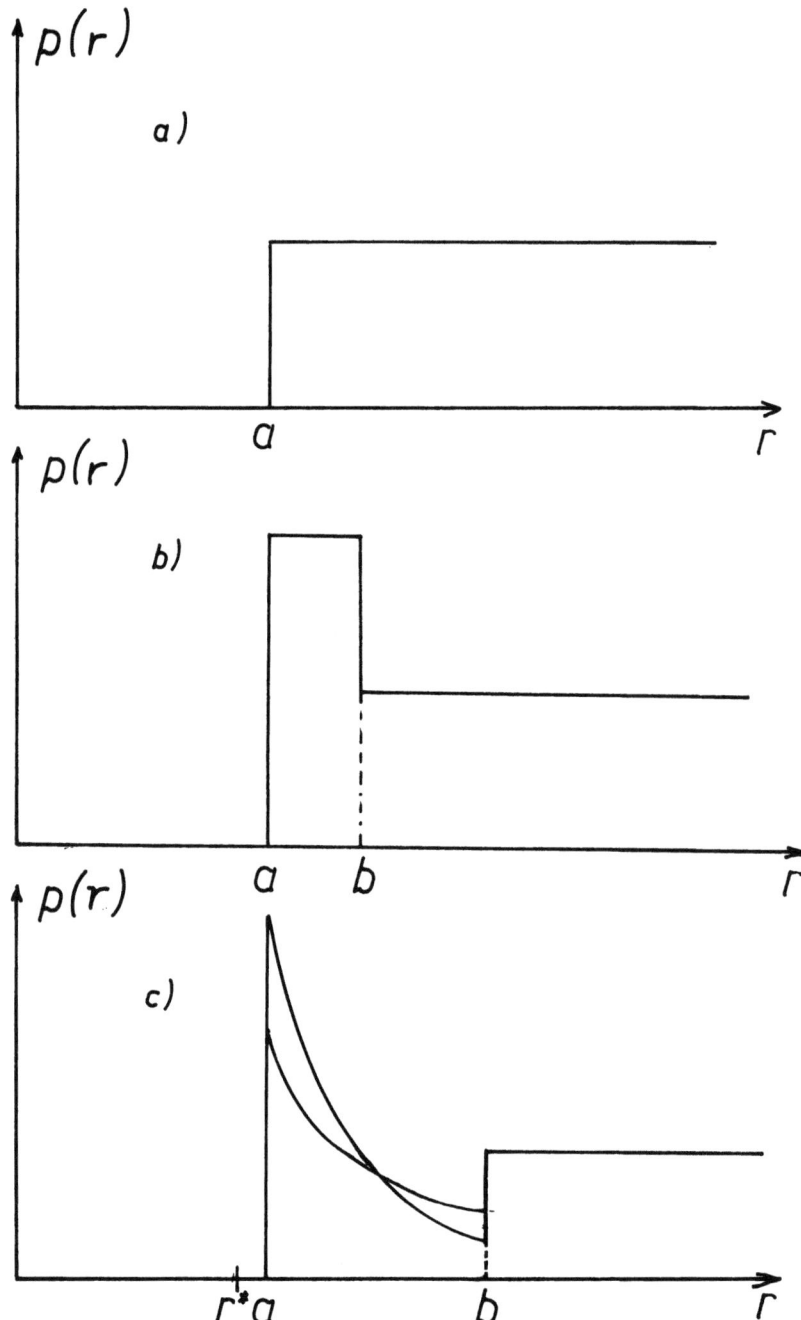

Fig. 2. Simple model pair distribution functions $p(r)$. The function of type c) is given for two different parameters n according to Equation (38a).

Let us discuss some possible approaches of the former type. The simplest choice is to use the propagator Equation (3) which is converted to regard relative motion by replacing D by $2D$. Next we quote a two parameter approach for the propagator which, due to its simplicity, has been used fairly often. This regards the jump model, we shall sketch this model in the particular form as introduced by Torrey [24]. The particles are at rest during a time τ_r. Then, during a very short time they get thermally agitated and come to rest with a displacement $\mathbf{r}_1$. The probability to find a given displacement $\mathbf{r}_1$ after one jump is

$$P(\mathbf{r}_1) = (\tfrac{3}{2}\pi \langle r_1^2 \rangle r)^{-1} \exp\left\{\frac{-r}{(\tfrac{1}{6}\langle r_1^2\rangle)^{1/2}}\right\} \tag{25}$$

with $\langle \mathbf{r}_1^2 \rangle$ = mean square displacement caused by one jump. At the same time the model is such that

$$\langle \mathbf{r}_1^2 \rangle = 6\tau_r D. \tag{26}$$

With a Poisson distribution of the jumping times Torrey computes the propagator as:

$$P(\mathbf{r}_0, \mathbf{r}, t) = (2\pi)^{-3} \int e^{-2\bar{D}x^2 t/(1+(1/6)\langle r_1^2\rangle x^2)} e^{-i\varkappa(\mathbf{r}-\mathbf{r}_0)}\, d\varkappa \tag{27}$$

$\bar{D}$ = mean self-diffusion coefficient of interaction partners. Insertion of Equation (27) in Equation (24) yields with Equation (21) (in the special case of equal spins, $\omega_I = \omega_S$, $\tilde{a}=0$)

$$\left(\frac{1}{T_1}\right)_{\text{inter}} = \frac{3\pi}{5} \gamma^4 \hbar^2 \frac{c'}{a^3 \omega_I}\{f(\alpha, x) + 2f(\alpha, \sqrt{2}\, x)\} \tag{28}$$

with

$$f(\alpha, x) = \frac{2}{x^2}\left\{v\left(1 - \frac{1}{u^2+v^2}\right) + e^{-2v}\left[v\left(1 + \frac{1}{u^2+v^2}\right) + 2\right] \times \right.$$
$$\left. \times \cos(2u) + e^{-2v} u\left(1 - \frac{1}{u^2+v^2}\right)\sin(2u)\right\}$$

and

$$u = \frac{1}{2}\sqrt{\frac{q(1-q)}{\alpha}}$$

$$v = \frac{1}{2}\sqrt{\frac{q(1+q)}{\alpha}}$$

$$q = \frac{\alpha x^2}{\sqrt{1+\alpha^2 x^4}}$$

$$\alpha = \langle r_1^2\rangle/12 a^2$$

$$x = \sqrt{\frac{\omega_I a^2}{\bar{D}}}, \quad c' = N/V \quad (V = \text{volume}),$$

where $p(\mathbf{r}_0)$ as given in Figure 2a has been applied. a is the closest distance of approach between the two interacting spins.

Is it possible to observe dynamical details by this method? To answer this question we have to find the smallest observation time $(\varkappa^2 D)^{-1} = \tau_{obs}$ which enters in the measurement of $(1/T_1)_{inter}$. The correlation function $g(t)$ has essentially decayed when the spherical harmonics $Y_2(\theta, \phi)$ have changed sign. The smallest displacement by which this can be accomplished is of the order of a. Thus the maximum $\varkappa^2 \approx a^{-2}$ and the shortest observation time $\tau_{obs} \approx a^2/D$.

With the exponential in Equation (27)

$$e^{-2t/(\tau_{obs} + (1/6)\langle r_1^2 \rangle/D)} \tag{29}$$

we see that the dynamical detail $\langle \mathbf{r}_1^2 \rangle$, or with Equation (26) τ_t, can be observed if $\langle \mathbf{r}_1^2 \rangle \gg a^2$. If $\langle \mathbf{r}_1^2 \rangle \ll a^2$ then Equation (27) represents the usual 'purely diffusional' propagator

$$P(\mathbf{r}_0, \mathbf{r}, t) = (2\pi)^{-3} \int e^{-2tD\varkappa^2} e^{-i\varkappa(\mathbf{r}-\mathbf{r}_0)} \, d\varkappa =$$

$$= \frac{1}{(8\pi Dt)^{3/2}} e^{-r^2/8\pi Dt}. \tag{3a}$$

In this event in Equation (28) $\alpha \to 0$.

The next type of a propagator one might like to test is of the general form of Equation (18).

$$P(\mathbf{r}_0, \mathbf{r}, t) = (2\pi)^{-3} \int \left(\sum \mathscr{P}_i'(\varkappa) e^{-D'_{irel}\varkappa^2 t} \right) e^{-i\varkappa(\mathbf{r}-\mathbf{r}_0)} \, d\varkappa'. \tag{30}$$

Consider the special case $i = 1, 2$ (see above, page 340–341). This model is similar to a jump model, however somewhat more realistic because it avoids infinitely short times of particle displacement. The model contains four adjustable parameters: $D_{A_{rel}} = 2D_A$, $D_{B_{rel}} = 2D_B$, the diffusion coefficients in the 'free' and 'bound' state, respectively, and τ_A and τ_B, the corresponding residence times. Again the shortest observation times are $\tau_{obs, A} \approx a^2/D_A$, $\tau_{obs, B} \approx a^2/D_B$. If $\tau_A, \tau_B \ll a^2/D_A, a^2/D_B$ we have 'fast exchange' the propagator reduces to Equation (3a). In the case of slow exchange $P(\mathbf{r}_0, \mathbf{r}, t)$ is of the form (18b). For the intermediate range of residence times τ_A, τ_B formulae for the intermolecular relaxation rate have not been derived yet.

Before going further with the analysis of Equation (24) it may be appropriate to say some words regarding the experimental situation. The most important condition for the separate determination of more than one parameter characterizing the propagator is the requirement $\tau_{obs}\omega_I \approx 1$. In nuclear magnetic resonance experiments with systems containing no paramagnetic species ω_I cannot be made higher than 10^9 s^{-1}. Thus, we find that the τ_{obs} are limited to be $\gtrsim 10^{-9}$ s. However, in all usual highly fluid liquids at room temperature we have $\tau_{obs} \approx \tau_c \lesssim 10^{-11}$ s. Consequently the method to investigate $P(\mathbf{r}_0, \mathbf{r}, t)$ is confined to those liquids (with molecules of moderate size) which can be cooled down sufficiently. Glycerol is a favourable example. In Figure 3 we present the intermolecular relaxation rate as a function of $v_I (v_I = \omega_I/2\pi)$ at two temperatures [25]. The data have been fitted to Equation (28). The resulting parameters are: $a = 2.5 \times 10^{-8}$ cm, $\alpha = 0$ for $T = 30\,°C$ ($D = 3.2 \times 10^{-8}$ cm^2 s^{-1}) and $a = 2.3 \times 10^{-8}$ cm $\alpha = 0.003$ at $T = 13\,°C$ ($D = 6.5 \times 10^{-9}$ cm^2 s^{-1}). The diffusion coefficients have been measured independently [26, 26a]. It may be seen from Figure that the fit is relatively poor. There are fast processes which are not

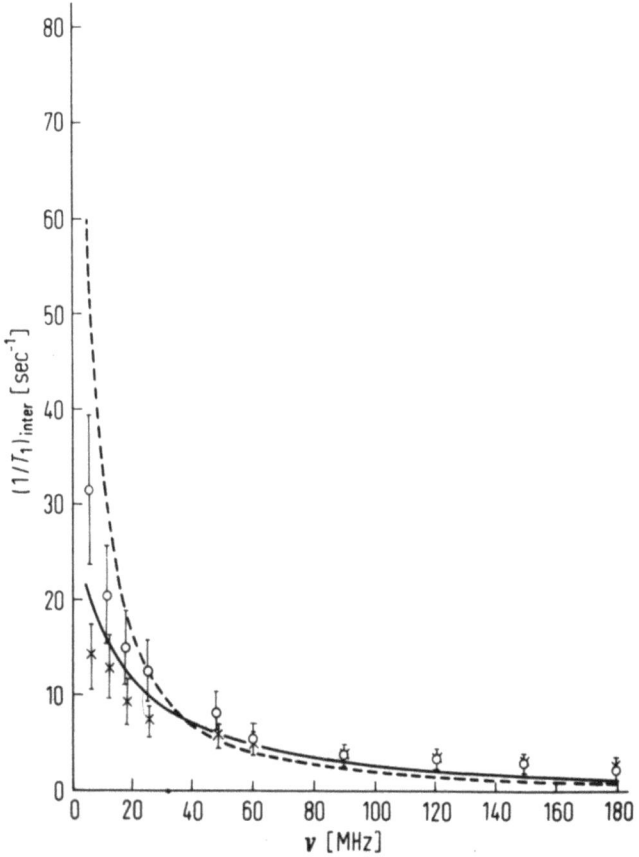

Fig. 3. The intermolecular proton relaxation rate of glycerol as a function of the resonance frequency [25]. $\times$: $1000/T = 3.3$; $\bigcirc$: $1000/T = 3.5$. Dashed and solid lines: Fits to the Torrey theory.

taken account of by Equation (28). It has not yet been tested whether a more satisfactory fit can be obtained by use of a propagator of the form Equation (30). A theoretical and experimental investigation regarding the intermolecular relaxation rate of ethane at low temperatures has been performed by Harmon and Muller [27]. In this work, a realistic pair distribution function has been used and an approximate frequency dependence of the relaxation rate is given. Three other points should be mentioned in connection with the interpretation of frequency dependent intermolecular relaxation rates in diamagnetic liquids.

(I) So far we applied a very primitive method when converting the absolute propagator to the relative one, we just took twice the diffusion coefficient. Actually however, $p(\mathbf{r}_0)$ and $P(\mathbf{r}_0, \mathbf{r}, t)$ are not independent. If $p(\mathbf{r}_0)$ has a very distinct maximum then one or more molecules are tightly coupled to the reference molecule. In this event the molecules stick together during a time of the order of or even longer than the rotational correlation time of the vector connecting the two interacting nuclei. Now, for these particles, $P(\mathbf{r}_0, \mathbf{r}, t)$ degenerates to a rotational propagator [1, 18, 32]

$$P(\mathbf{r}_0, \mathbf{r}, t) = \sum_{l, m} Y_l^{m*}(\phi_0, \theta_0) Y_l^m(\theta, \phi) e^{-l(l+1) D_r t} \tag{31}$$

with $P(\mathbf{r}_0, \mathbf{r}, t) = 0$ for $|\mathbf{r}| \neq |\mathbf{r}_0|$ and sufficiently short times. $D_r =$ rotational diffusion coefficient. This contribution may be combined in a suitable way with the usual $P(\mathbf{r}_0, \mathbf{r}, t)$. This, however, is not an easy matter [28]. Furthermore, even if the attractive potential is weak, the repulsive potential exists in any case and the volume occupied by the reference molecule is not accessible for the interaction partner. In order to take account of this situation, one may also add some rotational contribution to the translational propagator.

(II) This item is implicit in point (I) but it may be useful to mention it separately. The propagator $P(\mathbf{r}_0, \mathbf{r}, t)$ regards intermolecular pairs of interacting nuclei. Thus, for different nuclei on the same molecule the short time behaviour may be different. This comes about when the reference molecule and/or the partner molecule perform reorientational motion as a rigid body or as internal motions, and when the various nuclei are at non-equivalent positions. Propagators which represent a superposition of rotational and translational motions have been constructed by Zeidler [29] (see also Sears [30]):

$$P(\mathbf{r}_0, \mathbf{r}, t) = (2\pi)^{-3} \int e^{-2Dt\varkappa^2/(1 + D\tau_t \varkappa^2)} \left(\sum_l (2l+1) j_l^2(\varkappa d) \times \right.$$
$$\left. \times e^{-l(l+1) D_{reff} t} \right) e^{-i\varkappa(\mathbf{r}-\mathbf{r}_0)} d\varkappa, \tag{32}$$

$j_l(x) = \sqrt{\pi/2x} \, J_{l+1/2}$, $J_{l+1/2} =$ half-integer Bessel function, d is the distance nucleus – centre of mass. If both molecules perform tumbling motion, effective quantities $D_{reff} \approx 2D_r$ and $d_{eff} \approx 2d$ will be suitable approximations. The effect of the modulation of the magnetic interaction on the relaxation rate (in the case of extreme narrowing, see below) has been treated by Hubbard [31].

(III) In Equation (24) a factor N appears before the integral which means that the total correlation function is the sum of N independent correlation functions for the pairs: reference nucleus-interaction partner. But clearly the motions of the N particles are not strictly independent. Assume that n neighbours of the reference molecules have given positions. Then, at that instant for the $(n+1)$th particle these positions cannot be taken on. Furthermore, in principle there should also be slight cross-relaxation effects among the various interaction pairs [1, 32]. No attempt has as yet been undertaken to tackle these problems to the author's knowledge.

Summarizing, we see that in principle, detailed information regarding translational motion may be obtained from the frequency dependent intermolecular relaxation rate in pure diamagnetic liquids (and mixtures), however, the experimental as well as the computational effort needed is rather great.*)

4.2. SOLUTIONS OF PARAMAGNETIC PARTICLES

In solutions containing paramagnetic particles the resonance frequency ω_s of the electron spin acting as the interaction partner is about three orders of magnitudes larger than in those systems where the interaction partner is a nuclear spin. Now, the shortest observation time being relevant for the relaxation process of the (reference) nucleus is of the order of ω_s^{-1}. Thus, even with liquids of normal fluidity we have $\omega_s \tau_c \approx 1$. In the particular situation being valid in these solutions the theory yields the result [34–36, 18, 32]:

$$\left(\frac{1}{T_1}\right)_{\text{inter}} = \gamma_I^2 \gamma_s^2 t_1^2 S(S+1) \frac{16\pi \tau_d c'}{25 a^3} \{1 + \tfrac{7}{3} \hat{f}(\omega_s \cdot 6\tau_d)\} \qquad (33)$$

together with a similar expression for $(1/T_2)_{\text{inter}}$ which is not needed here. S is the electron spin, γ_s is the gyromagnetic ratio of the electron. $\tau_d = a^2/6\bar{D}$, $\bar{D}$ = mean self-diffusion coefficient. Again a is defined by Figure 2a. Since free radicals are fairly large molecules, the observation times are long. Thus, we quote $(1/T_1)_{\text{inter}}$ only in the form Equation (33), where the propagator Equation (3a) has been applied,

$$\hat{f}(\omega_s \cdot 6\tau_d) = \frac{15}{2x^2} \lim_{\alpha \to 0} f(\alpha, x),$$

where $f(\alpha, x)$ is given in Equation (28).

Studies with solutions of paramagnetic particles have the very great advantage that the intermolecular relaxation is the primary experimental quantity, no tedious isotopic substitution experiments are needed. Apart from this, all the remarks we have given in conjunction with diamagnetic solutions are valid as well, the only difference being that usually the concentration of electron paramagnetic interaction partners is relatively low which facilitates the consideration of relative and collective motions somewhat. Indeed, numerous experimental studies have been performed with solutions of free radicals and transition metal ions [37–46]. For the former systems often the electron-nucleus Overhauser effect has been utilized [1] which is another technique to study the frequency dependence of the spectral density $J(\omega)$ as given in Equation (23). The evaluation of the experimental results so far given in the literature has mostly been done on a qualitative level. It has been shown that Equation (33) (or the corresponding Overhauser coupling parameter) gives an approximate description of the experimental results. The order of magnitude of the self-diffusion coeffi-

---

* Burnett and Harmon [33] reported a method to obtain D directly from intermolecular relaxation measurements. It should be kept in mind that here as always relative translational motion is concerned, thus, in general no self-diffusion coefficients in the strict sense can be obtained.

cient which can be derived, is correct, some typical deviations from the expected behaviour at high frequencies have been observed.

In solutions of spherical metal ions [18, 32] in some regards the situation is simpler because there is no rotational motion of the particles carrying the unpaired electrons. On the other hand, due to the strong electrostatic forces exerted by the ion we have hydration (or solvation) which means that the rotational propagator Equation (31) is quite dominant. In this event, the spectral density is a simple Lorentz curve with minor alterations due to relative translational motion when the ions are farther apart from the relaxing nucleus. Probably, the simplest situation is given when the relaxing nucleus and the paramagnetic interaction partner are both spherical ions. In particular, if both ions are cations, then the ions keep far apart from one another, the closest distance of approach, a, becomes large, i.e. the shortest observation times become long, the diffusion equation should be reasonably valid, at least rotational contributions to the propagator due to attractive forces should be absent. From the frequency dependence of $1/T_1$ the two parameters $\bar{D}$ and a may be obtained. Such measurements have been made, the systems were aqueous solutions of $Li^+ + Mn^{2+}$ ($a = 4.8$ Å, $D = 0.8 \times 10^{-5}$ cm^2 s^{-1}), $Na^+ + Mn^{2+}$ ($a = 4.6$ Å, $D = 0.8 \times 10^{-5}$ cm^2 s^{-1}) [47], $Rb^+ + Mn^{2+}$ ($a = 4.3$ Å, $D = 0.8 \times 10^{-5}$ cm^2 s^{-1}), $Rb^+ + Cu^{2+}$ ($a = 2.7$ Å, $D = 0.35 \times 10^{-5}$ cm s^{-1}) [48]. The numbers in parentheses are the experimental results, the concentration was 1 M in both cations. The diffusion coefficient obtained for $Li^+ + Mn^{2+}$ and $Na^+ + Mn^{2+}$ are essentially correct, the figures given for Rb^+, in particular together with Cu^{2+}, are too small. Probably, here some relaxation by scalar interaction interferes.

4.3. The Situation of Extreme Narrowing

In ordinary fluid liquids at room temperature the relaxation rates $1/T_1 = 1/T_2$ are frequency independent because now $\omega\tau_c \ll 1$ (extreme narrowing situation). Now use of the propagator Equation (27) gives

$$\left(\frac{1}{T_1}\right)_{inter} = \pi\gamma_I^4\hbar^2 \frac{c'}{a^3} \tau_t \left(1 + \frac{12}{5} \frac{a^2}{\langle r^2\rangle}\right)$$

$$I = \tfrac{1}{2}, \qquad \tau_t = \langle r_1^2\rangle/6D, \qquad c' = N/V \tag{34}$$

for equal spins. If the interacting spins are different, this expression has to be multiplied by a factor $S(S+1)\cdot\tfrac{8}{9}$ ($I-S$ cross-coupling effects neglected). For 'true' or microstep diffusion $\langle r_1^2\rangle \ll a^2$ and Equation (34) reduces to

$$\left(\frac{1}{T_1}\right)_{inter} = \frac{2\pi}{5} \gamma^4\hbar^2 \frac{c'}{aD}. \tag{35}$$

It is clear that now the 'zero frequency' expression (34) does not give detailed information regarding the translational motion, that is, D and τ_t cannot be determined separately. Usually $(1/T_1)_{inter}$ according to Equation (35) comes out smaller than experimentally observed when the otherwise measured D is used. In this situation it is not correct to calculate τ_t from Equation (34) and from the experimental data because both Equation (34) and (35) do not represent rigorous relations. Some reasons for this being so have already been given, one other reason is that Equations (34) and (35) are only derived for systems with spherical symmetry with regard to the relaxing nucleus. Otherwise, a is not well-defined. Hubbard has improved the theory by taking account of the fact that the relaxing nucleus does not reside in the centre of the molecule [31]. Also, as was mentioned above, he calculated the effect of molecular rotation on the intermolecular relaxation rate.

What are the other shortcomings which are responsible for the lack of agreement between theory and experiment? We have already mentioned both of them: neglect of the appropriate effective intermolecular potential, i.e. the correct pair distribution function, and the neglect of collective motions. Indeed a more realistic radial distribution function has been incorporated in the theory by several authors [27, 49, 50], the effect on the relaxation rate was small. Still it would be convenient to have a direct means to study the influence of the pair distribution in a simple and qualitative way. This, in general, turns out to be mathematically difficult because of the particular functional dependence of the propagator on r_0 and r. In order to simplify the computation we may say that the decay of the correlation function is caused primarily by the rotation of the vector connecting the reference nucleus with the interaction partner. Superimposed on the rotational motion there is a radial displacement of the molecule which also contributes to the decay of the correlation function $g(t)$, but the radial displacement does not change the sign of the magnetic dipole interaction, thus it may be considered to be less effective. Finally we have to take into account the rotation of the individual molecule, the 'tumbling', both of the reference and the partner molecule.

These arguments lead us to the use of the rotational propagator Equation (31) alone in Equation (24). Then, after integration over all orientations we obtain from Equation (24)

$$g(t) = 4\pi N\, e^{-t/\tau_c^*} \int_0^\infty \frac{p(r_0)\, r_0^2\, dr_0}{r_0^6}, \qquad (36)$$

where τ_c^* is the correlation time which governs the decay of $g(t)$ and which now we have to consider. The time after which a partner molecule has moved from one side of the reference molecule to the other is

$$\tau_r \approx \frac{(2r_m)^2}{2\cdot 2\bar{D}} = \frac{r_m^2}{\bar{D}}. \qquad (37)$$

r_m = separation between the centre of mass of the reference and partner molecule. In order to take account of the relative motion we have taken $2\bar{D}$. We are interested in the correation time of the spherical harmonics of second order, thus

$$\tau_m = \frac{r_m^2}{3\bar{D}}$$

(which defines τ_m). But essentially Equation (37) corresponds to a projection of the motion in one dimension. In fact, during this motion the magnitude of r_m varies which causes an additional decrease of the correlation function, as indicated above. The time constant of this contribution is of the order of τ_b, the life time of a molecule in a given coordination sphere, we may set $\tau_b \approx \frac{1}{2}\tau_m$. Finally, the molecular tumbling modulates the magnetic interaction, in effect it reduces the correlation time of the interaction, we describe this contribution by a time $\tau_{mod} \approx \tau_m$. With these approximations we get

$$\frac{1}{\tau_c^*} = \frac{1}{\tau_m} + \frac{2}{\tau_m} + \frac{1}{\tau_m} = \frac{4}{\tau_m}.$$

Since the closet distance of approach between the protons very often is $a \approx \frac{1}{2}r_m$ and since the greatest contribution comes from the first coordination sphere, we may write

$$\tau_c^* = \frac{a^2}{3\bar{D}}. \tag{38}$$

One further comment is in place here, Equation (38) does not give the correct correlation time when the reference molecule and the partner molecule are closely connected so as to give effectively a 'single molecule'. Now, the experimental τ_c^* is shorter by about a factor $\frac{1}{2}$ than predicted by Equation (38). Now, due to the strong attraction potential the vector connecting interaction partners changes its direction on a direct way, not via a detour through regions where r is large. Since, as already mentioned, variations of r never change the sign of Y_2/r^3 this process is less effective for the decay of $g(t)$. Thus, the condition for Equation (38) to be applicable is that essentially equal attraction forces act from all sides of the partner molecule.

Now any radial distribution function may be introduced in Equation (36). Let us choose the three functions shown in Figure 2. The distribution function c may be written

$$Np(r_0) = \frac{p_0}{(r_0 - r^*)^n} \quad \text{for} \quad a \leq r_0 \leq b$$
$$Np(r_0) = c' \quad \text{for} \quad r_0 > b; \tag{38a}$$

as before c' is the spin concentration in the bulk liquid, p_0 is given by the relation

$$4\pi N \int_a^b p(r_0) r_0^2 \, dr_0 = n_c,$$

where n_c is the first coordination number around the molecular group carrying the reference nucleus. For the parameters r^* and b we set:

$$r^* = ka$$
$$b = la$$

we set $k = 0.9$.

With this distribution function [51] one obtains the intermolecular relaxation rate (like spins, $I=\frac{1}{2}$)

$$\left(\frac{1}{T_1}\right)_{\text{inter}} = \frac{1}{2}\gamma^4\hbar^2 \frac{n_c}{a^4\bar{D}} g(n, k, l) + \delta^*, \qquad (39)$$

where the function $g(r, k, l)$ is depicted in Figure 4. On the other hand, with the

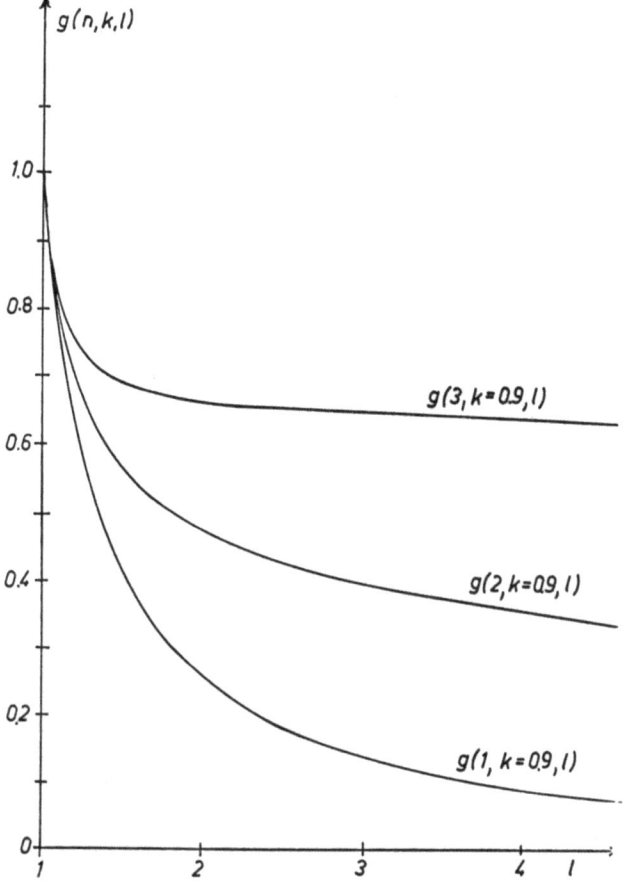

Fig. 4. The function $g(n, k, l)$ occurring in Equation (39).

distribution function b of Figure 2 the result is

$$\left(\frac{1}{T_1}\right)_{\text{inter}} = \frac{1}{2}\gamma^4\hbar^2 \frac{n_c}{a^4\bar{D}} + \delta^*,$$

in both cases

$$\delta^* = \frac{2\pi}{3} \gamma^4\hbar^2 \frac{c'}{b D}. \qquad (40)$$

Equation (39) contains three parameters: the steepness of the pair distribution function, i.e. the depth of the attractive potential, characterized by n, the coordination number n_c of the group or molecule containing the relaxing nucleus, and b, the radius of the second coordination sphere. It may be shown that satisfactory agreement between experimental results and Equation (39) can be achieved if the experimental diffusion coefficients are used and if the parameters n, n_c and b are suitably chosen. For a wide range of common organic liquids closest distances of approach $a = 2.5 - -4$ Å are found.

References

1. Abragam, A.: *The Principles of Nuclear Magnetization*, Oxford, 1961.
2. Slichter, C. P.: *Principles of Magnetic Resonance*, New York, Evanston and London, 1967.
3. Farrar, T. C. and Becker, E. D.: *Pulse and Fourier Transform NMR*, New York and London, 1971.
4. Carr, H. Y. and Maxwell, D. E.: *Phys. Rev.* **94**, 630 (1954).
5. McCall, D. W., Douglass, D. C., and Anderson, E. W.: *Ber Bunsenges. physik. Chem.* **67**, 336 (1963).
6. See e.g. Hertz, H. G.: *Ber. Bunsenges. phys. Chem.* **77**, 531 (1973).
7. Stejskal, E. O.: *J. Chem. Phys.* **43**, 3597 (1965).
8. Tanner, J. E.: *Rev. Sci. Instr.* **36**, 1086 (1965).
9. Tanner, J. E.: *J. Chem. Phys.* **52**, 2523 (1970).
10. Gross, B. and Kosfeld, R.: *Meßtechnik* **77**, 171 (1969).
11. Breuer, H.: *Rev. Sci. Instr.* **36**, 1666 (1965).
12. Tanner, J. E. and Stejskal, E. O.: *J. Chem. Phys.* **49**, 1768 (1968).
13. Kärger, J., Seyd, W., Pfeifer, H., and Geschke, D.: *Bull.* XVI, Coll. Ampere, Bukarest, 1970.
14. Kärger, J.: Dissertation, Leipzig, 1970.
15. Kärger, J.: *Ann. Physik.* 7. Flg. **24**, 1 (1969); **27**, 107 (1971).
16. See e.g. W. M. Lomer and G. G. Low in P. A. Egelstaff (ed.), *Thermal Neutron Scattering*, London, 1965.
17. Zimmerman, J. R. and Brittin, W. E.: *J. Phys. Chem.* **61**, 1328 (1957).
18. Hertz, H. G.: in F. Franks (ed.), *Water, a Comprehensive Treatise*, Vol. 3, Plenum Press New York, London, 1973, p. 355 and 389.
19. Bonera, G. and Rigamonti, A.: *J. Chem. Phys.* **42**, 171 (1965).
20. Zeidler, M. D.: *Ber. Bunsenges. physik. Chem.* **69**, 659 (1965).
21. Von Goldammer, E. and Zeidler, M. D.: *Ber. Bunsenges. physik. Chem.* **73**, 4 (1969).
22. Engelsmann, K., Hertz, H. G., and Zeidler, M. D.: *Z. physik. Chem.*, Frankfurt, in press.
23. Powles, J. G., Rhodes, M., and Strange, J. H.: *Mol. Phys.* **11**, 515 (1966).
24. Torrey, H. C.: *Phys. Rev.* **92**, 962 (1953).
25. Kintzinger, J. P. and Zeidler, M. D.: *Ber. Bunsenges. physik. Chem.* **77**, 98 (1973).
26. Tomlinson, D. J.: *Mol. Phys.* **25**, 735 (1972).
26a. Preissing, G., Noack, F., Kosfeld, R., and Gross, B.: *Z. Physik* **246**, 84 (1971).
27. Harmon, J. F. and Muller, B. H.: *Phys. Rev.* **182**, 400 (1969).
28. Hertz, H. G.: *Ber. Bunsenges. physik. Chem.* **71**, 999 (1967).
29. Zeidler, M. D.: *Ber. Bunsenges. physik. Chem.* **75**, 769 (1971).
30. Sears, V. F.: *Can. J. Phys.* **45**, 237 (1967).
31. Hubbard, P. S.: *Phys. Rev.* **131**, 275 (1963).
32. See e.g. Hertz, H. G.: *Progress in Nuclear Magnetic Resonance Spectroscopy*, Vol. 3, Pergamon Press, 1967, p. 180.
33. Burnett, L. J. and Harmon, J. F.: *J. Chem. Phys.* **57**, 1293 (1972).
34. Pfeifer, H.: *Annal. Physik* 7, Flg. 8, 1 (1961).
35. Pfeifer, H.: *Z. Naturforsch.* **17a**, 279 (1962).
36. Hubbard, P. S.: *Proc. Roy. Soc. London* **A291**, 537 (1966).
37. Hausser, R. and Noack, F.: *Z. Physik* **182**, 93 (1964).
38. Hausser, K. H., Krüger, G. J., and Noack, F.: *Z. Naturforsch.* **20a**, 91 (1965).

39. Hausser, R. and Noack, F.: *Z. Naturforsch.* **20a**, 1668 (1965).
40. Müller-Warmuth, W., Van Steenwinkel, R., and Yalciner, Aytac: *Mol. Phys.* **21**, 449 (1971).
41. Kramer, K. D., Müller-Warmuth, W., and Roth, N.: *Z. Naturforsch.* **20a**, 1391 (1965).
42. Grützediek, H., Müller-Warmuth, W., and Van Steenwinkel, R.: *Z. Naturforsch.* **25a**, 1703 (1970).
43. Kramer, K. D. and Müller-Warmuth, W.: *Z. Naturforsch.* **19a**, 375 (1964).
44. Müller-Warmuth, W., Van Steenwinkel, R., and Noack, F.: *Z. Naturforsch.* **23a**, 506 (1968).
45. Müller-Warmuth, W., Öztekin, Erol, Vilhjalmsson, Reynir, and Yalciner, Aytac: *Z. Naturforsch.* **25a**, 1688 (1970).
46. Günther, K. and Pfeifer, H.: *Zhurn. strukt. Khimi* **5**, 193 (1964); English translation, *J. Struct. Chem.* **5**, 177 (1964).
47. Göller, R., Hertz, H. G., and Tutsch, R.: *Pure Appl. Chemistry* **32**, 149 (1972).
48. Handschamann, W.: Diplomarbeit, Karlsruhe, 1973.
49. Oppenheim, I. and Bloom, M.: *Can. J. Phys.* **39**, 845 (1961).
50. Krynicki, K.: *Physica* **32**, 167 (1966).
51. Hertz, H. G. and Rädle, C.: *Ber. Bunsenges. physik. Chem.* **77**, 521 (1973).

DISCUSSION

Yip: (1) What is the theoretical basis for writing the intermediate scattering function as a sum of exponentials?

(2) It looks like in the past expression you showed for the propagator you have neglected the effects of translation – rotation coupling. Is this correct?

(3) Can you say again what are the precise relations between T_1, T_2 and your propagators?

Hertz: (1) This approach corresponds to the usual method which is applied when we have a superposition of a 'relaxation' and a (chemical) exchange process among various environments.

(2) Yes, the formula regards a model with superposition of independent translational and rotational diffusion.

(3) The propagator enters in the same form, the only difference between $1/T_1$ and $1/T_2$ being that these two quantities are given by different linear combinations of the spectral densities $J(\omega_i)$.

Friedman: I do not agree that one is free to exercise his fancy in constructing the propagator. To be self-consistent, the theory must allow for the fact that the same forces which determine the pair correlation function in the systems you described also contribute to the form of the propagator.

Hertz: I fully agree, and so far it is indeed left to the art of the scientist to use a pair distribution function and a propagator which do not contradict one another too much.

Kneubühl: You mentioned the measurement of propagators of glycerine by NMR-measurements. Glycerine was used for this study because it is a reasonably large molecule and it fits therefore into the gamut of the NMR method. I should like to ask you about the effect of the internal motion of the glycerine on the experimental result?

Hertz: You are right, in the strict sense, one has also to take account of internal motion of the molecule.

Comment by M. R. Hoare

May I just add a note of caution about the interpretation of Gaussian-type long time behaviour in any spatial transport process. If evolution to a Gaussian occurs it will be the result of some form of central-limit theorem reflecting the *large number* of steps which have occurred, not particularly the fact that these are of very small size. One is not justified in concluding that behaviour in any time-regime is described by Brownian-type diffusion with a Langevin friction coefficient.

However, as Prof. Rahman will certainly confirm, there seem to be no microscopic models for which computation can be carried out which do give purely Gaussian behaviour in the Van Hove self-correlation function as time goes to infinity.

NUCLEAR MAGNETIC RESONANCE RELAXATION STUDIES OF REORIENTATIONAL MOTIONS IN LIQUIDS AT HIGH PRESSURE

JIRI JONAS*, J. DEZWAAN, and J. H. CAMPBELL

Dept. of Chemistry, School of Chemical Sciences and Materials Research Laboratory, University of Illinois, Urbana, Ill., 61801, U.S.A.

Abstract. Several recent NMR relaxation studies in liquids at high pressure are discussed to illustrate the advantages of using pressure as an experimental variable in investigations of the dynamic structure of liquids. The first study deals with the reorientation and angular momentum correlation times in fluorobenzene-d_5. The second study follows the effects of pressure on molecular reorientation in liquid chloroform using analysis of both Raman lineshapes in terms of reorientational correlation functions and corresponding NMR relaxation data. The third study is devoted to the problem of pressure effects on the internal motion of groups within a molecule in the liquid state.

Résumé. Plusieurs études récentes des temps de relaxation de la Résonance Magnétique Nucléaire pour des liquides sous haute pression sont présentés pour illustrer les avantages de l'utilisation de la pression comme paramètre expérimental dans les études de la structure dynamique des liquides. Le sujet de la première étude est les temps de corrélation de réorientation moléculaire et de moment angulaire pour le fluorobenzène-d_5. La deuxième étude concerne l'effet de la pression sur la réorientation moléculaire du chloroforme liquide. Nous comparons une analyse de la forme des bandes de Raman en termes des fonctions de corrélation de réorientation avec une analyse basée sur des observations des temps de relaxation de la Résonance Magnétique Nucléaire.

La troisième étude est consacrée au problème de effets de la pression sur le mouvement interne des groupes à l'intérieur d'une molécule à l'état liquide.

1. Introduction

In a great majority of the NMR relaxation studies only temperature has been used as an experimental variable in spite of the fact that the volume changes have a major effect on molecular motions in liquids. If one uses both pressure and temperature as experimental variables in NMR experiments one can then separate the effects of density and temperature on molecular motions in liquids. We discuss several NMR relaxation studies [1] in liquids at high pressure performed recently in our laboratory. The first study [2] dealing with reorientation and angular momentum correlation times in liquid fluorobenzene-d_5 illustrates two aspects of high pressure measurements. Firstly, the experiment carried out at high pressure simply enables one to study the dynamics of molecular motions well above the normal boiling point of a liquid and also permits measurements in supercritical dense fluids. The simultaneous determination of both the reorientation, τ_θ, and angular momentum, τ_J, correlation times in a liquid provides detailed information about the mechanism of molecular reorientation. Secondly, we were able to separate the effect of density and temperature on the correlation times [3]. The purpose of the second study [4] was to follow the effect of

* J. S. Guggenheim Fellow, 1973.

pressure on the molecular reorientation of a symmetric top molecule, chloroform, in the liquid state using both the NMR and Raman techniques. The reorientational correlation times, τ_θ, were calculated from NMR deuteron spin-lattice relaxation time of chloroform-d_1 and compared with those calculated from corrleation functions obtained by the Raman experiment. The reorientational correlation functions were calculated using the method of spectral Fourier deconvolution introduced by Bratos [5]. In this specific example we emphasize the advantages of using the Raman technique to follow the effect of density on the detailed time dependence of the reorientational correlation functions in liquids. It has been shown that applying pressure to a liquid can have a large effect on the rate of molecular reorientation. In connection with our other studies [1] we became also interested in the problem of pressure effects on the internal rotation of groups within a molecule in the liquid state. The effect of pressure on the rate of overall and internal rotation in liquid benzylcyanide is followed in the final study [6].

2. Reorientation and Angular Momentum Correlation Times in Fluorobenzene-d_5

Two useful parameters characterizing a molecule's rotational motion in the liquid state are the angular position correlation time, τ_θ, and the angular momentum correlation time, τ_J. When the molecular reorientation takes place through small angular steps, the Debye rotational diffusion model [7] describes well the reorientation. Gordon [8] removed the restriction of small angular steps and proposed a generalized diffusion model for linear molecules which allows angular steps of arbitrarily large size. Two rotational diffusion processes designated as J-diffusion and M-diffusion were considered. These extended diffusion models have been applied to spherical molecules independently by Fixman and Rider [9], and by McClung [10].

Several authors [11–13] have recently determined the accurate values of τ_θ and τ_J for the same molecule in the liquid state. Maryott et al. [11] studied ClO_3F; Spiess et al. [12] measured $^{13}CS_2$ and Gillen et al. [13] investigated CCl_3F. They found that at low temperatures, Hubbard's relation [14]

$$\tau_\theta \tau_J = \frac{I}{6kT}, \qquad (1)$$

was obeyed, where I is the moment of inertia of the molecule. Hubbard's relation assumes a spherical molecule undergoing small step diffusional motion. At higher temperatures the molecule obeyed McClung's extended J-diffusion model where the rotational step size is allowed to increase arbitrarily. The molecule we chose was fluorobenzene-d_5. Since both the deuteron quadrupole coupling constant [2] is known and the diagonal elements of the spin rotation interaction tensor have been measured by the molecular beam method [15], it is relatively straightforward to obtain τ_θ from the deuteron spin-lattice relaxation data and τ_J from the fluorine spin-lattice relaxation data. Analysis of the relaxation data is given in detail in our original work [2].

The results of the analysis of the experimental relaxation data in fluorobenzene-d_5 at different temperatures and pressures are summarized in Figure 1 which shows the relationship between reduced correlation times τ_θ^* and τ_J^*. These are defined as $\tau_\theta^* = \tau_\theta \varpi$ and $\tau_J^* = \tau_J \varpi$, where $\varpi = (kT/I)^{1/2}$. The theoretical curves for the different models are also given. In Figure 1 Hubbard's equation which was discussed earlier is represented by the straight diagonal line. McClung's extended diffusion models [10] are represented by the two curves labelled M-, and J-diffusion. In this model, τ_J^*, the reduced time between collisions, is used as a parameter. Between collisions the molecules rotate

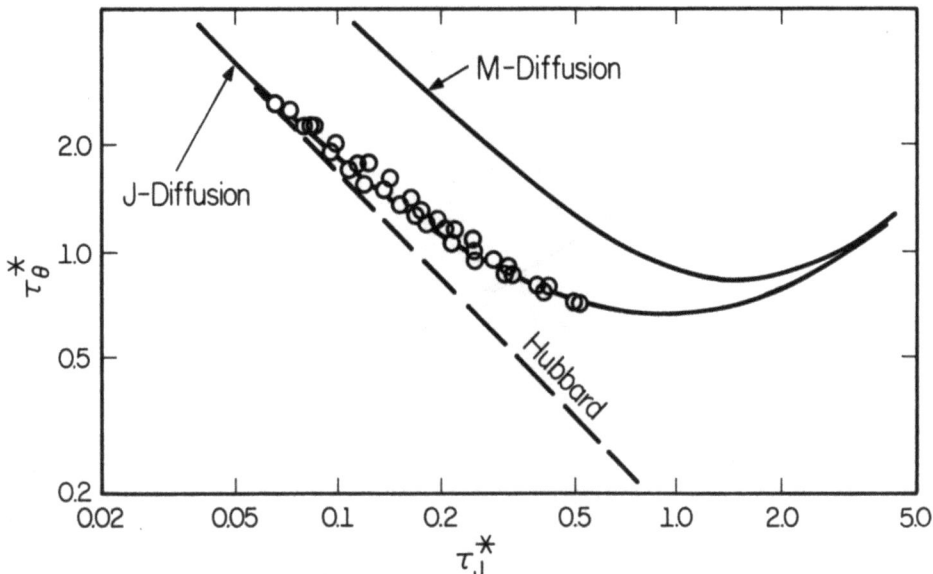

Fig. 1. The experimental dependence of τ_θ^* on τ_J^* in fluorobenzene-d_5 ($C_{\text{eff}} = 3.66$ kHz). Curves predicted by the various models discussed in the text are also shown.

with an angular velocity determined by the Boltzmann factor. M-diffusion means that only the direction of a molecule's angular momentum is randomized by a collision while J-diffusion means that both the direction and magnitude of a molecule's angular momentum is randomized. As expected, the J-diffusion model gives a shorter angular position correlation time for a given collision frequency. Figure 1 shows that the experimental data follow the J-diffusion model.

In the original study [2] of the τ_J, τ_θ relationship, the highest temperature measured was 309 °C and the range of τ_J was limited due to our equipment as it was not possible to accommodate a large change in the sample volume during the experiment. Therefore, we constructed [3] a new experimental setup which permits volume changes by a factor of 7 and also the high temperature limit is approximately 380 °C. We also measured the density of fluorobenzene-d_5 over a wide range of temperatures and pressures and thus we were able to separate the effect of temperature and density on the correlation times. Some of the results dealing with the angular momentum correlation time, τ_J, are given in Figures 2–4. We briefly discuss constant temperature, constant pressure and constant density experiments.

Bartoli and Litovitz [16] have recently used the 'moveable wall' cell model to predict the time between collisions τ_{BC}

$$\tau_{BC} = \tau_J = v^{-1}\left[(V/N_A)^{1/3} - \sigma\right], \tag{2}$$

where V/N_A is the volume per molecule, $(V/N_A)^{1/3}$ is the nearest neighbor distance for

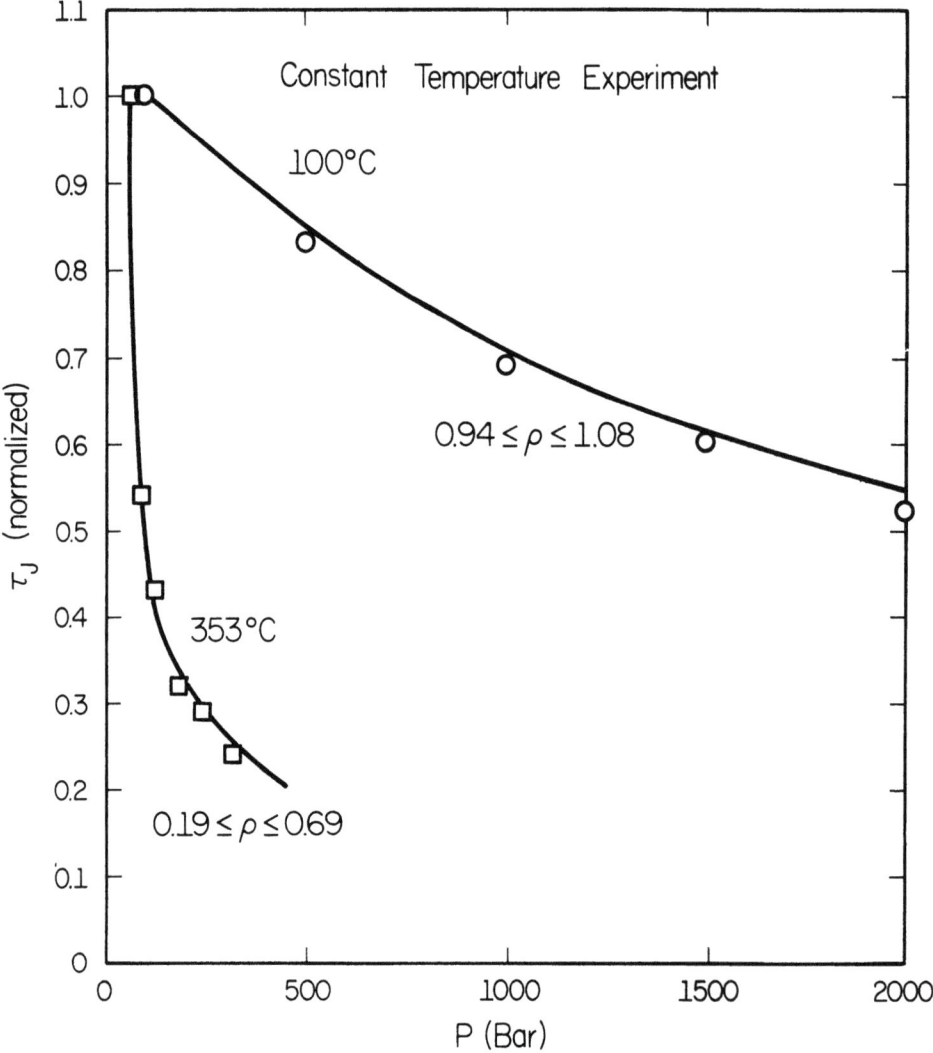

Fig. 2. The experimental and theoretical (Equation (2)) pressure dependence of normalized τ_J at constaot temperature in fluorobenzene-d_5. The density range for $t = 100\,°C$ was $0.94 \sim 1.08$ g cc^{-1}; for $t = 353\,°C$ $0.19 \sim 0.69$ g cc^{-1}.

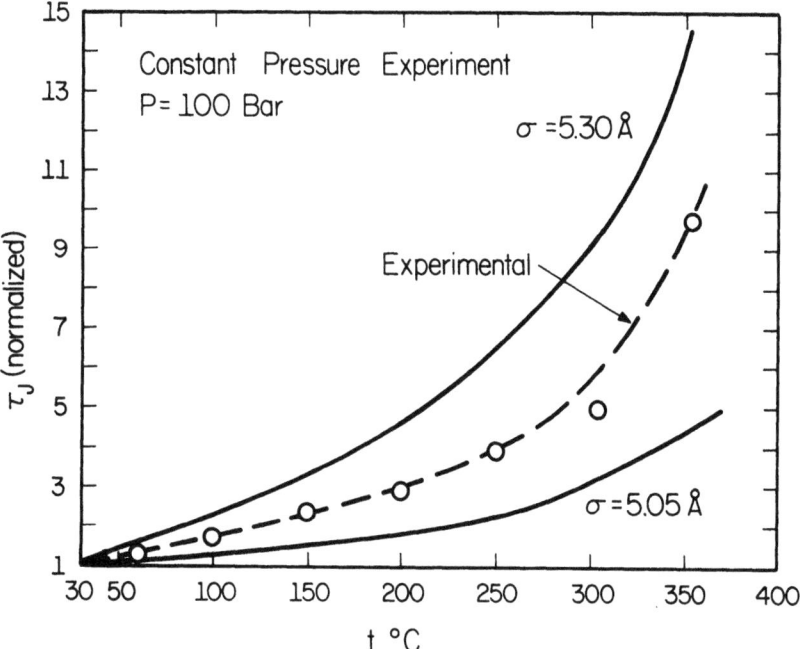

Fig. 3. The experimental and theoretical temperature dependence of normalized τ_J at constant pressure $P = 100$ bar in fluorobenzene-d_5.

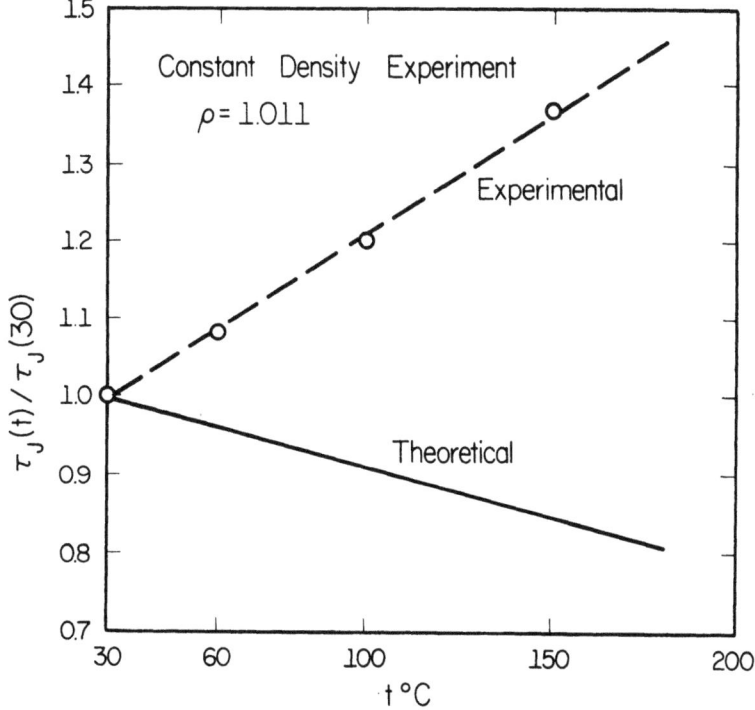

Fig. 4. The experimental and theoretical temperature dependence of normalized τ_J at constant density $\varrho = 1.011$ g cc^{-1}.

a simple cubic lattice, σ is the molecular collision diameter, and $\bar{v} = (8kT/\pi m)^{1/2}$ is the average particle velocity. Using the experimental data for τ_J in liquid CS_2 as obtained by Spiess et al. [12] Bartoli and Litovitz [16] were able to predict τ_J using Equation (2).

In our constant temperature experiment (see Figure 2) we use a normalized τ_J [defined as $\tau_J(P)/\tau_J(50$ bar$)$ or $\tau_J(P)/\tau_J(30$ bar$)$ for temperatures 100 °C and 353 °C, respectively]. The simple reason for using normalized quantities and also the pressure plot (instead of density) was to reduce the number of figures and show the general trend in the τ_J behavior (see also Figures 3, 4). In the case of $t = 353$ °C the density range covered was $0.19 \leqslant \varrho \leqslant 0.69$ and it appears that Equation (2) predicts well the density dependence of the normalized τ_J values for various constant temperatures. In general agreement with Equation (2) we found that the quantity $\tau_J(T)^{1/2}$ varies approximately as $V^{1/3}$ over the temperature range 30–353 °C (critical temperature 287 °C) at constant pressure of 100 bar. The predicted τ_J's differ from the experimental values by a factor of approximately 3 but the calculated value of $\sigma = 5.31$ Å is reasonable. It is quite clear (see Figure 3) that the choice of σ has a large effect on the actual predicted temperature dependence of τ_J and that using a suitable value of σ one could reproduce the experimental data.

The importance of separating the effect of temperature and density is well illustrated in Figure 4 which shows both the experimental temperature depedence of τ_J at constant density $\varrho = 1.011$ g cc^{-1} and also the theoretical prediction using Equation (2). At constant density the experimental angular momentum correlation time, τ_J, increases with temperature whereas Equation (2) predicts the opposite behavior. If one uses the usual simple interpretation of τ_J as time between collisions then to reproduce the experimental data by Equation (2) one has to allow for the temperature dependence of σ. It has been observed by Armstrong and collaborators [17, 18] that in dilute gases the effective cross section for the transfer of angular momentum during collisions decreases with increasing temperature. However, we prefer to obtain more experimental data on fluorobenzene-d_5 and other molecules before we attempt theoretical interpretation of our results. The experimental finding that τ_J increases with temperature at constant density is independent of any theoretical model. The purpose of these measurements was to show the importance of constant density experiments in liquids.

3. Raman and NMR Study of Reorientation in Liquid Chloroform

The pressure dependence of the reorientational correlation function for chloroform [4] has been measured by analysis of the Raman 3019 cm^{-1} A_1 C–H stretching line shape at 1, 1000, and 2000 bar and 23 °C. There reorientational correlation functions were obtained using the method of spectral Fourier deconvolution introduced by Bratos [5]. This method allows for determination of the rotational contribution to the vibrational Raman band in terms of the pure rotational correlation function. Since molecular reorientation contributes only to the depolarized scattering, then the spectral distribution of the polarized part is used as a basis for removing all other effects on the band shape. Details of the analysis are given in our original work [4].

The Fourier transforms calculated in this study were evaluated by direct numerical integration of the corresponding polarized and depolarized band transform integrals for each time point. The deconvolution to obtain the true rotational correlation function was then accomplished by division of the depolarized transform by the polarized transform at each time point. The errors in calculating the reorientational correlation functions were evaluated using standard statistical methods. The results of the calculations are given in Figure 5 which shows the time dependence of the rotational correlation functions at several pressures in liquid $CHCl_3$ together with the predicted short time behavior of the correlation function which was calculated using the Gaussian function introduced by Bratos and Maiechal [5].

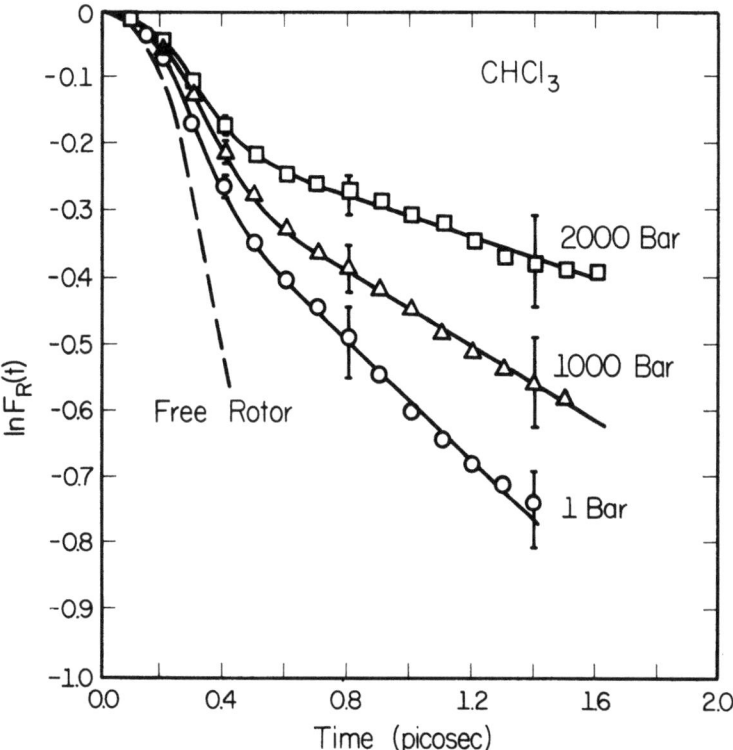

Fig. 5. The reorientational correlation function for chloroform at 23 °C and several high pressures. Theoretical free rotor behavior (---).

It is interesting to note that the chloroform rotational correlation functions at short times (less than 0.5 ps) are only moderately influenced by increasing density whereas the correlation functions at longer times experience a much greater change in the rate of decay with increasing pressure. Using the Raman frequency moments computed by Gordon [19] for a symmetric top we were able to express the short time behavior of the Raman rotational correlation function and estimate the molecular torques (see original work [4].

The Raman reorientational correlation times, τ_θ, have been calculated numerically from the Raman correlation functions and are listed in Table I. The integration was carried out after first extrapolating the correlation functions exponentially to zero at large times. These τ_θ's were compared with the NMR reorientation correlation times calculated from the deuteron spin-lattice relaxation times measured in liquid chloroform-d_1 under the same experimental conditions.

Assuming the rotational diffusion mechanisms, a theoretical τ_θ value has been calculated using the microviscosity approach of Gierer and Wirtz [20] and Bridgeman's [21] high pressure viscosity results for $CHCl_3$. These results are also given in Table I. Although the $\tau_{\theta\,\text{theoretical}}$ values are approximately 1.3 times larger than the $\tau_{\theta\,(\text{Raman})}$

TABLE I

Reorientational correlation times

Pressure (bar)	$\tau_\theta \times 10^{12}$ (s)		
	Theoret.	Expt. (Raman) (23 °C)	Expt. (NMR) (25 °C)
1	2.60	1.97	1.74
1000	4.26	3.14	2.60
2000	6.36	5.42	3.10

values, they both show approximately the same pressure dependence. This behavior was not observed in the NMR times.

The values of τ_θ calculated from Raman and NMR show good agreement at 1 bar, however, they appear to deviate substantially at higher pressures. Changes in the deuteron quadrupole coupling constant with density could cause such discrepancies. However, whether the well known self-association in chloroform does cause a change in the deuteron quadrupole coupling constant with increasing density remains to be determined experimentally.

4. Overall and Internal Rotation in Liquid Benzylcyanide

In this study [6] the deuterium spin-lattice relaxation times of the two selectively deuterated benzylcyanides, benzylcyanide-4-d_1(I) and α, α-dideuterobenzylcyanide(II) were measured as a function of pressure up to 3 kbar and 30° and 150 °C. The relaxation time of (I) gives a measure of the overall reorientation of the molecule, while the relaxation time of (II) reflects both the overall molecular reorientation and the internal rotation of the CD_2CN group. The choice of benzylcyanide was based on a study by Wallach [22] in which it has been shown that the internal motion of CD_2CN in methylbenzylcyanide proceeds at a rate comparable to the overall molecular motion and, thus, may be described by a rotational diffusion model. Using reported deuteron quadrupole coupling constants, the experimental spin-lattice relaxation time, T_1, can be analyzed in terms of the rotational diffusion constants for overall molecular, D_M, and internal, D_I, rotation.

The deuteron quadrupole coupling constants of 180 kHz were used for the ring and CD_2CN deuterons, respectively. In both cases the asymmetry parameter was neglected. The details of the analysis of the experimental relaxation data are again given in the original work [6]. In addition to relaxation measurements the pressure dependence of viscosity has also been determined. This enabled us to calculate the theoretical rotational diffusion constant $(D_M)_{thoer}$ for overall molecular rotation using the microviscosity approach [20] by Gierer and Wirtz [21]. These theoretical rotational diffusion constants are in reasonably good agreement with the experimental values as is shown in Figure 6. This supports the validity of our assumption that the overall

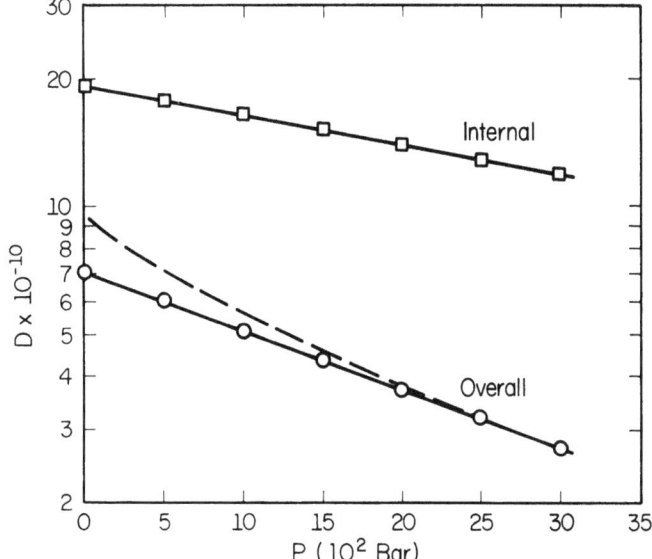

Fig. 6. The pressure dependence of the rotational diffusion constants $D_M(\bigcirc)$, $D_I(\square)$ and $(D_M)_{theor}$ (---) in liquid benzylcyanide at 150 °C.

molecular reorientation of benzylcyanide can be regarded as a small-step diffusion process.

In our earlier study [23] we have introduced the difference between the activation volumes for experimental quantity reflecting the reorientational motion and η/T as a qualitative measure of the degree of coupling between the rotational and translational motion. The value of $\Delta V^{\neq}$ for $D_M [\Delta V^{\neq}(D_M) = -RT (\partial \ln D_M/\partial P)_T]$ and D_I are the same at 30 °C $\Delta V^{\neq}(D_M) = \Delta V^{\neq}(D_I) = 14.4$ cc mole^{-1} and quite close to the value of $\Delta V^{\neq}$ for η/T; $\Delta V^{\neq}(\eta/T) = 16.2$ cc mole^{-1} $(\Delta V^{\neq}(\eta/T) = RT [\partial \ln (\eta/T)/\partial P]_T)$. This reflects a strong coupling between the rotational and translational motion of the benzylcyanide molecule. This finding also indicates that the assumption of a simple rotational diffusion is reasonable at least at 30 °C. However, at 150 °C the $\Delta V^{\neq}(D_M) = 11.2$ mole cc^{-1} whereas the $\Delta V^{\neq}(D_I)$ is only 5.5 cc mole^{-1} (see Figure 6). This is analogous to our earlier finding [24] in toluene-d_8 where at 100 °C the $\Delta V^{\neq}(D_M)$ was three times as large as $\Delta V^{\neq}(D_I)$. This result reflects the difference in frictional torques connected with the overall and internal rotation and also indicates that inertial effects influence the internal rotation at higher temperatures. The assumption of rotational diffusion is no longer adequate for describing the internal rotation of the CD_2CN group at higher temperatures.

Acknowledgements

The work discussed in this paper was partially supported by the U.S. Air Force Office of Scientific Research under Contract AF 72-2286 and by the National Science Foundation under Grant 28268X.

References

1. Jonas, J.: *Adv. Magn. Resonance* **6**, 73 (1973).
2. Assink, R. A. and Jonas, J.: *J. Chem. Phys.* **57**, 3329 (1972).
3. DeZwaan, J. and Jonas, J.: unpublished results.
4. Campbell, J. H. and Jonas, J.: *Chem. Phys. Letters.* **18**, 441 (1973).
5. Bratos, S. and Marechal, E.: *Phys. Rev.* **A4**, 1078 (1971).
6. DeZwaan, J. and Jonas, J.: *J. Phys. Chem.* **77**, 1768 (1973).
7. Debye, P.: *Polar Molecules* Dover, New York, 1928.
8. Gordon, R. G.: *J. Chem. Phys.* **44**, 1830 (1966).
9. Fixman, M. and Rider, K.: *J. Chem. Phys.* **51**, 2425 (1969).
10. McClung, R. E. D.: *J. Chem. Phys.* **51**, 3842 (1969).
11. Maryott, A. A., Farrar, T. C., and Malmberg, M. S.: *K. Chem. Phys.* **54**, 64 (1971).
12. Spiess, H. W., Schweitzer, D., Haeberlen, U., and Hausser, K. H.: *J. Magn. Resonance* **5**, 101 (1971).
13. Gillen, K. T., Douglass, D. C., Malmberg, M. S., and Maryott, A. A.: *J. Chem. Phys.* **57**, 5170 (1972).
14. Hubbard, P. S.: *Phys. Rev.* **131**, 1155 (1963).
15. Chan, S. I. and Dubin, A. S.: *J. Chem. Phys.* **46**, 1745 (1967).
16. Bartoli, F. J. and Litovitz, T. A.: *J. Chem. Phys.* **56**, 413 (1972).
17. Tward, E. and Armstrong, R. L.: *J. Chem. Phys.* **47**, 4068 (1967).
18. Armstrong, R. L. and Hanrahan, T. A. J.: *J. Chem. Phys.* **49**, 4777 (1968).
19. Gordon, R. G.: *J. Chem. Phys.* **40**, 1973 (1964).
20. Gierer, A. and Wirtz, K.: *Z. Naturforsch* **89**, 532 (1953).
21. Bridgeman, P. W.: *Proc. Am. Acad. Arts Sci.* **61**, 57 (1926).
22. Wallach, D.: *J. Chem. Phys.* **47**, 5258 (1967).
23. Assink, R. A., DeZwaan, J., and Jonas, J.: *J. Chem. Phys.* **56**, 4975 (1972).
24. Wilbur, D. and Jonas, J.: *J. Chem. Phys.* **55**, 5840 (1971).

DISCUSSION

Litovitz: I must agree that the assumption of Bartoli and myself that $\tau_J \propto \tau_{BC}$ is probably not true for many (if not most) liquids. I would however like to emphasize that when one plots τ_J vs $[(V/N)^{1/3} - 6]$ at constant density one is not testing the cell model which purports to predict τ_{BC} but rather one tests whether or not τ_J is proportional to τ_{BC}. I mean to suggest that the cell model might still work for τ_{BC} even if the above plot at constant density behaves 'uncorrectly'.

Jonas: The purpose of our experiments at constant density was twofold. Firstly, to determine the temperature dependence of the angular momentum correlation time, τ_J, at constant density, secondly, in a more general sense, to show on the example of the simple relationship discussed that constant density experiments can provide decisive evidence on the validity of some theoretical model. The important point is that such evidence is not available from measurements at constant pressure. The simple relationship for $\tau_J \simeq \tau_{BC}$ discussed in our communication provided a good example in this sense. At present time we do not attempt to prove or disprove the general validity of the cell model. We simply find that τ_J, which is the integral of the angular momentum correlation function, increases with temperature at constant density.

Friedman: Since this morning there was some mention of the question of the meaning of the activation energies of these relaxation processes, and since you have again referred to this question, it seems interesting to observe that from your results we learn that the process whose role is τ_J^{-1} has a negative activation energy. Perhaps this is sufficient to show that we need not discuss the physical significance of these parameters.

Jonas: I am glad to hear your comment which also reflects my view about the physical significance of activated process describing the molecular motions in liquids.

Gershon: (1) What is the error in assuming in NMR for the formula of $1/T_1$ that the orientational correlation function is exponential over the whole time range?

(2) In relation to the time between collisions mentioned in the lecture, what is a collision in a liquid?

Jonas: (1) I believe that the formulas for $1/T_1$ are all right if one takes τ_θ as the area under the correlation function. Similarly, we calculated the Raman correlation times from the experimental correlation functions by integrating the area. Of course, we assumed exponential decay of the functions at long times.

(2) Your comment on the physical meaning of a collision in a dense liquid poses a difficult question. At this point, I leave the answer to the theoreticians in this audience and particularly to those involved in molecular dynamics calculations.

Papoular: What is your general expectation as to the behavior of rotational relaxation times near a critical point? In particular, if you start out with a dense, rotationally hindered liquid, should you, or not, expect a more or less spectacular variation of τ near T_c?

Jonas: First, I would like to point out that any serious attempt to study the behavior of τ_θ near the critical point will require very different experimental set-up than we use. My opinion is that one will not find any large variation in τ_θ near T_c. At this point, I am unable to elaborate on the theoretical reasons for my statement but I may add that experimental evidence so far available in the literature supports my view.

Brot: Any measurement under constant volume condition is very interesting for obvious theoretical reason. Can you comment on the Arrhenius plots of $\tau_{\theta 2}$ under this condition? Were they reasonably linear and what was the experimental preexponential factor? It is expected that this factor should show more clear significance than in the usual constant pressure experiments (Brot, C.: *Chem. Phys. Letters* 3, 319, 1969). I refer here, of course, to cases where activated processes, not simple *J*-diffusion, occur.

Jonas: Our experiments on several molecules show non-Arrhenius behaviour of $\tau_{\theta 2}$ under constant volume condition. If one still goes ahead and calculates the 'activation' energies one obtains values close to kT. I do not recall the value of the experimental preexponential factor but I shall send you later a preprint of our paper on bromobenzene-d_5.

Rothschild: Have you checked whether, in your experiments on chloroform, you obtain the best literature value of the degree of depolarization of the band? I think this is very important when one attempts to compare Raman with nuclear magnetic relaxation correlation times.

Jonas: Yes, we devoted a lot of attention to this problem and found the value of the degree of depolarization in agreement with the literature value. When we started our experiments we tried to use sapphire windows in our high-pressure Raman cell but found this unsatisfactory due to scrambling of the polarization of the scattered and incident light. The use of fused quartz windows solved this problem.

Griffiths: Did I understand you to say that you used sapphire windows in your pressure cell.
Jonas: No.

ELECTRON SPIN RELAXATION IN THE STUDY OF THE MOTIONS OF PARAMAGNETIC PROBES IN LIQUIDS

G. MARTINI, M. ROMANELLI, and L. BURLAMACCHI

Institute of Physical Chemistry, University of Florence, Via G.Capponi 9, 50121 Florence, Italy

Abstract. The dynamics of the solvation sphere of several Mn(II) and Fe(III) complexes have been investigated with the Electron Spin Resonance (ESR) technique. The correlation times for the fluctuation of the Zero-Field Splitting components have been derived from the temperature and frequency dependence (X-, Q-, and S-band) of the electron spin relaxation time. Comparison between computed and experimental spectra suggests that the molecular motions in the second solvation sphere is slower than in the bulk solvent.

Résumé. La dynamique de la sphère de solvatation de quelques complexes de Mn(II) et Fe(III) a été étudiée par la technique de Resonance Paramagnetique Electronique (RPE). Les temps de corrélation pour la variation des composantes du Zero-Field Splitting ont été dérivées de la dependance du temps de relaxation du spin électronique en fonction de la température et du champ magnetique (bands X, Q, et S). Une comparaison entre le spectre calculés et ceux expérementaux suggére que le mouvements moléculaire dans le deuxième couche de solvatation sont plus lents que dans le solvent massif.

1. Introduction

Ions in the ^6S spin state ($3d^5$) occupy a particular position as paramagnetic probes for structural and dynamic studies in the field of Electron Spin Resonance (ESR). Their essential feature is that the g-factor and the hyperfine coupling of a solvated ion are essentially isotropic while the spin-spin intramolecular interaction, the so-called zero-field splitting (ZFS), is a second rank traceless tensor [1-3]. Its anisotropy is zero for a perfectly cubic crystal field on the ion while it increases with increasing deviation from the cubic symmetry of the ionic environment. In solution the anisotropy is normally averaged out by rapid rotation, however, the resulting modulation of the spin levels induces spin relaxation with consequent line broadening (see below).

Lineshape studies thus yield combined information about the ionic structure and its time dependence. With respect to the widely used $S=\frac{1}{2}$ compounds (vanadyl complexes, ClO_2, etc), which posses anisotropic g-factors, hyperfine coupling and spin-rotational coupling, all of which are almost unaffected by solvation, the $S=\frac{5}{2}$ ions appear interesting for solute-solvent interaction studies.

In this report, an investigation on the solvation sphere of Mn(II) and Fe(III) complexes by ESR is presented. The usual quantum-mechanical approach is used to evaluate the transition fields, and the general technique of the Redfield relaxation matrix [4] is employed to calculate the ESR linewidth. Most of the information is derived from comparison between the calculated and experimental spectra for three operation frequencies (S-, X-, and Q-band).

Although the general treatment could in principle be extended to Gd(III) ($4f^7$) and to Cr(III) ($3d^3$), we do not treat this argument here.

2. Experimental

Manganous solutions were prepared by dissolving anhydrous Mn(II) perchlorate in appropriate solvents or in a solvent mixture. $FeCl_4^-$ solution in *tris-n-butyl-phosphate* (TBP) was obtained from $FeCl_3$ with addition of dried LiCl up to saturation.

X-band (9200 Mc s^{-1}) ESR spectra were registered with a Varian V 4502 spectrometer. The S-band spectra were obtained by using a microwave bridge operating at 2950 Mc s^{-1}. Q-band spectra (34350 Mc s^{-1}) were carried out with a Varian V 4651 microwave bridge, courtesy of the Institute of Physics, University of Parma. With the use of the appropriate dewar cell, the V 4557 variable temperature assembly was used for temperature changes at all operational frequencies. The external magnetic field was calibrated by using a Varian Model F 8 nuclear fluxmeter from 1000 G upward. From 0 to 1000 G, a linear scanning from the Varian Mark I Fieledial was assumed.

3. Lineshape Analysis

The complete lineshape for a theoretical spectrum is interpreted in terms of the spin Hamiltonian:

$$\mathcal{H} = g\beta \mathbf{H}\cdot\mathbf{S} + A\mathbf{I}\cdot\mathbf{S} + \mathbf{S}_i\cdot\mathbf{D}\cdot\mathbf{S}_j. \tag{1}$$

The first two terms are almost isotropic and give the line positions. The third term whose explicit form is given by:

$$\mathcal{H}_{ZFS} = D[S_z^2 - S(S+1)/3] + E(S_x^2 - S_y^2) \tag{2}$$

is the only anisotropic term which contributes to the line broadening. The widths and the intensities of the five spin transitions for a ^6S spin state ion are given by the eigenvalues and the eigenvectors of the relaxation matrix (transverse relaxation) [3]:

$$-(D:D)/5 \begin{array}{c} \\ \langle 2| \\ \langle 1| \\ \langle 0| \\ \langle -1| \\ \langle -2| \end{array} \begin{array}{|ccccc|} |2\rangle & |1\rangle & |0\rangle & |-1\rangle & |-2\rangle \\ \hline A & D & E & 0 & 0 \\ D & B & 0 & F & 0 \\ E & 0 & C & 0 & E \\ 0 & F & 0 & B & D \\ 0 & 0 & E & D & A \end{array} \tag{3}$$

with

$$A = (24J_0 + 48J_1 + 28J_2)$$
$$B = (6J_0 + 36J_1 + 46J_2)$$
$$C = (16J_1 + 56J_2) \tag{4}$$
$$D = -(8\sqrt{10}\,J_1)$$
$$E = -(12\sqrt{5}\,J_2)$$
$$F = -36J_2$$

and

$$J_n = \frac{\tau_c}{1 + n^2\omega_0^2\tau_c^2}. \tag{5}$$

Here $(D:D) = \frac{2}{3}D^2 + 2E^2$ is the inner product of the ZFS tensor, including axial (D) and rhombic (E) parameters, ω_0 is the angular frequency of the radiation field, τ_c is the correlation time for the modulation, and the $|2\rangle$, $|1\rangle$... stand for $m_s = \frac{5}{2} \rightleftharpoons \frac{3}{2}$, $\frac{3}{2} \rightleftharpoons \frac{1}{2}$ ... transitions.

There is no direct correspondence of the linewidth against τ_c. It is found that, with good approximation, the linewidth is dominated by the $|0\rangle$ transition, whose matrix elements contain only J_n $(n \neq 0)$ spectral densities. The linewidth is then expected to pass through a maximum close to $\tau_c = 1/\omega_0$. The expected linewidths as a function of τ_c at the Q-, X-, and S-bands are shown in Figure 1. It is noteworthy that when $\omega_0^2 \tau_c^2 \ll 1$, the linewidth would be expected to be independent of the frequency and linearly dependent on τ_c. The experimental determination of the point of maximum linewidth at various frequencies allows the determination of τ_c as well as its temperature dependence [5].

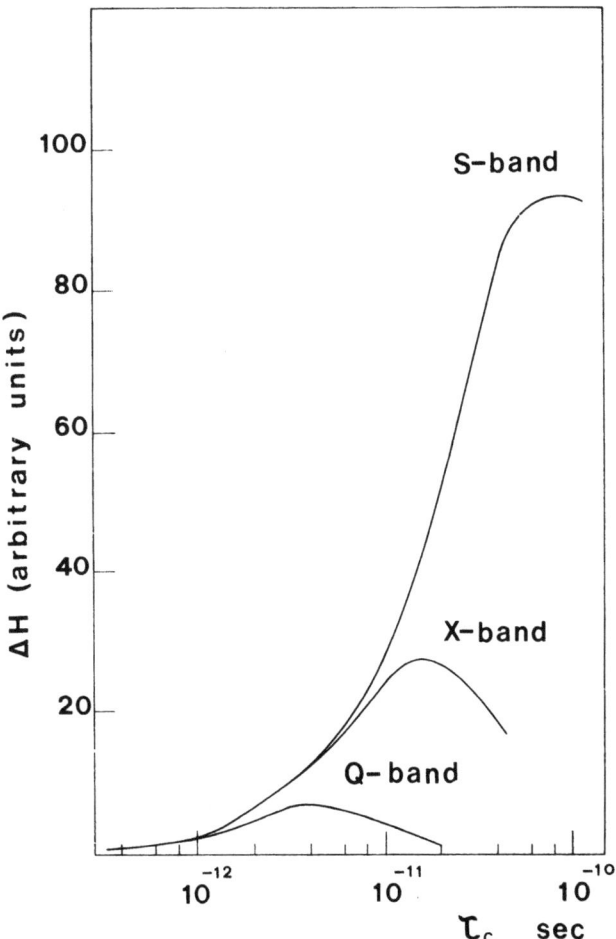

Fig. 1. Theoretical linewidths for a ^6S ion as a function of τ_c at 34,350 Mc s^{-1} (Q-band), 9200 Mc s^{-1} (X-band), and 2950 Mc s^{-1} (S-band).

For Mn(II) one must also account for the coupling of the electron spin with the nuclear spin of ^{55}Mn ($I=\frac{5}{2}$, $A\sim 95$ G). This has an important effect in two particular cases:

(i) if $A/\hbar\omega_0$ is not negligible, the degeneracy of the fine splitting transitions within each hyperfine line is removed by non-secular terms in the spin Hamiltonian. Since the resulting inhomogeneous broadening, together with overlapping of adjacent hyperfine components affect the observed linewidth, the problem of finding the true linewidth was resolved by computing the theoretical lineshape at the frequencies of interest and plotting the intrinsic/observed linewidths [6]. To this end, the line position were calculated to second order from the Hamiltonian (1). In order to overcome the difficulties due to the non-degenerate spin levels in the Redfield matrix, a complex Hermitian matrix, which includes the Redfield matrix and an imaginary contribution from the resonance frequency, is inverted at each point of the spectrum [7]. An example is shown in Figure 2, which reports the experimental and calculated S-band spectrum of Mn(II) in aqueous solution.

(ii) If $\omega_0^2\tau_c^2 > 1$, the $|\pm 1\rangle$ and $|\pm 2\rangle$ transitions are broadened out and second order effects in the remaining $|0\rangle$ transition affect the linewidth with the term [8]:

$$T_2^{-1} = \frac{(D:D)}{5}\left\{16 J_1 + 56 J_2 + 64\left(\frac{A}{\omega_0}\right)^2 [I(I+1) - m_I^2] J_0\right\}. \qquad (6)$$

Thus, as opposed to the fast motion narrowed spectrum, the central hyperfine lines become broader than the outer ones. The appearance of this inversion in the linewidth dependence of the hyperfine components is sometimes helpful in the evaluation of τ_c when unaveraged ZFS anisotropies complicate the lineshape.

Since the first coordination shell around the ion is actually octahedral (or tetrahedral for $FeCl_4^-$) with cubic symmetry, we must allow for a dynamic distortion from cubic symmetry to account for the ESR lineshape. It is well documented that unsymmetrical coordination in the first solvation sphere would induce ZFS too large to account for the observed linewidth, while the rate of ligand exchange ($\sim 10^7$ s^{-1}) is normally to slow to influence the lineshape [9]. Solvent fluctuations around the first coordination sphere are expected to induce ZFS values of the correct order of magnitude. Thus, the dynamics of the ZFS tensor may be associated with the ligand dynamics in the second solvation sphere. We will henceforth indicate by τ_f the mean lifetime of a distorted state, which, in principle, should correspond to the reciprocal of the rate of ligand exchange into the second solvation sphere.

τ_c, the correlation time for the electron spin relaxation, may be identified either with τ_f or with the re-orientational correlation time for Brownian motion [3, 10], τ_R, since both modulate the ZFS components along the quantization axis. Experimental results on Gd(III) complexes [11] show that the rotational correlation time described by the modified Debye expression [12]

$$\tau_R = \frac{4\pi\eta\kappa a^3}{3kT},$$

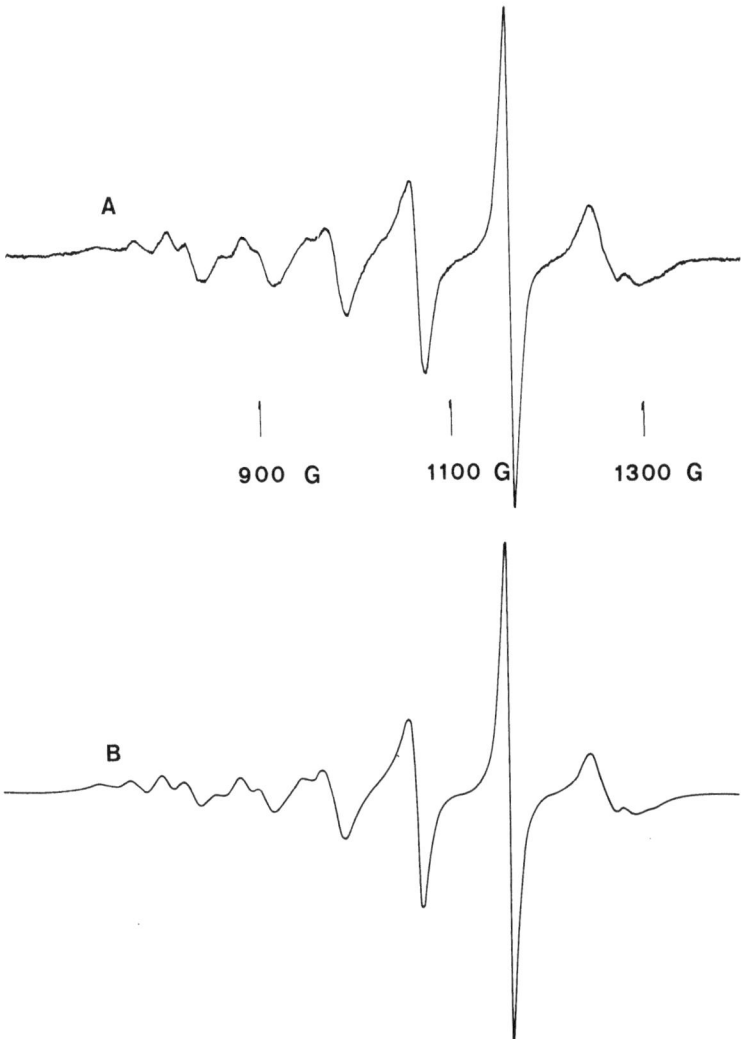

Fig. 2. Experimental (A) and computed (B) ESR spectrum of Mn(II)-water solution at S-band ($T=115\,°C$).

(where a^3 is the structural radius of the complex, and κ is an empirical factor) is of the correct order of magnitude to dominate the relaxation process. The observed dependence of τ_c on the size of the complex provides further support for the rotational model [5, 11]. This 'complex' dependence of τ_c could also have been attributed to changes in the dynamics of the environment. However, in the extreme case in which manganous ions are bound to very large molecules (macromolecules) in solution, even if these ions bind a large number of solvent molecules, a number of ESR spectra closely resemble the glassy patterns characterized by a wide distribution of completely *static* ZFS sites [13].

If $\tau_c = \tau_R$, then we have also $\tau_f > \tau_R$, that is, the rate of fluctuation is slower than the rate of reorientation. It appears that only limited information is available on the dynamics in the outer-solvation sphere, which however, would seem to show that the rate of ligand exchange may be slower than the usually quoted encounter lifetime in the bulk solvent [14]. This, which appears to be a very plausible assumption, further supports the rotational model. It has also been proposed that the modulation of the ZFS components could arise from collisions with solvent molecules not directly interacting with the ionic system (i.e. from the bulk solvent) [2]. If τ_c were dominated by this perturbation, then it should depend only on the nature of the solvent and not on the particular complex, which is not verified experimentally.

4. Results and Discussion

Figure 3a reports the linewidth as a function of the temperature, at the three operational frequencies, for 0.1 M $FeCl_4^-$ in TBP. The linewidth passes through a maximum both in the X-, and in the Q-bands. This provides an excellent test of the temperature dependence of τ_c. The calculated product $\eta\omega_0/T$ at the maximum linewidth ($\eta = 17.5$ cP and $\eta = 5.2$ cP at $T = -25\,°C$ and $T = 10\,°C$ respectively) is constant in the limit of the experimental errors, and agrees with the linear dependence predicted for τ_c. In the S-band, maximum linewidth is predicted around $-50\,°C$. However, below

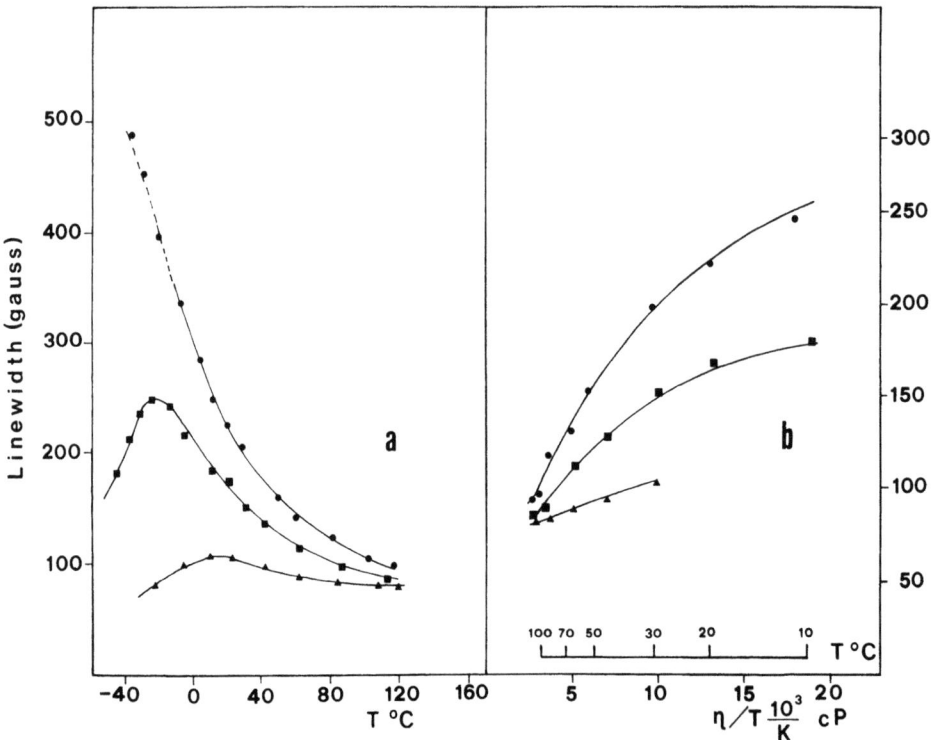

Fig. 3. Linewidth of 0.1 M $FeCl_4^-$ in TBP: (a) as a function of temperature; (b) as a function of η/T. ● – S-band; ■ - X-band; ▲ – Q-band

−15 °C, the lineshape is not symmetrical and does not allow an exact evaluation of the width. Figure 4 shows the observed S-band spectra of $FeCl_4^-$ in TBP in the range −140−+20 °C. The analysis of the glassy spectrum at X-band for this system has been previously reported [15]. The appearance of low field transitions ($g_{eff}=4.3$ and 6) in the S-band spectrum (Figure 4a) confirms the existence of a wide distribution of unaveraged ZFS sites. The combined analysis at X-, and S-band shows that the ZFS sites are mostly in the range from 500 to 2000 G, with only a small fraction off these limits.

In a rapidly tumbling system the anisotropy arising from a traceless interaction is averaged out when the variation of the line position $|\delta\omega|$ times the correlation time for the motion τ_c is $\ll 1$. With increasing temperature, the distribution of the sites is averaged out by fluctuations of the ZFS magnitude and the anisotropy of a single site is averaged out by rotation and/or fluctuations. At −40 °C (spectrum c of Figure 4) the almost equal distribution of sites as in the glassy state is maintained. At −15 °C ($\eta=10.5$ cP, $\tau_c=1\times 10^{-11}$ s) the spectrum is still unsymmetrical, which means that the distribution is still only partially averaged. Provided $\tau_c=\tau_f$, then the rate of fluctuation of the ZFS sites is $\sim 10^{11}$ s^{-1}. However, it is difficult to believe that the electron spin, whose relaxation time T_2 is $\sim 10^{-10}$ s, can distinguish among sites whose lifetime is $\sim 10^{-11}$ s.

At higher temperatures, the ZFS anisotropy over each ion is completely averaged and the spectrum is indeed symmetrical. The assumption of slow fluctuations now imposes that the spectrum consists of a sum of individual spectra with different widths. Unfortunately, the detection of superimposed spectra, especially in the region of partial averaging, is not as easy as it might seen. Computer analysis shows that the linewidth is dominated by the narrowest fraction, while slightly broader spectra contribute only in broading the wings and spectra with linewidth greater than a factor of three (with comparable intensity) practically do not affect the lineshape. Thus the lineshape, which obviously depends on the distribution function of the ZFS sites, closely resembles the 'pure' theoretical spectrum in the center, being broader in the wings. For $FeCl_4^-$ in TBP the line is lorentzian at 0 °C. Most probably the distribution is completely averaged at higher temperatures. The condition of averaging of degenerate spectra with different widths is that $T_2 \approx \tau_f$ [16], which gives $\tau_f \sim 10^{-10}$ s at 0 °C.

The mean value of the ZFS in solution is expected to be susceptible to change with temperature. This is better observed in Figure 3b, which shows the ESR linewidth as a function of η/T at the different microwave frequencies. From theory, assuming a constant $(D:D)$ value, the curves should be linear from the origin with slopes proportional to $(D:D)$ and independent of the frequency. Experimentally, the curves are not linear and largely field dependent. The slope increases with increasing temperature. Since no other relaxation processes, which are effective in line broadening with increasing temperature, can account for this system [17], it seems safe to assume that $(D:D)$ increases with the temperature.

There is no indication at this time about the field dependence, which is not predicted from general theory. Since D is of the order of magnitude of the Zeeman term,

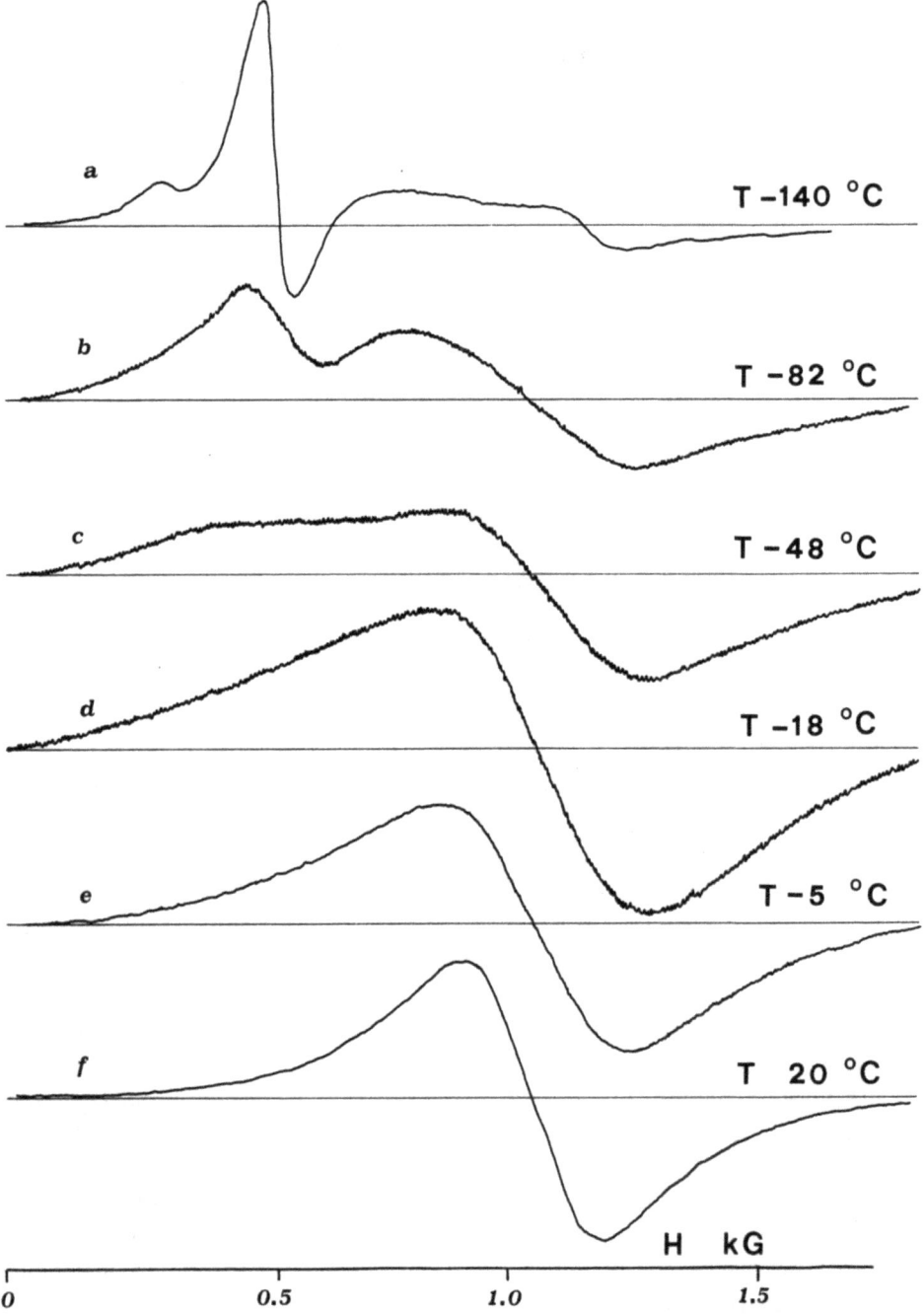

Fig. 4. S-band spectra of $FeCl_4^-$-TBP at various temperatures. Spectrum (e) has relative spectrometer gain = 2; spectra (b), (c), and (d) have relative spectrometer gain = 4.

at least in the S-band, it is probable that admixing of the two terms in the time dependent spin Hamiltonian should influence the total anisotropy of the spin levels. The important point, however, is that τ_c cannot be determined by simply fitting the theoretical frequency dependence at a single temperature.

The behaviour of the manganous ion in many solvents is very much the same as the $FeCl_4^-$-TBP system. Figure 5 a and b show the ESR linewidth as a function of temperature and η/T in aqueous solution. Unfortunately, the maximum linewidth is not observed because of the obvious limitation of the temperature range available. In methanol, Figure 6a and b, the maximum is clearly observed at X-band. Assuming linear dependence of τ_c on η/T, at Q-band the maximum is predicted at $\sim 10\,°C$. Indeed below this temperature, the spectrum shows the inversion of the hyperfine components predicted by Equation (6), which ensures that $\omega_0^2 \tau_c^2 > 1$. However, the linewidth increases with decreasing temperature, instead of decreasing. Indeed, for this system, the anisotropies of the single ZFS terms are expected to be much smaller than in $FeCl_4^-$-TBP, while τ_c in the range from 10 to $-20\,°C$ is between 4.6×10^{-12} and 9×10^{-12} s. Thus all anisotropies are certainly averaged out and the distribution of the cyrstal-field sites is expected to lead to degenerate spectra centered at $g_{eff} = 2$. Detailed analysis of the lineshape shows that the wings are indeed broader than predicted. Both the unexpected linewidth increase and the broadening of the wings can be attributed to superposition of spectra with different widths.

Spectra registered in mixed water-acetonitrile solution provide additional evidence in support of the proposed model. The general feature of the Mn(II) spectra in mixed solvents have been studied in detail [6]. It can be shown that in 5% water, the first coordination shell is built up of only water molecules, while presumably the second coordination sphere admixes acetonitrile and water molecules with progressively lower concentration of water with increasing distance from the central ion. The mixed coordination is expected to induce larger ZFS distribution. The linewidth behaviour reported in Figure 7a and b shows that the field dependence is larger as compared with pure liquid solution. The Q-band spectrum at $-15\,°C$ deviates appreciably from the theoretical lineshape and pronounced broadening in the wings of the hyperfine components is observed. At this temperature $\tau_c \simeq 6 \times 10^{-12}$ s is certainly able to average out all anysotropies, and we must assume $\tau_f \sim 10^{-9}$–10^{-10} s, in order to observe a ZFS distribution.

So far we have analyzed the Q-band spectra at low temperature, from which we derive that τ_f should be $\sim 10^{-9}$ s, a rate which is certainly slower than the collisional lifetime in the bulk solvent. At this temperature, X-band spectra are generally too broad to give accurate results.

At higher temperatures, the superposition of degenerate spectra is not easily detectable since the resulting lineshape tends to approach the single $(D:D)$ valued theoretical spectrum. Slight deviations are still observed at room temperature in the Q-band. At X-band, the relative heights of the hyperfine components deviate substantially from theory [6]. S-band spectra also deviate strongly but the theoretical spectra are extremely sensitive to small uncertainty in the line positions. Only at very

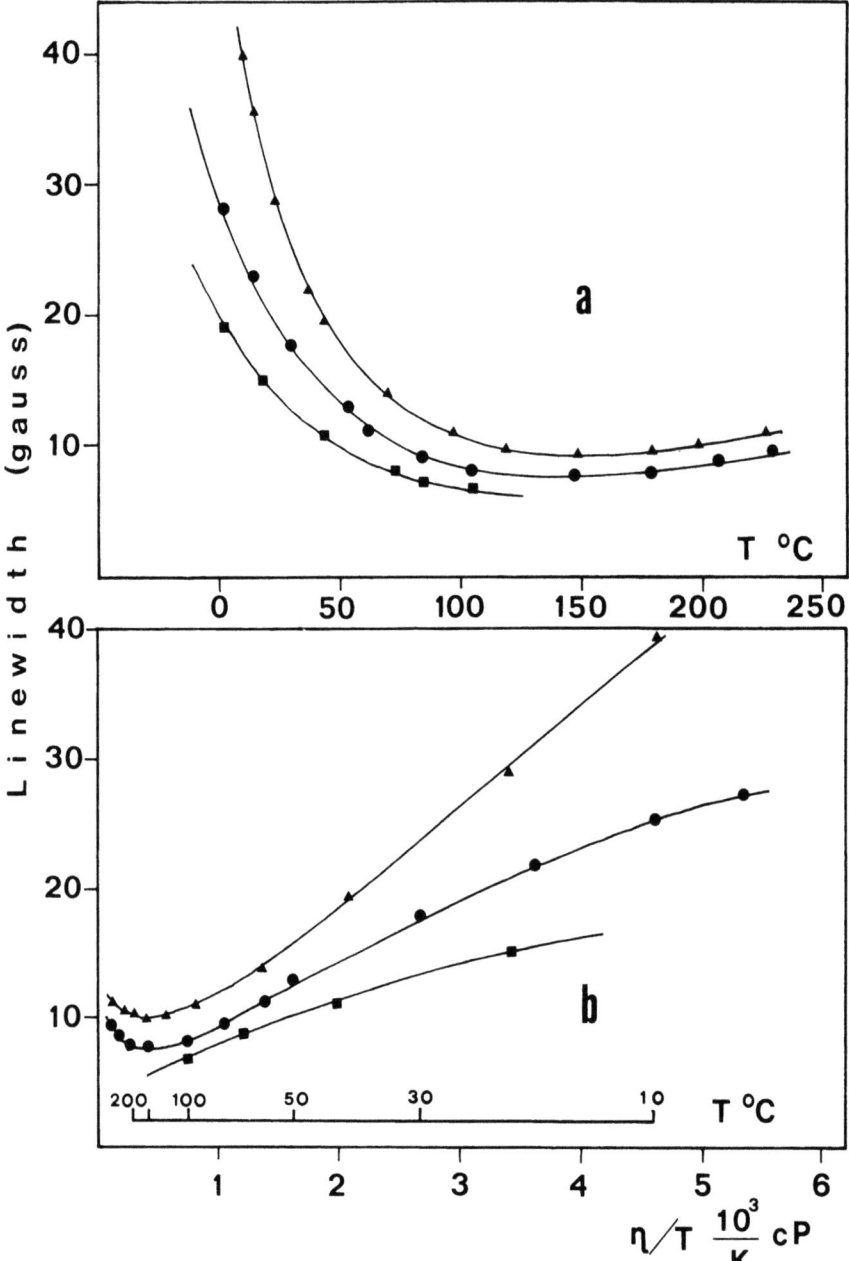

Fig. 5. Linewidth of 0.01 M Mn(II) in water solution: (a) as function of temperature, (b) as a function of η/T. ▲ – S-band, ● – X-band, ■ – Q-band.

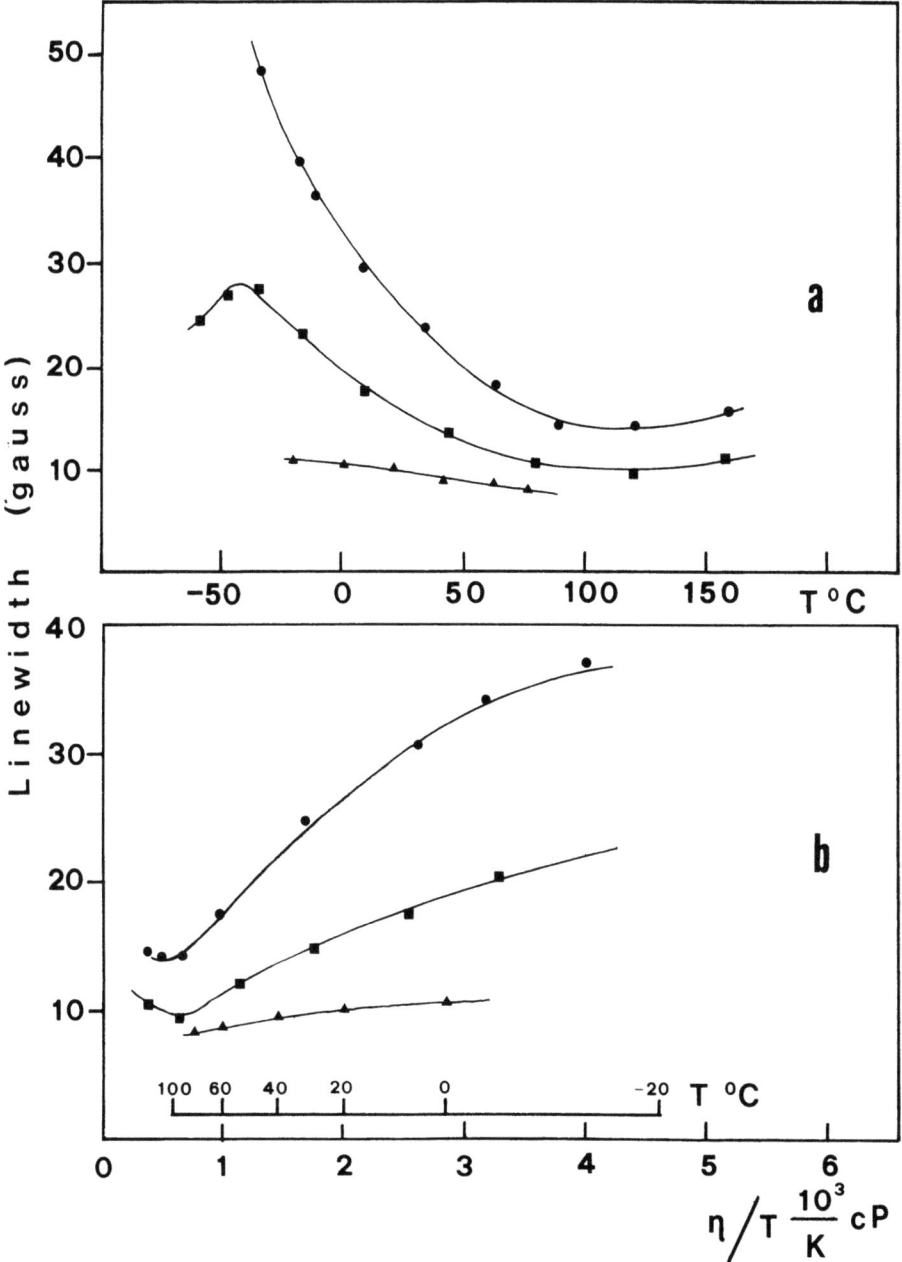

Fig. 6. Linewidth of 0.01 M Mn(II) in methanol: (a) as a function of temperature; (b) as a function of η/T. ● – S-band; ■ – X-band; ▲ – Q-band.

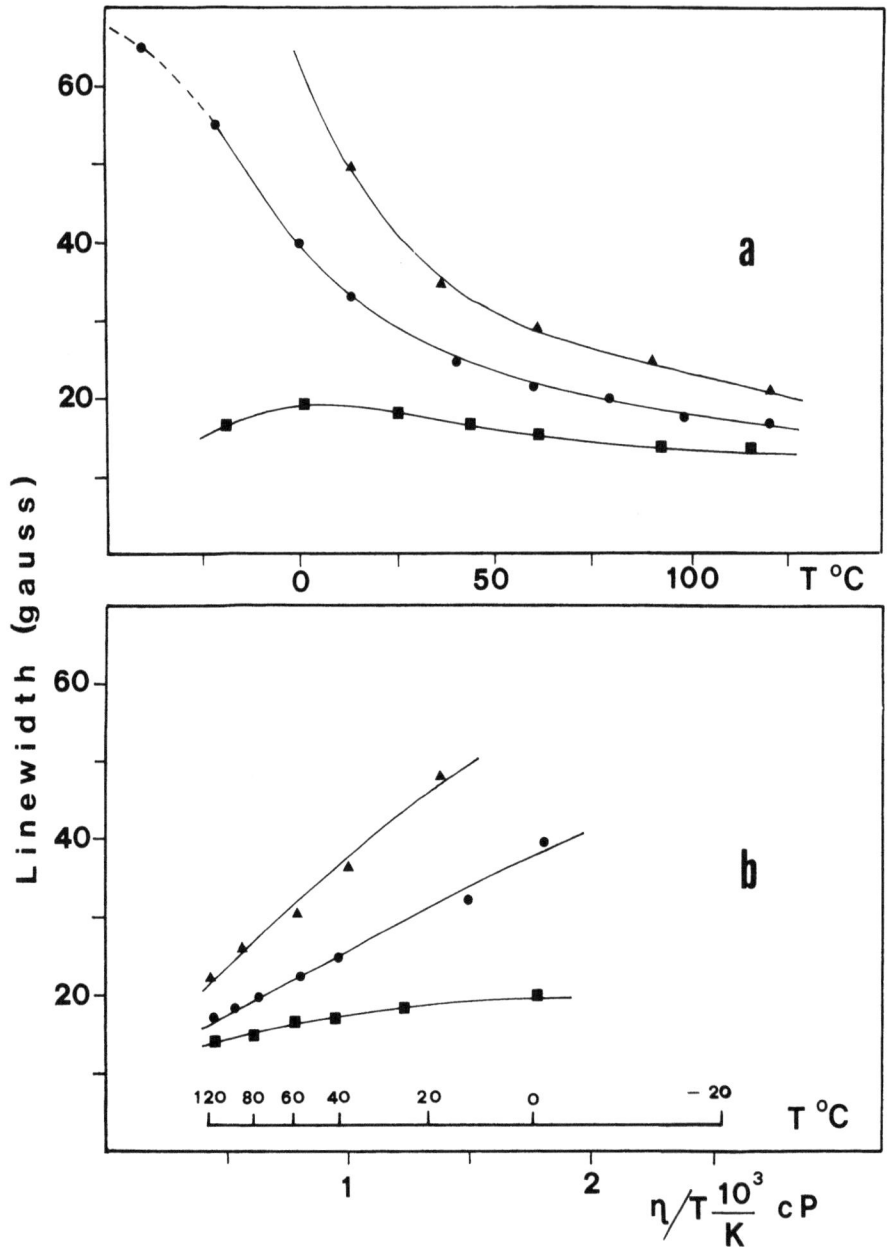

Fig. 7. Linewidth of 0.01 M Mn(II) in acetonitrile-water (10% volume) solution: (a) as a function of temperature; (b) as a function of η/T. ▲ – S-band; ● – X-band; ■ – Q-band.

high temperatures ($\gtrsim 80-100\,°C$) the agreement between theory and experiment is exceedingly good. To what extent the room temperature deviations are indicative of a residual ZFS distribution is a very subtle argument which requires more sophisticated investigation.

Acknowledgements

The authors are indebt to prof. C. Giori, Institute of Physics, University of Parma, for Q-band measurements, and to Mr E. Bragadin for the assembling of the S-band microwave bridge. Financial support was provided by the italian C.N.R.

References

1. McGarvey, B. R.: *J. Phys. Chem.* **61**, 1232 (1957).
2. Bloenbergen, N. and Morgan, L. O.: *J. Chem. Phys.* **34**, 842 (1961).
3. Hudson, A. and Luckhurst, G. R.: *Mol. Phys.* **16**, 395 (1969).
4. Redfield, A. G.: *IBM J. Res. Devel.* **1**, 19 (1957).
5. Burlamacchi, L.: *J. Chem. Phys.* **55**, 1205 (1971).
6. Burlamacchi, L., Martini, G., and Romanelli, M.: *J. Chem. Phys.* **59**, 3008 (1973).
7. Luckhurst, G. R. and Pedullli, G. F.: *Mol. Phys.* **22**, 931 (1971).
8. Luckhurst, G. R. and Pedulli, G. F.: *Chem. Phys. Letters* **7**, 49 (1970).
9. Stengle, T. R. and Langford, C. H.: *Coord. Chem. Rev.* **2**, 349 (1967).
10. Rubinstein, M., Baram, A., and Luz, Z.: *Mol. Phys.* **20**, 67 (1971).
11. Burlamacchi, L., Martini, G., and Romanelli, M.: *Mol. Phys.* **24**, 227 (1972).
12. McClung, R. E. D. and Kivelson, D.: *J. Chem. Phys.* **49**, 3380 (1968).
13. (a) Reed, G. H., Leigh, J. S., and Pearson, J. E.: *J. Chem. Phys.* **55**, 3311 (1971); (b) Reed, G. H. and Cohn, M.: *J. Biol. Chem.* **245**, 662 (1970); (c) Reed, G. H. and Cohn, M.: *J. Biol. Chem.* **247** 3073 (1972).
14. Langford, C. H.: in S. Petrucci (ed.), *Ionic Interactions*, Vol. II, Academic Press, New York, 1971, p. 2.
15. Burlamacchi, L. and Romanelli, M.: *J. Chem. Phys.* **58**, 3609 (1973).
16. McConnell, H. M.: *J. Chem. Phys.* **28**, 430 (1958).
17. Burlamacchi, L., Martini, G., and Tiezzi, E.: *J. Phys. Chem.* **74**, 3980 (1970).

DISCUSSION

Kneubühl: You studied the rotational diffusion of the complexes of S-state ions by observing the esr line broadening due to the statistical crystal-field splitting. However, a severe problem arises because the spin-Hamiltonians of the S-state ion complexes are isotropic. Thus the broadening observed is caused not only by the rotational diffusion, but also by other perturbations. It may be of interest to study complexes of ions with anisotropic spin-Hamiltonians, e.g. Cu^{2+}-ions. As an example I should like to mention a study of rotational diffusion of a complex formed by dissolving $CuCl_2$ in glycerine [1].

Hertz: (1) I would like to point that even in the first hydration sphere of Mn^{++} there is no strict cubic symmetry. From the studies of Mg^{++} we know that there is rotation of the water molecules about the dipole axis. This rotation together with the tumbling of the axes $Mn^{++}-OH_2$ will cause a fluctuating electric field. Perhaps one should consider this effect as well.

(2) I have the feeling that the residence times for the water in the first hydration sphere you have for Mn^{++} are a bit too long. The electron spin relaxation time and the residence time are both of the same order of magnitude, $\simeq 10^{-8}-10^{-9}$ s.

(3) What happens if you add diamagnetic electrolytes to your Mn^{++} solutions?

Martini: (1) Of course the first solvation sphere has temporary deviations from cubic symmetry. What we say is that they are induced by the molecular motions outside this sphere. We do not know anything about the effect of rotation about the dipolar axis, which is probably very fast. Experimentally, we observe some contribution to the line broadening which is not fully explained by the theory and which could be due to motions of this kind.

(2) To my knowledge, the residence time in the first solvation sphere is long in the time scale of EPR. A reassuring experimental evidence is given, for instance, from Mn^{++} in mixed acetonitrile-water solution where, at the appropriate solvent composition, one can clearly distinguish two superimposed spectra, one for the hexa-aquo and one for the hexa-ACTN comple [2].

(3) The argument is widely treated in the literature. Usually we observe line broadening and signal intensity decrease which are generally associated with outer and inner-sphere complex formation [3].

References

1. Kneubühl, F. K.: *J. Chem. Phys.* **33**, 1074 (1960).
2. Burlamacchi, L., Martini, G., and Romanelli, M.: *J. Chem. Phys.* **59**, 3008 (1973).
3. Burlamacchi, L., Martini, G., and Tiezzi, E.: *J. Phys. Chem.* **74**, 3980 (1970).

ASSOCIATION AND ANISOTROPIC MOLECULAR REORIENTATION IN LIQUIDS AS STUDIED BY NMR

R. ECKERT, G. LOOS, and H. SILLESCU,
Institut für Physikalische Chemie der Universität Frankfurt(M), F.R.G.

Abstract. The deuterium and chlorine relaxation times of CDCl₃ have been measured in solutions of diethyl ether, n-dibutyl ether, and various polyethers. The data have been analysed assuming anisotropic rotational diffusion of CDCl₃ in 2 possible states, a 'free' state, and a 'bound' state where CDCl₃ forms a 1:1 complex with any R–O–R' unit. By extrapolating to the limit of dilute ether solutions, rotational correlation times are obtained for motions of the complex in the chloroform solvent.

Résumé. Les temps de relaxation du deutérium et du chlore de CDCl₃ ont été mesurés dans des solutions d'éther éthylique, d'éther butylique normal et de divers polyéthers. Les valeurs mesurées ont été analysées en supposant qu'il y a une diffusion de rotation anisotrope de CDCl₃ dans deux états possibles, l'état 'libre' et l'état 'lié' dans lequel CDCl₃ forme un complexe 1:1 avec quelque élément R–O–R'. En extrapolant à la limite des solutions d'éther diluées on obtient les temps de corrélation de rotation pour les mouvements du complexe dans le solvant, le chloroforme.

1. Introduction

The molecular motion in liquid mixtures is influenced by possible intermolecular association phenomena. The dynamics of these phenomena can be studied by measuring spin lattice relaxation times that depend upon molecular motion. In the present investigation, we apply a simple 2 state model of molecular association defined as follows for a dilute solution [1–4]: (a) The solvent A and the solute B form a complex AB with a formation constant of

$$K = x_{AB}/x_A x_B, \qquad (1)$$

where x_A, x_B, and x_{AB} are mole fractions.

(b) The solution is an ideal mixture of its components A, B, and AB [5].

(c) The life time τ_g of the complex is very long compared with the rotational correlation time of the solvent molecule A being investigated by NMR.

(d) The rotational motion of A can be described by a combined process of anisotropic rotational diffusion *and* exchange between a 'free' state f and a 'bound' state g. We refer to References 3 and 4 for a discussion of this stochastic process which is characterized by rotational diffusion constants $D_f^\perp$, $D_f^\parallel$, $D_g^\perp$, and $D_g^\parallel$ for anisotropic molecular reorientation, and by the life times τ_f and τ_g of A in the states f and g, respectively.

If we denote by p_f and p_g the probabilities to find A in f or g, we have the relations $p_f + p_g = 1$ and $p_f/p_g = \tau_f/\tau_g$. Thus, 2 parameters (e.g., τ_g and p_g) are sufficient to describe the exchange between f and g. In dilute solutions, the theory of spin-lattice relaxation in 2 state systems yields for the case of nuclear quadrupole relaxation [4]:

$$T_{1f}/T_1 = 1 + p_g(\tau/\tau_{2f} - 1) \tag{2}$$

$$p_g = \tfrac{1}{2}(1 + \alpha)\left\{1 - \left[1 - \frac{4K}{K+1}\frac{\alpha}{(1+\alpha)^2}\right]^{1/2}\right\} \tag{3}$$

$$\alpha = n_B^0/n_A^0$$

$$\tau = \tfrac{1}{4}(3\cos^2\beta - 1)^2 \frac{1}{6D_g^\perp + \tau_g^{-1}} + 3\cos^2\beta \sin^2\beta \frac{1}{5D_g^\perp + D_g^\parallel + \tau_g^{-1}} +$$

$$+ \tfrac{3}{4}\sin^4\beta \frac{1}{2D_g^\perp + 4D_g^\parallel + \tau_g^{-1}}. \tag{4}$$

T_{1f} and T_1 are the spin-lattice relaxation times in the pure solvent and the solution, respectively. τ_{2f} is the rotational correlation time of A in the pure solvent [4]. n_A^0 and n_B^0 are the analytical mole numbers of A and B. β is the angle between the principal axes of the field gradient tensor and the rotational diffusion tensor; both tensors are assumed axially symmetric. Furthermore, we neglect the difference between the field gradients in the states f and g. Below, Equations (2)–(4) are applied to D- and Cl-relaxation of $CDCl_3$ where $\beta = 0$ for deuterium, and $\beta = 107.5°$ for chlorine.

The association constant K defined in Equation (1) cannot be determined from the relaxation results given below. We have assumed a value of $K = 1.5$ mf^{-1} for all ether chloroform complexes. In polyethers, where B stands for one segment, any cooperative effects between neighboring segments are neglected. The value of $K \approx 1.5$ is in harmony with the heat of mixture experiments of Becker et al. [5, 6]. In the limit of low ether concentrations the authors obtain K-values of 1.51 and 1.69 at 25 °C for diethyl ether and n-dibutyl ether, respectively. The diethyl ether value compares with the value of 1.46 obtained by Howard et al. [7] from NMR solvent shift measurements in solutions of diethyl ether and chloroform in carbon tetrachloride. Since p_g depends upon $K/(K+1)$ [see Equation (3)], the relaxation results become independent of K in the limit of 'complete' association ($K \gg 1$) which is approximated at low temperature [10] for the exothermic [5] ether chloroform complexes.

2. Experimental

The deuterium spin-lattice relaxation times have been measured at 8 MHz with a BRUKER B-KR 322s pulse spectrometer using the $90° - 90°$ pulse sequence. Some relaxation times of polymer protons are determined at 10 MHz and 55 MHz. The ^{35}Cl relaxation times are obtained from NMR line widths measured with a VARIAN VF 16 wide line NMR spectrometer at 5.6 MHz. The distance ΔH between maximum and minimum slope of the absorption line is related with the spin relaxation time $T_2 = T_1$ by $\Delta H = 2(\sqrt{3}\,\gamma T_2)^{-1}$. The accuracy of all relaxation times is estimated to be better than 10%.

Commercial substances have been used throughout. In particular, LUTONAL M40 and LUTONAL A50 from the BASF-Company were used for the polyvinyl methyl ether and polyvinyl ethyl ether samples, respectively. The samples have been used without further purification. However, oxygen was removed from the samples for proton relaxation measurements using the usual freeze-pump-thaw technique.

3. Results and Discussion

Some typical relaxation results are listed in Figure 1. The relaxation times T_{1f} in pure chloroform are 1.49 s for deuterium and 27.0 μs for chlorine. The quadrupole coupling constants e^2Qq/h of 168 kHz and 77 MHz yield the rotational correlation times τ_{2f} of 1.61×10^{-12} s and 1.58×10^{-12} s for D and Cl, respectively [8]. The τ-values ($\tau^{(D)}$ and $\tau^{(Cl)}$) given in Table I have been obtained from figures where T_{1f}/T_1 is drawn as a function of p_g which is evaluated from Equation (2) assuming $K=1.5$.

The rotational diffusion constants $D_g^\perp$ and $D_g^\parallel$ have been evaluated from Equation (4) assuming

$$\tau_g \gg (6D_g)^{-1} \tag{5}$$

which implies that $\tau^{(D)}$ equals the rotational correlation time $\tau_{2g}^{(D)} \equiv (6D_g^\perp)^{-1}$ for the reorientation of CDCl$_3$ in the bound state. The assumption of Equation (5) is justified a posteriori since the $D_g^\perp$-values of Table I are typical for rotational motions of the ether chloroform complex.

TABLE I

Reorientation of ether chloroform complexes. ($T = 22\,°C$, $K = 1.5$ mf^{-1})

	$\tau^{(D)} \times 10^{12}$ (s)	$\tau^{(Cl)} \times 10^{12}$ (s)	$D_g^\perp \times 10^{-10}$ (Hz)	$D_g^\parallel \times 10^{-10}$ (Hz)	$D_f^\perp/D_g^\perp$
diethyl ether	5.23	2.2	3.19 (4.1 a)	18	3.2 (3.0 a)
n-dibutyl ether	7.31	3.4	2.28	11	4.5
triethylene glycol dimethyl ether	8.91	5.4	1.87	5	5.5
tetraethylene glycol dimethyl ether	11.4	5.7	1.46	6	7.1
polyethylene glycol monomethyl ether (MW 1500)	14.6	5.7	1.14 (4.3 b, 3.3 c)	8	9.0
polyvinyl methyl ether	16.9	7.8	0.99	5	10
polyvinyl ethyl ether	31.5	8.7	0.53 ($\sim$0.2 c)	7	19
polyvinyl pyrrolidone d	(84)	(22)	(0.20)	(3)	(52)
chloroform-d		$D_f^\perp = 10.3 \times 10^{10}$ Hz	$D_f^\parallel = 11 \times 10^{10}$ Hz		

a From dielectric relaxation times of Reference 9.
b From dielectric relaxation times of Reference 12.
c From proton relaxation times for the polymer (see text).
d For comparison from Reference 4, evaluated for $K = 1.5$ mf^{-1}.

For the complexes of diethyl ether and n-dibutyl ether, the ratio $D_f^\perp/D_g^\perp$ reflects the increased size of the complex molecules as compared with the chloroform molecule. The ratio $D_f^\perp/D_g^\perp = 3.2$ for diethyl ether agrees with the corresponding ratio of 3.0 from dielectric relaxation in dilute solutions of ether and chloroform in cyclohexane [9]. If the Stokes-Einstein relation ($D = kT/8\pi\eta r^3$) applies, the ratio $D_f^\perp/D_g^\perp$ should

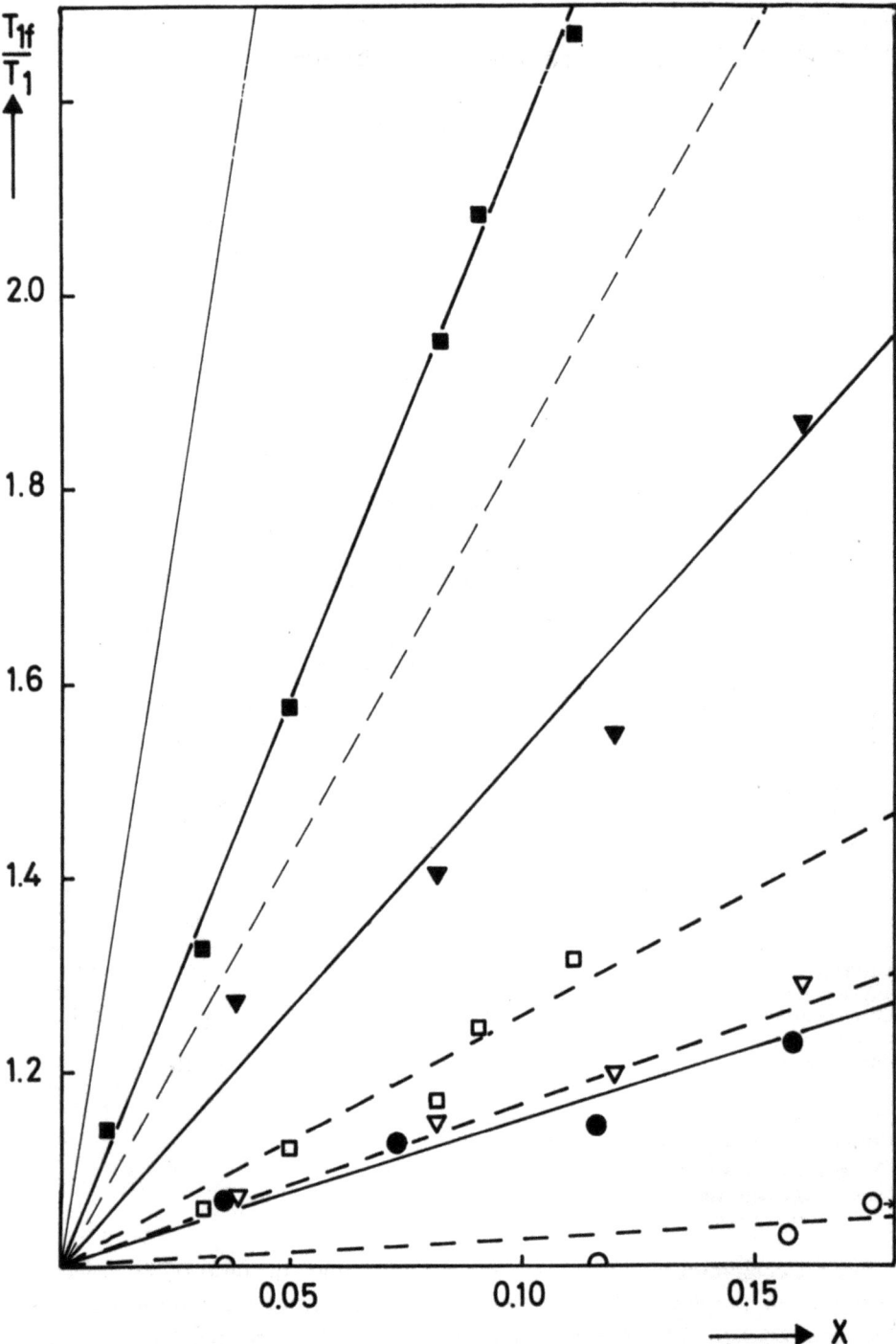

Fig. 1. Ratio of spin relaxation times T_{1f}/T_1, for $CDCl_3$ solutions of some ethers. x: analytical mole fraction of the solutes. Full lines: ^2H-relaxation. Dotted lines: ^{35}Cl-relaxation. The circles, triangles, and squares refer to diethyl ether, polyethylene glycol monomethyl ether 1500, and polyvinyl ethyl ether, respectively. For comparison, the polyvinyl pyrrolidone results of Reference 4 are indicated by thin lines.

(for spherical molecules) be equal to the volume ratio v (complex)$/v$ (CDCl$_3$), and this ratio should be temperature independent [10].

In polyethylene glycol ether solutions, $D_g^\perp$ is of the same order as the corresponding values from proton relaxation times [11] and dielectric relaxation times [12] of the polymer molecule in solution. We have also measured the proton relaxation time in 4.4 wt.% solutions of polyethylene glycol monomethyl ether 1500 in CDCl$_3$. The value of $T_1 = 0.74$ s at 10 MHz is in fair agreement with the values obtained by Liu and Ullman [11] in other solvents. If one assumes a rotational diffusion model for the motion of a proton proton vector in polyethylene glycol, one obtains the diffusion constant of 3.3×10^{10} Hz given in Table I. Of course, this is a very crude estimate since the segmental motion of the polymer molecule is more complicated [11]. As a matter of fact, we find a different value of $T_1 = 1.04$ s at 55 MHz which indicates that the 'extreme narrowing' condition does not apply. It should also be noted that the assumption $K = 1.5$ is only a crude estimate for the association in polyether solutions. Nevertheless, the numbers given in Table I indicate that the deuterium relaxation of CDCl$_3$ in the bound state reflects the segmental motion of the polymer rather than the motion of isolated chloroform molecules.

In polyvinyl ether solutions, $D_g^\perp$ is a measure of the side group rotation within the polymer molecule. This view is supported by proton relaxation experiments in 3.5 wt.% solutions of polyvinyl ethyl ether in CDCl$_3$. We find a non exponential time dependence of the nuclear magnetization which can be decomposed into 2 exponentials with a short relaxation time (0.13 s) typical for the backdone segmental motion [13], and a longer relaxation time of 0.33 s at 10 MHz that belongs to the protons of the ethyl side group. The rotational diffusion model yields an estimate of $D \sim 0.2 \times 10^{10}$ Hz for the side group rotation. The corresponding value $D_g^\perp = 0.53 \times 10^{10}$ Hz of CDCl$_3$ in the bound state is of the same order of magnitude. For comparison, we have listed the results of polyvinyl pyrrolidone solutions [4] where the segmental motion of the polymer is very slow compared with the value of $D_g^\perp$ for CDCl$_3$. Thus, the assumption of Equation (5) is not valid, and $\tau^{(D)} = 0.84 \times 10^{10}$ s is the life time of the hydrogen bond rather than the rotational correlation time for the complex. The largest $\tau^{(D)}$-value (0.32×10^{-10} s) found for ether chloroform complexes provides a lower limit for the life time of an ether chloroform hydrogen bond.

The values of $D_g^\parallel$ given in Table I are obtained from Equation (4) by combining the D and Cl relaxation results. It should be noted that the values for diethyl ether are not accurate since the increase of the chlorine NMR line widths at low ether concentrations is within the experimental error limits. Thus we had to use higher ether concentrations ($x \lesssim 0.4$) in order to determine $\tau^{(Cl)}$. In all ether chloroform complexes, $D_g^\parallel$ differs by less than a factor of 3 from the corresponding value $D_f^\parallel$ in pure CDCl$_3$. Thus, the complex formation has little influence upon the C_3-rotation of the chloroform molecule.

Acknowledgement

The support of the Deutsche Forschungsgemeinschaft is gratefully acknowledged.

References

1. Zimmerman, R. and Brittin, W. E.: *J. Phys. Chem.* **61**, 1328 (1957).
2. Beckert, D. and Pfeifer, H.: *Ann. Physik (Leipzig)* **16**, 262 (1965).
3. Sillescu, H.: *Adv. Mol. Relaxation Processes* **3**, 91 (1972).
4. Brussau, R. G. and Sillescu, H.: *Ber. Bunsenges. Physik. Chem.* **76**, 31 (1972).
5. Becker, F., Pflug, H. D. and Kiefer, M.: *Z. Naturforsch.* **23a**, 1805 (1968).
6. Kiefer, M.: Dissertation, Saarbrücken, 1969.
7. Howard, B. B., Jumper, C. F., and Emerson, M. T.: *J. Mol. Spectrosc.* **10**, 117 (1963).
8. See Reference 4 for a discussion of the $CDCl_3$ reorientation in pure $CDCl_3$.
9. Magee, M. D. and Walker, S.: *J. Chem. Phys.* **50**, 1019 (1969).
10. Eckert, R.: Dissertation, Frankfurt (M), 1972. (Further low temperature experiments are in progress.)
11. Liu, K.-J. and Ullman, R.: *J. Chem. Phys.* **48**, 1158 (1968).
12. Davies, M., Williams, G., and Loveluck, G. D.: *Ber. Bunsenges. physik. Chem.* **64**, 575 (1960).
13. Rockelmann, H. and Sillescu, H.: to be published.

DISCUSSION

Rivail: I would like to ask two questions. (1) Did you manage to take into account the variation of the quadrupole coupling constant of deuterium in chloroform when this molecule is bonded to the ether molecule? (2) It appears that you studied solution with high chloroform contents. What do you think of the possibility of making 1 ether-2 chloroform complexes?

Loewenstein: What is the error introduced by possible change in $(e^2 gQ)_D$ in the complexed state?

Sillescu: By comparing deuterium quadrupole coupling constants in gases with those in solids and liquids one can estimate that the change between the states f and g will probably not exceed 15-20%. However, this would change appreciably the numerical values of τ that vary proportionally, to the square of the quadrupole coupling.

The second question cannot be answered from our NMR experiments. However, the heat of mixture experiments of Becker *et al.* give an excellent fit with a 1:1 complex for diethylether and dibutylether whereas 1:2 complexes are also considered in tetrahydrofurane mixtures with chloroform.

Hertz: (1) I feel that the increase of the Cl^{35} relaxation rate is remarkably small when compared with that of the deuteron relaxation. It might be difficult to account for this difference in terms of anisotropic rotation. Could this not be an indication for a change of the quadrupole coupling constant of the deuteron when $DCCl_3$ interacts with the ether oxygen?

(2) I would recommend the study of the self-diffusion coefficient of $HCCl_3$ in the presence of the ethers. D should decrease because now we have the translational analogy to the rotational case you have studied.

(3) If in a mixture you add a large molecule to a small one, you get an increase of $1/T_1$ of the small species which is due to the repulsive and weak attraction part of the intermolecular potential (background effect). Did you correct for this background effect?

Sillescu: (1) A change in the deuterium quadrupole coupling constant alters the numerical values of $D_g^\perp$ and $D_g^\parallel$. However, the statement that complex formation has little influence upon the C_3-rotation of chloroform remains valid if an up to 20% change in the quadrupole coupling is considered (Compare Table I where $\tau^{(Cl)}$ is equal but $\tau^{(D)}$ differs for tetra- and polyethylene glycol, respectively).

(2) Self-diffusion experiments are in progress in these systems.

(3) From experiments in the mixed solvents $CDCl_3/C_6H_{12}$ and $CHCl_3/C_6D_{12}$ (H. Rockelmann and H. Sillescu, to be published) we conclude that the 'background effect' will give corrections of less than 10%.

Rothschild: Would it be correct to generalize that your experiments truly not so much indicate a 'complex' but that the rotation of chloroform around the symmetry axis does not need a significant rearrangement of the complexing molecules whereas the tilting of the symmetry axis of the chloroform molecules in the neighborhood of the complexing molecules requires that these jump out of the way?

Sillescu: Our results could also be interpreted in terms of jump or cell models where complex formation would show up the way you indicate. I should note that it is very probably the chloroform molecule that 'jumps out of the way' when it is bound to a large ether molecule.

SELF-DIFFUSION MEASUREMENTS IN SIMPLE AND COMPLEX LIQUIDS

R. MILLS

Diffusion Research Unit, Research School of Physical Sciences, Australian National University, Canberra, Australia

Abstract. The isotopic method for measuring self-diffusion in liquids is still more accurate than newer methods such as nmr and neutron diffraction and this is illustrated by our recent study of liquid water. The self-diffusion results for the isotopic forms of water in H_2O and D_2O are shown to agree with a mass-dependence equation derived from the kinetic theory of gases. A diaphragm-cell apparatus for measuring self-diffusion in cryogenic liquids is described. Our recent molecular dynamics calculations of self-diffusion in argon using the Barker-Bobetic potential are discussed also.

Résumé. La méthode isotopique de mesure de la self-diffusion dans les liquides reste plus précise que des méthodes plus récentes telles que la RMN et la diffraction des neutrons: ceci est illustré par notre étude récente de l'eau liquide. On montre que les valeurs de self-diffusion des formes isotopiques de l'eau dans H_2O et D_2O sont en accord avec une équation dépendant de la masse et dérivée de la théorie cinétique des gaz. On décrit une cellule à diaphragme pour la mesure de la self-diffusion dans les liquides cryogéniques. On discute également nos calculs récents de dynamique moléculaire sur la self-diffusion dans l'argon, qui utilisent le potentiel de Barker-Bobetic.

Self-diffusion coefficients which are related to the simple translational movements of species within a liquid, uncomplicated by concentration gradients, are an important measurable property of liquids. Such data are important for two reasons. In the first instance they are needed to check the calculations of both pure theory and of machine simulation techniques such as molecular dynamics. In the latter case, because the calculations themselves are rigorous what is being tested is the correctness of the assumed intermolecular potential between molecules. Secondly, self-diffusion measurements in conjunction with other known properties can be used to provide direct information on the structure and nature of liquids. An example of how knowledge of the microscopic nature of liquids can be obtained from such studies is illustrated in the water case below.

Diffusion measurements by classical techniques are difficult to make accurately and with precision, primarily because of the necessity to eliminate completely the effect of convective flow. This difficulty is accentuated in the case of isotopic self-diffusion where one has the absence of stabilizing density gradients and in addition there are errors of analysis associated with radioactive counting or mass spectrometric techniques. The agreement amongst various workers has consequently not always been good. The situation is now further complicated by the development over the last decade of the nuclear magnetic resonance, neutron diffraction and molecular dynamics methods of measuring self-diffusion. None of these methods are as precise as the isotopic method and there are also serious discrepancies between the data reported by the latter and these newer methods. Unfortunately this has not deterred a number of workers in the nmr field who have reported large amounts of data for a series of liquids over ranges of temperature and pressure, which because of their uncertain accuracy are of very limited use. The only saving grace of certain of these sets of data is that they are internally self-consistent so that one can get some information on the temperature or pressure dependence of self-diffusion.

To try and rationalize the situation, we have in this Unit brought together three of the above methods, isotopic tracer, nmr and molecular dynamics with the object of beginning an integrated programme of self-diffusion measurements. In this programme we are endeavouring not only to increase the accuracy and precision of the individual methods but also to ensure that there is agreement among them. Two liquids were chosen for concentrated study a complex one, water, and a simple one, argon. This paper summarizes the successes and difficulties arising from this project.

1. Liquid Water

Accurate and precise self-diffusion coefficients for liquid water were a first priority. Any information that such data can provide about the interactions and structure of water is of obvious importance. In addition ordinary water with its two protons is the most-studied of all liquids by the nmr technique and so reliable isotopic data are needed as a basis for comparison. In the literature, self-diffusion values for water at 25°C for example for both the isotopic and nmr methods have varied from about 2.2 to 2.6×10^{-9} $m^2 s^{-1}$ a total variation of almost 20% [1]. In the isotopic method a complicating factor is that the isotopically-substituted tracers differ appreciably in mass both from each other and from the solvent water. In the nmr case the main source of systematic error appears to be in the evaluation of the strength of the magnetic field gradient which enters into the equations which give the self-diffusion coefficient.

To try and resolve these anomalies, we have recently studied the self-diffusion of tritiated normal water (HTO) and heavy water (DTO) in solvent water (H_2O). Diaphragm cells were used in conjunction with sophisticated counting techniques. The experiments were carried out in the temperature range 1 to 45°C. The results which have recently been published are summarized in Table I.

In Table I the error quoted ($\sim \pm 0.2\%$) is the rms deviation of the data for several runs at each temperature. Longsworth [2] whose data are shown in columns 3 and 5 used a Rayleigh interferometer to measure the mutual diffusion in H_2O and D_2O to a precision of about $\pm 0.1\%$. The limiting mutual coefficients of this system are equivalent to the tracer diffusion coefficients of HDO in H_2O and D_2O. In Figure 1 is shown a combined plot of Longsworth's HDO and our HTO and DTO data. The parallelism both between the two tracer species in H_2O and D_2O and between different tracers in the same solvent is a marked feature. It is also remarkable that the distance between the curves are almost exactly proportional to the mass differences and this fact has encouraged us to extrapolate these results to obtain the values of pure H_2O and D_2O which are shown in the last two columns and illustrated by broken lines in the figure.

TABLE I
Self-diffusion of isotopic species of water

T (°C)	HTO*/H$_2$O ($D^* \times 10^{-9}$ m^2 s^{-1})	HDO*/H$_2$O [a]	DTO*/D$_2$O	HDO*/D$_2$O [a]	H$_2$O*/H$_2$O	D$_2$O*/D$_2$O
1	1.113 ± .002	1.128			1.149	
4	1.236 ± .003				1.276	
5	1.272 ± .002	1.295	1.001 ± .001	1.034	1.313	1.015
15	1.724 ± .003				1.777	
25	2.236 ± .004	2.272	1.849 ± .001	1.902	2.299	1.872
35	2.838 ± .002				2.919	
45	3.474 ± .003	3.532	2.939 ± .005	3.027	3.575	2.979

[a] Longsworth's data [1, 2] (precision ±0.1%).

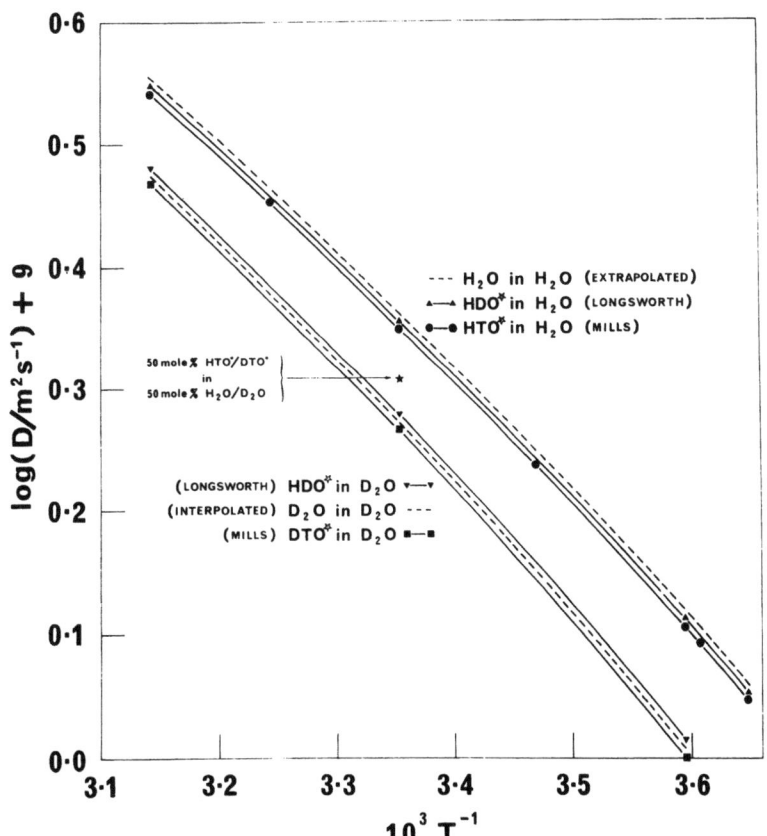

Fig. 1. Self-diffusion coefficients of isotopic species of water in H$_2$O and D$_2$O in the temperature range 1–45 °C.

As stated earlier, one of the objects of this water study was to obtain such values for pure water self-diffusion (H_2O/H_2O) which can be compared directly with molecular dynamics and nmr studies. It will be realized that both these techniques determine this type of coefficient and are not concerned with isotopic effects. The only current molecular dynamics results for water are those of Rahman and Stillinger [3] who in their original paper obtain a value for D of 4.2×10^{-9} m^2 s^{-1} at 34.3° which corresponds with our value at that temperature of 2.84. Considering the difficulties of simulating the transport properties of a complex liquid like water on the computer the agreement is encouraging. Nmr results for water as was mentioned previously are widespread and fall above and below the extrapolated data. However, three very recent determinations, those of Murday and Cotts [4], O'Reilly and Peterson [5] and Gillen, Douglass and Hoch [6] agree very well with our results. These nmr values however are quoted with errors of $\pm 2-5\%$ which tends to make the agreement less meaningful. The magnetic field gradient g enters as a constant in the self-diffusion equation for a pulsed gradient machine as under

$$R = \exp\left[-\gamma^2 \delta^2 g^2 \left(\Delta - \tfrac{1}{3}\delta\right) D\right], \tag{1}$$

where R is the attenuation of the spin-echo amplitude, γ is the gyromagnetic ratio, δ the duration of the magnetic field gradient pulse, Δ is the time interval between gradient pulses and D is the self-diffusion coefficient. The g factor is difficult to measure with accuracy and it would seem that most of the error in nmr measurements is associated with it. We consider that, in view of these facts, one should determine this machine calibration factor by using the isotopic self-diffusion values now available for pure water. A pulsed nmr machine in these laboratories has been calibrated in this way and the same calibration factor (45.3 ± 0.1 g cm^{-1} A^{-1}) has been obtained over the temperature range 1-45° indicating at least the self-consistency of the water data. The machine calibration factor as calculated from an analysis of the spin-echo shape was however 48.52 g cm^{-1} A^{-1}. Evidently there is a discrepancy here and it is undergoing further investigation. An associated study in this laboratory is the determination of mass effects in the diffusion of hydrocarbons by use of deuterated and normal species as tracers. Suitable extrapolation will give diffusion data which should provide further calibration points for the nmr machine.

The question of the mass dependence of isotopic species in liquids has been a subject of some controversy. A recent example is afforded by two studies on the self-diffusion of benzene at 25°C. Eppstein and Albright [7] in their work found relatively large differences between the diffusion of benzene samples labelled with one and two C^{14} atoms respectively. Allen and Dunlop [8] on the other hand using the same tracer species found very small differences. Again Miller [9] has found recently in one system a reverse dependence on mass in which the heavier isotope diffuses faster than the lighter one. As shown in Table I and Figure 1 there are very clear differences between self-diffusion coefficients of the tracer species in D_2O and H_2O as measured by Longsworth and ourselves and these differences are almost exactly proportional to the mass changes. We have therefore tested these data for mass dependence by using a kinetic equation for the analogous case in gases. This has been given by LeClaire [10] as

$$D_1 = \frac{2(2kT)^{1/2}}{3\pi^{3/2}\varrho S^2(1+\delta)} \cdot \left(\frac{m_1+m_2}{m_1 m_2}\right)^{1/2}, \qquad (2)$$

where D_1 is the tracer diffusion of species 1, ϱ is the number density, S a cross-section for collision of two molecular species, δ a slowly-varying function of m_1/m_2 and m_1 and m_2 the molecular masses of the tracer and the host liquid respectively. The mass effect can then be defined as the ratio of the D's for two isotopic species having masses $m_1 = m^\alpha$ or m^β, i.e.,

$$\frac{D^\alpha}{D^\beta} = \left(\frac{m^\beta}{m^\alpha}\right)^{1/2} \left(\frac{m^\alpha+m_2}{m^\beta+m_2}\right)^{1/2} \qquad (3)$$

the ratio of the δ terms being very close to unity in our case. In Table II the data from Table I are tested against Equation (3).

TABLE II

Ratios of isotopic self-diffusion coefficients in a common solvent

T (°C)	Trace species	Solvent	D^α/D^β	$\left[\left(\frac{m^\beta}{m^\alpha}\right)^{1/2} \cdot \left(\frac{m^\alpha+m_2}{m^\beta+m_2}\right)^{1/2}\right]$
5	HDO/HTO (α/β)	H_2O	1.018	1.013
25	HDO/HTO (α/β)	H_2O	1.016	1.013
45	HDO/HTO (α/β)	H_2O	1.017	1.013
5	HDO/DTO (α/β)	D_2O	1.033	1.024
25	HDO/DTO (α/β)	D_2O	1.029	1.024
45	HDO/DTO (α/β)	D_2O	1.030	1.024

It is evident that the isotopic differences can be almost completely accounted for in terms of a mass dependence. It is a little surprising in fact that Equation (3) which is derived from kinetic theory and involving concepts such as mean free paths, binary collisions and the persistence of velocities should fit the data of a complex liquid like water so effectively.

It will be observed also that the ratios of the experimental values are slightly higher than the RHS of Equation (3) in a fairly systematic manner. These differences are greater than the combined experimental error and would appear to be real. One can speculate on how these differences arise and two plausible explanations are suggested. In deriving Equation (3) from (2) it has been assumed that the collisional terms for the isotopic species are equal and this may not be strictly correct. The S factors are related to the intermolecular potential between species and in the case of gases this is believed to vary for isotopic species where the mass differences are proportionately large [11]. In this regard Steele [12] in discussing the intermolecular potentials for normal and deuterated benzenes, concludes that the potential of the latter was about

0.5% less than that of the former. Such a variation is approximately of the right order to account for the differences in Table II. An alternative explanation is that the differences may arise from a change in strength between the O—D···O and O—T···O hydrogen-type bonds. If the bond involving tritium were stronger then again qualitatively it would be in the right direction to explain the data. However, the fact that the differences are substantially the same for the 5 and 45° data and that one would expect more H-bonds at the lower temperature, the explanation in terms of a change in bond strength does not seem valid. The preceding section on isotopic mass effects emphasizes that once self-diffusion data reach a certain degree of precision, they can be used to provide information on the microscopic nature of liquids.

2. Liquid Argon

The situation with regard to self-diffusion measurements in argon has been reviewed recently by several authors including van Loef [13], Ricci [14] and Mills [1] and it is not intended to repeat that information here. However, aspects of the subject which are pertinent to the present study need to be discussed briefly. In Figure 2 we summarize the present status of experimental self-diffusion values for liquid argon, at or near atmospheric pressure and particularly around 85 K.

It will be noted that the three values determined by the open-ended capillary method show a large degree of variation. However, as was argued in an earlier paper [1],

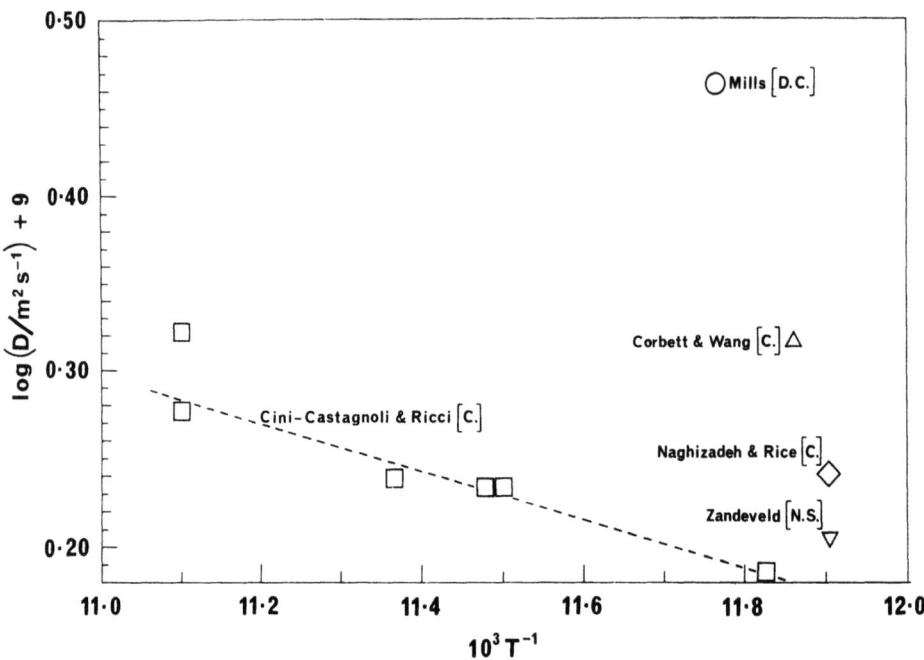

Fig. 2. Experimental self-diffusion results for liquid argon in the temperature range 84.5–90 K and near atmospheric pressure.

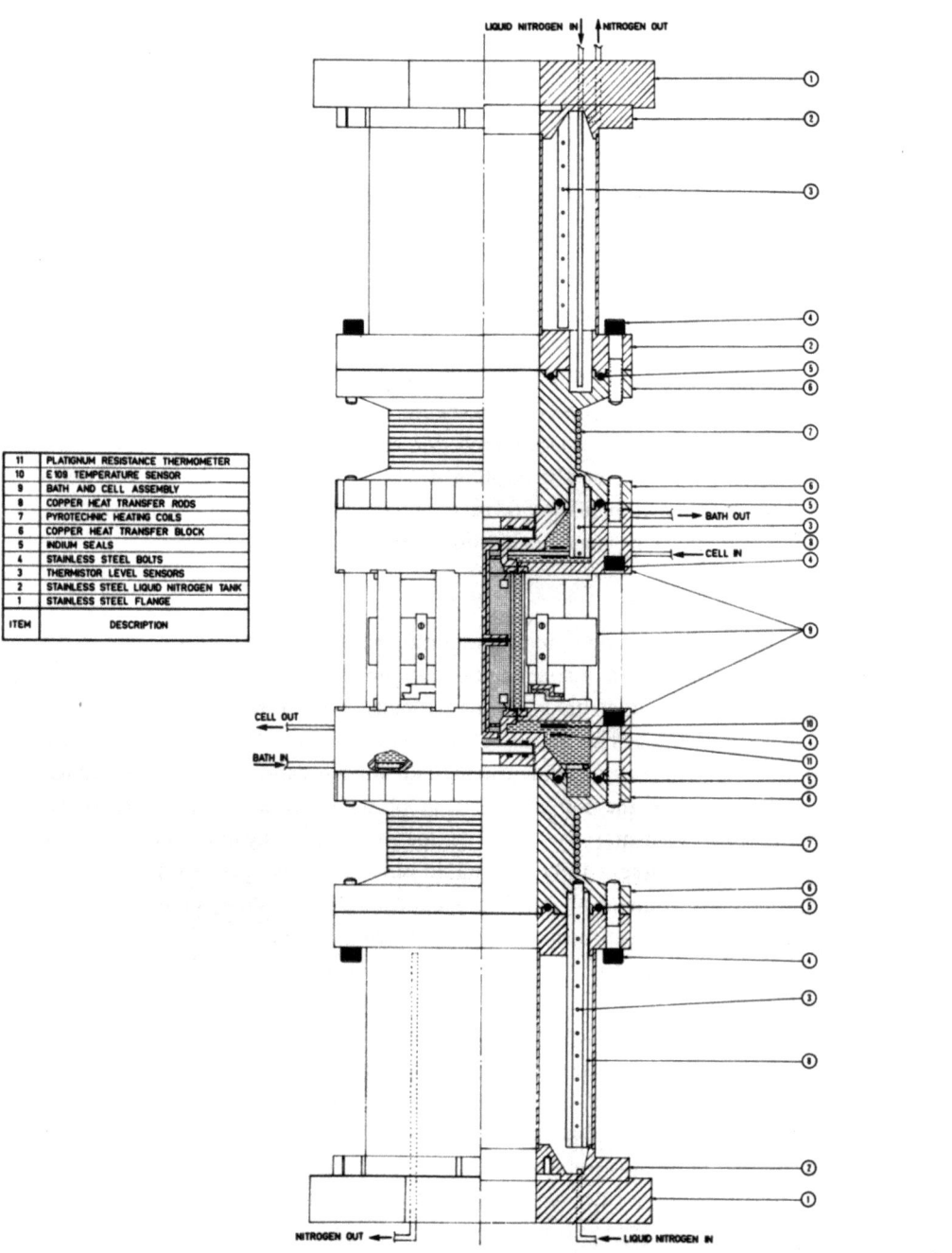

Fig. 3. Scale drawing of cryogenic diaphragm-cell apparatus for self-diffusion measurements.

the values of Cini-Castagnoli and Ricci [15] are most likely to be the correct ones. In this laboratory, we have attempted to confirm the latter results by using a different isotopic diffusion method, the diaphragm-cell technique which has not previously been used in the low-temperature region. The cryogenic diffusion cell is shown in Figure 3 below. A detailed description of the apparatus will eventually be published elsewhere. However, a few of the main features may be of interest.

2.1. Nature of cell

Diaphragm cells need to be calibrated with a system whose diffusion characteristics are known accurately. As there are no confirmed data at low temperature it was necessary to calibrate the diaphragm at 298 K using the standard system of 0.5 M KCl diffusing into water. It was necessary therefore to use a quartz cell as this material has a minimal temperature coefficient of expansion. If a metal diaphragm had been used its pore volume which determines the cell constant would be altered by the 210° change in temperature. Severe problems in sealing between the quartz cell and outer bath and the metal frame were encountered due to the differential shrinkage of the quartz and stainless steel during the cooling down period. Stirring of the cell compartments also presented difficulties and several modifications were tried; eventally fixed stainless steel blades driven by rotating magnets proved successful.

2.2. Temperature control

The cryostat is suspended in a vacuum tank maintained at about 10^{-6} torr. The upper and lower tanks contain liquid nitrogen and these abut on to copper blocks provided with heating coils. The copper blocks form the ends of an argon thermostatic bath which also surrounds the cell by means of a quartz cylinder. Two Artronix bridge-type temperature controllers independently regulate the temperature through platinum resistance elements set in the top and bottom of the bath. Another platinum resistance thermometer measures the actual temperature. As diffusion runs may take from 1–3 days the apparatus is fully automated and liquid nitrogen is replenished in the tanks through electronic level controllers connected to a logic system. Control is at present ± 0.1 K but can be improved when needed.

2.3. Counting technique

In the original version of the cryostat, the cell compartments were closed at their ends by quartz windows and were monitored by photomultipliers placed in the hollow centre of the apparatus. Most cryogenic fluids scintillate under the influence of radiation and in this case the radiation emanated from the A^{37} dissolved in the normal argon. Although many diffusion runs were attempted with this procedure, the fluctuations in the counting rates were too large to give meaningful results. The present procedure is, at the termination of a run, to evaporate off small samples from the two compartments and use gas proportional counters.

The cryogenic diffusion apparatus described above was brought into operation a little over a year ago. The self-diffusion coefficients obtained were all within a few per cent of 2.9×10^{-9} m^2 s^{-1} and as Figure 2 illustrates this is almost a factor of two greater than the best capillary results. As high apparent diffusion rates can usually be attributed to bulk flow, various experimental parameters were changed, and in particular much thicker diaphragms were used and the stirring method changed. The new results still centred on the above figure to within $\pm 5\%$. At this stage it was decided to replace the diaphragm cell portion with an open-ended capillary apparatus. The new cell will be in operation shortly.

3. Molecular Dynamics Calculations

A programme of molecular dynamics calculations is being carried out in this laboratory by Watts and co-workers [16]. Their recent results for argon are shown in Table III below.

TABLE III

Self-diffusion coefficients of argon
(All values of the coefficients are expressed in units of 10^{-9} m^2 s^{-1})

V_M (cm^3)	T (K)	D_{va}	D_{ms}	D_{LJ}	D_{exp}
26.26	89.6	0.71	0.80	1.01	0.14
27.04	87.9	1.16	1.28	1.38	0.42
	94.1	1.28	1.37 (1.39)[a]	1.52	0.28
	105.0	1.46	1.56	1.80	0.19
27.85	83.6	1.78	1.67	1.65	0.93
	89.7	1.64	1.84	1.82	1.12
	103.2	2.08	2.04	2.19	0.53
28.73	91.6	–	2.28	2.30	1.70
	96.5	2.26	–	2.44	1.50
	108.9	2.72	2.76	2.80	1.65
57.45	156.9	21.3	20.4	–	–
65.66	168.9	26.3	27.6	–	–
70.71	159.1	25.9	25.8	–	–
76.60	158.7	27.0	25.6	29.0	27–30
91.92	159.1	33.5	29.6 (26.8)[a]	–	–

[a] The values in parentheses were calculated from Equation (5) for t in the range 5×10^{-12} to 9×10^{-12} s. They are a check on the accuracy of the D_{ms} values.

The Barker-Bobetic two-body potential [17] was used to model the argon interactions on which these calculations were based. Because of the excessive computer time that would have been needed the two-body potential was truncated at 2.5 R_{min} (where R_{min} is the separation of the atoms at the minimum of the potential). The three-body potential developed by Bobetic and Barker [17] for equilibrium studies on argon was omitted also.

There are several features of interest in Table III. The first is that the comparable esults obtained by Levesque and Verlet [18], who used a Lennard-Jones potential, agree reasonably well with those obtained here. This agreement would indicate that the molecular dynamics calculations are not greatly affected by a change in potential parameters. The only extensive experimental results for the self-diffusion of argon under pressure are those of Naghizadeh and Rice [19]. It will be seen that both sets of molecular dynamics data differ from the experimental results both in magnitude and in their dependence on temperature. The experimental data have been extrapolated on the assumption that $\ln D_{exp}$ is a linear function of the pressure and either this procedure or the original data are incorrect. The value calculated from Watts' results for the experimental conditions given in Figure 2 is 1.75×10^{-9} m^2 s^{-1} which is in reasonable agreement with open-ended capillary and neutron diffraction results.

4. Conclusions

In summarising the results of self-diffusion in liquids it is evident that the situation in water is very satisfactory and there are now reliable and precise data over the range 1 to 45 °C. Following on from this, it should now be possible to calibrate spin-echo nmr machines with this data and so considerably reduce the error limits on self-diffusion coefficients from this source. The availability of accurate self-diffusion data for water has allowed also study of isotopic mass effects in liquids. The situation in argon is far from satisfactory however. The data at atmospheric pressure needs to be confirmed and its precision increased. The pressure dependence of the self-diffusion coefficient needs then to be studied accurately up to pressures of a few thousand atmospheres. Once data of this kind are available meaningful comparisons can be made with molecular dynamics results for the corresponding ranges.

References

1. Mills, R.: *Ber. Bunsenges. physik. Chem.* **75**, 195 (1971).
2. Longsworth, L. G.: *J. Phys. Chem.* **64**, 1914 (1960).
3. Rahman, A. and Stillinger, F. H.: *J. Chem. Phys.* **55**, 3336 (1971).
4. Murday, J. S. and Cotts, R. M.: *J. Chem. Phys.* **53**, 4724 (1970).
5. O'Reilly, D. E. and Peterson, E. M.: *J. Chem. Phys.* **55**, 2155 (1971).
6. Gillen, K. T., Douglass, D. C., and Hoch, M. J. R.: *J. Chem. Phys.* **57**, 5117 (1972).
7. Eppstein, L. B. and Albright, J. G.: *J. Phys. Chem.* **75**, 1315 (1971).
8. Allen, G. G. and Dunlop, P. J.: *Phys. Rev. Letters* **30**, 316 (1973).
9. Miller, L.: *Nature* **243**, 32 (1973).
10. LeClaire, A. D.: AERE Report No. R6313, 1970.
11. Harris, K. R., Bell, T. N., and Dunlop, P. J.: *Can. J. Phys.* **50**, 1644 (1972).
12. Steele, W. A.: *J. Chem. Phys.* **33**, 1619 (1960).
13. Van Loef, J. J.: *Physica* **62**, 345 (1972).
14. Ricci, F. P.: *Phys. Rev.* **156**, 184 (1967).
15. Cini-Castagnoli, G. and Ricci, F. P.: *Nuovo Cimento* **15**, 795 (1960).
16. Fisher, R. A. and Watts, R. O.: *Australian J. Phys.* **25**, 529 (1972).
17. Bobetic, M. V. and Barker, J. A.: *Phys. Rev.* **B2**, 4169 (1970).
18. Levesque, D. and Verlet, L.: *Phys. Rev.* **A2**, 2514 (1971).
19. Naghizadeh, J. and Rice, S. A.: *J. Chem. Phys.* **36**, 2710 (1962).

DISCUSSION

Rahman: In more recent calculations with an improved interaction for water, Stillinger and I have obtained a value of 1.9×10^{-5} cm^2 s^{-1} at 10 °C. This is probably closer to the correct value. Do you have any comments?

Mills: The value for the self-diffusion coefficient of water at 10 °C obtained in our experiments was about 1.5×10^{-5} cm^2 s^{-1}. In view of the difficulties involved in simulating the properties of a complex molecule like water, I think that the agreement is very good.

Janik: I think that the accuracy of the neutron method as applied to diffusion coefficient determination is poorer than was mentioned in the lecture. If we limit ourselves only to molecular liquids, we must admit that in all large neutron momentum transfer experiments the neutron scattering pattern contains both the translational and rotational contributions. It is often very difficult to make a proper separation. There exists nowadays only one instrument for very low momentum transfer measurements (in Prof. Springer's laboratory) which leads to diffusion coefficients without any rotational contribution. But even in this case the accuracy is, I think, 10–20%.

Dorfmüller: The fact you mentioned, that the isotope effect in the diffusion coefficient can be described in terms of a gas-like model, is to be expected since the collective motions might contribute only a little to the mass transport and will contribute even less to the isotope effect. This is so because the reduced mass of the ensemble of the molecules participating in a collective mode is rather high. This result shows how useful the study of isotope effects can be for separating gas-like individual from solid-like collective modes in a liquid.

Schofield: Did you compare the velocity autocorrelation functions for the Barker-Bobetic and Lennard-Jones potentials? Your results for the self-diffusion coefficient would appear to support the arguments I give in my paper that although different repulsive potentials give different velocity autocorrelation functions, nevertheless the contribution to the time integral (giving the coefficient) is extremely insensitive to details of the potential and depends only on quantities which are determined from fitting the potential to the thermodynamic data.

Mills: The molecular dynamics calculations of the self-diffusion coefficients of liquid argon were carried out by Dr R. O. Watts and R. A. Fisher who were members of my group in Canberra. To the best of my knowledge they did not compare their velocity autocorrelation functions with those of Levesque and Verlet directly. However, as you say, the agreement between the final results for the two potentials supports your contention.

Friedman: I believe that your experimental results are not inconsistent with the theory I described because, when you vary the solute molecule, its moment of inertia changes by a different ratio than its total mass. On the other hand, I am not sure whether the theory you described includes either rotation-translation coupling or the quantum effect which might also be important here.

Mills: Quantum effects and rotational-translational coupling could well be incorporated in the collisional term, S, of the first kinetic theory equation.

Hertz: I would like to defend a little bit the self-diffusion measurements made by NMR (with accuracy 5–10%). In mixtures the concentration dependence of D sometimes is much more important than the precise absolute value. For instance, if some association or complex formation is claimed to exist, this may easily be checked by the study of the concentration dependence of D: D must decrease as the association partner is added.

De Graaf: I would like to stress that up to now the neutron scattering technique has not been used in an optimal way to extract center-of-mass diffusion coefficients. When transforming the neutron scattering data, obtained in (K, ω) space, to the intermediate scattering function in the (K, t) domain, for long times the time dependence of the rotational motion effect will have decayed and the diffusion coefficient can be obtained from the long-time behaviour. When necessary, experiments could be set up along these lines with an improvement in the accuracy of the D values.

STUDY OF MOLECULAR MOTIONS IN LIQUID METHYLAMINES BY NUCLEAR MAGNETIC RESONANCE, II

A. LOEWENSTEIN, E. GLASER, and R. ADER

Dept. of Chemistry, Technion – Israel Institute of Technology, Haifa, Israel

Abstract. Measurements are reported for the self diffusion coefficients, D, of protons neat liquid $(CH_3)_3N$ and CH_3ND_2 at various temperatures. From these results the molecular radii and the parameter κ are evaluated. Also measured were spin-lattice relaxation times, T_1, of the protons in neat $(CH_3)_3N$ and in a dilute solution of $(CH_3)_3N$ in $(CD_3)_3N$ as a function of the temperature. The inter- and intramolecular contributions to T_1 were calculated assuming dipolar mechanism and using data taken from the D and T_1 of nitrogen and deuterium measurements. Agreement between the calculated and measured T_1 values is satisfactory. The temperature dependence of D and T_1 are examined in light of a model which assumes unconditional reorientational relaxation to vacant sites in the liquid.

Résumé. Les coefficients d'auto-diffusion D des protons dans les liquides purs $(CH_3)_3N$ et CH_3ND_2 ont été mesurés à différentes températures. A l'aide de ces résultats on a calculé les rayons moléculaires et le paramètre κ. On a mesuré également en fonction de la température les temps de relaxation spin-réseau T_1 des protons dans le liquide pur $(CH_3)_3N$, et dans une solution diluée de $(CH_3)_3N$ dans $(CD_3)_3N$. Les contributions inter-moléculaires et intra-moléculaires à T_1 ont été calculées dans l'hypothèse d'un mecanisme dipolaire, à l'aide des mesures de D et T_1 sur l'azote et sur le deutérium. L'accord entre les valeurs de T_1 calculées et mesurées ets satisfaisant. Les variations de T_1 et de D en fonction de la température sont étudiées à l'aide d'un modèle supposant une relaxation de réorientation inconditionnelle vers les sites vacants dans le liquide.

1. Introduction

In a previous paper [1] (henceforth denoted as I) we reported the deuterium and nitrogen magnetic relaxation times in liquid methyl-, dimethyl-, and trimethyl-amine. The results have been interpreted in terms of the quasilattice random-flight (QLRF) model introduced by O'Reilly et al. [2]. In this model the correlation function for large-step rotational diffusion, as derived by Ivanov [3], was applied, assuming isotropic motion of the whole molecule and jumps (of $2\pi/3$ rad) for the internal motion of methyl groups. O'Reilly further assumes [4] that vacancies exist in the liquid and that there is an unconditional relation between translational and rotational motions, i.e. – every translation of the molecule to a vacant site is associated with rotational reorientation. These assumptions lead to the expression: $w = E_D - E_R$ where w is the energy needed to create a vacancy and E_D, E_R are the activation energies for translational and rotational diffusion respectively.

This report extends the measurements given in I with the purpose of examining the reliability of the QLRF model. This is done through measurements of translational diffusion constants, proton spin-lattice relaxation times and an attempt to fit these measurements into the scheme used previously.

2. Experimental

2.1. Materials

CH_3ND_2 was produced by distillation from CH_3ND_3Cl and KOD. $(CH_3)_3N$ and $(CD_3)_3N$ were commercial products. All samples were carefully dried over sodalime, sodium and thaw-freeze degassed before sealing in 10 mm o.d. pyrex tubes.

2.2. NMR measurements

All measurements were performed on a Bruker B-KR 322S pulse spectrometer equipped with the B-KR 300 Z 18 field gradient unit for self diffusion measurements and a B-ST 100/700 temperature controller. T_1 was evaluated from τ null using the $\pi - \pi/2$ pulse method. D was measured from the intensity of the echoes after a $\pi/2 - \pi$ pulses sequence in the presence of a continuous magnetic field gradient. The accuracy of the measurements is indicated together with the results.

3. Results and Discussion

The self diffusion coefficients of liquid methyl- and trimethylamine are shown as a function of the reciprocal absolute temperature in Figures 1 and 2. CH_3ND_2 was used to avoid the complexity of the spectra which would result from the presence of two non-equivalent sets of protons.

The molecular radii, r_0 have been calculated from a plot of $D\eta$ vs T using the modified Stokes-Einstein equation [5] $D = kT/4\pi r_0 \eta$. The viscosities of the liquids have been reported in I. The values for r_0 thus obtained are: 2.56 and 2.75 Å for methyl- and trimethylamine respectively. These values agree quite well with the corresponding values of 2.7 and 2.9 Å, calculated from the densities, assuming hexagonal close packing (cf. Table 3 in I). The assumption of large coefficient of sliding friction between the particle and its medium which amounts to the application of the conventional Stokes-Einstein equation $(D = kT/6\pi \eta r_0)$ would give rise to rather low values of r_0. O'Reilly [6] has reached the same conclusion for the case of liquid ammonia. The values of r_0 can also be related to the hydrodynamic radii, σ, obtained from the rotational Stokes-Debye relation: $\tau_c = 4\pi\sigma^3\eta/3kT$ where the reorientational correlation times, τ_c, are most conveniently obtained from the quadupolar relaxation time of deuterium or nitrogen (Table 3 of I). The relation is [7]: $\sigma = \kappa^{1/3} r_0$ where κ is a temperature independent parameter which takes the values of $0 < \kappa < 1$. For methyl- and trimethylamine the values of κ are 0.11 to 0.28 and 0.10 to 0.29 respectively. The limits depend on the choice of τ_c (cf. Table 3 of I) from either $T_1(N)$ or $T_1(D)$ measurements. For comparison we calculated the value of κ for liquid NH_3 from the τ_c values evaluated from T_1 measurements of the nitrogen [8, 8a] and obtained $\kappa = 0.26$. These values compare reasonably well with those obtained for several other liquids. [7]

The Arrhenius activation energies for the self diffusion constants in methyl- and trimethylamine are 2.20 ± 0.03 and 1.83 ± 0.03 kcal mole^{-1} respectively. The activa-

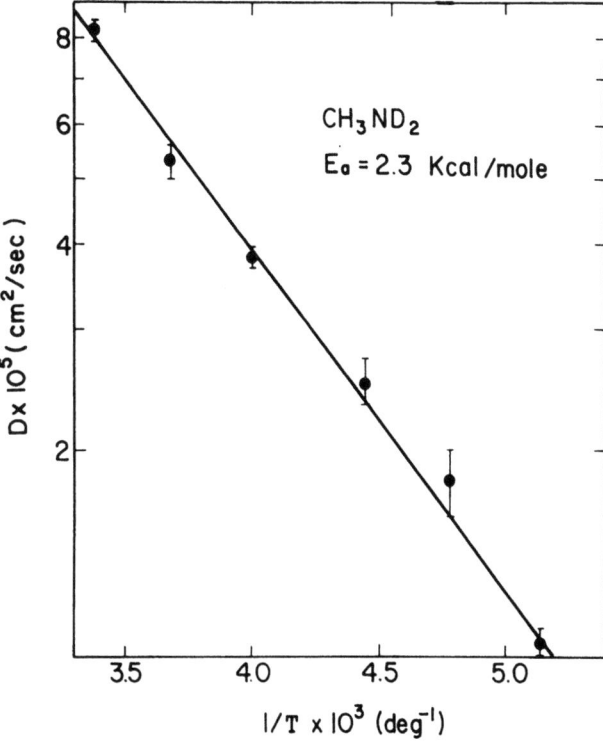

Fig. 1. The self diffusion coefficient of liquid methylamine as a function of the reciprocal absolute temperature.

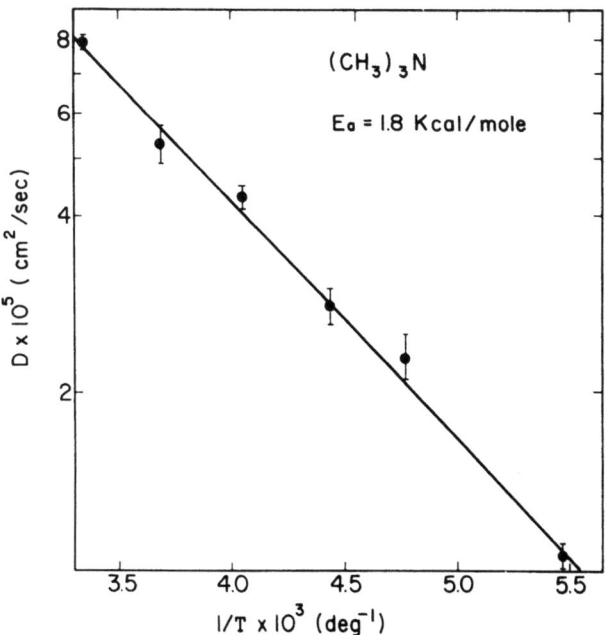

Fig. 2. The self diffusion coefficient of liquid trimethylamine as a function of the reciprocal absolute temperature.

tion energies for molecular reorientation for these liquids, assuming the QLRF model, are 2.8 and 1.2 kcal mole^{-1} respectively (Table 2 of I). Hence the values of the energies needed to create a vacancy, w, in methyl- and trimethylamine are -0.6 and $+0.6$ kcal mole^{-1} respectively. The negative value obtained for w in methylamine is, of course, physically unacceptable and means that such an approach to the interpretation of the data fails in this case. We have calculated w for trimethylamine from the scaled-particle (SP) model and also from the temperature dependence of the density, as suggested by O'Reilly and Peterson [2], and obtained the values of 0.5 and 1.4 kcal mole^{-1} respectively. The densities were taken from the work of Swift [9] and the density of the solid was obtained by extrapolation. Obviously the two procedures for calculating w give different results and only one of them (the SP model) gives reasonable agreement with the experiment. These discrepancies were also observed in the case of several other liquids, studied by O'Reilly and coworkers [2, 4, 10]. These results lead to the conclusion that this rather oversimplified approach to the relation between the activation parameters of molecular translations and rotations is not adquate.

The proton spin-lattice relaxation rates, T_1^{-1}, in neat $(CH_3)_3N$ as a function of the reciprocal temperature are shown in Figure 3. If we assume that the contributions to T_1^{-1} from the spin-rotation interaction and from the anisotropy of the chemical shift can be neglected, we may write:

$$T_1^{-1}(\text{observed}) = T_1^{-1}(\text{intra}, d) + T_1^{-1}(\text{inter}, d),$$

where d indicates that the relaxation mechanism is due to dipolar interaction. The intra-molecular contribution to T_1^{-1} may be calculated using the approximate expression [13]:

$$T_1^{-1}(\text{intra}, d) = \tfrac{3}{2}\gamma^4\hbar^2 \left(\frac{2}{n}\right) \sum_{i<j} dij^{-6} \tau_2,$$

where dij is the distance between the hydrogens, τ_2 is the reorientational correlation time and n is the number of nuclei in a group interacting with one another. We further assumed that $(2/n) \sum dij^{-6} = 2d'^{-6} + 6\langle d''^{-6}\rangle$ where $d'(=1.78\,\text{Å})$ is the proton-proton distance [12] in a methyl group and $\langle d''^{-6}\rangle$ is the average inverse sixth power of the proton-proton distance between two different methyl groups $((\langle d''^{-6}\rangle)^{+1/6}=2.78\,\text{Å})$. The values for τ_2 at different temperatures were taken from I (Figure 4) where they were derived from $T_1(N)$ and $T_1(D)$ measurements through the use of the QLRF model. The calculated values of $T_1^{-1}(\text{intra}, d)$ are shown in Figure 3.

The interamolecular contribution to T_1^{-1} was calculated from the following expression which takes into account both translational and rotational motions [14]

$$T_1^{-1}(\text{inter}, d) = \tfrac{9}{5}\pi\hbar^2\gamma^4 \frac{N}{rD}[1 + 0.233\,(b/r)^2 + 0.15\,(b/r)^4],$$

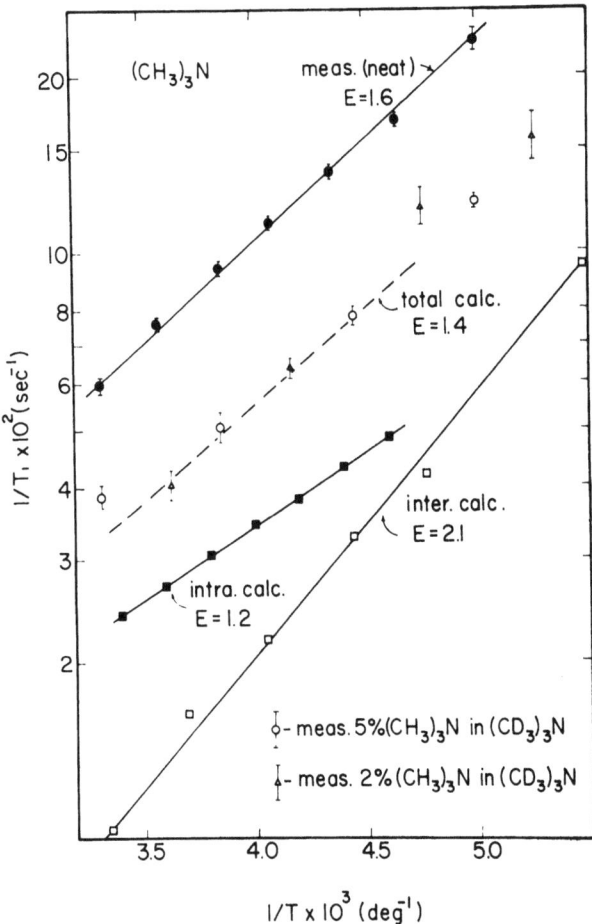

Fig. 3. Measured values of T_1^{-1} for protons in $(CH_3)_3N$ as a function of the reciprocal absolute temperature (●). Also given are values of T_1^{-1} for protons in a dilute solution of $(CH_3)_3N$ in $(CD_3)_3N$ (△ and ○) and the calculated values for the inter- (□) and intra- (■) molecular contributions to T_1^{-1}. The sum of the calculated T_1^{-1} (intra) and T_1^{-1} (inter) is shown by the dashed line. The activation energies, E, are indicated near the lines.

where N is the number of molecules per cubic centimetre, D the self diffusion constant, r is the effective radius of the molecule and b is the distance of each spin from the center of the molecule. This expression has been derived for spherical molecules and therefore does not strictly apply in our case. Furthermore, the choice of b and r is somewhat arbitrary. We have chosen $r = 2.9$ Å (hexagonal close packing value) and $b = 2.09$ Å (distance between N and H atoms). The densities at different temperature were taken from the work of Swift [9] and the self diffusion constants from Figure 2. The results for the calculated values of T_1^{-1} (inter, d) are shown in Figure 3. The sum of the calculated T_1^{-1} inter- and intra-values is given by the broken line in Figure 3 and the calculated values are about 40% lower than the experimental results but have almost the same temperature dependence. We believe that for the rough model used this agreement is fair. Furthermore, the discrepancy may arise from two other sources: (a) the values for τ_2 computed from the QLRF model are lower than the values of $D_\perp$ and $D_\parallel$ obtained by the application of the small-step diffusional model (cf. Figure 3 of I). These values might have been used in the calculation of T_1^{-1} (intra) [5, 16]. (b) The value chosen for r in the calculation of T_1^{-1} (inter) is probably an upper limit as has been indicated above by comparison with the values of r obtained from translational or rotational data. Both factors tend to reduce the calculated values of T_1^{-1} (intra) and slight changes in τ_2 or r would bring the experimental and calculated values of T_1^{-1} to close agreement. It should be noted that a precise calculation of T_1^{-1} (dipolar) should take into account $D_\perp$, $D_\parallel$ and in addition the rate of methyl group internal reorientation and molecular geometry data [17].

To check the calculations of T_1^{-1} (intra) we have measured the T_1^{-1} of dilute solutions (ca. 5% and 2%) of $(CH_3)_3N$ in $(CD_3)_3N$. The results are given in Figure 3. As indicated above, the calculated values of T_1^{-1} (intra) are evidently too low. It is also seen that the temperature dependence of T_1^{-1} (inter) is somewhat bigger than that of T_1^{-1} (intra).

In conclusion it might be said that the QLRF model is only partly satisfactory for the interpretation of our nmr results. Extension of the measurements both by nmr and by other techniques would be needed to obtain a more complete picture of the molecular dynamics in liquid methylamines.

References

1. Loewenstein, A. and Waiman, R.: *Mol. Phys.* **25**, 49 (1973).
2. O'Reilly, D. E. and Peterson, E. M.: *J. Chem. Phys.* **55**, 2155 (1971).
3. Ivanov, E. N.: *Soviet Phys. - JETP* **18**, 1041 (1964).
4. O'Reilly, D. E.: *J. Chem. Phys.* **49**, 5416 (1968).
5. McLaughlin, E.: *Trans. Faraday Soc.* **55**, 28 (1959).
6. O'Reilly, D. E.: *J. Chem. Phys.* **50**, 5379 (1966).
7. McClung, R. E. D. and Kivelson, D.: *J. Chem. Phys.* **49**, 3380 (1968): Kivelson, D., Kivelson, M. G., and Oppenheim, I.: *J. Chem. Phys.* **52**, 1810 (1970); Hwang, J., Kivelson, D., and Plachy, W.: *J. Chem. Phys.* **58**, 1753 (1973).
8. Atkins, P. W., Loewenstein, A., and Margalit, Y.: *Mol. Phys.* **17**, 329 (1969).
8a. Self-diffusion measurements for liquid NH3 are reported by:
    McCall, D. W., Douglass, D. C., and Anderson, E. W. *Phys. Fuids* **4**, 1317 (1961);
    O'Reilly, D. E. and Peterson, E. M.: *J. Phys. Chem.* **74**, 3280 (1970);
    Dorfmueller, T.: *Ber. Kernforschanlage Jülich*, 238 (1965).
9. Swift, E., Jr.: *J. Am. Chem. Soc.* **64**, 115 (1972).
10. O'Reilly, D. E., Peterson, E. M., and Hogenboom, D. L.: *J. Chem. Phys.* **57**, 3969 (1972).
11. Abragam, A.: *The Principle of Nuclear Magnetism*, Oxford university Press, 1961, Ch. VIII.
12. Wollrab, J. E. and Laurie, V. W.: *J. Chem. Phys.* **51**, 1580 (1969). We wish to thank Dr O. Kafri for his assistance in the computation.
13. cf. Smith, D. W. G. and Powles, J. G.: *Mol. Phys.* **10**, 451 (1966).
14. Hubbard, P. S.: *Phys. Rev.* **131**, 275 (1963). It might be noted that the use of an alternate 'hard sphere' formula which relates T_1^{-1} (inter, d) to D (cf. Torrey, H. C.: *Phys. Rev.* **92**, 962 (1953) or Reference 11) would reduce the calculated values of T_1^{-1} by just about 15%.
15. Zeidler, M. D.: *Ber. Bunsenges. physika Chem.* **75**, 770 (1971).
16. Kitchlew, A. and Nageswara Rao, B. D.: *Chem. Phys. Letters* **18**, 123 (1973).
17. Hubbard, P. S.: *J. Chem. Phys.* **52**, 563 (1970).

DISCUSSION

Hertz: I would like to point out that perhaps it is advisable to avoid the formulation: reorientation time of the molecule and reorientation time of the methyl (or other) group. Sometimes it is difficult to tell which part of the molecule is 'the molecule' and which is only a group of the molecule. Perhaps it may be better to speak only of the reorientation of certain groups.

Loewenstein: I agree to call a certain group of atoms which includes all the atoms in the chemical unit 'a molecule' and then accept Prof. Hertz's suggestion.

Rothschild: To answer Prof. Hertz's concern, this question could, hopefully, be answered by doing a Raman experiment and obtaining the correlation function. For instance, if one chooses a convenient mode, the vibrational transition moment of this mode would, by symmetry considerations, not be dependent on the internal rotation of the methyl groups but only on the end-over-end orientational motion of the whole molecule.

Bovee: You used the QLRF model which assumes isotropic overall reorientation of the molecules The activation energy for overall reorientation increases from mono- to dimethylamine, and decreases. from di- to trimethylamine, which can be caused by the fact in trimethylamine that there is no $N-H$ left to form hydrogen bonds. Would it in view of this not be more reasonable to take into account anisotropic overall reorientation?

Loewenstein: The answer is given in Reference 1 (part I).

INVESTIGATION OF MOLECULAR LIQUIDS BY NEUTRON SPECTROSCOPY

TASSO SPRINGER

Institut für Festkörperforschung, der Kernforschungsanlage Jülich, F.R.G.

Abstract. It will be discussed what physical information on molecular liquids can be obtained from neutron scattering experiments in the region of small energy transfers, in particular referring to rotational and diffusive motions. Neutron spectroscopy will be compared with other methods, and a few experimental examples will be presented.

Résumé. On discute, spécialement par rapport aux mouvements rotationnels et à la diffusion, les informations qui peuvent etre obtenues sur les liquides moléculaires par la diffusion des neutrons à bas transfert d'énergie. La spectroscopie de neutrons est comparée à d'autres méthodes. Quelques exemples expérimentaux sont présentés.

It will be reviewed what kind of information on the dynamics of molecular liquids can be obtained by means of neutron spectroscopy. We restrict ourselves to the discussion of rotational and diffusive motions, and we will not discuss the vibrations of the molecule. As a basis of our discussion we have to start with molecular motions in *solids*; in this field, neutron spectroscopy is well established, and it is going to become a standard technique in physical chemistry, as nmr, dielectric relaxation experiments, or optical spectroscopy. Then we extend our discussion to molecular *liquids*; here, the interpretation of the spectral data is somewhat difficult since one necessarily gets a superposition of rotational and translational contributions and we have to consider under what conditions an unambiguous interpretation is feasible at all.

In the frame of the Van Hove theory [1], a neutron scattering experiment is nothing else but a Fourier analysis of the density distribution of the scattering particles in time and in space. In this scattering process, the energy transfer determines the frequency of this Fourier component, ω, whereas the momentum transfer defines its wave number, κ, and one has

$$\hbar\omega_0 = E_0 - E_1 \quad \text{and} \quad \hbar\kappa = \hbar(\mathbf{k}_0 - \mathbf{k}_1), \tag{1}$$

where E_0, E_1, and $\mathbf{k}_0$, $\mathbf{k}_1$ are the incident or the scattered energies and wave vectors of the neutron, respectively.

The region of κ which can be covered in typical cases is between 10^{-2} and several Å^{-1}. The energy transfers reach from about 10^{-4} to several 0.1 eV for conventional time-of-flight and for typical triple axis spectrometers. Recently, the energy region has been extended by more than two orders of magnitude, using the so-called 'backscattering spectrometer': To separate and analyse the energies, this spectrometer applies Bragg reflection at an angle of 90°. In this backscattering situation, the diffracted wave-length depends on the Bragg angle only in second order. As a consequence, the resulting energy resolution becomes extremely high [2]. For the instrument at the Jülich FRJ-2 reactor it is 0.7 µeV. For this type of spectrometer, the energy transfer cannot be changed by the Bragg angle; therefore, a Doppler motion of the diffraction crystals has to be achieved. At present, a region of energy transfers of about ± 10 µeV can be covered by this method. According to these parameters, the typical time scale which can be studied by neutron spectrometers is between several 10^{-13} and 10^{-8} s. In terms of self-diffusion constants, this means values above $\omega/\kappa^2 \simeq 10^{-7}$ cm^2 s^{-1}.

In the following we will mainly deal with scattering on hydrogen compounds. Therefore, we restrict the discussion to the incoherent scattering intensity which is the dominating contribution for the neutron-proton interaction.* The scattering intensity for the hydrogen is proportional to the incoherent scattering law $S_{inc}(\kappa, \omega)$ which can be written as

$$S_{inc}(\kappa, \omega) = \frac{1}{2\pi} \iint e^{i(\kappa r - \omega t)} G_s(r, t) \, d^3r \, dt, \qquad (2)$$

where $G_s(r, t)$ is the probability to find the scattering particle at r for time t if it has been at $r = 0$ for an earlier time $t = 0$.

In the frame of this concept we discuss now $G_s(r, t)$ for a rotating molecule.** First, consider a hydrogen atom (or proton) bound to a rotating molecule which is fixed in a *solid*. In this case, the molecular rotation distributes the proton over a spherical shell. Consider a volume element on this sphere close to the (arbitrary) origin $r = 0$ where G_s has the following behaviour: For small times, $G_s(r, t)$ is large because, per definition, it is *certain* to find the scattering proton at the origin for $t = 0$. With increasing time, $G_s(r, t)$ decays: In the average, the scattering protons are going to 'forget' their original position. If time goes to infinity, G_s approaches a finite and constant value, G_∞, because the proton is confined to the surface of the sphere on which it rotates ('stationary' probability distribution). Therefore one has (Figure 1) [3]

$$G_s(r, t) = G_\infty(r) + G'_s(r, t). \qquad (3)$$

According to (2) and (3), the scattering law for the rotating proton $S^R_{inc}(\kappa, \omega)$ is a superposition of a *quasielastic* part, S_{qe}, which is approximately Lorentzian-shaped (due to the decaying part G'_s) and a *delta function* from G_∞, namely

$$S^R_{inc}(\kappa, \omega) = S_0(\kappa) \delta(\omega) + S_{qe}(\kappa, \omega). \qquad (4)$$

This shape of the spectrum holds quite generally for all kinds of random molecular rotations in solids. The intensity S_0 of the sharp elastic line is the Fourier transform of G_∞ in space. Physically this means the diffraction of the neutron wave on the single proton which is distributed over a finite region in space by virtue of the molecular motion. It should be noticed that such a sharp line in the neutron spectrum has the same origin as the Mössbauer line in gamma ray interactions: It corresponds to

---
* For reviews or books dealing with neutron scattering theory and experiments see [3]–[7], and the conference reports [8].
** We exclude the case of a completely periodic rotation.

processes where only momentum is transferred to the lattice as a whole, and no energy. Now consider a *diffusing* molecule, so that the center of gravity is not fixed. Here, the probability of finding the particle in a certain volume element practically decays to zero as time goes to infinity ($G_\infty \to 1/V_0$ for a sample volume V_0). As a consequence, the delta function in (4) disappears and is replaced by a Lorentzian-shaped curve (Figure 1).

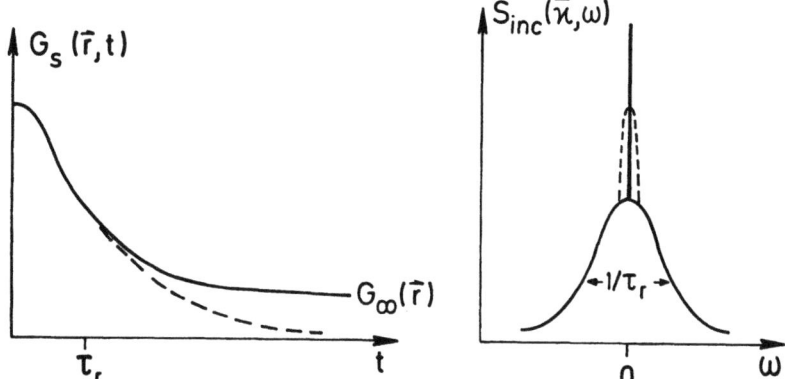

Fig. 1. Self-correlation function and incoherent scattering law in the region of small energy transfers $\hbar\omega$ (quasielastic spectrum). Solid line: molecule in a solid; dashed line: molecule in a liquid.

How compares the physical information according to (4) with the information which one gets from alternative methods? [9] It is well-known that infrared spectroscopy yields the Fourier spectrum $\tilde{F}_1(\omega)$ of the time-dependent dipole correlation function $F_1(t) = \langle \cos\beta(t) \rangle$, with $\cos\beta(t) = \mathbf{p}(0)\,\mathbf{p}(t)$, where $\mathbf{p}(t)$ is the dipole vector of the molecule. Raman spectroscopy gives the Fourier spectrum $\tilde{F}_2(\omega)$ of the correlation function $F_2 = \langle P_2[(\cos\beta(t)] \rangle$ (with $\cos\beta(t) = \mathbf{e}(0)\,\mathbf{e}(t)$, where the vector $\mathbf{e}$ is connected with the orientation of the polarization ellipsoid). Also the spin-lattice relaxation time T_1 is related to the Fourier transform of this function [9]. On the other hand, for neutrons one gets a superposition of Fourier spectra $\tilde{F}_l(\omega)$ in all orders l, namely [10], for the quasi-elastic part in (4):

$$S_{qe}(\kappa, \omega) = \sum_{l=1}^{\infty} (2l+1)\, j_l^2(\kappa d)\, \tilde{F}_l(\omega). \tag{5}$$

j_l is a spherical Bessel function of order l and $\tilde{F}_l(\omega)$ is the Fourier transform of $F_l(t) = \langle P_l[\cos\beta(t)] \rangle$ with $\cos\beta(t) = \mathbf{d}(0)\,\mathbf{d}(t)$, where $\mathbf{d}$ is the position vector of the rotating proton relative to the center of rotation (for a diatomic polar molecule, like HCl, $\mathbf{d}$ and $\mathbf{p}$ would coincide). Considering equations (4) and (5) one might expect to get more information from neutron spectroscopy than from the other methods. Practically, this is not the case: For small κ, the term $l=1$ is dominating and the $l=2$ term is difficult to evaluate. For large κ, one obtains a superposition of numerous terms in (5), and their separation would be difficult and inaccurate. As a consequence, neutron scattering essentially yields the 'dipole' term $\tilde{F}_1(\omega)$.

The *translational* motion of the molecule can be described by a scattering law $S_{inc}^D(\kappa, \omega)$ which is a Lorentzian in ω for small κ. Its half-width is $\Gamma_{1/2}^D = 2\kappa^2 D$, where D is the macroscopic self-diffusion constant of the molecule. Therefore, D can be determined. At larger κ, the widths $\Gamma_{1/2}^D(\kappa)$ deviates from this κ^2-relation, and it is related to the parameters of the single diffusive step; e.g. in the model of a jump diffusion, one obtains τ_0, the mean-time between jumps (see e.g. [5]).

Neutron spectroscopy has been successfully applied to study rotations in molecular solids, and diffusion in *simple* liquids like liquid metals or nobel gases. For a *molecular* liquid, the scattering law depends on the combined rotational and translational motion of the scattering proton. If both motions would be uncorrelated, the resulting scattering law is a convolution between $S_{inc}^D(\kappa, \omega)$ and $S_{inc}^R(\kappa, \omega)$. Let us consider now under what conditions one can expect to obtain separate information on both contributions. In principle, the criterium would be approximately $\Gamma_{1/2}^R \gtrsim \Gamma_{1/2}^D$ or $2D_r \gtrsim \kappa^2 D$, where D_r is the rotational diffusion constant.

It seems that small κ-values could help, provided that the available energy resolution is sufficiently high. However, if κ is small, the rotational spectrum is too weak to be observable (due to the decrease of the ratio j_1/j_0 with increasing κ, Equation (5)). A reasonable intensity requires $\kappa d \simeq 1$, therefore one gets

$$2d^2 D_r \gtrsim D. \tag{6}$$

This can only be fulfilled for liquids with relatively high viscosity (e.g. liquid crystals). For liquids whose molecules have a small momentum of inertia $\mathfrak{J}$ and a weak mutual interaction, the rotations are weakly hindered (e.g. for liquid CH_4). Here, this criterium should be written as $d^2(kT/\mathfrak{J})^{1/2} \gtrsim D$ (see [7]).

Complications for the interpretation of the neutron spectra occur also if there are several protons with different position vectors **d**, following the same rotation. Furthermore, a molecule may perform rotations around different molecular axes. To solve such problems one could scatter on molecules which are partly deuterated. In this way, the contribution of certain protons can be suppressed, or determined separately.

Finally, we briefly present a few *experimental results* as examples for the foregoing discussion. Figure 2 shows the neutron spectrum at very small energy transfers for a molecular *solid*, namely adamantane ($C_{10}H_{16}$) in its plastic phase [11]. The purely elastic line and the spectrum due to rotational motions can clearly be separated (see Equation (4)); its width gives a typical rotational correlation time of 1.5×10^{-8} s. Figure 3 presents neutron spectra for a nematic liquid crystal (PAA) [12]. At relatively small κ one gets only a Lorentzian-shaped quasielastic spectrum which is interpreted in terms of purely translational motions. Its half-width as a function of κ yields the self-diffusion constant $D = (4.1 \pm 0.3) \times 10^{-6}$ cm^2 s^{-1} in good agreement with tracer experiments. On the other hand, experiments at much larger $\kappa (\gtrsim 2$ Å$^{-1})$ [13] gave a width which should be interpreted in terms of molecular rotations. These two experiments [11, 12] have been carried out by means of the backscattering spectrometer. Figure 4 presents time-of-flight spectra on liquid NH_3 [14]. The spectrum includes translational and rotational contributions. A fit has been performed on the basis of the following assumptions: (i) rotations and translations are uncoupled which is assumed to be a good approximation for such a molecule, (ii) translations are treated by diffusion, using the macroscopic self-diffusion constant, and (iii) the rotations are calculated in terms of the model described above; the quasielastic part Equation (5) has essentially been calculated on the basis of the 'dipole' function

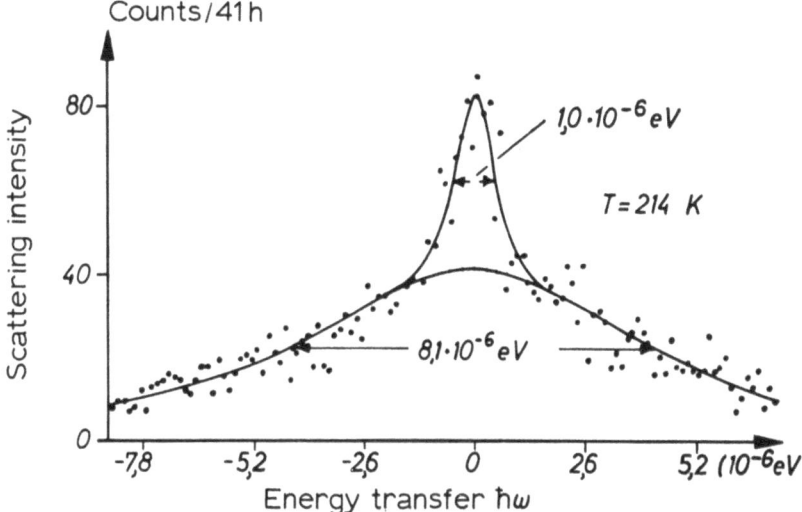

Fig. 2. Spectrum of neutrons scattered on plastic adamantane at 205 K as a typical example for rotations in a solid. The width of the line at energy transfer $\hbar\omega = 0$ is determined by the instrumental resolution (Alefeld and Stockmeyer [11]).

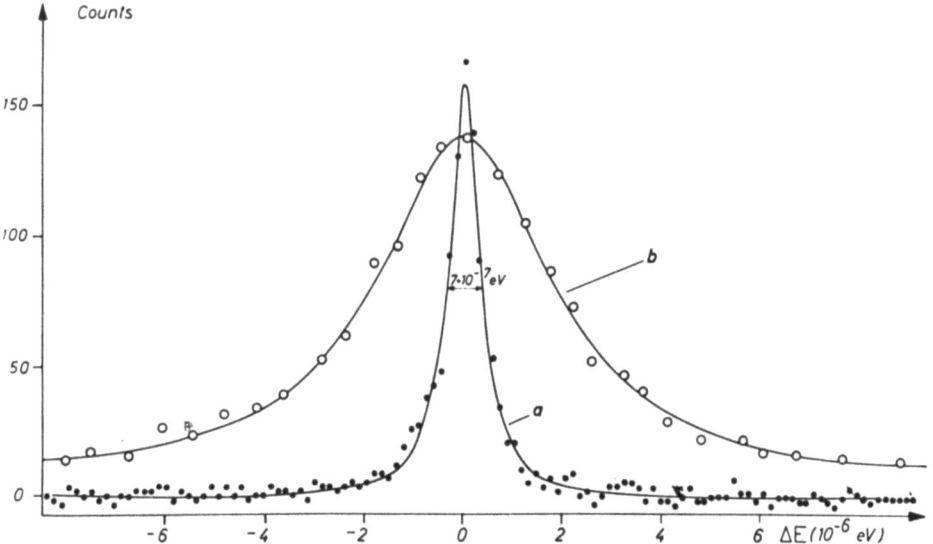

Fig. 3. Quasielastic spectrum on nematic PAA (122°C) for a scattering vector $\kappa = 0.27$ Å^{-1} as a function of energy transfer $\hbar\omega$ (curve b. Curve a: spectrum of solid PAA which is practically the resolution spectrum). (Töpler et al. [12]).

$F_1(t)$. A good fit with the experiments has been obtained under the assumption that $F_1(t)$ behaves, for small times $(t \leqslant \tau^*)$, as for free rotations, and like rotational diffusion for large times. It turns out that the experiments are able to yield two characteristic times which were found to be $\tau^* = 5 \times 10^{-14}$ s and $1/D_r = 2 \times 10^{-13}$ s.

The *coherent* scattering law depends on correlations existing between the motion

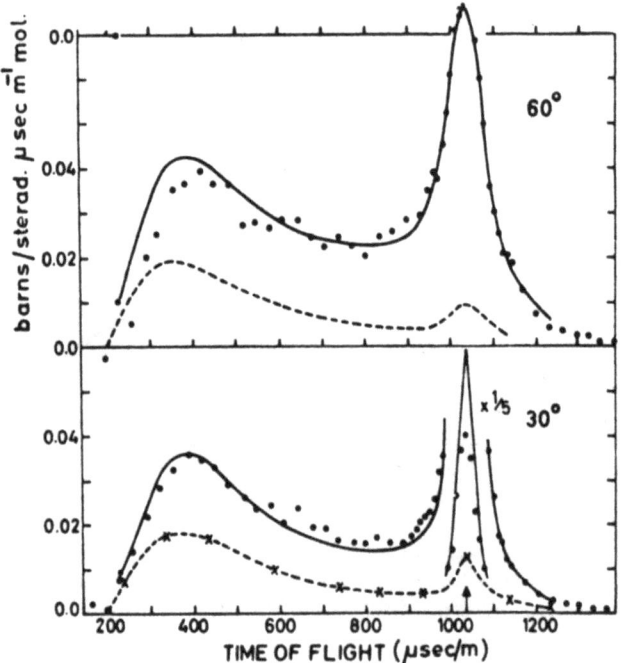

Fig. 4. Time-of-flight spectrum of liquid NH₃ at 218 K. ● = experiment; solid curve: theory (see text). Dashed curve: multiple scattering correction (Dasannacharya et al. [14]).

of *pairs* of the scattering atoms. Therefore, this would allow to study orientational correlations in liquids. However, the evaluation and interpretation of the spectra will be quite complex because one gets, in addition to the self-terms, a superposition of *inter-* and *intra*molecular interferences. So far, this possibility has not yet been exploited for molecular liquids.

Acknowledgements

The author wishes to thank Dr B. Alefeld and his colleagues for many valuable discussions and for allowing him to reproduce his unpublished data.

References

1. Van Hove, L.: *Phys. Rev.* **95**, 249 (1954).
2. Alefeld, B., Birr, M., and Heidemann, A.: *Naturwissenschaften* **56**, 410 (1969).

3. Turchin, V. F.: *Slow neutrons,* Jerusalem: Israel Program for Sci. Transl., 1965.
4. Gurevich, I. I. and Tarasov, L. V.: *Low Energy Neutron Physics,* North-Holland Publ. Comp., Amsterdam, 1968.
5. Egelstaff, P. A. (ed.): *Thermal Neutron Scattering,* Academic Press, London, 1965.
6. Egelstaff, P. A.: *An Introduction to the Liquid State,* Academic Press, London, 1967.
7. Springer, T.: *Springer Tracts in Modern Phys.* **64** (1972).
8. Neutron inelastic scattering; International Atomic Energy Agency, Conference proceedings: Vienna, 1965 (Bombay Symposium); Vienna, 1968 (Copenhagen Symposium); Vienna, 1972 (Grenoble Symposium).
9. Gordon, R. G.: *J. Chem. Phys.* **42**, 3658 (1965): *J. Chem. Phys.* **44**, 1830 (1966); *J. Chem. Phys.* **43**, 1307 (1965).
10. Sears, V. F.: *Can. J. Phys.* **45**, 237 (1967); **44**, 1279 (1966).
11. Alefeld, B. and Stockmeyer, R.: unpublished, 1973.
12. Töpler, J., Alefeld, B. and Springer, T.: *J. Mol. Cryst. Liquid Cryst.*, to be printed (1974).
13. Janik, J. A., Janik, J. M., Otnes, K., and Riste, T.: *Mol. Cryst. and Liq. Cryst.* **15**, 189 (1971).
14. Dasannacharya, B. A., Thaper, C. L., and Goyal, P. S.: *Neutron inelastic scattering;* International Atomic Energy Agency. Proceedings of the Grenoble Conference, Vienna, 1972, p. 477.

DISCUSSION

Hertz: I may make a comment regarding the possibility to separate the 'rotational' motion of chemically non-equivalent protons. This can be done by replacing those protons by deuterons in which we are not interested. But it may be so that the amount of substance needed is too large in many cases.

You told us that it is difficult to separate the translational from the rotational part of the Van Hove self-correlation function. Is this always possible in principle? Assume the molecule in question is in an associated state or in a free state. The effective (instantaneous) centre of mass will differ in both cases and thus the 'd' occuring in the corresponding formulae is not well defined.

Springer: I imagine that this effect might be studied as such by the scattering experiment.

Rahman: The dependence of the width of $S_{\text{inc}}(\kappa, \omega)$ on K^2 turns away from the K^2D line due to non-Gaussian shape of the Van Hove $G_s(\mathbf{r}, t)$; interpretation of this turning away in terms of jumps may not be quite correct.

Springer: I agree completely. Interpretation in terms of jumps has just been mentioned as an example for a very simple explanation of such data.

Schofield: It is not nearly as simple as Rahman implies. One reason is that the neutron scattering from molecular (as opposed to monoatomic) liquids does not give the mass density fluctuation. (There is no correlation between nuclear mass and neutron scattering length). Hence the theory is more complicated than a simple generalisation of the formalism for a monoatomic liquid, based on mass and momentum conservation considerations.

Springer: This refers to Rahman's comment about my statement that interpretation of data is more complicated in the case of coherent scattering. Coherent scattering spectra from monoatomic liquids, like liquid argon, can be interpreted in terms of existing theories. However for molecular liquids, where you have various kinds of atoms in one molecule and different kinds of rotations, I believe that the interpretation is very complex.

Yip: One difficulty in analyzing coherent scattering spectra of molecular liquids lies in the proper treatment of rotational effects. Weinrub (Ph.D. Thesis, Harvard University, 1971) has found that the partial wave expansion method of Sears provides a reasonable basis for calculating the rotational effects but one needs to keep several terms (up to five in some cases) in the series. Generally speaking, the structural effects depending on the relative separation of different scattering centers in the same molecule or different molecules can not be simply separated from the truly dynamical effects as expressed by the various rotational correlation functions.

Dorfmüller: I wonder if you did observe any difference in the neutron spectrum in going from the nematic phase of PAA to the isotropic phase. Such measurements in both phases could provide useful information about the effect of the onset of rotation about the two others axes on the spectra. There has been some evidence of pre- and post-transition phenomena in the phase change of PAA. I would suggest that neutron spectroscopy in the vicinity of the transition temperature might shed some light into the problem of these phenomena.

Springer: Quasi-elastic scattering at large κ values would, in fact, give this kind of information. I think, Prof. Janik could comment on such kind of experiments.

Janik: We made a large κ quasi-elastic neutron scattering experiment also for isotropic liquid PAA. The broadening, we believe, is due predominantly to rotations, not essentially different from those in nematic PAA (molecular rotations around the long molecular axis). A certain contribution of translational diffusion is however possible.

Davies: I should like to ask Dr Springer whether his data show signs of any special interaction equivalent to association between the ammonia molecules? And what assessment can he provide of the local field in the PAA? When such anisotropy can be measured it would be interesting to follow its reduction as the temperature is raised or as the system is diluted by solvent.

It seems worth emphasizing that when relaxation times can be measured, one of the most revealing factors is the temperature dependence that gives the activation enthalpy for the process. The chemist is well aware that absolute rates (i.e. reciprocals of relaxation times) are very difficult to calculate and are compounded of *at least* a frequency factor and an energy term. The latter term is often of particular significance to our understanding of the molecular process involved.

Springer: To your question: Incoherent neutron scattering on the protons in the molecule depends on the *individual* particle motion. Therefore, no *direct* information can be obtained on association, or more generally, on the relative motion of different molecules.

I would like to clarify that the data presented on liquid ammonia were by Dasannacharya *et al.* (*Proceedings of the Conference on Neutron Inelastic Scattering* 1972, IAEA Vienna, p. 477).

Chen: What is the ultimate accuracy of diffusion coefficient measurement by neutrons in liquids? What is the role of neutron scattering in the study of dense gases?

Springer: If you clearly separate the diffusive part of the spectrum from the rotational contributions you may reach an accuracy in D of about $\pm 10\%$. This needs experiments at relatively small scattering vectors in connection with a correspondingly high energy resolution. Concerning dense gases, this is a very promising field of investigations, since the theoretical interpretation can start from a more first-principle basis as in the case of liquids. Experimentally, there are difficulties because of the scattering contributions from the sample containers.

De Graaf: I would consider NH_3 a bad example because one would expect difficulties there in explaining the results. The center of mass in this compound is not coincident with the center of rotation, and the moments of inertia in this symmetrical top are different for different rotation axes.

In discussing the neutron data it is advantageous to use the k, t domain, because there one clearly sees that the contribution of the rotational motions decays in time as sketched in the figure.

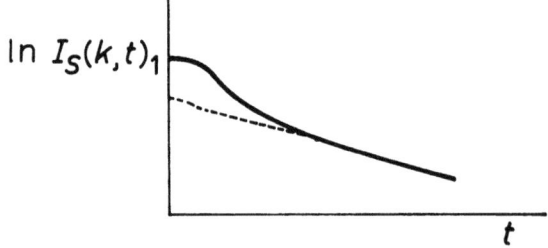

Springer: A very good example for the study of rotations is methane. This substance has been investigated quite long ago by Venkataraman (see [10]).

Janik: I would like to give a more detailed explanation of the history of the situation in PAA in connection with Prof. Davies' question. The measurements for large κ, Prof. Springer is referring to, were made with samples oriented by either electric or magnetic fields. Originally we interpreted them as if the broadening were caused by translational diffusion only. That gave the simplest explanation of the anisotropy observed. Later experiments of Springer *et al.* (and also our dielectric and IR broadening data) led us to a revision of that interpretation. In fact one observes pure diffusion anisotropy only in Springer's low κ experiments, whereas in our large κ experiment there is a mixture of translational diffusion and rotations. The anisotropy arises from a complicated competition of all three motions.

Carneiro: I would like to draw your attention to the Marx- (Multi Angle Reflecting Crystal)-

spectrometer, that with a resolution at 60 μeV seems to fill out the gap between a conventional spectrometer and the back-scattering spectrometer.

Springer: A certain gap in the energy resolution is still in the 5–50 μeV region.

Schofield: I would like to make a general remark based on my experience of the last few days in seeing the experimental results of measurements on time correlation functions. If the data are as good as appears (which I doubt, as they are based in many cases on numerical Fourier transformation of spectral line shapes) then it should be possible to extract the 'memory function' (Equation (26) of my paper in these proceedings). The memory function is essentially the autocorrelation function of the random force, and as such gives a much clearer distinction of time scales than appears in the spectrum. Since a rather simple memory function can give complex correlation functions, it could be that conclusions drawn from direct study of the correlation function turn out to be misleading. However, one should give a word of warning: the computation of the memory function can be very difficult numerically and requires very good data. (An excellent account of memory functions and the numerical techniques have been given by Berne and Harp, Reference 12 of my paper).

COMPARATIVE STUDY OF MOLECULAR MOTION IN LIQUIDS BY NMR AND NEUTRON SPECTROSCOPY

MANFRED D. ZEIDLER

Institut für Physikalische Chemie und Elektrochemie der Universität Karlsruhe, F.R.G.

Abstract. Translational and rotational motions of molecules in the liquid phase are investigated using quasielastic neutron scattering and nuclear magnetic relaxation methods.

The common theoretical basis which was reviewed in a previous article of the author is extended to include internal rotational motions.

Neutron scattering data and deuteron relaxation rates were obtained for benzene, cyclohexane and methanol at different temperatures. Spin-echo self-diffusion constants were measured for cyclohexane. The experiments are evaluated on the basis of the presented theory to yield information on absolute and relative translational motion and on the rotational contribution to the quasielastic neutron scattering peak.

Résumé. Les mouvements de translation et de rotation des molécules en phase liquide sont étudiées par les méthodes de la diffusion quasi-élastique des neutrons et de la relaxation magnétique nucléaire.

La base théorique commune qui a été élaborée dans un article antérieur de l'auteur est étendue pour y inclure les mouvements de rotation interne.

Les données de diffusion de neutrons et de vitesses de relaxation des deutons ont été obtenues pour le benzène, le cyclohexane et le méthanol à différents températures. Les constantes de diffusion par spin-écho ont été mesurées pour le cyclohexane. Les expériences ont été dépouillées sur la base de la théorie présentée ci-dessus, pour obtenir des renseignements sur les mouvements de translation relative et absolue et sur la contribution rotationelle à la raie de diffusion quasi-élastique des neutrons.

1. Introduction

Information on the (single particle) molecular motion in liquids can be obtained from different experimental techniques among which (incoherent) neutron scattering and nuclear magnetic relaxation are of prime importance. Typical characteristics of these two methods are compared in Table I. As can be seen both methods have their shortcomings regarding separability of translational and rotational motions or time scale of the experiment (in normal mobile liquids characteristic times are of the order of 10^{-12} s). Therefore a combination of both techniques is necessary and the following procedure may be used: Through intramolecular relaxation data we get information on the rotational motion, and these results allow us to make an estimate of the rotational contribution to the quasielastic neutron scattering. After taking care of this effect the remaining translational contribution is found which may be analysed in terms of macroscopic self-diffusion data available from independent experiments. On the other hand the intermolecular relaxation is related to the relative translational motion between neighbouring molecules and thus a comparison with the single particle translation gives an idea about the collective motions in the liquid.

Both types of experimental data, i.e. the neutron scattering data finally expressed as the incoherent scattering law and the nuclear magnetic relaxation data, can be related to Fourier-transforms of certain time-correlation functions. These time-

TABLE I

Comparison of typical characteristics of neutron scattering and nuclear magnetic relaxation in the study of molecular motion in liquids

	Neutron scattering	Nuclear magnetic relaxation
Type of motion	Translation Rotation Vibration	Relative translation Rotation

Separability of different motions	(a) Quasielastic region:
	  Translation and rotation
	(b) Inelastic region:
	  Vibration
		(a) Intermolecular relaxation:
		  Relative translation and relative rotation
		(b) Intramolecular relaxation:
		  Rotation

Time scale	Subthermal neutrons: 10^{-12}–10^{-11} s	Strongest magnetic fields: 10^{-8} s

correlation functions thus provide the important theoretical link between the different kinds of experimental data, for whatever time-dependent variables are contained in these functions we must presume that their time evolution is identical. Unfortunately, mainly because of the inadequate time-scale of the nmr experiment, we have to employ a model in order to describe the time-dependence of the correlation functions. This was chosen to be the 'random walk' model or even its limiting case, the diffusion model, in connnection with the rotational motion. Anyhow it should be noted that the neutron scattering and nmr relaxation experiments could be analyzed in terms of the same model which is a necessary requirement in order to make meaningful comparisons.

In a previous paper concerning this topic [1] the theoretical foundations for a common analysis of experimental results have been pointed out and experimental data for the acetonitril system were presented. In the present publication the theoretical expression for the rotational intermediate scattering law is extended to include the effect of intramolecular rotation, and experimental results for three more systems – benzene, cyclohexane and methanol – will be given.

2. Theory

The theory was treated extensively in the previous paper [1]. Except for the rotational intermediate scattering law, for which a more general formula is derived, the final expressions of that paper are repeated here.

2.1. INCOHERENT INTERMEDIATE SCATTERING LAW

The intermediate scattering law is obtaind from the scattering law, which is closely related to the experimental double differential scattering cross section, by a Fourier-transformation (note that always the incoherent contributions are merely considered since they are alone of importance in the scattering experiments on hydrogenous samples) [2]:

$$I(\kappa, t) = \int_{-\infty}^{+\infty} S(\kappa, \omega) e^{i\omega t} \, d\omega. \qquad (1)$$

The intermediate scattering law $I(\kappa, t)$ may be considered as a product of the translational, rotational and vibrational contributions, the latter contribution affecting the intensity but not the width of the quasielastic scattering peak. Therefore Fourier-transformation of the quasielastic peak alone yields the product $I_{trans} \cdot I_{rot}$. We will consider the two parts separately.

2.1.1. Rotational Contribution

The rotational intermediate scattering law is represented by the following time-correlation function:

$$I_{rot}(\kappa, t) = \langle e^{-i\kappa \mathbf{d}(0)} \cdot e^{i\kappa \mathbf{d}(t)} \rangle. \qquad (2)$$

Here **d** is the vector from the center of mass of the molecule to the scattering nucleus (proton). Equation (2) may be evaluated by using the relation

$$\langle e^{-i\kappa d(0)} e^{i\kappa d(t)} \rangle = \iint P(\Omega, \Omega_0, t) e^{-i\kappa d(\Omega_0)} e^{i\kappa d(\Omega)} p(\Omega_0) d\Omega_0 d\Omega, \qquad (3)$$

where $P(\Omega, \Omega_0, t)$ is the conditional probability density (so-called propagator) that the orientation of the vector **d** is specified by Ω at time t if it was specified by Ω_0 at time zero, and Ω, Ω_0 are sets of Eulerian angles describing the orientation relative to the laboratory coordinate system. $p(\Omega_0)$ is the a priori-probability density for the orientation to be specified by Ω_0 at time zero. In contrast to the previous treatment [1] we will assume that the scattering nucleus is part of a group within the molecule, for example a methyl group which might undergo internal rotational motion superimposed on the rotational motion of the other part of the molecule. This internal motion, however, should not change the length of the vector **d**. Defining two molecular coordinate systems, one attached to the internally rotating group (the z-axis of this system should coincide with the internal rotation axis) and the other fixed to the rigid part of the molecule (the z-axis of the latter system should coincide with the symmetry axis of the rigid part which therefore must be describable as a symmetric rotator), we can expand the exponential in the following way:

$$e^{\pm i\kappa d} = \sum_{l=0}^{\infty} \sqrt{4\pi(2l+1)} (\pm i)^l j_l(\kappa d) \sum_{m=-l}^{+l} D_{m0}^{(l)}(\Omega) \times$$

$$\times \sum_{n=-l}^{+l} D_{nm}^{(l)}(\Omega') Y_l^n(\theta'', \phi'') \qquad (4)$$

which is an extension of Equation (16) in reference [1]. j_l is a spherical Bessel function, $D_{mn}^{(l)}$ an element of a Wigner rotation matrix and Y_l^n a spherical harmonic. The double primed polar coordinates fix the vector **d** within the internally rotating system, Ω' specifies the orientation of this system relative to the rigd molecular system whereas Ω specifies the orientation of the latter relative to the laboratory system. If both rotational motions can be described by diffusione quations and occur independently of each other the propagator in Equation (3) is the product of the following two propagators [3]:

$$P_1(\Omega, \Omega_0, t) = \sum_{L=0}^{\infty} \sum_{M'=-L}^{+L} \sum_{M=-L}^{+L} \frac{2L+1}{8\pi^2} D_{M'M}^{(L)}(\Omega_0)^* D_{M'M}^{(L)}(\Omega) \times$$

$$\times e^{-\{L(L+1)D_\perp + M'^2(D_\parallel - D_\perp)\} \cdot t} \qquad (5)$$

$$P_2(\alpha', \alpha'_0, t) = \frac{1}{2\pi} \sum_{N=-\infty}^{+\infty} e^{-iN(\alpha'-\alpha'_0) - N^2 D_{\text{int}} t}. \qquad (6)$$

Furthermore we have for $p(\Omega_0)$ in Equation (3)

$$p(\Omega_0) = p_1(\Omega_0) \cdot p_2(\alpha'_0) = \frac{1}{8\pi^2} \cdot \frac{1}{2\pi}. \qquad (7)$$

Inserting Equations (4) to (7) into (3) and taking advantage of orthogonality relations gives the result:

$$I_{rot}(\kappa, t) = \sum_{l=0}^{\infty} (2l+1) j_l(\kappa d)^2 e^{-l(l+1)D_\perp t} \sum_{m=-l}^{+l} e^{-m^2(D_\parallel - D_\perp)t} \times$$

$$\times \sum_{n=-l}^{+l} d_{nm}^{(l)}(\beta')^2 d_{no}^{(l)}(\theta'')^2 e^{-n^2 D_{int} \cdot t}, \qquad (8)$$

where the diffusion constants $D_\perp$, $D_\parallel$ and D_{int} describe the rotation perpendicular to the symmetry axis of the rigid part of the molecule, parallel to this symmetry axis and around the internal rotation axis respectively. θ'' is the angle between the vector **d** and the internal rotation axis, β' is the angle between this axis and the symmetry axis of the rigid part of the molecule. The definition of the function $\mathbf{d}_{nm}^{(l)}$ as found in reference [4], for example, is:

$$d_{nm}^{(l)}(\beta) = \sqrt{\frac{(l+n)!(l-n)!}{(l+m)!(l-m)!}} \sum_s (-1)^{l-n-s} \binom{l+m}{l-n-s}\binom{l-m}{s} \times$$

$$\times \left(\cos\frac{\beta}{2}\right)^{2s+n+m} \left(\sin\frac{\beta}{2}\right)^{2l-2s-n-m}. \qquad (9)$$

Equation (8) which is an extension of Equation (20) in Reference [1] reduces to the latter if $D_{int}=0$, i.e. $\theta''=0$.

2.1.2. *Translational Contribution*

The final result for the translational intermediate scattering law on the basis of the random walk model is

$$I_{trans}(\kappa, t) = e^{-D\kappa^2/(1+D\tau\kappa^2)t}, \qquad (10)$$

where D is the translational diffusion coefficient and τ is the mean time between jumps.

2.2. NUCLEAR MAGNETIC RELAXATION RATES

Relaxation experiments on hydrogenous samples may be performed with two different isotopes: light hydrogen (^{1}H) or deuterium (^{2}H). Thus either proton or deuteron resonances are investigated. The relaxation also is dominated by different types of spin-lattice interactions: magnetic dipole-dipole interactions for protons and electric quadrupole-field gradient interactions for deuterons. Depending on whether the two interaction partners belong to the same or to different molecules we differentiate between intra- and intermolecular contributions. Now quadrupolar interactions are dominated by intramolecular relaxation, whereas in dipolar interactions both contributions are of comparable magnitude. Their experimental separation, however, is possible in many cases by use of the isotopic dilution method [5].

Intramolecular relaxation is determined completely by the rotational motion of the molecule. Since in intermolecular relaxation the interacting protons belong to different molecules, their relative translational as well as their relative rotational motions contribute to the relaxation. Theoretical estimates seem to indicate that the rotational contribution to the intermolecular dipolar relaxation is of minor importance [6], therefore we will equate the translational contribution with the intermolecular relaxation.

Note that due to the inadequate time-scale of the nmr experiment the relaxation rates are related to integrals ($\omega = 0$ Fourier-transform) over appropriate time-correlation functions [1].

2.2.1. Rotational Contribution

As in the case of the intermediate scattering law we consider a motion where an internal diffusional rotation is superimposed on the rotation of the rigid part of the molecule with the latter being considered a symmetric rotator. The result is [7]:

$$\left(\frac{1}{T_1}\right)_{intra} = C \sum_{m=-2}^{+2} \sum_{n=-2}^{+2} d_{nm}^{(2)}(\beta')^2 \, d_{n0}^{(2)}(\theta'')^2 \times $$

$$\times \frac{1}{6D_\perp + m^2(D_\parallel - D_\perp) + n^2 D_{int}}, \quad (11)$$

where

$$C = \begin{cases} 2I(I+1)\gamma^4\hbar^2 r^{-6} & \text{dipolar interaction} \\ \frac{3}{8}\cdot(eqQ/\hbar)^2 & \text{quadrupolar interaction } (\eta = 0, I = 1). \end{cases}$$

Equation (11) is completely equivalent to Equation (8). Here also θ'' is the angle between the dipole-dipole vector or the field gradient symmetry axis (the asymmetry parameter η was set equal to zero in this derivation and therefore corresponds to cylindrical symmetry of the field gradient), depending on the interaction type, and the internal rotation axis, β' is the angle between this axis and the symmetry axis of the rigid part of the molecule. The field gradient axis usually coincides with the direction of the chemical bond. The new symbols have the meaning: I nuclear spin, γ magnetogyric ratio, r intramolecular dipole-dipole separation, eqQ/h quadrupole coupling constant.

Equation (11) which is an extension of Equations (33) and (34) in Reference [1] reduces to the latter if $D_{int} = 0$, i.e. $\theta'' = 0$.

2.2.2. Translational Contribution

In order to take the relative translational motion of the molecules into account it was assumed that they perform independent random walks, uninfluenced by the presence of other molecules. The result for the intermolecular dipolar relaxation rate is:

$$\left(\frac{1}{T_1}\right)_{inter} = \frac{4\pi}{3} I(I+1)\gamma^4\hbar^2 \frac{n\tau}{a^3}\left\{1 + \frac{2}{5}\frac{a^2}{D\tau}\right\}, \quad (12)$$

where the new symbols have the following meaning: n nuclear spin concentration (i.e. protons per cm^3), a distance of closest approach of the molecular centers of mass. Equation (12) is completely equivalent to Equation (10).

3. Experimental

The neutron scattering experiments were performed with the cyrstal time-of-flight spectrometer at the Karlsruhe research reactor FR-2. Some details were given in the previous paper [1].

Deuteron relaxation rates were measured by the 90°–90° pulse sequence using a nmr pulse spectrometer operating at 12 MHz.

The translational self-diffusion constants were determined by the nmr spin-echo method, observing the proton resonance at 25.5 MHz, with a steady field-gradient.

4. Results

4.1. Benzene (C_6H_6)

The measured deuteron relaxation rates are plotted in Figure 1 together with results from other authors [8, 9]. The agreement is quite satisfactory over the total temperature range 10° to 75°C.

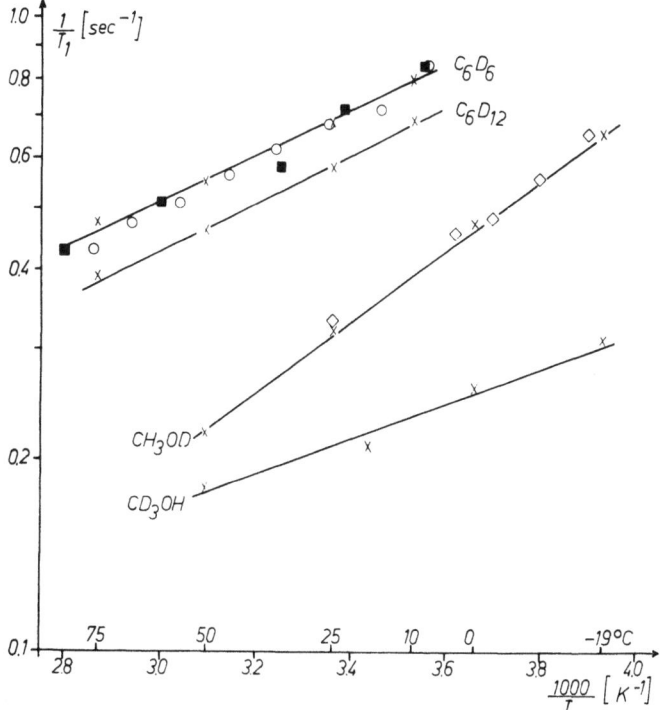

Fig. 1. Deuteron relaxation rates. × own measurements; ○ data from Reference 8; ■ data from Reference 9; ◇ data from Reference 17. The values for CH$_3$OD are reduced by the factor 10.

Since benzene is a symmetric rotator Equation (11) with $D_{int} = 0$ should be applied to evaluate the rotational diffusion constants $D_\perp$ and $D_\parallel$. Of course from one experimental quantity alone both constants cannot be obtained. Recently Gillen and Griffiths [10] employed the Raman line shape method to obtain $D_\perp$ and also combined these results with deuteron relaxation rates (their experimental deuteron rates completely agree with the results in Figure 1) to obtain $D_\parallel$. The constant C in Equation (11) could be obtained from the known quadrupole coupling constant of 193 kHz [11], β' is equal to $90°$ since the C_6-axis which is the symmetry axis and the C—D bonds are perpendicular to each other.

Intra- and intermolecular proton relaxation rates were separately determined by Powles and Figgins [12] over a wide temperature range. From the intramolecular rate no new information is obtained because all proton-proton vectors again are perpendicular to the symmetry axis, however, the above results from the deuteron rates are confirmed using the known intramolecular proton-proton distances [5]. The intermolecular rate is evaluated on the basis of Equation (12). For this purpose the distance of closest approach between molecular centers, a, was estimated from the molar volume [5] and a molecular model, an average value of 6 Å was found. With this figure the values for τ, the mean time between jumps, were calculated using available literature data for the translational self-diffusion coefficient D. The latter obtained from nmr spin-echo experiments [13] and radioactive tracer studies [14] are shown in Figure 2. It should be noted that calculation of a from Equation (12) by assuming $\tau = 0$ yields unreasonably small values between 3.4 and 2.6 Å in the temperature range from 10 °C to 75 °C.

Now with the rotational diffusion constants obtained from the deuteron relaxation rates the rotational contribution to the intermediate scattering law can be calculated numerically from Equation (8) with $D_{int} = 0$. The distance d from the molecular center of mass to the protons equals 2.44 Å, for β' we find $90°$. The sum over l converged quite rapidly and not more than 2 to 6 terms (depending on κ) had to be included. For the calculation of the translational contribution to the intermediate scattering law Equation (10) was used, again the translational self-diffusion constants as found from spin-echo and tracer measurements were inserted. On the other hand the time τ was treated as a fit parameter and adjusted to the 'experimental' data of $I_{trans} \cdot I_{rot}$ which were obtained as described in detail in the previous paper [1]. The calculated product functions $I_{trans} \cdot I_{rot}$ are compared with the experimental data in Figure 3 at the temperatures 10, 25, 50, and 75 °C. The unbroken lines represent the calculated product $I_{trans} \cdot I_{rot}$, the dashed lines give the calculated I_{trans} alone while the experimental data are represented by different symbols. $I(\kappa, t)$ is plotted as a function of time, different curves with κ as parameter are shown. The included error bars were obtained by drawing extreme smoothed curves through the experimental points of the scattering law and Fourier-transforming these. A comparison between theoretical and measured curves shows that only for longer times, usually above 3×10^{-12} s, a satisfactory fit could be obtained. Therefore curves with a large κ-value cannot be fitted at all. This behaviour was expected since the diffusion equations

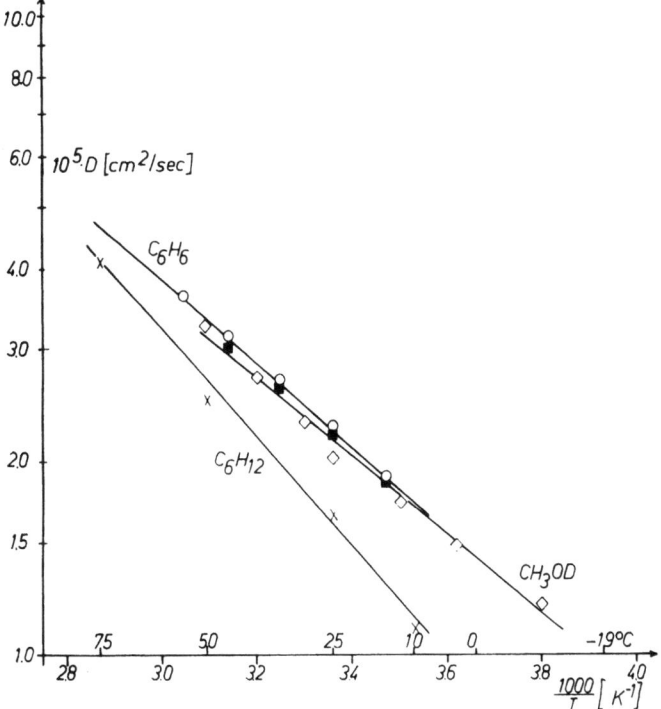

Fig. 2. Translational self-diffusion coefficients. × own measurements; ○ data from Reference 13; ■ data from Reference 14; ◇ data from Reference 20.

do not hold for short times. The most important result is that the rotational contribution is rather large as may be seen from a comparison with the I_{trans}-curves at the same κ-values in the figure. This has the effect of making the total intermediate scattering law non-exponential (or the scattering law non-Lorentzian). The adjusted τ-values are close to zero, i.e. the translational motion is described quite satisfactorily by a diffusion equation. In contrary the τ-values derived from the intermolecular relaxation rates are larger by one order of magnitude. This result leads to the conclusion that the translational motion is a collective effect, i.e. the whole neighbourhood of a selected molecule moves in the same direction for a certain time thereby reducing the relative translational diffusion coefficient (which corresponds to an increase of τ). The short-range dependence of the dipolar interaction justifies the statement that only the nearest neighbourhood is affected in this collective process.

All experimental nmr data and self-diffusion values, calculated rotational diffusion constants and ajusted τ-parameters for benzene are collected in Table II.

4.2. Cyclohexane (C_6H_{12})

The same procedure as described for benzene was used to evaluate the experimental results for cyclohexane.

The measured deuteron relaxation rates are included in Figure 1. From these data

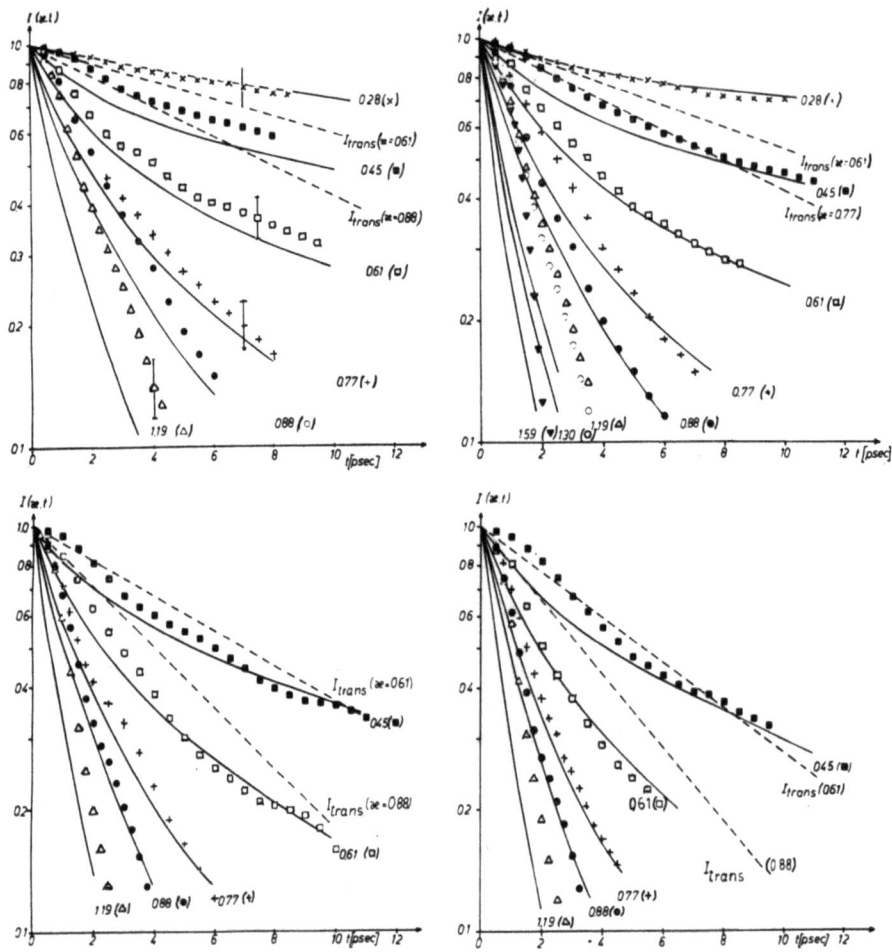

Fig. 3. Intermediate scattering law for benzene at 10°C, 25°C, 50°C and 75°C (from upper left to lower right).

the isotropic rotational diffusion constant $D_\parallel = D_\perp$ was calculated using Equation (11) again with $D_{int} = 0$. In this case Equation (11) reduces to the well-known formula

$$\left(\frac{1}{T_1}\right)_{intra} = C \cdot \frac{1}{6D_\perp}. \tag{13}$$

The quadrupole coupling constant was taken as 174 kHz [11]. Separate intra- and intermolecular proton relaxation rates are available only at 25°C from a previous publication [5]. Translational self-diffusion constants were measured and are shown in Figure 2.

The rotational contribution to the intermediate scattering law is found from Equation (8) with $D_\parallel = D_\perp$ and $D_{int} = 0$. In this case Equation (8) reduces to

$$I_{rot}(\kappa, t) = \sum_{l=0}^{\infty} (2l + 1) j_l(\kappa d)^2 e^{-l(l+1) D_\perp \cdot t}. \tag{14}$$

TABLE II
Experimental and calculated values for benzene

T (°C)	Experimental relaxation rate $1/T_1$ (s^{-1})			Experimental translational self-diffusion constant D (cm^2 s^{-1}) References 13 and 14	Calculated rotational diffusion constant Reference 10		Calculated mean time between jumps τ (s)	
	Deuterons	Protons (intra) Reference 12	Protons (inter) Reference 12		$D_\perp$ (s^{-1})	$D_\parallel$ (s^{-1})	Neutron scattering	NMR
10	0.80	1.09×10^{-2}	5.0×10^{-2}	1.7×10^{-5}	0.042×10^{12}	0.33×10^{12}	4×10^{-12}	65×10^{-12}
25	0.70	0.97	3.9	2.2	0.053	0.35	4	53
50	0.56	0.85	2.9	3.3	0.074	0.37	2	47
75	0.46	0.74	2.2	4.7	0.099	0.40	2	41

The value to be used for d is an average of the two sets of protons in cyclohexane, which has the D_{3d} symmetry (chair form), and distances of 1.98 Å and 2.49 Å (average 2.24 Å) occur [5]. The calculated total intermediate scattering law again can be fitted satisfactorily to the experimental data, as shown in Figure 4 for four different temperatures. The importance of the rotational contribution is manifested also in this case. The best fit was achieved for $\tau = 2 \times 10^{-12}$ s, again rather close to the diffusion limit.

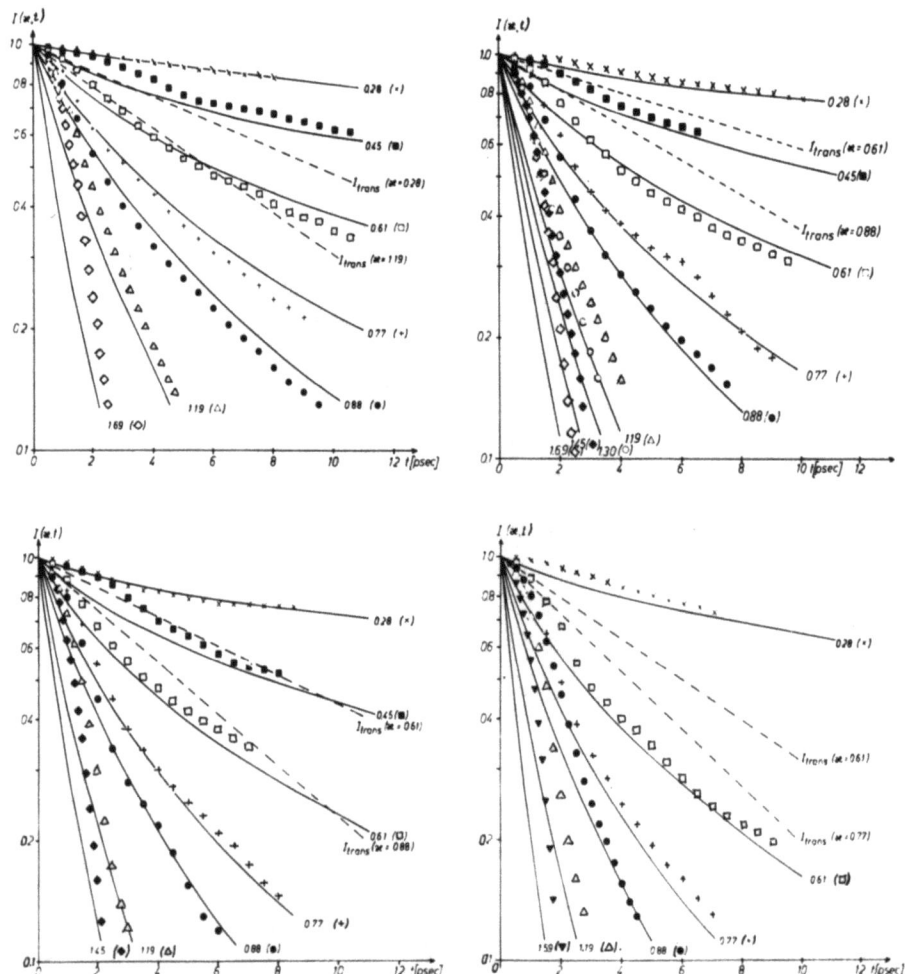

Fig. 4. Intermediate scattering law for cyclohexane at 10 °C, 25 °C, 50 °C and 75 °C (from upper left to lower right).

From the nmr intermolecular proton rate we get 100×10^{-12} s with $a = 6.4$ Å which is considerably larger and thus indicates collective translational motion also in liquid cyclohexane.

Table III summarizes all numerical data.

TABLE III
Experimental and calculated values for cyclohexane

T (°C)	Experimental relaxation rate $1/T_1$ (s^{-1})			Experimental translational self-diffusion constant D (cm^2 s^{-1})	Calculated rotational diffusion constant $D_\perp$ (s^{-1})	Calculated mean time between jumps τ (s)	
	Deuterons	Protons (intra) Reference 5	Protons (inter) Reference 5			Neutron scattering	NMR
10	0.68	—	—	1.1×10^{-5}	0.11×10^{12}	2×10^{-12}	—
25	0.58	5.1×10^{-2}	8.9×10^{-2}	1.6	0.13	2	100×10^{-12}
50	0.46	—	—	2.7	0.16	2	—
75	0.38	—	—	4.1	0.20	2	—

4.3. Methanol (CD$_3$OH and CH$_3$OD)

From nmr relaxation measurements it was concluded that the methyl group in methanol performs fast internal rotational motion [15, 16, 17]. Quasielastic neutron scattering should lead to the same conclusion if scattering at the methyl group using CH$_3$OD and scattering at the hydroxyl group using CD$_3$OH are compared.

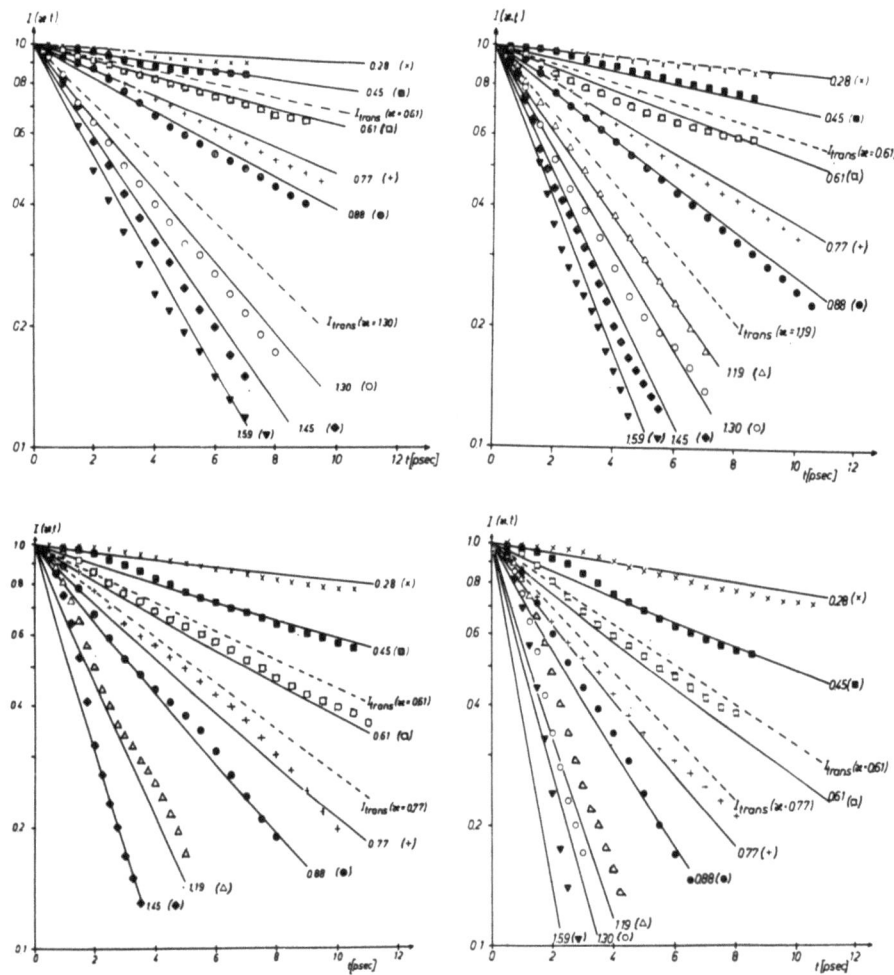

Fig. 5. Intermediate scattering law for methanol (CD$_3$OH) at $-19\,°C$, $0\,°C$, $25\,°C$ and $50\,°C$ (from upper left to lower right).

Previous experiments of this kind [18] which however were evaluated in a different way than described here seem to indicate indeed larger broadening of the quasi-elastic peak for CH$_3$OD.

Deuteron relaxation rates were measured for both alcohols, the data are found in Figure 1 together with values for CH$_3$OD from the literature [17]. The rates for

TABLE IV
Experimental and calculated values for methanol

T (°C)	Experimental deuteron relaxation rate $1/T_1$ (s^{-1})		Experimental translational self-diffusion constant D (cm^2 s^{-1})	Calculated rotational diffusion constant $D_\perp$ (s^{-1})	Calculated rotational diffussion constant D_{int} (s^{-1})
	CH$_3$OD	CD$_3$OH			
−19	6.6	0.30	1.0×10^{-5}	0.02×10^{12}	9.7×10^{12}
0	4.6	0.26	1.4	0.03	3.7
25	3.1	0.21	2.2	0.05	2.6
50	2.2	0.18	3.1	0.07	

CH$_3$OD are larger by one order of magnitude and also the activation energy amounts to 2.6 kcal mole^{-1} as compared to 1.3 kcal mole^{-1} for CD$_3$OH. Other authors give similar values for the activation energies [16, 17]. Thus the easier reorientation of the methyl group shows up also in the smaller activation energy. For further evaluation of the relaxation rates the following quadrupole coupling constants were chosen: 170 kHz for C—D [19] and 248 kHz for O—D the latter value being the coupling constant in D$_2$O [9, 15]. For evaluation of the rotational diffusion constants we will consider the rotation of the hydroxyl group to be isotropic, thus Equation (13) is used for CH$_3$OD, and Equation (11) with $D_\| = D_\perp$ and $\beta' = 0$ is used for CD$_3$OH. θ'' the angle between the C—D bond and the internal rotation axis is tetrahedral. Except at the lowest measured temperature acceptable values for D_{int} were calculated, the failure at $-19\,°C$ may be due to an inappropriate coupling constant used in the calculation.

Translational self-diffusion constants are available from the literature [5, 20]. They are plotted in Figure 2.

The rotational contribution to the intermediate scattering law in the case of CD$_3$OH (isotropic rotation of the hydroxyl group) is calculated from Equation (14). d is taken as 1.4 Å. The calculated and experimental products $I_{trans} \cdot I_{rot}$ for CD$_3$OH are reproduced in Figure 5. The best fit was achieved for $\tau = 0$ at all temperatures. Because of the small rotational diffusion constant the rotational contribution is rather small and the functions are very nearly exponential.

By combining the scattering data for CH$_3$OD and CD$_3$OH the ratio $I_{rot}^{CH_3OD}/I_{rot}^{CD_3OH}$ could be found since the translational contributions are equal and thus cancel. Theoretically this ratio can be calculated from a knowledge of $D_\perp$ and D_{int}, with $d = 1.5$ Å and $\theta'' = 42°$ for the methyl protons, using Equations (8) (with $D_\perp = D_\|$) and (14). We expect this ratio to be less than one and qualitatively this is found experimentally, i.e. the quasielastic line is additionally broadened for CH$_3$OD due to the internal rotation. However, the experimental data are not accurate enough (the resolution is not sufficient) to justify quantitative evaluations.

Table IV contains all numerical data for methanol.

5. General Conclusions

From the results presented in the previous section we may derive the following general conclusions:

(1) NMR relaxation times and quasielastic neutron scattering line widths are comparable on the basis of the random-walk theory.

(2) The broadening of the quasielastic neutron scattering line results from translational *and* rotational processes.

(3) The rotational motion renders the quasielastic line non-Lorentzian.

(4) For the mean time between translational jumps a few pico-seconds are calculated and often the limiting case of pure diffusion is approached.

(5) Collective translational motion is present so that the relative diffusion coef-

ficient between neighbouring molecules is smaller than in the case of independent motion.

Acknowledgements

The author is grateful to Dr W. Gläser of the Kernforschungszentrum Karlsruhe for the possibility to use the time-of-flight spectrometer at the Research Reactor FR 2.

References

1. Zeidler, M. D.: *Ber. Bunsenges. physik. Chem.* **75**, 769 (1971).
2. Springer, T.: *Quasielastic Neutron Scattering for the Investigation of Diffusive Motions in Solids and Liquids,* Springer-Verlag, Berlin, 1972.
3. Favro, D.: R .E. Burgess (ed.), *Fluctuation Phenomena in Solids,* Academic Press, New York, 1965.
4. Edmonds, A. R.: *Drehimpulse in der Quantenmechanik,* Bibliographisches Institut, Mannheim, 1964.
5. Zeidler, M. D.: *Ber. Bunsenges. physik. Chem.* **69**, 659 (1965).
6. Hubbard, P. S.: *Physic. Rev.* **131**, 275 (1963).
7. Versmold, H.: *Z. Naturforsch.* **25a**, 367 (1970).
8. Woessner, D. E.: *J. Chem. Phys.* **40**, 2341 (1964).
9. Powles, J. G., Rhodes, M., and Strange, J. H.: *Mol. Phys.* **11**, 515 (1966).
10. Gillen, K. T. and Griffiths, J. E.: *Chem. Phys. Letters* **17**, 359 (1972).
11. Rowell, J. C., Phillips, W. D., Melby, L. R., and Panar, M.: *J. Chem. Phys.* **43**, 3442 (1965).
12. Powles, J. G. and Figgins, R.: *Mol. Phys.* **10**, 155 (1966).
13. Falcone, D. R., Douglass, D. C., and McCall, D. W.: *J. Phys. Chem.* **71**, 2754 (1967).
14. Collings, A. F. and Mills, R.: *Trans. Faraday Soc.* **66**, 2761 (1970).
15. von Goldammer, E. and Zeidler, M. D.: *Ber. Bunsenges. physik. Chem.* **73**, 4 (1969).
16. von Goldammer, E. and Hertz, H. G.: *J. Phys. Chem.* **74**, 3734 (1970).
17. O'Reilly, D. E. and Peterson, E. M.: *J. Chem. Phys.* **55**, 2155 (1971).
18. Aldred, B. K., Eden, R. C., and White, J. W.: *Disc. Faraday Soc.* **43**, 169 (1967).
19. Lucken, E. A. C.: *Nuclear Quadrupole Coupling Constants,* Academic Press, New York, 1969.
20. Nothnagel, K. H. and Weiss, A.: *Ber. Bunsenges. physik. Chem.* **73**, 1033 (1969).

DISCUSSION

Friedman: In your talk you implied that there is some appreciable error in transforming the experimental data to get the $I(\kappa, t)$ function. I wonder, then, why you do calculate the effective scattering function from your model and the instrumental factors to get something which can be compared more directly with experiment.

Zeidler: Fourier transformation from the frequency to the time domain has to be performed anyhow in order to deconvolute the measured spectra from the instrumental line width. Therefore one might as well do the comparison with the theoretical data, computed directly as $I(\kappa, t)$, in this representation. Of course one could turn around and calculate the effective scattering law, as you suggest, but I prefer to have experimental data, in this case the time correlation function $I(\kappa, t)$, which are typical of the investigated liquid only.

Springer: You have found a much larger intermolecular characteristic relaxation time from NMR as from neutron scattering experiments. Can this be interpreted in terms of collective or 'coordinated' motions?

Zeidler: The mean time between jumps, τ, should actually come out to be the same from the two types of experiments, if the underlying theory is correct. The observed discrepancy is therefore due to an unjustified assumption in the derivations, which if traced back, most probably arises from the calculation of the relative translational propagator (appearing in the intermolecular dipolar relaxation rate) according to Equation (25) in Reference 1. This equation is valid only for independent translational motion of the molecules. In order to obtain from the intermolecular relaxation rate through Equation (12) the same τ as from neutron scattering, the translational diffusion coefficient D must be smaller than the spin-echo self-diffusion coefficient which was inserted to calculate τ before. This smaller value of D indicates that the relative diffusion, especially between neighbouring molecules since the dipolar interaction is rapidly decreasing with internuclear distance, is delayed, i.e. neighbouring molecules tend to move in the same direction.

White: Could you make an estimate of the relative contribution of rotational motion to the neutron quasielastic scattering in CH_3OD and CD_3OH? Is there any evidence that the centre of rotation in CH_3OH is near to the OH group and away from the free molecule centre of mass?

Zeidler: The rotational contribution to the quasielastic peak is larger for CH_3OD due to the fast internal rotation of the methyl group.

The value for d in Equation (8) which corresponds to the distance from the centre of mass of the molecule to the scattering nucleus was not fitted to the experimental data but was calculated from the molecular model. I therefore have no knowledge about a possible shift of the centre of mass.

Volino: Did you plot the values of the translational correlation time τ deduced from neutron data, as a function of reciprocal temperature, in order to deduce some activation energy and how can it be compared with the similar activation energy deduced from T_1 NMR data?

Zeidler: The τ values deduced from neutron scattering were obtained by first estimating the rotational contribution to the intermediate scattering law and then fitting the residuary intermediate scattering law to the translational contribution as given by Equation (10). Any error in estimating the rotational contribution produces a large error in τ. Probably ± 2 ps is a reasonable estimate for the error. Therefore it is not justified to calculate activation energies. On the other hand the temperature dependence is as expected, the values of τ decrease with increasing temperature.

Hertz: It should be clearly pointed out that these very long τ's indicate that the theory is not good enough and they should not be taken as an evidence for a long sticking together of two benzene molecules.

Zeidler: As pointed out in the answer to Dr Springer's question the large τ values obtained from the intermolecular dipolar relaxation rates have no direct physical meaning.

Litovitz: How model dependent are the results in benzene which indicate that the relative diffusion coefficient is much less than the self-diffusion coefficient?

Zeidler: I think that the difference observed in the τ values deduced from the different experiments is too large to be due alone to an inadequate model. However it should be mentioned that in the expression for the intermolecular relaxation rate Equation (12) two simplifications were introduced: the relative rotational motion is neglected and the pair distribution function is approximated by a step function. Theoretical arguments indicate that these neglects are not very serious (see for example Harmon, J. F. and Muller, B. H.: *Phys. Rev.* **182**, 400 (1969)).

INELASTIC NEUTRON SCATTERING BY CHEMICAL REACTIONS

I: *Structure and High Frequency Dynamics of Trifluoroacetic Acids*

J. C. LASSEGUES

Laboratoire de Spectroscopie Infrarouge, Université de Bordeaux I, Bordeaux, France

and

J. W. WHITE

Physical Chemistry Laboratory, South Parks Road, Oxford, England

Abstract. In this paper, we analyse the possibilites of application of inelastic neutron scattering to the study of chemical reactions.
In the first part of the work on the reacting system trifluoroacetic acid/water, we present the results obtained for the pure liquid and solid CF_3COOH and CF_3COOD. The neutron diffraction, time of flight and energy loss experiments provide new information on the structure and vibrational assignment of these acids in both phases. The quasielastic broadening in the neutron time of flight spectra of the liquids and the NMR spin-spin and deuteron spin lattice relaxation time measurements are then applied to the study of molecular dynamics. These techniques give upper limits of 1.7×10^{-5} cm^2 s^{-1} and 1.6×10^{10} s^{-1} respectively for the centre of mass and rotational diffusion constants.

Résumé. Les possibilités d'application de la diffusion inélastique des neutrons à l'étude des réactions chimiques sont analysées. Dans le cadre d'une étude du système acide trifluoroacétique/eau, nous présentons tout d'abord les résultats obtenus pour les acides CF_3COOH et CF_3COOD liquides et solides purs. Des mesures de diffraction, de temps de vol et de perte d'énergie avec des neutrons froids, permettent d'obtenir des informations nouvelles sur la structure et l'attribution vibrationnelle de ces acides. L'analyse de l'élargissement du pic quasiélastique et la détermination des temps de relaxation spin-spin et deueteron-spin réseau par RMN permettent ensuite d'évaluer les limites supérieures des constantes de diffusion du centre de masse et de diffusion rotationnelle (1.7×10^{-5} cm^2 s^{-1} et 1.6×10^{10} s^{-1} respectivement).

1. Introduction

Many features of the method of inelastic neutron scattering spectroscopy indicate that this technique may provide novel information on the kinetics of very fast, diffusion limited chemical reactions. The method gives unique information on the microscopic aspects of atomic and molecular diffusion in simple and molecular liquids [1, 2, 3] because of the large and controllable momentum transfer which attends slow neutron scattering by molecules. Typically this is about 1 Å^{-1} and so processes of correlation length near to intra and intermolecular distances are 'observable'. The dependence on momentum transfer is studied by observing scattering spectra at many scattering angles and the approach to breakdown of hydrodynamic equations of motion at short times can be thus directly observed. In addition the scattering cross-section may be strongly weighted in favour of scattering by chosen nuclei in a molecule and this selectivity has great potential for separating the contributions from diverse processes [3, 4]. This initial paper sketches some of theoretical possibilities for neutron scattering by reactions, and presents the quasielastic and inelastic neutron scattering spectra of solid and liquid trifluoroacetic acids.

Recently considerable attention has been given by three groups to the spectrum of light scattered by a chemical reaction [5, 6, 7]. The formalism of Berne and Pecora is most close to that usually encountered in neutron scattering theories so it will be the more often referred to here. They write the intensity of light scattered by polarisability fluctuations into unit volume and for frequency shift ω as

$$I(Q, \omega) = \left(\frac{k_f^4 |E_0|^2}{16\pi^2 \varepsilon^2}\right) \cdot \bar{C}(Q, \omega), \tag{1}$$

where the momentum transfer, $\mathbf{Q}$, in scattering from initial, $\mathbf{k}_0$, to final wavevector $\mathbf{k}_f$ is given by

$$\mathbf{Q} = \mathbf{k}_f - \mathbf{k}_0. \tag{2}$$

E_0 is the incident electric field vector (directed along the z-axis), ε is the dielectric constant of the scattering medium and $\bar{C}(Q, \omega)$ is the symmetrical scattering law defined by the time Fourier transform

$$\bar{C}(Q, \omega) = (2\pi v)^{-1} \int_{-\infty}^{\infty} dt \exp[-i\omega t] \cdot \left\langle \sum_{i=1, j=1}^{N} \alpha_{yz}^{j*}(0) \alpha_{yz}^{i}(t) \times \right.$$
$$\left. \times \exp - i\mathbf{Q} \cdot \mathbf{r}_i(0) \cdot \exp i\mathbf{Q} \cdot \mathbf{r}_j(t) \right\rangle, \tag{3}$$

where $\alpha_{yz}^{i}(t)$ is the yz component of the polarisability of molecule i at time t. The analogous expression for the symmetrical neutron scattering law $S(Q, \omega)$ is [8]

$$S(Q, \omega) = (2\pi v)^{-1} \int_{-\infty}^{\infty} dt \exp[-i\omega t] \left\langle \sum_{i=1, j=1}^{N} b^{*i}(0) b^{j}(t) \times \right.$$
$$\left. \times \exp - i\mathbf{Q} \cdot \mathbf{r}_i(0) \exp i\mathbf{Q} \cdot \mathbf{r}_j(t) \right\rangle, \tag{4}$$

where b^i, b^j are the neutron scattering lengths of the *nuclei* i, j respectively at positions $\mathbf{r}_i(0)$, $\mathbf{r}_j(t)$.

The close similarity between (3) and (4) is obvious but so also are the first major differences namely that the neutron method picks out either correlated or individual nucleir motions (depending on whether the distinct or self part of (4) is examined). Unlike light scattering where the reaction modulates a molecular polarisability, neutron scattering can be arranged to 'observe' the reacting or exchanging particle. Because Q is much larger for neutrons the spacial extent of the observable dynamical processes may also be smaller by a factor of about 10^{-3} than in light scattering. Among the interesting consequences of the first point are that very small changes in molecular structure may be detectable; exchange of, for example, a single proton may not significantly change the molecular inertia tensor thence the scattering from rotational diffusion, and the isotopic substitution method [4] may play a valuable role.

To obtain working expressions for the light scattering cross-section Berne and Pecora make two important assumptions, namely that a continuum equation of motion for the probability distribution function $P_{\gamma\alpha}(\mathbf{r}, \Omega, t)$ [5c] can be set up and that this adequately represents the microscopic motion of the system. Secondly, with Blum and Salzburg [6a] they distinguish the relaxation produced by transport processes such as diffusion of the centre of mass (which are controlled by spacial 'concentration gradients') from relaxation by chemical reaction whose rate is deemed to be controlled by point gradients (of the free energy ΔG).

The optical scattering law derived from this basis in the limit of small Q contains Lorenzian terms representing the rotational and centre of mass diffusion of the unreacted molecule, and a simple Lorenzian of fixed width from the reaction. This sum of Lorenzians is centred around zero frequency. The centre of mass diffusion has a momentum dependent width $\Delta E = 2\hbar DQ^2$ where D is the Fick's law diffusion coefficient. In the weak hindering approximation the tensor correlations in Equation (3) select rotational relaxation functions of order $l=2$ from a spherical harmonic expansion of the solutions to the rotational diffusion equation [9]. Further simplification arises for cylindircal molecules since then only one m value contributes to the cross-section. Such simplifications to the neutron scattering cross-section naturally arise in the same way as can be seen for example by a group theoretical expansion of the double sum in Equation (4). Further information on diffusion functions (e.g. at higher Q) may be obtained from studies of the inelastic part of the spectrum because of the necessary convolution of vibrational and diffusive spectra.

Hydrodynamic models of the molecular diffusive motions are probably satisfactory at small Q values but already neutron experiments indicate some limits to their applicability at short times and ranges [1, 10] even for monatomic liquids. Similar departures from simple diffusive behaviour occur for molecular liquids but perhaps the most interesting aspect of neutron scattering from a reacting fluid at momentum transfers between 0.1 Å^{-1} and 10 Å^{-1} arises from the possibility that the reactive part of the cross-section becomes Q dependent. Examples of this behaviour arise with coupling between reactive and 'normal' transport processes [11] and if the 'reacting' moeity moves in a free energy gradient during the course of the reaction.

The region of energy-momentum phase space in which scattering from reactons should be observable with present neutron spectrometers lies in the area $\Delta E = 10^8$–10^{11} s^{-1}, $Q = 0.01$ Å^{-1}–10 Å^{-1} and this covers a number of diffusion limited proton transfer reactions and intramolecular rearrangements. In case of Q independent reactive scattering there are considerable advantages to be gained by working at low Q so that centre of mass and rotational diffusion effects are suppressed. Although there is some doubt about the interpretation of these results it seems that amongst the fastest of the proton transfer reactions is that between hydronium and trifluoroacetate ions [12]

$$CF_3COOH + H_2O \rightleftharpoons CF_3COO^- + H_3O^+$$

which will be studied as part of the present work.

From studies of molecular dynamics in liquids it is clear that to interpret the neutron scattering from reacting systems at this stage, full use must be made of all information on molecular diffusion and chemical kinetics from other techniques. Thus for the trifluoroacetic acid and trifluoroacetic acid/water systems we will draw on nuclear magnetic resonance spin-spin relaxation time (T_2) measurements in pulsed field gradients for the macroscopic centre of mass diffusion constants of ^{1}H and ^{19}F [13]. From deuteron spin lattice relaxation times [14] and depolarised Raman [15] and Infra-red [16] line broadening we obtain estimates of the relaxation times for the tensor and vector rotational correlation functions $F_2(t)$, $F_1(t)$ [9] respectively.

2. The Trifluoroacetic Acid System

In most pure carboxylic acids and their solutions there is extensive molecular association [17] and rates of processes connecting dimers, monomers and chains must first be shown to lie outside the observed frequency region of neutron scattering.

2.1. Thermodynamics and Structure

Table I [18] summarizes some of the thermodynamic data for the dimerization of formic, acetic and trifluoroacetic acids.

TABLE I

Thermodynamics of association – carboxylic acids

Acid	Solution in benzene at 30°C	Vapour phase at 100°C			
	K_{30}	K_{100}	ΔF_{100} (cal mole^{-1})	ΔH_{100} (cal mole^{-1})	ΔS_{100} (cal mole^{-1} deg^{-1})
HCOOH	7.1	301	686	14.110	36.0
CH$_3$COOH	2.4	92.0	1570	15.270	36.7
CF$_3$COOH	–	435	415	14.050	36.5
CF$_3$COOD	–	461	370	13.920	36.3

Although the hydrogen bonded dimers in CF$_3$OOCH are weaker than for CH$_3$COOH the tendency is still strong and persists in solutions. Association in non and slightly polar solutions has been extensively studied [19, 20] with the conclusion that for CF$_3$COOH the dimer is cyclic and probably the only associated species [20]. In aqueous solutions CF$_3$COOH is a strong electrolyte [21] (Acidity constant K_{25} between 4 and 9 mole l^{-1}) compared to acetic acid $K_{25} = 1.8 \times 10^{-5}$.

The high enthalpy of association for CF$_3$COOH undoubtedly affects the structures of the crystal and pure liquid phases. No crystal structure of solid phases of CF$_3$COOH has been published, but a recent study of solid monofluoroacetic acid [22] shows that the cyclic dimers persist in the solid. The space group $P2_{1/c}$ derives from isolated dimers in contrast to the situation in acetic and formic acids which have space groups

$P_{na}2_1$ based on chains [23, 24]. For CF_3COOH, Berney has recently shown [25] that there is a specific heat anomaly at 220. K, possibly due to a structural change. In all phases the weigth of spectroscopic evidence [26] suggests a centrosymmetrical crystal structure involving cyclic dimers, but it is not yet possible to decide if these cyclic dimers are isolated [25] like in monofluoroacetic acid [22] or strongly bonded to form chains [25].

When the pure liquid warms up from the melting point. 257.9 K ($-15°C$) to room temperature a large increase in dielectric constant occurs [27], which indicates large structural changes, possibly dissociation. Tentatively therefore we may conclude that the pure liquid at 25 °C. is an intermediate situation between formic acid which forms polymeric chains and acetic acid mainly composed of cyclic dimers.

2.2. Association Kinetics

Extensive measurements by ultrasonic relaxation have been made on pure carboxylic acids [28]. Two relaxation intervals around 10^6 and 10^9 s^{-1} have been found. The first one has been assigned to an intramolecular equilibrium between cyclic and open dimers and the second to the equilibrium between monomer and dimers.

$$2R-C\begin{matrix}O\\O-H\end{matrix} \underset{\sim 10^9 s^{-1}}{\overset{\sim 10^9 s^{-1}}{\rightleftarrows}} \begin{matrix} R-C\begin{matrix}O\cdots H-O\\O-H\cdots O\end{matrix}C-R \\ \updownarrow \sim 10^6 s^{-1} \\ R-C\begin{matrix}O\quad H-O\\O-H\cdots O\end{matrix}C-R \end{matrix}$$

No such measurement has been made on trifluoroacetic acid, but if the CF_3COOH system resembles other carboxylic acids as seems likely from the arguments above, only the faster rate process ($\sim 10^9$ s^{-1}) is likely to come near to the rate of process contributing to neutron inelastic scattering with a spectrometer resolution of $\sim 10^{10}$ s^{-1}.

3. Experimental Conditions

Inelastic neutron scattering measurements by time of flight were made with the 6H [29] and 4H5 cold neutron spectrometers on the DIDO reactor and with the Beryllium filter spectrometer [30] on the PLUTO reactor at HARWELL. The incident neutron energies E_0 were 37 cm^{-1} and 25 cm^{-1} respectively and the chopper configuration used gave resolutions of about $\Delta E/E = 10\%$. In some cases a higher resolution was achieved by running the sample with a vertical orientation to the beam [31]. The resolution of the Beryllium filter spectrometer varied from 34 cm^{-1} at 300 cm^{-1} to 28 cm^{-1} at 1100 cm^{-1}.

Some preliminary powder neutron diffraction measurements were made on the

Badger diffractometer at DIDO. The samples were held in cylindrical vanadium containers (12 mm diameter). These studies help to interpret the coherent features in the quasielastic scattering studies but are also interesting from the point of view of the solid and liquid structures.

The CF_3COOH and CF_3COOD were respectively 'pure' and 'puriss' KOCH-LIGHT products. They were enclosed in very thin lead 'sachets' (0.014 cm thickness) sealed with solder and always handled in a very dry atmosphere. The aqueous solutions of these acids have been studied in a polytetrafluoroethylene/container. The neutron spectra were corrected for counter efficiency by the 'CIRCA' and 'ROUNDABOUT' suite of programs [3].

4. Results

4.1. Diffraction from CF_3COOH and CF_3COOD

Figures 1 and 2 show the neutron diffraction from solid and liquid phases of CF_3COOH and CF_3COOD at various temperatures, all taken in a counting time of about 3 min each angular position. For lack of detailed crystal structures, nett plane distances have been calculated for each peak.

The pronounced effect of substituting D for H in the structure is obvious for the low temperature patterns. Since length of the cyclic dimer is about 7.3 Å, it seems possible that this species contributes strongly to the structure factor for the '3.65 Å' peak. Also, the peaks at $2\theta = 28°$; 2.3 Å, $Q = 2.72$ Å^{-1} in CF_3COOH may be related by a factor of two in the order of diffraction and arise from the sideways packing of the dimers or monomers in an orthorhombic or near orthorhombic structure. None of these crystallographic spacings agree with the model of strongly bonded cyclic dimers arranged in chains to form a $P_{2/m}$ monoclinic system [25]. The hypothesis of isolated cyclic dimers [26] is much more likely.

Even at the low resolution used here ($\Delta Q/Q \simeq 5\%$), the change in diffraction pattern between 70 K and 240 K is sufficiently large to confirm that some kind of disorder phenomenon occurs before melting. A shift and progressive attenuation of the '3.65 Å' peak occurs with higher temperatures, continuing into the liquid phase as shown by the 273 and 298 K results for CF_3COOD.

4.2. Trifluoroacetic acids – solids

The time of flight spectra of CF_3COOH and CF_3COOD at 213 K represented on Figure 3 for three different angles of scattering, show a well structured inelastic region and a narrow elastic peak.

4.2.1. *Elastic Region*

A careful comparison of the width of the elastic peak with the resolution width of the spectrometer indicates no detectable broadening in the range of momentum transfer from 0.25 to 2 Å^{-1}. On the other hand, the peak height is strongly increased for some particular values of the momentum transfer due to Bragg diffraction. The positions of these features agree with the data of §4.1 (Figures 1, 2 and 8).

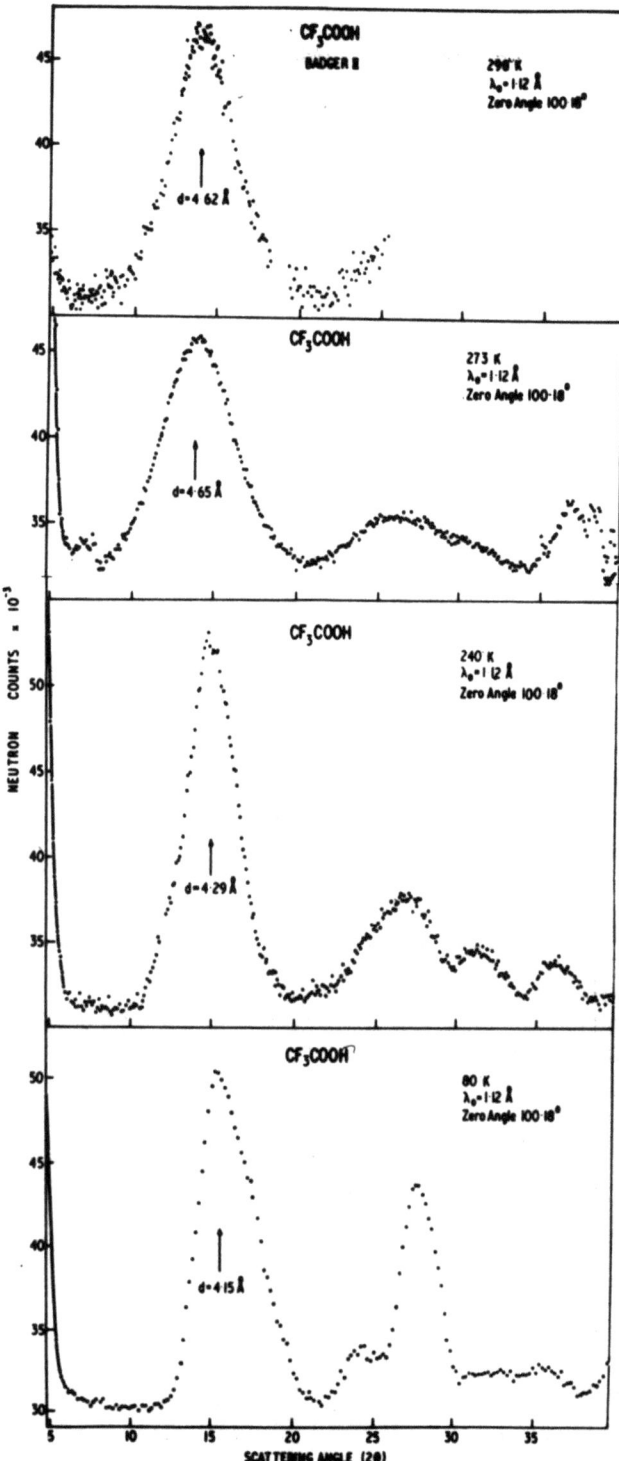

Fig. 1. Diffraction of 1.12 Å neutrons from liquid and polycrystalline CF_3COOH at 298, 273 K and 240, 80 K.

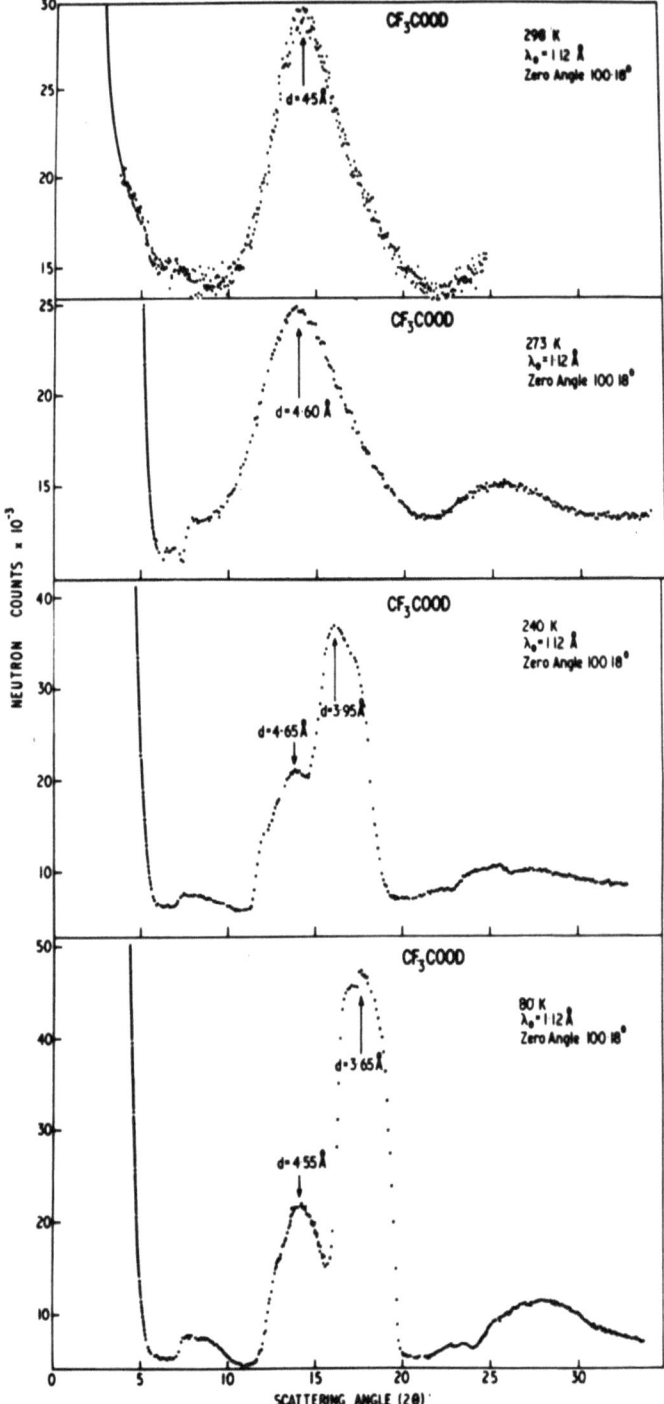

Fig. 2. Diffraction of 1.12 Å neutrons from liquid and polycrystalline CF₃COOD at 298, 273 and 240, 80 K.

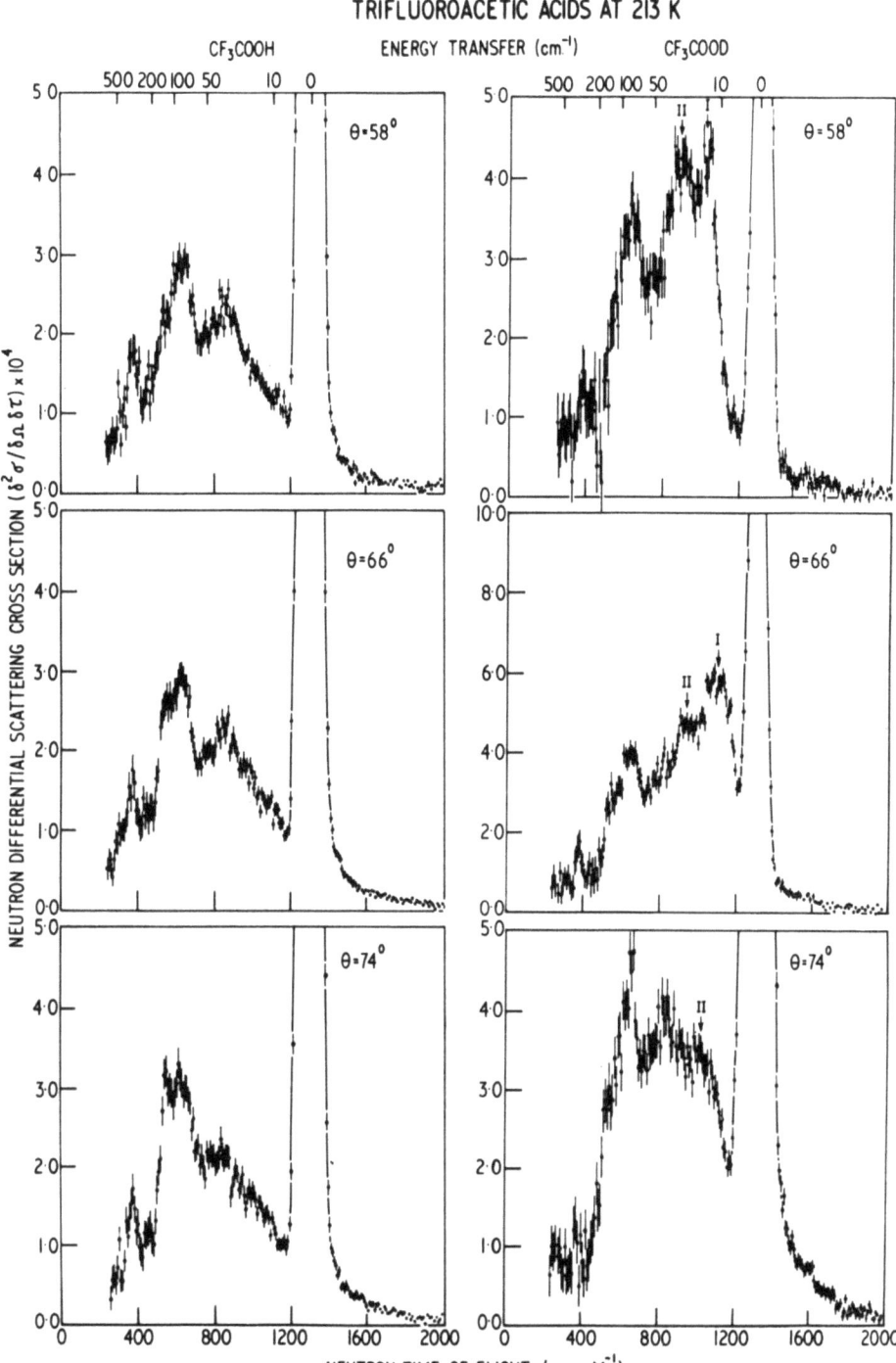

Fig. 3. Inelastic scattering spectra of 5.2 Å neutrons from solid CF₃COOH and CF₃COOD at 213 K taken at scattering angles $\theta = 58$, 66 and 74° to the incident beam direction and with the sample inclined at 45° to the beam.

4.2.2. Inelastic Scattering

The vibrational analysis of a solid composed of isolated cyclic dimers can be made with good approximation using group frequencies.

The comparison with previous infrared and Raman work [26] is interesting. For Solid CF_3COOH, CF_3COOD the infrared and Raman bands have quite different

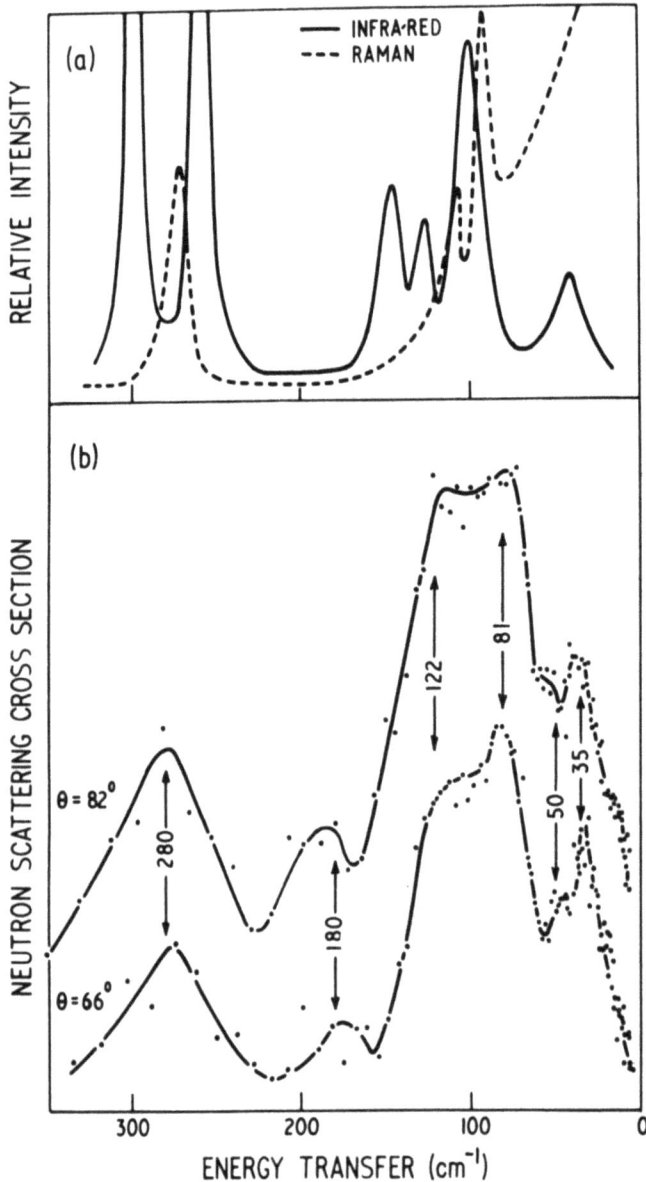

Fig. 4. Comparison of low frequency region (a) in the Infra-red, Raman (26) and (b) in the neutron spectra of solid CF_3COOH taken at two different angles of scattering.

frequencies because of selection rules. The very weak polarizability of the OH group explains the absence of scattering from the OH intramolecular modex. In the inelastic neutron scattering no such selection rules occur and the intensity of a peak is simply related to the cross section of the atoms involved in the vibration and the to amplitude of this vibration. Due to the very large hydrogen cross section, the vibrations involving large proton amplitudes appear very distinctly.

This allows a useful and complementary vibrational assignment to be made. As shown in Figure 4, where the infrared and Raman spectra of solid CF_3COOD [26] between 0 and 300 cm^{-1}, have been compared to the corresponding neutron spectra.

Between 30 and 130 cm^{-1}, very intense peaks appear in the neutron spectra. We think that they are due to the torsion t (OH$\cdots$0) (35 cm^{-1}), the in-plane deformation

TABLE II

Vibrational assignment of CF_3COOH and CF_3COOD

Assignment	Infrared and Raman				Inelastic neutron scattering	
	Liquid [36, 37]		Solid [26]		Liquid	Solid
	IR	R	IR	R		
t (OH$\cdots$0)			41			35
β (OH$\cdots$0)						50 (?)
γ (OH$\cdots$0)			100	92		81.122
t (CF$_3$)		120	100 (?)	108?	80–110	
comb. $\gamma + t$ (?)			127			
ν(OH$\cdots$0)			146			180
ϱ(CF$_3$)	261.284	258.268	258	267	270	280
			300	272		
δ(CCO)	401	400	395	423	380	420*
δs(CF$_3$)	435.449	420.437	458	438	430.460 +	
γ(CCO)	515	525	518	522	530*.529 +	525*
δa(AF$_3$)	598	600	607	599	594 +	615*
δ(COO)	675	689	719	701	730*.710 +	710*
ν(C$-$C)	810	815	828	819	780*.772 +	81υ
$\delta^1 s$(CF$_3$)	893		890			
γ(OH)	886		902		875.905* +	895*
δ(OH) U	1207	1241 (?)	1305	1330		
ν(C$-$O)U	1456	1445	1477	1482		
t (OD$\cdots$0)			41			40
β (OD$\cdots$0)						50 (?)
γ (OD$\cdots$0)			100 (?)			80.116
t (CF$_3$)			100 (?)			
comb. $\gamma + t$ (?)			127			
ν(OD$\cdots$0)			146			
γ(OD)			642			650*
δ(OD)	1039		1037	1076		
ν(C$-$D)	1430	1427	1442	1444		

We have recorded the Raman spectrum of CF_3COOH liquid between 1000 and 50 cm^{-1}. The other experimental values for the liquids come from Reference 36 and 37 and have been sometimes reassigned using the neutron results. Part of these neutron results have been obtained with the beryllium filter spectrometer, by ourselves (*) or by M. F. Collins and B. C. Haywood (+).

$\beta(\text{OH}\cdots\text{O})$ (50 cm^{-1}) and the two components of the out of plane deformation $\gamma(\text{OH}\cdots\text{O})$ (81 and 122 cm^{-1}). These vibrations involve effectively large proton amplitudes.

A new weaker band is seen in the neutron spectra at about 180 cm^{-1} and assigned to the stretching vibration $\nu(\text{OH}\cdots\text{O})$ of A_g symmetry because of the expected low amplitude of the proton in this motion.

At higher energy transfers, the resolution decreases quickly and we can see for example that the different infrared and Raman bands due to the CF$_3$ rocking: $\varrho(\text{CF}_3)$, merge into a single broad band in the neutron spectra.

Table II summarises the vibrational analysis of CF$_3$COOH and CF$_3$COOH in all our time of flight and Beryllium filter data.

In addition to the molecular vibrational density of states, new peaks appear in the 0–60 cm^{-1} region for CF$_3$COOD (Figure 3). Their frequency varies with scattering angle (moving to lower energies as θ increases) and they may be ascribed to average longitudinal phonons (32). Two separate excitations (I) and (II) can be distinguished and their dispersion curves are shown in Figure 5. The lines extrapolate to Q values

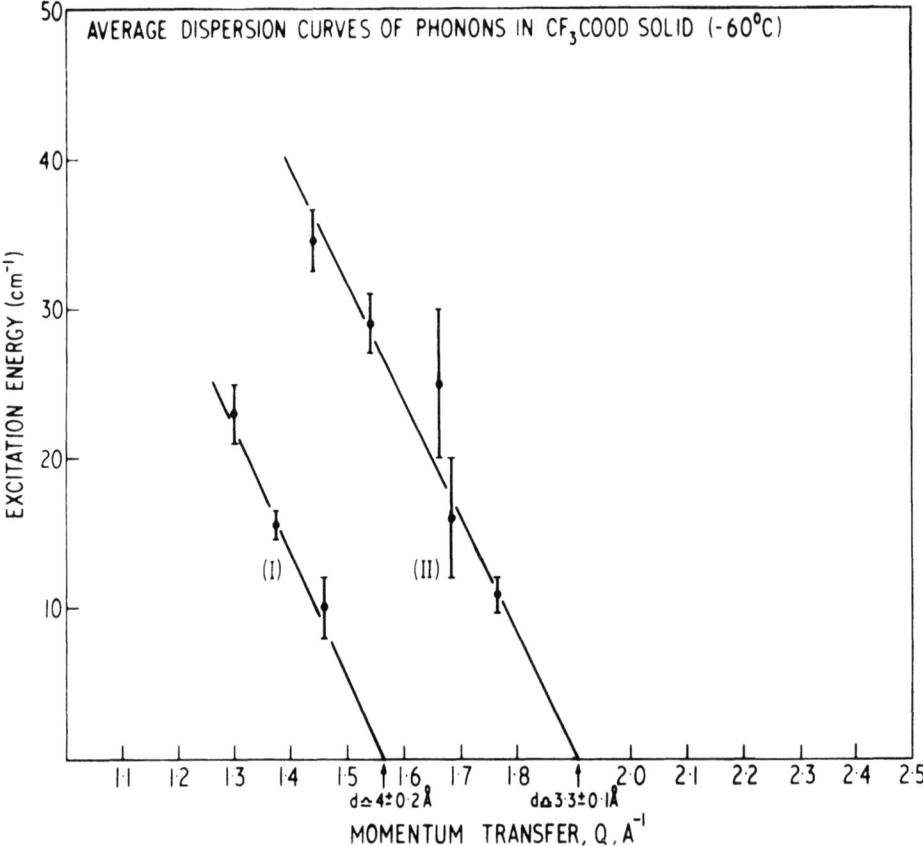

Fig. 5. Mean dispersion curves for the two phonons occurring in the time of flight spectra (Figure 3) of solid CF$_3$COOD.

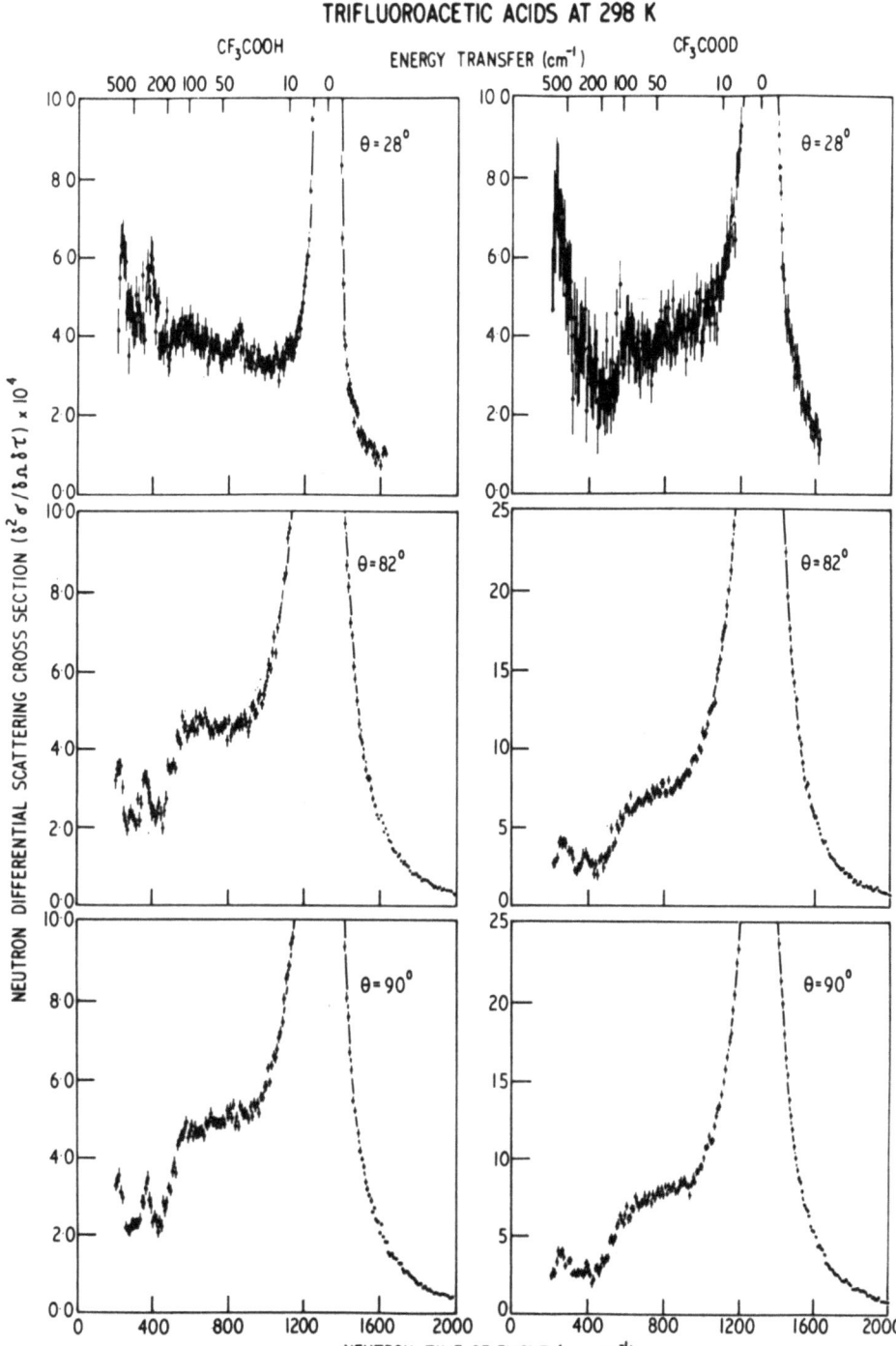

Fig. 6. Inelastic scattering spectra of 5.2 Å neutrons from liquid CF_3COOH and CF_3COOD at 298 K taken at scattering angles $\theta = 28°, 82°, 90°$ to the beam and with the sample inclined at 45° to the beam.

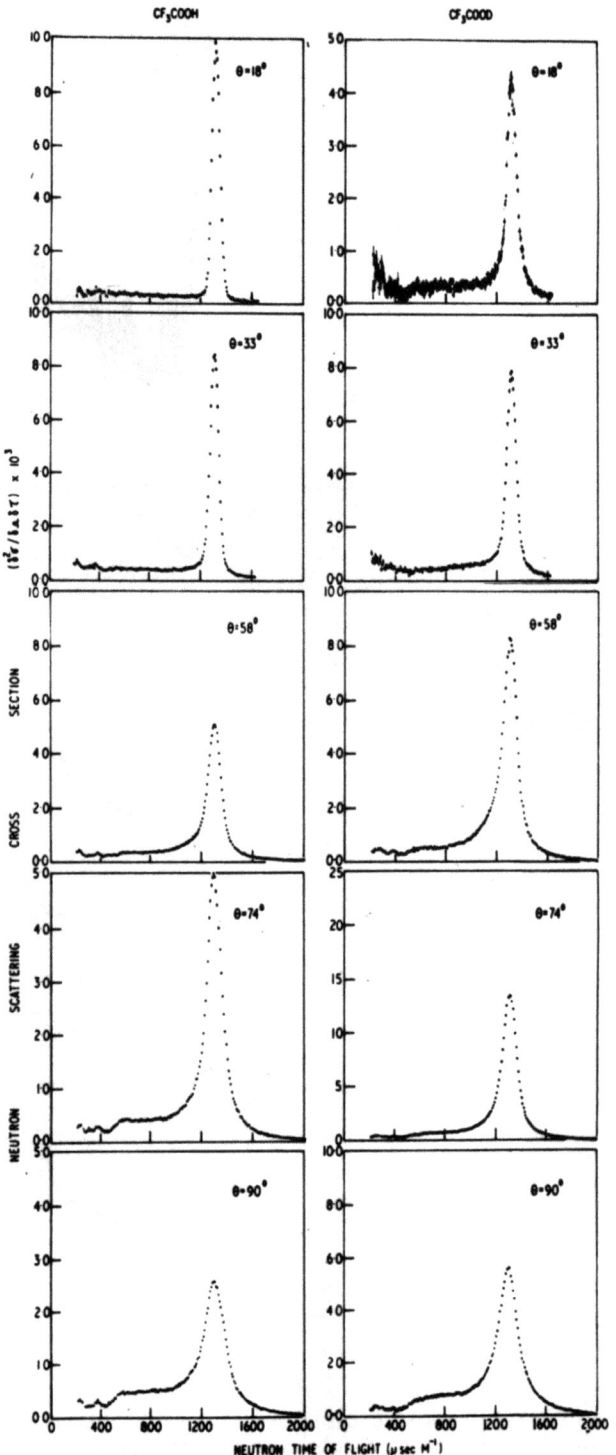

Fig. 7. Emphasis of the quasielastic scattering region in liquid CF₃COOH and CF₃COOD at 298 K and scattering angles $\theta = 18°, 33°, 58°, 79°$ and $90°$.

corresponding to diffraction maxima in Figure 2 at $2\theta = 16$ and $19°$, i.e. in the broad main diffraction peak. The mean sound velocity is of the order of 10^5 cm s^{-1} for both curves suggesting that the same excitation is seen in two Brillioun zones.

4.3. Trifluoroacetic acids – liquids

From 200 cm^{-1} up to 800 cm^{-1}, several well defined peaks may be observed on the time of flight spectra of liquid CF_3COOH and CF_3COOD (Figure 6). Under 200 cm^{-1} the inelastic region is less well structured than for the solid state, particularly because of a great wing on the quasielastic peak.

4.3.1. Quasielastic Region

The peak width and height of the two acids are very dependent on the momentum transfer as shown in Figure 7. This experimental profile is the convolution of the machine resolution by the function describing the motions of the scattering centres.

To study the single particle diffusive motions the quasielastic broadening should be observed away from Bragg scattering. The diffraction results and the analysis of Figure 8 show that serious interference occurs at Q^2 values near 0.5 and 2 Å^{-2}, where the peak widths are much smaller than expected. The regions between $Q = 0$ and $Q^2 = 0.4$, $Q^2 = 0.6$ and $Q^2 = 2.0$ are relatively free of diffraction effects and have been given most weight in the diffusion analysis. At low Q values the inelastic 'wing' between 0–30 cm^{-1} is of low intensity. The scattering law [1, 4] $S(Q, \omega)$ is a nearly symmetrical function which approximates quite well to the expected convolution of a Lorenzian line (due to the diffusive motions) convoluted with the instrumental Gaussian resolution function which is now well known for the 4H5 nad 6H spectrometers [31].

Two methods of deconvolution were used to find the Lorenzian width. The first used the height, halfwidth and area of the convoluted functions (Area or A method), the other used tabulated values of the Lorenzian-Gaussian convolution (T method). The broadening of E vs Q^2 for CF_3COOH in three different experiment conditions (Figure 9) and for both acids in the same experimental conditions (Figure 8) are plotted for both methods. It can be seen that of the two approaches the Table (T) method is more reproducible for this system.

Figure 9 illustrates an important experimental point when working with coherent

TABLE III

Translational diffusion coefficients measured by Neutron scattering and macroscopic methods

Substance	Temperature	Method	D cm^2 s^{-1}
CF_3COOH	22 °C	^{1}H NMR	1.35×10^{-5}
CF_3COOH	22 °C	^{19}F NMR	1.40×10^{-5}
CF_3COOD	22 °C	^{19}F NMR	1.40×10^{-5}
CF_3COOH	23 °C	Neutron	1.7×10^{-5}
CF_3COOD	23 °C	Neutron	1.6×10^{-5}
CF_3COOH	23 °C	Stokes-Einstein	$3–7 \times 10^{-6}$

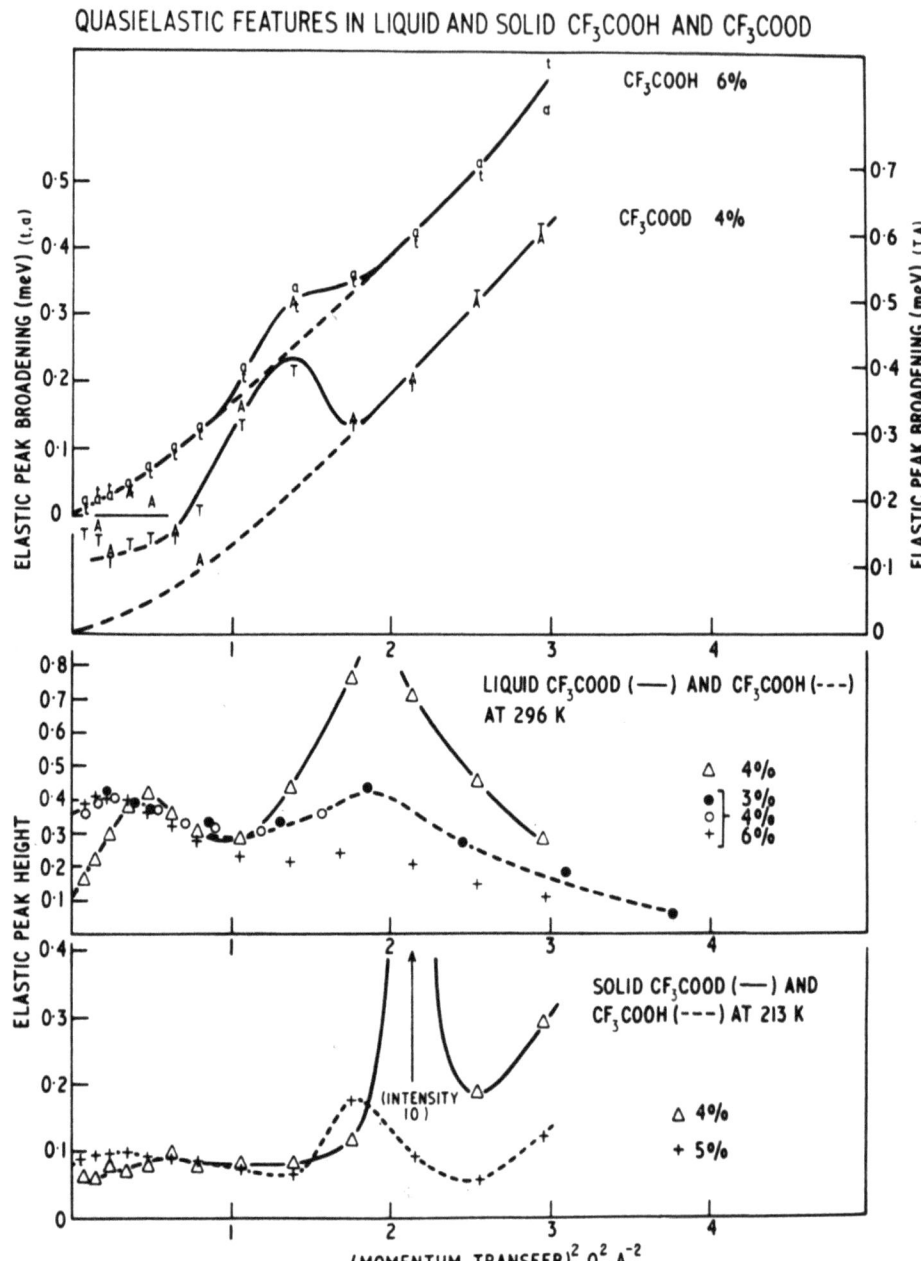

Fig. 8. (a) Quasielastic peak full width at half height $\triangle E$ vs Q^2 for liquid CF_3COOH and CF_3COOD. Variation of the peak height with Q^2 for the liquid, (b) and solid (c) acids. The percentage of scattering of the neutron beam is specified for each sample.

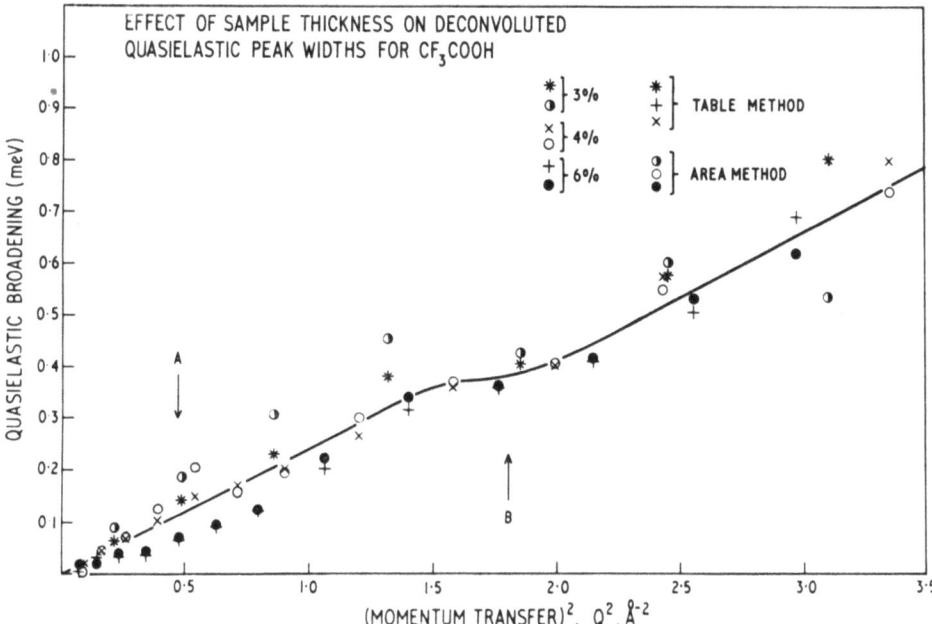

Fig. 9. Deconvoluted quasielastic peak widths computed by the Area ◐, ○, ●, and table *, +, × methods for liquid CF₃COOH at 298 K and samples scattering 3, 4 and 6% of the nuetron beam.

scattering samples. The largest broadening was always observable with the thinnest sample of CF_3COOH (3.0% total scattering). The other two samples were only marginally thicker 4% and 6% scattering, but already for them the tails of the diffraction peaks were sufficient to cause narrowing in the Q region of the quasielastic observations. Since the Bragg maximum is intense, multiple Bragg effects may be important near 2 Å^{-1}. Table III summarises the diffusion constants obtained from these neutron scattering experiments in contrast with those from ^{19}F and ^1H NMR pulsed field gradient studies and from a Stokes Einstein Law estimate based on mean values of the radius (3–5 Å) and of the viscosity (1–1.5 cp).

5. Discussion

Before interpreting the diffusion coefficients found by neutron and macroscopic methods, it is necessary to discuss the possible structure of the liquid and then to put upper limits to the quasielastic effects arising from rotational diffusion using the Sears model [9] as an approximation and the rotational diffusion constant from NMR relaxation.

5.1. STRUCTURE OF THE LIQUID

Studies of liquid CH_3COOH [34, 35] show that infrared and raman spectroscopy together may distinguish and assign the vibrations of the monomers, cyclic and open dimers and polymers, particularly in the $v_{(C=O)}$ region. Consequently the equilibrium constants can be evaluated.

The first infrared and Raman spectra obtained for CF_3COOH [36, 37] are not sufficiently precise for such an analysis. They only show that the amount of monomers is very small. However, we can assume from the dielectric measurements [27] that the pure liquid is composed of a mixture of cyclic and open dimers and polymers.

The principal peak in the diffraction pattern of CF_3COOH and CF_3COOD liquid at 298 K and 273 K is broad and the position varies slightly with temperature. The maximum corresponds approximately to spacings of 4.5 and 4.6 Å at 298 K and 273 K in the case of CF_3COOH. This broad peak most probably has contributions from dimer and chain structures, the percentage of which varies with temperature. A similar diffraction pattern has been obtained in this laboratory for the polymeric chains of methyl alcohol in the liquid phase.

From structural studies of the CF_3COOH cyclic dimer in the gas phase [40] and of the CH_2FCOOH cyclic dimer [22] and CH_3COOH chains [24] in the solid state, it is possible to make estimates of the structural parameters if the associated forms of CF_3COOH in the/liquid state.

For the cyclic dimer, the principal moments of inertia are:

I_X 435 10^{-40} g cm^{-2}
I_Y 3850 10^{-40} g cm^{-2}
I_Z 4000 10^{-40} g cm^{-2}.

The OX, OY and OZ axes are respectively drawn from the centra of symmetry O of the cyclic dimer, through the C—C bonds, perpendicular to OX in the plane of the cycle and perpendicular to the plane of the cycle.

5.2. ROTATIONAL DIFFUSION

5.2.1. *Internal Rotation of the CF_3 Group*

The microwave data obtained for the gas phase [40] indicate a very low barrier to internal rotation of the CF_3 group. On the other hand the moment of inertia for this group is high $IX(CF_3) \sim 150 \times 10^{-40}$ g cm^{-2}, so the CF_3 torsion never observed by optical methods, certainly occurs at a very low frequency. Because of the free rotation and the low CF_3 scattering cross section, little contribution from this group is expected in the quasielastic region.

5.2.2. *Rotational Diffusion of the Different Species*

A measurement of the spin lattice relaxation time of deuterium in pure CF_3COOD at 22°C has been made by Dr G. Tiddy (Unilever Port Sunlight Laboratories) for us to get an estimate of this quantity averaged over the species present. Deuterium is relaxed by the electric quadrupole interaction mostly [38] and the spin lattice relaxation rate $(1/T_1)$ in the absence of other than rotational modulation of $\partial^2 V/\partial Z'^2$ is related to τ_2

by the expression

$$\frac{1}{T_1} = \frac{3}{8} \cdot \left(\frac{eQ}{\hbar} \cdot \frac{\partial^2 V}{\partial Z'^2}\right)^2 \tau_2, \tag{5}$$

where $(eQ/\hbar) \cdot (\partial^2 V/\partial Z'^2)$ is the quadrupole coupling constant. To calculate τ_2 the value of the coupling constant assumed was that for hydrogen bonded deuterium in water (237 kHz). The value of τ_2 found was 9.1×10^{-12} s. In the approximation of rotational diffusion this τ_2 gives a diffusion constant $D_r = 1.6 \times 10^{10}$ s^{-1}.

5.2.3. Sears Model Estimate

The incoherent scattering due to rotational diffusion may be expressed by the following relation

$$\frac{\delta^2 \sigma}{\delta\Omega\delta\omega} = \frac{N}{\pi} \frac{k}{k_0} \exp\left(\frac{\hbar\omega}{2kT}\right) \sum_{l=0}^{\infty} (2l+1) b_i^2 j_i^2 (Qr) \frac{l(l+1)Dr}{[l(l+1)Dr]^2 + \omega^2} \tag{6}$$

which gives Lorentzian profiles about zero energy transfer. $j_i(Qr)$ is a spherical Bessel function, and b_i the incoherent scattering length of atoms arranged spherically at distance r from the centre of mass. We can estimate upper limits for the intensity contributions of the s, p and d partial wave.

Amongst the previously discussed species, the cyclic dimer has the lowest available moment of inertia (about the OX-axis). Thus the highest frequency diffusive component is likely to come from motion about this axis. The H atoms are distant 1.125 Å approximately from this axis and Table IV summarizes the different values of $(2l+1) \times j_l^2(Qr)$ when Q varies from 0 to 2.0 Å^{-1} and for l values of 0, 1 and 2.

TABLE IV

Q	Qr	$j_0^2(Qr)$	$3j_1^2(Qr)$	$5j_2^2(Qr)$
0	0	1.00	0	0
0.178	0.2	9.87×10^{-1}	1.32×10^{-2}	3.5×10^{-5}
0.355	0.4	9.48×10^{-1}	5.17×10^{-2}	5.5×10^{-4}
0.534	0.6	8.86×10^{-1}	1.12×10^{-1}	2.74×10^{-3}
0.710	0.8	8.04×10^{-1}	1.875×10^{-1}	8.30×10^{-3}
0.888	1.0	7.08×10^{-1}	2.72×10^{-1}	1.925×10^{-2}
1.068	1.2	6.03×10^{-1}	3.576×10^{-1}	3.74×10^{-2}
1.243	1.4	4.955×10^{-1}	4.362×10^{-1}	6.43×10^{-2}
1.440	1.6	3.903×10^{-1}	5.01×10^{-1}	1.00×10^{-1}
1.598	1.8	2.927×10^{-1}	5.466×10^{-1}	1.45×10^{-1}
1.778	2.0	2.067×10^{-1}	5.70×10^{-1}	0.97×10^{-1}
1.955	2.2	1.351×10^{-1}	5.664×10^{-1}	2.58×10^{-1}

These values show that the contribution of the $l=1$ (and certainly the $l=2$) terms is negligible up to Q values of about 0.8 Å^{-1}. This is in any case the useful region for analysing quasielastic broadening vs momentum transfer (Figure 8) because of the increasing importance of coherent effects at higher Q values. From § 5.22 the width of the $l=1$ Lorentzian is about 3.2×10^{10} Hz or about 300 μeV which would be detectable by its contribution to the total quasielastic peak at Q's about 1 Å^{-1} with our energy and momentum resolution but barely perceptible at lower Q values because of the small intensity.

5.3. Comparison of the Macroscopic and Microscopic Diffusion Constants

The macroscopic diffusion constants for ^{1}H and ^{19}F measured by Dr Tiddy (Table III) were made relative to pure water at 22 °C. D values for water at this temperature vary between 2.5×10^{-5} and 2.15×10^{-5} cm^2 s^{-1} [39]. The value of 2.3×10^{-5} cm^2 s^{-1} corrected for temperature from the preferred value of 2.36×10^{-5} cm^2 s^{-1} at 23 °C was used to calculate D in the table.

The D values from neutron scattering are consistently a little higher than those from the NMR measurements. Only further measurements will show whether this can be explained by the rotational contribution being higher than the estimates above or due to chemical exchange kinetics of the proton arising from traces of water in the system. It is important to note that the neutron D values for CF$_3$COOH and CF$_3$COOD are within experimental error the same, which suggests that in the pure acids only centre of mass translations of the molecules contribute to the mass transport on the observation time scales of neutron scattering.

References

1. Larsson, K. E.: 'Neutron Inelastic Scattering', *Proc. Copenhagen Symposium*, I.A.E.A., Vienna, 1968, p. 397.
2. Egelstaff, P. A.: *An Introduction to the Liquid State*, Academic Press, 1967.
3. Aldred, B. K., Stirling, G. C., and White, J. W.: *Faraday Symp. Chem. Soc.* **6**, 135 (1972).
4. Aldred, B. K., Eden, R. C., and White, J. W.: *Disc. Faraday Soc.* **43**, 169 (1967).
5. (a) Berne, B. J. and Frisch, H. L.: *J. Chem. Phys.* **47**, 3675 (1967);
   (b) Berne, B. J., Deutch, J. W., Hynes, J. T., and Frisch, H. L.: *J. Chem. Phys.* **49**, 2864 (1968).
   (c) Berne, B. J. and Pecora, R.: *J. Chem. Phys.* **50**, 783 (1969).
6. (a) Blum, L. and Salsburg, Z. W.: *J. Chem. Phys.* **48**, 2292 (1968).
   (b) Blum, L. and Salsburg, Z. W.: *J. Chem. Phys.* **50**, 1654 (1969).
7. (a) Yeh, Y. and Keeler, R. N.: *J. Chem. Phys.* **51**, 1120 (1969).
   (b) Yeh, Y.: *J. Chem. Phys.* **52**, 6218 (1970).
8. Lomer, W. M. and Law, G. G. in P. A. Egelstaff (ed.), *Thermal Neutron Scattering*, Academic, 1965, Ch. I.
9. Sears, V. F.: *Can. J. Phys.* **45**, 239 (1967).
10. Buyers, W. J. L., Sears, V. F., Lonnig, P. A., and Lonnig, D. A.: 'Neutron Inelastic Scattering', *Proc. Grenoble Symposium*, I.A.E.A., Vienna, 1972, p. 399.
11. Taylor, A. D.: private communication.
12. Kreevoy, M. M. and Mead, C. A.: *Disc. Faraday Soc.* **39**, 166 (1965).
13. See for example McCall, D. W. and Douglass, D. C.: *J. Phys. Chem.* **71**, 987 (1967).
14. See for example, Gordon, R. G.: *Adv. Magn. Resonance* **3**, 1 (1968).
15. (a) Gordon, R. G.: *J. Chem. Phys.* **43**, 1307 (1965).
    (b) Marechal, E.: *Ber. Bunsenges.* **75**, 343 (1971).
16. Bratos, S., Rios, J., and Guissani, Y.: *J. Chem. Phys.* **52**, 439 (1970)
17. Allen, G. and Caldin, E. F.: *Quart. Rev.* **7**, 255 (1953).
18. Taylor, M. D. and Templeman, M. B.: *J. Am. Chem. Soc.* **78**, 2950 (1956).
19. Murty, T. S. S. R. and Pitzer, K. S.: *J. Phys. Chem.* **73**, 1426 (1969).
20. Kirzenbaum, M., Corset, J., and Josien, M. L.: *J. Phys. Chem.* **75**, 1327 (1971).

21. Covington, A. K., Freeman, J. G., and Lilley, T. H.: *J. Phys. Chem.* **74**, 3773 (1970),
22. Kanters, J. A. and Kroon, J.: *Acta Cryst.* **28B**, 1946 (1972).
23. Holtzberg, F., Post, B., and Fankucken, J.: *Acta Cryst.* **6**, 127 (1953).
24. (a) Nahringbauer, I.: *Acta Chem. Scand.* **24**, 453 (1970).
    (b) Jönsson, P. G.: *Acta Cryst.* **27B**, 893 (1971).
25. Berney, C. V.: *J. Am. Chem. Soc.* **95**, 708 (1973).
26. Clague, D. and Novak, A.: *J. Chim. Phys.* **67**, 1126 (1970).
27. Simons, J. H. and Lorentzen, K. E.: *J. Am. Chem. Soc.* **72**, 1426 (1950).
28. See for example (a) Bader, F. and Plaß, K. G.: *Ber. Bunsenges. physik. Chem.* **75**, 553 (1971). (b) Lanshina, L. V., Lupina, M. I., and Khabibullaev, P. K.: *Soviet Phys. Acoust.* **16**, 343 (1971).
29. Bunce, L. J., Harris, D. H. C., and Stirling, G. C.: U.K.A.E. Report. AERE R.6246 (H.M.S.O., 1970).
30. Collins, M. F., Haywood, B. C., and Stirling, G. C.: *J. Chem. Phys.* **52**, 1828, (1970).
31. Harryman, M. B. M. and Hayter, J. B.: U.K.A.E.A. Report. AERE.
32. Guner, Z. and Cocking, S. J.: 'Inelastic Scattering of Neutrons', *Proc. Symp. Chalk River 1962*, I.A.E.A., Vienna, **1**, 237 (1963)
33. Collins, M. F. and Haywood, B. C.: *J. Chem. Phys.* **52**, 5740 (1970).
34. Haurie, M. and Novak, A.: *Compt. Rend.* **264**, 694 (1967).
35. Haurie, M.: Thesis, Bordeaux, 1966.
36. Fuson, N., Josien, M. L., Jones, E. A., and Lawson, J. R.: *J. Chem. Phys.* **20**, 1627 (1952).
37. Barcelo, J. R. and Otero, C.: *Spectrochim. Acta* **18**, 1231 (1962).
38. Abragam, A.: *The Principles of Nuclear Magnetism*, Oxford, 1961.
39. Trappeniers, N. J., Gerritsma, C. J., and Oosting, P. H.: *Phys. Letters* **18**, 256 (1965).
40. Costain, C. C. and Srivastava, G. P.: *J. Chem. Phys.* **41**, 1620 (1964).

DISCUSSION

Hertz: You have given us a definite structure of the pure liquid CF_3COOH: cyclic dimers. There is some evidence from NMR studies that in the pure liquid acetic acid other association forms are present or may ever prevail. I would like to know on which experimental facts your proposal of the structure is based?

White: The composition of the pure liquid has been studied extensively by infrared and Raman scattering (see references in the paper; see [2]). Because of the separate appearance of bands in Raman and IR the presence of cyclic dimers is strongly suggested. Additionally the high dielectric constant indicates more open chain structures than in acetic acid. The diffraction data we presented certainly need to be followed up especially just above the melting point.

COLLECTIVE EXCITATIONS IN LIQUID HYDROGEN OBSERVED BY COHERENT NEUTRON SCATTERING

K. CARNEIRO and M. NIELSEN

Danish Atomic Energy Commission Research Establishment Risø, DK-4000 Roskilde, Denmark

and

J. P. McTAGUE*

Danish Atomic Energy Commission Research Establishment Risø, DK-4000 Roskilde, Denmark

and

University of California Los Angeles, Calif. 90024, U.S.A.

Abstract. Coherent scattering of neutrons by liquid parahydrogen shows the existence of well-defined collective excitations in this liquid. Similarity with the scattering from both liquid helium and classical liquids is found. Furthermore in the range of observed wavevectors, $0.7 \text{ Å}^{-1} \leq \kappa \leq 3.1 \text{ Å}^{-1}$, extending from the first through the third Brillouin zones in the solid, $S(\kappa, \omega)$ is remarkably similar to the scattering law expected from a polycrystal. For $\kappa \leq 2.3 \text{ Å}^{-1}$ the observed spectra satisfy the one-phonon sum rule with a mean square displacement $\langle u^2 \rangle = 1.0 \text{ Å}^2$.

Résumé. La diffusion cohérente des neutrons par le parahydrogène liquide indique l'existence d' excitations collectives bien definies dans ce liquide. On a constaté une ressemblance avec la diffusion, et par l'helium liquide, et par les liquides classiques. Par ailleurs, pour les vecteurs d'onde κ entre 0.7 Å^{-1} et 3.1 Å^{-1}, vecteurs qui correspondent à la première, deuxième et troisième zone de Brillouin en phase solide, $S(\kappa, \omega)$ ressemble remarquablement à la loi de diffusion caracterisant un polycristal. Quand $\kappa \leq 2.3 \text{ Å}^{-1}$ les spectres observés obéissent à la règle d'un phonon avec un déplacement carré moyen de $\langle u^2 \rangle = 1.0 \text{ Å}^2$.

1. Introduction

The intermolecular collective excitation in liquids can be observed most directly by two experimental methods, Brillouin scattering of light, and coherent neutron scattering, which in principle both measure the scattering law $S(\kappa, \omega)$. This quantity is the Fourier transform of the total van Hove correlation function $G(\mathbf{r}, t)$. But whereas light scattering samples the excitations in the long wavelength or hydrodynamic region, neutron scattering is sensitive when the wavelength is comparable to the intermolecular spacing, corresponding to the phonon region in the solid phase.

Liquid hydrogen has properties that make it a good candidate for a neutron scattering study. This is due in part to the existence of the para and ortho modifications of molecular hydrogen, $p\text{-}H_2$ and $o\text{-}H_2$ respectively. In Table I is shown how the symmetry of the molecular wavefunction Ψ divides the rotational states into two groups, the para states where the rotational quantum number J is even and the total spin $I = 0$, and the ortho states where J is odd and $I = 1$. Because of the small moment of inertia of the H_2 molecule and because the anisotropic interactions are small, J is a good quantum number. Further, since the energy of the first rotational state $\Delta E = E_{(J=1)}$

---

* Alfred P. Sloan Research Fellow. Research partially supported by the National Science Foundation.

TABLE I

Symmetries of the molecular wavefunction Ψ, in the case of H_2, when the two protons are interchanged. S = symmetric, AS = antisymmetric. At the right is shown the resulting spins of the protons corresponding to the symmetry of the spinfunction e_I

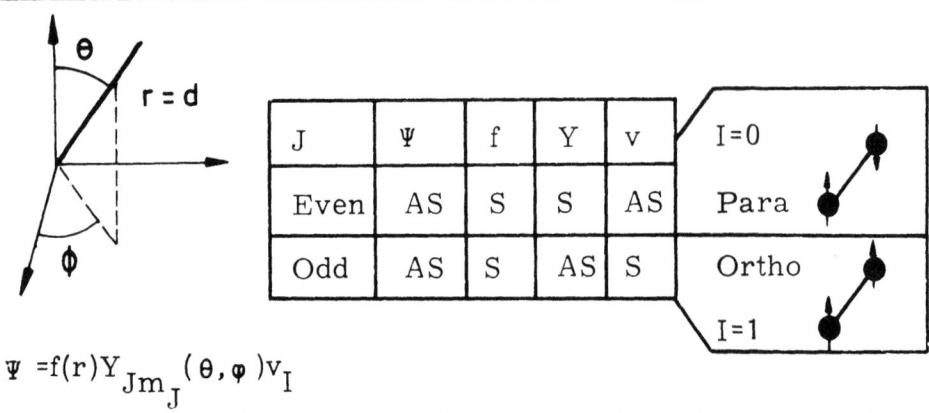

$\Psi = f(r) Y_{Jm_J}(\theta, \varphi) v_I$

$-E_{(J=0)} = 14.6$ meV $= 170$ K, only p-H_2 with $J=0$ and o-H_2 with $J=1$ are present below the boiling point $T_B = 20.4$ K.

The fact that the $J=0$ state for p-H_2 is spherically symmetric, makes this liquids a simple rare gas-like van der Waals liquid. In Table II we compare some characteristics of H_2 with the liquids Ar, Ne and He. It is seen that H_2 has Lennard-Jones parameters similar to Ne, but because of the smaller mass, the quantum nature of H_2 is much more pronounced. A measure of this is the de Boer parameter $\Lambda^* = (h^2/m\sigma^2\varepsilon)^{1/2}$, which may be taken as the square root of the energy of the zero point motion relative to the potential energy, or as the de Broglie wavelength relative to the interparticle spacing. Also closely related to the quantum nature of a liquid is the degree of thermal excitation, qualitatively characterized by the ratio T/θ_D, where θ_D is the Debye temperature. From the data in Table II we see that the high frequency, short wavelength modes in both H_2 and He are essentially unpopulated at liquid temperatures, whereas for Ne and Ar this is not true. Both Λ^* and the ratio T_B/θ_D indicate that liquid H_2 is intermediate between He and Ne and has little similarity with the classical liquid Ar.

TABLE II

Some characteristics of the van der Waals liquids Ar, Ne, H_2, and He

Liquid	Ar	Ne	H_2	^{4}He
σ [Å]	3.41	2.79	2.93	2.58
ε/k_B [K]	120	36	37	10
Λ^*	0.196	0.593	1.73	2.68
T_B	87.5	27.1	21.4	4.2
θ_D	87	64	103	25
T_B/θ_D	1.05	0.42	0.20	0.17

Although the above mentioned is only strictly true for p-H_2, it holds well also for o-H_2 since the importance of the electrical quadrupole-quadrupole (EQQ) forces is small in the liquid phase [1].

2. Neutron Scattering Properties of H_2

The other important consequence of Table I is that p-H_2 has $I=0$ i.e. a spin singlet eigenstate. This makes the neutron scattering within the $J=0$ state purely coherent, so that p-H_2 acts as a scatterer with a coherent scattering length per molecule of $2 \cdot b_{coh} \cdot j_0(\kappa d/2)$. b_{coh} is the coherent scattering length for the proton, $d=0.746$ Å is the internuclear bond length, and j_0 is the zero order spherical Bessel function. κ is the wavevector transferred to the liquid by a neutron.

Sarma [2] has worked out the full expressions for the neutron scattering from all the combinations of p-H_2 and o-H_2. The result is summarized on Figure 1, as seen in the

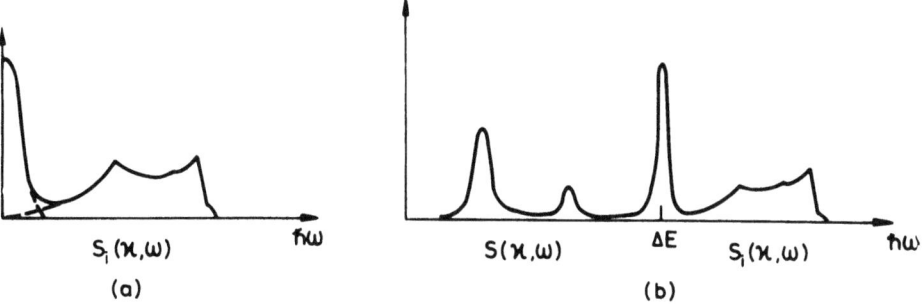

Fig. 1. Neutron scattering from molecular hydrogen, obtained by a constant wavevector experiment. (a) Scattering from o-H_2 yielding $S_i(\kappa, \omega)$; (b) Scattering from p-H_2 yielding $S(\kappa, \omega)$ and $S_i(\kappa, \omega)$.

solid and the liquid phases. Figure 1a shows the scattering from pure o-H_2 yielding the incoherent scattering law $S_i(\kappa, \omega)$ consisting of a quasielastic part and an inelastic part due to density of states. Figure 1b shows that S_i, when pure p-H_2 is investigated, is translated by ΔE, and below this energy, $S(\kappa, \omega)$ can be investigated. Because of the principle of detailed balance:

$$S(\kappa, -\omega) = e^{-\hbar\omega/k_B T} S(\kappa, \omega) \qquad (1)$$

all scattering laws are essentially one-sided below T_B.

The fact that o-H_2 can be prepared at a sufficient purity and stored enough long for a neutron scattering experiment has been used by Egelstaff et al. [1] in the liquid phase, and by Stein et al. [3] in the solid phase. The equilibrium composition of 99.8% p-H_2 has been investigated by Nielsen [4] in the solid and by Carneiro et al [5] in the liquid.

3. Results and Discussion

In preparing the sample we carefully catalyzed the hydrogen gas before letting it condence in a 30 mm i.e. aluminum container with a ferric hydroxide gel ortho/para catalyst at the top and at the bottom. The experiments were performed on the Risø triple axis spectrometers. Second order contamination was reduced by use of either a graphite filter or a germanium monochromator.

Scans were performed in the region $0.7 \text{ Å}^{-1} \leqslant \kappa \leqslant 3.1 \text{ Å}^{-1}$ at 14.7 K and at saturated vapour pressure. The following corrections have been considered and taking into account where significant: scattering from the empty can (uniformly $\sim 15\%$), incoherent scattering from the small amount of $o\text{-}H_2$ ($<2\%$), instrumental sensitivity (essentially a k^3 correction [6]), and multiple scattering ($<2\%$ in this region [7]).

The resulting $S(\kappa, \omega)$ is presented in Figure 2 for the wavevector values $\kappa = 0.7$, 1.1, 1.5, 1.9, 2.3, 2.7, and 3.1 Å^{-1}. It demonstrates that a well-defined peak exists in $S(\kappa, \omega)$ for $\omega = 0$, signifying the existence of a collective excitation in the density-density correlation function. In both Ar [8] and Ne [9] peaks are not seen in $S(\kappa, \omega)$ but only in the velocity correlation function $\omega^2 S(\kappa, \omega)$, which of course has a peak at finite energy even when the density correlation function is overdamped.

The total spectra have been tested for the ACB (Ambegaokar, Conway and Baym) one-phonon sum rule [10], which is a specialization of the universal Placzek sum rule, valid for a cubic harmonic solid. The Placzek sum rule states that:

$$I_P \equiv \int_{-\infty}^{\infty} d\omega \, \omega S(\kappa, \omega) = \kappa^2/(2M). \tag{2a}$$

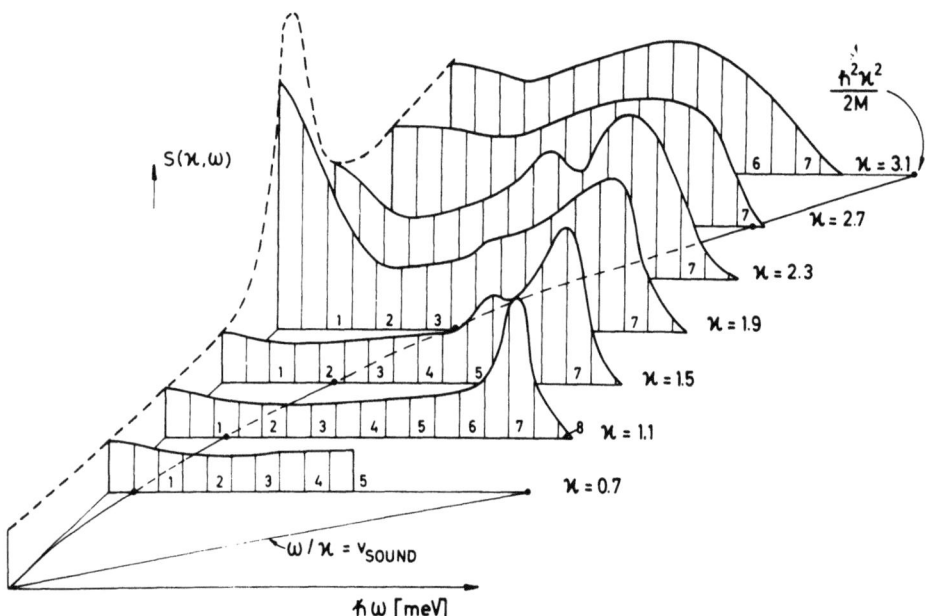

Fig. 2. Scattering law $S(\kappa, \omega)$ for liquid parahydrogen at $T = 14.7$ K and saturated vapour pressure. In the base plane the recoil energy curve is shown, and the extension of the position of the Brilloun peak is indicated.

Applying (1), (2a) becomes:

$$I_P \equiv \int_0^\infty d\omega (1 - \exp(-\hbar\omega/k_B T)) \omega S(\kappa, \omega) = \kappa^2/(2M). \tag{2b}$$

The ABC sum rule then states that the one-phonon part S^1 of S obeys the relation similar to (2):

$$I_{ACB} \equiv \int_0^\infty d\omega (1 - \exp(-\hbar\omega/k_B T)) \omega S^1(\kappa, \omega) = \frac{\kappa^2}{2M} \exp(-\tfrac{1}{3}\kappa^2 \langle u^2 \rangle). \tag{3}$$

In Figure 3 we show $I_{ACB}(2M/\kappa^2)$ obtained from the spectra shown in Figure 2. For $\kappa \leqslant 2.3 \, \text{Å}^{-1}$ our results are consistent with one-phonon scattering, with a mean

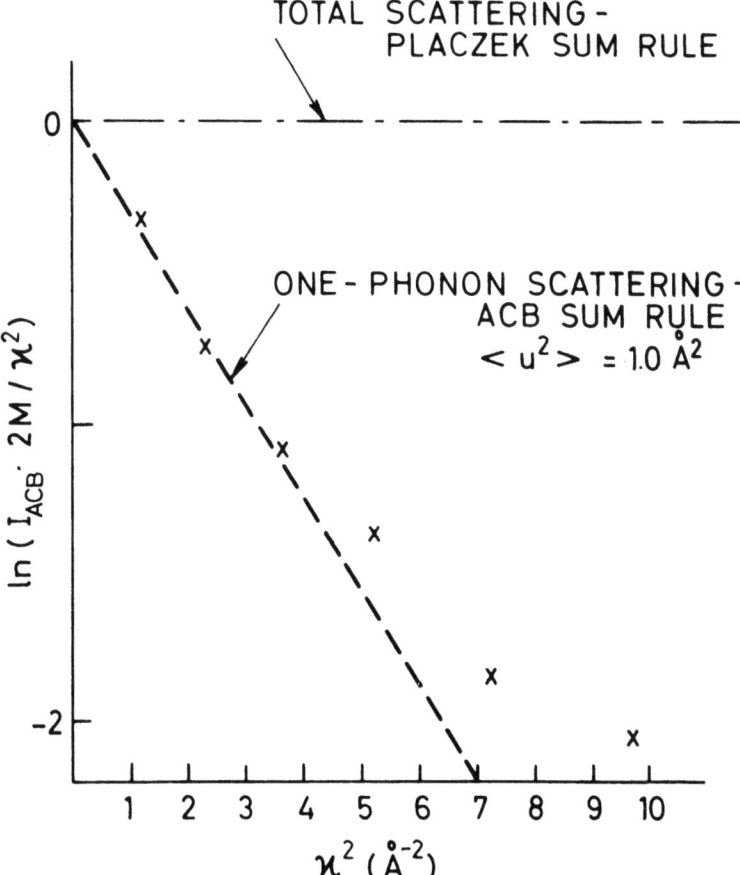

Fig. 3. The ACB one-phonon sum rule applied to the measured spectra. The dash-dotted line corresponds to the Placzek sum rule for the total scattering. The finite slope of the dashed line through the measured points corresponds to a mean square displacement $\langle u^2 \rangle = 1.0 \, \text{Å}^2$.

square displacement $\langle u^2 \rangle = 1.0$ Å^2. At larger κ's, however, $S(\kappa, \omega)$ shows a significant multiphonon contribution. The magnitude of $\langle u^2 \rangle$ indicates that multiphonon scattering should predominate for $\kappa > 1.5$ Å^{-1}, but this scattering seems to be concentrated mainly above 8 meV, a region not investigated in this experiment.

The excitations seem solidlike in other respects also. For instance the longitudinal phonon energy at the first zone boundary in the solid at a density equal to that of liquid H$_2$ would be at $\kappa = 1$ Å^{-1} and $\hbar\omega = 7.5$ meV, as estimated by the Grüneisen relation and the data of Reference 4. This compares favourably with the peak in our $S(\kappa, \omega)$ at the same κ but with $\hbar\omega = 7.2$ meV.

At higher κ's, corresponding to the second and third Brillouin zones in the solid intensity appears in the liquid at lower energy, reminiscent of the neutron scattering law calculated for a polycrystalline powder [11]. In the latter case this additional feature is related to modes which possess appreciable transverse character, suggesting that transverse modes may contribute to the scattering in the liquid, although it is not entirely clear how the neutron can couple to transverse modes in the absence of a reciprocal lattice. Such a coupling would imply that the liquid as a whole, similar to a crystal lattice, could take up momentum of $\hbar\kappa_0$, where κ_0 is the wavevector at which the major peak in the structure factor occurs. In our case $\kappa_0 = 2$ Å^{-1}, which is close to an average of the first three Bragg peaks in the solid.

4. Comparison with Other Van der Waals Liquids

Compared with the coherent neutron scattering patterns obtained from the other liquids shown on Table II, $S(\kappa, \omega)$ for p-H$_2$ indeed seems to be intermediate between the extreme quantum case He [12] and the classical liquids. Both Ar and Ne show a quasielastic peak at κ_0, but in neither case do phononlike excitations seem to exist. The free-particle recoil effect, observed in Ne for $\kappa > 4$ Å^{-1} is not seen in Ar and H$_2$, in the regions investigated.

H$_2$ and He are similar because of the exsitence of well-defined propagating modes. In He no elastic scattering occurs, and the exictations are very similar to the longitudinal part of the polycrystalline phonon spectrum discussed above.

We conclude that thermal population, rather that the lack of structural order, appears to be the main factor determining the behaviour of the collective excitations in simple van der Waals liquids. Furthermore, although many models of liquids characterize the mode damping process in terms of self-diffusion, such an interpretation is in contradiction with experimental results. Finally by incorporating the 'low temperature liquid' H$_2$ among the van der Waals liquids we feel that a more consistent picture of the nature of the dynamics of liquids will emerge.

References

1. Egelstaff, P. A., Haywood, B. C., and Webb, F. J.: *Proc. Phys. Soc.* **90**, 681 (1967).
2. Sarma, G.: in *Inelastic Scattering of Neutrons in Solids and Liquids*, IAEA, Vienna, 1961, p. 397.
3. Stein, H., Stiller, H., and Stockmeyer, R.: *J. Chem. Phys.* **57**, 1726 (1972).

4. Nielsen, M.: *Phys. Rev.* **B7**, 1626 (1973).
5. Carneiro, K., Nielsen, M., and McTague, J. P.: *Phys. Rev. Letters* **30**, 481 (1973).
6. Bjerrum Møller, H. and Nielsen, M.: in *Instrumentation for Neutron Inelastic Scattering Research*, IAEA, Vienna, 1970, p. 49.
7. Blech, I. A. and Averbach, B. L.: *Phys. Rev.* **137**, A 1113 81965).
8. Sköld, K., Rowe, I. M., Ostrowski, G., and Randolph, P. D.: *Phys. Rev.* **A6** 1107 (1972).
9. Kerr, W. C. and Singwi, K. S.: *Phys. Rev.* **A7** 1043 (1973).
10. Ambegaokar, V., Conway, J. M., and Baym, G.: in R. F. Wallis (ed.), *Lattice Dynamics*, Pergamon, New York, 1965.
11. de Wette, F. J. and Rahman, A.: *Phys. Rev.* **176**, 786 (1968).
12. Cowley, R. A. and Woods, A. D. B.: *Can. J. Phys.* **49**, 177 (1971).
13. Hubbard, J. and Beeby, J. L.: *J. Phys.* **C2** 2262 (1969).

DISCUSSION

Springer: I was impressed by the very pronounced peaks due to collective excitations in liquid H_2. Very recently, we found also such collective excitations in liquid neon, but not so pronounced. Is the displacement of these peak positions compatible with the sound velocity? Are your measurements at K-values within the first 'Brillouin zone' of the liquid?

Carneiro: I find the recent measurement on Ne very encouraging. Perhaps it would be possible to see the sound waves in Ar at the small wave vectors that you have now achieved although Sköld and al. were not able to see them. In H_2 it is not possible to do Brillouin scattering because of the high sound velocity.

Concerning your second question, we do not know sound velocity accurately enough to compare in detail with our spectra. We compared with the solid as shown in the text. The range of wave vectors is also described in the text.

Rahman: Recent calculations (M.D.) and neutron experiments show very clear 'phonon' peaks even up to about 0.8 Å^{-1}. One would like to understand what is the property of the pair potential which determines this phenomenon? Or is it just $(C_p - C_v)/C_v$?

Carneiro: I cannot answer your question but as I have tried to point out, the tendency in the dynamics of Van der Waals liquids seems clear. In liquid metals the difference between Na, where 'phonons' are not seen by neutron scattering, and Rb, where they have recently been measured, may lie in the non-negligible incoherent scattering in Na, since you see no difference. Generally speaking neutron Brillouin scattering is a difficult experiment and one has to select the liquid with great care.

Dorfmüller: Is there any chance of measuring the changes in the phonon spectrum in liquids like CH_3, C_2H_6 on melting?

Carneiro: I do not know enough about the molecular liquids to answer your question.

Yip: In answer to the previous question, I would like to say that there exist in the neutron scattering literature several studies of molecular liquids such as CH_4 and CD_4 (e.g. B. A. Dasannachyra, and G. Venkataraman: *Phys. Rev.*, 1967). One does not talk about phonons in these systems, but one can compare the effective frequency spectrum in the case of liquid CH_4 with that observed in the solid. Generally speaking, the rotational effects appear to be similar in the liquid and solid phases which is perhaps not surprising. Moreover, away from the very low frequency region the translational effects also are not too different. Such a comparison has been discussed by A. K. Agrawal and S. Yip: *Nuclear Science and Engineering*, 1969.

MOLECULAR DYNAMICS: COMPUTER 'EXPERIMENTS' ON SIMPLE LIQUIDS

LOUP VERLET

Laboratoire de Physique Théorique et Hautes Energies*, Université de Paris-Sud, 91405 Orsay, France

Abstract. We show that computer 'experiments' yield informations on collective motions involving a limited number of atoms in simple liquids. Three examples are given: The study of the velocity autocorrelation function demonstrates the existence of vortex-like motions at intermediate densities and of a quasi-elastic behavior near solidification. This type of solid-like behavior is also apparent in the frequency-dependant shear-viscosity near solidification. The last example, that of reorientational motion of diatomics is of a more negative nature: in that case no appreciable collective behavior is apparent.

Résumé. La simulation sur ordinateur renseigne sur les mouvements collectifs dans les liquides impliquant un nombre limité d'atomes. On donne trois exemples: l'étude de la fonction d'autocorrélation des vitesses montre l'existence de mouvements tourbillonaires aux densités intermédiaires et d'un comportement quasi-élastique près de la solidification. Ce type de comportement du genre solide est aussi apparent dans le viscosité dépendant de la fréquence au voisinage de la solidification. Le dernier exemple, le mouvement de réorientation de molécules diatomiques, est en fait contre-exemple: dans ce cas les mouvements collectifs provoquent des effets qui se situent à un niveau non mesurable.

1. Introduction

At the macroscopic level, liquids and dense gases can be described, in the linear regime, through the linearized hydrodynamics. The constants entering the set of Navier-Stokes equations – thermodynamical constants and transport coefficients – can be obtained, for a given force law, in a computer experiment. At the microscopic level where wavelengths are of the order of interparticle separation, and inverse frequencies of the order of the mean collision time, the collective motions are no longer seen and a description of the liquid can be made knowing the behavior of the various time correlation function for small times. This situation corresponds to the usual neutron scattering experiment and can be further studied by molecular dynamics simulation.

This paper is mostly concerned with the intermediate situation where the wavelengths are of the order of several interparticle distances and the inverse frequencies of the order of several mean collision times. There, we learn about collective motions at the microscopic scale, which are most conveniently described through the formalism of generalized hydrodynamics [1]. We shall give three examples of the applications of computer 'experiments' to the study of these collective motions.

The first case is the well-known velocity autocorrelation function (v.a.f.) which exhibits two kinds of collective motions according to the density and temperature. In the moderately dense gas region the motion of a given particle entails the formation of a surrounding vortex pattern which gives rise to a long positive tail in the v.a.f.

---

* Laboratoire associé au Centre National de la Recherche Scientifique.

Another type of behavior is observed near solidification, for instance near the triple point where the v.a.f. becomes negative after one or two collisions and stays negative for a duration involving several relaxation times.

The second example of these collective modes is seen through the study of the time correlation of the xz component of the stress tensor. This time correlation is the integrand of the Kubo formula for the shear viscosity. For densities smaller, by 10%, than the solidification density, this function has a range of the order of the mean collision time. It can be approximated by an exponential behavior: this description corresponds to the Maxwell viscoelastic theory. Near solidification, molecular dynamics 'experiments' [2, 3] show the presence of a tail at large time in the Kubo integrand for the shear viscosity. This function can be adequately approximated by a sum of two exponentials. This leads to generalized viscoelastic theory with two relaxation times. One is the usual molecular relaxation time, the other is a much longer time connected with a collective mode. It has been shown [3] that using a model due to Zwanzig and Bixon [4] and this viscoelastic theory one can explain the negative plateau of the v.a.f. near solidification. Also, this collective mode should be included in the description of the results obtained by molecular dynamics near the triple point for the dynamical structure factor $S(k, \omega)$.

The last example given will be that of the orientational correlation in diatomics. There one would think of a collective behavior [5] which could be cranked by the rotational motion. The analysis of computer experiments [6] show that, within the rather smaller errors in these 'experiments' no such collective behavior is seen.

2. The Velocity Autocorrelation Function

2.1. Theoretical considerations

The v.a.f. normalized to one for $t=0$ is defined

$$f(t) = \langle \mathbf{v}_i(0) \mathbf{v}_i(t) \rangle \frac{m}{3k_B T}. \tag{1}$$

There, the label refers to a distinguished particle whose self motion is studied and the bracket implies an average over an equilibrium ensemble of initial times.

When the time t is much larger than the mean collision time [7], the hydrodynamical regime is reached which corresponds here to a simple diffusion process. In this limit:

$$\langle (\mathbf{r}_i(t) - \mathbf{r}_i(0))^2 \rangle = 6Dt, \tag{2}$$

where D is the self-diffusion constant.

From there it is easily deduced that:

$$D = \frac{kT}{m} \int_0^\infty f(t) \, dt. \tag{3}$$

This is the Kubo formula for the self diffusion constant.

Let us suppose, as a first approximation, that the motion of the labeled particle can be described by a Langevin equation. It follows the equation for the v.a.f.

$$\frac{\partial f}{\partial t} + \gamma f(t) = 0, \tag{4}$$

where γ is the friction constant.

Taking the Laplace-Fourier transform:

$$\tilde{f}(\omega) = \int_0^\infty e^{-i\omega t} f(t)\, dt, \tag{5}$$

we get

$$\tilde{f}(\omega) = \frac{1}{-i\omega + \gamma}. \tag{6}$$

Using (3), we get in order to satisfy the hydrodynamic limit, the Stokes-Einstein relation

$$\gamma = \frac{kT}{mD}. \tag{7}$$

The v.a.f. corresponding to the Langevin equation is of the form $e^{-\gamma t}$. It is inconsistent with the behavior of the v.a.f. at small times which can be deduced from the equations of the motion:

$$f(t) = 1 - \frac{\Omega_0^2 t^2}{2} + \cdots, \tag{8}$$

where

$$\Omega_0^2 = \frac{\varrho}{3m} \int \nabla^2 V(r)\, g(r)\, d\mathbf{r}. \tag{9}$$

There ϱ is the particle density, $V(r)$ the two-body interaction potential, $g(r)$ the radial distribution function.

In order to make out for this inconsistency, one introduces a frequency dependent friction coefficient $\tilde{\gamma}(\omega)$, which at zero frequency reduces to the friction constant γ.

Equation (6), with γ replaced by $\tilde{\gamma}(\omega)$ is then valid for all frequencies (it is, in fact, a definition of $\tilde{\gamma}(\omega)$).

This corresponds to the generalized Langevin equation [8]:

$$\frac{\partial f}{\partial t} + \int_0^\infty \gamma(t - \tau) f(\tau)\, d\tau = 0. \tag{10}$$

It involves the knowledge of the system at times prior to the time t. This is why the function $\gamma(t)$ is often referred to as the memory function of the system.

The known behavior of $f(t)$ for small times is easily incorporated in this formalism by choosing a memory function of the form

$$\gamma(t) = \Omega_0^2 n(t) \tag{11}$$

where $n(0)=1$.

If the self motion did not induce an appreciable collective motion, then the memory function would extend only over times of the order of the mean collision time. One could try for instance a relaxation time approximation [9]

$$n(t) = e^{-t/\tau} \tag{12}$$

for which $f(t)$ is easily obtained analytically. Although this v.a.f. is much better than the simple Langevin exponential, it fails to reproduce even qualitatively, the behavior at large time of the v.a.f. as obtained from computer 'experiments' [10, 11].

2.2. THE VELOCITY AUTOCORRELATION IN THE DENSE GAS REGION

Alder and collaborators have shown [12] by computer experiment that the v.a.f. of moderately dense gas (density about half of the solidification density) behaves approximately as $1/t$ for very large times, i.e. times larger than about 20 mean collision times. This slow decay can be understood as caused by the collective motion accompanying the motion of the labeled particle: this motion initiates the formation of a vortex which closes up and pushes at later time the particle in its initial direction of motion. This would correspond to a negative memory function at large times (negative friction).

One can use the Navier-Stokes equation [12, 13] to see how an initial disturbance caused by the motion of the self particle decays at large times. This gives a behavior as $1/t$ in two dimensions and $1/t^{3/2}$ at three dimensions. The argument is admittedly weak in two dimensions where the Navier-Stokes equations have been shown not to exist [14]. It is yet convincing in three dimensions. In that case Alder has not checked that the $1/t^{3/2}$ behavior was reached, because the systems of 108 particles he uses is small enough so that by the time when the asymptotic region of the v.a.f. could be reached a sound wave could have propagated through the periodic boundaries. We report here preliminary results obtained in Orsay for a system of 4000 particles. The program, set up by Levesque, uses a speeding up technique very similar to that described by Schofield [15]. The particles interact through the part of the Lennard-Jones potential that gives rise to the repulsive force, i.e. through the potential

$$v_0(r) = v_{LJ}(r) + \varepsilon, \tag{13}$$

where $v_{LJ}(r) = 4\varepsilon[(\sigma/r)^{12} - (\sigma/r)^6]$.

The temperature is $T = 2\varepsilon/k_B$, the density $\varrho = 0.45\,\sigma^{-3}$.

At the time of writing the equations of motion were integrated for times [16] of the order of 600 τ_0.

If we fit the results for

$$12\,\tau_0 < t < 24\,\tau_0$$

by an $At^{-\alpha}$, we get

$$1.35 \leqslant \alpha \leqslant 1.65.$$

For that state we intend to calculate the thermodynamic and transport properties of the liquid in order to test the existing theories for the amplitude A of the tail of the v.a.f.

2.3. THE VELOCITY AUTOCORRELATION FUNCTION NEAR THE SOLIDIFICATION DENSITY

We have recalled in the introduction that the vaf near solidification presents a negative region which extends at fairly large times. This was observed by Rahman [10] in his classical study of the Lennard-Jones fluid near its triple point. The same structure is observed at higher temperature along the solidification line [17] and in the hard sphere system. This clearly corresponds to a retarded friction which can be understood in the following way: as in the less dense case the labeled particle tends to push and drag its neighbors in its forward motion. In the very dense case this forward motion tends to be reversed because of the barrier opposed by the repulsive cores. In a first step the motion of the labeled particle itself is reversed. At a later stage the motion of the group of accompanying particles is itself reversed. It is this backward collective motion which creates this persisting negative friction.

The spatial extent of this correlated motion can be studied by studying the Fourier transform of the self current-current correlation function (whose $k=0$ component is the v.a.f.). Then it is seen [11] that the two regimes considered above have a quite different character. The spatial extent of the vortex motion appears to be much larger than the extent of the group of particles involved in the collective motion near the triple point.

3. Shear Waves and Phonons in Simple Liquids

3.1. SHEAR VISCOSITY AND SHEAR WAVES

The shear viscosity can be calculated through the well-known Kubo relation

$$\eta = \int_0^\infty \eta(t)\, dt \qquad (15)$$

with

$$\eta(t) = \frac{\varrho}{k_B T} \langle \tau^{xz}(t)\, \tau^{xz}(0) \rangle, \qquad (16)$$

where $\tau^{xz}(t)$ is the xz component of the stress tensor at time t.

Using the microscopic expression for the stress tensor [18]

$$\tau^{xz}(t) = \sum_{i=1}^{i=N} \left(m v_i^x v_i^z - \tfrac{1}{2} \sum_{j \neq i} \frac{x_{ij} z_{ij}}{r_{ij}} \frac{\partial v(r)}{\partial r_{ij}} \right) \qquad (17)$$

one can calculate the function $\eta(t)$ in the course of a molecular dynamics experiment.

Such a computation has been done by Alder's group [3] for the hard sphere fluid at various densities and in Orsay [2] for the Lennard-Jones fluid near the triple point. Further computations are presently under way in Orsay on large Lennard-Jones systems.

As seen in the hard sphere system, the Kubo correlation for the shear viscosity is of short range in time for densities more than 10% lower than the solidification density. For simple fluids with continuous potentials one may then use the Maxwell viscoelastic theory and approximate $\eta(t)$ by a simple exponential

$$\eta(t) = G_\infty e^{-t/\tau}, \tag{18}$$

where G_∞ is the infinite frequency shear modulus, which according to (16) is an equilibrium quantity expressed in terms of the potential and the r.d.f. τ is a relaxation time of the order of the mean collision time.

Let us introduce the transverse current correlation function

$$C_t(k, t) = k^2 \left\langle \sum_{i=1}^{i=N} v_i^x(t) e^{-ikz_i(t)} \sum_{j=1}^{j=N} v_j^x(0) e^{-ikz_j(0)} \right\rangle. \tag{19}$$

In the long wavelength limit

$$\tilde{C}_t(k, \omega) = \frac{k^2 k_B T/m}{-i\omega + \tilde{\eta}(\omega) \dfrac{k^2}{m\varrho}}. \tag{20}$$

The hydrodynamical equation for the damping of shear waves is recovered in the small ω limit where, according to (15),

$$\tilde{\eta}(0) = \eta.$$

Replacing in (20), $\tilde{\eta}(\omega)$ by its relaxation time approximation

$$\tilde{\eta}(\omega) = G_\infty/(-i\omega + 1/\tau) \tag{21}$$

one easily sees that at low k one gets a diffusion equation, and a high k damped shear waves reminiscent of those present in a solid for all k's.

Molecular dynamics experiments show that near solidification there appears a long tail in $\eta(t)$. For the Lennard-Jones fluid near triple point ($kT/\varepsilon = 0.722$, $\varrho\sigma^3 = 0.8442$) one can fit the function $\eta(t)$ by a sum of two exponentials:

$$\tilde{\eta}(\omega) = G_\infty [(1 - \alpha)/(-i\omega + 1/\tau_<) + \alpha/(-i\omega + 1/\tau_>)] \tag{22}$$

with

$\alpha = 0.128$
$\tau_< = 0.63\, \tau_0$
$\tau_> = 4.72\, \tau_0$.

We see that $\tau_<$ is of the order of the mean collision time, but that $\tau_>$ is much larger.

With the above numbers it is clear that a large part of the total viscosity is contributed to by the long-time tail. This implies an enhancement of the shear viscosity (as compared to the values that can be obtained from the Enskog theory applied to the equivalent hard sphere gas) when one approaches solidification. We see from (22) that solid like modes in $\tilde{\eta}(\omega)$ appear at much lower frequency than in the one relaxation time approximation. This viscoelastic theory with two relaxation times leads to shear waves less damped and appearing at a lower value of k than with the usual one time model. For finite k's one can introduce a generalized viscosity $\tilde{\eta}_k(\omega)$ by using Equation (20) as a definition for this quantity. For this generalized viscosity we extend Equation (22) by using k dependent quantities. The k dependence of the infinite frequency shear modulus is known explicitly. $\tau_>$ is kept as a constant. For each k the value of α and τ_c is fitted so as to reproduce the computer experiments for $\tilde{C}_t(k, \omega)$. This fit shows that $\alpha(k)$ decreases very rapidly with k. This shows that the phenomenon giving rise to the tail involves a sizable group of atoms.

Direct computations of the shear viscosity have been performed for various states of the Lennard-Jones potential. The results can be compared with the present one near the triple point. Gosling, McDonald and Singer [19] have produces an actual shear of a sinusoidal form in the z direction of the enclosing box. When a small system (256 particles) is used, the long wavelengths are missed and the contribution of the long time tail to the shear viscosity is underestimated. To remedy this defect, these authors double the size of the system in the z direction, and find an excellent agreement with our results. Ashurst and Hoover [20] consider a fluid placed between moving planes with stick boundary conditions. For high velocities of the moving planes (of the order of σ/τ_0) the shear viscosity is strongly underestimated. The computation is repeated by lower velocities: its extrapolation to zero agrees very well with the values obtained through the Kubo formula.

3.2. VISCOELASTIC THEORY OF THE V.A.F.

Zwanzig and Bixon [21] have introduced a model which realizes the coupling between the motion of the labeled particle and the others: As they are interested in the behavior at large times, the system is described through the Navier-Stokes equation. The coupling is realized by approximating the labeled particle by hard spheres coupled to the medium through suitable boundary conditions, as in the derivation of the Stokes law. The coupling to thermal diffusion is altogether neglected and the viscosities are treated in the viscoelastic approximation with a single relaxation time. Then one obtains a v.a.f. which after a negative minimum becomes again positive in contrast with the results obtained from computer 'experiments'. If we introduce now the viscoelastic theory with two times we obtain a v.a.f. which stays negative in a large time region in qualitative agreement with the computer results.

3.3. LONGITUDINAL CURRENT CORRELATION FUNCTION

The longitudinal-current correlation is given by

$$C_l(k, t) = k^2 \left\langle \sum_{i=1}^{i=N} v_i^z(t) e^{-ikz_i(t)} \sum_{j=1}^{j=N} v_j^z(0) e^{ikz_j(0)} \right\rangle. \tag{23}$$

Its Fourier-Laplace transform can be written as

$$\tilde{C}_l(k, \omega) = \frac{k^2 k_B T/m}{-i\omega + \dfrac{k^2 k_B T}{-i\omega m S(k)} + \tilde{N}_l(k, \omega)}, \tag{24}$$

where $S(k)$ is the structure factor and $N_l(k, t)$ the appropriate memory function.

From this longitudinal current correlation function, one easily gets the dynamical structure factor $S(k, \omega)$ which is reached experimentally by studying the coherent neutron scattering:

$$S(k, \omega) = \frac{1}{\pi} \operatorname{Re} \frac{\tilde{C}_l(k, \omega)}{\omega^2}. \tag{25}$$

This function has also been studied by molecular dynamics [3, 22]. For the Lennard-Jones fluid near its triple point one observes clear phonon peaks for those wave vectors smaller than $1/\sigma$ that can be reached with a finite system ($N=864$, a system of 4000 particles is now under study).

In the hydrodynamic limit, these peaks are present (they are known as the Brillouin doublets). They are obtained by studying the poles of $\tilde{C}_l(k, \omega)$ in the low k and ω limit with

$$\tilde{N}_l(k, \omega) = \frac{\tfrac{4}{3}\eta + \zeta}{m\varrho} + \frac{k^2 k_B T}{mS(k)} \frac{\gamma - 1}{-i\omega + ak^2}. \tag{26}$$

There ζ is the bulk viscosity and a the heat conductivity K divided by ϱC_V. The hydrodynamic expression for $S(k, \omega)$ does not provide any fit for the molecular dynamic results even for the lowest value of k ($k=0.62\,\sigma^{-1}$): in particular it does not show any phonon peaks which disappear at a much lower value of k. An excellent fit is obtained by introducing for the generalized shear viscosity the same memory function with two relaxation times as in the transverse case. The bulk viscosity correlation has no tail, and gives a rather small contribution, because $\zeta/\eta \simeq 1/4$. $\gamma = C_V/C_P$ is replaced by $\gamma(k)$ which tends rapidly to 1 as k is increased: the coupling with thermal modes disappears at large wave vectors. $a = K/\varrho C_V$ is simply replaced by $a(k)$ which becomes negligible when k becomes of the order of σ^{-1}. We note that this analysis can be successfully carried out because we have made use of the frequency dependent shear viscosity with a long relaxation time. Careful experiments made with high flux reactors which make possible the investigation of $S(k, \omega)$ at low k and ω should permit the direct verification of these predictions.

4. Reorientational Correlations in Diatomics

We now consider the case of somewhat elongated molecules, namely diatomics such as nitrogen, which can be simulated by tying two Lennard-Jones potentials together [6]. A priori we expect the following: when one of these molecules rotates, it cranks the surrounding molecules into motion. If we describe the surrounding medium through Navier-Stokes equation (with frequency-dependent viscosities) and if we couple the rotating particle to the medium through suitable boundary conditions analogously to the Zwanzig-Bixon model we may hope in this case also to reach the collective behavior studied above. This question has been studied theoretically by Berne [5] and by Pomeau [23] in the density region where vortex rings are important. The molecular dynamics study of Levesque et al. concerns mainly the dense liquid region.

These authors calculate the correlation function

$$F(t) = \langle \mathbf{U}(0) \cdot \mathbf{U}(t) \rangle,$$

where $\mathbf{U}(t)$ is a unit vector attached to the molecule. This function is reached experimentally through the study of the infra-red spectra.

In order to exploit the analogy with the preceding sections, we introduce the function

$$C(t) = \langle \dot{U}(t) \dot{U}(0) \rangle$$

from which $F(t)$ is easily obtained by integration.

It is easily seen that the memory function representation for $C(t)$, which exhibits its known behavior at small times, is

$$\tilde{C}(\omega) = \frac{\omega_0^2}{-i\omega + \dfrac{\omega_0^2}{-i\omega} + (\Omega_0^2 - \omega_0^2)\tilde{\eta}(\omega)}$$

with

$$\omega_0^2 = 2k_B T/I.$$

(I is the moment of inertia) and

$$\Omega_0^2 = 2\omega_0^2 + \frac{\langle N^2 \rangle}{I^2},$$

where $\langle N^2 \rangle$ is the mean square torque which is an equilibrium quantity.

An excellent fit to the data for $C(t)$ and $F(t)$ is obtained by taking for the memory function $n(t)$ a single relaxation time approximation of the exponential type. A perfect agreement within the statistical error (around 3%) is obtained with a Gaussian memory function. This means that diatomics are not elongated enough to induce collective motions at an appreciable level.

References

1. Kadanoff, L. P. and Martin, P. C.: *Ann. Phys. (N.Y.)* **24**, 419 (1969).
2. Alder, B. J., Gass, D. M., and Wainwright, T. E.: *J. Chem. Phys.* **53**, 3813 (1970).
3. Levesque, D., Verlet, L., and Kurkijarvi, J.: *Phys. Rev.* **A7**, 1690 (1973).
4. Zwanzig, R. Z. and Bixon, M.: *Phys. Rev.* **A2**, 2906 (1970).
5. Berne, B. J. *J. Chem. Phys.* **56**, 2164 (1972).
6. Barojas, J., Levesque, D., and Quentrec, B.: *Phys. Rev.* **A7**, 1092 (1973).

7. We refer somewhat loosely to a mean collision time, which represents the typical relaxation time of the two-body processes. It can be given a precise meaning by introducing an equivalent hard-sphere gas. See for instance Reference 5.
8. Mori, H.: *Prog. Theoret. Phys. (Kyoto)* **33**, 423 (1965).
9. Berne, B. J., Boon, J. P., and Rice, S. A.: *J. Chem. Phys.* **45**, 1086 (1966).
10. Rahman, A.: *Phys. Rev.* **136**, A405 (1964).
11. Levesque, D. and Verlet, L.: *Phys. Rev.* **A2**, 2514 (1970).
12. Alder, B. J. and Wainwright, T. E.: *Phys. Rev.* **A1**, 18 (1970).
13. Dorfman, J. R. and Cohen, E. G. D.: *Phys. Rev. Letters* **25**, 1257 (1970). Ernst, M. H., Hauge, E. H., and Van Leeuwen, J. M. J.: *Phys. Rev. Letters* **25**, 1254 (1970).
14. Pomeau, Y.: *Phys. Rev.* **A5**, 2569 (1972).
15. Schofield, P.: *Computer Phys. Commun.* **5**, 17 (1973).
16. We choose as a time unit τ_0 equal to $(m\sigma/48\varepsilon)^{1/2}$. For Argon, with the usual values for σ and ε ($\sigma = 3.405$ Å, $\varepsilon = 119.8° \, k_B$) one has $\tau_0 = 3.112 \times 10^{-13}$ s. It may be of interest to note that for Argon near its triple point (to be specific, for $kT/\varepsilon = 0.722$ and $\varrho\sigma^3 = 0.8442$), one has for the equivalent hard sphere mean collision time $\tau = 0.226 \, \tau_0$.
17. Unpublished results obtained in our laboratory.
18. Schofield, P.: *Proc. Phys. Soc. (London)* **88**, 149 (1966).
19. Goshing, E. M., McDonald, I. R., and Singer, K.: to be published in *Mol. Phys*.
20. Ashurst, W. T. and Hoover, W. G.: to be published in *Phys. Rev. Letters*.
21. Zwanzig, R. and Bixon, M.: *Phys. Rev.* **A2**, 2005 (1970).
22. Rahman, A. *Proc. IUPAP Conf.*, Chicago, 1971.
23. Pomeau Y.: to be published.

MOLECULAR DYNAMICS CALCULATION OF NEUTRON INELASTIC SCATTERING FROM WATER*

F. H. STILLINGER

Bell Laboratories, Murray Hill, N.J. 07974, U.S.A.

and

A. RAHMAN

Argonne National Laboratory, Argonne, Ill. 60439, U.S.A.

Abstract. The data generated by a molecular dynamics calculation on liquid water at 1 gm cm^{-3} and 10°C have been used to simulate a neutron inelastic scattering experiment on water. It is shown that already at $\kappa \sim 1$ Å^{-1} the function $\omega^2 S_{\text{inc}}(\kappa, \omega)/\kappa^2$ is a good replica of the spectrum of proton velocity autocorrelation. It is emphasized that the separation of $S_{\text{inc}}(\kappa, \omega)$ into quasielastic and inelastic parts or the use of a phonon expansion in analyzing the data are both invalid procedures.

Résumé. La diffusion inélastique des neutrons par l'eau est simulée, utilisant des trajectoires produites par un calcul de dynamique moléculaire sur l'eau de densité 1 gm cm^{-3} et à une température 10°C. On a trouvé qu' une valur de $\kappa \sim 1$ Å^{-1} est déjà suffisamment petite pour que $\omega^2 S_{\text{inc}}(\kappa, \omega)/\kappa^2$ soit une bonne représentation du spectre de corrélation des vitesses de protons. On constate que (i) la séparation de $S_{\text{inc}}(\kappa, \omega)$ en une partie quasiélastique et une partie inélastique et (ii) l'usage d'une expansion phononique pour l'analyse des données d'une expérience de cette sorte sont, tous les deux, des procédés non-valables.

1. Introduction

In recent years computer simulation of liquids has led to an increasing degree of insight into the structural and kinetic properties of liquids at the molecular level [1]. Firstly, by solving a classical *N*-body problem under completely well defined mathematical conditions one can provide data for the development of the theory of liquids; the best example of this is the creation of the perturbation theory of monatomic liquids which would have been impossible otherwise [2]. Secondly, by drawing attention to rather unexpected aspects of the structure and kinetics of liquids, molecular dynamics data is able to provide guidelines for the analysis of data obtained on real materials in the laboratory [3]; it is worth recalling that the quasi-crystalline model for the analysis of liquid state properties has been abandoned almost universally because of the overwhelming evidence from molecular dynamics that such a model is quite inappropriate. Thirdly, by comparing experimental and molecular dynamics data of comparable accuracy, it has become possible to throw light on various interparticle interactions that are considered appropriate in certain liquids; a beginning in this direction has been made by studying liquid alkali metals experimentally and by molecular dynamics [4]. The recent work on water also falls into this category [5, 6].

---

* Work performed in part under the auspices of the U.S. Atomic Energy Commission.

It is the purpose of the following contribution to give an example of the second feature of molecular dynamics mentioned above. We have shown that valuable insight into the structural and kinetic properties of liquid water can be obtained by making molecular dynamics calculation on water. On the other hand neutron inelastic scattering is a valuable tool for the study of proton motions in water (in fact in any hydrogenous material). In this paper we shall present a mock-up of a neutron experiment using the trajectories of the protons generated in a molecular dynamics calculation on water.

In section 2 we give a short summary of the various ingredients which make up the molecular dynamics project on water which has been underway now for about three years. In Section 3 we present the results from one such molecular dynamics run; keeping in view the theme of this paper, only those aspects of the results are presented which are germane to the theme, namely the time behavior of proton trajectories; the detailed presentation of the results in their totality will be made elsewhere [6]. Section 4 describes the manner in which neutron inelastic scattering would analyze the motion of protons generated in the molecular dynamics calculation. In Section 5 the results are presented; where appropriate an effort is made to contrast the conclusions one can draw from these results with the manner in which neutron data has been analyzed and interpreted in the literature. In the final Section 6 a short discussion is given for the manner in which questions of a quantum mechanical nature related to our problem might be resolved. The difficulties of correcting neutron inelastic scattering data for multiple scattering are also mentioned there. Lastly, brief remarks regarding the vibratory modes of the molecules have been made in that section.

2. Molecular Dynamics Model for Water

The total potential energy used in the calculation being reported here consists of a sum of effective pair potentials. The present version of this effective pair potential uses a four-point-charge model for each molecule; the molecule itself is considered to be a rigid structure. Specifically

$$V(1, 2) = V_{LJ}(r_{12}) + S(r_{12}) V_{el}(1, 2), \qquad (2.1)$$

where r_{12} is the separation between oxygen nuclei. V_{LJ} is a central interaction of the Lennard-Jones type:

$$V_{LJ}(r) = 4\varepsilon [(\sigma/r)^{12} - (\sigma/r)^6], \qquad (2.2)$$

with $\varepsilon = 5.2605 \times 10^{-15}$ erg and $\sigma = 3.1$ Å. The four point charges, two $+q$ and two $-q$ on each molecule, contribute sixteen Coulombic interactions gathered in V_{el}. The charges have magnitudes 0.2357 e or 1.13194×10^{-10} esu. The two $+q$'s are at 1 Å distance from the oxygen nucleus at the positions occupied by the protons. The $-q$'s are 0.8 Å distance from the oxygen nucleus. The four lines joining the oxygen to these charges form precise tetrahedral angles (109°28′) with one another. Finally

$$S(r) = 0 \quad 0 \leqslant r < R_L,$$
$$= \frac{(r - R_L)^2 (3R_U - R_L - 2r)}{(R_U - R_L)^3} \quad R_L \leqslant r < R_U,$$
$$= 1 \quad R_U \leqslant r,$$

with $R_L = 2.0160$ Å and $R_U = 3.1287$ Å.

The molecular dynamics run was made with 216 water molecules at a density 1 g cm^{-3} and $T = 10\,°$C. The total run time was 38100 Δt; the integration step, Δt, was $10^{-4}\,\tau$ where $\tau = (M\sigma^2/\varepsilon)^{1/2} = 2.126 \times 10^{-12}$ s, M being the mass of the molecule.

This potential seems to give a moderately good account of the properties of liquid water. In particular the constant of self-diffusion is found to be 1.9×10^{-5} cm^2 s^{-1}. Details will be published elsewhere [6].

3. Characteristics of Proton Motions

Each proton participates in the translational motion of the center of mass of the parent molecule and in the rotational motion of the molecule around the center of mass. We will be interested in studying the details of this motion for microscopic times of the order of 10^{-12} s and hence over distances of the order of 10^{-8} cm. A neutron inelastic scattering experiment on water gives a composite picture of proton motions whereas in molecular dynamics one has the ability to study each aspect separately.

The mean square displacement of the center of mass of a molecule is defined in terms of its position $\mathbf{r}^{CM}(t)$ as

$$\langle \mathbf{r}^2 \rangle = \left\langle \frac{1}{N} \sum_{j=1}^{N} (\mathbf{r}_j^{CM}(t+\tau) - \mathbf{r}_j^{CM}(\tau))^2 \right\rangle_\tau. \tag{3.1}$$

In principle $\langle \mathbf{r}^2 \rangle$ contains all pertinent information about the process of self-diffusion. In practice however it is more profitable to consider its second derivative, the velocity autocorrelation function,

$$\langle \mathbf{V}(0) \cdot \mathbf{V}(t) \rangle = \left\langle \frac{1}{N} \sum_{j=1}^{N} \mathbf{V}_j^{CM}(t+\tau) \cdot \mathbf{V}_j^{CM}(\tau) \right\rangle_\tau. \tag{3.2}$$

The frequency spectrum of $\langle \mathbf{V}(0) \cdot \mathbf{V}(t) \rangle$ is denoted by $f_{CM}(\omega)$ and is given by

$$f_{CM}(\omega) = \int_0^\infty dt \cos \omega t \langle \mathbf{V}(0) \cdot \mathbf{V}(t) \rangle / \langle \mathbf{V}^2(0) \rangle. \tag{3.3}$$

Note that, as a matter of convenience, we have used $\langle \mathbf{V}^2(0) \rangle$ for normalizing the autocorrelation to unity at $t=0$.

The function $f_{CM}(\omega)$ is displayed in Figure 1. Note the clarity with which two regions of frequency stand out. There is a low frequency region at $\omega\tau=20$ (or 50 cm^{-1}) and a high frequency one at $\omega\tau=90$ (or 225 cm^{-1}). From Equation (3.3) we get $f_{CM}(0)=MD/k_B T$. From Figure 1, $f_{CM}(0)=0.0069\,\tau$ and hence $D=1.9\times 10^{-5}$ cm^2 s^{-1}. The region around $\omega\tau\sim 20$ corresponds to the motion of a molecule together with its immediate neighbors moving roughly as a 'cluster' while the region around $\omega\tau\sim 100$ corresponds to the oscillation of a molecule against its immediate neighbors.

The degree of damping however is so large (i.e., the width is comparable to the value of the frequency itself) that there is no tendency whatsoever for a molecule to

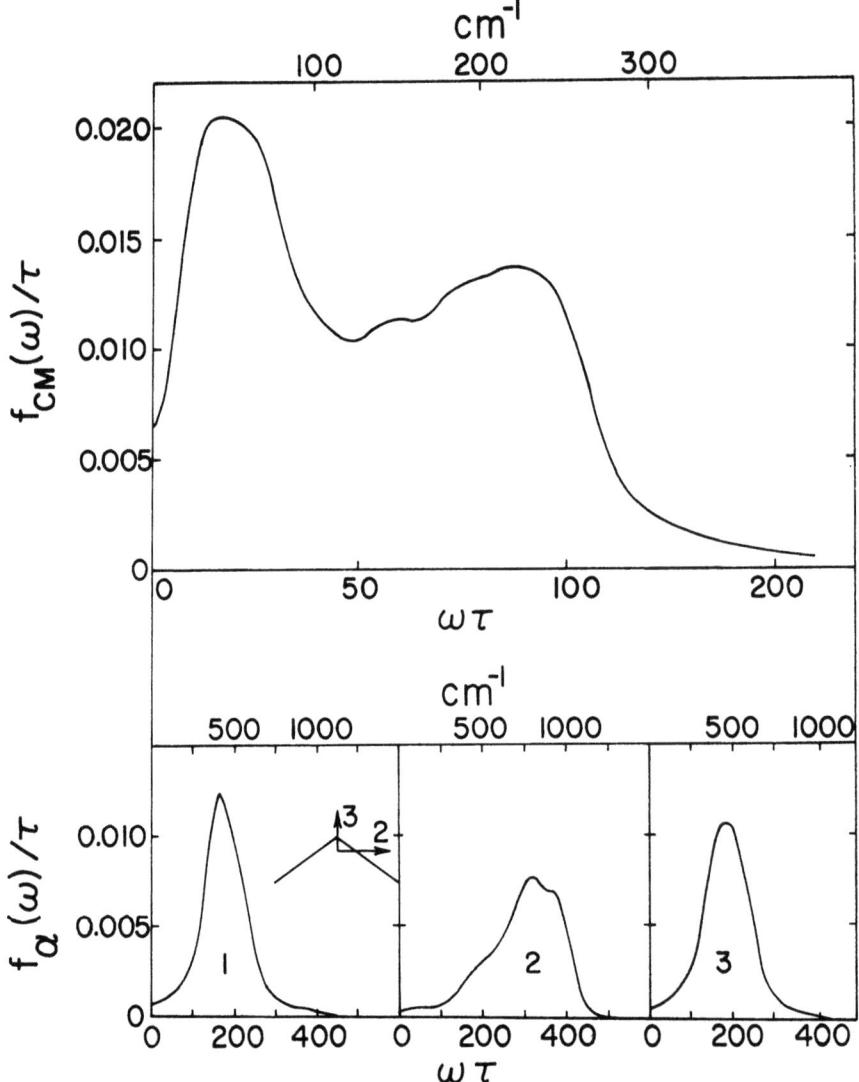

Fig. 1. Spectra $f_{CM}(\omega)$ [Equation (3.3)] and $f_\alpha(\omega)$ [Equation (3.4)]. Area under each curve is $\pi/2$. Unit of time $\tau=2.126$ ps.

be localized around its momentary position for any recognizable length of time. In other words there is *no* evidence of the so-called 'solid-like' behavior.

If $\omega_\alpha(t)$ is the angular velocity of the molecule around a principal axis α, the autocorrelations $\langle \omega_\alpha(0) \omega_\alpha(t) \rangle$, $\alpha = 1, 2, 3$ are of interest in studying the librational characteristics of molecular motion. The frequency spectrum is denoted by $f_\alpha(\omega)$ and is given by

$$f_\alpha(\omega) = \int_0^\infty dt \cos \omega t \langle \omega_\alpha(0) \omega_\alpha(t) \rangle / \langle \omega_\alpha^2(0) \rangle. \tag{3.4}$$

A slow time dependence of $\langle \omega_\alpha(0) \omega_\alpha(t) \rangle$ indicates relatively free rotations whereas a rapidly oscillating behavior indicates the trapping of a molecule in a certain orientation by its neighbors. The lower part of Figure 1 shows the three functions $f_\alpha(\omega)$; the choice of axes is also shown there. The moments of inertia are such that $I_1 > I_3 > I_2$. The frequency at which the spectra $f_\alpha(\omega)$ show maxima are of course in the reverse order. The values of these frequencies are $\omega \tau = 165$, 320–370 and 180 (or ~ 410 cm^{-1}, 800–925 cm^{-1}, and 450 cm^{-1}). It was shown in our previous work [5] that the localization of the molecules around their momentary orientations is only marginal.

We refer the reader to Eisenberg and Kauzmann [7] where a summary of the experimentally observed characteristics of molecular motions in water is given. The five characteristic frequencies we have found in the molecular dynamics model of water (50 cm^{-1}, 225 cm^{-1}, 410 cm^{-1}, 450 cm^{-1}, 800–925 cm^{-1}) are close enough to the experimental values for us to believe that the analysis and discussions in the following section do throw useful light on the neutron inelastic scattering technique for investigating the motion of molecules in water.

4. Theory of Neutron Inelastic Scattering from Water [8]

In the present context the relevant feature of the interaction between protons and neutrons is the fact that the neutron scattering cross section is dominated by the interference of waves scattered from the same proton at different times. Thus the neutron, when scattered from a hydrogenous material, samples the quantity $\mathbf{r}_j(t+\tau) - \mathbf{r}_j(\tau)$ where $\mathbf{r}_j$ denotes the position of proton j. Thus we will be interested in the function

$$F_s(\kappa, t) = \left\langle \frac{1}{N} \sum_{j=1}^N \exp\{i\mathbf{\kappa} \cdot (\mathbf{r}_j(t+\tau) - \mathbf{r}_j(\tau))\} \right\rangle_\tau, \tag{4.1}$$

for wave vector $\mathbf{\kappa}$, whose magnitude is denoted by κ. We shall further consider the Fourier transform of $F_s(\kappa, t)$; namely $S_{\text{inc}}(\kappa, \omega)$, given by

$$S_{\text{inc}}(\kappa, \omega) = \int_0^\infty dt \cos \omega t F_s(\kappa, t). \tag{4.2}$$

In the case of hydrogenous materials the cross section for scattering of a neutron, with momentum change $\hbar\kappa$ and energy change $\hbar\omega$, is directly related to $S_{\mathrm{inc}}(\kappa, \omega)$.

We shall also consider the autocorrelation $C_s(\kappa, t)$ defined by

$$C_s(\kappa, t) = \frac{1}{\langle(\kappa\cdot\mathbf{V})^2\rangle}\left\langle\frac{1}{N}\sum_{j=1}^{N}(\kappa\cdot\mathbf{V}_j(t+\tau))(\kappa\cdot\mathbf{V}_j(\tau)) \times \exp\{i\kappa\cdot(\mathbf{r}_j(t+\tau)-\mathbf{r}_j(\tau))\}\right\rangle_\tau, \qquad (4.3)$$

where

$$\langle(\kappa\cdot\mathbf{V})^2\rangle = \left\langle\frac{1}{N}\sum_{j=1}^{N}(\kappa\cdot\mathbf{V}_j)^2\right\rangle.$$

It is easy to show that

$$\frac{\mathrm{d}^2 F_s(\kappa, t)}{\mathrm{d}t^2} = -\langle(\kappa\cdot\mathbf{V})^2\rangle\, C_s(\kappa, t). \qquad (4.4)$$

The Fourier transform of $C_s(\kappa, t)$ is of course $\omega^2 S(\kappa, \omega)/\langle(\kappa\cdot\mathbf{V})^2\rangle$. In the limit $\kappa \to 0$ we find that $\omega^2 S_{\mathrm{inc}}(\kappa, \omega)/\langle(\kappa\cdot\mathbf{V})^2\rangle$ becomes the Fourier transform, $f_p(\omega)$, of the proton velocity autocorrelation function; i.e.,

$$f_p(\omega) = \lim_{\kappa\to 0}(\omega^2 S_{\mathrm{inc}}(\kappa, \omega)/\langle(\kappa\cdot\mathbf{V})^2\rangle) = \int_0^\infty \mathrm{d}t\,\cos\omega t\,\frac{\langle\mathbf{V}_p(0)\cdot\mathbf{V}_p(t)\rangle}{\langle\mathbf{V}_p^2(0)\rangle}. \qquad (4.5)$$

The suffix p makes it explicit that we are concerned with the velocity of the proton.

As stated above $\mathbf{V}_p$ is a composite of the translational and rotational motion of the molecules. There is no way, except through approximations in the interpretation of $S_{\mathrm{inc}}(\kappa, \omega)$ or of $f_p(\omega)$ to separate the various components of molecular motion embedded therein.

In the limit $\kappa \to 0$ another property of $S_{\mathrm{inc}}(\kappa, \omega)$ is of interest. For small κ only the large time behavior of $F_s(\kappa, t)$ remains relevant. In this limit $F_s(\kappa, t) \to \exp(-\kappa^2 \times (Dt+C))$ where D is the constant of self diffusion and C a constant with dimensionality $(\text{length})^2$. Hence the half width of $S_{\mathrm{inc}}(\kappa, \omega)$ at half height becomes $\kappa^2 D$.

In the limit $\kappa \to \infty$ the phase interference in the exponent of the right side of Equation (4.1) is so rapid that only the small time behavior of $F_s(\kappa, t)$ is relevant. In this limit we can write $\mathbf{r}_j(t+\tau)-\mathbf{r}_j(\tau)=t\mathbf{V}_j(\tau)$ and hence $F_s(\kappa, t)=\exp(-\langle(\kappa\cdot\mathbf{V})^2\rangle \times t^2/2)$.

It is necessary to emphasize here that in general, except in the two limiting cases $\kappa \to 0$ and $\kappa \to \infty$, $F_s(\kappa, t)$ *cannot* be written as $\exp(-\kappa^2 W(t))$. An exception is provided by a harmonically vibrating solid; this exception is also peculiar in that $W(t)$ goes to a constant value for $t\to\infty$ thus allowing an expansion of $F_s(\kappa, t)$ in powers of $W(t)-W(\infty)$. This is the well known phonon expansion.

Before proceeding to the presentation of the results in the next section it is worthwhile to recall the manner in which experimental data for scattering of neutrons from water has been treated. Unfortunately the *only* unambigous treatment of data is that of Sakamoto et al. [9]. These authors used the 'constant Q' technique to get the $S_{inc}(\kappa, \omega)$ directly and then transformed their data first to get $F_s(\kappa, t)$ and then the Van Hove function $G_s(r, t)$ for the protons. These results are in overall accord with the molecular dynamics results already reported [5, 6]. No other experiment along these lines has been performed over the last ten years for improving the accuracy of the results in the ω wings of $S_{inc}(\kappa, \omega)$; i.e., for large energy transfers to the neutron. In our opinion the analysis of all other experiments is fraught with ambiguities and with assumptions that cannot be justified. Firstly, there has always been an attempt to separate out the so-called quasi-elastic scattering from the inelastic part; secondly, it has been almost invariably assumed that $F_s(\kappa, t)$ has the form $\exp(-\kappa^2 \langle \mathbf{r}^2 \rangle / 6)$; thirdly, by combining the above two notions the scattering is expressed in terms of a phonon expansion involving a Debye-Waller factor which is not a definable quantity for motion of particles in a liquid. Another type of approach is one in which the proton is supposed to diffuse by a recognizable jump process between neighboring sites after a measurable stay at a site. Molecular dynamics data already published shows that none of the ideas mentioned above is valid for the purpose of describing the motion of protons in water. Also, the unambigous analysis of their experimental data by Sakamoto et al. [9], in spite of the limited accuracy available at that time, does not give any hint that assumptions like the ones described above have any validity.

Once a molecular dynamics run of sufficient length has been made and the chronological sequence of positions and velocities of all the protons recorded, the calculation of $F_s(\kappa, t)$ or $C_s(\kappa, t)$, Equations (4.1) and (4.3), respectively, is quite a simple matter.

5. Molecular Dynamics Results

5.1. $F_s(\kappa, t)$

In Figure 2 we have plotted the function $\log F_s(\kappa, t)$ for a few values of κ, to bring out an important feature of its dependence on κ and t. It is quite clear that $F_s(\kappa, t)$ *cannot* be written in the form $\exp(-\kappa^2 \langle \mathbf{r}^2 \rangle / 6)$. Hence any analysis of the data based on such a form would lead to erroneous conclusions regarding the behavior of $\langle \mathbf{r}^2 \rangle$. To obtain the correct asymptotic behavior of $\langle \mathbf{r}^2 \rangle$ for large t, and hence the constant of self diffusion, it is necessary to extract the small κ behavior of the function by a suitable extrapolation procedure. Figure 3 shows a plot of the same function as in Figure 2 but as a function of κ^2 for various values of t on the right end of Figure 2. Extrapolating to the value $\kappa = 0$ one gets the values marked off on the ordinate of Figure 3. Using these values and plotting them back on Figure 2 one gets the dashed line shown there which therefore indicates the asymptotic behavior in time for $\kappa = 0$. This gives a value of $D = 2.1 \times 10^{-5}$ cm^2 s^{-1}. The difference between this value and the value 1.9×10^{-5} cm^2 s^{-1} obtained from the motion of the center of mass and given in Section 3 is due to the fact that in Figure 2 the data needs to be taken further out

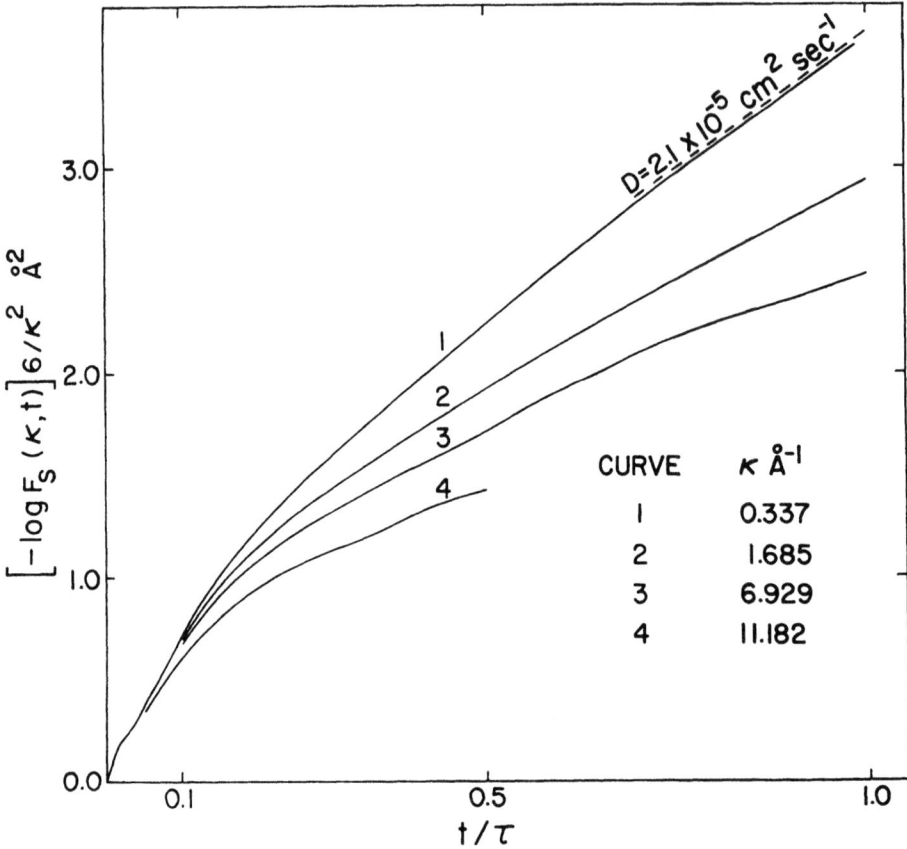

Fig. 2. Dependence of $F_s(\kappa, t)$ [Equation (4.1)] on t for various κ. A non-Gaussian distribution of proton displacements is clearly indicated; in other words $F_s(\kappa, t) \neq \exp(-\kappa^2 \langle r^2 \rangle/6)$.

in time to allow for the long time that the molecule takes to complete one whole diffusive rotation.

Note that in Figure 2 the curve for $\kappa = 0.337 \text{ Å}^{-1}$ indeed comes quite close to the curve for $\kappa = 0$. In other words an experiment made in the region of $\kappa \sim 0.3 \text{ Å}^{-1}$ to obtain $F_s(\kappa, t)$ from $S_{\text{inc}}(\kappa, \omega)$ is in principle capable of giving at least the large time diffusive behavior of $F_s(\kappa, t)$ correctly. The small time behavior will be considered in subsection 5.4. below. Notice that at $t = 0.025 \, \tau$ there is an interesting structured region in $F_s(\kappa, t)$ and also that the overall curvature extends to about $0.3 \, \tau$. This information will be contained in the high frequency region of $S_{\text{inc}}(\kappa, \omega)$.

5.2. $S_{\text{inc}}(\kappa, \omega)$

The function $S_{\text{inc}}(\kappa, \omega)$ is shown in Figure 4 for $\kappa = 0.337 \text{ Å}^{-1}$. Notice that the intensity is almost entirely in the region $\omega\tau < 0.2$. The interesting region (Figure 1) beyond $\omega\tau \sim 10$ is very difficult to observe at present even with the best equipment of the neutron inelastic scattering method. In Figure 4 we have also shown $S_{\text{inc}}(\kappa, \omega)$ for

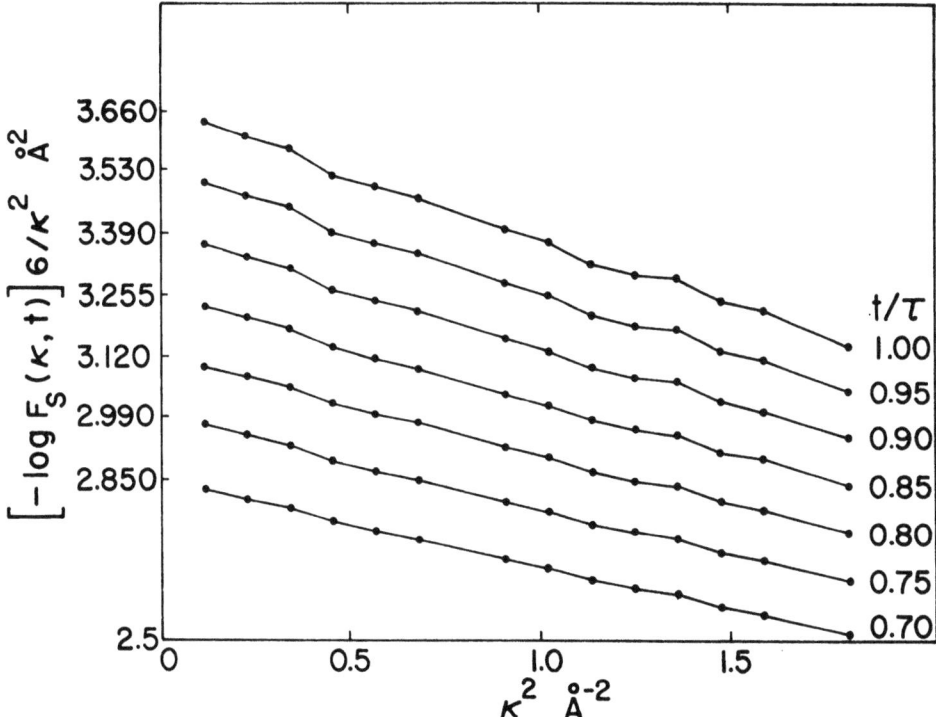

Fig. 3. Dependence of $F_s(\kappa, t)$ on κ^2 for various t between $0.7\,\tau$ and τ. Extrapolation to the left gives the values indicated. These values are shown in Figure 2 as a dashed line.

$\kappa = 0.674$ Å^{-1}. This curve will be reconsidered further below in this subsection. The upper part of Figure 4 shows the half width at half height of $S_{\text{inc}}(\kappa, \omega)$ as a function of κ^2. The discussion in Section 4 on the small κ behavior of $F_s(\kappa, t)$ shows that the half width at half height of $S_{\text{inc}}(\kappa, \omega)$ should go to zero as $\kappa^2 D$ when $\kappa \to 0$. From the upper graph in Figure 4 we see that to extract this information from $S_{\text{inc}}(\kappa, \omega)$ the region $\kappa \lesssim 0.5$ Å^{-1} is a region of small enough κ. The diffusion constant derived from the graph in Figure 4 is indicated on the graph.

From the point of view of the main theme of this paper it is necessary to emphasize that the bending of the half-width-at-half-height curve away from the straight line has no simple explanation in terms of the behavior of the mean square displacement alone of the protons. It arises from the fact that the Van Hove function $G_s(r, t)$ for the protons departs appreciably from a Gaussian shape. This has already been shown explicitly in Figure 2. We therefore believe that the construction of models [10] for $\langle \mathbf{r}_p^2 \rangle$, the mean square displacement of the protons, to account for an observed departure from the value $\kappa^2 D$ of the half-width-at-half-height of $S_{\text{inc}}(\kappa, \omega)$ is not useful; in fact it is an erroneous procedure. We have already drawn attention to this in previous publications [5, 11].

$S_{\text{inc}}(\kappa, \omega)$ for $\kappa = 0.674$ Å^{-1} is shown in Figure 5 for large values of ω. The frequency scale in Figure 5 is 100 times that in Figure 4. The values of $S_{\text{inc}}(\kappa, \omega)$ fall by

3 orders of magnitude or more in going from $\omega\tau=0$ to $\omega\tau=50$ so that the numerical transformation procedure (to get $S_{\text{inc}}(\kappa,\omega)$ from $F_s(\kappa,t)$) starts to generate noise which is seen on the right hand part of Figure 5. However, even in such difficult circumstances the values shown contain some useful information (apart from the width information shown in Figure 4). Multiplying by ω^2 one gets the values shown on the top part of Figure 5. Note that a maximum at $\omega\tau=20$ is now discernible. This corresponds to the low frequency element of diffusive motion of a molecule shown in Figure 1. Apart from this no further information can be obtained from $S_{\text{inc}}(\kappa,\omega)$ for small κ due to the practical difficulty of calculating for large ω the low intensity wings of this function when κ is in the region $<1\,\text{Å}^{-1}$. A method of getting around this practical difficulty of numerical transformation is dealt with in subsection 5.4 below.

Regarding the overall shape of $S_{\text{inc}}(\kappa,\omega)$ we notice that there is no logical way of separating out the so-called quasi-elastic part from the inelastic part. The curve of

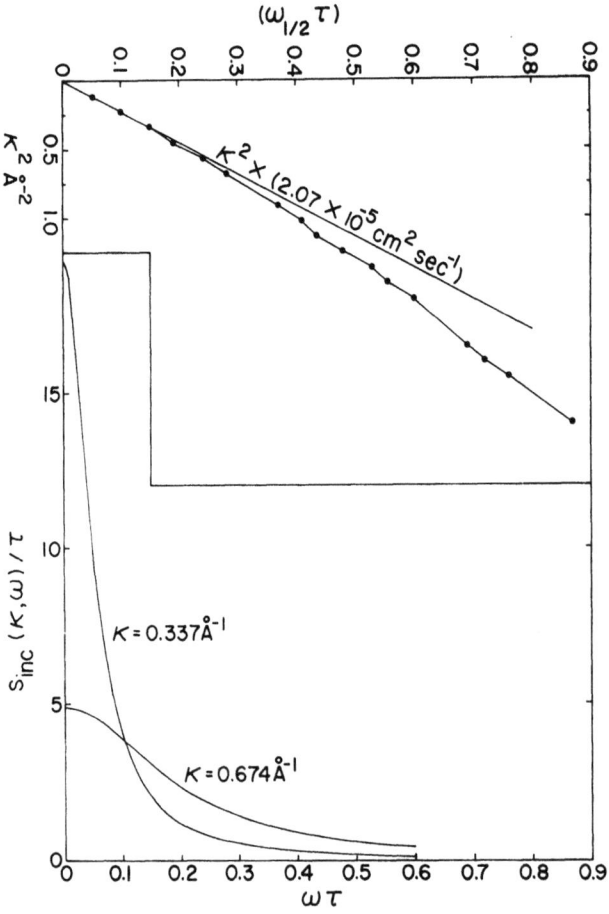

Fig. 4. Lower plot shows $S_{\text{inc}}(\kappa,\omega)$ for two small values of κ. The upper plot gives the half-width-at-half-height of $S_{\text{inc}}(\kappa,\omega)$ as a function of κ^2; the straight line behavior for small κ corresponds to $D = 2.07 \times 10^{-5}\,\text{cm}^2\,\text{s}^{-1}$.

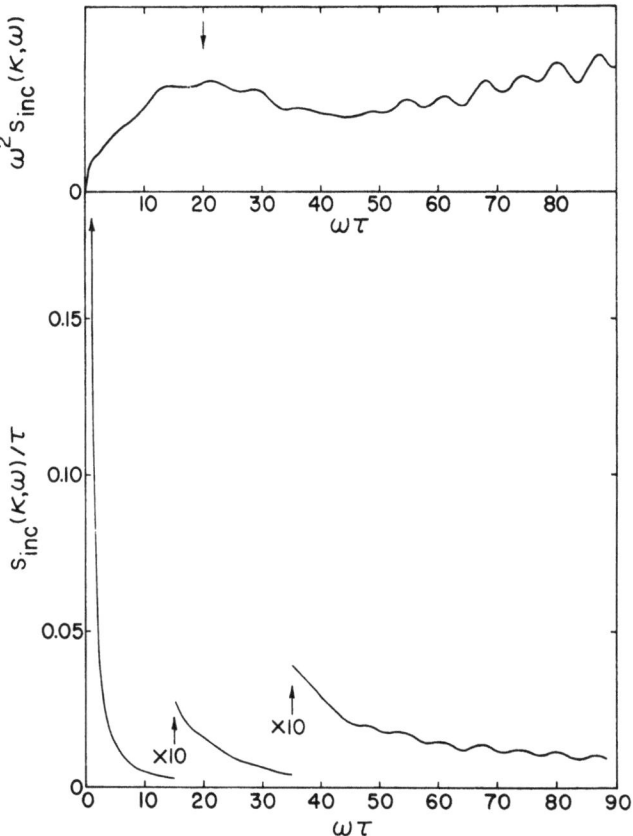

Fig. 5. $S_{\text{inc}}(\kappa, \omega)$ for $\kappa = 0.674 \text{Å}^{-1}$ (same as in Figure 4) is shown on an expanded scale. Note the large drop in intensity. The upper part shows $\omega^2 S_{\text{inc}}(\kappa, \omega)$ on an arbitrary scale. The arrow indicates the position of the first maximum in Figure 1.

$S_{\text{inc}}(\kappa, \omega)$ is one composite representation of the dynamics of the protons and has to be considered and analyzed in its entirety.

5.3. $\langle \mathbf{V}_p(0) \cdot \mathbf{V}_p(t) \rangle$

It is now clear that $S_{\text{inc}}(\kappa, \omega)$, in the small κ region, due to practical difficulties alluded to above, is just barely capable of bringing out a part of the interesting frequency characteristics of molecular motions shown in Figure 1. At this point therefore we revert to the primitive function itself; namely, $\langle \mathbf{V}_p(0) \cdot \mathbf{V}_p(t) \rangle$, the velocity autocorrelation function of the protons. The transform $f_p(\omega)$ defined in Equation (4.5) is shown in Figure 6. The frequency regions of importance taken from Figure 1 are indicated by arrows; as expected the proton gives a composite picture of the translational and rotational characteristics of the motion of water molecules. The shape of $f_p(\omega)$ is in overall accord with the available results from neutron inelastic scattering. However, in view of the experimental difficulties and methodological errors involved in extracting this spectrum from experimental data not much significance can be given to this accord.

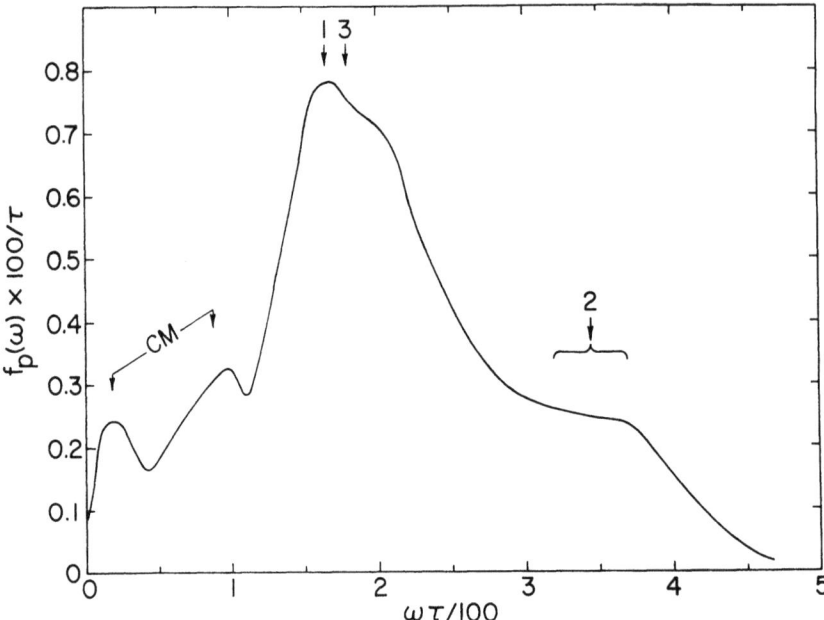

Fig. 6. Spectrum of the velocity autocorrelation of the proton. Arrows indicate expected positions of maxima from Figure 1. Librations around axes 1 and 3 are not separated. Their width makes the librations frequency around axis 2 and the high frequency component of center of mass motion appear only as shoulders in the spectrum.

In Figure 6 the only doubtful region is that of $\omega\tau \lesssim 2$. This is because the time a molecule takes to turn through a right angle is the Debye relaxation time ($\sim 2\,\tau$) and since $\langle \mathbf{V}_p(0) \cdot \mathbf{V}_p(t) \rangle$ has been calculated only up to $t = \tau/2$ there is a, residual, small negative tail of magnitude <0.01 in the correlation function which, if included, would reduce the values calculated for $f_p(\omega)$ in the region $\omega\tau < 2$ by about 10%. This region of frequency is of no significance for the following discussion.

The peak at $\omega\tau = 20$ stands out quite clearly (compare Figures 1 and 6). However the maxima at $\omega\tau \sim 180$ are so broad that they overshadow the structure which one would have liked to see more clearly at $\omega\tau \sim 90$ (intermolecular vibrations) and $\omega\tau \sim 350$ (librations around principal axis #2).

5.4. $f_p(\omega)$ AND $\omega^2 S(\kappa, \omega)/\langle(\boldsymbol{\kappa} \cdot \mathbf{V})^2\rangle$

As mentioned in Section 4, $\langle \mathbf{V}_p(0) \cdot \mathbf{V}_p(t) \rangle$ is the limiting form, for $\kappa \to 0$, of the autocorrelation function $C_s(\kappa, t)$ of Equation (4.3). It is interesting to know the degree to which the transforms of the two differ for $\kappa \sim 1\,\text{Å}^{-1}$, since this is a convenient region of κ for experimental work with neutrons. The transform of $C_s(\kappa, t)$ is shown in Figure 7; the values obtained are very close to $f_p(\omega)$ shown in Figure 6.

The conclusion to be drawn from Figure 7 is that in the region of $\kappa \sim 1\,\text{Å}^{-1}$, for $\omega\tau > 2$, $f_p(\omega)$ is extremely well represented by $\omega^2 S(\kappa, \omega)/\langle(\boldsymbol{\kappa} \cdot \mathbf{V})^2\rangle$ which is the transform of $C_s(\kappa, t)$. The significance of this is that a neutron inelastic scattering experi-

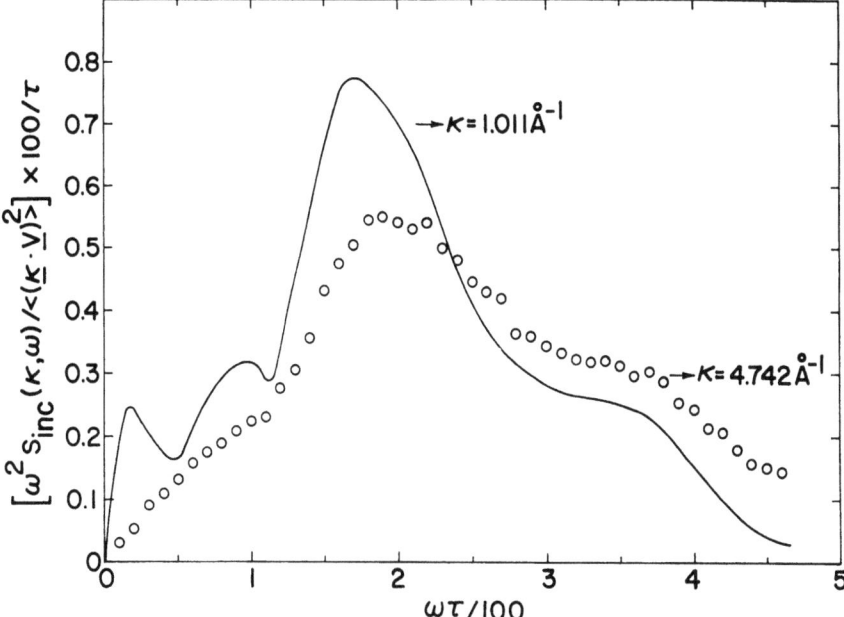

Fig. 7. Spectrum of $C_s(\kappa, t)$ [Equation (4.3)] for two values of κ. For $\kappa \sim 1$ Å^{-1} this is a very good representation of $f_p(\omega)$ of Figure 6. For $\kappa \sim 5$ Å^{-1} the departure from $f_p(\omega)$ is quite large.

ment made at $\kappa \sim 1$ Å^{-1} to determine $S_{\text{inc}}(\kappa, \omega)$ will give $f_p(\omega)$ by simply taking $\omega^2 S_{\text{inc}}(\kappa, \omega)$ and normalizing the area to correspond to a convenient value. In this presentation the definitions of all the transforms (Equations (4.2), (4.5)) are such that the area under the transforms from $\omega = 0$ to $\omega = \infty$ is $\pi/2$.

For large κ the difference between $f_p(\omega)$ and the transform of $C_s(\kappa, t)$ becomes more pronounced. For $\kappa = 4.742$ Å^{-1} the results are shown in Figure 7. Here we have only a rough, qualitatively similar, behavior, with respect to $f_p(\omega)$.

5.5. Conclusions

It thus appears that up to $\kappa \sim 1$ Å^{-1} an experimentally determined $S_{\text{inc}}(\kappa, \omega)$ simply multiplied by ω^2 (and the area $\int_0^\infty \omega^2 S(\kappa, \omega) \, d\omega$ suitably normalized) will give a reasonably faithful description of $f_p(\omega)$ and hence of the dynamics of the protons in the system. There is no question at all of a possible separation between the so-called quasi-elastic region and the inelastic region. Further, the notion of a phonon expansion is not only unnecessary but in fact it is incorrect for analyzing the data from liquid water.

6. Concluding Remarks

The outstanding difficulty in making neutron inelastic scattering experiments on water arises from multiple scattering in the sample. A good example of this is provided by the work of Blanckenhagen [12]; Figure 2 of Reference 12 shows the ratio of single to multiple scattering. It is seen there that the contribution from multiple scattering is not a 'correction' to, but a major part of, what is observed. We hope to investigate this problem in the near future by using the $S_{inc}(\kappa, \omega)$ generated by molecular dynamics to simulate the multiple scattering phenomenon in an experiment with thin plate geometry with 90% transmission and an incident neutron beam of wave length 1 Å [13].

All our calculations have been based on a system obeying classical statistical mechanics. However, the nature of the conclusions drawn here about the difficulties inherent in the interpretation of neutron inelastic scattering data are still valid conclusions. Of course a direct comparison between actual data and our classically calculated results needs much greater care and justification. The first step in making such a direct comparison will be to symmetrize the experimental data by writing

$$S_{inc}^{exp}(\kappa, \omega) = \exp(-\hbar\omega/2k_BT) S_{inc}^{sym}(\kappa, \omega),$$

where $\hbar\omega$ denotes energy gain by the neutron. This ensures compatibility with the requirements of detailed balance [14]. The comparison will then have to be made between $S_{inc}^{sym}(\kappa, \omega)$ and our calculated results. To what quantitative extent this is sufficient to account for quantum mechanical corrections is a question which needs further study.

The presence of molecular vibrations gives a further contribution to the neutron inelastic scattering; this is excluded from our calculations since we have treated the molecules as rigid. This contribution occurs [15] at $\omega\tau \approx 600$ and at $\omega\tau \approx 1200$ (see ω scale on Figure 6). However, the main argument in this presentation will not be affected by this extension in the spectrum.

References

1. The literature on molecular dynamics studies of various liquids has already become quite extensive. Apart from the Lennard-Jones liquid which has been most thoroughly investigated, work on liquid alkali metals, diatomic liquids and water is now available in the literature.
2. Verlet, L. and Weis, J. J.: *Phys. Rev.* **A5**, 939 (1972).
3. Rowe, J. M. and Sköld, K.: *Proc. Symposium on Neutron Inelastic Scattering at Grenoble*, I.A.E.A. Vienna, 1972, p. 413.
4. Rahman, A.: 'Number and Kinetic Energy Density Fluctuations in Classical Liquids' in S. A. Rice, K. F. Freed, and J. C. Light (eds.), *Statistical Mechanics New Concepts, Problems Applications*, University of Chicago Press, Chicago, 1972, p. 177. A similar study of liquid rubidium is under way.
5. Rahman, A. and Stillinger, F. H.: *J. Chem. Phys.* **55**, 3336 (1971); and Stillinger, F. H. and Rahman, A.: *J. Chem. Phys.* **57**, 1281 (1972).
6. Stillinger, F. H. and Rahman, A.: Results similar to those in Reference 5 have been obtained with a new potential (see Section 2 below); the new results, which are in better overall accord with experiment than the results presented in Reference 5, are being prepared for publication.
7. Eisenberg, D. and Kauzmann, W.: *The Structure and Properties of Water*, Oxford U.P., New York, 1969, Chap. 4, Table 4.10.
8. Egelstaff, P. A.: *An Introducton to the Liquid State*, Academic Press, London, 1967.
9. Sakamoto, M., Brockhouse, B. N., Johnson, R. G., and Pope, N. K.: *J. Phys. Soc. Japan* **17**, Supp. B-II, 370 (1962).
10. Rahman, A., Singwi, K. S., and Sjölander, A.: *Phys. Rev.* **126**, 997 (1962).
11. Rahman, A.: 'Molecular Dynamics Studies of Liquids', in P. C. Gehlen, J. R. Beeler, and R. I. Jaffee (eds.), Plenum Publishing Corp., New York, 1970, p. 233.

12. Von Blanckenhagen, P.: *Ber. Bunsenses. physik. Chem.* **76**, 891 (1972).
13. For a recent treatment of the data reduction and multiple scattering problem: Copley, J. R. D., Price, D. L., and Rowe, J. M.: *Nucl. Instr. and Methods* **107**, 501 (1973); and Copley, J. R. D.: *Computer Phys. Commun.*, to be published.
14. Schofield, P.: *Phys. Rev. Letters* **4**, 239 (1960).
15. Harling, O. K.: *Proc. Symposium on Neutron Inelastic Scattering*, I.A.E.A. Vienna, 1968, p. 507.

DISCUSSION

Schofield: Referring to the historic past of ten years ago, may I point out that jump diffusion may have been a feature of the Rahman, Singwi and Sjölander model. It was not a feature of the Schofield-Egelstaff model! May I point out, however, that the jump diffusion model does not give a Gaussian $G_s(\mathbf{r}, t)$.

Rahman: I did mention during the talk that I was a participant in the development of some of the wrong approaches of ten years ago. May I recall that the Rahman-Singwi-Sjölander approach used a Gaussian quasi-crystalline model whereas the earlier Singwi-Sjölander model used a jump model. It is true that jump diffusion model leads to a non-Gaussian $G_s(\mathbf{r}, t)$ but the trouble with the jump model is that the consequent mean square displacement is not correct at all.

Sillescu: Can you calculate the propagator $P(\mathbf{r}_1(0)/\mathbf{r}_0(t))$ for protons of 2 water molecule 1 and 2 that have their closest distance of approach at time zero, and compare it with $P(\mathbf{r}_1(0)/\mathbf{r}_1(t))$ in order to check whether there is collective translational motion of water molecules?

Rahman: We are going to calculate the propagator of interest to NMR in the near future.

Brot: (1) Looking into your comparison of classical vs quantal correlation functions for a variety of simple systems, what is the better approximation for the classical CF: the real part of the quantal one or its modulus? The first choice corresponds to the symmetrized CF of Kubo of course, the second one to an assumption frequently made by experimentalists. There exists also a third possibility which is to take the geometric mean between the real and the imaginary part, as suggested by Schofield a few years ago.

(2) If your intermediate scattering function F_s is for the protons and not for the centers of gravity, much of its non-Gaussian character must arise from its rotational part. Supposing a Gaussian rotational F_s amounts to truncating its Sears expansion at the first spherical harmonic, a procedure which is frequently incorrect.

Rahman: Up to now we have only compared the real part of the QM result with the classical result. The other alternatives have not yet been considered. Regarding your second question, the F_s we calculated was of course for protons and the non-Gaussian behavior does arise mostly out of rotational motion.

Springer: You have described the H_2O intermolecular interaction simply in terms of Coulomb forces between a tetrahedral arrangement of electric charges. Does this imply that the 'hydrogen bonding' can be understood essentially in these terms?

Rahman: I have a very meagre knowledge of the varieties of ways in which chemists have found it useful to think of hydrogen bonds. One way is a simple coulombic approach. We have shown by MD calculations that a coulombic picture, using a distribution of a few point charges in the molecule, is quite a useful approach.

Yip: (1) Does your effective frequency distribution agree with those derived directly from neutron inelastic scattering measurements?

(2) Would you care to comment in the context of your results on the assumption that the validity of rotation-translation coupling effects can be ignored?

Rahman: The agreement is really quite good; reference to the recent work of Blankenhagen (see list of references in our paper) will show this quite clearly. Such a comparison was purposely left out of this presentation in favour of a focus on the difficulties that arise in the analysis of data obtained in a neutron experiment.

The separability, to a fairly good approximation, of the translational and rotational motions of the molecules was mentioned in our first paper on water. In view of the interest shown here in this regard we will try to give more attention to it in our future work and, if necessary, we will publish numerical details as well.

Chen: Regarding the quantum correction of the Van-Hove selfcorrelation function, if I calculate the correlation function classically, what is wrong by just hang on the detail balance factor?

Rahman: That is just what we are doing at present as you will see in the text of the paper. However, in the literature alternative, and essentially arbitrary, recipes are available. We would like to consider them as well.

Schofield: The relation

$$S_{\text{Cl}} \approx \exp\left(-\frac{\hbar\omega}{2kT}\right) S_Q \bigg/ \cosh\left(\frac{\hbar\omega}{2kT}\right)$$

comes from the quantum and classical versions of the Kubo-Wosi relaxation function theory. I think this particular choice is mainly an aesthetic one. In relation to the question of quantum vs. classical spectral functions, Berne has done calculations for rotational motion and has shown recurrences in the time correlation function which do not occur classically.

Litovitz: Do the hindered translational peaks depend strongly on temperature? Is your result consistent with that of Walrafen who uses the intensity of these peaks to estimate the extent of hydrogen bonding?

Rahman: The peaks do depend on temperature though I do not recall the details. I do not know how Walrafen connects this spectrum with the number of bonds. We shall surely look into it.

Magat: Il y a, à mon avis, deux questions qui sont probablement liées et pour lesquelles il n'y a pas eu jusqu'à présent des réponses entièrement satisfaisantes. (1) Pourquoi l'intensité des bandes Raman intermoléculaires 160, 170, 500 et 700 cm^{-1} décroit-elle très rapidement entre 20 et 60°C, comme je l'avais observé il y a 40 ans et comme l'a confirmé Walrafen, tandis que l'intensité de ces mêmes bandes en infrarouge est indépendante de la température? (2) La constante diélectrique complexe de l'eau peut être parfaitement représentée par un demi-cercle de Cole et Cole à condition de prendre $\varepsilon_\infty = 5 \pm 0{,}5$. Or, d'une part n^2_{opt} de l'eau liquide est $\sim 1{,}7$, la décroissance de ε_∞ de 5 à 1,7 se faisant au voisinage de 170 cm^{-1} (fréquence de vibration-élongation qui ne saurait être responsable de la variation de ε_∞); d'autre part ε_∞ de la glace est de 3,2. Je crains que, tant que nous n'aurons pas d'explication entièrement satisfaisante de ces observations, nous ne pourrons pas dire que nous 'comprenons' l'eau liquide et ceci malgré les progrès énormes (dus en grande partie à Rahman et Stillinger) qui ont été faits ces dernières années.

Rahman: Qualitatively speaking the intensity of an IR band should be less dependent upon the surroundings of a molecule than that of the corresponding Raman band; with increase in temperature the immediate surroundings of a molecule become less ordered, the hydrogen bonds become more ill defined and hence the derivative of the polarizability of the cluster consisting of a molecule and its immediate neighbors should vary rapidly with temperature. A quantitative study of this question will involve the very difficult task of calculating the appropriate quantum mechanical transition matrix elements as a function of local disorder.

As regards ε_∞, (future) molecular dynamics calculations incorporating flexible water molecules will probably throw some useful light on the dielectric problem.

Friedman: Concerning the so-called translation-rotation coupling, one could imagine that your MD calculations were repeated for different proton masses, to see whether you get the large observed isotope effect (a factor of 1,2) for the change in diffusion coefficient or viscosity in going from H_2O to D_2O. If you do, then there must be translation-rotation coupling in your model fluid. If not, then in view of the excellence of your model, we could conclude that this particular isotope effect has a quantum mechanical origin. But the biggest quantal effect presumably comes from the librations (~ 500 cm^{-1}) so even the quantal effect would be a kind of translation-rotation coupling.

Rahman: These are very interesting points which we will pursue in further analysis of our data. It is possible that a small degree of coupling is enough to bring about the change you have mentioned.

Comment by M. R. Hoare

It would be a pity if no one were to mention the work of Zarar, Hasted and Chamberlain on submillimetre dielectric dispersion in water which will have appeared too recently for most participants to have seen it. (*Nature Phys. Sci.* **243**, 106 (1973)).

Interpreting their measurements of complex refractive index in the wave-number range 20–100 cm^{-1} they conclude that there is a second relaxation process with characteristic time 0.53×10^{-13} s, which they suggest arises from the rotation of whole molecules (or OH-groups) not breaking a hydrogen bond in the process.

It would be interesting to know whether such movements can be 'seen' in the latest molecular dynamics results.

COMPUTATION AND INTERPRETATION OF THE SPECTROSCOPIC PROPERTIES OF LIQUIDS

J. H. R. CLARKE* and S. MILLER

Dept. of Chemistry, Southampton University, Southampton, England

and

L. V. WOODCOCK

Dept. of Physical Chemistry, Leeds University, Leeds LS2 9JT, England

Abstract. Molecular dynamics calculations for spectroscopic properties of some simple liquids are compared with experiment, previous MD results, and physical approximations of analytical theories. Correlation functions obtained from MD data are used to examine models for intermolecular light scattering form liquid argon and molten KCL. MD computations based upon an inter-molecular diatomic (neon-neon) potential plus an intramolecular Morse potential have been carried out for liquids CO and N_2, from which angular correlation functions $\phi_{P_1}(t)$, $\phi_{P_1}(t)$, $\phi_J(t)$ and $\phi_{T_q}(t)$ are computed. The pair potentials employed for CO and N_2 lead to negative pressures (~ -0.5 kbar) at the experimental liquid densities. The Raman spectrum of molten KCN is reported and interpreted with the aid of MD computations using modified Tosi-Fumi potentials.

Résumé. Les calculs de dynamique moléculaire (DM) des propriétés spectroscopiques de quelques liquides simples sont comparés à l'expérience, à des résultats de DM antérieurs et aux approximations physiques des théories analytiques. Les fonctions de corrélation obtenues à partir des données de DM servent à examiner des modèles de diffusion intermoléculaire de la lumière par l'argon liquide et le KCL fondu. Les calculs de DM fondés sur un potentiel intermoléculaire diatomique (néon-néon) plus un potentiel de Morse intramoléculaire ont été conduits pour CO et N_2 liquides, ce qui pérmet de calculer les fonctions de corrélation angulaire $\phi_{P_2}(t)$, $\phi_{P_2}(t)$, $\phi_J(t)$ et $\phi_{T_q}(t)$. Les potenties lemployés pour CO et N_2 conduisent à des pressions négatives (env. $-0,5$ kbar) aux densités liquides expériementales. On présente le spectre Raman de KCN fondu, qui est interprété par des calculs de DM utilisant des potentiels de Tosi-Fumi modifiés.

1. Introduction

During the past decade there have been major developments in progress towards the quantitative theoretical interpretation of the optical properties of liquids at the molecular level. Gordon [1], for instance, has shown that the intensity profiles of infrared absorption and Raman scattering of light are related to time-correlation functions of the dipole moment and the polarisability tensor respectively. The advent of molecular dynamics (MD) computations [2], moreover, has given rise to the prospects of calculating these time-correlation functions, and hence the power spectra, without resort to mathematical truncation or semi-empirical parametrisation. Although the techniques of computer simulation are now widely employed to interpret thermodynamic and certain transport properties, optical properties have received comparatively little attention.

MD computations of high-frequency spectral data can be applied in a number of

---

* Present address: Dept. of Chemistry, University of Manchester Institute of Science and Technology, Manchester.

capacities in the study either of intermolecular forces in liquids or in the quantitative interpretation of the optical phenomena *per se*. Whereas thermodynamic and transport properties may be largely insensitive to the orientation dependence of intermolecular potentials, or polarisation forces in ionic and dipolar media, these are manifested in certain spectroscopic properties. Comparisons between computed and experimental spectra, therefore, should lead to deductions regarding the nature of the potential surfaces in the real liquid. These comparisons can also be used to test proposed mechanisms for the interactions between the scattering media and the optical field.

There have been many attempts to approximate the relevant time-correlation functions, and their corresponding memory functions, in terms of simplified models [3]. With the availability of essentially exact (MD) results, applications remain obscure, in a practical sense, but the computed correlation functions, nevertheless, may be used to illustrate the shortcomings with regard to quantitative interpretation. Not only can a particular spectroscopic property be computed, but the information required to understand the behaviour may be obtained by resolving the computed function to which it is related into theoretically or conceptually meaningful components.

A number of previous authors have reported time-correlation functions for model liquids which relate to those obtained by Fourier transformation of experimental spectra. Of notable interest in the present context are the simulations of liquids CO and N_2 using Stockmayer type potentials [3, 4], the calculations reported recently for a non-dipolar diatomic potential representation of liquid N_2 [5], calculations of the integrated intensity and spectral band shape for intermolecular light scattering from liquid argon [6, 7], and a preliminary report of a computation of high-frequency polarised light scattering from molten KCl [8].

In the present report, experimental and MD data for liquids N_2, CO, A, KCl and KCN are studied. In the case of argon and KCl, attention is confined to the intermolecular light scattering spectra. For CO, N_2 and KCN, IR absorption and Raman scattering due to angular rotations are considered. Further preliminary MD data on these liquids are reported, together with the experimentally measured Raman spectrum of molten KCN.

2. Intermolecular Light-Scattering

The spectrum of light scattered inelastically from simple atomic liquids, observed as a continuum in the wave number range 5–100 cm^{-1}, has aroused considerable interest as a potential source of information on polarisation forces and atomic motions.

The most notable features of the spectrum measured for liquid argon [9] (Figure 1) are a virtual absence of polarised scattering (the depolarisation ratio $\varrho = 0.72$) and an integrated intensity per atom much less than the vapour phase spectrum. The most plausible model that has been proposed to explain the spectrum is the dipole-induced-dipole (DID) mechanism. The induced dipole moment responsible for scattering at the ith atom is due partly to the incident optical field ($\mathbf{E}_0$) and partly to the local field ($\mathbf{E}_l$) produced by dipoles induced in surrounding atoms. The instantaneous value of

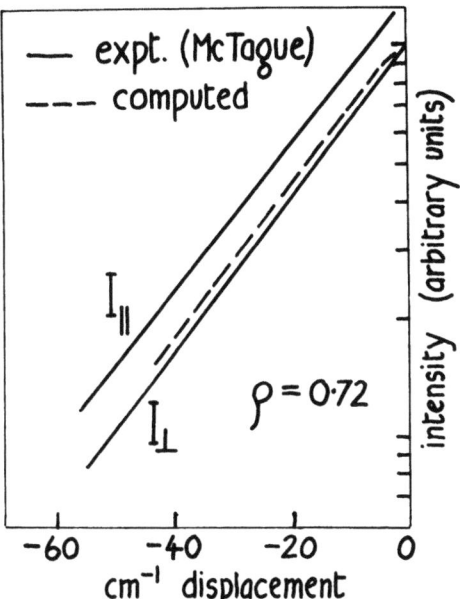

Fig. 1. Experimental intermolecular light scattering spectra for liquid argon (from Reference 9) together with a computed spectral density. $I_\parallel$ and $I_\perp$ are, respectively, the scattering observed with electric vector, parallel to and perpendicular to that of the incident light. ϱ is the depolarisation ratio.

E_l is given by the expression

$$\mathbf{E}_l(t) = \mathbf{E}_0(t) \cdot \sum_j^N \mathbf{T}_{ij}(t)$$

in which

$$\mathbf{T}_{ij} = \frac{1}{r_{ij}^3}\left[1 - \frac{3\mathbf{r}_{ij} \cdot \mathbf{r}_{ij}}{r_{ij}^2}\right]$$

E_l fluctuates according to the local particle distribution within the liquid and results in depolarised scattering. In the present work only the single particle correlation function

$$\left\langle \sum_j^N \mathbf{T}_{ij}(0) \sum_j^N \mathbf{T}_{ij}(t) \right\rangle$$

has so far been computed but the Fourier transform shows close agreement with the shape of the experimental spectrum (Figure 1). Alder et al. have computed the more general function

$$\left\langle \sum_{i,j}^N \mathbf{T}_{ij}(0) \sum_{i,j}^N \mathbf{T}_{ij}(t) \right\rangle$$

and the corresponding mean-squared quantity for a number of model potentials over a range of densities [6, 7]. They find that although the DID model alone does not account for the observed density dependence of intensity [6], it does nevertheless predict the correct form of the spectrum of liquid argon [7].

The inelastic scattering spectrum of liquid potassium chloride at 1073 K is shown in Figure 2. It has a similar exponential form and an integrated intensity comparable to that of liquid argon but the scattering in this case is predominantly *polarised* ($\varrho = 0.31$).

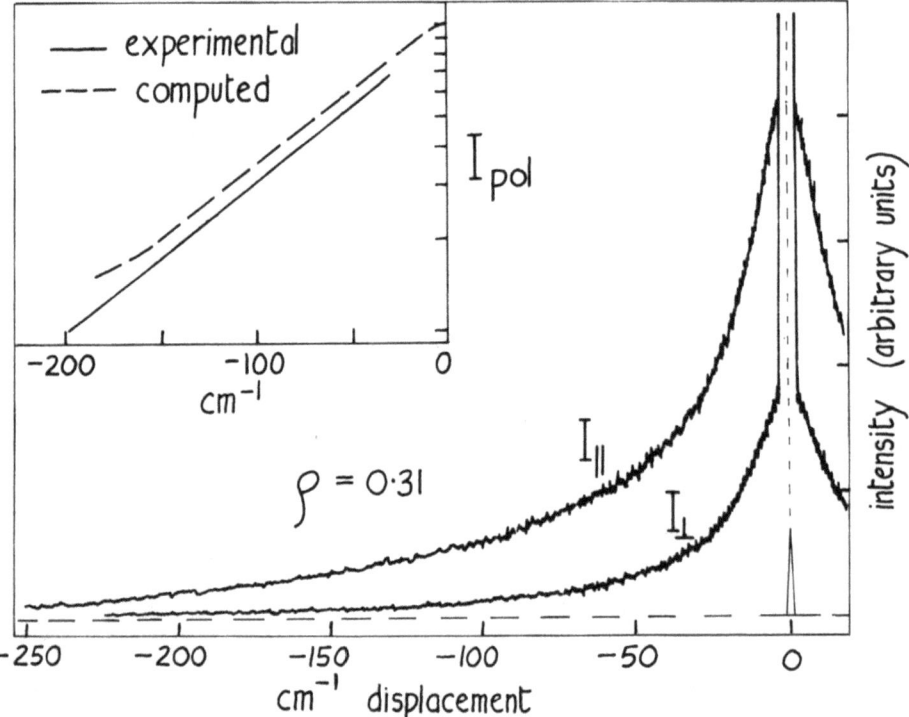

Fig. 2. Intermolecular light scattering spectra for molten potassium chloride at 1073 K. I_p ( ) is the trace scattering ($= I_\parallel(\omega) - \frac{4}{3} I_\perp(\omega)$).

Since the DID model does not give rise to polarised scattering (for an isotropic system, the trace of the induced anisotropy is zero), an alternative model has been suggested [8] in which the scattering arises predominantly from charge-induced polarisability anisotropy (CIA) due to the fluctuating electrostatic field in ionic liquids.

The instantaneous electric field ϕ in an ionic liquid is order-of-magnitude greater than in a non-polar liquid. In these circumstances, the total polarisability of a spherical ion can be expanded as a Taylor series [13] and the dipole moment induced by the optical field at the ith ion is given by

$$\mu(t) = (\alpha_i + \tfrac{1}{2}\gamma_i \psi_i^2(t) + \cdots) \mathbf{E}_0(t).$$

This model suggests that the polarised inelastic scattering should be related to the correlation function

$$\langle \mathrm{Tr}\, \psi_i^2(0)\, \psi_i^2(t) \rangle$$

and furthermore that the scattered intensity should be proportional to the square of the hyperpolarisability, γ_i. In Figure 3, relative integrated intensities are plotted for a series of sodium halide melts as a function of the estimated hyperpolarisabilities of the halide ions. (The hyperpolarisabilities are taken as proportional to the optical values for the isoelectronic rare gas atoms [16]. The much less polarisable sodium ions

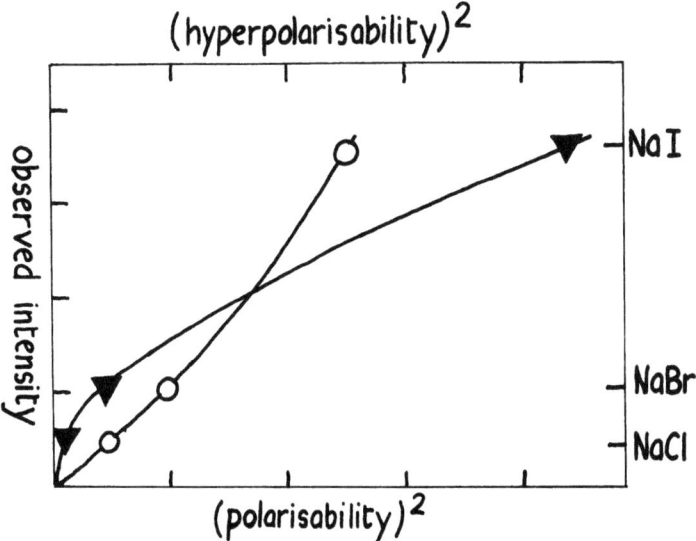

Fig. 3. Dependence of integrated light scattering intensities for liquid sodium halides at 1073 K on polarisabilities and hyperpolarisabilities of the anions. ○ polarisability (α); ▼ hyperpolarisability (γ).

are not considered.) There are large deviations from linearity, sufficient to suggest that this model is not satisfactory.

It is interesting to note from Figure 3 that the observed intensities are approximately proportional to the square of the leading polarisability, α_i, of the anions. An alternative description of the CIA mechanism is that the polarisability distortion is proportional to the magnitude of the intermolecular force [14]. A polarised spectrum computed on this basis [8] has a very similar profile to the experimental data (Figure 2). Quantitatively one can understand that polarised scattering would result from the appreciable symmetrical component in the motions of clusters of ions that are known to be a characteristic feature of ionic liquid microstructure [12].

3. Rotational-Vibrational Spectra

For linear molecules, Gordon [1] has shown that the infrared and Raman band profiles in liquids may be described in terms of the orientational correlation functions. For IR absorption, we have

$$A(\omega) \propto \frac{1}{2\pi} \int_0^\infty \langle \cos\theta(0) \cos\theta(t) \rangle \cos\omega t \, dt \tag{4}$$

and for Raman scattering

$$I_\perp(\omega) \propto \frac{1}{2\pi} \int_0^\infty \langle [\cos^2\theta(0) - \tfrac{1}{3}][\cos^2\theta(t) - \tfrac{1}{3}] \rangle \cos\omega t \, dt. \qquad (5)$$

$A(\omega)$ and $I_\perp(\omega)$ are, respectively, the absorbance and depolarised scattering intensity at a frequency displacement ω from the vibrational band origin. θ is the orientational displacement.

The autocorrelation functions in Equations (4) and (5), hereafter referred to as $\phi_{P_1}(t)$ and $\phi_{P_2}(t)$ respectively, have been computed from MD data on N_2 and CO liquids, both in the present and earlier work [3–5], for various models. Unfortunately, the available experimental spectra on both liquid CO (IR) [17] and N_2 (Raman) [18] probably are not sufficiently accurate for detailed comparisons. However, in Figure 4

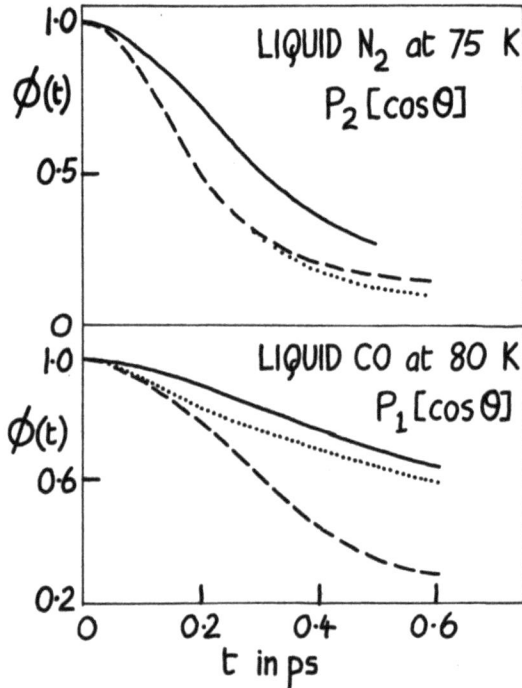

Fig. 4. Orientational correlation functions for liquids nitrogen and carbon monoxide. — computed in present work; – – – experimental data from Reference 18 for N_2, from Reference 17 for CO; ..... computed curves from Reference 5 for N_2, Reference 3 for CO.

the experimental correlation functions [19, 20], $\phi_{P_1}(t)$ for CO and $\phi_{P_2}(t)$ for N_2, are plotted together with the present results using a neon-neon diatomic potential (see Appendix) and with results for other models [4, 5]. In the case of the Raman spectra of liquid N_2, general qualitative agreement is obtained, but quantitative comparisons should be possible when accurate experimental data become available. In the case of liquid CO it is significant that two very different models yield similar correlation functions both of which are in large disagreement with the experimental curve. These and other angular correlation functions descriptive of rotational diffusion, namely the angular momentum $\phi_J(t)$ and the torque $\phi_{T_q}(t)$, calculated for liquids CO and N_2 from the present models (see Appendix), are shown in Figures 5 and 6. The torque correlation functions, exhibit some characteristics of a Brownian type rotational diffusion, i.e. rapid initial decay due to the random torque followed by the slow decay in the negative region associated with a frictional torque. The absence of a negative region in $\phi_J(t)$ for N_2 probably reflects the fact that on 'collision' there is an interchange between translational and rotational energy. The dipole and quadrupole moments in the modified Stockmayer models [3, 4] lead to appreciably more rotational

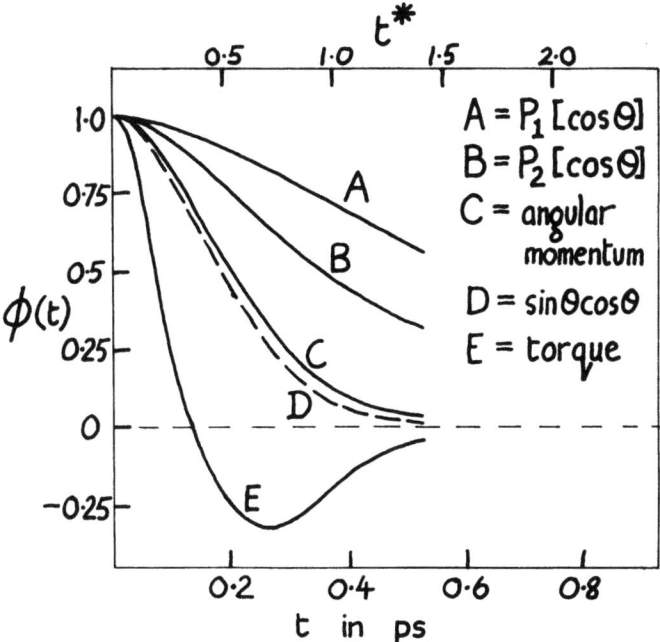

Fig. 5. Angular autocorrelation functions for the present liquid nitrogen model (see Appendix) $t*$ is the reduced time $(= t \times \neq kT/I)$.

backscattering despite the similarities between $\phi_{P_1}(t)$ and $\phi_{P_2}(t)$ in the two models.
The correlation function

$$\phi_{cs}(t) = \langle \cos\theta \sin\theta(0) \cos\theta \sin\theta(t) \rangle$$

has also been computed for the present model of liquid N_2 (Figure 5). It has been suggested [23] that this function decays much faster than $\phi_J(t)$ so that the following approximation is valid

$$\frac{d^2\phi_{P_2}(t)}{dt^2} \approx \phi_{\ddot{P}_2}(t) = \langle \cos\theta \sin\theta\dot\theta(0) \cos\theta \sin\theta\dot\theta(t) \rangle$$

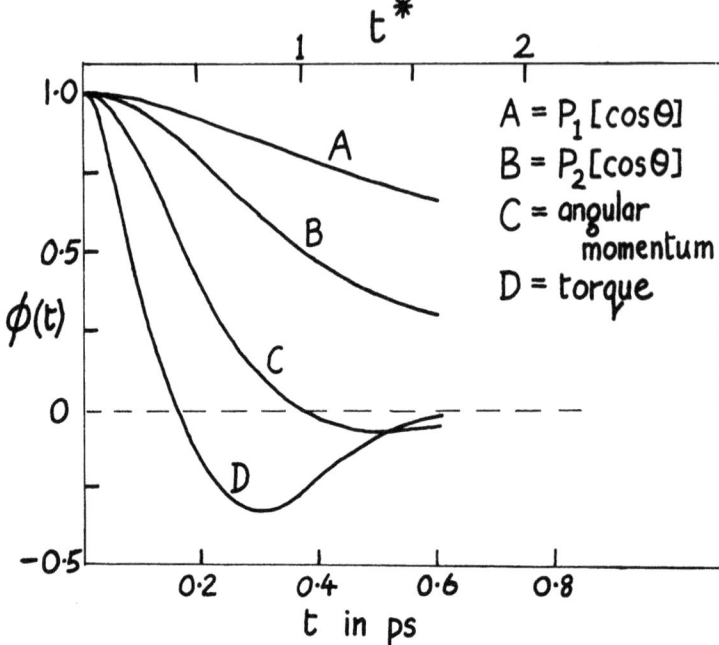

Fig. 6. Angular autocorrelation functions for the present liquid carbon monoxide model (see Appendix). Symbols as for Figure 5.

and that $\phi_{cs}(t)$ can be decoupled from the correlation function on the right-hand side, giving

$$\frac{d^2 \phi_{P_2}(t)}{dt^2} \approx \phi_J(t).$$

This, however, appears not to be the case, at least in simple diatomic liquids. Figure 5 shows that $\phi_J(t)$ is close to $\phi_{cs}(t)$ at all times. Both the computed and the experimental $d^2\phi_{P_2}(t)/dt^2$ function decay rapidly and go negative (c.f. the curve for benzene in Reference 23), but this type of behaviour is not exhibited by $\phi_J(t)$ in the diatomic model. If these observations are to any extent generally applicable, the 'mean collision times' in Reference 23 will tend to be underestimated.

The Raman spectrum of molten KCN has been measured and is shown in Figure 7. The orientational correlation function was derived from the spectrum by a previously described method [24]. MD calculations have been carried out using modified Tosi-Fumi potentials and a diatomic model for the CN^- ion. The computed $\phi_{P_2}(t)$ function agrees quantitatively with the experimental data over the whole time range of significance (this comparison is also shown in Figure 7). In Figure 8 all the computed angular correlation functions are shown.

When the time scale is reduced by the free rotor frequency factor $\sqrt{kT/I}$, where I is the moment of inertia, and T is the temperature, both $\phi_J(t)$ and $\phi_{T_q}(t)$ for KCN closely resemble those for CO and N_2. The orientation correlation functions $\phi_{P_1}(t)$

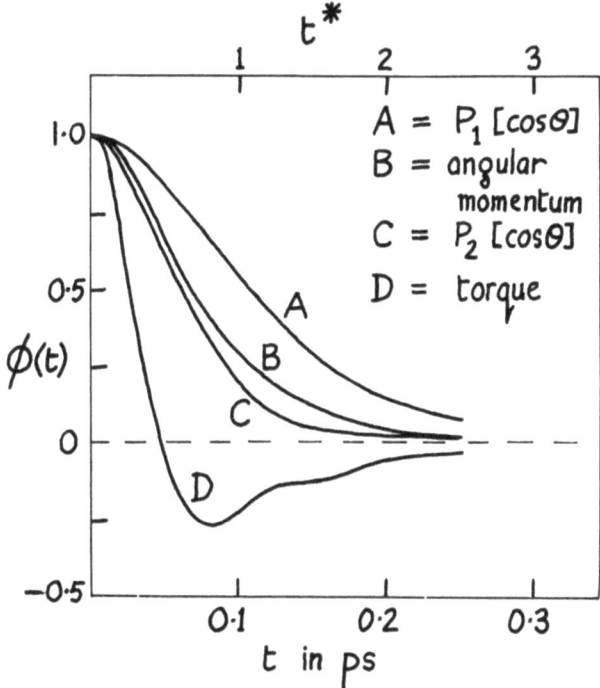

Fig. 7. Raman scattering from the 2065 cm^{-1} CN$^-$ stretching vibration in molten KCN at 1000 K. Inset shows a comparison of the Fourier transform of the orientational contribution with the MD computed $\phi_P(t)$ correlation function for CN$^-$.

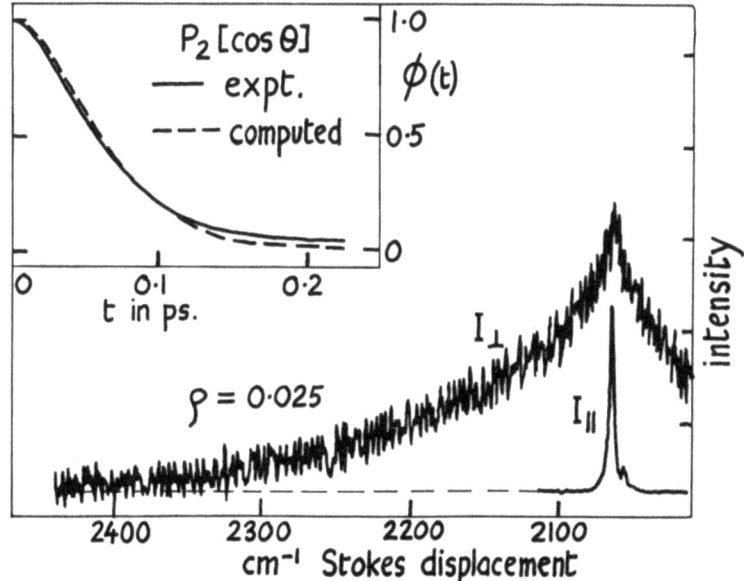

Fig. 8. Angular autocorrelation functions for CN$^-$ in molten KCN from MD computations (see Appendix). Symbols as for Figure 5.

and $\phi_{P_2}(t)$, however, display a relatively more rapid decay for CN$^-$. Cross-correlations

$$\left\langle \sum_{\substack{i,j \\ j \neq i}}^{N} [\cos^2 \theta_i(0) - \tfrac{1}{3}][\cos^2 \theta_j(t) - \tfrac{1}{3}] \right\rangle$$

have also been computed for CN$^-$ ions in molten KCN. The equilibrium covariance is found to be around 20% of the variance

$$\langle [\cos^2 \theta_i(t) - \tfrac{1}{3}]^2 \rangle$$

and negative. The sign reflects the presence, an average of K$^+$ ions directly between neighbouring CN$^-$ ions, as the coulombic field will tend to align the neighbouring CN$^-$ dipoles in opposite directions.

A few general observations are worth noting from Figures 5, 6 and 8 regarding approximate analytical models. The small-angle diffusion model [21] predicts that $\tau_{p_1}/\tau_{p_2} = 3$ and a jump diffusion model [22] predicts that this ratio should be unity (τ_{p_1} and τ_{p_2} are relaxation times assuming an exponential decay of the correlation functions). The former model also implies that $\phi_J(t)$ should decay much faster than either $\phi_{P_1}(t)$ or $\phi_{P_2}(t)$. The most favourable situation for the small angle diffusion model is found for the weak dipolar CO model for which $\tau_{p_1}/\tau_{p_2} \to 3$. However, the situation is progressively less favourable in the cases of the N$_2$ and KCN models. For liquid KCN, in fact, $\phi_{P_2}(t)$ is always close to $\phi_J(t)$ implying a predominance of large angle jumps although $\phi_{P_2}(t)$ and $\phi_{P_1}(t)$ are well separated. Alternative models due to Gordon allowing rotational jumps of varying magnitudes and include the effects of free rotation, i.e. that $\phi_{Tq}(t)$ decays well in advance of the free rotation time. In each of the above cases, $\phi_{Tq}(t)$ decays to zero in approximately one third of $\sqrt{I/kT}$ thus lending some support for these models.

Acknowledgements

We wish to thank the C.E.G.B. for a research fellowship (to J.H.R.C.) and a research studentship (to S.M.), and the Ramsay Memorial Trust for a (General British) fellowship (to L.V.W.). Financial assistance from N.A.T.O. grant no. 379 is gratefully acknowledged.

Appendix

MD computations are based upon the algorithm of Verlet [26] and have been carried out for liquid argon ($\varrho = 1.417$ g cm^{-3}, $\bar{T} = 83.7$ K), liquid N$_2$ ($\varrho = 0.9345$ g cm^{-3}, $\bar{T} = 75$ K), liquid CO ($\varrho = 0.9345$ g cm^{-3}, $\bar{T} = 75$ K), liquid KCl ($\varrho = 1.530$ g cm^{-3}, $\bar{T} = 1045$ K), and liquid KCN ($\varrho = 1.373$ g cm^{-3}, $\bar{T} = 1000$ K) using natural atomic masses. The pair potentials constituting the MD models all incorporate the EXP-6 function.

In the case of argon a simple EXP-6 potential is used with the parameters given by van Thiel and Alder [27], and the number of atoms in the MD cube (N) is 108. For

KCl, the Tosi-Fumi [28] EXP-6-8 plus ion-ion coulombic potentials are used and $N=216$ ions.

Diatomic pair potentials are used for N_2 and CO of the modified Buckingham form with the parameters corresponding to the neon-neon potential taken from Hirchfelder et al. [29]. The dipole moment of CO is represented by point charges on the atomic nuclei. An intramolecular Morse potential is incorporated into the model with the parameters for the dissociation energy, the internuclear distance and the force constant all as given in Moore [30]. All the MD runs for N_2 are with $N=216$; computations using both $N=216$ and $N=64$ have been carried out for CO.

The Tosi-Fumi potentials for KCl have been modified for KCN. The K^+-K^+ interaction is as KCl but the distance parameters (r_i+r_j in reference C) between K-C and K-N are 2.498 Å to conform with the CN^- ionic radii as given by X-ray data on the crystal. A diatomic model is also used for the CN^- ion incorporating an intraionic Morse potential and a separation of the charge to give the dipole moment. Computations with $N=324$ and $N=96$ have been carried out for KCN but only the $N=324$ results are used here.

Thermodynamic averages are monitored to test the adequacy of the models. For A, KCl and KCN, satisfactory agreement is obtained at the present level of comparison with experiment. The pressures calculated in the N_2 and CO simulations, however, are -0.41 ± 0.1 and -0.79 ± 0.1 kbar respectively. Whilst the depth of the neon-neon potentials are close to the semi-empirical effective diatomic potential derived from crystal data [31], an unexpectedly large value of σ (in the LJ diatomic potential) is required, and which is in fact comparable with the argon-argon distance.

The Raman spectrum of molten KCN was determined using a Varian-Cary 82 Raman Spectrometer. The high temperature sample cell has been described previously [32].

References

1. Gordon, R. G.: *Adv. Magn. Resonance* **3**, 1 (1968).
2. Rahman, A.: *Phys. Rev.* **A136**, 405 (1964); for reference to MD computations see the previous two papers and References 3-6 below.
3. Berne, B. J. and Harp, G. D.: *Adv. Chem. Phys.* **17**, 63 (1970).
4. Harp, G. D. and Berne, B. J.: *Phys. Rev.* **A2**, 2514 (1970).
5. Barajas, J., Levesque, D., and Quantrie, B.: *Phys. Rev.* **A7**, 1092 (1973).
6. Alder, B. J., Weis, J. J., and Strauss, H. L.: *Phys Rev.* **A7**, 282 (1973).
7. Alder, B. J., Weis, J. J., and Strauss, H. L.: in press.
8. Clarke, J. H. R. and Woodcock, L. V.: *J. Chem. Phys.* **57**, 1006 (1972).
9. McTague, J. P., Fleury, P. A., and Dupre, D.: *Phys. Rev.* **188**, 303 (1969).
10. Buckingham, A. D. and Stephen, M. J.: *Trans. Faraday Soc.* **53**, 884 (1957).
11. Kirkwood, J. G.: *J. Chem. Phys.* **4**, 592 (1936).
12. Woodcock, L. V.: *Proc. Roy. Soc. (London)* **A328**, 83 (1972).
13. Buckingham, A. D. and Pople, J. A.: *Proc. Phys. Soc.* **A68**, 905 (1955).
14. Lewis, J. C. and van Kranendonk, J.: *Phys. Rev. Letters* **24**, 802 (1970).
15. Pauling, L.: *Proc. Roy. Soc. (London)* **A114**, 181 (1972).
16. Buckingham, A. D. and Dunmur, D. A.: *Trans. Faraday Soc.* **64**, 1776 (1968).
17. Ewing, G.: *J. Chem. Phys.* **37**, 2250 (1962).
18. Grawford, M. F., Welsh, H. L., and Harold, J. H.: *Can. J. Phys.* **30**, 81 (1952).
19. Gordon, R. G.: *J. Chem. Phys.* **43**, 1307 (1965).

20. Gordon, R. G.: *J. Chem. Phys.* **42**, 3658 (1965).
21. Debye, P.: *Polar Molecules*, Dover Publications Inc., 1929, Ch. 5.
22. Ivanov, E. N.: *Soviet JETP* **18**, 1041 (1964).
23. Hardy, H., Volterra, V., and Litovitz, T.: *Faraday Soc. Symposum*, 1972.
24. Clarke, J. H. R. and Miller, S.: *Chem. Phys. Letters* **13**, 97 (1972).
25. Gordon, R. G.: *J. Chem. Phys.* **44**, 1830 (1966).
26. Verlet, L.: *Phys. Rev.* **159**, 98 (1967).
27. van Thiel, M. and Alder, B. J.: *J. Chem. Phys.* **44**, 1056 (1966).
28. Tosi M. P. and Fumi, F. G.: *J. Phys. Chem. Solids* (1964).
29. Hirchfelder, J. O., Curtiss, C. F., and Bird, R. B.: *Molecular Theory of Gases and Liquids*, John Wiley and Sons, New York, 1954 p. 180.
30. Moore, W. J.: *Physical Chemistry*, 4th ed., Longmans, London, 1962, pp. 592 and 605.
31. Sweet, J. R. and Steele, W. A.: *J. Chem. Phys.* **47**, 3022 (1967); see also Reference 5.
32. Clarke, J. H. R., Hartley, P. J., and Kuroda, Y.: *J. Phys. Chem.* **78**, 1831 (1972).

DISCUSSION

Brot: Nous avons calculé, pour le potentiel utilisé par Levesque et Quentrec, pour N_2, les courbes équipotentielles d'interaction entre une molécule et un atome supposé indépendant (équipotentielle de $U(\mathbf{r} - \mathbf{r}_A) + U(\mathbf{r} - \mathbf{r}_A')$ où $\mathbf{r}_A$ et $\mathbf{r}_A'$ sont les vecteurs position des deux atomes de la molécule). Nous pouvons définir la 'forme' de la molécule comme étant l'enveloppe intérieure de la famille de cercles de rayon $\sigma/2$ dont le centre se déplace le long de l'équipotentielle zéro. Cette 'forme' coincide d'une facon quasi-miraculeurse avec ce qui parait être la meilleure définition physique de la forme d'une molécule, à savoir, la courbe d'équiprobabilité de présence des électrons correspondant à la densité $1,35 \times 10^{-2}$ électron par angström au cube (Raich, J. C. and Mills, R. L.: *J. Chem. Phys.* **55** (4), 1811 (1971)). De plus, l'amincissement en haltère que l'on pourrait reprocher à ce modèle comme n'étant pas physiquement satisfaisant n'apparît que pour des valeurs très élevées fu potentiel, valeurs qui ne seraient jamais atteintes dans les zones de température où nous travaillons (liquide dense).

Woodcock: I agree. The diatomic potential contains two adjustable parameters which may be semi-empirically fitted either advertently or possibly fortuitously. As you imply, the evidence from *ab initio* calculation of electron density contours, from previous studies on crystalline phases, and from the combined MD computations to date, suggests that the diatomic potential is an inadequate representation. Of course, our ultimate objective in comparing calculated and experimental spectra is to obtain information *inter alia* on the orientation dependence of the intermolecular potential which may not be forthcoming from thermodynamic and bulk transport data. From the practical point of view, however, using the diatomic representation in MD computations has the advantage of amenability to direct use in the cartesian coordinate system without having to compute Eulerian angles.

De Graaf: The rotational motions in N_2 and of the CN^- ion in KCN are quite similar in the liquid and in the high-temperature solid phase. By comparing MD calculations with lattice dynamics results (from dispersion relation measurements) would one not be able to get more information about the interatomic potentials?

Woodcock: As far as I am aware, no one has yet reported MD results for molecular crystals but when these are forthcoming, comparisons with experimental dynamic properties should certainly provide additional criteria for parametrising model pair potentials.

Kneubühl: Constant and others found a relation between the orientational correlation functions and the related angular correlation functions if there are frequent collisions. This relation can be described by the following statement. The orientational memory function is an approximation of the angular momentum correlation function. Do you know of any recent and more precise statement on this relation?

Woodcock: I am not aware or the work to which you refer of the relationship itself. Molecules in the liquid phase move cooperatively on very complex fluctuating potential energy surfaces. Unless 'collision' is defined in a strict statistical mechanical sense, I would be sceptical of any approximate relationship obtained on that basis. For polar liquids the angular momentum correlation function has a negative region unlike the orientational memory function.

PRESSURE DEPENDENCE OF THE INCOHERENT SCATTERING LAW AND THE TIME-DEPENDENT TRIPLET CORRELATION FUNCTION IN DENSE HYDROGEN GAS*

S. H. CHEN, Y. LEFEVRE, and G. MAZENKO**

Nuclear Engineering Dept., Massachusetts Institute of Technology Cambridge, Mass. 02139 U.S.A.

and

P. A. EGELSTAFF

Physics Dept., University of Guelph, Guelph, Ontario, Canada

Abstract. The incoherent scattering laws (or the van Hove self correlation functions) of hydrogen gas are measured at 85 K and pressures of 110, 120 and 130 atm. From these data we are able to determine the pressure derivative of the scattering law for the wave vector transfer $Q = 0.2, 0.4$ and 0.6 Å^{-1}. The mean density of the gas is $n = 8.56 \times 10^{21}$ cm^{-3} which is close to the critical density of hydrogen $n_c = 9.34 \times 10^{21}$ cm^{-3}. We relate the pressure derivative to a certain integral of the time-dependent triplet self correlation function according to Egelstaff *et al.* [1] We then derive an exact expression for the pressure derivative in terms of the pressure derivative of the memory function associated with the phase-space self-correlation function [2] and evaluate it in the one-relaxation time approximation. We then discuss the qualitative features of the experimental results in terms of the approximation.

Résumé. On mesure à 85 K sous des pressions de 110, 120 et 130 atm la diffusion incohérente (ou les fonctions d'autocorrélation de van Hove) pour l'hydrogène gazeux. On en déduit la variation en fonction de la pression de la loi de diffusion pour des vecteurs de transfert $Q = 0.2$–0.4 et 0.6 Å^{-1}. La densité moyenne du gaz (8.56×10^{21} cm^{-3}) est proche de la densité critique de l'hydrogène qui est 9.34×10^{21} cm^{-3}. On relie la variation avec la pression à une certaine intégrale de la fonction triplet d'autocorrélation dépendant du temps (Egelstaff *et al.*) [1]. On en tire une expression exacte de la variation avec la pression: elle est exprimée par une fonction mémoire associée à la fonction d'autocorrélation dans l'espace des phases (référence 2) et évaluée dans l'approximation à un seul temps de relaxation. Les résultats expérimentaux qualitatifs sont discutés en fonction de cette approximation.

1. Introduction

It is well known [3] that the double differential incoherent scattering crossection of neutrons by a dense fluid is directly related to the spectral density of the classical test particle correlation function $G_s(r, t)$. i.e.,

$$S_s(k\omega) = \frac{1}{2\pi} \int_{-\infty}^{+\infty} dt \int d^3r \, e^{+i(\mathbf{k}\cdot\mathbf{r} - \omega t)} G_s(r, t), \qquad (1)$$

---

* Work supported by contract AEC AT(11-1) 3352
** Also Dept. of Physics, Harvard University, Cambridge, Mass. 02138, U.S.A.

where

$$G_s(\mathbf{r}_1 - \mathbf{r}_2, t_1 - t_2) = \frac{1}{n} \langle N\delta(\mathbf{r}_1 - \mathbf{R}(t_1))\delta(\mathbf{r}_2 - \mathbf{R}(t_2))\rangle. \quad (2)$$

= probability of finding the position of the test particle at $\mathbf{R}(t_2)$ at t_2 given that it is at $\mathbf{R}(t_1)$ at t_1.

In practice, however, measurement of $S_s(k\omega)$ as a function of ω at constant k gives too much redundant information since the line shape of $S_s(k\omega)$ is severely restricted by the known sum rules such as

$$\langle \omega^0 \rangle \equiv \int_{-\infty}^{+\infty} d\omega \, S_s(k\omega) = 1 \quad (3)$$

$$\langle \omega^2 \rangle \equiv \int_{-\infty}^{+\infty} d\omega \, \omega^2 S_s(k\omega) = \frac{k_B T}{m} k^2, \quad (4)$$

where m is the mass of the fluid particles. Since $S_s(k\omega)$ is also a function of the thermodynamic state of the fluid namely a function of temperature T and pressure P, what is really relevant to the theory of dense fluids is changes in $S_s(k\omega)$ induced by changes in T and P. In this connection Egelstaff et al. [1, 4] pointed out recently that a particularly simple relation exists between the isothermal pressure derivative of $S_s(k\omega)$ and Fourier integrals of a certain time-dependent three-particle self correlation function $K(\mathbf{r}, \mathbf{s}, t)$. Following their notation we have:

$$A_T(k\omega) = \left.\frac{\partial S_s(k\omega)}{\partial P}\right|_T = \frac{1}{k_B T} \int dt \, d^3r \, e^{+i(\mathbf{k}\cdot\mathbf{r} - \omega t)} \int d^3s K(\mathbf{r}, \mathbf{s}, t), \quad (5)$$

where

$$K(\mathbf{r}, \mathbf{s}, t) \equiv G_{3c}(\mathbf{r}, \mathbf{s}, t) - ng(s) G_s(r, t) \quad (6)$$

$G_{3c}(\mathbf{r}, \mathbf{s}, t)$ = Given that at $t=0$ there is a test particle at position 0 and another particle at position $\mathbf{s}$, it is the probability of finding the test particle propagating to a new position $\mathbf{r}$ at t under the influence of the other particle (whatever the motion of it).

If one switches off interaction between the test particle and the particle at $\mathbf{s}$, $G_{3c}(\mathbf{r}, \mathbf{s}, t)$ obviously reduces to $ng(s) G_s(r, t)$. $K(\mathbf{r}, \mathbf{s}, t)$ therefore is the interacting part of G_{3c} that takes into account propagation of the test particle to $\mathbf{r}$ having scattered by the particle initially at $\mathbf{s}$. In the so-called Vinyards model of liquids [5] $K(\mathbf{r}, \mathbf{s}, t)$ would be identically zero and so also would $A_T(k, \omega)$ be zero. Therefore the experimental fact (we shall see later) that at certain ranges of k the pressure derivatives are appreciably non-zero would immediately rule out the validity of the model at these ranges of k. Although the integration over $\mathbf{s}$ in Equation (5) largely removes the utility of using $A_T(k\omega)$ to obtain the detail information on the three particle function $K(\mathbf{r}, \mathbf{s}, t)$, however, it is true that the relation such as (5) gives some information about the three particle correlation in dense fluids by being able to rule out some unrealistic models of the fluid such as the example given above.

The sum rules (3) and (4) together with (5) immediately gives us sum rules for $A_T(k\omega)$:

$$\int d\omega\, A_T(k, \omega) = 0 \tag{7}$$

$$\int d\omega\, \omega^2 A_T(k\omega) = 0. \tag{8}$$

These two sum rules taken together tell us that at $k = $ const. the spectrum of $A_T(k\omega)$ has both positive and negative parts and furthermore the curve has to have zero crossings at least twice. We shall see this feature later.

In this paper we present a method by which $A_T(k\omega)$ can in principle be rigorously calculated. We shall use the generalized kinetic theory approach of Mazenko [2] and we shall relate $A_T(k\omega)$ directly to the pressure derivative of the memory function associated with the phase-space self-correlation function. Since the memory function is in the heart of the physics governing dynamics of the microscopic single particle motions in fluids, the experimental curves of $A_T(k\omega)$ should in the true sense contain the most relevant information with regard to theory of liquids. Since the measurements to be presented are still not extensive enough and in order to understand the qualitative features of the result we shall take the one relaxation time model [6] approximation of the memory function and calculate $A_T(k\omega)$. We shall show that $A_T(k\omega)$ indeed is directly proportional to the pressure derivative of the binary collision frequency which is the only relevant parameter of the theory in this approximation.

2. Theory

In this section we shall use exclusively the notation of Reference 2 which we shall call FRKT. Readers are recommended to consult the reference for detailed definitions of various quantities. We shall begin by writing down the van Hove self-correlation function, in terms of the phase-space self correlation function.

$$S_s(k\omega) = -\frac{1}{\pi n}\,\mathrm{Im}\int d^3p\, d^3p'\, C_s(k, pp', \omega + i\varepsilon). \tag{9}$$

From which we have

$$\left[\frac{\partial}{\partial P} S_s(k\omega)\right]_\beta = -\frac{1}{\pi}\,\mathrm{Im}\int d^3p\, d^3p'\,\frac{\partial}{\partial P}\left[\frac{1}{n} C_s(k, pp', \omega + i\varepsilon)\right]_\beta. \tag{10}$$

Consider first an identity

$$\left[\frac{\partial}{\partial P} C_s(k, pp', z)\right]_\beta \equiv \int d^3\bar{p}\, d^3\bar{p}'\, C_s(k, p\bar{p}, z)\left[-\frac{\partial}{\partial P} C_s^{-1}(k, \bar{p}\bar{p}', z)\right]_\beta \times \\ \times C_s(k, \bar{p}'p', z), \tag{11}$$

where $C_s^{-1}(k, pp', z)$ can be obtained in terms of the generalized kinetic equation [2]

$$\int d^3\bar{p}\, C^{-1}(k, p\bar{p}, z)\, C(k, \bar{p}p', z) = \delta(p - p') \tag{12a}$$

$$C_s^{-1}(k, pp', z) = \left(z - \frac{k \cdot p}{m}\right)\frac{\delta(p - p')}{nf_0(p)} - \frac{1}{nf_0(p)} \phi_s(k, pp', z) \tag{12b}$$

Clearly

$$\frac{\partial}{\partial P} C_s^{-1}(k, pp', z) = \left[\frac{\partial}{\partial P}\left(\frac{1}{n}\right)\right]_\beta nC_s^{-1}(k, pp', z) - \frac{1}{nf_0(p)} \times$$

$$\times \left[\frac{\partial}{\partial P} \phi_s(k, pp', z)\right]_\beta \tag{13}$$

so that by putting (13) into (11) we have

$$\left[\frac{\partial}{\partial P} C_s(k, pp', z)\right]_\beta = -\left[\frac{\partial}{\partial P} n^{-1}\right]_\beta nC_s(k, pp', z) + \int d^3\bar{p}\, d^3\bar{p}' \times$$

$$\times C_s(k, p\bar{p}, z)\frac{1}{nf_0(\bar{p})}\left[\frac{\partial}{\partial P}\phi_s(k, \bar{p}\bar{p}', z)\right]_\beta C_s(k, \bar{p}'p', z). \tag{14}$$

We then put (14) into (10) to obtain finally

$$\left[\frac{\partial S_s(k\omega)}{\partial P}\right]_\beta = -\frac{1}{\pi n^2} \mathrm{Im} \int d^3p\, d^3p'\, d^3\bar{p}\, d^3\bar{p}'\, \frac{1}{f_0(\bar{p})} C_s(k, p\bar{p}, z) \times$$

$$\times \left[\frac{\partial}{\partial P} \phi_s(k, \bar{p}\bar{p}', z)\right]_\beta C_s(k, \bar{p}'p', z). \tag{15}$$

The above expression is an exact result. We see that the pressure derivative of the van Hove self correlation function is directly related to the pressure derivative of the memory function. To evaluate (15) we have to actually solve the kinetic Equation (12) for $C_s(k, pp', z)$. This is numerically possible for moderate density fluids with hard sphere potential since in this case the memory function is considerably simpler [7]. We shall not do so here because we are only attempting to understand the qualitative features of the experimental result here. We shall, however, use the one-relaxation time approximation to the kinetic equation which is known to give quite favorable results for the line width of $S_s(k\omega)$ at this range of hydrogen gas density, [6].

In this approximation we have ϕ_s independent of k and z, and given by an expression

$$\phi_s(pp') = -i\alpha[\delta(p - p') - f_0(p)], \tag{16}$$

where α is the typical collision frequency of molecules in the gas. The α is related to the self diffusion coefficient D and the thermal speed $v_0 = (k_B T/m)^{1/2}$ by $\alpha = v_0^2/D$. Using (16) in (15) we have

$$\left[\frac{\partial S_s(k\omega)}{\partial P}\right]_\beta = \frac{1}{\pi n^2}\left(\frac{\partial \alpha}{\partial P}\right)_\beta \mathrm{Im}\left\{i\int d^3\bar{p}\, d^3p\, d^3p'\left[C_s(k, p\bar{p}, z) \times \right.\right.$$

$$\times C_s(k, \bar{p}p', z) \frac{1}{f_0(\bar{p})} - \int d^3\bar{p}\, d^3\bar{p}'\, d_3^3 p\, d^3 p' \times$$
$$\times C_s(k, p\bar{p}, z)\, C_s(k, \bar{p}'p', z)\Big]\Big\}. \qquad (17)$$

The solution $C_s(k, pp', z)$ in this case is also well known [6] and we have

$$\int d^3 p\, C_s(k, pp', z) = \frac{1}{1 - i\alpha D(kz)} \frac{nf_0(p')}{z - \frac{k\cdot p'}{m} + i\alpha} \qquad (18)$$

$$C_s(kz) = \int d^3 p\, d^3 p'\, C_s(k, pp', z) = \frac{nD(kz)}{1 - i\alpha D(kz)}, \qquad (19)$$

where

$$D(kz) = \int d^3 p \, \frac{f_0(p)}{z - \frac{k\cdot p}{m} + i\alpha}. \qquad (20)$$

Putting (17), (18) and (19) together we have finally

$$\left[\frac{\partial S_s(k\omega)}{\partial P}\right]_\beta = \frac{1}{\pi}\left(\frac{\partial \alpha}{\partial P}\right)_\beta \mathrm{Im}\Big\{i\, \frac{1}{(1 - i\alpha D(kz))^2} \times$$
$$\times \left[\int \frac{d^2\bar{p} f_0(\bar{p})}{\left(z - \frac{k\cdot \bar{p}}{m} + i\alpha\right)^2} - D^3(kz)\right]\Big\}. \qquad (21)$$

Equation (21) can be put into analytic form at the special point $\omega = 0$. It is not hard to show that in this case

$$\left[\frac{\partial S_s(k0)}{\partial P}\right]_\beta = \frac{1}{\pi\alpha^2}\left(\frac{\partial \alpha}{\partial P}\right)_\beta \frac{D(y) + d^2(y)}{(1 - d(y))^2}, \qquad (22)$$

where

$$d(y) = \sqrt{\frac{\pi}{2}}\, y\, e^{y^2/2}\, \mathrm{erfc}\, \frac{y}{\sqrt{2}} \qquad (23)$$

$$D(y) = \sqrt{\frac{\pi}{2}}\, y^2\left[e^{y^2}\, \mathrm{erfc}\, \frac{y}{\sqrt{2}}\left(\frac{z}{y} - y\right) + \sqrt{\frac{2}{\pi}}\right] \qquad (24)$$

and the $\mathrm{erfc}(x)$ is defined in [8].

A point to note in (21) and (22) is that it is proportional to a factor $(\partial\alpha/\partial P)_T$ which is the fundamental quantity of physical interest in this binary collision model. One knows that any collision operator such as (16) which does not violate the property of particle number conservation would automatically produce the correct limiting behavior

$$S_s(k\omega) = (\sqrt{\pi} k v_0)^{-1} \exp[-\omega^2/k^2 v_0^2], \quad y \ll 1 \tag{25}$$

$$S_s(k\omega) = \frac{1}{\pi} \frac{Dk^2}{\omega^2 + (Dk^2)^2}, \quad y \gg 1, \tag{26}$$

where the collision parameter y is defined as $\alpha/\sqrt{2}kv_0$, a measure of the ratio of the wave length of fluctuations to the collision mean free path in the fluid.

3. Experiment

The low-energy neutron scattering cross section of hydrogen molecules can be expressed in terms of contributions corresponding to definite rotational transitions [9]. At a temperature of 85 K, the sample temperature of our experiment, we need only consider the $J=0$ (para) and $J=1$ (ortho) states since population of the $J>2$ levels is negligible. Even though para-hydrogen scattering is purely coherent we can neglect it because the orthohydrogen cross section is predominantly incoherent and is some 30 times that of parahydrogen. If we further restrict our measurement and analysis to the quasi-elastic region of the scattering, we can also neglect the $1 \to 0$ transition because of the presence of a form factor $j_1^2(Qb)$, where j_1 is the first-order spherical Bessel function and $2b = 0.75$ Å is the internuclear distance, and $Q \leqslant 0.6$ Å^{-1} is the wavenumber transfer in the experiment. Consequently the observed line shape may be analyzed in terms of the double differential cross section [6]

$$\frac{d^2\sigma}{d\Omega\, d\omega} = a_{\text{inc}}^2 \frac{k_f}{k_0} [j_0^2(Qb) + 2j_2^2(Qb)] e^{-\hbar\omega/2k_BT} e^{-\hbar^2 Q^2/8mk_BT} S_s(Q, \omega), \tag{27}$$

where a_{inc} is the incoherent scattering length (2.53×10^{12} cm), k_0 and k_f are the incident and scattered neutron wave numbers and $\hbar\omega$, $\hbar Q$ are the energy and momentum transfers of neutrons to the sample. The exponential factors appear because we are using a classical time correlation function which does not preserve the property of detailed balance or describe the recoil effects. According to (27) the energy distribution of quasi-elastically scattered neutrons is determined, aside from the known factors, by the spectral distribution of the van Hove self correlation function.

The experiment was carried out with a fixed incident energy, $E_0 = 12.6$ meV ($\lambda = 2.55$ Å) and energy analysis of the quasi-elastic peak was made at constant $Q = 0.2$, 0.4 and 0.6 Å^{-1}. The sample holder was made of 6061-T6 aluminum with cylindrical bore of $1\frac{1}{2}''$ high $\times \frac{1}{4}''$ diameter. The pressure of the hydrogen gas was changed at every (Q, ω) point from 110 atm to 130 atm. The measured $S_s(Q, \omega)$ for these two pressures are shown in Figure 1 and Figure 2 with the background already subtracted and the detail balance factor removed. In order to take the derivative of the curves we did the following: We fit the curves with the analytic solution of the one

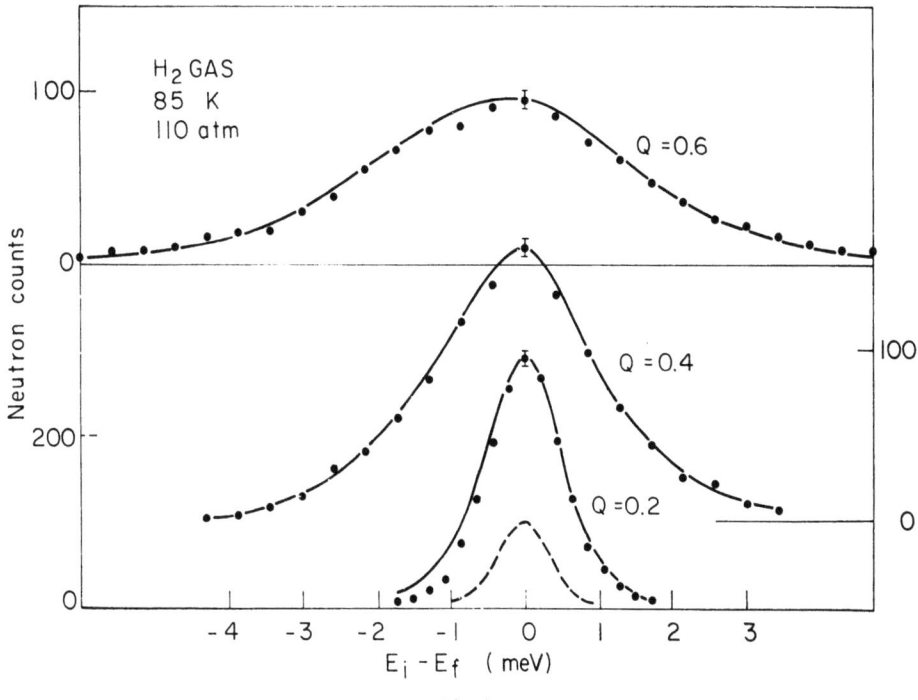

Fig. 1.

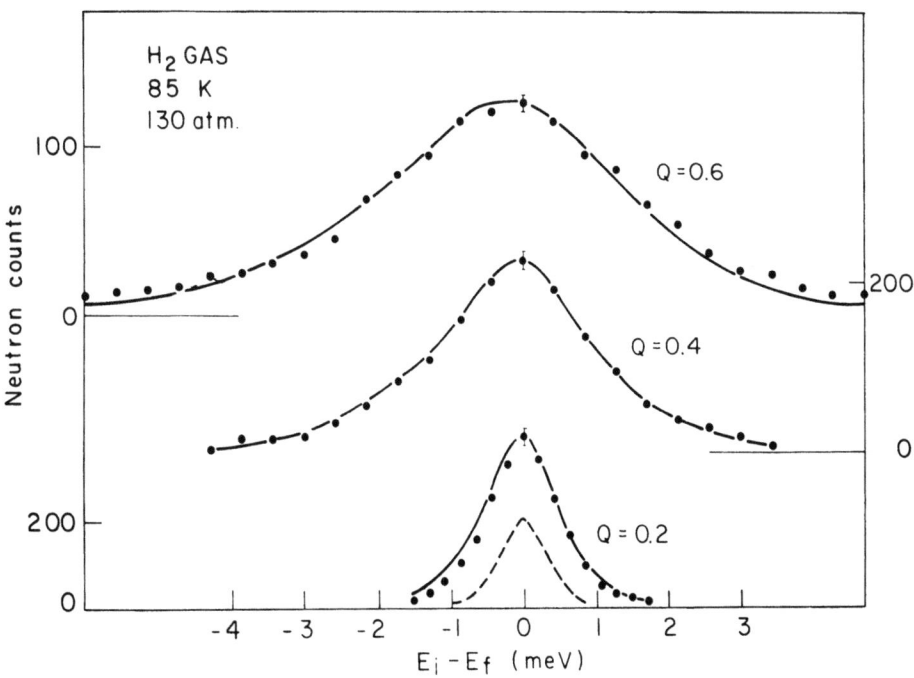

Fig. 2.

relaxation time model [6] by varying the parameter α. The best fit was shown by the solid lines in Figure 1 and 2. We then took the difference of the fitted curves to obtain the derivatives. This procedure is consistent with the discussion of the result with Equations (21) and (22).

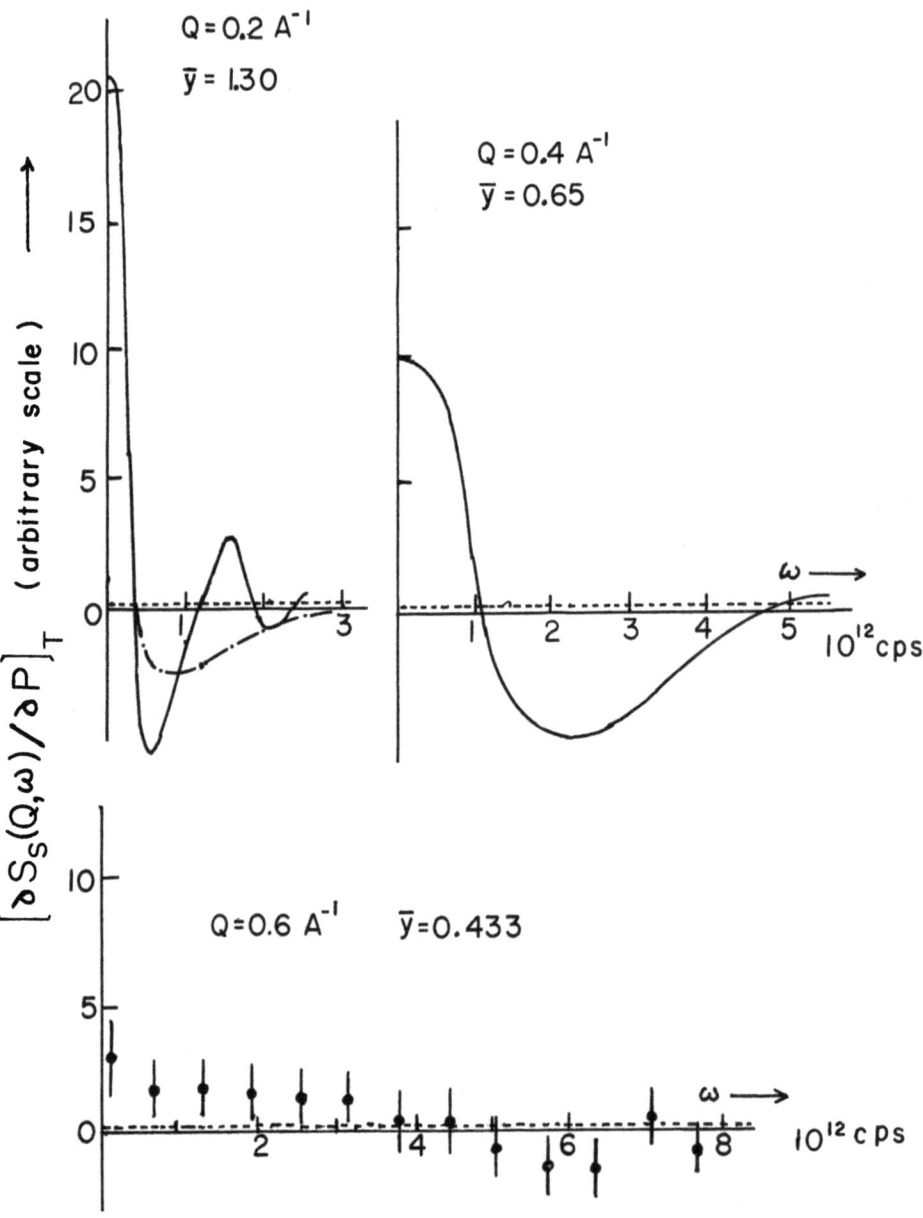

Fig. 3.

4. Results and Discussion

The pressure derivatives for $Q = 0.2$, 0.4 and 0.6 Å^{-1} data are shown in Figure 3 in solid lines and black points. These indicate the difference between data taken at $P = 130$ and 110 atm. The mean y values are also indicated. The dotted lines indicate that the derivative would be identically zero if the Vineyard's model is correct. We see that for $Q = 0.2$ and 0.4 Å^{-1}, the model is far from correct. We summarize here the qualitative features of the curves:

(i) The curves are oscillatory with zero area and more than two zero crossings.

(ii) $\omega = 0$ values decrease rapidly and the curves damp out faster as Q values increase.

(iii) At small Q such that $\bar{y} > 1$ the first zero crossing occurs at ω corresponding to the half width of $S_s(Q, \omega)$.

We can easily understand (i) from two sum rules (7) and (8). A rough understanding of (ii) and (iii) comes from looking at the two limiting behaviors in (25) and (26). At large Q such that $y < 0.15$, (25) is valid and the pressure derivative should be zero because v_0 in (25) does not depend on pressure. At small Q such that $y > 1.3$, (26) is approximately valid for ω which is less than the half width. From (26) we have

$$\left[\frac{\partial S_s(Q, \omega)}{\partial P}\right]_T = \left(\frac{\partial D}{\partial n}\right)_T \left(\frac{\partial n}{\partial P}\right)_T \frac{Q^2(\omega^2 - Q^4 D^2)}{\pi(\omega^2 + Q^4 D^2)^2}. \tag{28}$$

We see that the first zero crossing happens at $\omega = DQ^2$ which is the half width of (26). The expression (28) is plotted in the first graph of Figure 3 as a dot-dash line. The agreement with experiment is quite good up to the first zero crossing. However the (28) fails at $\omega > DQ^2$ because it fails to produce further zero crossings. In order to have more quantitative comparisons with curves for $Q = 0.4$ and $Q = 0.6$ Å^{-1} we have to evaluate numerically (21) and (22). A rough estimate shows that they are qualitatively correct, however we haven't yet carried out the numerical integrations.

In conclusion, we emphasize that a rigorous formula (15) exists which is tractable in the case of a moderately dense hard sphere fluid. The significance of the pressure derivative measurement is that it is related directly to the pressure derivative of the memory function which is a central quantity in the kinetic theory of fluids.

Acknowledgements

We gratefully acknowledge the generous help of P. Walsh and H. C. Teh during the course of the experiment and also for the data analysis.

References

1. Egelstaff, P. A., Gray, C. G., and Gubbins, K. E.: *Phys Letters* **37A**, 321 (1971).
2. Mazenko, G. F.: *Phys. Rev.* **A7**, 209 (1973).
3. Van Hove, L.: *Phys. Rev.* **95**, 249 (1954); Aamodt, R., et al.: *Phys. Rev.* **126**, 1165 (1962).

4. Egelstaff, P. A.: *Neutron Inelastic Scattering*, IAEA, Vienna, 1972, p. 383.
5. Vineyard, G. H.: *Phys. Rev.* **110**, 999 (1958).
6. Nelkin, M. and Ghatak, A.: *Phys. Rev.* **A135**, 4 (1964): Lefevre, Y., Chen, S. H., and Yip, S.: *Neutron Inelastic Scattering*, IAEA, Vienna 1972, p. 445.
7. Mazenko, G. F., Wei, T., and Yip, S.: *Phys. Rev.* **A6**, 1981 (1972).
8. Abranowitz, and Stegun, (eds.): in *Handbook of Mathematical Functions*, Dover, New York, p. 297.
9. Sears, V. F.: *Proc. Phys. Soc.* **86**, 965 (1965).

DISCUSSION

Carneiro: You mentioned that your dense gas was 'almost a liquid'. Is it true that a gas and a liquid are the same substances at the same density?

Chen: The density of the hydrogen gas used is $n = 8.56 \times 10^{21}$ cm^{-3} while the critical density of hydrogen is $n_c = 9.34 \times 10^{21}$ cm^{-3}. Effect of raising temperature in a fluid is to increase the mean thermal velocity by $v_0 = \sqrt{K_0 T/m}$ and as a result the collision is 'harder', making the gas behaves like a hard sphere system. On the other hand the increase in density has a much more important effect in dynamics of the fluid because the collective motions are much more sensitive to density increase. It is perhaps fair to say that from a theoretical point of view a high density high temperature fluid is the same as a hard sphere liquid at the same temperature.

Ailawadi: From your experiments, were you able to get some information about the correlation function $K(\mathbf{r}, \mathbf{s}, t)$ for dense orthohydrogen gas?

Chen: From its definition

$$K(\mathbf{r}, \mathbf{s}, t) = G_{3c}(\mathbf{r}, \mathbf{s}, t) - g(s) G_S(\mathbf{r}, t).$$

For example, if we take Vineyard's picture of liquids

$$G_{3c}(\mathbf{r}, \mathbf{s}, t) \simeq g(s) G_S(\mathbf{r}, t)$$

therefore $K(\mathbf{r}, \mathbf{s}, t) \simeq 0$ at all r and t. The experiment shows that this is not the case for $Q = 0.2 + 0.4$ Å^{-1}. At $Q = 0.6$ it perhaps is a good approximation. When the experimental data are fully analysed by the single-relaxation model in the paper, we should know the Q-dependence of

$$\int d\mathbf{S} \int e^{i\mathbf{Q}\cdot\mathbf{r}} K(\mathbf{r}, \mathbf{s}, t) \, d\mathbf{r}.$$

The experiment, whatever accurate, would never give detailed information of $K(\mathbf{r}, \mathbf{s}, t)$ because one of the coordinate **s** is integrated out.

Laulicht: Could you apply your theory for Raman scattering experiments? I understand that you are dealing with translational diffusion but could not that approach be applied also for rotational diffusion?

Chen: First of all the rotational Raman scattering band reflects the translational motion of molecules only at low density. If one does Raman scattering at high density one only observes the 'pressure broadening', an entirely different effect from 'pressure narrowing' observed by the neutron incoherent scattering. One can perhaps describe the pressure broadening of Raman line by a correlation function. But it will be a correlation function very different from what I described.

ETUDE DES MOUVEMENTS MOLECULAIRES PAR DIFFUSION DE LA LUMIERE

PIERRE LALLEMAND

Laboratoire de Spectroscopie Hertzienne de l'Ecole Normale Supérieure,
24 rue Lhomond, 75005 Paris, France

Résumé. On présente une mise au point sur les applications de la diffusion de la lumière à l'etude des liquides. On rappelle d'abord ce qu'on peut obtenir comme information à partir du spectre Rayleigh-Brillouin, en particulier en ce qui concerne les processus de relaxation. On discute ensuite brièvement la partie dépolarisée donnant la raie Rayleigh dépolarisée. On montre par quelques exemples que le centre des raies n'est pas lorentzien comme on le suppose d'habitude. On présente enfin quelques idées sur la partie dépolarisée due aux mouvements translationnels.

Abstract. This paper reviews the main applications of light scattering to the study of liquids. We first show what kind of information can be obtained from the polarized Rayleigh-Brillouin spectrum in the case of a simple fluid, concerning thermodynamic properties and transport coefficients. We then discuss the case where there are relaxation processes. We show how the often used frequency dependent transport coefficients lead to observable changes in the spectrum, thus providing a way to determine relaxation times. We show that light scattering can be used to extend the ultrasonic techniques towards both shorter and longer relaxation times.

We discuss the depolarized part of the spectrum in the simpler case of uncorrelated molecules. We present some data showing that the center of the line is not lorentzian as is usually claimed, and deduce that a Cole-Davidson distribution of relaxation times leads to very good agreement with experiment in the case of low temperature glycerol. This indicates that there is a need to improve the resolution used in making depolarized Rayleigh measurements.

We then present a discussion of the depolarized spectrum in the case of optically isotropic molecules. We first recall the two possible origins of depolarization: dipole induced dipole and electronic distortion of the atoms during collisions. Then we review some results obtained in gases, and describe some recent results derived from molecular dynamics calculations, which show that one cannot use a binary collision model in liquids.

On se propose de présenter une discussion des principaux domaines d'étude des liquides pour lesquels la diffusion de la lumière peut être un outil d'étude expérimentale utile. Pour cela, nous indiquerons d'abord les traits généraux du phénomène de diffusion de la lumière, puis nous discuterons un certain nombre d'applications particulières.

Lorsque l'on étudie l'interaction d'un faisceau lumineux, que l'on suppose monochromatique, avec un milieu matériel transparent, on trouve qu'on peut séparer en deux contributions la réponse des atomes ou molécules qui constituent le milieu. On a d'une part une réponse cohérente qui est responsable de l'indice de réfraction du milieu, et d'autre part une réponse incohérente. Cette réponse incohérente se traduit par des changements dans la direction de propagation du faisceau et dans la fréquence de l'onde. Ces changements de fréquence peuvent s'étendre sur une très vaste gamme allant de 0 à plusieurs milliers de cm^{-1}. On classe traditionnellement dans la diffusion de la lumière tous les processus au cours desquels il n'y a pas de modification de l'état interne des molécules. Ainsi, la diffusion de la lumière est-elle complémentaire de la diffusion Raman.

Nous allons d'abord rappeler brièvement la théorie moléculaire de la diffusion [1, 2]. Considérons pour cela un ensemble de molécules identiques situées au point $\mathbf{r}_i(t)$, et qui possèdent un tenseur de polarisabilité $\boldsymbol{\alpha}_i$ qu'on suppose symétrique. Appliquons un champ électrique incident sous forme d'une onde plane $\mathbf{E}_{\text{inc}} = \mathbf{E} \exp(i\mathbf{k}_i \cdot \mathbf{r} - i\omega t)$. Ce champ crée un dipôle oscillant $\mathbf{P}(\mathbf{r}_i) = \boldsymbol{\alpha}_i \mathbf{E}_{\text{loc}}(\mathbf{r}_i)$ où $\mathbf{E}_{\text{loc}}(\mathbf{r}_i)$ est le champ local au point $\mathbf{r}_i$. Nous supposons pour le moment que $\mathbf{E}_{\text{loc}}$ est proportionnel à $\mathbf{E}_{\text{inc}}$ (avec une constante de proportionnalité qu'on peut prendre de l'ordre du facteur de Lorentz-Lorenz $(n^2+2)/3$). Avant de calculer le champ diffusé, on commence par décomposer le tenseur de polarisabilité $\boldsymbol{\alpha}$ en deux parties

$$\boldsymbol{\alpha} = \tfrac{1}{3} \operatorname{Tr}(\alpha) \boldsymbol{I} + (\boldsymbol{\alpha} - \tfrac{1}{3} \operatorname{Tr}(\alpha) \boldsymbol{I}).$$

Le premier terme est proportionnel à la matrice unité $\boldsymbol{I}$, tandis que le second terme est un tenseur symétrique du second ordre à trace nulle. L'utilité de cette décomposition provient du fait qu'on peut montrer que les deux contributions de $\boldsymbol{\alpha}$ au champ diffusé sont statistiquement indépendantes pour un ensemble de molécules dont les axes sont distribués au hasard. Le terme proportionnel à $\operatorname{Tr}(\alpha)$ donne un spectre polarisé tandis que l'autre terme donne un spectre dépolarisé.

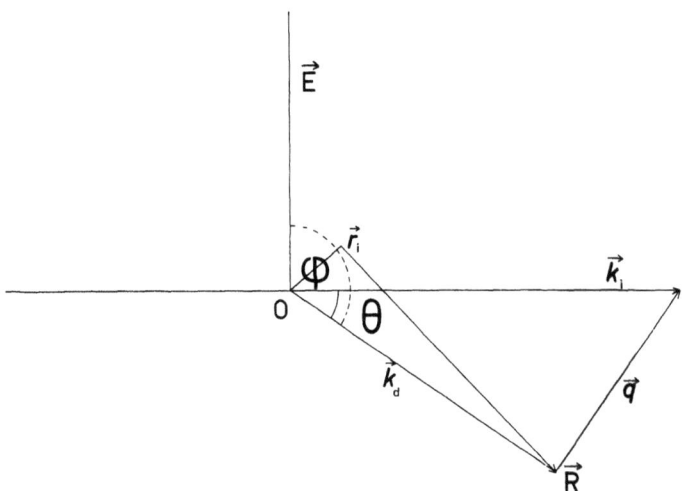

Fig. 1. Schéma du calcul du champ diffusé au point éloigné **R**.

On trouve alors que le terme $\bar{\alpha} = \tfrac{1}{3} \operatorname{Tr}(\alpha)$ donne comme champ diffusé au point éloigné **R** (figure 1)

$$\mathbf{E}_d(\mathbf{R}) = \frac{\bar{\alpha}}{R} \mathbf{k}_d \wedge \left(\mathbf{k}_d \wedge \sum_i \boldsymbol{\varepsilon}_{\text{loc}} e^{i(\mathbf{k}_i - \mathbf{k}_d) \cdot \mathbf{r}_i} \right),$$

où $\mathbf{k}_d$ est un vecteur d'onde dans la direction de **R**. On en déduit la section efficace de diffusion

$$\frac{\partial \sigma}{\partial \Omega} = \bar{\alpha}^2 k^4 \sin^2 \varphi \left(\frac{n^2 + 2}{3} \right)^2 |\varrho_\mathbf{q}|^2,$$

où φ est l'angle entre la direction de diffusion et la direction de polarisation du champ incident, $\mathbf{q} = \mathbf{k}_i - \mathbf{k}_d$ le vecteur d'onde de diffusion, et où on a introduit la transformée de Fourier spatiale de la densité de particules:

$$\varrho_\mathbf{q} = \sum_i \int \delta(\mathbf{r} - \mathbf{r}_i) e^{i\mathbf{q}\cdot\mathbf{r}} d\mathbf{r}.$$

La section efficace differentielle s'obtient en prenant la transformée de Fourier de la fonction de corrélation du champ diffusé, d'où

$$\frac{\partial^2 \sigma}{\partial\Omega\partial\omega} = \bar{\alpha}^2 k_d^4 \sin^2\varphi \left(\frac{n^2 + 2}{3}\right)^2 \int \langle \varrho_\mathbf{q}(t) \varrho_\mathbf{q}^*(0) \rangle e^{i\omega\tau} dt.$$

Les seules hypothèses faites jusqu'ici concernent le facteur de Lorentz-Lorenz, mais ce n'est pas très important car il est très rare que l'on fasse des mesures absolues de section efficace de diffusion.

On trouve ainsi un résultat analogue à celui de la diffusion des neutrons, la seule différence importante étant due au domaine de vecteurs d'onde $\mathbf{q}$ accessibles. En effet, si θ est l'angle de diffusion, on a $q = 2nk_i \sin(\theta/2)$ où n est l'indice de réfraction. On ne pourra donc explorer que les vecteurs d'onde très petits devant l'inverse des dimensions atomiques typiques.

Le calcul du spectre ou, ce qui revient au même, de la fonction de corrélation de la densité $C(\tau) = \langle \varrho_\mathbf{q}(\tau) \varrho_\mathbf{q}^*(0) \rangle$ peut être fait de nombreuses manières. Pour montrer le lien avec les études ultrasonores, nous décrirons les grandes lignes de la méthode de Landau [3] selon laquelle la fonction de corrélation $C(\tau)$ a la même dépendance temporelle que la densité macroscopique du milieu. C'est à dire qu'on peut prendre les équations macroscopiques de l'évolution du système en vue de trouver $C(\tau)$.

Soit ϱ, v, T les variables thermodynamiques du système. Elles sont reliées entre elles par les équations différentielles de l'hydrodynamique généralisée qui considère des coefficients de transport dépendant de $\mathbf{q}$ et de ω (essentiellement $\eta_v(\mathbf{q}, \omega)$). On linéarise [4, 5] ces équations pour obtenir

$$\frac{\partial \varrho}{\partial t} = -\varrho_0 \frac{\partial v}{\partial z}$$

$$\frac{\partial v}{\partial t} = -\frac{C_0^2}{\varrho_0 \gamma} \frac{\partial \varrho}{\partial z} - \beta \frac{C_0^2}{\gamma} \frac{\partial T}{\partial z} + \frac{4\eta_s}{3\varrho_0} \frac{\partial^2 v}{\partial z^2} + \frac{1}{\varrho_0} \int_0^t \eta_v(t-t') \frac{\partial^2 \dot{v}(t')}{\partial z^2} dt'$$

$$\frac{\partial T}{\partial t} = \frac{\lambda}{\varrho_0 C_v} \frac{\partial^2 T}{\partial z^2} - \frac{\gamma - 1}{\varrho_0 \beta} \frac{\partial \varrho}{\partial t},$$

où les indices '0' indiquent les valeurs à l'équilibre.

C_0 est la vitesse du son à basse fréquence, $\gamma = C_p/C_v$ est le rapport des chaleurs spécifiques, $\beta = (1/V)(\partial V/\partial T)_P$ est le coefficient de dilatation thermique à pression constante, λ la conductibilité thermique, η_s la viscosité de cisaillement et η_v la viscosité de volume. On élimine $\partial v/\partial z$ puis on prend la transformée de Fourier spatiale pour le vecteur d'onde q. On prend ensuite la transformée de Laplace en vue de tenir compte des conditions initiales qui sont données par la thermodynamique statistique [3] : à savoir

$$\left\langle \frac{\delta \varrho^2}{\varrho_0} \right\rangle = \varrho_0^2 k T_{\chi_T}; \quad \langle \delta \varrho \cdot \delta T \rangle = 0 \quad \text{et} \quad \langle \delta T^2 \rangle = \frac{kT^2}{C_v},$$

où χ_T est le coefficient de compressibilité isotherme. On obtient alors $\varrho_q(s)$ en fonction de $\varrho_q(0)$, d'où l'on déduit $G(s) = \langle \varrho_q(s) \varrho_q(0) \rangle$. On détermine alors le spectre en prenant $S(\omega) = 2 \text{ Re} \{G(s=i\omega)\}$. $G(s)$ peut se mettre sous la forme du rapport de deux fonctions $G(s) = N(s)/D(s)$. On obtient les modes propres du système en résolvant l'équation de dispersion $D(s) = 0$. En vue de poursuivre le calcul, on doit trouver les racines de

$$D(s) = 0 = \begin{vmatrix} s^2 + \dfrac{c_0^2 q^2}{\gamma} + \left(\dfrac{4 \eta_s}{3 \varrho_0} + \dfrac{\eta_v}{\varrho_0} + \dfrac{\eta_v(s)}{\varrho_0} \right) q^2 s & \beta \dfrac{c_0^2}{\gamma} \varrho_0 q^2 \\ -s(\gamma - 1) \dfrac{1}{\varrho_0 \beta} & s + \dfrac{\lambda q^2}{\varrho_0 C_v} \end{vmatrix}.$$

Lorsque $\eta_v(s) \equiv 0$, on trouve trois modes : les deux modes acoustiques qui donnent lieu au doublet Brillouin, et le mode thermique qui donne lieu à la raie Rayleigh. On a :

$$s_{\text{Brillouin}} = -\tfrac{1}{2} \Gamma q^2 \pm iCq$$

où

$$C = C_0 \left[1 - \frac{1}{2} \frac{\Gamma^2}{c_0^2} q^2 \right] \quad \text{et} \quad \Gamma = \frac{4 \eta_s}{3 \varrho_0} + \frac{\eta_v}{\varrho_0} + \frac{\lambda}{\varrho_0} \left(\frac{1}{C_v} - \frac{1}{C_p} \right)$$

et

$$s_{\text{Rayleigh}} = - \frac{\lambda q^2}{\varrho_0 C_p}.$$

Le rapport des intensités $I_{\text{Ray}}/2 I_{\text{Bri}} = \gamma - 1$ s'appelle rapport de Landau-Placzek. [6].

On voit que l'étude du spectre donne de nombreux renseignements : $\lambda/\varrho_0 C_p$; $\tfrac{4}{3}\eta_s + \eta_v$; C_0 ; C_p/C_v et χ_T en mesurant l'intensité diffusée totale. Ces différents résultats sont résumés sur la figure 2, qui présente trois raies lorentziennes. (Cela n'est vrai que si la largeur des raies Brillouin est faible devant le déplacement Brillouin, car autrement il y a des composantes antisymétriques centrées sur les fréquences Brillouin.)

On voit donc que l'étude des raies Brillouin, qui fournit essentiellement C_0 permet d'étendre les mesures acoustiques jusqu'à des fréquences de quelques GHz. Cependant, il faut remarquer que bien qu'on ait la même équation de dispersion que celle qu'on utilise en acoustique, on la résoud ici pour q réel, alors qu'on la résoud pour ω réel en acoustique. Cette distinction entre le domaine temps et le domaine espace ne joue de rôle que lorsque l'atténuation est forte.

La situation est nettement plus complexe dans le cas plus intéressant où $\eta_v(s) \not\equiv 0$. Rappelons que l'on introduit une viscosité de volume dépendant du temps [7] lorsque le système présente un processus de relaxation. L'emploi de coefficients de transport dépendant du temps peut s'interpréter dans le cadre de l'équation de Langevin généralisée où on a introduit des termes dissipatifs non locaux dans le temps. Nous ne cherchons pas ici à les justifier, mais à voir si on peut les déterminer expérimentalement, car leur connaissance pourra éventuellement donner des informations sur les mouvements moléculaires dans les liquides.

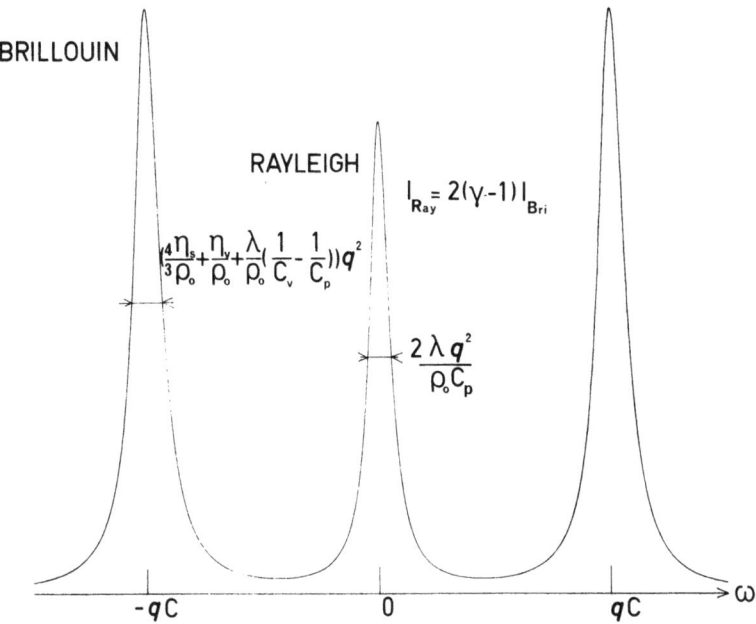

Fig. 2. Résumé des renseignements contenus dans le spectre Rayleigh-Brillouin. I_{Ray} et I_{Bri} représentent respectivement l'intensité intégrée de la raie centrale et celle d'une des composantes du doublet Brillouin.

On prend donc l'équation de dispersion $D(s)=0$, et on essaye de la résoudre lorsque le terme $\eta_v(s)$ est présent. Mountain a donné une solution lorsque $\eta_v(s)=a/(1+s\tau)$, c'est à dire lorsqu'il n'y a qu'un seul temps de relaxation τ. Il a montré qu'il apparaît un quatrième mode dans le spectre correspondant à la racine $s \approx -1/\tau$. Cela entraîne la présence d'une raie centrée à la fréquence incidente de largeur totale à mi-hauteur de l'ordre de $2/\tau$. On l'appelle raie de Mountain. Elle a été observée dans un certain nombre de cas [8, 9, 10]. Plus récemment, Demoulin et al. [11] ont étudié la contribution de cette racine au spectre dans la glycérine à basse température où le temps de relaxation devient très long. Dans ces conditions, on peut mesurer la dépendance temporelle de $\eta_v(t)$ avec une précision remarquable.

A part le cas spectaculaire des liquides visqueux tels que la glycérine, la raie de Mountain n'est pas très intense. On étudie donc plutôt son influence sur le spectre Brillouin qui est profondément modifié lorsque $C_0 q \tau \approx 1$. En effet, si l'on fait varier $C_0 q \tau$ (par changement de q en variant l'angle de diffusion ou la longueur d'onde du laser, ou en changeant τ qui dépend souvent de la température), on trouve que C varie entre C_0 et C_∞ (les vitesses du son à basse et haute fréquence) et que la largeur Γ passe par un maximum. Les études ainsi réalisées sont voisines des mesures acoustiques [12]. On peut donner des expressions littérales approchées [5, 9] des détails des spectres, mais en pratique il est souvent plus simple d'utiliser directement la fonction de corrélation d'autant plus que $\eta_v(s)$ est rarement une fonction aussi simple que celle que l'on a indiquée plus haut. En effet, on a en général une distribution [13] de temps de relaxation telle que

$$\eta_v(s) = (c_\infty^2 - c_0^2)\varrho_0 \int g(\tau) \frac{\tau}{1+s\tau} d\tau,$$

où la distribution est normalisée à l'unité. On peut en principe déduire $g(\tau)$ de la forme des spectres, mais la précision des mesures expérimentales n'est pas assez bonne en pratique.

Montrons maintenant quelques applications de cette technique expérimentale pour la connaissance des liquides.

1. Intensité diffusée

On peut en principe étudier $\langle \varrho_\mathbf{q}(0) \varrho_\mathbf{q}^*(0) \rangle$ en faisant varier l'angle de diffusion, mais il faut se rappeler que les vecteurs d'onde accessibles sont très petits. Grâce à des expériences extrèmement soignées, Cannell et Lunacek [14] ont réussi à mesurer des longueurs de corrélation spatiale aussi petites que 20 Å avec un faisceau laser à 6328 Å. On voit donc que la diffusion de la lumière n'est utilisable que pour des systèmes présentant des longueurs de corrélation assez grandes: points critiques, cristaux liquides. La mesure de l'intensité diffusée totale n'est jamais très précise, mais on peut faire de très bonnes mesures relatives en vue de déterminer la variation de la compressibilité isotherme X_T, par exemple au voisinage d'un point critique.

2. Spectre

La diffusion de la lumière est très utilisée. On peut déterminer avec une assez bonne precision la diffusivité thermique (à 5% près environ). Dans ce but, on peut utiliser un grand angle de diffusion, pour lequel la largeur de la raie Rayleigh est de l'ordre de quelques dizaines de MHz. On fait les mesures avec un interféromètre Fabry-Pérot de très haute résolution [15, 16]. On préfère souvent diffuser la lumière à petit angle où la raie est très étroite (domaine du kHz). On doit alors utiliser les techniques spectroscopiques des battements lumineux [17, 18] dont nous parlerons plus loin. Signalons aussi que de nombreuses expériences ont été faites pour mesurer des coefficients de diffusion dans des mélanges binaires [19].

Les résultats les plus intéressants obtenus par diffusion de la lumière concernent la partie Brillouin du spectre. En effet, comme on l'a indiqué plus haut, l'analyse des spectres expérimentaux permet de déduire la vitesse et l'atténuation des ondes sonores de haute fréquence (jusqu'à 6 à 10 GHz). On a pu ainsi étudier en détail les processus de relaxation structurale et vibrationnelle dans de nombreux liquides [20] où les temps de relaxation sont supérieurs à environ quelques 10^{-11} s. On en trouvera un exemple d'application dans l'exposé de Munch et Candau [21]. Quels sont les temps de relaxation les plus longs que l'on puisse mesurer par ces techniques ? En principe il n'y a pas de limitation, mais d'un point de vue expérimental, on peut dire qu'il y a deux gammes de temps suivant les instruments utilisés. Le Fabry-Pérot permet d'étudier les ondes sonores de fréquences comprises entre quelques GHz et une centaine de MHz, soit τ entre 10^{-11} et 10^{-8} s environ. Les méthodes par battements lumineux ne marchent bien que pour des temps supérieurs à 10^{-6} s environ, la limite supérieure étant fixée par la stabilité des instruments : laser et électronique, et par la patience de l'expérimentateur. On voit donc qu'il y a un trou, mais heureusement, il se trouve dans la zone favorable pour les études ultrasonores.

Il ne semble pas que l'on ait mis en évidence dans le cas des liquides ordinaires le fait que les coefficients de transport puissent dépendre du vecteur d'onde q. Cette dépendance doit probablement se produire pour des vecteurs d'onde beaucoup plus grands que ceux qu'on peut atteindre par diffusion de la lumière. Il n'en est pas de même dans les gaz où l'on obtient de très grandes dispersions pour $ql \sim 1$ où l est le libre parcours moyen. Cette condition se présentant pour des conditions expérimentales aisément accessibles, on a pu étudier en détail la dispersion ultrasonore dans les gaz ainsi que la dépendance en q et en ω de la viscosité de volume [22, 23, 24].

Nous pouvons résumer cette discussion en disant que l'étude de la diffusion polarisée de la lumière permet de déterminer plusieurs propriétés thermodynamiques des liquides, et d'étendre vers les hautes et les basses fréquences les techniques ultrasonores en vue de mesurer des temps de relaxation, sources d'information pour l'étude des mouvements moléculaires dans les liquides.

Nous allons maintenant discuter le spectre dû à la partie symétrique de trace nulle du tenseur de polarisabilité moléculaire $\boldsymbol{\alpha}$. Considérons pour simplifier un ensemble de molécules optiquement anisotropes présentant un axe de révolution. La partie symétrique de trace nulle du tenseur $\boldsymbol{\alpha}$ prend alors une forme particulièrement simple si on l'exprime dans un système de coordonnées lié à la molécule

$$\boldsymbol{\alpha} = \bar{\alpha} \begin{pmatrix} 1 & 0 & 0 \\ 0 & 1 & 0 \\ 0 & 0 & 1 \end{pmatrix} + \begin{pmatrix} \alpha_\perp - \bar{\alpha} & 0 & 0 \\ 0 & \alpha_\perp - \bar{\alpha} & 0 \\ 0 & 0 & \alpha_\parallel - \bar{\alpha} \end{pmatrix}.$$

Considérons le cas usuel où l'on observe la lumière diffusée à 90° du vecteur d'onde incident (figure 3). Le champ résultant des dipôles induits sur chaque molécule par le champ incident présente alors deux composantes v et h selon deux axes perpendiculaires (v pour vertical et h pour horizontal par rapport au plan de diffusion). On précise le dispositif expérimental en indiquant la polarisation incidente qu'on prend comme étant soit V soit H. On trouve en faisant les hypothèses restrictives suivantes :
– absence de corrélations angulaires entre molécules voisines, – molécules de dimensions petites devant la longueur d'onde de la lumière, que

$$\tfrac{4}{3} I_{Vv} = I_{Vh} = I_{Hv} = I_{Hh}.$$

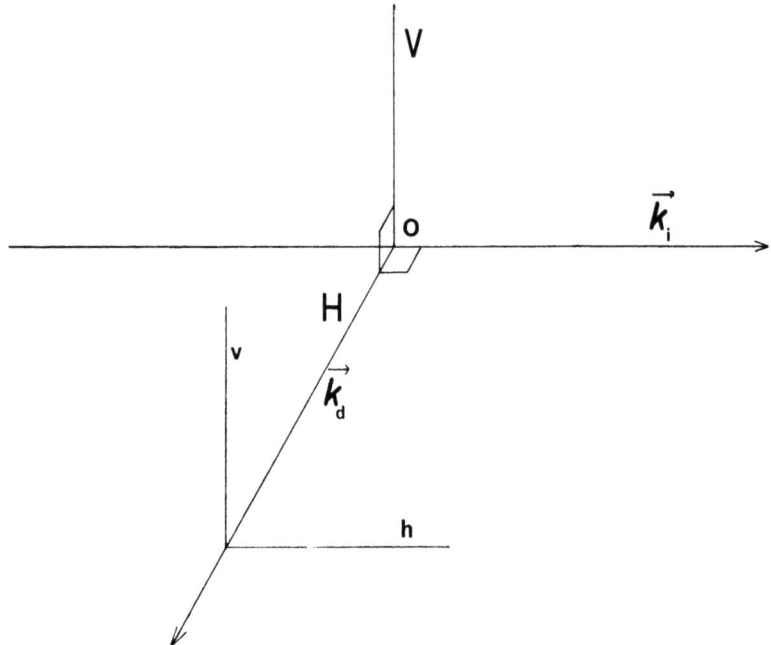

Fig. 3. Convention de polarisation pour la diffusion à 90°.

En supposant qu'il n'y a pas de corrélation entre l'orientation des molécules et les fluctuations de densité, on trouve que les quatre situations expérimentales conduisent à la même fonction de corrélation pour le champ diffusé, soit pour la géométrie Vv:

$$\langle E_d(t+\tau) E_d^*(t) \rangle = \tfrac{4}{45}(\alpha_\parallel - \alpha_\perp)^2 \left(\frac{n^2+2}{3}\right)^2 \frac{k^4}{R^2} \langle P_2(\cos\theta(\tau)) \rangle,$$

où P_2 est un polynôme de Legendre et $\theta(\tau)$ l'angle de rotation de la molécule entre les instants t et $t+\tau$, et où on a introduit le facteur de Lorentz-Lorenz dont la valeur exacte est en doute surtout ici où les molécules sont optiquement anisotropes. Il s'agit d'un problème important car d'assez nombreuses études ont été faites non pas de l'intensité diffusée absolue, mais du rapport des intensités polarisées et dépolarisées [25]. En effet, on suppose que les composantes du tenseur de polarisabilité sont connues (par exemple à partir de mesures sur des gaz) et on compare le rapport des intensités polarisées et dépolarisées calculé au rapport des mesures. On interprète alors les différences entre ces deux rapports comme provenant de corrélations instantanées entre molécules voisines.

Mais l'essentiel des études consiste à mesurer la distribution spectrale de la raie dépolarisée qu'on appelle raie Rayleigh dépolarisée. Cette raie s'observe le plus aisément dans une géométrie où la partie polarisée ne contribue pas. Les temps de corrélation sont en général très courts, il en résulte qu'il est plus facile de mesurer le spectre qui s'étend sur une très large gamme à l'aide d'interféromètres Fabry-Pérot ou de monochromateurs. Vu le très grand nombre d'études qui ont été faites, il n'est pas possible de donner une idée complète du sujet. Nous nous contenterons donc de faire quelques remarques sur l'utilisation des spectres expérimentaux. Il faut dire auparavant que ces études sont très voisines de celles qu'on fait au moyen de la diffusion Raman. Rappelons qu'on mesure en Raman les fonctions de corrélation C_{vib} pour la partie polarisée et $C_{vib} \cdot C_{rot}$ pour la partie dépolarisée si l'on suppose la rotation et la vibration incorrélées [26]. L'avantage de l'effet Raman est qu'on dispose de plusieurs raies correspondant à des axes principaux qui peuvent être différents [27]. En outre, les vibrations de molécules voisines étant très probablement incorrélées [28], il n'y a plus de problème dû aux corrélations angulaires entre molécules voisines. En revanche, la séparation des contributions C_{vib} et C_{rot} n'est pas toujours aisée à faire, ce qui est particulièrement important quand on s'intéresse à la partie centrale des raies (comportement aux temps longs).

La figure 4 montre un exemple de spectre dépolarisé pris dans un cas favorable, celui du sulfure de carbone. Que pouvons-nous déduire de données expérimentales aussi détaillées? Si l'on trace le spectre avec une échelle linéaire, on obtient une raie approximativement lorentzienne dont on peut déterminer la largeur à mi-hauteur, soit ici $\Delta\omega = 5{,}87$ cm^{-1} Ce genre de mesure est fait très couramment en fonction de la température [29, 30]. On vérifie alors que $\tau_{or} = 1/\Delta\omega = A \exp - H/kT$ où H est l'énergie d'activation pour les mouvements rotationnels. On compare ensuite au temps de re-

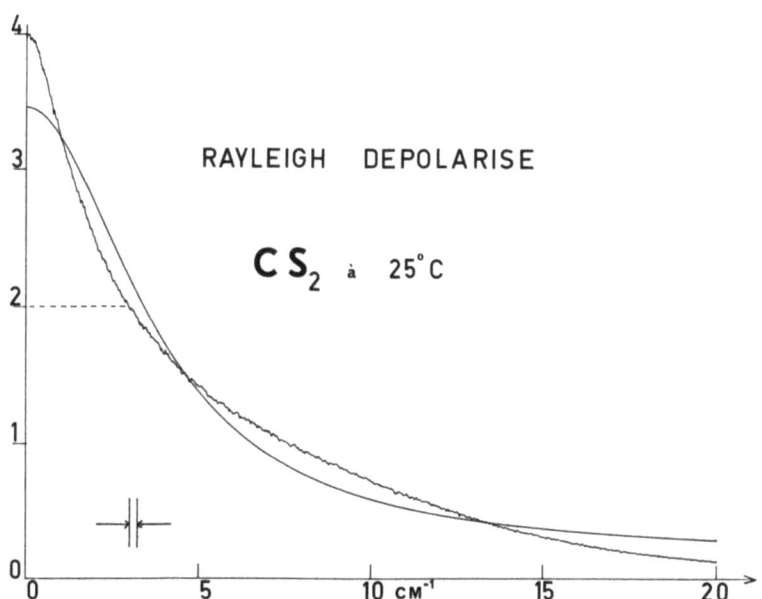

Fig. 4. Spectre de la lumière dépolarisée diffusée par du sulfure de carbone à 25°C. La largeur totale à mi-hauteur de la fonction d'appareil est de 0,2 cm^{-1}. Il est clair que la raie n'est pas lorentzienne comme on peut le voir en la comparant à une courbe lorentzienne dont la largeur a été choisie pour avoir le meilleur accord avec la courbe expérimentale. La largeur à mi-hauteur de la courbe expérimentale est de 5,87 cm^{-1}, tandis que celle de la lorentzienne est de 9,46 cm^{-1}.

laxation diélectrique, et suivant la valeur du rapport $x = \tau_{or}/\tau_{\text{diélectrique}}$, on dit que le mouvement est du type diffusif si $x \approx \frac{1}{3}$ ou du type par sauts brusques si $x \approx 1$. L'utilisation de différents modèles [30, 31] pour la rotation permet d'évaluer x en comparant τ_{or} au temps entre collisions successives de deux molécules dans le cas où l'on suppose qu'il se produit des collisions binaires dans le liquide. Il serait important de vérifier que le centre de la raie correspond bien à une raie lorentzienne car on compare souvent ces données à celles qu'on obtient en RQN ou RMN où d'habitude on suppose explicitement que les fonctions de corrélation sont exponentielles aux temps longs.

On peut ensuite examiner les ailes de la raie. On trouve que très loin du centre, la forme est approximativement exponentielle, ce qui rappelle les formes spectrales observées dans les gaz rares où le processus de diffusion est du type translationnel. Comme nous le verrons plus loin, le spectre translationnel a exactement les mêmes caractéristiques que le spectre Rayleigh dépolarisé, en ce qui concerne sa polarisation. Il n'y a donc pas moyen de le séparer expérimentalement. Peut-on l'estimer théoriquement? oui pensent certains auteurs [32, 33]. On peut alors le soustraire, et déterminer les 'vraies' ailes Rayleigh, qui sont dues au mouvement rotationnel aux temps courts. Il est alors commode de considérer non pas le spectre mais sa transformée de Fourier, soit la fonction de corrélation $c(\tau)$. On fait alors un développement de $c(\tau)$ en série de Taylor

$$c(\tau) = 1 - \frac{\gamma_2}{2!}\tau^2 + \frac{\gamma_4}{4!}\tau^4 + \cdots,$$

où γ_2 et γ_4 sont directement reliés aux moments de la distribution spectrale par

$$\gamma_{2n} = \int_{-\infty}^{\infty} s(\omega)\omega^{2n} d\omega.$$

Ces moments ont été étudiés en grand détail entre autres par Gordon [34] qui trouve que γ_2 est relié aux effets inertiels:

$$\gamma_2 = \frac{3kT}{I}.$$

(I est le moment d'inertie de la molécule par rapport à un axe perpendiculaire à l'axe de révolution de la molécule). γ_4 lui, dépend du couple moyen subi par la molécule en raison de ses interactions avec ses voisines.

Une autre façon d'analyser la fonction de corrélation consiste à considérer la fonction de corrélation de la vitesse angulaire de la molécule. En effet, considérons des temps assez courts pour que $\theta(\tau) \ll 1$. On a alors

$$c(\tau) \approx 1 - \tfrac{3}{4}\theta^2(\tau).$$

Prenons alors la dérivée seconde par rapport au temps, et réarrangeons les termes, on trouve:

$$G(\tau) = \langle \Omega(0)\Omega(\tau) \rangle \propto \frac{d^2}{dt^2} C(\tau)$$

soit en revenant aux spectres

$$G(\omega) \simeq \omega^2 C(\omega)$$

valable seulement pour ω très grand.

Cette méthode due à Dardy et Litovitz [35], a été appliquée à plusieurs liquides. Elle donne des idées précises sur le temps au bout duquel la vitesse angulaire de la molécule s'annule. Nous voyons malgré tout qu'elle suppose que l'on sache retrancher la partie du spectre due aux effets translationnels, ce qui est difficile.

Revenons à l'étude du centre de la raie ou à la partie de la fonction de corrélation aux temps longs. Quelle est la forme exacte de la fonction de corrélation? Les expériences usuelles ne permettent pas de répondre avec certitude à cette question à cause de la résolution limitée des appareils utilisés. La plupart des auteurs affirment malgré tout qu'une forme lorentzienne est une bonne approximation pour les liquides simples. Nous allons maintenant décrire des mesures directes de la fonction de corrélation du champ diffusé dépolarisé faites avec une très bonne résolution sur de la glycérine à basse température. La raison du choix de ce liquide est que la méthode expérimentale utilisée est limitée aux temps relativement longs ($\tau > 10^{-6}$ s environ). En voici le principe: On sait qu'il existe plusieurs méthodes de détermination directe des fonctions de corrélation d'un champ optique, soit

$$g^{(1)}(\tau) = \langle E_d(\tau) E_d^*(0) \rangle / \langle E_d(0) E_d^*(0) \rangle.$$

On peut introduire le retard τ soit par un chemin optique $L=c\tau$ dans un interféromètre de Michelson, et mesurer directement $g^{(1)}(\tau)$ comme le fait Connes [36]. On peut aussi introduire le retard τ dans le courant photoélectrique délivré par le photomultiplicateur, et déterminer la fonction

$$g^{(2)}(\tau) = \langle E_d(\tau) E_d^*(\tau) E_d(0) E_d^*(0) \rangle / \langle E_d(0) E_d^*(0) E_d(0) E_d^*(0) \rangle$$

que l'on obtient en faisant l'intégrale du produit du courant photoélectrique $i(\tau)$ par le courant $i(0)$ à l'aide d'un appareil électronique appelé corrélateur. On peut montrer que lorsque E_d est une variable aléatoire gaussienne stationnaire, on a

$$g^{(2)}(\tau) = 1 + |g^{(1)}(\tau)|^2.$$

On perd l'information sur la phase de $g^{(1)}(\tau)$, mais ceci n'est en général pas gênant. Cette méthode dite des battements lumineux dérive directement de l'expérience de Hanbury-Brown et Twiss [37]. Elle est très utilisée car elle permet d'obtenir la fonction $g^{(1)}(\tau)$ avec une très bonne précision pour $\tau > 10^{-7}$ s. Les détails des mesures dépendent du corrélateur utilisé pour calculer la fonction de corrélation $\langle i(\tau) i(0) \rangle$. En général on détermine en même temps une centaine de valeurs de τ pour $\tau = \theta$;

$2\theta; \ldots; 100\theta$. Il est clair que cette méthode susceptible de donner une résolution de 10^{15} n'est utilisable que si la source de lumière incidente est un laser stable. La figure 5 montre un exemple de fonction de corrélation $g^{(2)}(\tau)$ obtenue en raccordant plusieurs courbes expérimentales prises pour des valeurs de θ différentes, à l'aide d'un corrélateur digital. On peut alors comparer la fonction $g^{(2)}(\tau)$ mesurée expérimentalement au carré d'une exponentielle. On trouve qu'il n'est pas possible d'avoir un bon accord.

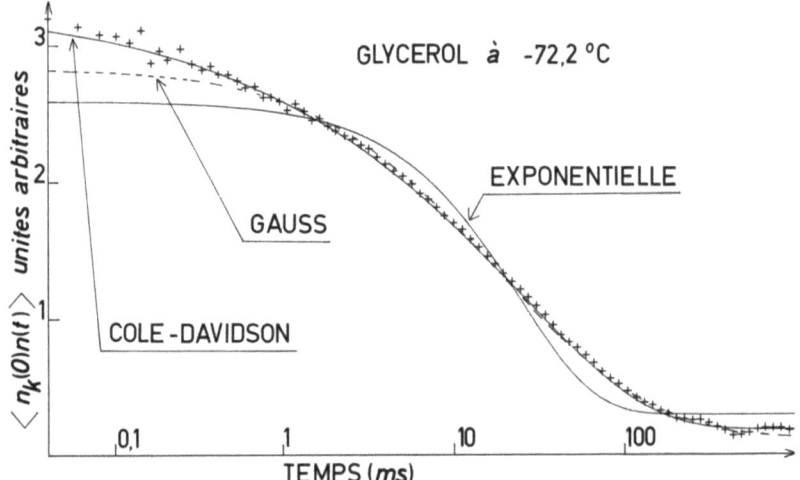

Fig. 5. Fonction d'autocorrélation du courant photoélectrique dû au champ électrique dépolarisé diffusé par du glycérol à $-72,2\,°C$. Les données expérimentales sont représentées par les croix. La courbe 'exponentielle' a été obtenue en cherchant la meilleure détermination d'un temps de relaxation unique. Les courbes Cole-Davidson et Gauss ont été obtenues en cherchant les meilleures valeurs de β et τ_0 pour chacune de ces distributions: à savoir la distribution de Cole-Davidson et la distribution log-gaussienne [13].

En revanche, si l'on considère qu'il y a une distribution $g(\theta)$ de temps de relaxation, telle que

$$C(\tau) = \int_0^\infty g(\theta) \exp -\tau/\theta \, d\theta$$

on trouve un très bon accord avec les données expérimentales pour

$$g(\theta) = \frac{\tau_0}{\theta} \left(\frac{\theta/\tau_0}{1 - \theta/\tau_0} \right)^\beta, \quad 0 < \theta < \tau_0$$

qui est la fonction de distribution de Cole-Davidson si $\beta=0,4$ et $\tau_0=0,23$ s. On a trouvé que la même distribution permettait de très bien rendre compte de la fonction de corrélation de la partie polarisée du spectre, à condition de prendre une échelle de temps légèrement différente conduisant à un temps de relaxation moyen $\bar{\tau}=\beta\tau/1,5$ fois plus grand pour la partie polarisée que pour la partie dépolarisée. Il est utile de rappeler ici que des études de relaxation diélectrique [38] dans la glycérine conduisent aussi à une distribution de temps de relaxation de Cole-Davidson pour $\beta = 0,55$ (figure 6).

Ces résultats montrent qu'on peut obtenir des données très précises sur la fonction de corrélation $c(\tau)$ pour des temps assez longs. On en déduit que les mouvements rotationnels dans un liquide associé tel que la glycérine se produisent par sauts angu-

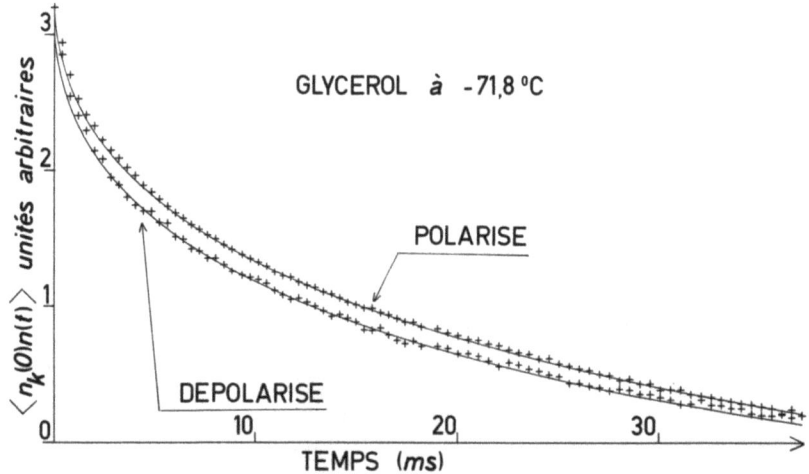

Fig. 6. Comparaison des fonctions d'autocorrélation du courant photoélectrique dû au champ diffusé polarisé et dépolarisé par du glycérol à $-71,8\,°C$. Les données expérimentales sont représentées par les croix. Les courbes ont été tracées pour les meilleures valeurs des paramètres β et τ_0 de la distribution de Cole-Davidson.

laires importants qui se produisent lors des réarrangements structuraux du liquide. Ceci confirme les résultats obtenus à haute température [30].

On voit donc que l'étude du spectre dépolarisé apporte une très grande quantité d'informations sur les mouvements rotationnels au sein des liquides, qui mises en commun avec celles qui proviennent d'autres techniques expérimentales, permettent de préciser les paramètres intervenant dans telle ou telle théorie.

Nous discuterons finalement une contribution aux ailes de la raie Rayleigh due aux mouvements translationnels des molécules. Considérons un fluide composé de molécules optiquement isotropes tel que l'argon. D'après ce que nous avons dit au début, la diffusion devrait être complètement polarisée, car chaque atome possède un tenseur de polarisabilité proportionnel au tenseur unité. En réalité, on trouve qu'il y a une légère dépolarisation que l'on peut interpréter simplement de la manière suivante. Considérons l'atome i situé au point $\mathbf{r}_i$, ainsi que le champ local qui le polarise. Ce champ diffère du champ incident car il contient aussi les contributions dues à la polarisation de tous les autres atomes du milieu [39]. L'effet des atomes situés à grande distance se traduit par un changement du vecteur d'onde de la lumière $\mathbf{k}_i \to n\mathbf{k}_i$, où n est l'indice de réfraction. L'effet des atomes voisins est de produire une contribution proportionnelle à $\sum_{\text{voisins}} T^{\alpha\beta}(\mathbf{r}_{ij}) \mathbf{E}_j$ où $T^{\alpha\beta}(r_{ij})$ est le propagateur du champ dipolaire entre l'atome i et l'atome j. Cette contribution dépend de la structure précise de l'environnement. Elle est donc reliée aux fluctuations de densité au voisinage de l'atome i, et s'annule pour un milieu homogène. On trouve que si le champ incident est polarisé dans la direction z, on a une contribution

$$\langle E_z^2 \rangle \propto \sum_{i \neq j} \sum_{k \neq l} \alpha^2 \, T^{zz}(\mathbf{r}_{ij}) \, T^{zz}(\mathbf{r}_{kl})$$

et

$$\langle E_x^2 \rangle \propto \sum_{i \neq j} \sum_{k \neq l} \alpha^2 \, T^{xz}(\mathbf{r}_{ij}) \, T^{xz}(\mathbf{r}_{kl}),$$

où

$$T^{zz}(\mathbf{r}_{ij}) = \frac{1}{r_{ij}^3} \left(1 - \frac{3 z_{ij} z_{ij}}{r_{ij}^2} \right)$$

et

$$T^{xz}(\mathbf{r}_{ij}) = \frac{1}{r_{ij}^3} \left(-\frac{3 x_{ij} z_{ij}}{r_{ij}^2} \right),$$

où x_{ij}, y_{ij}, z_{ij} sont les coordonnées cartésiennes du vecteur $\mathbf{r}_{ij} = \mathbf{r}_i - \mathbf{r}_j$.

Il existe un autre mécanisme de dépolarisation plus subtil. Il se peut que les fonctions d'onde de l'atome soient déformées par ses voisins et que par conséquent son tenseur de polarisabilité ait perdu la symétrie sphérique. Cet effet doit se produire lorsque la distance entre atomes est faible et qu'il y a recouvrement partiel des orbitales électroniques. On voit donc que la diffusion translationnelle peut avoir deux origines différentes.

On peut montrer que le terme dû aux fluctuations de densité conduit à un coefficient de dépolarisation de $\frac{3}{4}$, comme pour la raie Rayleigh due à la rotation des molécules. On pense qu'il doit en être de même pour l'autre contribution. Quant au spectre, on le calcule à partir de la fonction de corrélation du champ diffusé. On obtient par exemple pour la polarisation z

$$C(\tau) = \langle E_d(\tau) E_d^*(0) \rangle \propto \left\langle \sum_{ijkl} T^{zz}(\mathbf{r}_{ij}(\tau)) \, T^{zz}(\mathbf{r}_{kl}(0)) \right\rangle.$$

Grâce à la forme particulièrement simple de T, on peut montrer que

$$C(\tau) \propto \left\langle \sum_{i \neq j} \sum_{k \neq l} \beta(r_{ij}(\tau)) \, \beta(r_{kl}(0)) \, P_2(\hat{r}_{ij}(\tau) \cdot \hat{r}_{kl}(0)) \right\rangle,$$

où l'on a supposé que la déformation des orbitales est additive et conduit à une partie additionnelle $\gamma(r)$ sous forme de tenseur symétrique de trace nulle d'axe de révolution parallèle au rayon vecteur $\mathbf{r}$, de sorte que

$$\beta(r) = 6\alpha^2/r^3 + \gamma(r),$$

où α est la polarisabilité de l'atome isolé.

Avant d'utiliser cette théorie dans le cas des liquides, nous indiquerons quelques résultats obtenus dans les gaz de faible densité où on pourra négliger les termes à plus de deux particules. On aura ainsi un modèle où seules les paires jouent un rôle, ce qui simplifie énormément l'analyse. On trouve une composante dépolarisée dont l'intensité est proportionnelle au carré de la densité, donc au nombre de paires d'atomes. La comparaison avec le modèle décrit plus haut montre qu'il est nécessaire de tenir compte à la fois des fluctuations de densité et des distortions électroniques au cours des collisions. On a essayé de trouver la dépendance en r du terme $\gamma(r)$ pour l'argon à partir des données expérimentales [40, 41]. On a trouvé que le terme $\gamma(r)$ doit avoir une dépendance en r relativement lente ($\exp(-r/1,43)$ ou $\exp(-r/1,25)$), ce qui semble surprenant pour un effet dont l'origine réside dans une déformation de nuages électroniques par suite de leur recouvrement. On s'attendrait plutôt à trouver un terme variant à peu près comme les forces répulsives entre atomes, soit environ [42] $\exp(-r/0,28)$ où r est toujours en Å. Comme on ne peut pas mesurer cette fonction par une autre méthode, il faut avoir recours à un calcul quantique de la polarisabilité en fonction de la distance r. Cela a été fait pour He–He [43] et pour Ar–Ar [44] en utilisant la méthode SCF en présence d'un champ électrique et en développant sur une base de fonctions gaussiennes à un électron, mais sans tenir compte des corrélations. Les résultats sont évidemment beaucoup plus précis dans le cas de He–He car il n'y a que quatre électrons. On trouve que $\gamma(r)$ est de la forme $\exp(-r/0,74)$ ce qui de nouveau est à plus grande portée que le potentiel répulsif [42] $\exp(-r/0,25)$. Les résultats des calculs dans l'argon ne sont pas assez précis pour qu'on puisse en tirer de telles conclusions. Il n'est pas possible d'entrer ici dans le détail de l'analyse des données expérimentales, mais il faut dire que les méthodes de détermination du terme $\gamma(r)$ ne donnent réellement que son comportement au voisinage de $r \approx \sigma$, distance de Lennard-Jones entre atomes. On pourrait certainement trouver d'autres types de fonction permettant d'obtenir le même accord entre les données expérimentales et le calcul théorique. Il est probable que des mesures très précises faites à diverses températures pourraient permettre de déduire la fonction $\gamma(r)$ avec plus de précision.

Nous allons maintenant passer au cas beaucoup plus complexe des gaz denses ou des liquides. La situation est assez développée sur le plan expérimental. Il existe d'une part des mesures du coefficient de dépolarisation dans l'argon à température ordinaire entre 0 et 600 amagats [45]. On trouve que l'intensité dépolarisée croit d'abord comme le carré de la densité, mais que très vite l'effet des corrélations entre paires réduit fortement l'intensité diffusée [46]. On peut dire que le milieu devient de plus en plus symétrique lorsque la densité augmente. D'autre part, Fleury *et al.* [47] ont fait des études très complètes des ailes du spectre dépolarisé dans Ar et Ne à des températures variées et à des pressions élevées. Il est clair que même si l'on prend le modèle des fluctuations de densité où l'on néglige la variation de la polarisabilité, le calcul théorique est très complexe car il faut calculer une expression à quatre particules, c'est à dire qu'il faut considérer des contributions dépendant de

$$n_{12}(\mathbf{r}_{12}), n_{123}(\mathbf{r}_{12}, \mathbf{r}_{13}) \quad \text{et} \quad n_{1234}(\mathbf{r}_{12}, \mathbf{r}_{13}, \mathbf{r}_{34}).$$

Fleury *et al.* [48] ont essayé de traiter les fluctuations de densité comme une diffusion Raman du second ordre, mais cela présente des difficultés dans un liquide. En effet la fonction T qui apparaît dans l'expression du champ électrique diffusé varie comme $1/r^3$, par conséquent, il faut considérer des fluctuations de densité du milieu de vecteurs d'onde très grands $q \approx 1/\sigma$, qui sont mal définis en raison de l'amortissement. En outre, il faut faire des hypothèses hardies sur les fonctions n_{12}, n_{123} et n_{1234}.

En vue de faire un calcul quantitatif, il a fallu recourir aux méthodes de simulation par ordinateur. Alder et al. [42] ont d'abord calculé le coefficient de dépolarisation en ne considérant que les fluctuations de densité. Ils trouvent qu'ils peuvent rendre compte des caractéristiques générales des données expérimentales, mais l'accord n'est pas quantitatif. Il est probable qu'il faudrait prendre une polarisabilité effective moyenne dépendant de la densité. En ce qui concerne le spectre, la situation se présente beaucoup mieux. Berne et al. [50] ont calculé la fonction de corrélation du champ diffusé. Ils ont trouvé que l'essentiel de la dépendance temporelle de $c(\tau)$ provient du terme $\langle P_2(\hat{r}_{ij}(\tau) \cdot \hat{r}_{kl}(0)) \rangle$. Il en résulte que la forme exacte de $\gamma(r)$ ne joue qu'un rôle secondaire dans le calcul de la forme du spectre. Ce résultat contredit certains modèles de chocs binaires dans les liquides où l'on isole deux molécules sous prétexte que $\gamma(r)$ est de la forme $1/r^n$ avec $n \approx 15$ et que donc seuls des chocs très énergétiques peuvent contribuer. Plus récemment, Alder et al. [51] ont fait des calculs très détaillés de $c(\tau)$ pour divers potentiels interatomiques. Ils obtiennent des résultats en très bon accord avec les formes spectrales mesurées par Fleury et al. Il ne semble pas d'après les calculs qu'il y ait de comportement particulier de $c(\tau)$ pour les grands τ comme on en trouve pour la viscosité ou la fonction de corrélation des vitesses. On peut donc dire que l'on comprend assez bien maintenant le mécanisme de cette diffusion dépolarisée translationnelle, mais qu'il reste beaucoup à faire en vue de pouvoir calculer l'amplitude de cette diffusion afin de pouvoir soustraire avec certitude sa contribution aux ailes de la raie Rayleigh en vue de déterminer avec encore plus de précision les fonctions de corrélation dépendant de la partie anisotrope du tenseur de polarisabilité moléculaire.

Nous avons vu au cours de cette introduction à la diffusion de la lumière qu'il s'agit d'une méthode très puissante permettant d'obtenir de nombreux renseignements sur la dynamique des fluides. En effet, l'interaction de la lumière avec le milieu permet de transférer la partie, de vecteur d'onde sélectionné, de l'information contenue dans le bruit de basse fréquence entretenu par l'agitation thermique vers le domaine des fréquences optiques où l'on dispose de méthodes d'analyse très fines et très sensibles. Bien que les vecteurs d'onde accessibles soient très petits, on peut obtenir des données précises sur certaines propriétés thermodynamiques et sur les coefficients de transport et leur dépendance en fonction de la fréquence, sur les mouvements rotationnels et sur les mouvements translationnels.

Bibliographie

1. Fabelinskii, I. L.: *Molecular Scattering of Light*, Plenum Press, N.Y., 1968.
2. Benedek, G. B.: 'Thermal Fluctuations and The Scattering of Light', dans M. Chrétien, E. P. Gross, et S. Deser (eds.), *1966 Brandeis University Summer Institute in Theoretical Physics*, Gordon and Breach, N.Y., 1968.

3. Landau, L. et Lifshitz, I.: *Statistical Physics*, Pergamon Press, London.
4. Mountain, R.: *Rev. Mod. Phys.* **38**, 205 (1968).
5. Mountain, R.: *J. Res. N.B.S.* **70A** 207 (1966).
6. Cummins, H. Z. et Gamon, R. W.: *J. Chem. Phys.* **44**, 2785 (1966).
7. Zwanzig, R.: *J. Chem. Phys.* **43**, 714 (1965).
8. Gornall, W. S., Stegeman, G. I. A., Stoicheff, B., Stolen, R. H., et Volterra, V.: *Phys. Rev. Letters* **17**, 297 (1966).
9. Montrose, C. J., Solovyev, V. A., et Litovitz, T. A.: *J. Acous. Soc. Am.* **43**, 117 (1968).
10. Nichols, W. H. et Carome, E. F.: *J. Chem. Phys.* **49**, 1000 et 1013 (1968).
11. Demoulin, C., Montrose, C. J., et Ostrowski, N.: ce volume, p. 575.
12. Herzfeld, K. F. et Litovitz, T. A.: *Absorption and Disperson of Ultrasonic Waves*, Academic Press Inc., N.Y., 1959.
13. Litovitz, T. A. et Davis, C. M.: dans W. P. Mason (ed.), *Physical Acoustics*, Academic Press Inc., 1965, Vol. 2A.
14. Cannell, D. S. et Lunacek, J. H.: *J. Physique* **33C** (1), 91 (1972).
15. Oliver, C. J. et Pike, E. R.: *Phys. Letters* **31A**, 373 (1970).
16. Beysens, D.: Thèse, Université Paris-6, 1973, non publié.
17. Lastovska, J. B. et Benedek, G. B.: *Phys. Rev. Letters* **17**, 1035 (1966).
18. Cummins, H. Z. et Swinney, H. L.: dans E. Wolf (ed.), *Light Beating Spectroscopy* dans *Progress in Optics*, Vol. VIII, 1970.
19. Dubois, M. et Bergé, P.: *J. Physique* **33C** (1), 37 (1972).
20. Chiao, R. Y. et Fleury, P. A.: dans P. L. Kelley, B. Lax, et P. E. Tannenwald (eds.), *Physics of Quantum Electronics*, McGraw Hill, N.Y., 1966.
21. Munch, J. P. et Candau, S.: ce volume, p. 535.
22. May, A. D. et Hara, E. H.: *Can. J. Phys.* **49**, 421 (1971).
23. Cazabat-Longequeue, A. M. et Lallemand, P.: *J. Phys.* **33C** (1), 57 (1972).
24. Clark, N. A., Mellman, G. R., et Greytak, T. J.: *Phys. Rev. Letters* **29**, 150 (1972).
25. Zamir, E. et Ben-Reuven, A.: *J. Phys.* **33C** (1), 237 (1972).
26. Bratos, S. et Maréchal, A.: *Phys. Rev.* **A4**, 1078 (1971).
27. Goldberg, H. S. et Pershan, P. S.: *J. Chem. Phys.* **58**, 3816 (1973).
28. Bartoli, F. J. et Litovitz, T. A.: *J. Chem. Phys.* **56**, 404 et 413 (1972).
29. Shapiro, S. L. et Broida, H. P.: *Phys. Rev.* **154**, 129 (1967).
30. Pinnow, D. A., Candau, S. J., et Litovitz, T. A.: *J. Chem. Phys.* **49**, 347 (1968).
31. Ivanov, E. N.: *JEPT* **18**, 1041 (1964).
32. Bucaro, J. A. et Litovitz, T. A.: *J. Chem. Phys.* **54**, 3846 (1971).
33. Shin, Hyung Kyu: *J. Chem. Phys.* **56**, 2617 (1972).
34. Gordon, R. G.: *J. Chem. Phys.* **43**, 1307 (1965).
35. Dardy, H. D., Volterra, V., et Litovitz, T. A.: *J. Chem. Phys.* **59**, 4491, (1973).
36. Connes, P.: 'Fourier-Transform Spectroscopy', dans C. H. Townes et P. A. Miles (eds.), *Quantum Electronics and Coherent Light*, Academic Press Inc., N.Y., 1964.
37. Hanbury-Brown, R. et Twiss, R.: *Proc. Roy. Soc. London* **243A**, 29 (1958).
38. Davidson, D. W. et Cole, R. H.: *J. Chem. Phys.* **19**, 1484 (1951).
39. Buckingham, A. D. et Stephen, M. J.: *Trans. Faraday Soc.* **53**, 884 (1957).
40. McTague, J. P., Ellenson, W. D., et Hall, L. H.: *J. Phys.* **33C** (1) 241 (1972).
41. Lallemand, P.: *J. Phys.* **33C** (1), 257 (1972).
42. Hirschfelder, J. O., Curtiss, C. F., et Bird, R. B.: *Molecular Theory of Gases and Liquids*, John Wiley, N.Y., 1954, p. 181.
43. O'Brien, E. F., Gutschick, V. P., McKoy, V., et McTague, J. P.: *Phys. Rev.* **A8**, 690 (1973).
44. Lallemand, P., David, D. J., et Bigot, B.: *Molecular Physics*, 1974.
45. Thibeau, M., Oksengorn, B., et Vodar, B.: *J. Phys.* **29**, 287 (1968).
46. Gelbart, W. M.: *J. Chem. Phys.* **57**, 699 (1972).
47. Fleury, P. A., Daniels, W. B., et Worlock, J. M.: *Phys. Rev. Letters* **27**, 1493 (1971).
48. McTague, J. P., Fleury, P. A., et DuPré, D. B.: *Phys. Rev.* **188**, 303 (1969).
49. Alder, B. J., Weis, J. J., et Strauss, H. L.: *Phys. Rev.* **A7**, 281 (1973).
50. Berne, B. J., Bishop. M., et Rahman, A.: *J. Chem. Phys.* **58**, 2696 (1973).
51. Alder, B. J., Strauss, H. L., et Weis, J. J.: *J. Chem. Phys.* **59**, 1002 (1973).

DISCUSSION

Weiss: The spectral band shape is probably most sensitive to the angular part of the distortion anisotropy and less to the precise radial dependence.

Gershon: For the fitting of the depolarized scattering from CS_2 was only one Lorentzian used?

Lallemand: The actual spectrum looks like the sum of two Lorentzian curves, a very broad one and a sharp one that was shown. The ratio of widths is quite large ($\sim$ factor 20) so that one could use a flat background to fit the sharp curve. Now, I did not use more than one Lorentzian in the center, and I did not vary the spectral range over which the central line would be Lorentzian. Clearly, this data was just shown to warn against the simple use of half widths to deduce relaxation times.

Jackson: In the spectrum shown of glycerol the polarized and depolarized correlation times appear to be the same, is this the case? Don't you have both a Mountain line and the spectral feature associated with molecular reorientation?

Lallemand: We did not show on the last slide the asymptote of the spectrum. Thus the 1.5 factor between τ_{pol} and τ_{depol} does not show up clearly here. The ratio of polarized over depolarized is roughly a factor 10, so that the amount of depolarized in the polarized is fairly small. Furthermore we use homodyne spectroscopy so that we have to compare the square of the ratio of polarized and depolarized components.

Janik: Can the formula you mentioned

$$\frac{1}{\tau_{eff}} = \frac{1}{\tau} + D^* k^2$$

be practically applied for diffusion coefficient determination?

Lallemand: This has not been used to determine D, but rather to interpret some of the spectra in gases. Light scattering can be used to study mutual diffusion coefficients in mixtures of liquids that have different indices of refraction (see Reference 19 of the paper).

ETUDE DES MECANISMES DE RELAXATION DE VIBRATION DANS LE TETRACHLORURE DE CARBONE ET LE CHLOROFORME LIQUIDES PAR ANALYSE DES SPECTRES DE DIFFUSION BRILLOUIN

J. P. MUNCH et S. CANDAU

*Laboratoire d'Acoustique Moléculaire, E.R.A. au C.N.R.S. Institut de Physique,
Université Louis Pasteur, 4, rue Blaise Pascal, Strasbourg, France*

Résumé. Nous avons mesuré les spectres de diffusion Brillouin pour le tétrachlorure de carbone et le chloroforme liquides à différentes températures. Les spectres obtenus sont comparés à des spectres calculés dans l'hypothèse d'une relaxation à un ou deux temps de la chaleur spécifique de vibration. Nous avons également analysé les résultats de mesures ultrasonores réalisées par d'autres auteurs. Les résultats sont discutés à partir de la théorie quantique de Slawsky, Schwartz et Herzfeld.

Abstract. Brillouin spectra have been measured in liquid carbone tetrachloride and chloroform as a function of temperature. Experimental spectra have been compared with calculated ones assuming a single or double relaxation of the vibrational specific heat. Acoustical data obtained by other authors have been similarly analysed. Experimental results relative to carbone tetrachloride can be accounted for by assuming a single relaxation process involving the specific heat of the three lowest vibrational modes of the molecules. The same process has been observed in the whole investigated temperature range. These results disagree with previous litterature studies. In chloroform the best fit between experimental and theoretical spectra is obtained with the assumption of double relaxation, the order of magnitude of the ratio of relaxation times being 5.

To explain the experimentally observed processes, we have used a quantum theory which has been successfully developed for gases by Slawsky, Schwartz, Herzfeld and extended to liquids by Litovitz. This theory assumes that the transfer of energy between vibrational and translational degrees of freedom is due to highly energetic binary collisions, the collision efficiency being the same in the liquid as in the gas except for a factor related to the effect of attractive forces in the collision efficiency.

Acoustical dispersion curves have been calculated according to this model. From this calculation, a dispersion zone corresponding to the simultaneous relaxation of the three lowest vibrational modes only, should appear in the high frequency domain for carbone tetrachloride. This is in agreement with our observations as stated above. As for chloroform, two dispersion zones show up in the same frequency range. The high frequency one involves the specific heat of the two lowest vibrational modes. The low frequency one involves the specific heat of the three other modes.

1. Introduction

L'existence de processus de relaxation vibrationnelle dans les liquides a été mise en évidence par Lamb *et al* qui ont établi par des mesures de dispersion ultrasonore que toute la chaleur spécifique de vibration relaxait avec un temps caractéristique unique [1]. Des mesures ultérieures de Litovitz *et al* ont montré que ce temps de relaxation était considérablement affecté par une variation de pression ou l'addition d'impuretés [2]. L'analyse de ces résultats a conduit Litovitz à considérer que les transferts d'énergie entre degrés de translation et de vibration ne s'effectuent pas de façon coopérative mais plutôt selon un processus de collisions intermoléculaires comme c'est le cas en milieu gazeux.

Dans les gaz, les mécanismes de relaxation vibrationnelle sont mieux connus et peuvent être interprétés au moyen d'une théorie quantique due à Slawsky, Schwartz et Herzfeld (SSH) et Tanczos [3, 4], théorie fondée sur l'hypothèse que les transferts d'énergie ne se produisent que lors de collisions bimoléculaires très énergétiques. La cinétique de ces transferts est décrite au meyon de vitesses de transition $k_{ij}^{kl}(a, b)$ qui, dans cette hypothèse peuvent s'écrire sous la forme:

$$k_{ij}^{kl}(a, b) = P_{ij}^{kl}(a, b) M,$$

où i et j représentent les états du mode vibrationnel (a) avant et après collision, k et l, ceux du mode (b). $P_{ij}^{kl}(a, b)$ est la probabilité qu'une collision induise un transfert entre énergies de vibration et de translation. Cette probabilité dépend essentiellement du potentiel intermoléculaire, des nombres d'onde des modes vibrationnels et de leur amplitude de vibration dans le système de coordonnées normales. M est la fréquence de collisions.

Dans les liquides où les libres parcours moyens des molécules sont beaucoup plus faibles et par suite les fréquences de collisions plus élevées, l'hypothèse de collisions bimoléculaires paraît moins fondée. Cependant Herzfeld et Litovitz ont admis, qu'étant données les énergies de transition mises en cause, la probabilité est faible qu'une collision entre 3 (ou plus de 3) molécules soit efficace [2, 4]. Il s'ensuit que le processus de relaxation observé à l'état gazeux doit être conservé à l'état liquide et simplement déplacé dans le temps. Le rapport des vitesses de transition: (k_{lig}/k_{gaz}) n'est cependant pas simplement égal au rapport des fréquences de collisions M_{liq}/M_{gaz}; il faut en effet selon Litovitz [2] modifier également l'expression de la probabilité de transition, les forces attractives à longue distance ne jouant pas de rôle dans les transferts d'énergie pour les liquides.

L'expression de la vitesse de transition pour un liquide s'écrit alors:

$$k_{liq} = k_{gaz} \frac{M(\sigma_s)_{liq}}{M_{gaz}} \frac{1 + C/T}{e^{\varepsilon/kT}} = M(\sigma_s)_{liq} P_{gaz} \frac{1 + C/T}{e^{\varepsilon/kT}}. \quad (1)$$

M_{gaz} se déduit des mesures de viscosité [5]. C est la constante de Sutherland. ε est la profondeur du puits de potentiel de Lennard-Jones. Pour évaluer M_{liq}, Litovitz a proposé un modèle de cellule à parois mobiles dans lequel les molécules sont représentées par des sphères rigides de diamètre σ_s (diamètre de cœur dur de Sutherland) disposées aux différents nœuds d'un réseau [2].

Plus récemment, Davis et Oppenheim ont développé une nouvelle théorie de la relaxation de vibration à partir de la mécanique statistique et ont obtenu, moyennant certaines approximations, l'expression suivante [6].

$$k_{liq} = k_{gaz} \frac{g(R^*)}{g_0(R^*)} \quad (2)$$

$g(R)$ étant la fonction de corrélation de paires du liquide et $g_0(R)$ sa valeur extrapolée aux faibles densités. R^* est une distance indéterminée mais toujours inférieure au diamètre de cœur dur de Sutherland σ_s.

Le rapport $g(R^*)/g_0(R^*)$ peut être considéré comme le rapport des fréquences de collisions, ces dernières étant alors définies comme des événements au cours desquels deux molécules s'approchent à une distance inférieure à R^*. Davis a montré que ces fréquences de collisions sont nettement inférieures à celles qu'à obtenues Litovitz à partir du modèle de cellule [6].

Cependant la relation (2) implique, comme la relation (1), proportionnalité entre les fréquences de relaxation du liquide et celles du gaz correspondant, ces dernières pouvant être calculées à partir de la théorie quantique de SSH.

Les résultats expérimentaux relatifs à deux liquides triatomiques (CO_2 et CS_2) présentant une relaxation unique de toute la chaleur spécifique de vibration se sont révélés en bon accord avec le modèle de Litovitz [2]. Plus récemment, des mesures de diffusion Brillouin ont permis de montrer que le processus de relaxation observé dans le dichlorométhane peut être interprété à partir de la théorie de collisions bimoléculaires [7].

Dans ce travail nous présentons des résultats relatifs au tétrachlorure de carbone et au chloroforme. Des spectres de diffusion Brillouin ont été enregistrés dans le domaine de températures $-20\,°C$, $+55\,°C$. L'analyse de ces spectres ainsi que des données ultrasonores obtenues par d'autres auteurs nous a permis de déterminer les courbes de dispersion acoustique et de les comparer à des courbes calculées à partir de la théorie de SSH Tanczos.

2. Calcul des courbes de dispersion acoustique

Ces courbes ont été calculées pour le tétrachlorure de carbone et le chloroforme gazeux à la température de $25\,°C$ par Tanczos [4]. Pour les détails de ce calcul, le lecteur peut se référer aux manuscrits originaux [3, 4].

Les courbes de dispersion de la vitesse V et de l'absorption ultrasonore α s'écrivent:

$$\left(\frac{V}{V_0}\right)^2 = 1 + \sum_i \frac{4\pi^2 f^2 \tau_i^2 D_i}{1 + 4\pi^2 f^2 \tau_i^2}$$

$$\left(\frac{\alpha}{f^2}\right) = 2\pi^2 \frac{V_0^2}{V^3} \sum_i \frac{D_i \tau_i}{1 + 4\pi^2 f^2 \tau_i^2} + \left(\frac{\alpha}{f^2}\right)_{Cl}, \quad (3)$$

où f est la fréquence de l'onde sonore, V_0 la vitesse extrapolée à fréquence nulle. $(\alpha/f^2)_{Cl}$ représente les pertes d'origine visqueuse:

$$\left(\frac{\alpha}{f^2}\right)_{Cl} = \frac{2\pi^2}{\varrho_0 V_\infty^3}(\eta_v + \tfrac{4}{3}\eta_s)$$

η_s étant la viscosité de cisaillement, η_v la viscosité de volume résiduelle provenant de réarrangements structuraux.

Les amplitudes et temps de relaxation D_i et τ_i sont fonction essentiellement des vitesses de transition individuelles $k_{ij}^{kl}(a, b)$ et des chaleurs spécifiques de vibration $C'(v)$ des modes mis en jeu.

$C'(v)$ se calcule à partir de la relation de Planck-Einstein

$$C'(v) = nR \left(\frac{hv}{kT}\right) \frac{\exp(-hv/kT)}{|1 - \exp(-hv/kT)|^2}$$

n étant la dégénérescence de la vibration de nombre d'onde v, h la constante de Planck, k la constante de Boltzmann et R celle des gaz parfaits.

Nous avons repris le calcul de Tanczos pour différentes températures comprises entre $-20\,°C$ et $+55\,°C$. Nous avons ainsi pu à partir de la relation (1) calculer des courbes de dispersion acoustique pour le tétrachlorure de carbone et le chloroforme liquides. Dans ce calcul, nous avons utilisé les valeurs des fréquences de collisions déterminées par Bartoli et Litovitz à partir de l'élargissement des raies Raman [8].

Le calcul fournit les résultats suivants:

2.1. Tetrachlorure de carbone

La molécule de CCl_4 possède quatre modes fondamentaux de vibration (tableau I. Dans le tableau II sont reportées les valeurs calculées des vitesses de transition en fonc-

TABLEAU I
Tétrachlorure de carbone

Mode	v cm^{-1}	$C'(v)$ (cal mole^{-1} °C^{-1})		
		$T = -20\,°C$	$T = 30\,°C$	$T = 55°$
1	218	3,5024	3,6369	3,685
2	314	4,602	4,9690	5,1027
3	458	1,1625	1,3598	1,435
4	776	1,443	2,1407	2,457

TABLEAU II
Tétrachlorure de carbone

Nature de la transition		$k_{ij}^{kl}(a, b)\ 10^{-9}\ s^{-1}$		
$ij\,(a)$	$kl\,(b)$	$T = -20\,°C$	$T = 30\,°C$	$T = 55\,°C$
10 (1)		5,2	5,12	5,26
10 (1)	01 (2)	6,1	5,13	5,22
20 (2)	01 (3)	0,233	0,172	0,1647
10 (2)		0,063	0,078	0,093
10 (2)	01 (1)	9,637	7,81	7,820
10 (2)	01 (3)	0,093	0,093	0,1016
20 (2)	01 (4)	—	—	—
10 (3)		…	0,00024	—
10 (3)	02 (1)	2,18	1,57	1,518
10 (3)	01 (2)	0,79	0,72	0,768
10 (3)	01 (4)	—	—	—
10 (4)		—	$0,35 \times 10^{-8}$	—
10 (4)	01 (3)	—	—	—
10 (4)	02 (2)	0,00065	0,000585	0,00068

tion de la température. On peut constater que les probabilités de désactivation simple définies par $k_{10}(i)$ deviennent négligeables pour des niveaux énergétiques supérieurs à $h\nu_1$.

A partir de ces vitesses de transition, nous avons calculé les fréquences et amplitudes de relaxation décrivant la dispersion en vitesse (tableau III). L'examen des valeurs de $\{f_i, D_i\}$ à chaque température fait apparaître l'existence de trois zones de relaxation.

– La première, dont la fréquence est de l'ordre de 2,5 GHz possède une amplitude négligeable.

– La deuxième zone est caractérisée par deux fréquences de relaxation très proches. Pour $T=-20\,°C$ leur rapport est 1,03. Cette zone peut être considérée comme une simple relaxation dont l'amplitude est la somme algébrique des deux amplitudes et

TABLEAU III

Tétrachlorure de carbone

$T=-20\,°C$		$T=30\,°C$		$T=55\,°C$	
f_i (Hz)	D_i	f_i (Hz)	D_i	f_i (Hz)	D_i
$0,29 \times 10^{10}$	0,00034	$0,25 \times 10^{10}$	0,0000	$0,27 \times 10^{10}$	0,0005
$0,82 \times 10^{9}$	0,029	$0,78 \times 10^{9}$	0,341	$0,66 \times 10^{9}$	0,007
$0,79 \times 10^{9}$	0,33	$0,67 \times 10^{9}$	0,0015	$0,879 \times 10^{9}$	0,4
$0,112 \times 10^{6}$	0,023	$0,104 \times 10^{6}$	0,039	$0,124 \times 10^{6}$	0,04

la fréquence une moyenne pondérée des deux fréquences. Lorsque la température augmente, le rapport de ces deux fréquences de relaxation augmente également, mais l'amplitude de l'une devient négligeable.

– En troisième lieu, nous observons une zone de relaxation à basse fréquence avec une amplitude environ dix fois plus faible que la précédente.

La courbe de dispersion en vitesse peut donc s'interpréter en termes de deux relaxations possédant deux fréquences très différentes. On peut remarquer en comparant les tableaux I et IV que les chaleurs spécifiques associées aux relaxations respectivement basse fréquence et haute fréquence sont celles des vibrations (ν_4) et $(\nu_1 + \nu_2 + \nu_3)$.

2.2. Chloroforme

La molécule de $CHCl_3$ possède cinq modes normaux de vibration (tableau V). Pour ce liquide également, le calcul prévoit deux zones de dispersion d'amplitudes non négligeables, les fréquences de relaxation étant dans un rapport d'environ 50 (tableau VI).

La chaleur spécifique impliquée dans le processus haute fréquence est celle des modes ν_4 et ν_5, la relaxation des trois autres modes s'effectuant par le processus basse fréquence (cf. comparaison des tableaux V et VI).

TABLEAU IV
Tétrachlorure de carbone

$T = -20°C$			$T = 30°C$			$T = 55°C$		
f_i (Hz)	D_i	C'_i (cal mole^{-1} °C^{-1})	f_i (Hz)	D_i	C'_i (cal mole^{-1} °C^{-1})	f_i (Hz)	D_i	C'_i (cal mole^{-1} °C^{-1})
$0{,}79 \times 10^9$	0,3	9,3	$0{,}78 \times 10^9$	0,34	10	$0{,}87 \times 10^9$	0,4	10,2
$0{,}112 \times 10^6$	0,023	1,44	$0{,}104 \times 10^6$	0,033	2,14	$0{,}124 \times 10^6$	0,04	2,4

TABLEAU V

Chloroforme ($T = 20\,°C$)

Mode (a)	v (cm^{-1})	$C'_a(v)$ (cal mole^{-1} °C^{-1})	$\sum\limits_{a=n}^{a=n'} C'_a$ (cal mole^{-1} °C^{-1})	
1	261	3,5022	$n=1$	
2	363	1,5614		5,0636
3	670	0,9102	$n'=2$	
4	760	1,4835	$n=3$	
5	1211	0,3875	$n'=5$	2,7812

TABLEAU VI

Chloroforme ($T = 20\,°C$)

$10^{-9} f_i$ (Hz)	D_i	C'_i (cal mole^{-1} °C^{-1})
5,3	0,0001	–
1,9	0,175	5,0
0,81	0,00002	–
0,45	0,0012	–
0,04	0,055	2,7

3. Caractérisation des processus de relaxation

3.1. Dispersion acoustique et spectroscopie Brillouin

Les équations (3) du paragraphe précédent permettent de relier les mécanismes de base de relaxation à la dispersion de la vitesse de propagation et de l'absorption spatiale des ondes ultrasonores. Lorsque la fréquence de relaxation devient plus grande que 10^9 Hz, il est utile de faire appel à la diffusion Brillouin pour compléter les données ultrasonores. La diffusion Brillouin permet d'analyser l'évolution temporelle des fluctuations collectives de densité, fluctuations dont la fréquence et le temps de vie sont affectés par la relaxation des degrés de liberté internes.

Mountain [9] reprenant une procédure due à Landau [10] a calculé la densité spectrale associée à ces fluctuations. Dans le cas d'un liquide dispersif présentant un processus de relaxation unique, cette densité spectrale s'écrit:

$$\sigma(k,\omega) = (1 - 1/\gamma) \frac{2\lambda k^2/\varrho_0 c_p}{(\lambda k^2/\varrho_0 c_p)^2 + \omega^2} +$$

$$+ \frac{2\varrho_0 V_0^2}{\gamma} \frac{\eta_0 + \varrho_0 (V_\infty^2 - V_0^2)/(1 + \omega^2 \tau^2)}{|\omega\eta_0 + \omega\varrho_0\tau(V_\infty^2 - V_0^2)/(1 + \omega^2\tau^2)|^2 + |\varrho_0\omega^2/k^2 - \varrho_0(V_0^2 - V_\infty^2 \omega^2\tau^2)/(1 + \omega^2\tau^2)|^2}$$

k est le vecteur d'onde de la fluctuation, relié à l'angle de diffusion θ par la loi de Bragg

$$k = 2n k_0 \sin\theta/2$$

n étant l'indice de réfraction du milieu, k_0 le vecteur d'onde de la lumière incidente. ω est le changement de fréquence circulaire par rapport à celle de la lumière incidente. γ est le rapport des chaleurs spécifiques à pression constante et à volume constant, ϱ_0 la densité du milieu. λ est la conductibilité thermique. V_∞ qui représente la limite haute fréquence de la vitesse de propagation ultrasonore est donné par:

$$\left(\frac{V_\infty}{V_0}\right)^2 = \frac{c_v}{c_p}\frac{c_p - c'}{c_v - c'}, \qquad (4)$$

où c' est la quantité de chaleur spécifique qui relaxe.

L'équation (3) peut être facilement généralisée au cas où un processus de relaxation multiple est présent dans le milieu [11, 12].

Ainsi, à partir du spectre de lumière diffusée par un liquide, il est théoriquement possible de déterminer les paramètres de relaxation. Nous donnons dans le paragraphe suivant la procédure que nous avons adoptée pour l'analyse de ces spectres.

3.2. Procédure d'analyse

Pour comparer directement les spectres expérimentaux et calculés, il faut tout d'abord effectuer le produit de composition de ces derniers avec le profil instrumental. Celui-ci est obtenu en enregistrant un spectre de lumière diffusée sans changement de fréquence.

On commence tout d'abord par se donner une hypothèse sur le processus de relaxation et on calcule $\sigma(k, \omega)$ en fonction des paramètres associés à ce processus, paramètres que l'on fait varier afin d'établir une compatibilité éventuelle avec l'ensemble du spectre Brillouin observé. Lors de la convolution, on tient compte de la présence dans le spectre expérimental d'une faible composante de lumière parasite dont la distribution spectrale est celle du profil instrumental.

Pratiquement, cette méthode de comparaison directe ne peut conduire à des conclusions sûres que si le processus de relaxation est caractérisé par un seul temps ou deux temps très différents. On procède donc en plusieurs étapes en commençant par l'hypothèse la plus simple.

1ère hypothèse: relaxation unique de toute la chaleur spécifique de vibration.

La densité spectrale est dans ce cas donnée par la relation 3, V_∞ étant calculé pour $c' = \sum_a c'_a$ (équation (4)). V_0 est fourni par des mesures ultrasonores à basse fréquence. Les paramètres variables sont alors le temps de relaxation et la viscosité de volume η_v. La valeur de η_v qui a pu être mesurée pour de nombreux liquides a toujours été trouvée comprise entre 0 et $(4/3)\eta_s$ [2]. On se limite donc à ce domaine de variation et on cherche le couple de valeurs $\{\tau, \eta_v\}$ donnant un accord satisfaisant avec l'expérience. Si aucune solution n'est obtenue, on passe à l'analyse en double relaxation.

2ème hypothèse: double relaxation caractérisée par deux temps de relaxation très différents $(\tau_2/\tau_1 > 20)$.

Lorsque cette condition est réalisée, il est généralement possible de caractériser un des deux processus, en amplitude et en fréquence au moyen des techniques ultrasonores [13, 7]. Si tel n'est par le cas, si par exemple une partie de la chaleur spécifique relaxe à des fréquences inférieures au domaine exploré par les techniques ultrasonores conventionnelles (c'est le cas prévu par le calcul pour le CCl_4), on est ramené à une analyse de simple relaxation d'une fraction de la chaleur spécifique de vibration totale. Pour différentes valeurs de c' comprises entre 0 et $\sum_a c'_a$, on cherche les conditions de compatibilité du temps de relaxation et de la viscosité de volume résiduelle avec les spectres observés. Pour éviter de tracer tous les spectres associés aux différentes combinaisons possibles des paramètres, on opère de la manière suivante [14].

On sélectionne tout d'abord parmi les combinaisons (τ, η_v, c') celles qui conduisent à une coïncidence des fréquences ω_B des sommets des raies Brillouin des spectres expérimentaux et calculés. On peut alors comparer les largeurs à mi-hauteur calculées et observées de ces composantes (la fonction calculée ayant été préalablement normée sur les hauteurs relatives des composantes Brillouin). On ne conserve que les solutions donnant un accord sur ces largeurs.

3ème hypothèse: double relaxation caractérisée par un rapport des deux temps inférieur à 20.

Dans ce cas une analyse par superposition des spectres expérimentaux et calculés ne peut conduire à des conclusions précises et il est nécessaire de recourir à l'analyse des courbes de dispersion acoustique (équations (3)). Les spectres de diffusion Brillouin permettent de compléter ces courbes de dispersion dans le domaine des hautes fréquences. En effet la vitesse de propagation à la fréquence ω_B caractérisant la position du sommet des raies Brillouin est donnée en première approximation par la relation :

$$V(\omega_B) = \omega_B/k.$$

Lorsque l'amplitude de dispersion est importante, il convient d'effectuer des corrections [11]. Pour le chloroforme et le tétrachlorure de carbone, elles sont inférieures à 0,6%.

Signalons d'autre part que pour chaque hypothèse de travail il est très important de corroborer les conclusions déduites de la comparaison sur les spectres Brillouin par une analyse des courbes de dispersion acoustique.

3.3. Dispositif expérimental

La technique de diffusion Brillouin est décrite en détail dans de nombreux ouvrages [15]. La source lumineuse est un laser monomode longitudinal à hélium-néon. La lumière diffusée est recueillie par un cône réflecteur (l'angle de diffusion étant 90°).

Le spectre est analysé au moyen d'un interféromètre de Pérot et Fabry plan de finesse 30 environ. La largeur du profil instrumental était dans nos expériences de l'ordre de 300 MHz soit sensiblement plus faible que dans les expériences précédemment réalisées sur le tétrachlorure de carbone [14, 16].

4. Résultats expérimentaux

4.1. Tétrachlorure de carbone

De nombreuses études ont été réalisées sur le tétrachlorure de carbone. C'est ainsi que les mesures de dispersion ultrasonore réalisées par Plass [17] ainsi que par Berdyev et al. [18] ont mis en évidence l'amorce d'une relaxation dans le domaine de 10^9 Hz. D'autre part, des analyses de spectres Brillouin effectuées par trois groupes différents ont donné des résultats compatibles avec une relaxation unique de toute la chaleur spécifique de vibration à température ambiante [14, 16, 19]; cependant les divers auteurs obtiennent des paramètres de relaxation qui diffèrent assez sensiblement entre eux. Par ailleurs une de ces études, réalisée par diffusion Brillouin stimulée conduit à un mécanisme de relaxation variable avec la température, ce qui est en désaccord avec les conclusions de notre calcul du paragraphe 2 [19].

Compte tenu de l'ensemble de ces résultats, nous avons analysé selon la procédure décrite dans le paragraphe précédent, des spectres Brillouin enregistrés à diverses températures comprises entre $-5\,°C$ et $+60\,°C$.

La figure 1 représente la variation thermique de ω_B/k. Dans le domaine de température étudié, ω_B/k prend des valeurs intermédiaires entre V_0 et V_∞, traduisant ainsi l'existence de relaxations à des fréquences de l'ordre de 10^9 Hz.

Envisageons tout d'abord la première hypothèse de la procédure d'analyse.

(a) Hypothèse d'une relaxation unique de toute la chaleur spécifique de vibration.

Sur le tableau VII on a reporté les valeurs de la différence relative Δ entre la largeur à mi-hauteur des raies Brillouin, calculée pour une fréquence de relaxation f_r donnant une superposition des positions des raies Brillouin et la largeur à mi-hauteur mesurée. Ces valeurs sont pour toutes les températures, positives et supérieures aux erreurs expérimentales ($\sim 5\%$).

Nous avons par ailleurs effectué le calcul des courbes de dispersion (à 20°C) pour les valeurs de la fréquence de relaxation et de la viscosité de volume donnant le meilleur accord en diffusion Brillouin ($f_r = 2{,}8 \times 10^9$ Hz et $\eta_v = 0$). Les valeurs expérimentales de l'absorption ultrasonore obtenues par Berdyev et al. [18] se placent nettement au-dessous de la courbe calculée; ce résultat est cohérent avec la valeur positive de Δ à $3{,}385 \times 10^9$ Hz (figure 2).

(b) Hypothèse d'une double relaxation ($\tau_2/\tau_1 > 20$).

Ainsi qu'on l'a indiqué, en l'absence de processus mis en évidence par les techniques ultrasonores, la première étape de l'analyse consiste à admettre une simple relaxation d'une fraction de la chaleur spécifique de vibration totale.

Le tableau VIII donne les valeurs des paramètres variables donnant l'accord le plus satisfaisant entre spectres calculés et observés.

La lecture de ce tableau permet de constater que dans tout le domaine de température étudié, les spectres Brillouin peuvent être interprétés en termes d'une relaxation unique d'environ 80% de la chaleur spécifique de vibration. Cette fraction est du même ordre de grandeur que la contribution c_{123} des trois premiers modes de vibration. La viscosité de volume résiduelle est de l'ordre de $(\frac{4}{3})\eta_s$.

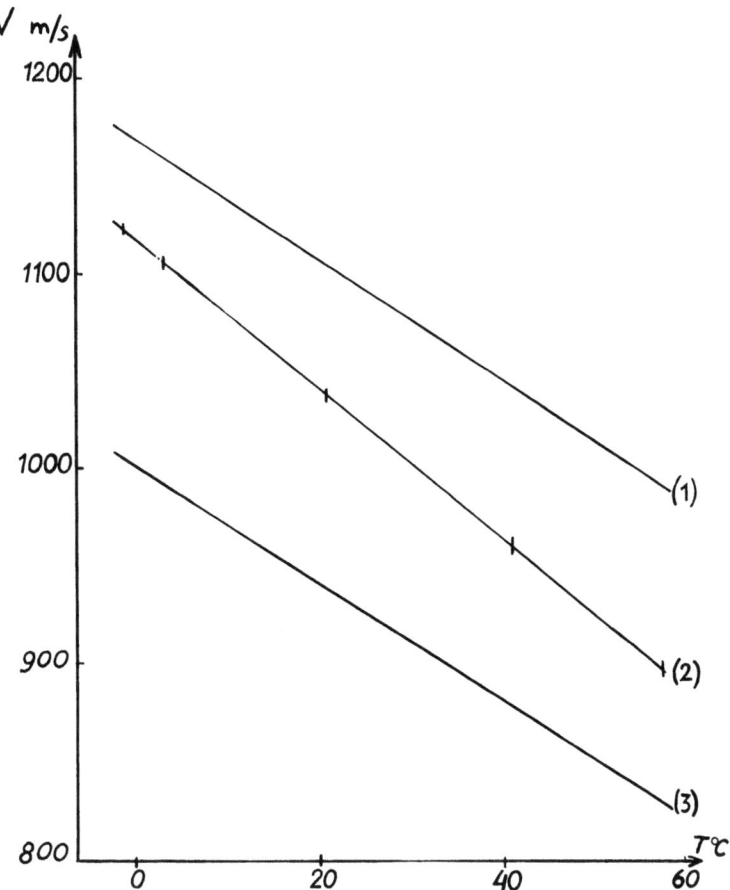

Fig. 1. Tétrachlorure de carbone. Variations thermiques de $V_\infty(1)$, $\omega_B/k(2)$ et $V_0(3)$.

TABLEAU VII

Tetrachlorure de carbone

$T\,°C$	η_v	$10^{-9}f_r$ (Hz)	$\Delta\,\%$
−3	0	2,2	+12
	$\tfrac{4}{3}\eta_s$	2	+17
2	0	2,2	+10
	$\tfrac{4}{3}\eta_s$	2	+14
20	0	2,8	+12
	$\tfrac{4}{3}\eta_s$	2,6	+17
40	0	3	+9
	$\tfrac{4}{3}\eta_s$	2,9	+15
57,5	0	3,3	+14
	$\tfrac{4}{3}\eta_s$	3,2	+18

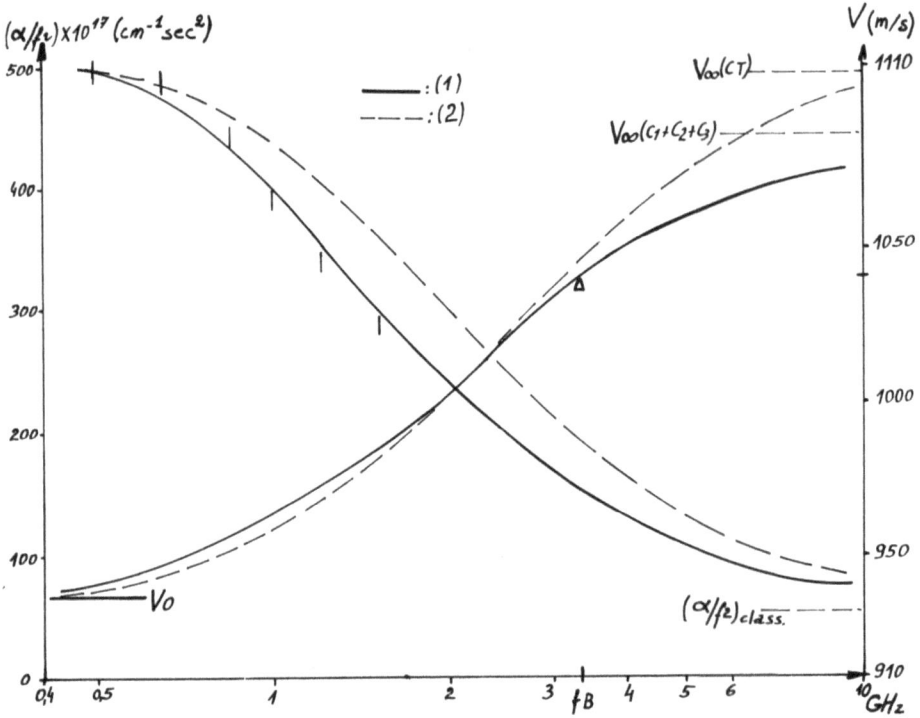

Fig. 2. Tétrachlorure de carbone. Courbes de dispersion calculées à 20°C dans les hypothèses suivantes: (1) Relaxation de la chaleur spécifique des trois premiers modes de vibration; (2) Relaxation unique de toute la chaleur spécifique de vibration. △ Mesure de ω_B/k réalisée à partir du spectre Brillouin. | Mesures ultrasonores de Berdyev et al. [18].

TABLEAU VIII

Tetrachlorure de carbone ($\eta_v = \frac{4}{3}\eta_s$)

$T\,°C$	$10^{-9} f_r$ (Hz)	C'/C_T	$\Delta \%$	C_{123}/C_T	$10^{-9}\,(f_r\text{ calc}^a)$ (Hz)
−3	2	0,82	+1	0,86	1,96
2	2	0,82	+0,8	0,84	2,05
20	2,2	0,80	+0,4	0,83	2,25
40	2,5	0,81	+1,5	0,80	2,56
57,5	2,7	0,80	+1,2	0,79	2,76

[a] (f_r calc) a été calculée à partir des valeurs expérimentales de $(\alpha/f^2)_0$, de V_0 et des valeurs de C' données dans la troisième colonne de ce tableau.

La figure 3 représente la comparaison entre spectres expérimental et calculé à 20 °C pour les valeurs de f_r et c' données dans le tableau VIII. Nous avons également calculé les courbes de dispersion acoustique (figure 2). L'accord entre résultats expérimentaux et courbes calculées peut être considéré comme satisfaisant dans tout le domaine de fréquences.

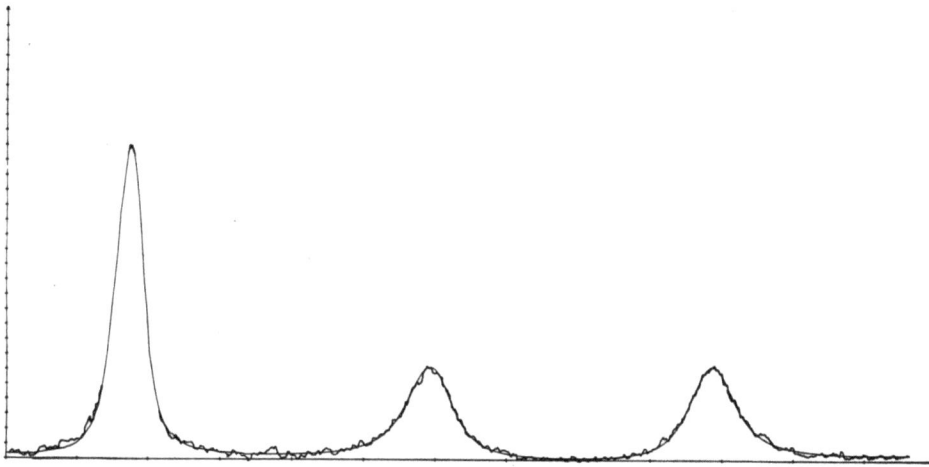

Fig. 3. Tétrachlorure de carbone. Comparaison des spectres enregistré et calculé (courbe en trait plein) à 20°C ($c' = c_1 + c_2 + c_3$), $\eta_v = \frac{4}{3}\eta_s$, $f_r = 2{,}2 \times 10^9$ Hz.

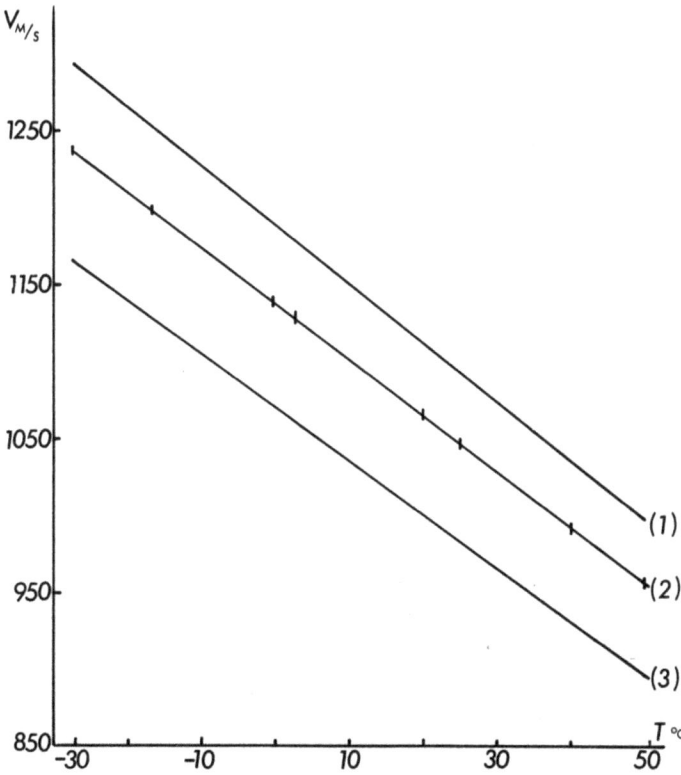

Fig. 4. Chloroforme. Variations thermiques de $V_\infty(1)$, $\omega_B/k(2)$, $V_0(3)$.

Ainsi l'analyse des résultats expérimentaux fournit un processus entièrement en accord avec le mécanisme prévu par le calcul du paragraphe 3 dans tout le domaine de température étudié. Les résultats expérimentaux précédents, notamment ceux de diffusion Brillouin stimulée avaient conduit à des conclusions différentes. On sait cependant que diverses causes d'erreur peuvent affecter les mesures de diffusion Brillouin stimulée [20]. Signalons également que l'étroitesse du profil instrumental dans nos expériences permettait une bonne précision sur les mesures d'élargissement des raies Brillouin, cette mesure fournissant le critère le plus sensible dans la procédure de superposition de spectres.

4.2. Chloroforme

L'existence de processus de relaxation dans le domaine de 10^9 Hz a été mise en évidence dans le chloroforme par Plass [17] et Berdyev et al. [18] au moyen de techniques ultrasonores.

Nous avons enregistré des spectres de diffusion Brillouin entre $-27\,°C$ et $+52\,°C$. La figure 4 représente la variation thermique de ω_B/k. Nous avons alors comparé les spectres expérimentaux avec des spectres calculés dans le cadre des deux premières hypothèses. Cette comparaison nous a permis d'exclure la possibilité d'une relaxation

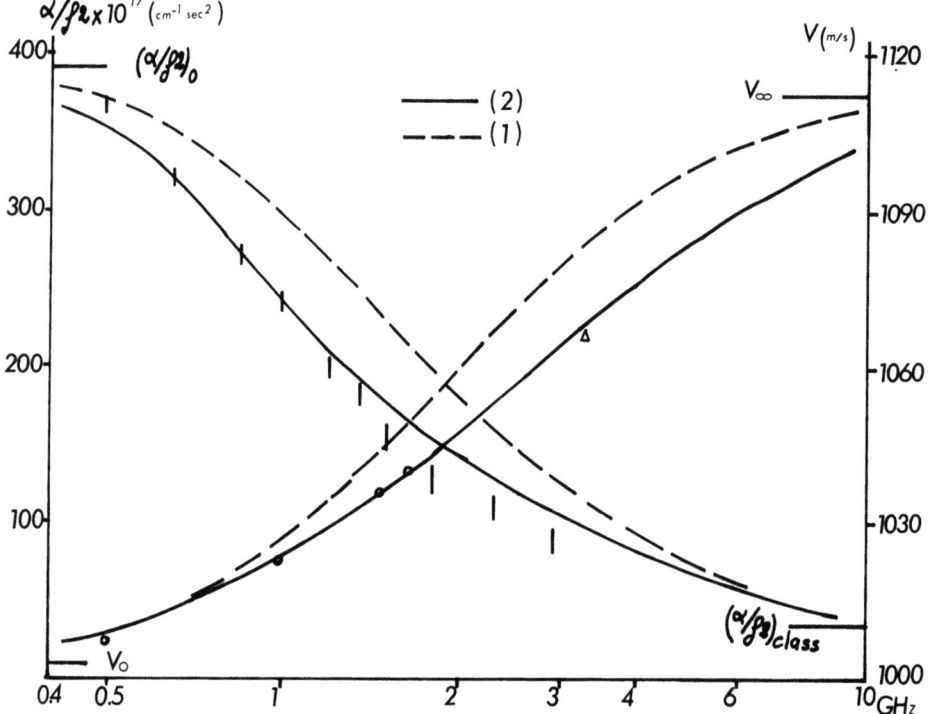

Fig. 5. Chloroforme. Courbes de dispersion calculées à 20°C dans les hypothèses suivantes: (1) Relaxation unique de toute la chaleur spécifique de vibration ($f_r = 1,96$ GHz); (2) Double relaxation ($f_{r1} = 0,78$ GHz, $f_{r2} = 3,5$ GHz). △ Mesure de ω_B/k réalisée à partir du spectre Brillouin. ○| Mesures ultrasonores de Berdyev et al. [18].

d'une fraction ou de la totalité de la chaleur spécifique dans tout le domaine de température exploré. Nous avons alors envisagé l'existence de deux relaxations avec des temps dont le rapport est inférieur à 20. Comme nous l'avons dit précédemment, on ne peut dans ce cas procéder qu'à l'analyse des résultats de vitesse depropagation et d'absorption ultrasonores. Nous ne disposions de ces derniers qu'à la température de 20 °C [18]. Nous sommes partis de l'hypothèse d'une double relaxation dont les amplitudes sont déterminées, en accord avec le mécanisme calculé, par (c_1+c_2) et $(c_3+c_4+c_5)$. Dès lors, les deux paramètres variables sont les deux fréquences de relaxation (η_v étant fixé à $\frac{4}{3}\eta_s$). Nous avons alors déterminé à l'aide d'un programme de minimisation de fonction (sur l'absorption) la valeur de ces deux fréquences donnant le meilleur accord avec les résultats expérimentaux.

Ces valeurs sont: $f_1=0{,}78$ GHz et $f_2=3{,}5$ GHz. Les amplitudes de relaxation correspondantes sont: $D_1=0{,}055$ et $D_2=0{,}175$.

Les courbes de dispersion calculées à partir de ces paramètres sont tracées dans la figure 5. L'accord entre la courbe de vitesse calculée et les résultats de Berdyev est excellent mais les résultats relatifs à l'absorption ultrasonore sont beaucoup moins satisfaisants.

5. Discussion

Avant de discuter des résultats obtenus dans les liquides, il est utile de rappeler quelques résultats des gaz. Pour la plupart des gaz étudiés, les fréquences de relaxation calculées à partir de la théorie de SSH diffèrent des fréquences déterminées expérimentalement. A l'origine de ce désaccord on peut incriminer des erreurs systématiques sur les probabilités de transition, provenant du choix du potentiel intermoléculaire plutôt que d'une évaluation erronée des fréquences de collisions. Un résultat est significatif à cet égard, c'est celui obtenu dans le dichlorométhane pour lequel on observe deux relaxations très séparées. Le rapport des deux fréquences de relaxation ne dépend pas de la fréquence de collisions; or les valeurs calculées et mesurées de ce rapport sont sensiblement différentes [4].

Il est probable que pour un processus de relaxation caractérisé par un seul temps ou deux temps très différents, l'erreur sur les probabilités de transition ne conduise qu'à un simple décalage des fréquences de relaxation. Par contre dans un cas comme celui du chloroforme, il est possible que ces valeurs différentes des probabilités de transition affectent le processus lui-même, la description en deux zones n'étant plus suffisante.

L'ensemble des restrictions mentionnées ci-dessus permet de comprendre que pour les liquides, il n'est guère possible de déduire des expériences ultrasonores ou de diffusion Brillouin des fréquences de collisions, ni de trancher entre le modèle de Litovitz et celui de Davis et Oppenheim. La seule comparaison possible porte sur les variations thermiques des fréquences de relaxation calculées et expérimentales qui se sont avérées moins sensibles à la forme du potentiel intermoléculaire [5].

Nous avons reporté sur le tableau IX les valeurs calculées et mesurées des fréquences de relaxation pour le tétrachlorure de carbone. Le rapport de ces valeurs demeure

TABLEAU IX
Tétrachlorure de carbone

$T(°C)$	$10^{-9} f_r$ calc (Hz)	$10^{-9} f_r$ exp (Hz)	$\dfrac{f_r \text{ exp}}{f_r \text{ calc}}$
0	0,72	2,06	2,8
20	0,78	2,2	2,9
40	0,84	2,5	2,9
57,5	0,90	2,7	3

constant en fonction de la température. Une conclusion identique ayant été préalablement obtenue pour le dichlorométhane [7], il semble possible d'interpréter les résultats expérimentaux à partir du modèle de Litovitz.

Il faut cependant remarquer que l'on fait une hypothèse implicite, à savoir que les fréquences de collisions relatives à la translation des molécules varient avec la température comme les fréquences de collision déterminées expérimentalement à partir de la fonction de corrélation pour les orientations des molécules. Des expériences sur des liquides plus simples pour lesquels on pourrait adopter comme valeurs de fréquences de collisions celles qu'on obtient par simulation sur ordinateur, paraissent souhaitables.

En ce qui concerne les hypothèses fondamentales de la théorie de SSH, il semble, et c'est la conclusion la plus importante de ce travail, qu'elles permettent d'interpréter généralement les processus de relaxation de vibration des liquides, comme le montrent les résultats obtenus pour le tétrachlorure de carbone.

Remerciements

Les auteurs tiennent à remercier M. Caloin pour l'aide apportée lors de la réalisation de ce travail ainsi que pour de fructueuses discussions.

Bibliographie

1. Heasel, E. et Lamb, J.: *Proc. Roy. Soc.* **B69**, 869 (1956).
2. Litovitz, T. A.: *Proc. Int. School of Physics, Enrico Fermi, Course* **XXVII**, Acad. Press, N.Y., 1963.
3. Slawsky, Z., Schwartz, R. N., et Herzfeld, K. F.: *J. Chem. Phys.* **20**, 1591 (1952).
4. Tanczos, F. I.: *J. Chem. Phys.* **25**, 439 (1956).
5. Herzfeld, K. F. et Litovitz, T. A.: *Absorption and Dispersion of Ultrasonic Waves*, Acad. Press, N.Y., 1959.
6. Davis, P. K. et Oppenheim, I.: *J. Chem. Phys.* 1972 **57**, 505 (1972).
   Davis, P. K.: *J. Chem. Phys.* **57**, 517 (1972).
7. Caloin, M. et Candau, S.: *J. Phys.* **33C** 1-7 (1972).
8. Bartoli, F. J. et Litovitz, T. A.: *J. Chem. Phys.* **56**, 314 (1972).
9. Mountain, R. D.: *J. Res. Nat. Bur. Std* **72A**, 95 (1968).
10. Landau, L. et Lifshitz, E. M.: *Fluid Mechanics*, Pergamon Press, Inc. N.Y.
11. Montrose, C. J., Solovyev, V. A., et Litovitz, T. A.: *J. Acad. Soc. Am.* **43**, 117 (1968).
12. Nichols, W. H. et Carome, E. F.: *J. Chem. Phys.* **49**, 1000 (1968).

13. Andreae, J. H.: *Proc. Phys. Soc. (London)* **B70**, 71 (1957).
    Andreae, J. H., Joyce, P. L., et Oliver, R. J.: *Proc. Phys. Soc. (London)* **75**, 82 (1960).
14. Une procédure analogue est détaillée dans la référence: Stegeman, G. J. A., Gornall, W. S., Volterra, V., et Stoicheff, B. P.: *J. Acad. Soc. Am.* **49**, 979 (1971).
15. Voir par exemple: Sette, D.: dans Temperley, Rawlinson, and Rushbroke (eds.), *Physics of Simple Liquids*, North Holland Publishing Company, Amsterdam, 1968.
16. Carome, E. F., Nichols, W. H., Kunsitis-Swyt, R. C., et Singal, S. P.: *J. Chem. Phys.* **49**, 1013 (1968).
17. Plass, K. G.: *Ber. Bunsenges. physik. Chem.* **4**, 74 (1970).
18. Berdyev, A. A., Lezhnev, N. B., Lapkin, V. V., et Shubina, M. G.: *Proc. Sixth Int. Congr. on Acoust.*, Tokyo, 1968, vol. V, J-5-3.
    Berdyev, A. A. et Lezhnev, N. B.: *Proc. Seventh Int. Congr. on Acoust.*, Budapest, 1971.
19. Barocchi, F. et Vallauri, R.: *J. Chem. Phys.* **51**, 10 (1969).
20. Goldblatt, N. R. et Litovitz, T. A.: *J. Acad. Soc. Am.* **41**, 1301 (1967).

DISCUSSION

Rouch: Que représente α/f^2? Avez-vous tenu compte de la viscosité de volume? Avez-vous fait varier l'angle de diffusion?

Munch: $(\alpha/f^2)_0$ représente les pertes d'origine visqueuse. Il inclut la viscosité de cisaillement mesurée et la viscosité de volume résiduelle comme paramètre pouvant varier entre 0 et $(\frac{4}{3}) \eta_s$.

Dans ces expériences l'angle de diffusion était constant.

MOLECULAR MOTION IN LIQUIDS FROM LOW FREQUENCY DEPOLARIZED LIGHT SCATTERING*

NAHUM D. GERSHON and IRWIN OPPENHEIM

Chemistry Dept., Ben-Gurion University of the Negev, Beer-Sheva, Israel

and

Chemistry Dept., Massachusetts Institute of Technology, Cambridge, Mass. 02139, U.S.A.

Abstract. In this work we present theories for low frequency depolarized light scattering from liquids. Expressions for the various line shapes are given which depend on the geometry of the experiments and on temperature. We show how details of the rotational, translational and vibrational molecular motion can be extracted from these experiments.

Résumé. Dans ce travail nous présentons des théories sur la diffusion dépolarisée de la lumière, à basse fréquence, par les liquides. Nous donnons des expressions pour les diverses formes de raies spectrales, dépendant de la géometrie expérimentale et de la température. On peut voir comment il est possible d'extraire de ces expériences des détails sur les mouvements moléculaires de rotation, de translation, et de vibration.

1. Introduction

Molecular motion in liquids is of vast interest. Many techniques are used to follow this motion. Among them are dielectric relaxation, NMR, spectroscopy, light scattering and hydrodynamical methods. These techniques measure the long time behaviour (not very long) of correlation functions of variables which depend upon the orientational, translational and vibrational motion of the molecules. For example, with dielectric relaxation, NMR and conventional depolarized light scattering one can obtain values for the characteristic times of the molecular reorientation in liquids. On the other hand, ultrasonic and Brillouin measurements yield details on the translational and sometimes on the vibrational motion of the molecules in the liquid state. The development of lasers and sensitive detection devices has made it possible to measure accurately the depolarized light scattering from liquids [1, 2, 3]. This means that now we can observe the fine structure of this spectrum.

In Section 2 we present a description of the depolarized spectra at low frequencies. In the 3rd section various forms of the spectrum in the VH geometry are given and in the 4th, an expression for the HH spectrum is presented. In the 5th section we consider the effects of the vibrational motion of the molecules on the depolarized spectra.

2. The Problem

The depolarized light scattering spectrum depends upon combinations of frequency dependent autocorrelation functions of various components of the polarizability

---

* A portion of this work was supported by the National Science Foundation.

density [4] which is defined by

$$\alpha'_k(t) = \sum_{j=1}^{N} \alpha'_j(t) \exp(i\mathbf{k}\cdot\mathbf{r}_j(t)), \tag{2.1}$$

where $\alpha'_j(t)$ is the polarizability tensor of the molecule j, and $\mathbf{r}_j(t)$ is its centre of mass position at time t; $\mathbf{k}$ is the wave vector. It is assumed that the $\alpha'_j(t)$ is an average quantity which depends only upon the orientation of molecule j in space. Effects of induced polarizability are on a much shorter time scale than the times considered here.

The active components of the polarizability density are selected by the geometry of the experiment. Two possible geometries are the VH and the HH. In the VH geometry the polarization of the incident wave is perpendicular to the scattering plane and that of the outgoing wave is in the scattering plane. The expression for the spectrum is given by

$$I_{VH} \propto \sin^2 \frac{\theta}{2} \operatorname{Re} \langle \alpha_{k\omega,12} \alpha_{-k,12} \rangle + \cos^2 \frac{\theta}{2} \operatorname{Re} \langle \alpha_{k\omega,31} \alpha_{-k,13} \rangle. \tag{2.2}$$

In the HH geometry both of the polarization vectors are in the scattering plane and the spectrum is given by

$$I_{HH} \propto (\cos^2 \tfrac{1}{2}\theta + \tfrac{1}{2}\sin^2 \tfrac{1}{2}\theta)^2 \operatorname{Re}\langle \alpha_{k\omega,33}\alpha_{-k,33}\rangle + \sin^4 \tfrac{1}{2}\theta \operatorname{Re}\langle \bar{\alpha}_{k\omega}\bar{\alpha}_{-k}\rangle + \cos^2 \theta \operatorname{Re}\langle a_{k\omega}a_{-k}\rangle, \tag{2.3}$$

where θ is the scattering angle; α_k is the anisotropic part of the polarizability density tensor; 1, 2, 3 are the x, y, z laboratory axes where z is chosen along $\mathbf{k}$, the wave vector of the particular experiment; $\bar{\alpha}_k \equiv \tfrac{1}{2}(\alpha'_{k,11} - \alpha'_{k,22})$; $a'_k = \tfrac{1}{3}\operatorname{Tr}\alpha'_k$; and ω is the difference between the scattered frequency and the incident frequency.

If the molecules are rigid (which is usually a good approximation at low frequencies – we shall see exceptions later) and if the reorientation process is much faster than the hydrodynamic ones, α_k relaxes mostly via the orientational motion of the molecules. Orientational relaxation times of ordinary molecular liquids at ordinary temperatures are in the region of 10^{-10}–10^{-12} s. If the only slow dynamical variables in the system are the components of α_k we would obtain Lorentzian line shapes for the depolarized spectra. This is in accord with the results of conventional light scattering measurements.

The use of modern techniques has revealed that in the VH spectrum far from the freezing point there is a dip in the centre while in the HH spectrum there are fine structures at the Brillouin frequencies [1, 2, 3]. This fine structure is due to the coupling with the slowly varying hydrodynamic variables. These are [5] the number density, n_k, the linear momentum density, $\mathbf{g}_k$ and the energy density, e_k. In addition, in systems of nonspherical molecules the internal angular momentum density, $\mathbf{s}_k$, might also be slowly varying. There has been some confusion in the past as to which are the appropriate slow variables of these systems. Some investigators [6, 7, 8, 9] considered the components of the stress tensor, τ_k, as slow variables and claimed a proportionality between the dielectric tensor fluctuations and τ_k. In liquids at temperatures well above the freezing point, the stress tensor varies quickly compared with hydrodynamical times. Therefore, the approaches of Ben-Reuven and Gershon [10], of Keyes and Kivelson [11] and of Andersen and Pecora [12] in which the hydrodynamical variables together with α_k or with the dielectric constat fluctuation tensor are taken as the slow variables are preferred. Keyes and Kivelson took the hydrodynamical variables together with the polarizability density as the slow variables. Their description is essentially correct except for the fact that their coupling parameter depends upon the static correlation between the stress tensor and the polarizability density which vanishes. Also the theory of Ben-Reuven and Gershon failed to give precise results for the fine structure. On the other hand, Andersen and Pecora obtained general expressions for the VH spectrum using a different technique. Their two variable theories can be reduced to the results of Section 3 is their tensor B is identified with the polarizability density and if both a calculation of the modes and a partial fraction analysis are done.

The theory that we present here uses the generalized hydrodynamical method of Selwyn and Oppenheim [13] extended to include non-conserved variables [14, 15]. If A_k is a vector containing all the system slow variables then we obtain [15, 16] successive approximations to the hydrodynamical matrix $M(k, \omega)$ which is defined by [13]:

$$\langle A_{k\omega} A^T_{-k} \rangle \equiv \frac{1}{-i\omega I + M(k, \omega)} \langle A_k A^T_{-k} \rangle, \qquad (2.4)$$

where I is the unit matrix and T denotes a transpose operation. When A_k includes all the slow dynamical variables of the system, $M(k, \omega)$ is independent of ω at low frequencies, and this limit is denoted by M. The expressions for M are given in References 15 and 16 in terms of correlation functions of molecular quantities. Given M it is possible to calculate from (2.4) the various correlation functions of $\langle \alpha_{k\omega} \alpha_{-k} \rangle$. This is done in the next sections for the depolarized spectra.

3. The Vertical Horizontal Geometry

It is evident from Equation (2.2) that the active components in the VH geometry are $\alpha_{k,12}$ and $\alpha_{k,31}$. Thus, in order to obtain the VH line shapes it is necessary to calculate the two autocorrelation functions: $\langle \alpha_{k\omega,12} \alpha_{-k,12} \rangle$ and $\langle \alpha_{k\omega,31} \alpha_{-k,13} \rangle$. This is done by using Equation (2.4) a procedure which is simplified considerably by the fact [15, 16] that M can be broken into blocks which are not connected to each other. The calculation of References 15 and 16 shows that $\alpha_{k,12}$ is not coupled to any other variables of the set, i.e. its autocorrelation function at low frequencies is a Lorentzian. On the other hand $\alpha_{k,31}$ is coupled to $g_{k,1}$ and to $S_{k,2}$. Therefore, this autocorrelation function can be formally written as

$$\langle \alpha_{k\omega,31} \alpha_{-k,13} \rangle = \frac{-(\omega - iR'_2)(\omega - iR'_3)}{i(\omega - iR_1)(\omega - iR_2)(\omega - iR_3)} \langle \alpha_{k,31} \alpha_{-k,13} \rangle, \qquad (3.1)$$

where R'_2 and R'_3 are the roots of the numerator while R_1, R_2, R_3 are the roots of the denominator and are the modes of this particular subsystem. The roots are calculated from the expression of the $\alpha_{31}\alpha_{31}$ element of $(-i\omega I + M)^{-1}$. If terms including time derivatives of $S_{k,2}(t)$ are large compared to the others, then it can be shown [17] that R'_3 is equal to R_3 up to terms of $O(k^2)$ and that

$$R'_2 = -\frac{k^2\eta}{mn} - M^{(1)}_{g_1\alpha_{31}} M^{(1)}_{\alpha_{31}g_1}/M^{(0)}_{\alpha_{31}\alpha_{31}}, \qquad (3.2)$$

where η is the shear viscosity, mn is the mass density and $M^{(l)}_{\alpha\beta}$ is the $\alpha\beta$ element of M which is of $O(k^l)$. The denominator is now a quadratic function of ω and its roots can be calculated exactly. They are:

$$R_{1,2} = -\tfrac{1}{2}\left(\frac{k^2\eta}{mn} + x + M_{\alpha_{31}\alpha_{31}}\right) \pm \tfrac{1}{2}\Bigg[\left(\frac{k^2\eta}{mn} + x - M_{\alpha_{31}\alpha_{31}}\right)^2 + $$
$$+ 4M^{(1)}_{g_1\alpha_{31}} M^{(1)}_{\alpha_{31}g_1}\Bigg]^{1/2}, \qquad (3.3)$$

where

$$x = M^{(1)}_{\alpha_{31}g_1} M^{(1)}_{g_1\alpha_{31}}/M^{(0)}_{\alpha_{31}\alpha_{31}} < 0 \qquad (3.4)$$

$$M_{\alpha_{31}\alpha_{31}} = M^{(0)}_{\alpha_{31}\alpha_{31}} + M^{(2)}_{\alpha_{31}\alpha_{31}}. \qquad (3.5)$$

The low frequency dependence of $\langle \alpha_{k\omega,31}\alpha_{-k,13}\rangle$ and consequently the line shape of the VH spectrum depends upon whether the discriminant

$$D = \left[\left(\frac{k^2\eta}{mn} + M^{(1)}_{g_1\alpha_{31}} M^{(1)}_{\alpha_{31}g_1}/M^{(0)}_{\alpha_{31}\alpha_{31}} - M_{\alpha_{31}\alpha_{31}}\right)^2 + 4M^{(1)}_{g_1\alpha_{31}} M^{(1)}_{\alpha_{31}g_1}\right] \qquad (3.6)$$

is positive, zero or negative.

3.1. The Discriminant is Positive

In this regime all the modes are dissipative and the VH spectrum has the following form:

$$I_{\mathrm{VH}} \propto \sin^2\frac{\theta}{2} \frac{M_{\alpha_{12}\alpha_{12}}}{\omega^2 + (M_{\alpha_{12}\alpha_{12}})^2} \langle \alpha_{k,12}\alpha_{-k,12}\rangle + \cos^2\frac{\theta}{2} \frac{1}{(R_1 - R_2)} \times$$
$$\times \left[(R_1 - R'_2)\frac{-R_1}{\omega^2 + R_1^2} - (R_2 - R'_2)\frac{-R_2}{\omega^2 + R_2^2}\right]\langle \alpha_{k,31}\alpha_{-k,31}\rangle, \qquad (3.7)$$

where $M_{\alpha_{12}\alpha_{12}}$ is $M^{(0)}_{\alpha_{12}\alpha_{12}} + M^{(2)}_{\alpha_{12}\alpha_{12}}$.

It is possible to distinguish between two regions:

(1) If $M_{\alpha_{31}\alpha_{31}} > (k^2\eta/mn) + x$, the spectrum consists of two positive Lorentzians with widths $M_{\alpha_{12}\alpha_{12}}$ and $-R_2$, and one negative narrow Lorentzian with width $-R_1$. In quinoline for example this is the region of room temperature which is far from the freezing point [3].

(2) If $(k^2\eta/mn) + x > M_{\alpha_{31}\alpha_{31}}$, then the spectrum has two positive Lorentzians with widths $M_{\alpha_{12}\alpha_{12}}$ and $-R_1$ and a broad negative Lorentzian with a width of $-R_2$.

3.2. THE DISCRIMINANT IS ZERO

When $D=0$, R_1 is equal to R_2 and the spectral line shape is:

$$I_{\text{VH}} \propto \sin^2\frac{\theta}{2} \frac{M_{\alpha_{12}\alpha_{12}}}{\omega^2 + (M_{\alpha_{12}\alpha_{12}})^2} \langle \alpha_{k,12}\alpha_{-k,12} \rangle +$$
$$+ \cos^2\frac{\theta}{2} \frac{\omega^2[R_2' - 2R_1] + R_2'R_1^2}{(\omega^2 + R_1^2)^2} \langle \alpha_{k,31}\alpha_{-k,13} \rangle. \tag{3.8}$$

The 1st term is Lorentzian while the second one is non-Lorentzian.

3.3. THE DISCRIMINANT IS NEGATIVE

In this regime R_1 and R_2 are complex. This means that in contrast to the previous cases the modes of $\langle \alpha_{k\omega,31}\alpha_{-k,13} \rangle$ are propagative rather than only dissipative. The spectral line shape is:

$$I_{\text{VH}} \propto \sin^2\tfrac{1}{2}\theta \frac{M_{\alpha_{12}\alpha_{12}}}{\omega^2 + (M_{\alpha_{12}\alpha_{12}})^2} \langle \alpha_{k,12}\alpha_{-k,12} \rangle +$$
$$+ \cos^2\frac{\theta}{2}\frac{1}{2\Omega}\left\{(-\Gamma - R_2')\frac{\omega+\Omega}{(\omega+\Omega)^2 + \Gamma^2} + \Omega\frac{\Gamma}{(\omega+\Omega)^2+\Gamma^2} + \right.$$
$$\left. + (\Gamma + R_2')\frac{\omega-\Omega}{(\omega-\Omega)^2 + \Gamma^2} + \Omega\frac{\Gamma}{(\omega-\Omega)^2+\Gamma^2}\right\} \langle \alpha_{k,31}\alpha_{-k,13} \rangle,$$

where $\tag{3.9}$

$$\Gamma = \tfrac{1}{2}\left(\frac{k^2\eta}{mn} + x + M_{\alpha_{31}\alpha_{31}}\right) \tag{3.10}$$

$$\Omega = \tfrac{1}{2}\left[-\left(\frac{k^2\eta}{mn} + x - M_{\alpha_{31}\alpha_{31}}\right)^2 - 4M^{(1)}_{g_1\alpha_{31}}M^{(1)}_{\alpha_{31}g_1}\right]^{1/2}. \tag{3.11}$$

This line shape is composed of a Lorentzian centered at 0, one Lorentzian and a dispersion curve centered at $+\Omega$ and another Lorentzian and dispersion curve centered at $-\Omega$.

When $M_{\alpha_{31}\alpha_{31}} > (k^2\eta/mn) + x$ the dispersion curves are negative in the region $\omega = -\Omega$ till $\omega = +\Omega$ while when $M_{\alpha_{31}\alpha_{31}} < (k^2\eta/mn) + x$ they are positive in this interval. At the point where $M_{\alpha_{31}\alpha_{31}} = (k^2\eta/mn) + x$ the dispersion curves disappear.

In this regime the total line shape would have the appearance of a triplet. This case may have been seen by Fabelinskii et al. [18] in salol and benzophenone at low temperatures.

4. The Horizontal-Horizontal Geometry

In this geometry the active components are $\alpha_{k,33}$, $\bar{\alpha}_k$ and a_k, as is clear from Equation (2.3). $\bar{\alpha}_k$ is not coupled to any of the other slow variables and contributes a Lorentzian to the spectrum. $\langle a_{k\omega}a_{-k}\rangle$ gives rise to the Rayleigh-Brillouin lines whose contribution disappears at scattering angles of 90 and 270°. On the other hand, $\alpha_{k,33}$ is coupled to n_k, $g_{k,3}$ and e_k as is seen from the structure of M calculated before [15, 16]. Using a similar procedure to that of the previous section, the form of the HH spectrum is obtained [16]. Under the condition that terms of $O(k^0)$ are much greater than terms of $O(k^2)$ the form is:

$$I_{HH} \propto (\cos^2 \tfrac{1}{2}\theta + \tfrac{1}{2}\sin^2 \tfrac{1}{2}\theta)^2 \left[\frac{M}{\omega^2 + M^2} + \tfrac{1}{2} \frac{MZ}{M^2 + R^2} \times \right.$$

$$\times \left(\frac{S}{(\omega - R)^2 + S^2} + \frac{S}{(\omega + R)^2 + S^2} \right) - \tfrac{1}{2} \frac{RZ}{M^2 + R^2} \times$$

$$\left. \times \left(\frac{\omega - R}{(\omega - R)^2 + S^2} - \frac{\omega + R}{(\omega + R)^2 + S^2} \right) \right] \langle \alpha_{k,33} \alpha_{-k,33} \rangle +$$

$$+ \sin^4 \tfrac{1}{2}\theta \frac{M_{\bar{\alpha}\bar{\alpha}}}{\omega^2 + (M_{\bar{\alpha}\bar{\alpha}})^2} \langle \bar{\alpha}_k \bar{\alpha}_{-k} \rangle + \cos^2 \theta \, \text{Re} \langle a_{k\omega} a_{-k} \rangle, \quad (4.1)$$

where

$$M = M^{(0)}_{\alpha_{33}\alpha_{33}} + M^{(2)}_{\alpha_{33}\alpha_{33}} + Z \quad (4.2)$$

$$Z = M^{(1)}_{\alpha_{33}g_3} M^{(1)}_{g_3\alpha_{33}} / M^{(0)}_{\alpha_{33}\alpha_{33}} \quad (4.3)$$

$$R = (-M^{(1)}_{ng_3} M^{(1)}_{g_3 n} - M^{(1)}_{eg_3} M^{(1)}_{g_3 e})^{1/2} \quad (4.4)$$

$$2S = k^2(\tfrac{4}{3}\eta + \eta_v) - (M^{(1)}_{ng_3} M^{(1)}_{g_3 n} M^{(2)}_{ee} - M^{(1)}_{ng_3} M^{(1)}_{g_3 e} M^{(2)}_{en}) \times$$
$$\times (M^{(1)}_{ng_3} M^{(1)}_{g_3 n} + M^{(1)}_{g_3 e} M^{(1)}_{eg_3})^{-1} + M^{(2)}_{ee}. \quad (4.5)$$

R and S are the sound wave frequency and width, respectively, and η_v is the bulk viscosity. In the vertical-vertical geometry far away from the freezing point, the Brillouin lines can be much more intense than the fine structures of $\langle \alpha_{k\omega,33} \alpha_{-k,33} \rangle$ which also appear in the polarized spectrum [19]. Thus in interpreting the polarized spectrum in the present regime one can obtain R, S and the widths of the broad lines which will appear as one Lorentzian. This can be compared with the results from the HH spectrum according to Equation (4.1).

5. The Effects of the Vibrational Degrees of Freedom on the Orientational Motion

The treatment so far has neglected the fact that the molecules have vibrational degrees of freedom. The vibrational motion is usually faster by orders of magnitude than the orientational and mean translational motion of the molecules so that in many cases the vibrations can be neglected. However, there can be a situation where the vibrational degrees of freedom exchange energy quickly but the exchange of the vibrational energy with the other degrees of freedom is as fast as or slower than the translational and rotational motions. In such cases the vibrational energy is an additional slow variable and must be taken into account when calculating the depolarized spectra.

As was done by Weinberg and Oppenheim [14], the vibrational energy $\tilde{U}_k$ is chosen to be orthogonal to the other variables. This choice simplifies M considerably.

It is clear from the structure of M [20z that $\alpha_{k,21}$ $\alpha_{k,31}$, $g_{k,1}$ and $S_{k,2}$ are not coupled to $\tilde{U}_k$ and therefore the VH spectrum is the same as in section 3. On the other hand, $\alpha_{k,33}$, n_k, $g_{k,3}$ and e_k are coupled to $\tilde{U}_k$ as was described before [14, 20]. We restrict the system to cases where the polarizability density varies much faster than the conserved variables, while the U_k can vary much faster, faster or as fast as the conserved variables. The order of magnitude of the non conserved part of the time derivative of $\tilde{U}_k$ is denoted by λ which depends upon the strength of the coupling between $\tilde{U}_k$ and the other degrees of freedom. Calculations show [20] that when $\lambda \sim 1 \gg \sigma k$ (where σk is the magnitude of the time derivatives of the conserved variables) and when $\lambda \sim \sigma k$ the form of the HH spectrum at 90° remains as in Equation (4.1). In the second case the shift and the width of the fine structure have the same k and λ dependence as in the Rayleigh-Brillouin spectrum [14]. In the intermediate case, $\lambda^2 \sim \sigma k$, the HH 90° spectrum contains only two Lorentzians without fine structures [20]. The widths of the two Lorentzians are similar. In all of the above mentioned results the roots are computed up to $O(k^2)$ or equivalent orders in λ and k.

6. Discussion

In the previous sections expressions for the depolarized spectra in the VH and the HH geometries were presented. The quantities which determine the spectra are orientational widths ($M_{\alpha_{31}\alpha_{31}}$, $M_{\alpha_{12}\alpha_{12}}$, $M_{\bar{\alpha}\bar{\alpha}}$, $M_{\alpha_{33}\alpha_{33}}$), coupling parameters ($M^{(1)}_{g_1\alpha_{31}}$ $M^{(1)}_{\alpha_{31}g_1}$, $M^{(1)}_{g_3\alpha_{33}}$, $M^{(1)}_{\alpha_{33}g_3}$), η, and the sound frequency R and width S (Equation (4.1)). These parameters can be calculated from the experimental results. The orientational widths yield the orientational relaxation times of second rank tensors in liquids. These values can be compared with the results of dielectric relaxation and spin resonance measurements. The differences between the values obtained from the depolarized spectra and dielectric relaxation might be due in part to the fact that they are relaxation times of tensors of different rank [21], and to uncertainty in the values deduced from dielectric measurements. The differences between values obtained for orientational characteristic times deduced from depolarized light scattering and spin relaxation might arise, in part, from pair correlations among the molecules in the liquids [4].

The coupling constants $M^{(1)}_{g_1\alpha_{31}}$ $M^{(1)}_{\alpha_{31}g_1}$ and $M^{(1)}_{g_3\alpha_{33}}$ $M^{(1)}_{\alpha_{33}g_3}$ give information on the strength and temperature dependence of the coupling between the orientational motion and the translational motion.

To sum up, the combination of theory and experiments on low frequency depolarized light scattering is a powerful technique for the study of orientational and translational motions in liquids.

References

1. Starunov, V. S., Tiganov, E. V., and Fabelinskii, I. L.: *Soviet Phys. JETP* **5**, 317 (1967).
2. Stegeman, G. I. A. and Stoicheff, B. P.: *Phys. Rev. Letters* **21**, 202 (1968);
   Stegeman, G. I. A.: Ph.D. Thesis, University of Toronto, 1969, unpublished;
   Stegeman, G. I. A. and Stoicheff, B. P.: *Phys. Rev.* **A7**, 1160 (1973).

3. Chabrat, J. P., Latemendia, L., Rouch, J., and Vaucamps, C.: unpublished results.
4. Ben-Reuven, A. and Gershon, N. D.: *J. Chem. Phys.* **51**, 893 (1969).
5. The definitions are (15).

$$n_k(t) = \sum_j \exp(i\mathbf{k}\cdot\mathbf{r}_j(t))$$

$$\mathbf{g}_k(t) = \sum_j \mathbf{p}_j(t) \exp(i\mathbf{k}\cdot\mathbf{r}_j(t))$$

$$e_k(t) = \sum_j \left\{\frac{\mathbf{p}_j^2}{2m} + \frac{\mathbf{s}_j^2}{2i} + \tfrac{1}{2}\sum_{l\neq j}[U(r_{lj}) + U'(r_{lj})Q(\zeta_{lj})]\right\} \exp(i\mathbf{k}\cdot\mathbf{r}_j(t))$$

$$\mathbf{S}_k(t) = \sum_j \mathbf{s}_j(t) \exp(i\mathbf{k}\cdot\mathbf{r}_j(t))$$

$\mathbf{s}_j$ is the angular momentum of molecule j around its center of mass.
$U(r_{lj})$ and $U(r_{lj})Q(\zeta_{lj})$ are the radial and the angular parts of the intermolecular potential, respectively, between molecules j and l.

6. Leontovitch, M. A.: *Bull. Acad. Sci. USSR, Phys. Ser.* **4**, 499 (1941).
7. Rytov, S. M.: *Soviet Phys. JETP* **6**, 130 (1953); **6**, 401 (1958); **6**, 513 (1959).
8. Volterra, V.: *Phys. Rev.* **180**, 156 (1969).
9. Chung, C. H. and Yip, S.: *Phys. Rev.* **A4**, 928 (1971).
10. Ben-Reuven, A. and Gershon, N. D.: *J. Chem. Phys.* **54**, 1049 (1971).
11. Keyes, and Kivelson, D.: *J. Chem. Phys.* **54**, 1786 (1971); **56**, 1876 (1972).
12. Andersen, H. C. and Pecora, R.: *J. Chem. Phys.* **54**, 2584 (1971).
13. Selwyn, P. A. and Oppenheim, I.: *Physica* **54**, 161 (1971).
14. Weinberg, M. and Oppenheim, I.: *Physica* **61**, 1 (1972).
15. Gershon, N. D. and Oppenheim, I.: *Physica* **62**, 198 (1972).
16. Gershon, N. D. and Oppenheim, I.: *Physica* **64**, 247 (1973).
17. Gershon, N. D. and Oppenheim, I.: *J. Chem. Phys.* **59**, 1337 (1973).
18. Sabviov, L. M., Starunov, V. S., Fabelinskii, I. L.: *Soviet Phys. JETP* **33**, 83 (1971).
19. Gershon, N. D. and Oppenheim, I.: *Physica* **65**, 625 (1973).
20. Gershon, N. D. and Oppenheim, I.: 'Coupling of Rotational, Vibrational and Translational Molecular Motions in Fluids and Low Frequency Depolarized Light Scattering', *Physica*, in press.
21. Ivanov, E. N.: *Soviet Phys. JETP* **18**, 1041 (1964).

DISCUSSION

Bratos: Your theory involves a number of variables such as number density, angular momentum density, etc. Is it possible for a experimental spectrum to assign a well determined variable to observed fine structure?

Gershon: The problem which are the slow dynamical variables of the system is fundamental in the theory. One has to consider the physics of the system and see what are the possible slow processes. Then the examination of the results compared with the experimental data should be done and the success of the particular choice should be considered. The symmetry of the variables allows certain couplings to occur and thus plays an important role in the determination of the line shapes.

Ailawadi: (1) Can you give us an argument as to why angular momentum variable decays faster? As far as I know, experiments like NMR or Raman scattering spectroscopy give information about single molecular motion only. To extract information about collective variables you talk about in your paper could be misleading.

(2) Since some of the experiments you reported here were done in supercooled state, I should think the collective angular momentum density, at least in this case, is not necessarily a fast decaying variable.

Gershon: (1) The angular momentum decays in normal liquids much faster than the reorientation. In liquids like SF_6 or CH_4 when the reorientation is almost free the angular momentum decays slower than the orientation as has been presented by Prof. McLung *et al.* in this conference. The values of reorientational correlation times of normal liquids deduced from depolarized light scattering are close to those from dielectric relaxation which also *include dynamic pair correlations*. These times are also similar to values obtained from NMR and Raman techniques which follow after the motion of single molecules. The similarity shows that the dynamical pair correlation do not contribute significantly to these correlation times. This similarity between single molecule and collective correlation times might be the case also for angular momentum relaxation times.

(2) Physically, when a liquid is cooled the molecular motion is slowed down. Thus its orientational fluctuations decay slower at low temperatures than at high temperatures. On the other hand, any angular momentum decays to zero faster at low temperatures than at high temperatures as its existence involves actual motion of the molecules which is slowed down. The same consideration applies to the collective motion.

Friedman: Does the shear wave which you mentioned correspond to the shear wave in Rytov's theory? That is, is it found for the same systems and does it lead to the same spectral features?

Gershon: The basic assumption of Rytov's theory is that fluctuations in the dielectric tensor are proportional to the stress tensor. In the liquids studied in these experiments this tensor is not a slow variable. The shear waves which we report are different from Rytov's, and his theory does not lead to the same spectral features as our work. If one neglects the physics of the systems, Rystov's theory fits the experimental contours of Stegeman and Stoicheff far from the freezing point. At low temperatures there are difficulties with the comparison of experimental data with Rytov's theory.

Jackson: Did you say that the fine structure would disappear if one approached or passed through a dispersion region? In liquids such as benzene one observes no evidence of fine structure. However it is possible to make Brillouin measurements below the relaxation frequency by performing the experiment with a small scattering angle and hence effectively at low frequency. Is it possible that the fine structure might be observed in a liquid such as benzene by performing a similar small scattering angle experiment?

Gershon: Each dispersion region has its particular line shape. In the VH geometry the magnitudes of M, X, $k^2\eta/mn$ determine in what region a system is in. The scattering angle, θ, influences the magnitude of k (through $\sin\frac{1}{2}\theta$) and on the other hand also the intensity of $\langle \alpha_{k\omega, 31}\alpha_{-k,13}\rangle$ term (through $\cos^2\theta/2$). It is possible that at low θ one can observe fine structure in systems which do not have it at $\theta = 90°$. However one should keep in mind that there is the experimental difficulty that, because of the noise, it is possible to separate two Lorentzians only if their widths are clearly different.

Rouch: Avez-vous entrepris le calcul du spectre HH et quel en est le résultat?

Gershon: Expressions for the HH geometry were derived. The theory predicts the fine structures observed at the Brillouin frequencies.

APPORT DES PROCESSUS DE REDISTRIBUTION ET DE REORIENTATION MOLECULAIRE DANS LES EFFETS ELECTRO-OPTIQUES NON LINEAIRES

B. KASPROWICZ-KIELICH et S. KIELICH

Institut de Physique de l'Université A. Mickiewicz, 60780 Poznań, Grunwaldzka 6, Pologne

et

J. R. LALANNE

Université de Bordeaux I, et Centre de Recherches Paul Pascal, 33 – Talence, France

Résumé. Nous présentons une théorie rendant compte de la relaxation moléculaire intervenant dans les fluides isotropes soumis aux effets électro-optiques non linéaires. Elle permet de recueillir des informations sur trois temps de relaxation, notés $\tau_1 : \tau_2 : \tau_3$ et tels que $\tau_1 = 3\tau_2 = 6\tau_3$. L'étude des effets non linéaires dans les milieux denses fournit des renseignements sur les mouvements moléculaires (translation et orientation) intervenant dans les processus de réorientation, de translation et de translation – orientation se produisant dans les régions d'ordre à courte distance existant dans les liquides. Les temps mis par ces processus pour atteindre l'état stationnaire sont suffisamment différents les uns des autres pour permettre leur séparation par les techniques laser impulsionnelles modernes.

Abstract. A relaxation – molecular theory of non-linear electro-optical effects in isotropic media is proposed and shown to provide information on the three successively shorter and shorter relaxation times $\tau_1 = 3\tau_2 = 6\tau_3$. The study of non-linear effects in condensed media provides information on molecular translational and orientational motion related with the processes of reorientation as well as translational, translational – orientational and orientational fluctuations in regions of near ordering in liquids. The times required by these processes to attain the steady state differ sufficiently from one another to permit their separation by modern laser pulsed techniques.

1. Introduction

L'étude du comportement non linéaire des propriétés électriques et optiques des corps isotropes, soumis à l'action d'un champ électrique extérieur intense, a permis de recueillir de nombreuses données présentant un grand intérêt et concernant la structure électrique et optique des atomes et des molécules ainsi que leurs interactions dans les états condensés [1–6]. Outre des effets déjà classiques, tels l'effet Kerr [1, 2] et la saturation diélectrique [3, 6], les techniques lasers ont révélé de nombreux et nouveaux effets électro-optiques et effets optiques non linéaires [6–9].

L'effet Kerr induit par un champ électrique oscillant intense, jadis étudié par Peterlin et Stuart [5], conduit, aux fréquences optiques, à l'effet Kerr optique [10] découvert par Mayer et Gires [11]. Des études approfondies de la biréfringence optique induite, dans les liquides, par un faisceau intense laser ont été effectuées par Paillette [12] et Martin et Lalanne [13]. Cet effet, couplé aux mesures de diffusion Rayleigh dépolarisée, conduit aux valeurs de la polarisabilité non linéaire d'ordre 3 pour les molécules faiblement anisotropes [14], ainsi qu'à de nombreux renseignements concernant les corrélations angulaires moléculaires [15, 16] et leurs fluctuations statistiques de translation [17-19]. La redistribution moléculaire [17, 20], due aux fluctuations de translation, étudiée par Yvon [21] en réfraction moléculaire et par Kirkwood [22] en polarisation électrique linéaire, n'exprime que les mouvements moléculaires dans les régions d'ordre à courte distance des liquides. Les fluctuations purement translationnelles ne se produisent que dans les liquides composés d'atomes [23]. Dans ceux constitués de molécules anisotropes, un rôle important échoit en plus aux fluctuations de translation-orientation [18, 24].

Un deuxième phénomène électro-optique nouveau consiste en la génération, induite par un champ électrique statique, du deuxième harmonique d'une onde laser. Observé dans les liquides [7], les gaz constitués de molécules polaires [25], et les gaz composés de molécules non polaires [26], cet effet fut présenté par Mayer [25] comme une méthode de détermination de la polarisabilité non linéaire d'ordre 2 des molécules polaires, et par Fin et Ward [26] comme un procédé permettant d'obtenir les polarisabilités non linéaires d'ordre 3 dans les atomes. La génération du deuxième harmonique, dans les corps isotropes, est liée non seulement à la polarisabilité électronique non linéaire, mais surtout à la réorientation électrique des dipôles permanents et induits [27]. Au contraire, la génération du troisième harmonique dans les liquides [28] et les gaz [29], dépend principalement de la polarisabilité électronique non linéaire des atomes et des molécules.

Dans ce travail, nous considérons ces deux phénomènes et les mécanismes microscopiques qu'ils on en commun ou qui les différencient. Les temps nécessaires aux différents processus pour atteindre l'état stationnaire sont, en général, différents. D'autre part, le traitement théorique de leur évolution dans le temps pose de très grandes difficultés surtout lorsqu'il s'agit des milieux denses. Aussi, notre effort principal portera sur la recherche d'un traitement, en cinétique statistique des processus non linéaires et grâce à la théorie simplifiée de la relaxation moléculaire de Debye [3, 5]. Dans le cas des milieux denses, nous nous contenterons de discuter quelques exemples simples qui, toutefois, distingueront les principaux processus statistiques de fluctuations liés au mouvement moléculaire dans les régions d'ordre à courte distance.

2. Théorie

Un traitement phénoménologique montre que la polarisation électrique non linéaire d'ordre 3, induite dans le milieu au point **r** et au temps t, prend la forme suivante [7]:

$$P_i^{(3)}(\mathbf{r}, t) = \chi_{ijkl}^{(3)} E_j(\mathbf{r}, t) E_k(\mathbf{r}, t) E_l(\mathbf{r}, t). \tag{1}$$

Le tenseur de rang 4, $\chi_{ijkl}^{(3)}$, traduit la susceptibilité électrique d'ordre 3 tenant compte de la structure et de l'état thermodynamique du milieu. L'équation (1) et celles utilisées par la suite respecteront la convention de sommation bien connue.

Le champ électrique de l'onde électromagnétique s'écrit:

$$\mathbf{E}(\mathbf{r}, t) = \sum_a \mathbf{E}(\omega_a) \exp\left[i\left(\mathbf{k}_a \cdot \mathbf{r} - \omega_a t\right)\right]. \tag{2}$$

La sommation s'étend sur toutes les valeurs, positives et négatives, des fréquences

ω_a et des vecteurs d'onde $\mathbf{k}_a$. L'indice inférieur distingue le mode. Dans un corps isotrope, la théorie statistique-moléculaire permet d'écrire la polarisation (1), à la fréquence ω_P de l'onde mesure, sous la forme:

$$P_i^{(3)}(-\omega_p) = \sum_{abc} \{Q_1(-\omega_p, \omega_a, \omega_b, \omega_c)\delta_{ij}E_k^{\omega_b}E_k^{\omega_c} +$$
$$+ \tfrac{1}{2}Q_2(-\omega_p, \omega_a, \omega_b, \omega_c)(3E_i^{\omega_b}E_j^{\omega_c} - \delta_{ij}E_k^{\omega_b}E_k^{\omega_c}) +$$
$$+ \tfrac{1}{2}Q_3(-\omega_p, \omega_a, \omega_b, \omega_c)(3E_j^{\omega_b}E_i^{\omega_c} - \delta_{ij}E_k^{\omega_b}E_k^{\omega_c})\} E_j^{\omega_a} \times$$
$$\times \exp\{i[(\mathbf{k}_a + \mathbf{k}_b + \mathbf{k}_c)\cdot\mathbf{r} - (\omega_a + \omega_b + \omega_c)t]\}, \qquad (3)$$

où Q_1, Q_2 et Q_3 sont des grandeurs caractérisant les micromécanismes des non-linéarités induites dans le milieu isotrope aux fréquences ω_a, ω_b et ω_c.

Dans les milieux statistiquement inhomogènes, on rencontre outre des processus non linéaires purement électroniques, différents processus moléculaires de fluctuations [6]. Les grandeurs Q_s ($s=1, 2, 3$) deviennent alors fonctions de la température absolue du système T:

$$Q_s(T) = \sum_{n=0}^{3} Q_s^{(n)} T^{-n}. \qquad (4)$$

Une telle dépendance avec la température résulte de l'application du calcul statistique des perturbations jusqu'à l'ordre $n=3$ inclus. Nous allons maintenant procéder à une discussion de ces approximations successives.

2.1. Approximation d'ordre 0

Dans l'approximation d'ordre 0, on admet qu'aucun processus d'ordre statistique ne se produit au sein du milieu atomique ou moléculaire ou que ces processus sont tout à fait négligeables. La non-linéarité du milieu n'est alors due qu'aux phénomènes purement électroniques, et les $Q_s^{(0)}$ prennent la forme:

$$Q_1^{(0)}(-\omega_p, \omega_a, \omega_b, \omega_c) = \frac{\varrho}{9} c_{\alpha\alpha\beta\beta}(-\omega_p, \omega_a, \omega_b, \omega_c)$$

$$Q_2^{(0)}(-\omega_p, \omega_a, \omega_b, \omega_c) = \frac{\varrho}{45} \{4c_{\alpha\beta\alpha\beta}(-\omega_p, \omega_a, \omega_b, \omega_c) +$$
$$- c_{\alpha\beta\beta\alpha}(-\omega_p, \omega_a, \omega_b, \omega_c) - c_{\alpha\alpha\beta\beta}(-\omega_p, \omega_a, \omega_b, \omega_c)\}$$

$$Q_3^{(0)}(-\omega_p, \omega_a, \omega_b, \omega_c) = \frac{\varrho}{45} \{4c_{\alpha\beta\beta\alpha}(-\omega_p, \omega_a, \omega_b, \omega_c) +$$
$$- c_{\alpha\beta\alpha\beta}(-\omega_p, \omega_a, \omega_b, \omega_c) - c_{\alpha\alpha\beta\beta}(-\omega_p, \omega_a, \omega_b, \omega_c)\}, \qquad (5)$$

où le tenseur $c_{\alpha\beta\gamma\delta}(-\omega_P, \omega_a, \omega_b, \omega_c)$ définit la polarisabilité non linéaire d'ordre 3 de l'atome ou de la molécule [14] et $\varrho = N/V$ est le nombre-densité du milieu.

En particulier, si le tenseur $c_{\alpha\beta\gamma\delta}$ est totalement symétrique ce qui est généralement le cas en absence de dispersion et d'absorption électronique, les trois $Q_s^{(0)}$ de l'équation (5) se réduisent à une seule grandeur indépendante:

$$2Q_1^{(0)} = 5Q_2^{(0)} = 5Q_3^{(0)} = \tfrac{1}{9}\varrho\, c_{\alpha\alpha\beta\beta}. \qquad (5a)$$

2.2. La première approximation

La première approximation $n=1$ du calcul statistique de perturbations permet de calculer les $Q_s^{(1)}$ du développement en série (4). Admettons que les variations des champs électriques dans le temps sont suffisamment lentes pour permettre aux molécules de les suivre en s'orientant grâce à leurs moments dipolaires et à leurs ellipsoides de polarisabilité. Afin de simplifier nos calculs, nous ne considérons que des molécules axialement symétriques. La méthode des processus de relaxation moléculaires de Debye [3, 5, 6] donne alors:

$$Q_1^{(1)}(-\omega_p, \omega_a, \omega_b, \omega_c) = \frac{\varrho}{3k} \{b(-\omega_p, \omega_b, \omega_c) m(\omega_a) D_1(\omega_a) +$$
$$+ m(-\omega_p) b(\omega_a, \omega_b, \omega_c) D_1(\omega_a + \omega_b + \omega_c)\} \quad (6)$$

$$Q_2^{(1)}(-\omega_p, \omega_a, \omega_b, \omega_c) = Q_3^{(1)}(-\omega_p, \omega_a, \omega_b, \omega_c) =$$
$$= \frac{2\varrho}{45k} \{3b(-\omega_p, \omega_b, \omega_c) m(\omega_a) D_1(\omega_a) +$$
$$+ \gamma(-\omega_p, \omega_a) \gamma(\omega_b, \omega_c) D_2(\omega_b + \omega_c) +$$
$$+ 3m(-\omega_p) b(\omega_a, \omega_b, \omega_c) D_1(\omega_a + \omega_b + \omega_c)\}. \quad (7)$$

Ici, k est la constante de Boltzmann, m le moment électrique dipolaire permanent de la molécule, $\gamma = a_{33} - a_{11}$ l'anisotropie électrique de sa polarisabilité linéaire, et $b = (b_{333} + 2b_{113})/3$ sa polarisabilité non linéaire moyenne d'ordre 2.

Les équations (6) et (7) font intervenir les dénominateurs suivants traduisant la relaxation:

$$D_1(\omega_a) = (1 - i\omega_a \tau_1)^{-1}$$
$$D_2(\omega_b + \omega_c) = \{1 - i(\omega_b + \omega_c)\tau_2\}^{-1} \quad (8)$$
$$D_1(\omega_a + \omega_b + \omega_c) = \{1 - i(\omega_a + \omega_b + \omega_c)\tau_1\}^{-1},$$

où $\tau_1 = \tau_D$ est le premier temps de relàxation temps de relaxation de Debye [3], tandis que $\tau_2 = \tau_D/3$ représente le deuxième temps de relaxation moléculaire [5, 6].

Le calcul des $Q_s^{(1)}$, si on veut tenir compte simultanément de la relaxation moléculaire dans des champs extérieurs et des corrélations moléculaires statistiques, pose de très grandes difficultés et n'est possible que dans des cas particuliers [27].

2.3. La deuxième approximation

La deuxième approximation $n=2$ du calcul statistique de perturbations enduit, grâce à la théorie de la relaxation, aux $Q_s^{(2)}$ suivants:

$$Q_1^{(2)}(-\omega_p, \omega_a, \omega_b, \omega_c) = \frac{2\varrho}{27k^2} m(-\omega_p) m(\omega_a) \gamma(\omega_b, \omega_c) G_1(\omega_a + \omega_b + \omega_c) \quad (9)$$

$$Q_2^{(2)}(-\omega_p, \omega_a, \omega_b, \omega_c) = Q_3^{(2)}(-\omega_p, \omega_a, \omega_b, \omega_c) =$$
$$= \frac{\varrho}{45k^2} \{\gamma(-\omega_p, \omega_a) m(\omega_b) m(\omega_c) F(\omega_b + \omega_c) +$$
$$+ 2m(-\omega_p) m(\omega_a) \gamma(\omega_b + \omega_c) G_3(\omega_a + \omega_b + \omega_c)\}, \quad (10)$$

où nous avons employé la notation:

$$F(\omega_b + \omega_c) = \tfrac{1}{2}[D_1(\omega_b) + D_1(\omega_c)] D_2(\omega_b + \omega_c) \tag{11}$$

$$G_1(\omega_a + \omega_b + \omega_c) = \{2[2D_1(\omega_a) + D_2(\omega_b + \omega_c)] D_3(\omega_a + \omega_b + \omega_c) + \\ - 3[D_1(\omega_a) + D_2(\omega_a + \omega_b)]\} D_1(\omega_a + \omega_b + \omega_c) \tag{12}$$

$$G_3(\omega_a + \omega_b + \omega_c) = \tfrac{1}{3}[2D_1(\omega_a) + D_2(\omega_a + \omega_b)] D_3(\omega_a + \omega_b + \omega_c). \tag{13}$$

Ici, nous avons, outre les dénominateurs traduisant la relaxation (8) le dénominateur suivant:

$$D_3(\omega_a + \omega_b + \omega_c) = \{1 - i(\omega_a + \omega_b + \omega_c)\tau_3\}^{-1}, \tag{14}$$

où apparaît le troisième temps de relaxation $\tau_3 = \tau_D/6$. Les formules (9) et (10) montrent que la deuxième approximation fait intervenir un processus de réorientation simultanée des moments électriques permanents et induits. Les corrélations moléculaires et les fluctuations statistiques modifient ces processus de façon assez compliquée.

2.4. La troisième approximation

La troisième approximation statistique $n=3$ n'est liée qu'à la réorientation des moments électriques permanents.

Il vient:

$$Q_1^{(3)}(-\omega_p, \omega_a, \omega_b, \omega_c) = -\frac{2\varrho}{54k^3} m(-\omega_p) m(\omega_a) m(\omega_b) m(\omega_c) H_1(\omega_a + \omega_b + \omega_c) \tag{15}$$

$$Q_2^{(3)}(-\omega_p, \omega_a, \omega_b, \omega_c) = Q_3^{(3)}(-\omega_p, \omega_a, \omega_b, \omega_c) = \\ = \frac{\varrho}{135k^3} m(-\omega_p) m(\omega_a) m(\omega_b) m(\omega_c) H_3(\omega_a + \omega_b + \omega_c) \tag{16}$$

avec:

$$H_1(\omega_a + \omega_b + \omega_c) = F(\omega_a + \omega_b)[3 - 2 D_3(\omega_a + \omega_b + \omega_c)] D_1(\omega_a + \omega_b + \omega_c) \tag{17}$$

$$H_3(\omega_a + \omega_b + \omega_c) = F(\omega_a + \omega_b) D_3(\omega_a + \omega_b + \omega_c). \tag{18}$$

Il est en fait possible et si tous les champs électriques sont statiques ou lentement variables, de tenir compte, dans cette approximation, des corrélations moléculaires. On parvient ainsi à la saturation diélectrique, qui dépend fortement des différentes corrélations angulaires [6, 30].

3. Applications et discussion

L'application de l'équation générale (3) permet la description des différents phénomènes non linéaires dans lesquels s'expriment les processus moléculaires-statistiques. La théorie précédente montre que, pour une symétrie bien définie des tenseurs des polarisabilités moléculaires non linéaires, on a en général $Q_1 \neq Q_2 = Q_3$ ce qui permet d'écrire l'équation (3) sous la forme suivante:

$$P_i^{(3)}(-\omega_p) = \sum_{abc} \{Q_1(-\omega_p, \omega_a, \omega_b, \omega_c) \delta_{ij} E_k^{\omega_b} E_k^{\omega_c} +$$
$$+ Q_2(-\omega_p, \omega_a, \omega_b, \omega_c) (\tfrac{3}{2} E_i^{\omega_b} E_j^{\omega_c} + \tfrac{3}{2} E_j^{\omega_b} E_i^{\omega_c} +$$
$$- \delta_{ij} E_k^{\omega_b} E_k^{\omega_c})\} E_j^{\omega_a} \exp\{i[(\mathbf{k}_a + \mathbf{k}_b + \mathbf{k}_c) \cdot \mathbf{r} +$$
$$- (\omega_a + \omega_b + \omega_c) t]\}. \qquad (19)$$

Q_1 est une grandeur rendant compte de la non-linéarité induite isotrope. $Q_2 = Q_3$ définissent l'anisotropie de la non-linéarité induite dans le corps isotrope. Les grandeurs Q_1 et Q_2 ont la forme donnée par le développement en série (4). Les perturbations successives traduisent les différents mécanismes moléculaires, tout particulièrement lorsqu'il s'agit d'un milieu condensé.

3.1. GENERATION DU TROISIEME HARMONIQUE

La polarisation non linéaire (3), ou (19), définit très généralement les processus de couplage de trois ondes dans un milieu isotrope et en particulier le triplement de la fréquence d'une onde lumineuse lorsque $\omega_a = \omega_b = \omega_c = \omega$. Dans ce cas seul le processus purement électronique intervient. Les constantes (5) permettent de déterminer les polarisabilités non linéaires d'ordre 3, $c_{\alpha\beta\gamma\delta}(-3\omega, \omega, \omega, \omega)$, pour des atomes et des molécules [28, 29, 31]. Dans les milieux denses, les $Q_s^{(0)}$ ont la forme (5) et ne sont que faiblement modifiés par les champs électriques locaux et la redistribution atomique et moléculaire [17, 27].

3.2. GENERATION DU DEUXIEME HARMONIQUE EN PRESENCE D'UN CHAMP ELECTRIQUE STATIQUE

Ce processus correspond au cas $\omega_a = 0$, $\omega_b = \omega_c = \omega$. Outre $Q_s^{(0)}(-2\omega, 0, \omega, \omega)$, il fait intervenir, pour les systèmes moléculaires, les grandeurs $Q_s^{(1)}(-2\omega, 0, \omega, \omega)$. Dans le cas où la polarisation du milieu est induite par un champ électrique variable de fréquence $\omega_a = \omega_0 \ll \omega = \omega_b = \omega_c$, la théorie de la relaxation fournit (formules (6) et (7)) l'apport suivant dépendant de la température :

$$2Q_1^{(1)}(-2\omega, \omega_0, \omega, \omega) = 5Q_2^{(1)}(-2\omega, \omega_0, \omega, \omega) =$$
$$= \frac{\varrho}{6k} b(-2\omega, \omega, \omega) m(\omega_0) D_1(\omega_0). \qquad (20)$$

Les deux autres termes des formules (6) et (7) s'annulent, puisque d'après (8) on a $D_2(2\omega) = 0$ et $D_1(\omega_0 + 2\omega) = 0$ si $\omega \to \infty$. Dans le cas du champ statique, l'apport (20) ne s'annule pas, puisque pour $\omega_0 = 0$ on trouve $D_1(0) = 1$.

On voit que l'étude de la génération du deuxième harmonique dans un champ électrique variable $|\omega_0 \ll \omega|$ permet la détermination de $D_1(\omega_0)$ et par suite, à partir de l'équation (8) celle du premier temps de relaxation de Debye $\tau_1 = \tau_D$.

Dans les gaz condensés et les liquides moléculaires, les corrélations du type dipolaire et multipolaire ainsi que les champs moléculaires ont une influence très prononcée sur la génération du deuxième harmonique dans un champ électrique statique. Le traitement semi-macroscopique de Kirkwood [32], généralisé pour inclure les effets électro-optiques non linéaires [33], permet d'en tenir compte facilement.

Dans le cas d'un gaz ou d'un liquide constitué de molécules centrosymétriques, l'apport (20) s'annule. En négligeant l'anisotropie de la polarisabilité, on obtient alors :

$$2Q_1^{2\omega} = 5Q_2^{2\omega} = \frac{5\varrho}{18} c_{2\omega} \left(1 + \frac{a}{kT} \langle F_m^2 \rangle \right), \quad (21)$$

où :

$$a = a_{\alpha\alpha}/3 \qquad c_{2\omega} = c_{\alpha\alpha\beta\beta}(-2\omega, 0, \omega, \omega)/5 \quad (22)$$

désignent, respectivement, la polarisabilité électrique linéaire moyenne et la polarisabilité non linéaire moyenne à la fréquence 2ω.

La formule (21) présente un très grand intérêt car elle permet, dès qu'on connaît a et $c_{2\omega}$, de déterminer la valeur de $\langle F_m^2 \rangle$, c'est-à-dire la moyenne statistique du carré du champ électrique moléculaire F_m existant dans la région d'ordre proche. Des expériences récentes ont montré qu'on peut déterminer $\langle F_m^2 \rangle$ à partir des mesures de la diffusion coopérative biharmonique dans les liquides constitués de molécules centrosymétriques [34].

3.3. Effet Kerr induit par un faisceau laser

La biréfringence optique, analysée avec une onde lumineuse de fréquence $\omega_a = \omega_A$, et induite par un faisceau intense laser à la fréquence $\omega_I = \omega_b = \omega_c$, dépend des Q_s et du deuxième terme de la grandeur anisotrope (7) :

$$Q_2^{(1)}(-\omega_A, \omega_A, \omega_I, -\omega_I) = \frac{\varrho}{45k} \gamma(-\omega_A, \omega_A) \gamma(\omega_I, -\omega_I). \quad (23)$$

Les autres termes de (6) et de (7) s'annulent aux fréquences optiques. A partir de la formule (23), valable pour les gaz moléculaires, on peut déterminer les anisotropies optiques des polarisabilités linéaires $\gamma_{\omega_A} = \gamma(-\omega_A, \omega_A)$ et $\gamma_{\omega_I} = \gamma(\omega_I, -\omega_I)$ pour des molécules asymétriques.

Pour les atomes et les molécules à polarisabilité linéaire isotrope p.ex. CCl_4, la constante d'anisotropie (23) résultant de la réorientation de Langevin [2, 10] s'annule, et la biréfringence du gaz n'est due qu'à l'effet non linéaire de Voigt [1, 4] décrit par les formules (5). Dans les états condensés, toutefois, les fluctuations de translation d'Yvon-Kirkwood font que la constante anisotrope ne s'annule pas. Elle est alors donnée par la formule suivante [17, 23] :

$$Q_2^{(1)}(\omega_A, \omega_I) = \frac{2\varrho}{5k} a^2{}_{\omega_A} a^2{}_{\omega_I} (\langle r_{12}^{-6} \rangle + 2 \langle r_{12}^{-3} r_{23}^{-3} \rangle + \cdots) \quad (24)$$

r_{12} est la distance entre les atomes 1 et 2. Le paramètre de corrélations radiales binaires $\langle r_{12}^{-6} \rangle$ est positif dans tous les cas, tandis que celui traduisant les corrélations radiales ternaires $\langle r_{12}^{-3} r_{23}^{-3} \rangle$ est positif ou négatif selon la structure d'ordre proche [18, 23].

Dans les liquides constitués de molécules anisotropes, outre les fluctuations de translation (24), existent des corrélations angulaires ainsi que des fluctuations de translation-orientation données par la formule :

$$Q_2^{(1)}(\omega_A, \omega_I) = \frac{\varrho}{45k} \{\gamma_{\omega_A}\gamma_{\omega_I}(1 + J_A) +$$
$$+ 2[\gamma_{\omega_A}a_{\omega_I}(3a_{\omega_I} - \gamma_{\omega_I}) + \gamma_{\omega_I}a_{\omega_A}(3a_{\omega_A} - \gamma_{\omega_A})]J_{RA} +$$
$$+ 2\gamma_{\omega_A}\gamma_{\omega_I}(3a_{\omega_A} + 3a_{\omega_I} + \gamma_{\omega_A} + \gamma_{\omega_I})K_{RA} + \cdots\}, \qquad (25)$$

où J_A désigne le paramètre de corrélations angulaires [17] et J_{RA} et K_{RA} ceux des corrélations radiales-angulaires [18, 24]. Ce sont des paramètres traduisant les fluctuations de translation-orientation. Le paramètre J_{RA} joue un rôle particulièrement important en effet Kerr et en diffusion Rayleigh anisotrope des solutions diluées, où il rend compte de corrélations existant entre la molécule anisotrope en solution et les molécules, considérées comme quasi-sphériques, du solvant [24, 35].

3.4. Effet Kerr

La biréfringence électro-optique dépend généralement des $Q_s^{(0)}$ et $Q_s^{(1)}$ que nous avons déjà discutés, mais aussi des Q_s donnés dans la théorie de la relaxation par les formules (9) et (10). La théorie de l'effet Kerr induit dans les gaz fut établie par Voigt [1], Langevin [2] et Born [4] dans leurs travaux déjà classiques. Elle est décrite par les formules (4)–(14) et (19), avec $\omega_a = \omega_A$ et $\omega_b = \omega_c = 0$. Si la biréfringence est causée par un champ électrique alternant oscillant à une fréquence $\omega_0 \ll \omega_A$, on est conduit, en posant $\omega_a = \omega_A$ et $\omega_b = \omega_c = 0$ dans (6)–(10), aux formules de Peterlin et Stuart [5] pour la dispersion diélectrique de l'effet Kerr. Cet effet permet de déterminer outre $\tau_1 = \tau_D$ le paramètre $\tau_2 = \tau_D/3$.

Dans les gaz aux molécules non dipolaires, les apports d'ordre 2, notamment (9) et (10), s'annulent. La situation dans les états condensés est radicalement différente. Dans ce cas outre les apports (24) et (25) (dans lesquels il faut faire $\omega_I = 0$) intervient également la constante anisotrope suivante d'ordre 2 due aux fluctuations des champs des quadrupôles électriques θ des molécules:

$$Q_2^{(2)}(\omega_A) = \frac{\varrho a^2}{105k^2}\gamma_{\omega_A}\theta^2 \langle r_{12}^{-8}\rangle. \qquad (26)$$

La valeur moyenne radiale $\langle r_{12}^{-8}\rangle$ peut être déterminée à partir de la diffusion biharmonique coopérative [34].

De même, d'autres processus statistiques de fluctuations liés au mouvement des molécules, peuvent entraîner, p.ex. la non-additivité de la constante de Kerr dans le cas des solutions [6, 36].

3.5. Saturation électrique

On mesure dans ce cas les variations de la permittivité électrique, induite par un champ électrique intense statique ou alternant, et révélées par un champ électrique lentement variable qui détecte toutes les composantes du développement en série (4). Les gaz dipolaires, outre les mécanismes précédemment discutés, présentent celui de la réorientation des dipôles électriques eux-mêmes. Dans le traitement de la théorie de la relaxation, ce mécanisme est traduit par les expressions (15)–(18). Les formules

(3)–(19), dont nous venons de donner la première démonstration, définissent la théorie moléculaire-relaxationnelle de la saturation électrique des substances dipolaires. L'étude de la dispersion diélectrique de l'effet Kerr et de la saturation électrique permet de déterminer les temps de relaxation τ_2 et τ_3. Il est à regretter que l'on ne dispose pas encore d'études expérimentales à ce sujet: toutefois, le premier essai vient d'être entrepris [37].

L'étude des variations non linéaires de la permittivité électrique des liquides et des solutions moléculaires permet de recueillir de nombreux renseignements d'un haut intérêt concernant les différentes corrélations moléculaires angulaires [30, 33] ainsi que les processus de fluctuations statistiques, surtout dans les substances non dipolaires [6].

4. Conclusion

Nous avons montré que les effets électro-optiques non linéaires sont plus sensibles aux processus moléculaires statistiques dans les gaz et les liquides (surtout aux mouvements moléculaires de translation et de rotation) que les effets optiques et électriques linéaires jusqu'à présent étudiés. Ces processus microscopiques réagissent individuellement et à des vitesses très différentes, aux impulsions électriques du champ. Ces 'temporal responses' ou réactions déclenchant un tel processus ont des durées allant de 10^{-16} s pour les processus électroniques jusqu'à 10^{-12} s pour la relaxation des molécules individuelles et 10^{-10} s pour les différents processus coopératifs de fluctuations dans les régions d'ordre proche. Les techniques laser dont on dispose actuellement rendent possible la séparation de ces processus, particulièrement s'il s'agit de l'effet Kerr optique [38–40] et de l'autoconvergence des faisceaux laser [41–46].

La théorie de la relaxation moléculaire de Debye, généralisée aux perturbations d'ordre élevé, nous a permis de donner une description uniforme de l'évolution dans le temps des différents mécanismes contribuant aux non-linéarités des substances polaires dans des champs électriques extérieurs oscillants.

Cette théorie fait intervenir trois temps de relaxation, τ_1, τ_2 et τ_3 liés par:

$$\frac{\tau_1}{\tau_2} : \frac{\tau_1}{\tau_3} = 3:6,$$

où $\tau_1 = \tau_D$. On voit que τ_2 et τ_3 ne valent qu'un tiers et un sixième du temps de relaxation de Debye, τ_D.

Le traitement non linéaire de la dispersion peut être déduit des méthodes quantiques [47, 48] ou de celle de Kubo [49]. Ces méthodes, appliquées à des milieux statistiquement inhomogènes, entraînent de grandes difficultés. Nous n'avons par conséquent, traité ici que de quelques résultats simples découlant de la statistique classique des processus en équilibre thermodynamique, formules (21), (24)–(26).

Dans les liquides atomiques, le rôle principal appartient aux fluctuations du nombre-densité ainsi qu'aux fluctuations de translation d'Yvon-Kirkwood (qui entraînent des regroupements spatiaux des atomes dans les régions d'ordre à courte distance). Par suite des interactions binaires, ternaires, et plus élevées, les atomes en corrélations se comportent momentanément comme des agrégats anisotropes subissant une réorientation dans le champ électrique extérieur.

Dans les liquides moléculaires, outre le processus de réorientation de Langevin et Debye, les fluctuations de translation-orientation et d'orientation commencent à jouer un rôle important. Dans les régions d'ordre proche, les molécules anisotropes en mouvement de translation se trouvent simultanément engagées dans des mouvements d'orientation. Le cas des liquides constitués de molécules non polaires est particulièrement intéressant: les fluctuations dans le temps et l'espace des champs moléculaires induits par les quadrupôles et les multipôles électriques font que les termes de la série (4), fonctions plus élevées de la température, ne s'annulent pas. C'est la raison pour laquelle le comportement thermique des Q_s des liquides non dipolaires est similaire à celui des Q_s des gaz dipolaires.

Nous venons de traiter rapidement de cinq phénomènes non linéaires actuellement étudiés expérimentalement. Nous les avons considéré dans l'ordre croissant de l'influence qu'exercent les processus moléculaires statistiques. L'étude simultanée de ces effets anisotropes non linéaires, effectuée sur des liquides atomiques ou moléculaires judicieusement choisis, ouvre la voie à une nouvelle étude des mouvements moléculaires et de la structure dynamique de l'ordre à courte distance régnant au sein du liquide. De tels renseignements ne semblent pas pouvoir être fournis par la seule étude des effets linéaires isotropes.

Enfin ces études conduisent à l'amélioration de nos connaissances relatives aux fonctions de corrélations intermoléculaires dans les fluides denses.

Bibliographie

1. Voigt, W.: *Magneto- und Electro-Optik*, Teubner, Leipzig, 1908.
2. Langevin, P.: *Radium* **7**, 249 (1910).
3. Debye, P.: *Polare Molekeln*, Leipzig, 1929.
4. Born, M.: *Optik*, J. Springer, Berlin, 1933.
5. Peterlin, A. et Stuart, H.: *Doppelbrechung insbesondere Künstliche Doppelbrechung*, Akademische Verlagsgesellschaft, Leipzig, 1943; Benoit, H.: *Ann. Phys.* **6**, 561 (1951).
6. Kielich, S.: dans M. Davies (ed.), *Dielectric and Related Molecular Processes*, Chem. Soc. London, 1972, Chapter 7.
7. Maker, P. D. et Terhune, R. W.: *Phys. Rev.* **137A**, 801 (1965).
8. Akhmanov, S. A. et Chirkin, A. S.: *Statistical Effects in Nonlinear Optics*, Moscow Univ. Press, 1971.
9. Kielich, S.: *Fundamentals of Molecular Nonlinear Optics*, A. Mickiewicz University Press, Poznań, Part I (1972) and Part II (1973).
10. Buckingham, A. D.: *Proc. Phys. Soc.* **B69**, 344 (1956).
11. Mayer, G. et Gires, F.: *Compt. Rend. Acad. Sci.* **258B**, 2039 (1964).
12. Paillette, M.: *Ann. Phys.* **4**, 671 (1969).
13. Martin, F. B. et Lalanne, J. R.: *Optics Communications* **2**, 219 (1970); *Phys. Rev.* **A4**, 1275 (1971); Martin, F. B.: Thèse, Bordeaux, 1971.
14. Kielich, S., Lalanne, J. R., et Martin, F. B.: *Compt. Rend. Acad. Sci.* **272B**, 731 (1971); *IEEE J. Quantum Electronics* **QE-9**, 601 (1973).
15. Kasprowicz, B. et Kielich, S.: *Acta Phys. Polonica* **33**, 495 (1968).
16. Lalanne, J. R.: *J. Phys.* **30**, 643 (1969).
17. Kielich, S.: *Acta Phys. Polonica* **22**, 299 (1962); **30** 683 (1966); **A41** 653 (1972).

18. Kielich, S., Lalanne, J. R., et Martin, F. B.: *J. Phys.* **33**, C1-191 (1972).
19. Lalanne, J. R., Martin, F. B., et Bothorel, P.: *J. Colloid. Interface Sci.* **39**, 601 (1972).
20. Hellwarth, R. W.: *Phys. Rev.* **152**, 156 (1966); *J. Chem. Phys.* **52**, 2128 (1970).
21. Yvon, J.: *Compt. Rend. Acad. Sci.* **202**, 35 (1936).
22. Kirkwood, J. G.: *J. Chem. Phys.* **4**, 592 (1936).
23. Kielich, S.: *Optics Communications* **4**, 135 (1971).
24. Kielich, S.: *Chem. Phys. Letters* **10**, 516 (1971); 'Erratum', *ibid.* **19**, 609 (1973).
    Kielich, S. et Woźniak, S.: *Acta Phys. Polonica* **A45**, 105 (1974).
25. Mayer, G.: *C.R. Acad. Sci.* **267B**, 54 (1968);
    Hauchecorne, G., Kervervé, F., et Mayer, G.: *J. Phys.* **32**, 47 (1971).
26. Fin, R. S. et Ward, J. F.: *Phys. Rev. Letters* **26**, 285 (1971).
27. Kielich, S.: *Chem. Phys. Letters* **2**, 569 (1968); *Acta Phys. Polonica* **37A**, 205 (1970).
28. Terhune, R. W., Maker, P. C., et Savage, C. M.: dans P. Grivet et N. Bloembergen, (eds.), *Third Conference on Quantum Electronics, Paris (1963)*, Columbia University Press, New York, 1964 p. 1559.
29. Ward, J. F. et New, G. H. C.: *Phys. Rev.* **185**, 57 (1969).
30. Kielich, S. et Piekara, A.: *Acta Phys. Polonica* **18**, 439 (1959);
    Piekara, A.: *J. Chem. Phys.* **36**, 2145 (1962);
    Davies, M.: *Ann. Reports* **A67**, 65 (1970).
31. Kielich, S.: *Opto-electronics* **2**, 125 (1970; *Ferroelectrics* **4**, 257 (1972).
32. Kirkwood, J. G.: *J. Chem. Phys.* **7**, 911 (1939).
33. Kielich, S.: *Acta Phys. Polonica* **17**, 239 (1958); **32**, 405 (1967); **33**, 89 (1968).
34. Kielich, S., Lalanne, J. R., et Martin, F. B.: *Phys. Rev. Letters* **26**, 1295 (1971); *Acta Phys. Polonica* **A41**, 479 (1972);
    J. Raman Spectroscopy **1**, 119 (1973).
35. Buckingham, A. D., Stiles, P. J., et Ritchie, G. L. D.: *Trans. Faraday Soc.* **67**, 577 (1971).
36. Kielich, S.: *Mol. Phys.* **6**, 49 (1963).
37. Gregson, M., Parry Jones, G., et Davies, M.: *Trans. Faraday Soc.* **67**, 1630 (1971).
38. Bloembergen, N.: *Am. J. Phys.* **35**, 989 (1967);
    Chiao, R. Y. and Godine, J.: *Phys. Rev.* **185**, 430 (1969).
39. Fabielinski, I. L.: *Usp. Fiz. Nauk* **104**, 77, (1971).
40. Hellwarth, R. W., Owyoung, A., et George, N.: *Phys. Rev.* **A4**, 2342 (1971).
41. Shen, Y. R. et Shaham, Y. J.: *Phys. Rev.* **163**, 224 (1967).
42. Brewer, R. G. et Lee, C. H.: *Phys. Rev. Letters* **21**, 267 (1968).
43. Cubeddu, R., Polloni, R., Sacchi, C. A., et Svelto, O.: *Phys. Rev.* **A2**, 1955 (1970.
44. Fleck, J. A., Jr. et Carman, R. L.: *Appl. Letters* **20**, 290 (1972).
45. Alfano, R. R., Hope, L. L., et Shapiro, S. L.: *Phys. Rev.* **A6**, 433 (1972).
46. Gustafson, T. K. et Townes, C. H.: *Phys. Rev.* **A6**, 1659 (1972).
47. Kielich, S.: *Acta Phys. Polonica* **26**, 135 (1964).
48. Takatsuji, M.: *Phys. Rev.* **155**, 980 (1967).
49. Pecora, R.: *J. Chem. Phys.* **50**, 2650 (1969).
50. Hill, N. E., Vaugham, W. E., Price, A. H., et Davies, M.: *Dielectric Properties and Molecular Behaviour*, Van Nostrand Reinhold, London, 1969.

DISCUSSION

Lallemand: Il paraît nécessaire de mentionner la diffusion non linéaire à 2 photons dans laquelle on absorbe 2 photons laser et on observe 1 photon diffusé à une fréquence voisine de $2\omega_{laser}$. Ceci peut permettre la mesure d'une fonction de corrélation angulaire $P_3(\cos\theta)$, ce qui peut servir à préciser les mouvements rotationnels dans les liquides.

Lalanne: Oui, bien sûr. Maker a montré que l'étude spectrale de la diffusion harmonique d'une onde laser intense par un ensemble de molécules sans centre de symétrie donnait des renseignements importants sur les mouvements de rotation des molécules (ordre trois) dans les liquides. Ce travail a été repris à la Ford Motor Company par Peterson (thèse de doctorat).

ETUDE DES PROPRIETES VISCOELASTIQUES D'UN LIQUIDE PAR DIFFUSION DE LA LUMIERE

C. DEMOULIN, C. J. MONTROSE*, et N. OSTROWSKY

Laboratoire de Physique de l'Ecole Normale Supérieure, Paris, France

Résumé. La diffusion Rayleigh par le glycerol en surfusion a été étudiée par une technique de corrélation de photons, qui permet de mesurer des temps de relaxation beaucoup plus longs que les méthodes usuelles (ultra sons) et donc de s'approcher beaucoup plus près de la température de transition de verre. On relie le signal observé à la fonction de corrélation des fluctuations de densité du liquide, qui, à l'échelle de temps observée, ne dépendent plus que de la relaxation structurale, à température et à pression constante. On donne une description du montage expérimental et de la méthode d'analyse des données. Les résultats sont comparés avec ceux obtenus à plus haute température par ultra sons et par mesures de spectres de la lumière diffusée.

Abstract. The Rayleigh scattering of light from supercooled liquid glycerol is studied by a photon counting technique which measures the clipped correlation product $\langle n_k(0) n(t) \rangle$, where $n(t)$ is the number of photoelectrons detected in a sampling time around the time t and $n_k(0)$ is one (or zero) if $n(0)$ is greater (or smaller) than an integer k. The relationship between this signal and the correlation function of the density fluctuations in the liquid is derived. Experimental conditions are such that the structural relaxation time is much longer than all the other characteristic times of the problem (period and relaxation time of sound waves, thermal relaxation time). On this time scale, the measured signal is therefore determined by the isothermal structural relaxation dynamics of the liquid. A description of the experimental set-up and the data analysis is given. The isothermal structural relaxation function was obtained at a dozen temperatures in the range $-80\,°C$ to $-50\,°C$, over which range the average relaxation time τ_{PT} was found to vary by a factor of nearly 10^5. The shape of the correlation function was well described by a distribution of relaxation times of the Cole Davidson type with a width parameter $\beta = 0.4$. This value is inferior to the one obtained by dielectric relaxation measurements in the same temperature region, but is consistent with the value obtained ultrasonically at higher temperatures.

Traditionnellement, les propriétés viscoélastiques des liquides ont été étudiées par des mesures de relaxation diélectrique [1, 2] et surtout par la mesure de l'absorption et de la dispersion des ultra sons [3]. Afin d'augmenter la limite supérieure des fréquences mesurables, on a fait appel plus récemment [4, 5] aux techniques de diffusion de la lumière. Nous avons utilisé une technique particulière de mesure de corrélation de photons qui nous a permis d'explorer la limite inférieure des fréquences mesurables et d'étudier des temps de corrélation très longs (allant jusqu'à quelques secondes). Après avoir rappelé les caractéristiques du problème étudié (§1) nous expliquerons le principe de cette méthode (§2); nous décrirons ensuite le montage expérimental (§3), puis l'analyse des données et les résultats obtenus (§4).

1. Problème étudié

Le spectre de la lumière diffusée par un liquide viscoélastique est caractérisé par un certain nombre de paramètres:

---

* Adresse permanente: Dept. of Physics, Catholic University of America, Washington, D. C. 20017 U.S.A.

(1) $(\omega_B)^{-1}$ période des ondes sonores spontanées, qui sont à l'origine des raies Brillouin

(2) $((\lambda/\varrho c_p) k^2)^{-1}$ temps de relaxation thermique des fluctuations spontanées de densité de vecteur d'onde k (λ est la conductivité thermique, ϱ la densité et c_p la chaleur spécifique à pression constante); ces fluctuations sont responsables de la raie Rayleigh "ordinaire".

(3) τ temps de relaxation structurale. Dans les liquides visqueux étudiés, ce temps varie énormément avec la température. Par exemple, pour la glycérine en surfusion, il passe de 10^{-8} s à $0\,°C$ à quelques secondes à $-80\,°C$, jusqu'à quelques dizaines de minutes à $-90\,°C$, température au dessous de laquelle on dit que la glycérine est devenue un verre.

Jusqu'à présent, toutes les études théoriques et expérimentales se sont intéressées au domaine où $\tau \ll ((\lambda/\varrho c_p) k^2)^{-1}$. On peut alors montrer que suivant la valeur de τ par rapport à $(\omega_B)^{-1}$ la relaxation structurale affecte soit la partie adiabatique ($\tau \lesssim (\omega_B)^{-1}$) soit la partie isotherme ($\tau \gg (\omega_B)^{-1}$) des fluctuations de densité, responsables de la lumière diffusée. Dans le premier cas, on observe une modification du spectre Brillouin, et dans le deuxième cas, la relaxation structurale se traduit par une nouvelle raie dans le spectre, dite raie de Mountain, centrée à la même fréquence que celle de l'onde incidente, et de largeur proportionnelle à $1/\tau_{PS}$, donc beaucoup plus large que la raie Rayleigh ordinaire. Les indices P et S de τ_{PS} traduisent le fait que les "réarrangements de structure" doivent se produire à pression constante (car $\tau \gg (\omega_B)^{-1}$) et à entropie constante (car $\tau \ll (\lambda/\varrho c_p) k^2)^{-1}$.

Suivant une idée du professeur T. A. Litovitz, nous nous sommes intéressés au domaine de température où τ est le temps le plus long du problème, c'est à dire où $\tau \gg ((\lambda/\varrho c_p) k^2)^{-1}$. On s'attend alors à ce que la composante centrale du spectre soit formée d'une raie très fine, associée à la relaxation structurale, superposée sur une raie beaucoup plus large, associée à la relaxation thermique. Notons que dans ce cas, les fluctuations de chaleur relaxent très vite par rapport à τ; les réarrangements de structure se font donc à température constante et le temps de relaxation structural mesuré doit s'écrire τ_{PT}.

A l'échelle de temps de notre expérience $(t \gg ((\lambda/\varrho c_p) k^2)^{-1})$ la fonction de corrélation des fluctuations de densité $\phi(t)$ ne provient plus que de la relaxation structurale et le temps moyen de relaxation est donc égal à:

$$\tau_{PT} = \int_0^\infty \phi(t)\,dt. \tag{1}$$

2. Méthode utilisée: corrélation de photons

Les expériences de diffusion de la lumière permettent d'obtenir $\phi(t)$ en mesurant la fonction d'autocorrélation du champ électrique $E(t)$ diffusé dans un angle donné*.

---

* $E(t)$ représente en fait le signal analytique complexe associé au champ électrique réel diffusé et on peut écrire l'intensité lumineuse diffusée $I(t) = E^*(t) E(t)$.

On a en effet:

$$g^{(1)}(t) = \frac{\langle E^*(0) E(t) \rangle}{\langle E^*(t) E(t) \rangle} = A\phi(t). \tag{2}$$

où A dépend des caractéristiques du milieu diffusant et de la longueur d'onde utilisée.

Les méthodes classiques de mesures de spectre (spectrographes, interféromètres) consistent à interposer sur le trajet du faisceau diffusé un filtre à haute résolution et à fréquence variable, puis à mesurer à l'aide d'un photomultiplicateur l'intensité $I(\nu)$ donnée par:

$$I(\nu) = \frac{I}{2\pi} \int_{-\infty}^{+\infty} g^{(1)}(t) e^{2\pi i \nu t} dt. \tag{3}$$

Ces méthodes ne peuvent cependant pas s'appliquer à notre problème car les raies étudiées sont beaucoup trop fines ($1/\tau \approx 10^5$ s^{-1}) et la résolution des filtres qui serait nécessaire est bien supérieure aux possibilités actuelles. Nous avons donc employé une méthode entièrement différente, dite de "battements lumineux" [6] qui consiste à recevoir directement sur la cathode du photomultiplicateur toute la lumière diffusée et à mesurer la fonction d'autocorrélation du signal de sortie. Montrons comment cette fonction d'autocorrélation est reliée au signal $\phi(t)$ que l'on cherche à mesurer. Un photomultiplicateur est un détecteur quadratique c'est à dire qu'il donne en sortie un signal proportionnel au carré du champ électrique incident. Ce signal est soit un courant photoélectrique soit, comme dans le cas qui nous intéresse, un nombre d'impulsions par unité de temps. Si donc on appelle $n(t)$ le nombre d'impulsions fournies par le tube pendant un temps d'échantillonage θ (on suppose $\theta \ll \tau$) on aura:

$$n(t) = \bar{n} \frac{E^*(t) E(t)}{\langle |E(t)|^2 \rangle},$$

où $\bar{n}$ est le nombre moyen d'impulsions reçues pendant le temps θ. La fonction d'autocorrélation du signal de sortie du photomultiplicateur s'écrit donc:

$$\langle n(0) n(t) \rangle = \bar{n}^2 \frac{\langle E^*(0) E(0) E^*(t) E(t) \rangle}{\langle |E(t)|^2 \rangle^2}. \tag{4}$$

En faisant l'hypothèse justifiée dans nos expériences que les fluctuations de densité observées sont en nombre suffisant pour que $E(t)$ soit une variable aléatoire gaussienne, on voit que cette fonction de corrélation peut être reliée à la quantité recherchée par la relation [7]:

$$\langle E^*(0) E(0) E^*(t) E(t) \rangle = \langle E^*(0) E(0) \rangle \langle E^*(t) E(t) \rangle +$$
$$+ \langle E^*(0) E(t) \rangle \langle E(0) E^*(t) \rangle$$

La fonction d'autocorrélation du signal de sortie du photomultiplicateur permet donc de déduire $g^{(1)}(t)$:

$$\langle n(0) n(t) \rangle = \bar{n}^2 [1 + |g^{(1)}(t)|^2]. \tag{5}$$

Le signal véritablement mesuré dans nos expériences est une forme simplifiée de cette expression dans laquelle on remplace $n(0)$ par

$$\left.\begin{array}{l} n_k(0) = 1 \quad \text{si} \quad n(0) > k \\ n_k(0) = 0 \quad \text{si} \quad n(0) \leqslant k \end{array}\right\} \text{où } k \text{ est un entier positif ou nul choisi à l'avance.}$$

On peut alors montrer [8] que:

$$\langle n_k(0) n(t) \rangle = \bar{n} \bar{n}_k \left[1 + \frac{1+k}{1+\bar{n}} |g^{(1)}(t)|^2 \right]. \tag{6}$$

On voit donc que ce produit de corrélation "clippé", plus simple à réaliser électroniquement, donne les mêmes informations que l'expression complète (5) avec un rapport signal sur bruit légèrement inférieur, mais qui peut être facilement compensé par une prise de données plus longue.

3. Montage expérimental

Nos mesures ont porté sur le glycérol en surfusion pour des températures comprises entre $-48\,°C$ et $-80\,°C$.

(1) L'échantillon de glycérine anhydre (teneur en eau inférieure à 0,02% en poids) était préparé de la manière suivante: Après avoir pompé sous vide et à $+80\,°C$ l'air dissout dans le liquide, on élimine les poussières par filtration en circuit continu, à l'aide d'un filtre millipore (diamètre des pores $=0{,}45\,\mu$).

(2) Le schéma de l'expérience est représenté sur la figure 1. La lumière provenant d'un laser à argon ionisé (800 mw pour $\lambda_0 = 4880$ Å) est focalisé sur l'échantillon de glycérine contenu dans un dewar possédant 4 doubles fenêtres en croix, et thermostaté à $0{,}1°$ près. La lumière diffusée à $90°$ est focalisée sur un trou d'épingle de $300\,\mu$ de diamètre, et un diaphragme réglable permet de limiter à quelques unités le nombre d'aires de cohérence vues par le photomultiplicateur (radiotechnique 56DVP 03) placé derrière le trou d'épingle.

(3) Les impulsions sortant du tube photoélectrique sont amplifiées, passent ensuite par un discriminateur puis entrent dans un corrélateur digital à 100 cannaux construit dans le laboratoire [9]. Les temps d'échantillonnage peuvent être choisis entre 100 ns et 0,1 s ce qui permet de mesurer des temps de corrélation compris entre 1 μs et 1s. Pour chaque condition expérimentale donnée (intensité du signal et valeur de θ fixée) il était important d'ajuster l'entier k de la formule (6) à sa valeur optimale, c'est à dire celle qui donne le meilleur rapport signal sur bruit. Ceci était obtenu en choisissant k de l'ordre de $\bar{n}$, nombre moyen d'impulsions comptées pendant le temps θ. La valeur maximale de k prévue sur notre appareil étant de 6, nous avons dû, pour les expériences à grands θ ($\bar{n} \gg 10$) utiliser une méthode inspirée de la technique de "scaling" [10] et qui consiste à diviser extérieurement par un nombre s le nombre d'impulsions qui vont contribuer à former (dans le corrélateur) le nombre $n_k(0)$. On peut montrer [11] que dans le cas où $s \gg k$, cette méthode revient à mesurer le signal $\langle n_{k \times s}(0) n(t) \rangle$ et donc à augmenter extérieurement la capacité de la "grille k".

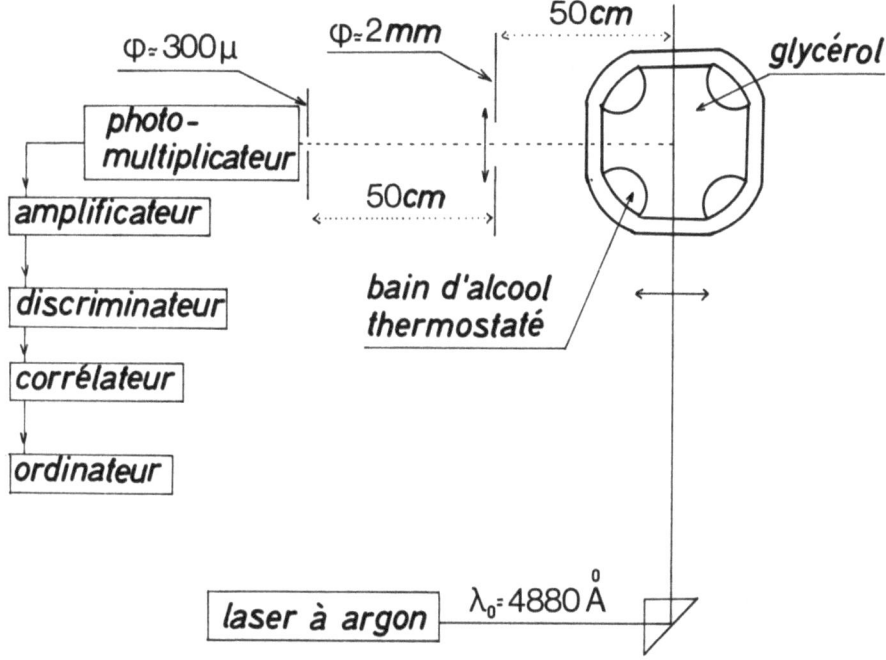

Fig. 1. Schéma du montage utilisé pour étudier la lumière diffusée à 90° par une technique de corrélation de photons.

4. Analyse des données – résultats

En faisant varier la température de $-48\,°C$ à $-80\,°C$ nous avons observé un temps de relaxation variant par un facteur de près de 10^5. On a reproduit sur la figure 2 un enregistrement typique des données contenues dans 90 des mémoires du corrélateur. Ces données sont analysées sur ordinateur par la méthode des moindres carrés, en prenant pour courbes théoriques

$$\langle n_k(0)\,n(t)\rangle_{\text{th}} = a + b\,|\phi(t)|^2. \tag{7}$$

On a observé qu'en prenant pour $\phi(t)$ une forme exponentielle $\exp(-t/\tau)$ (courbe en traits interrompus sur la figure 2) l'accord était mauvais et que de plus la meilleure valeur de τ choisie par l'ordinateur diminuait lorsqu'on réduisait l'intervalle θ entre chaque point expérimentaux, c'est à dire lorsqu'on diminuait le domaine de temps exploré expérimentalement. Ceci montre clairement que le phénomène de relaxation structurale ne peut être décrit par une seule constante de temps, mais fait intervenir une distribution de temps de relaxation avec une contribution non négligeable de temps beaucoup plus courts que le temps moyen τ_{PT}, ce qui est bien représenté par une distribution du type Cole-Davidson [1]. Ces auteurs ont utilisé, pour interpréter leurs résultats sur les mesures de relaxation diélectrique, une fonction de distribution de temps de relaxation à deux paramètres, τ_0 temps de relaxation maximum présent

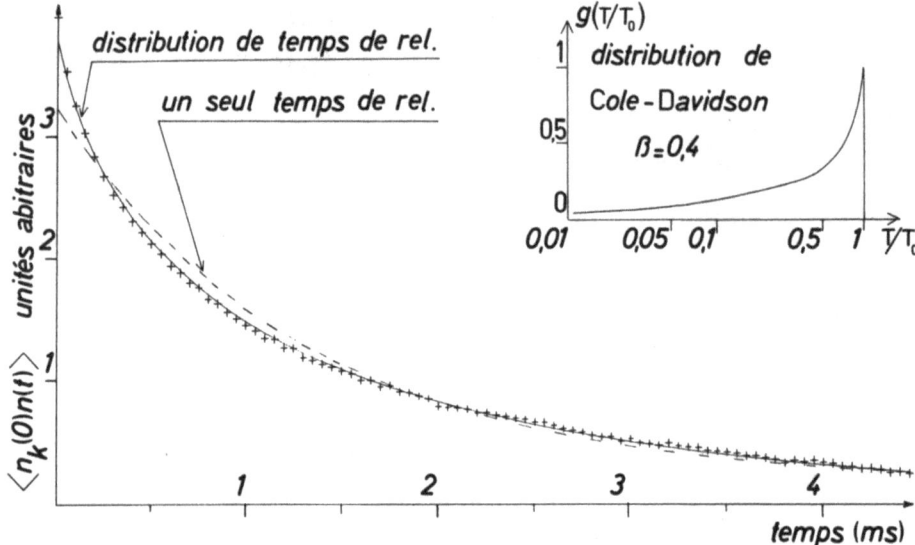

Fig. 2. Signal fourni par le corrélateur digital pour $\theta = 50~\mu$s. La hauteur de chaque croix est proportionnelle au nombre enregistré dans la mémoire correspondante. On a tracé deux courbes théoriques correspondant à une simple exponentielle (traits interrompus) et une superposition d'exponentielles (trait continu) correspondant à la distribution de temps de relaxation dessinée en haut et à droite.

dans la distribution et β paramètre compris entre 0 et 1 et décrivant la largeur de la distribution. Dans l'intervalle élémentaire d log (τ/τ_0) cette fonction de distribution est donnée par:

$$g(\tau/\tau_0) = \frac{\sin\beta\pi}{\pi}\left(\frac{\tau/\tau_0}{1-\tau/\tau_0}\right)^\beta \quad \text{si} \quad 0 \leq \frac{\tau}{\tau_0} < 1$$

$$= 0 \quad \text{si} \quad \frac{\tau}{\tau_0} \geq 1.$$

Cette fonction est représentée en haut et à droite sur la figure 2 pour $\beta = 0,4$. On peut montrer que si $\bar{\tau}$ est le temps de relaxation moyen, on a $\bar{\tau} = \beta\tau_0$. On a donc supposé que la fonction $\phi(t)$ que l'on cherche à mesurer était de la forme:

$$\phi(t) = \int_0^\infty d\log\left(\frac{\tau}{\tau_0}\right) \exp(-t/\tau)\, g(\tau/\tau_0).$$

Nous avons calculé numériquement cette fonction pour différentes valeurs du paramètre β et l'avons comparé (à l'aide de la relation (7)) aux points expérimentaux. Afin de rendre cette comparaison plus précise, nous avons cherché à connaître la fonction de corrélation expérimentale sur un très large domaine de temps. Nous avons donc enregistré, pour chaque température, un grand nombre de courbes avec des temps d'échantillonage variant entre $\tau/5000$ et $\tau/5$ et nous les avons raccordées par ordinateur. Un exemple d'une telle courbe est représenté sur la figure 3 en échelle

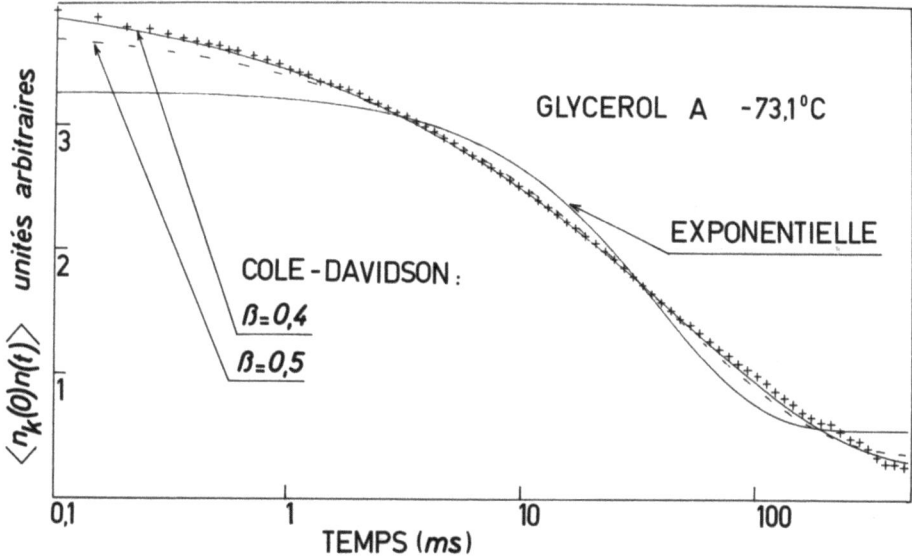

Fig. 3. Représentation en échelle de temps logarithmique de la fonction de corrélation obtenue en raccordant 7 courbes expérimentales avec θ variant de 50 μs à 5 ms. Les courbes théoriques correspondent à des distributions de largeur décroissante : $\beta = 0,4$ (trait continu), $\beta = 0,5$ (traits interrompus) et $\beta = 1$ (simple exponentielle, en trait continu).

de temps logarithmique. On a également porté sur cette figure les trois meilleures courbes théoriques correspondant à $\beta = 0,4$ $\beta = 0,5$ et $\beta = 1$ (simple exponentielle). Pour toutes les températures étudiées, on a trouvé que le meilleur accord était réalisé pour $\beta = 0,40 \pm 0,05$ ce qui est sensiblement inférieur à la valeur $\beta = 0,55$ trouvée par les mesures de relaxation diélectriques de Cole et Davidson [1] dans le même domaine de températures. On déduit de la meilleure valeur de β trouvée la valeur correspondante du temps de relaxation moyen τ_{PT}. Les résultats sont présentés sur la figure 4 (cercles noirs) et l'on voit que la dépendance en température est de la forme (loi d'Arrhénius :

$$\tau_{PT} \propto \exp(E^*/RT) \quad \text{avec } E^* \text{ de l'ordre de 32 kcal mole}^{-1}.$$

Ces résultats ont été comparés [12] avec les valeurs des temps de relaxation déduites des mesures à plus haute température par ultrasons [13] et qui sont représentées par des triangles sur la figure 4. Ces points expérimentaux se raccordent donc bien avec nos mesures.

Notons enfin que la valeur optimum trouvée pour le paramètre β de la distribution de Cole-Davidson est du même ordre que celui trouvé par les mesures d'ultrasons de Piccirelli *et al.* [13] pour des températures comprises entre $-50\,°C$ et $0\,°C$, et nettement inférieure à la valeur $\beta = 0,5$ déduite des mesures de spectre Brillouin de Pinnow *et al.* [4] pour des températures de l'ordre de 70 °C. Ceci semble donc confirmer le fait que la largeur de la distribution de temps de relaxation tend à augmenter lorsque la température diminue et semble atteindre une valeur limite $\beta = 0,4$ lorsqu'on s'approche de la température de transition de verre.

L'application de la technique de corrélation de photons à la mesure des propriétés viscoélastiques d'un liquide a donc permis d'étudier les phénomènes de relaxation structurale dans une échelle de temps inaccessible aux autres méthodes (ultrasons,

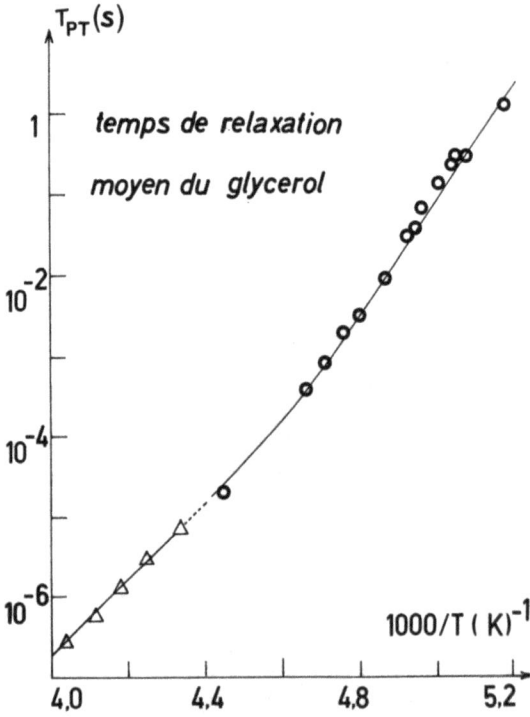

Fig. 4. Variation du temps de relaxation moyen $\tau_{PT} = \beta \tau_0$ en fonction de la température. Les cercles correspondent à nos mesures, et les triangles représentent les valeurs déduites des mesures par ultrasons de Piccirelli et al.

hypersons) et dans un régime où ces temps sont beaucoup plus longs que tous les autres temps caractéristiques du problème (période et temps de relaxation des ondes sonores, relaxation thermique) et concernent donc les phénomènes de réarrangements structuraux à température et à pression constante. En modifiant légèrement les conditions expérimentales, cette méthode peut en outre servir à étudier le couplage entre la relaxation structurale et la relaxation thermique et des expériences dans ce sens ont été entreprises.

Bibliographie

1. Davidson, D. W. and Cole, R. H.: *J. Chem. Phys.* **19**, 1484 (1951).
2. McDuffie Jr., G. E. and Litovitz, T. A.: *J. Chem. Phys.* **37**, 1699 (1962).
3. Voir par ex. Litovitz, T. A. and Davis, C. M.: dans W. P. Mason (ed.), *Physical Acoustics*, Academic Press, Inc., New York, 1965, Vol. 2, Part A Chap. 5.

4. Pinnow, D. A., Candau, S. J., Lamacchia, J. T., and Litovitz, T. A.: *J. Acoust. Soc. Am.* **43**, 131 (1968).
5. Katz, B.: Ph.D. Thesis, Catholic University of America, 1973.
6. Cummins, H. Z. and Swinney, H. L.: *Progress in Optics*, vol. VIII, North-Holland Publ. Co., Amsterdam.
7. Metha, C. L.: *Lectures on Theoretical Physics*, University of Colorado Press, Boulder, 1964, p. 345.
8. Jakeman, E. and Pike, E. R.: *J. Phys. A*, Ser. 2, **2**, 411 (1969).
9. Ce corrélateur a été conçu et réalisé par A. M. Berland, Ingénieur au C.N.R.S.
10. Jakeman, E., Oliver, C. J., Pike, E. R., and Pusey, N.: *J. Phys. A*, **5**, L93.
11. Demoulin, C.: Thèse de 3ème cycle, Université de Paris, Janvier 1974.
12. Demoulin, C., Montrose, C. J., and Ostrowsky, N.: *Phys Rev. A* à paraître.
13. Piccirelli, R. and Litovitz, T. A.: *J. Acoust. Soc. Am.* **29**, 1009 (1957).

DISCUSSION

Chen: I wonder whether you can comment on relative strength of light scattering from thermal fluctuation and structural fluctuation? If one considers that the structural relaxation occurs from reorientation of individual molecules which is more or less a single particle motion, while the thermal fluctuation is a collective motion, it seems that the strength of scattering would be much weaker.

Ostrowsky: The intensity of the 'mountain line' may be computed from the hydrodynamic equations with a frequency dependent viscosity and one finds that for glycerol, its intensity is about the same as the thermal Rayleigh line. Therefore, due to the presence of the Brillouin lines, the signal we are interested in represents a little less than a third of the total intensity.

Gershon: (1) How does the magnitude of β depend on temperature? (2) How are the values of β from polarized light scattering compared with the values of β obtained from the depolarized spectrum?

Ostrowsky: (1) β decreases from almost 1 at high temperature to 0.40 that we have found in our experiment. (2) They seem to be the same.

Balcou: In order to calculate the activation energy you used a diagram ($\log \tau$ vs $1000/T$) which seemed to present two distinct linear regions. There might be several explanations for this behaviour, e.g. a structural change, the inadequacy of the law in $1000/T$.... So did you try to plot your experimental points as a function of $1000/(T-\theta)$, θ being a constant, as is done in several works concerning viscoelastic or dielectric methods?

Ostrowsky: We have just tried to fit our data with the Arrhenius law, but in view of what was said at the conference, we shall also try a form $e^{B/(T-\theta)}$.

Cole: How do your relaxation times compare with dielectric relaxation times for glycerol?

Ostrowsky: They are of the same order of magnitude but as the width of the relaxation time distribution we find ($\beta = 0.40$) is different from the one you used ($\beta = 0.55$) it is difficult to make a precise and meaningful comparison.

THE SPECTRUM OF LIGHT SCATTERED FROM LIQUID n-ALKANES

J. V. CHAMPION

Physics Dept., City of London Polytechnic, England

and

D. A. JACKSON

The Physics Laboratories, University of Kent at Canterbury, England

Abstract. Using an argon ion laser with piezo-electrically and pressure scanning Fabry-Pérot interferometers, the polarised and depolarised spectra of some liquid n-alkanes has been determined over a wide range of temperatures. The relaxation times due to fluctuations in the orientational motions of the molecules are obtained from the low frequency depolarised spectra. These times are in close agreement with the corresponding times measured by flow birefringence. The hypersonic phonon velocity and absorption coefficients evaluated at ∼ 5 GHz from the polarised spectra show little dispersion in the frequency range 2 MHz to 5 GHz.

Résumé. Nous avons étudié les spectres polarisés et dépolarisés de quelques n-alkanes à l'état liquide dans un domaine étendu de température à l'aide d'un laser Arx et d'interféromètres de Fabry-Pérot balayés, soit par variation de pression, soit par un dispositif piézo-électrique. Les temps de relaxation d'orientation des molecules, obenus à partir de l'aile de basse fréquence du spectre dépolarisé, sont en bon accord avec ceux que l'on peut déduire des mesures de biréfringence d'écoulement. Les vitesses et les coefficients d'absorption des ondes hypersonores de fréquence 5 GHz que l'on peut tirer du spectre polarisé montrent une faible dispersion dans la bande de fréquence comprise entre 2 MHz et 5 GHz.

1. Introduction

When linearly polarized light is scattered by a liquid, the scattered light has a distribution in frequency about that of the incident beam. This spectrum contains a central Rayleigh line and a Brillouin doublet, both of which are polarized, and a symmetric broad line which is depolarized. The polarized and depolarized spectra can be studied separately using suitably oriented polarizers, a Fabry-Pérot interferometer and a laser acting as source.

The central Rayleigh line, caused by local isobaric entropy fluctuations in the liquid, has generally a line width of $\sim 10^6$ Hz which is much narrower than the instrumental line width when usual interferometer spacings (1 mm to 100 mm) are used.

The Brillouin frequency shift v_B typically of the order of a few GHz for 90° scattering is produced by adiabatic pressure fluctuations (hypersonic thermal phonons) and may be used to determine the phonon velocity V_S from the expression

$$V_S = \frac{v_B \lambda}{2n \sin \theta/2}, \qquad (1)$$

where λ is the wavelength in vacuo of the incident radiation, θ the scattering angle and n the liquid refractive index. The Brillouin linewidth Δv_B is related to the sound absorption coefficient α of the liquid by

$$\alpha = \frac{\Delta v_B}{V_S},\qquad(2)$$

where Δv_B is the frequency half width at half peak height of the Brillouin line. In a simple fluid the shear and volume viscosities govern the linewidth, however in a complicated system in which structural relaxations occur due to coupling of energy between internal and external molecular degrees of freedom the linewidth is usually dominated by this mechanism at frequencies where the energy interchange is possible.

The symmetric depolarized line is due to fluctuations of the local dielectric tensor. Generally the low frequency region of this spectrum is associated predominately with the orientational motion of anisotropic molecules [1]. A number of years ago Leontovich [1] showed that the relaxation time describing the fluctuation in orientation of an anisotropic molecule may also be related to the Maxwell constant of flow birefringence. In the case of light scattering, the spectrum is the Fourier transform of the time dependent molecular motion, the half width of the spectrum being reciprocally related to the relaxation time τ_s. In flow birefringence, the thermal randomising of the orientation of the molecule is opposed by the uniform flow tending to align the molecule in the direction of the streamlines. This results in a non-random orientational distribution which introduces a macroscopic anisotropy and consequently a birefringence. Leontovich using a phenomenological approach based on the fluctuations of the stress tensor has shown that the relaxation time τ_f of the anisotropy is related to the Maxwell constant M (induced birefringence/velocity gradient) by

$$\tau_f = \frac{16M^2 n^2}{\eta \beta_T \varrho \left(\dfrac{\partial \varepsilon}{\partial \varrho}\right)^2}\left[\frac{6 - 7\Delta u}{12\Delta u}\right],\qquad(3)$$

where η is the liquid viscosity, β_T the isothermal compressibility, Δu the depolarization ratio, n the mean refractive index, ϱ the density and ε the optical dielectric constant. The isothermal density derivative of the optical dielectric constant may be evaluated using the expression due to Meeten [2],

$$\varrho\left(\frac{\partial \varepsilon}{\partial \varrho}\right)_T = (n^2 - 1)(7n^2 + 23)/30.\qquad(4)$$

The aim of this work was to compare the values of the orientational relaxation times τ_s and τ_f in order to derive a fuller understanding of the problem of the interpretation of molecular orientational relaxation, illustrated by the wide divergence of current data dependent on the experimental techniques (e.g. NMR, dielectric relaxation, light scattering) used. The liquids chosen for this investigation were the homologous series of n-alkanes as the flow birefringence data was already available [3] over a range of temperatures. By a simple change of the instrumental polarization optics, it was also possible to investigate the internal molecular relaxation associated with conformational changes by determination of the dispersion of the hypersonic phonon velocities and absorption coefficients.

2. Results and Discussion

2.1. Depolarised Spectra

The relaxation times τ_s for the *n*-alkanes $C_{12}H_{26}$, $C_{14}H_{30}$, $C_{15}H_{32}$, $C_{16}H_{34}$ over the temperature range 20 °C to 140 °C have been determined directly from the low frequency region of the depolarised spectrum using a piezoelectrically scanned Fabry-Pérot interferometer with an argon ion laser in single frequency mode at 514.5 nm as source. The spectra were obtained at a fixed scattering angle of 90°. Typical spectra are shown in Figure 1. In order to cover the wide range of line widths (1 GHz for $C_{16}H_{34}$ at 20 °C, 15 GHz for $C_{12}H_{26}$ at 140 °C) two Fabry-Pérot etalon spacings 0.5 cm and 2.0 cm were used. The instrumental line shape was approximately Lorentzian and its width was at greatest 10% of the spectral line width ($C_{16}H_{34}$ at 20 °C and 2.0 cm spacer) and at smallest 3% ($C_{12}H_{26}$ at 140 °C and 0.5 cm spacer) and hence was simply subtracted from the measured line width as a first order deconvolution procedure. The line shapes were found to be single Lorentzians over the low frequency range of $\sim 4 \Delta v$, where Δv is the full width of the line at half peak height, when the contribution from overlapping spectra was taken into account. The relaxation times were obtained from the expression $\tau_s = (\pi \cdot \Delta v)^{-1}$ with a maximum error of 10%.

The flow birefringence relaxation times τ_f for the above *n*-alkanes over a more limited temperature range have been evaluated from the data of Champion and North [3] using Equations (3) and (4), and are compared with the values of τ_s in Figure 2.

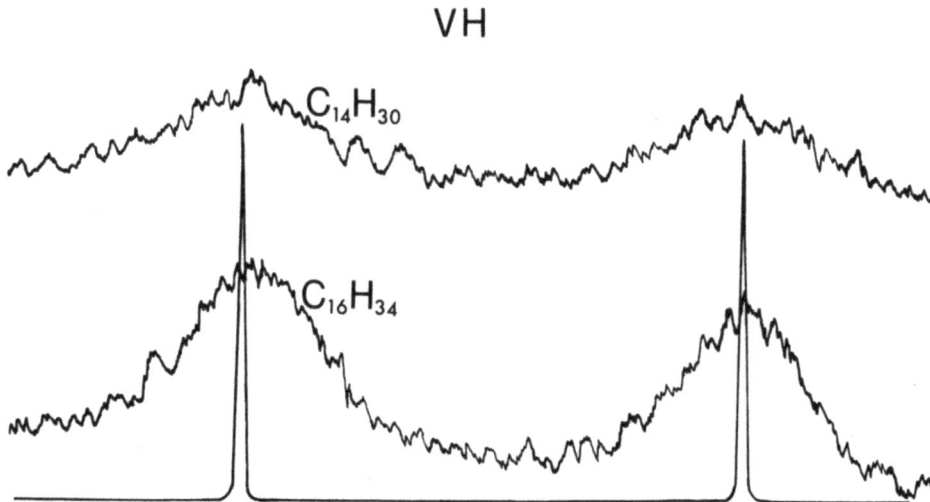

Fig. 1. Typical VH spectra at 60 °C for $C_{14}H_{30}$ and $C_{16}H_{34}$ with an etalon spacing of 2.0 cm. The instrument profile is also shown.

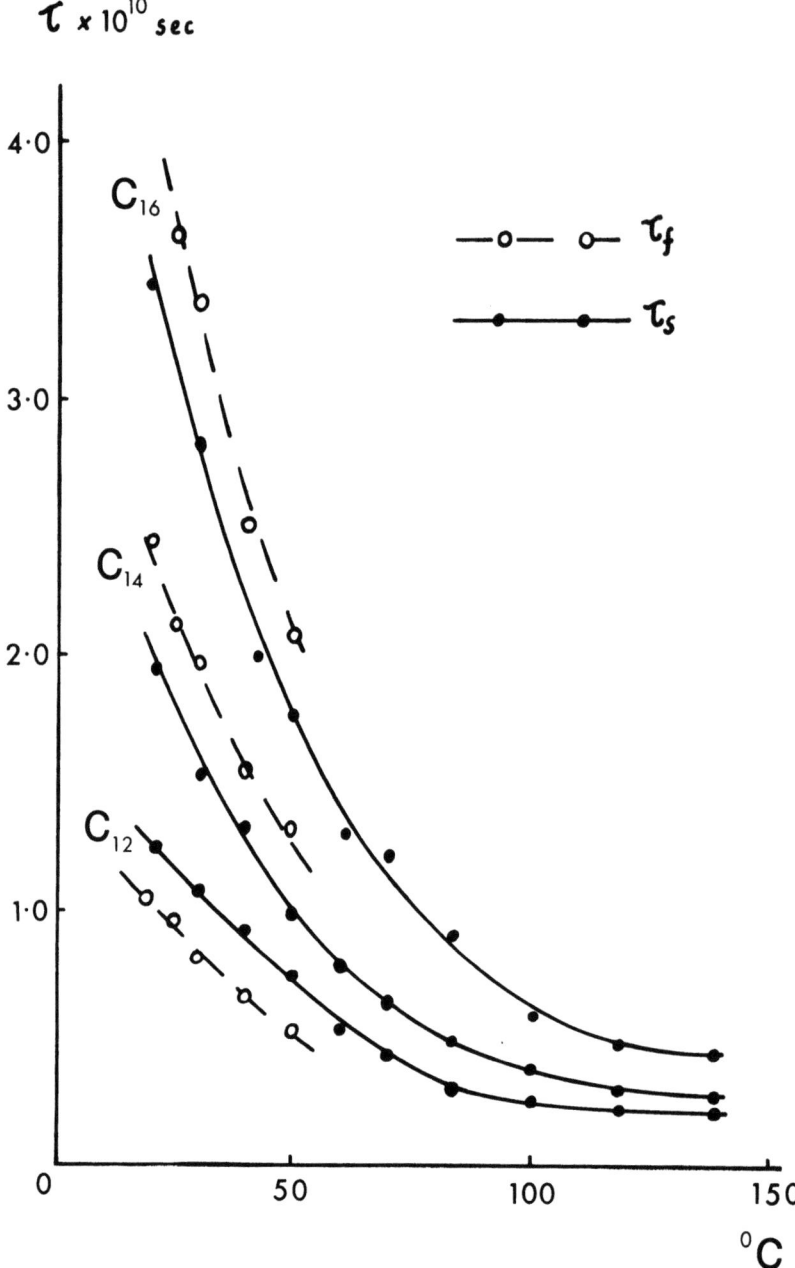

Fig. 2. The relaxation times determined from depolarised light scattering τ_s and flow birefringence τ_f, as a function of temperature. (Note $C_{15}H_{32}$ times not shown for clarity – data may be obtained from Figure 4).

The values of τ_f and τ_s agree to within 20% indicating that the same relaxation mechanism is being observed in the two experiments. This result is the first experimental verification of the Leontovich theory [1]. Also, the end to end distance of a n-alkane molecule depends not only on the number of atoms in the molecule but also on the distribution of rotational isomeric states of the molecule. When such a molecule is subjected to shear flow, this equilibrium distribution is perturbed and results in an increase in the molecular end to end distance. It has been suggested previously [3]

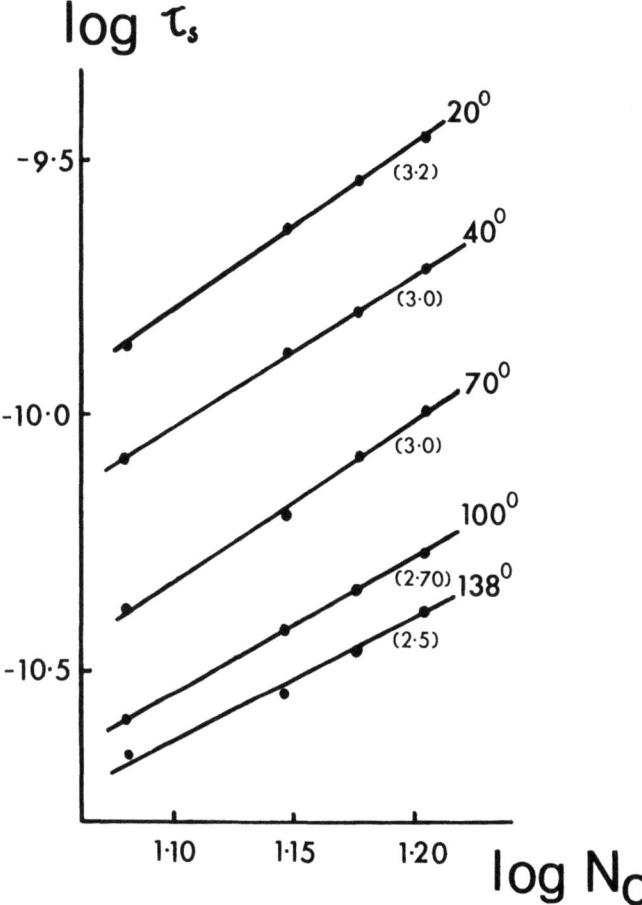

Fig. 3. The relaxation time determined from depolarised light scattering τ_s as a function of the number of carbon atoms N_c in the chain molecule, at various temperatures. The slopes of the lines are in brackets.

that the finite probability of a gauche isomer being present, introducing non-linearity in the molecular shape, increases abruptly on increasing the number of carbon atoms in the chain molecule from twelve to fourteen in the homologous series. As the reorientation time of the molecule is a function of the end to end distance, then the behaviour shown in Figure 2, that for $C_{12}H_{26}$, $\tau_f < \tau_s$, whilst for $C_{14}H_{30}$, $C_{15}H_{32}$, $C_{16}H_{34}$, $\tau_f > \tau_s$, is to be expected.

In Figure 3 τ_s is plotted as a function of the number of carbon atoms N_c in the chain molecule. For a rigid molecule, the orientational relaxation time is proportional to (length)^3 or $(N_c)^3$, [5], while for a long chain polymer molecule τ is proportional to $(N_c)^{3/2}$, [6]. The rigidity of the short chain polymer molecule decreases as its temperature increases, consequently it would be expected that the dependence of τ_s on carbon number would slowly change from a $(N_c)^3$ dependence towards a $(N_c)^{3/2}$ as the temperature is increased. Such behaviour is shown in Figure 3. Figure

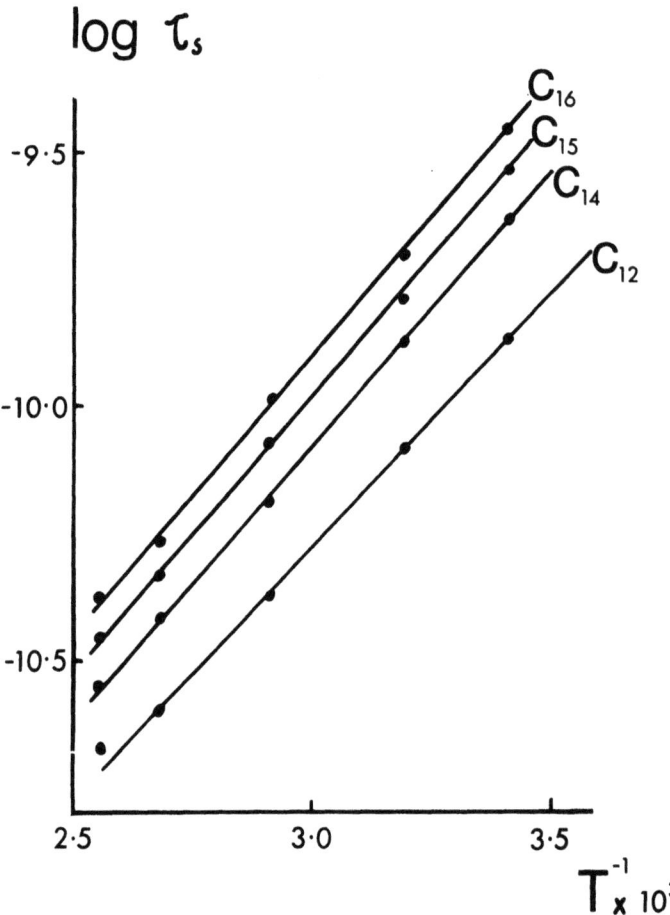

Fig. 4. The relaxation time determined from depolarised light scattering τ_s as a function of inverse temperature for the n-alkanes $C_{12}H_{26}$, $C_{14}H_{30}$, $C_{15}H_{32}$, $C_{16}H_{34}$.

Figure 4 shows that there is an Arrhenius temperature dependence of the relaxation time τ_s on temperature and illustrates, as in Figure 2, from the slopes of the lines the difference in behaviour of $C_{12}H_{26}$ compared with $C_{14}H_{30}$. The activation energies calculated from the slopes of the lines are $E_a(C_{12}H_{26}) = 4.5$ kcal mole^{-1}, $E_a(C_{14}H_{30}) = 4.9$ kcal mole^{-1}, $E_a(C_{15}H_{32}) = E_a(C_{16}H_{34}) = 5.0$ kcal mole^{-1}, showing an abrupt change occurs in E_a between molecules with twelve and fourteen carbon atoms in the homologous series. This may be interpreted as being due to the change in the probability of the presence of a gauche isomeric state, supporting the earlier interpretation of the change in the relative magnitudes of τ_s and τ_f shown in Figure 2. Furthermore this change in E_a (~ 0.5 kcal mole^{-1}) corresponds also to the difference in energy between the trans (linear) and gauche (kinked) isomeric states for rotation about a single carbon-carbon bond, indicating that the $C_{14}H_{30}$ molecule has on average one more gauche isomer, or kink in the chain, compared with the $C_{12}H_{26}$ molecule.

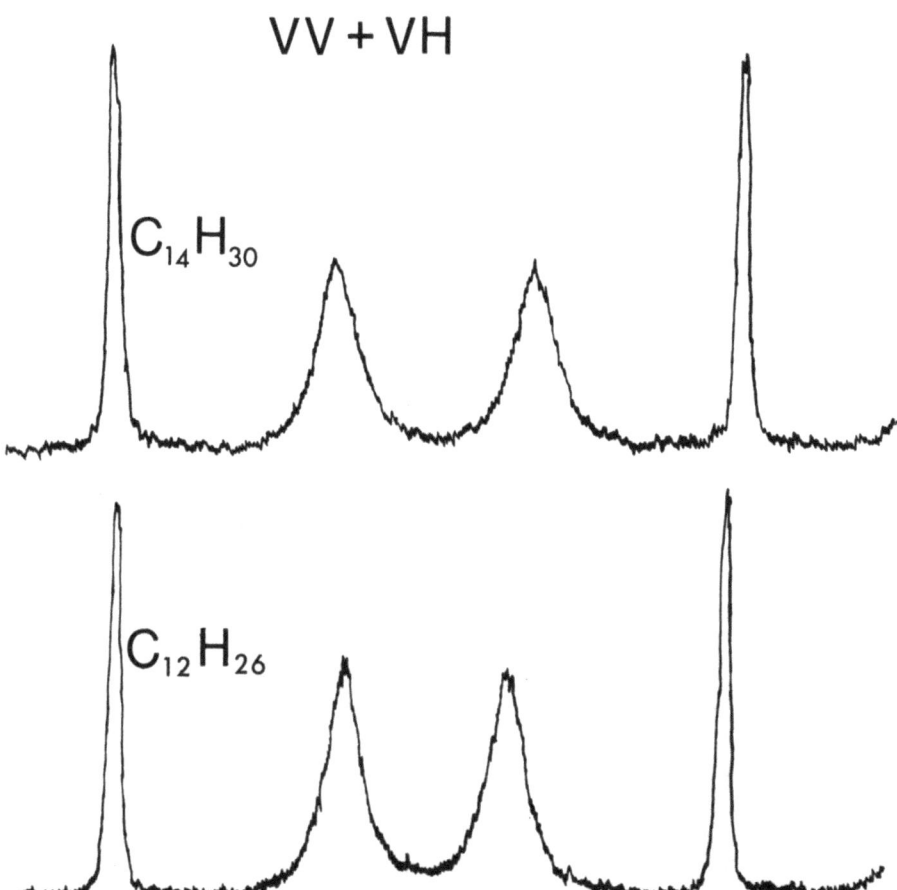

Fig. 5. Typical VH + VV spectra for $C_{12}H_{26}$, $C_{14}H_{30}$ at 50°C using an etalon spacing of 2.0 cm.

2.2. POLARISED SPECTRA

The polarised Brillouin spectra were obtained for the n-alkanes $C_{12}H_{26}$ to $C_{36}H_{74}$ over the temperature range 20°C to 140°C in the liquid state, using a pressure scanned Fabry-Pérot interferometer of 1.0 cm spacing as well as the piezo-electrically scanned instrument used above. Typical spectra (VV + VH) are shown in Figure 5. They were

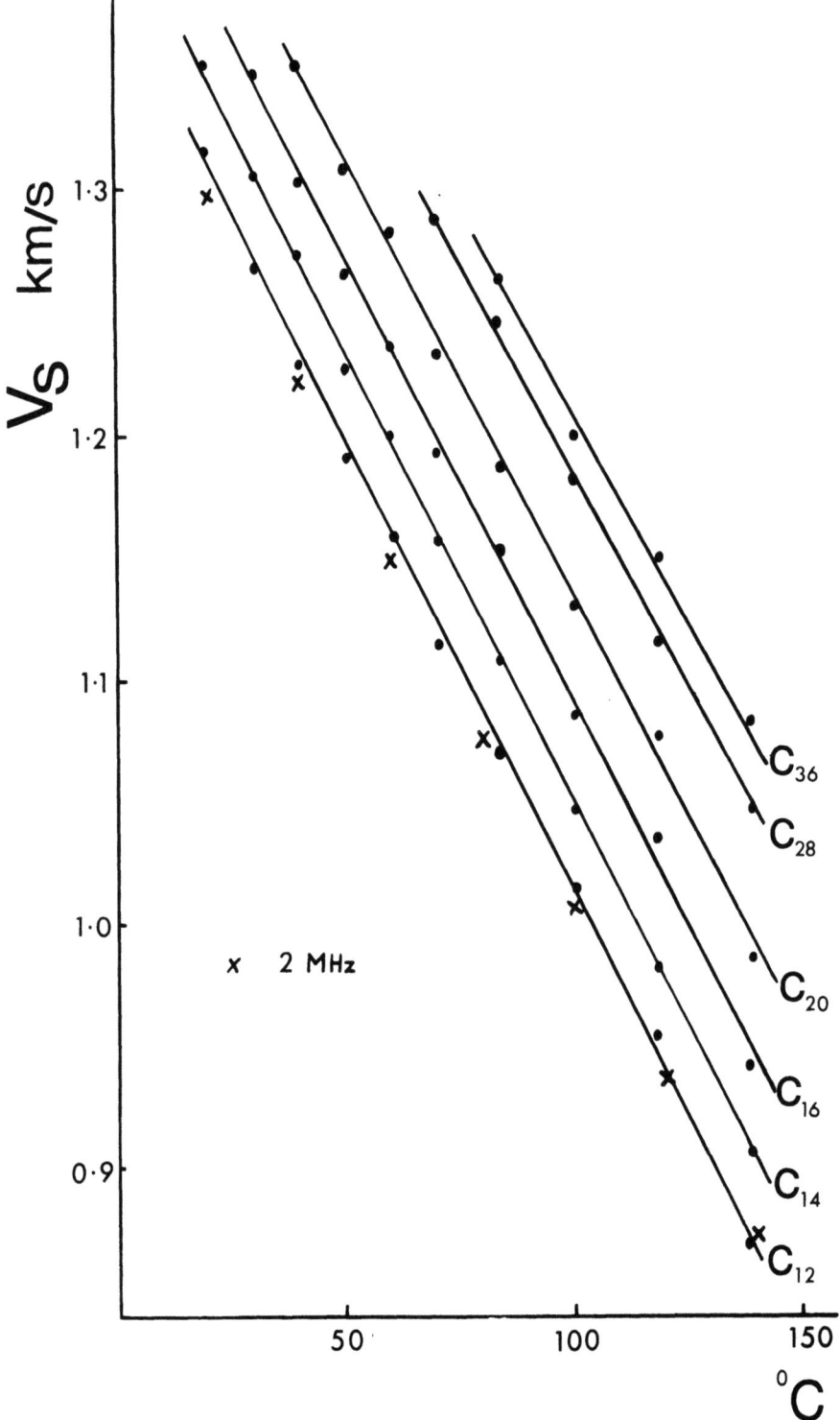

Fig. 6. The hypersonic phonon velocity V_S for the n-alkanes $C_{12}H_{26}$ to $C_{36}H_{74}$ as a function of temperature. The values of Boelhouwer [7] obtained at 2 MHz are also shown for $C_{12}H_{26}$.

measured with the sensitivity reduced by a factor of ~50 after removing the H polariser from the scattered beam. The Brillouin line shifts, widths and peak heights were obtained at a fixed scattering angle of 90°. The line shift measurements were made using the more-linear pressure-scanned instrument, as the position of the peak heights could be located with greater accuracy than the values of the peak widths. The hypersonic phonon velocities were obtained using Equation (1) and are shown in Figure 6. V_S decreases linearly with increasing temperature and has the same

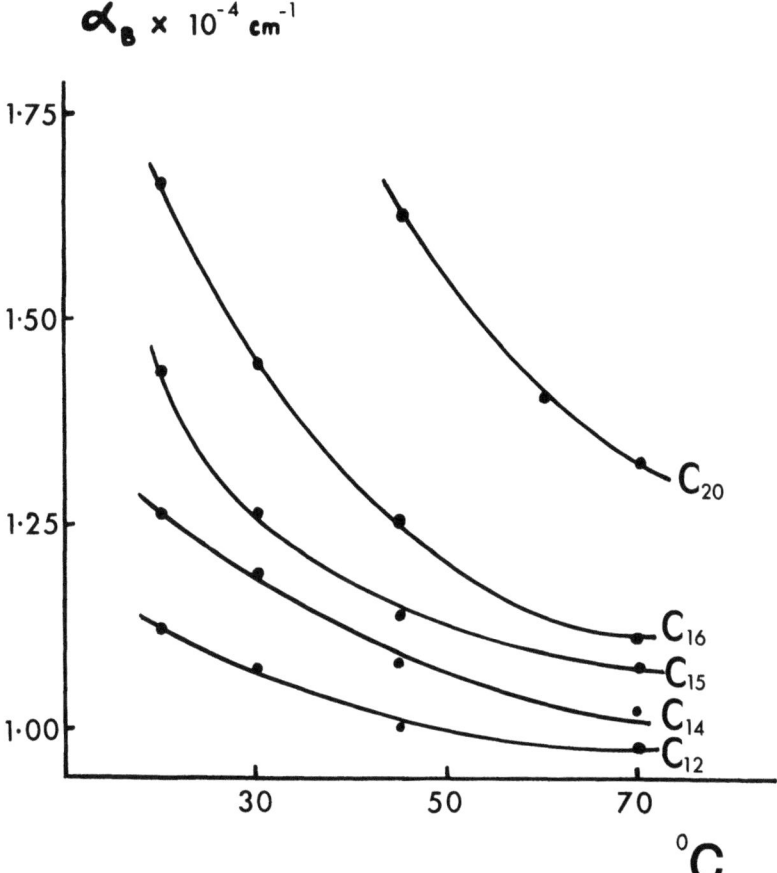

Fig. 7. The absorption coefficient of n-alkanes, derived from the width of the Brillouin line, as a function of temperature and at frequency ~5 GHz.

value to within 1% as that determined by Boelhouwer [7] using a pulsed echo technique at 2 MHz, indicating that little dispersion occurs in the frequency range 2 MHz to 5 GHz. The overall accuracy of the hypersonic velocity measurements was 0.5%.

Figure 7 shows the temperature dependence of the absorption coefficient α, determined using Equation (2). It decreases with increasing temperature, and figure 8 shows the marked increase of absorption with increasing number of carbon atoms,

N_c, in the chain molecule, similar behaviour has been reported by Micheles [8] in ultrasonic measurements performed at 7.96 MHz. The value of α/f^2 at this frequency for $C_{16}H_{34}$ is ~ 100 at 20 °C, the corresponding hypersonic value at 5 GHz is ~ 80. The value of $\alpha_{\text{class}}/f^2$ is ~ 50 indicating that at the hypersonic frequency of the

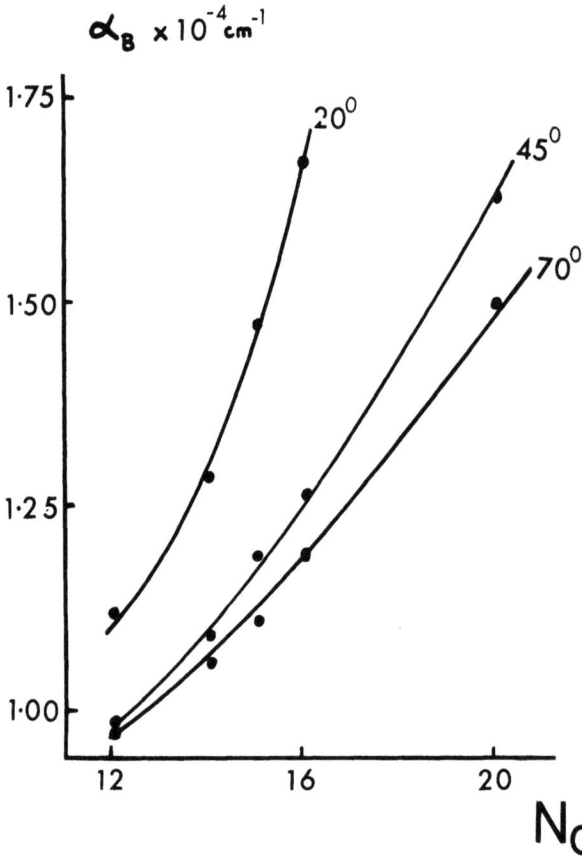

Fig. 8. The absorption coefficient of n-alkanes as a function of the number of carbon atoms N_c in the chain molecule at various temperatures.

measurements exchange of energy between internal and external degrees of freedom is possible. The variation of α with temperature shown in Figure 7 follows closely that predicted for α_{class} from the temperature dependence of the viscosity and density of the alkane and further substantiates the absence of any dispersion process.

References

1. Leontovich, M.: *J. Phys. U.S.S.R.* **4**, 499 (1941).
2. Meeten, G. H.: *Nature* **218**, 761 (1968).
3. Champion, J. V. and North, P. F.: *Trans. Faraday Soc.* **64**, 2287 (1968).
4. Meeten, G. H.: *J. Chim. Phys.* **66**, 1565 (1969).

5. Broersma, S.: *J. Chem. Phys.* **32**, 1626 (1960).
6. Isihara, A.: *J. Chem. Phys.* **47**, 3821 (1967).
7. Boelhouwer, J. W. M.: *Physica* **34**, 484 (1967).
8. Micheles, B.: *J. Chim. Phys.* **63**, 1123 (1966).

DISCUSSIONS

Candau: Could you give some details as to the way you obtain the flow birefringence relaxation time from the magnitude of the flow birefringence?

Champion: The flow birefringence relaxation time τ_f was obtained using the phenomenological theory of Leontovich. A dynamic equilibrium exists when a hydrodynamic field is applied to the liquid enabling the orientational relaxation time to be obtained from a steady state situation.

BRILLOUIN SCATTERING IN THE INERT-GAS LIQUIDS: ARGON AND KRYPTON

B. Y. BAHARUDIN*, D. A. JACKSON, and P. E. SCHOEN**

Physics Laboratory, The University, Canterbury, Kent. CT2 7NR, England

Abstract. We have observed the hypersonic velocities of sound in liquid argon and liquid krypton along their liquid-vapour coexistence curves from their respective triple points to their critical points. Measurements of the sound absorption coefficients were made over a more restricted temperature range. Within the limits of experimental error ($\sim 0.2\%$) we find no evidence of dispersion in the hypersonic velocity relative to the ultrasonic velocity in either liquid.

Résumé. Nous avons observé les vitesses hypersonores du son dans les liquides argon et krypton le long de leurs courbes de coexistence liquide-vapeur du point triple jusqu'au point critique. On a mesuré les coefficients de l'absorption du son dans une étendue de température plus restreinte. Nous n'avons éprouvé aucune dispersion de la vitesse hypersonore relatif à la vitesse ultrasonore dans les deux liquides dans la limite d'erreur expérimentale ($\sim 0,2\%$).

1. Introduction

We have observed Brillouin scattering in liquid argon and liquid krypton in equilibrium with the vapour from the triple point to the critical point. These are the first measurements of the hypersonic velocity in liquid krypton and of the hypersonic absorption in liquid argon and krypton.

Our data for argon were taken at Brillouin frequencies ranging from 400 MHz to 4 GHz. This large dynamic range has allowed us to search for dispersion in the sound velocity without having to rely upon intercomparisons between our data and that of other experiments performed using different apparatus under different conditions. The present data were taken over a large range of temperatures and frequencies in a single experiment, using the same sample.

Recent theoretical analyses [1, 2, 3] have suggested that at sufficiently high frequencies, approaching the inverse collision time, the propagation of sound in a liquid should show non-classical behaviour. In particular the velocity of sound should show a negative dispersion. The maximum frequency for Brillouin scattering in rare gas liquids is of order 10 GHz which is several orders of magnitude smaller than the collision frequency in simple liquids. Nevertheless it is possible that the onset of negative dispersion may be discernable. A model calculation by Boon and Deguent [3] predicts negative dispersion in liquid argon of $3 \times 10^{-3}\%$ for hypersound relative to ultrasound, but this effect is far too small to be detected by present Brillouin scattering techniques.

On the other hand Fleury and Boon [4] have reported measurements of the sound velocity in argon and neon by Brillouin scattering, over a frequency range of 1 GHz

---

* Supported by the National University of Malaysia.
** S.R.C. Research Fellow.

to 3 GHz, which in the case of argon were systematically ~0.4% lower than the values obtained previously by ultrasonic (~1 MHz) measurements. However, the overall accuracy of their measurements was ~0.5%. No values for the ultrasonic sound velocity in neon were then available to Fleury and Boon, but a subsequent experiment by Larson et al. [5] again indicated the ultrasonic velocity is higher than the hypersonic, in this case by about 1.5%. We have, therefore, performed measurements of the sound velocity of improved accuracy and extended the frequency range from hypersonic values downwards to reduce the gap between hypersonic and ultrasonic results as far as possible.

Additional information about a dispersion process may also be obtained from a study of the sound absorption coefficient as a function of temperature and frequency. In a simple atomic liquid in which there should be no internal relaxation process leading to dispersion effects the 'classical' sound attenuation coefficient α_c is simply related to the shear viscosity η_s and thermal conductivity κ of the liquid by [6]

$$\alpha_c = \tfrac{2}{3}[(2\pi f)^2/\varrho v^3]\left[\eta_s + \tfrac{3}{4}(\gamma - 1)\frac{\kappa}{C_p}\right],$$

where ϱ is the density of the liquid, γ the ratio of the specific heats and C_p the specific heat at constant pressure. The measured absorption coefficient α is larger than α_c and it is usual to associate this 'excess' absorption with a volume or bulk viscosity [6]

$$\eta_B = [2\varrho v^3/(2\pi)^2]\,[(\alpha-\alpha_c)/f^2].$$

The resolution of the Fleury and Boon experiment was insufficient to permit the determination of the sound absorption coefficient in the case of argon or neon. In our experiment we have a ten-fold greater instrumental resolution and this has enabled us to measure the sound absorption coefficients and compare the hypersonic η_B with ultrasonic η_B.

2. Experimental Technique

Light which is inelastically scattered from sound waves is shifted in frequency by an amount $f = qv/2\pi$. The sound velocity, v, is a function of temperature and the wave vector q is a function of the scattering angle θ which is given by

$$q = \frac{4\pi n}{\lambda}\sin\theta/2,$$

where n is the refractive index of the scattering medium and λ is the vacuum wavelength of the incident light. By choosing θ and the temperature one can vary f over a large range, although q is determined most accurately when θ is ~180°, due to its $\sin\theta/2$ dependence. Backscattering measurements therefore provide the best determination of the absolute value of the sound velocity, and the best data for comparison with ultrasonic measurements to check for velocity dispersion.

We have performed measurements at both back- and forward-scattering angles. This pair of angles yields the velocity of sound at frequencies differing by an order of magnitude or more, thus providing a test for dispersion independent of ultrasonic measurements. Additional measurements have also been made for 90° scattering, since at this angle the scattered light is very nearly free of stray light scattered from the cell windows.

Frequency shifts were determined by using two different Fabry-Pérot (FP) interferometers with Free Spectral Ranges (FSR) of 3 GHz and 5 GHz, and linewidths of 70 MHz and 110 MHz respectively. It was necessary to use two such instruments because in both liquid argon and krypton the sound velocity has a large temperature dependence, and since the FP is a periodic instrument the Brillouin frequency shift is some multiple of the FSR at certain temperatures. Consequently the Brillouin peaks tend to be lost in the strong zero frequency signal of the elastically scattered light. At other temperatures the peaks may overlap and be unresolved.

Measurements of the velocity of sound in argon at ultrasonic frequencies have been made by several authors [7, 8], and their measurements agree to $\sim 0.1\%$. To obtain an accuracy of that order by Brillouin scattering requires close attention to a number of experimental problems, particularly the accurate determination of the frequency shift, the measurement of scattering angles and the calculation of the refractive index. The most important source of error in our meauurements was laser frequency instability and non-linearity in the scan of our Fabry-Pérot interferometer. These combined to produce an uncertainty in the frequency shift of the Brillouin lines of about ± 5 MHz.

The scattering angles were defined by two laser beams crossing within the sample volume. By removing the sample cell, replacing it with a glass slide mounted on a goniometer and reflecting the beams back on each other we were able to measure scattering angles (after allowing for refraction effects in the cell) to $\sim 1'$.

The temperature of the scattering cell was stabilized to ± 0.003 deg for periods of > 30 min (the time required to obtain several orders of a spectrum) using a Fisher proportional controller and thermistor sensor. The temperature change across the sample region, determined using a pair of platinum resistance thermometers, was 0.005 deg. Absolute temperatures, measured by platinum thermometer (calibrated to ± 0.005 deg) and checked against the vapour pressure of the liquid argon, were determined with an accuracy better than 0.02 deg. The rate of change of hypersonic velocity in argon with temperature is about 1%/deg. Consequently the uncertainty in our temperature measurement produced an uncertainty in the measured velocity of less than 0.02%.

Our gases were supplied by the British Oxygen Company. (Argon: 99.9995% pure; krypton: 99.99% pure). The gases were checked by mass spectrometer to confirm their purity after use.

The overall accuracy of our velocity measurements for argon was better than $\pm 0.2\%$. The accuracy of the measurements for krypton was slightly less ($\pm 0.5\%$) due to a deterioration in the stability of our laser and uncertainty in the value of the refractive index.

The light source for both the argon and krypton experiments was an Argon-ion laser producing ~ 600 mW at 5145 Å.

The relation between Δf, the full width at half maximum of the Brillouin line, and the sound attenuation coefficient, α, is given by [9],

$$\alpha = \frac{\pi \Delta f}{v}.$$

Since Δf is proportional to q^2, the Brillouin linewidth as the Brillouin frequency shift was most accurately measured in a back-scattering experiment. In the rare gas liquids the linewidths are narrow. The predicted width using the ultrasonic value of α in the case of argon is ~ 50 MHz for a 180° scattering angle. The resolving powers of the FP's used for the velocity measurements were not sufficient to allow accurate linewidth determination and so a further experiment was performed using a confocal piezo-electrically scanned FP of ~ 25 MHz resolution. The nonlinearity of this device was $\sim 2-3\%$ so that it could not be used to measure velocity accurately.

The data thus obtained were deconvoluted using the method of La Macchia et al. [10]. Numerical calculations showed that corrections for aperture broadening were unnecessary because the collection solid angles used in the experiment were small ($\sim 3 \times 10^{-5}$ sr).

3. Results

In order to calculate the velocity from the Brillouin frequency shifts we must know both the wavelength and temperature dependence of the refractive indices of the liquids. In the case of liquid argon accurate values for the refractive index along the liquid-vapour coexistence line exist [11, 12], but for liquid krypton the data are scanty [12, 13, 14]. We have used the available information for krypton together with density data [15] to calculate the refractive index from the Clausius-Mossoti equation.

TABLE I

Sound velocity in liquid argon along the saturated vapour pressure curve (only representative points are tabulated)

T (K)	θ	f (GHz)	v (ms^{-1})	
			Hypersonic	Ultrasonic[a]
± 0.02	$\pm 1'$	± 0.006	± 2.0	± 0.08
85.25	13°35'	0.393	853.8	851.32
89.28	13°35'	0.379	824.5	823.54
110.12	13°35'	0.307	668.8	667.96
85.28	169°51'	4.062	850.2	851.16
89.32	169°51'	3.922	824.6	823.31
101.06	169°51'	3.474	737.5	738.53
84.77	93°48'	2.970	853.9	854.61
90.99	93°48'	2.803	810.8	811.59
102.99	102°47'	2.627	724.7	723.88
124.32	102°47'	1.930	544.9	543.99
143.19	102°47'	1.111	323.4	323.81

[a] Reference 8.

Tables I and II give a selection of our data for the Brillouin shifts and hypersonic velocities for liquid argon and liquid krypton.

TABLE II

Sound velocity in liquid krypton along the saturated vapour pressure curve (only representative points are tabulated)

T (K)	n	θ	f (GHz)	v (ms⁻¹)	
				Hypersonic	Ultrasonic[a]
±0.02	calc.	±1'	±0.010	±4.0	±0.06
117.12	1.3026	169°20'	3.526	693.3	695.53
125.26	1.2938	169°20'	3.307	659.1	662.61
129.07	1.2902	169°20'	3.256	650.8	646.80
117.97	1.3016	93°48'	2.536	691.4	692.15
126.53	1.2926	93°48'	2.383	654.1	657.37
132.35	1.2871	93°48'	2.289	631.0	632.99
145.35	1.2728	93°48'	2.072	577.3	576.11
162.19	1.2501	93°48'	1.732	491.1	495.55
197.19	1.1901	93°48'	0.916	272.6	272.53

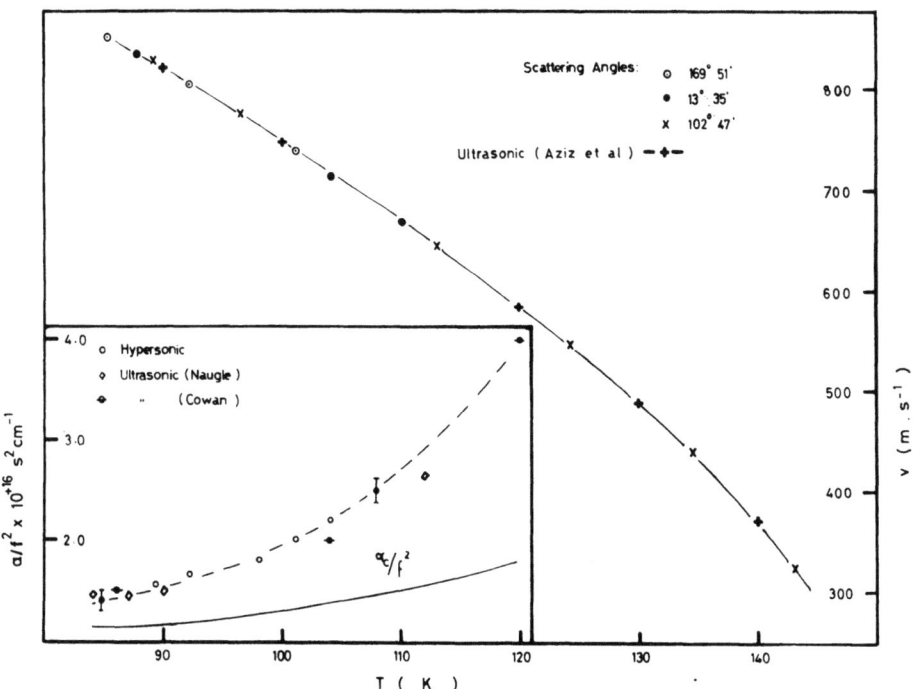

Fig. 1. *Argon:* Sound velocity vs temperature. Size of symbol is equal to error bar. ⎯⎯+⎯⎯ line is polynomial fit of Aziz *et al.* [8], to ultrasonic data. (Only representative points are shown.) *Inset:* Absorption vs temperature. Typical error bar is shown. Dashed line is 'least squares' fit to hypersonic data, solid line, calculated from classical absorption formula. Ultrasonic data are that of Naugle [16] and Cowan [17].

Because of the stray light problem encountered in forward- and back-scattering, measurements at these angles were taken over a temperature range of only 30 deg. As can be seen from Figures 1 and 2 the data for all angles agree with each other and with ultrasonic data to within $\pm 0.2\%$ for argon and $\pm 0.4\%$ for krypton for all temperatures. There is no sign of any systematic difference.

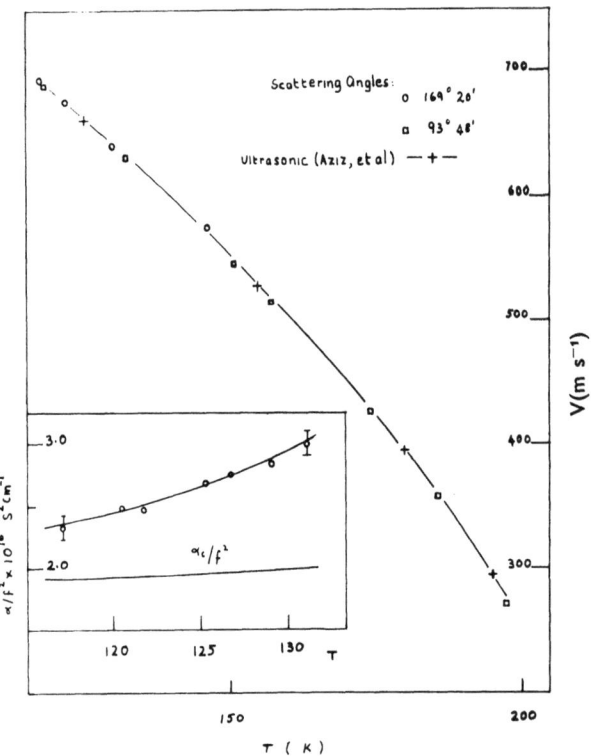

Fig. 2. *Krypton:* Sound velocity vs temperature. Size of symbol is equal to error bar. ──+── line is polynomial fit of Aziz *et al.* [8] to ultrasonic data. *Inset:* Absorption vs temperature. Dashed line is 'least squares' fit to hypersonic data, (typical error bar indicated). Solid line is calculated from classical absorption formula.

Figures 1 and 2 show the absorption coefficients for argon and krypton, respectively, expressed as α/f^2. Laser instability, Fabry-Pérot linewidth and deconvolution effects restricted the accuracy of these measurements to $\pm 10\%$. Figure 1 also presents the ultrasonic absorption data for argon (which is of comparable accuracy) [16, 17]. The agreement between ultrasonic and hypersonic absorption measurements is within experimental error.

Numerical values for the absorption coefficients at several temperatures are given in Table III for argon and krypton. Values of the apparent bulk viscosity, η_B, corresponding to the excess attenuation [18], $\alpha - \alpha_c$, are also presented in this table.

TABLE III

Smoothed sound attenuation coefficients and viscosities in liquid argon and krypton. η_s from Reference 18. η_B deduced from this experiment

	T (K)	$\alpha/f^2 \cdot 10^{16}$ (s^2 cm^{-1})	η_S (mP)	η_B (mP)	η_B/η_s
Argon	85	1.40	2.73	1.15	0.42
	90	1.57	2.32	1.53	0.66
	100	1.97	1.70	1.89	1.11
	110	2.74	1.30	2.32	1.79
Krypton	116	2.29	4.47	1.94	0.43
	120	2.44	4.04	2.27	0.56
	125	2.67	3.67	2.62	0.71
	130	2.93	3.37	2.93	0.87

4. Conclusions

We have demonstrated the versatility and accuracy of the Brillouin scattering technique in covering a large dynamic range in the measurement of sound velocity and attenuation. With better control of stray light and proper choice of FP spacers and scattering angles one could approach even closer the range of frequencies measured by ultrasonic techniques.

We have found no disagreement between our sound velocity measurements and those of ultrasonics and no difference between our own forward- and back-scattering measurements. We find no evidence of dispersion over the frequency range from 1 MHz to 4 GHz in either argon or krypton, and suggest that if dispersion exists, it will be necessary to turn to other experimental techniques and/or higher frequencies to find it.

Experiments are in progress now to measure hypersonic velocity and absorption in liquid xenon and neon. A more detailed report of our measurements in the four rare gas liquids, neon, argon, krypton and xenon, will be given in a future publication.

References

1. Gillis, N. S. and Puff, P. D.: *Phys. Rev. Letters* **16**, 606 (1966).
2. Berne, B. J., Boon, J. P., and Rice, S. A.: *J. Chem. Phys.* **47**, 2283 (1967).
3. Boon, J. P. and Deguent, P: *Phys. Rev.* **A2**, 2542 (1970).
4. Fleury, P. A. and Boon, J. P.: *Phys. Rev.* **186**, 244 (1966).
5. Larson, E. V., Naugle, D. G., and Adair, III T. W.: *J. Chem. Phys.* **54**, 2429 (1971).
6. Herzfeld, K. F. and Litovitz, T. A.: *Absorption and Dispersion of Ultrasonic Waves*, Academic Press, N.Y., London, 1959, p. 44.
7. Thoen, J., Vangeel, E., and Van Dael, W.: *Physica* **45**, 339 (1969).
8. Aziz, R. A., Bowman, D. H., Lim, C. C.: *Can. J. Chem.* **45**, 2079 (1967).
9. Fabelinskii, I. L.: *Molecular Scattering of Light*, Plenum Press, N. Y., 1968, and p. 89.
10. Leidecker, H. W., Jr. and La Macchia, J. T.: *J. Acad. Soc. Am.* **43**, 143 (1968).
11. Abbis, C. P., Knobler, C. M., Teague, D. K., and Pings, C. J.: *J. Chem. Phys.* **42**, 4145 (1965).
12. Sinnock, A. C. and Smith, B. L.: *Phys. Rev.* **181**, 1297 (1969).
13. Marcous, J.: *Can. J. Phys.* **48**, 243 (1970).
14. Marcous, J.: *Can. J. Phys.* **48**, 1947 (1970).

15. Cook, G. A.: *Argon, Helium and the Rare Gases*, Interscience Publications, N.Y., London, 1961, Ch. IX.
16. Naugle, D. G.: *J. Chem. Phys.* **44**, 741 (1966).
17. Ball, R. N. and Cowan, J. A.: *Phys. Can.* **27**, 68 (1971).
18. Parameters used to calculate the values of α_c all obtained from the literature.
    Boon, J. P. and Thomas, G.: *Physica* **29**, 208 (1963); Förster, S.: *Cryogenics* **3**, 176 (1963); Gladun, C. and Menzel, F.: *Cryogenics* **10** 210 (1971); also Reference 15 gives values of most of the parameters.

RAYLEIGH SCATTERING: DENSITY DEPENDENCE OF ORIENTATIONAL MOTIONS IN ACETONE AND BENZENE

J. DILL and T. A. LITOVITZ

Physics Dept., Catholic University of America, Washington, D.C., U.S.A.

Abstract. Results of a high pressure study of the depolarized scattering in acetone and benzene are presented. The density dependences of the angular velocity and angular position correlation functions are determined. In both liquids the AVCF show a negative overshoot, however the mechanism or reorientational motions appear to be significantly different.

Résumé. On présente les résultats de l'étude de la diffusion dépolarisée dans le benzène et l'acétone sous haute pression. On détermine l'influence de la densité sur les fonctions de corrélation de la vitesse angulaire et de la position angulaire. Dans les deux liquides on observe un dépassement de la fonction de corrélation de la vitesse angulaire, mais le mécanisme des mouvements réorientationnels semble différer de façon significative.

1. Introduction

One of the advantages of Rayleigh scattering for measurement of orientational motions in liquids is its ability to measure the very short time behavior ($\sim 10^{-13}$ s). To exploit this the orientational contributions to the spectrum must be separated from others such as collision induced scattering. In this research we attempt to do this and study the density dependence of the orientational motions in two liquids.

2. Theory

For single particle scattering the spectrum is given by [1]

$$I(\omega) \propto \int dt\, e^{-i\omega t} \langle (\hat{n}_i \cdot \boldsymbol{\alpha}(0) \cdot \hat{n}_s)(\hat{n}_i \cdot \boldsymbol{\alpha}(t) \cdot \hat{n}_s) \rangle, \qquad (1)$$

where $\hat{n}_i$ and $\hat{n}_s$ are unit vectors in the direction of polarization of the incident and scattered electromagnetic waves respectively, and $\boldsymbol{\alpha}$ is the polarizability tensor of a molecule in the liquid. In this work we studied depolarized light scattering at 90° which will be observed if $\boldsymbol{\alpha}$ is anisotropic. There are two common processes which give rise to a depolarized spectrum. If the molecule has a permanent anisotropy in $\boldsymbol{\alpha}$ a spectrum related to the reorientation of the molecules will be observed. Interactions of molecules with their neighbors also give rise to a depolarized spectrum, this related to the time dependence of the interactions. When both processes are present the polarizability can be written $\boldsymbol{\alpha}(t) = \boldsymbol{\alpha}_0 + \boldsymbol{\alpha}_t(t)$ where $\boldsymbol{\alpha}_0$ is the polarizability which a molecule would have if interactions were not present, and $\alpha_t(t)$ is the time dependent interaction induced polarizability. This polarizability leads to a spectrum of the form $I(\omega) = I_{OR}(\omega) + I_{INT}(\omega) + I_{OR\text{-}INT}(\omega)$ where $I_{OR}(\omega)$ is related to the reorientation of the molecule, (i.e. α_0), $I_{INT}(\omega)$ is related to $\alpha_t(t)$ and $I_{OR\text{-}INT}(\omega)$ is related to the cross correlation of the two processes.

In order to study the orientational spectrum one must separate $I_{OR}(\omega)$ from the measured spectrum. Such a process has been detailed by Dardy et al. [2] and will be briefly summarized here. There are two basic assumptions. First the cross correlation term $I_{OR\text{-}INT}$ is assumed to be negligible.

Then the interaction term $I_{INT}(\omega)$ is removed from the experimental spectrum as follows. The shape of $I_{INT}(\omega)$ is obtained from the slope of the experimental spectrum at high frequencies where the orientational contribution is small. At low frequencies where the shape is obscured by $I_{OR}(\omega)$ a shape calculated using a binary collision model is used. This calculated spectrum fits the observed spectrum well in liquids with spherical molecules where, due to the absence of a strong orientational component the collision spectrum can be followed to relatively low frequencies [3]. Since the collision contribution at low frequencies is a small part of our experimental spectrum (5% at most) the results are rather insensitive to the exact shape assumed there. The amplitude of the collision spectrum is obtained by using the fact that the second moment of an orientational spectrum, $M_2 = 6\,kT/I$. The experimental value of M_2 is calculated, and $I_{INT}(\omega)$ is adjusted until the second moment of the difference spectrum $I_{OR}(\omega) = I_{EXPT}(\omega) - I_{INT}(\omega)$ is equal to the theoretical orientational value.

From the orientational spectrum we obtain the angular position correlation function, APCF, and the angular velocity correlation function, AVCF. The relation between $I_{OR}(\omega)$ and the APCF is given by [4].

$$\langle P_2(x)\rangle = K\beta_0^2 \int d\omega\, e^{i\omega t} I_{OR}(\omega), \tag{2}$$

where $\langle P_2(x)\rangle = \tfrac{1}{2}(3\cos^2\theta(t) - 1)$ is the second order Legendre polynomial, K is a constant, $\beta_0 = \alpha_\| - \alpha_\perp$ is the anisotropy of the polarizability and $\theta(t)$ is the angle between the orientation at $t = 0$ and $t = t$. For times short compared to the decay time of $P_2(X)$ the AVCF can be obtained from $I_{OR}(\omega)$ using the relation [5, 6]

$$\langle \omega_\perp(0)\cdot\omega_\perp(t)\rangle = \frac{2}{l(l+1)} \int d\omega\, e^{i\omega t} \omega^2 I(\omega). \tag{3}$$

3. Analysis of Data

Using the above method $I_{OR}(\omega)$, the APCF and AVCF have been calculated in benzene and acetone. Due to the relatively weaker strength of the orientation spectrum at high frequencies in acetone, it should be noted that the shape of the AVCF is less accurate than in benzene. Data were taken as a function of pressure at room temperature ($23\,°C \pm 1\,°C$). Benzene was chosen because an extensive temperature study of $I_{OR}(\omega)$ was available for comparison [2]. Acetone was chosen because its 1 atm spectrum in contrast to that of benzene is very Lorentzian and shows no evidence of any shoulder in the high frequency wings indicating that the reorientational motions of the two liquids were quite different. Figures 1 and 2 present the APCF and AVCF in acetone. Figures 3 and 4 give the APCF and AVCF in benzene. A reduced time scale $[(t^* = (kT/I)^{1/2} t]$ has been used throughout.

In acetone at all pressures there is a fast initial loss of angular position correlation which goes over to an exponential at long times. The initial fast fall is generally associated with relatively free rotation of the molecule while the long time exponential is indicative of diffusional behavior. In the AVCF there is a cross over with a shallow negative peak which goes to zero rapidly.

The APCF in benzene is distinctly different from that in acetone. It starts to decay rapidly, hesitates, increases its slope again, and then slowly goes over to an exponential, taking over twice as long as acetone to become exponential. In the AVCF there is a cross over with a negative overshoot followed by a second positive overshoot.

In order to further discuss the details of the angular correlation functions we define a number of characteristic times. For the APCF these are: the average angular position correlation time, $\langle \tau_2 \rangle$ defined by the relation,

$$\langle \tau_2 \rangle = \int_0^\infty \langle P_2(x) \rangle \, dt \qquad (4)$$

the long time exponential slope τ_2, and the time required for the APCF to become exponential, τ_E. For the AVCF the only time which can be accurately determined experimentally is τ_{x0}, the time at which the AVCF changes sign for the first time. A

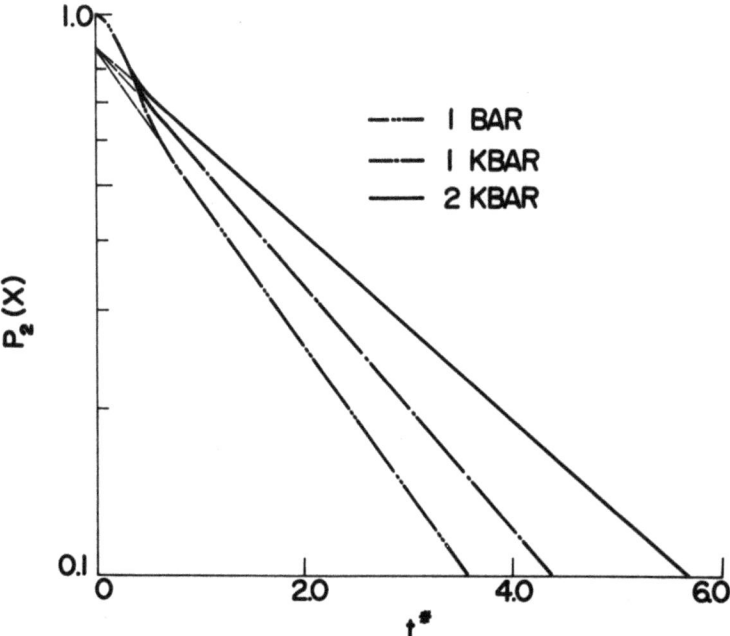

Fig. 1. Density dependence of APCF in acetone.

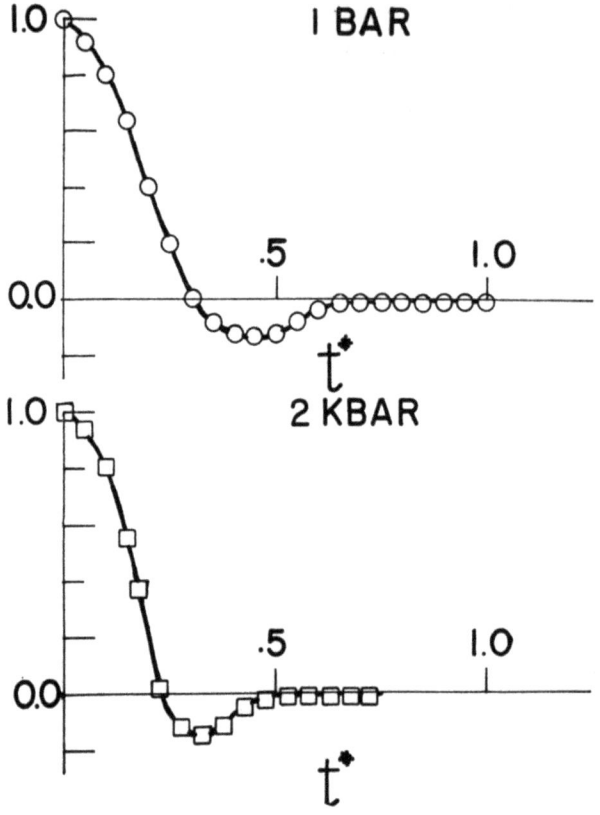

Fig. 2. Density dependence of AVCF in acetone.

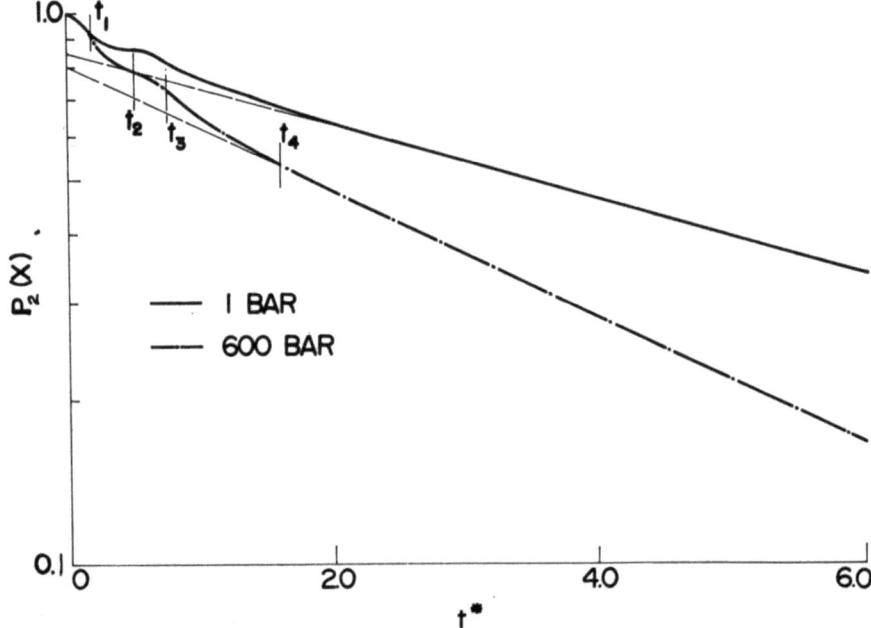

Fig. 3. Density dependence of APCF in benzene.

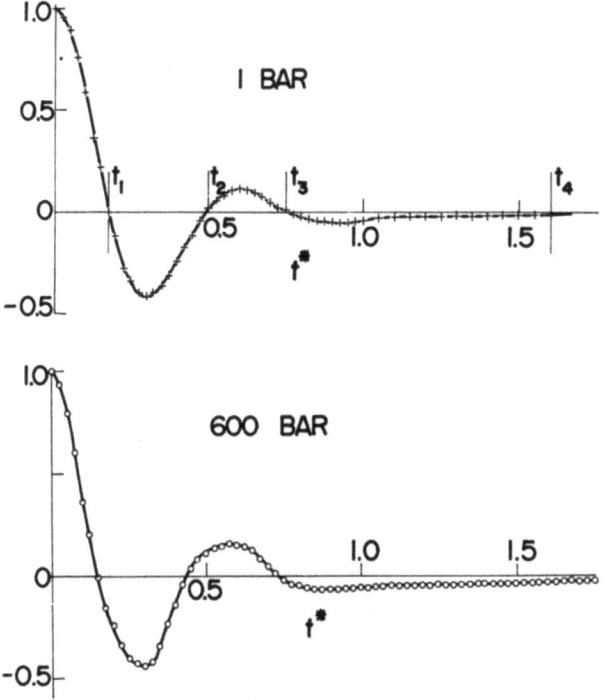

Fig. 4. Density dependence of AVCF in benzene.

second time related to the AVCF is $\langle \tau_J \rangle$, the average angular momentum correlation time defined by

$$\langle \tau_J \rangle = \int_0^\infty dt \langle A_J(t) \rangle, \qquad (5)$$

where $\langle A_J(t) \rangle$ is the normalized angular momentum function which is equivalent to the normalized AVCF. Unfortunately, since our experimentally obtained AVCF is valid only at short times we cannot perform the integral in Equation (5) to obtain $\langle \tau_J \rangle$. Instead we shall use the Hubbard relation [8].

$$\langle \tau_J \rangle = 6kT/I \langle \tau_2 \rangle \qquad (6)$$

to calculate $\langle \tau_J \rangle$ from $\langle \tau_2 \rangle$ which can be accurately determined. The Hubbard relation is applicable only for small angular step size with loss of angular velocity correlation after each collision. We shall see that although this model is not strictly applicable to our work its use can provide us with some insight into the actual interpretation of $\langle \tau_J \rangle$. Table I lists all the times defined above.

3.1. DENSITY DEPENDENCE: $\langle \tau_2 \rangle$ AND τ_2

In both liquids $\langle \tau_2 \rangle$ and τ_2 increase with increasing density as one would expect. The

TABLE I

Benzene

P (kbar)	$\varrho(p)/\varrho(1)$	$\langle\tau^*_2\rangle$	$\tau_2^*/\langle\tau^*_2\rangle$	τ_{x0}^*	τ_E^*	n_E
1	1.0	3.9	1.17	0.18	1.6	9
250	1.02	4.6	1.19	0.17	1.7	10
489	1.04	5.7	1.17	0.16	1.9	12
666	1.05	7.4	1.16	0.15	2.0	13

Acetone

P (kbar)	$\varrho(p)/\varrho(1)$	$\langle\tau^*_2\rangle$	$\tau_2^*/\langle\tau^*_2\rangle$	τ_{x0}^*	τ_E^*	n_E
1	1.0	1.47	1.30	0.32	0.65	2.0
500	1.04	1.65	1.33	0.29	0.61	2.1
1000	1.075	1.82	1.37	0.27	0.56	2.1
1500	1.104	2.04	1.32	0.25	0.50	2.0
2000	1.13	2.13	1.36	0.23	0.43	1.9

ratio $\langle\tau_2\rangle/\tau_2$ is independent of density. This result emphases the fact that whether one uses simply the Lorentzian half width (τ_2^{-1}) or first substracts $I_{INT}(\omega)$ before analyzing the data, the same low frequency information about the density dependence of the reorientational times is obtained.

3.2. DENSITY DEPENDENCE OF $\tau_{x0}(\tau_{BC})$

In Table I it can be seen that in both acetone and benzene the values of τ_{x0} decrease with increasing density. Since a strong interaction or collision causes a molecule to turn around (AVCF changing sign) τ_{x0} is proportional to τ_{BC}, the time between collisions which is of particular interest because it is a fundamental step in the reorientation process.

It has been fairly commonly noted [2, 9] that τ_{BC} scales as $\varrho^{-1/3}$. The cell model is a typical one which shows this density dependence. This model as derived by Madigowsky and Litovitz [10] to explain ultrasonic relaxation data predicts that

$$\tau_{BC} = (\bar{r} - \sigma)/\bar{v}, \qquad (10)$$

where $\bar{v} = (8kT/\pi m)^{1/2}$ is the average velocity of a molecule, k is the gas constant, T the absolute temperature, m the mass of the molecule, $\bar{r}$, the average distance between neighbors, σ the diameter of the molecule and τ_{BC} the time between collisions. For benzene, X-ray data indicates that the molecules tend to pack in a face centered cubic structure giving for the average distance between molecules $\bar{r} = (\sqrt{2}V/N_A)^{1/3}$ where V is the molar volume and N_A is Avogadro's For acetone we assumed simple cubic structure (i.e. $\bar{r} = (V/N_A)^{1/3}$ τ_{x0} is plotted vs $\bar{r}$ in Figure 5. For both liquids it is clear that τ_{x0}^* scales as $\varrho^{-1/3}$. Plotted along with the pressure data are the temperature data of Dardy et al. [2]. The agreement between the two sets of data show the cell model can yield both the correct density and temperature dependence. In the cell model the intercepts yield the effective molecular diameter. In benzene a value 5.28 Å is found which is in good agreement with values obtained from gas viscosity data (5.27 Å). In acetone a gas viscosity value is not available however the intercept value

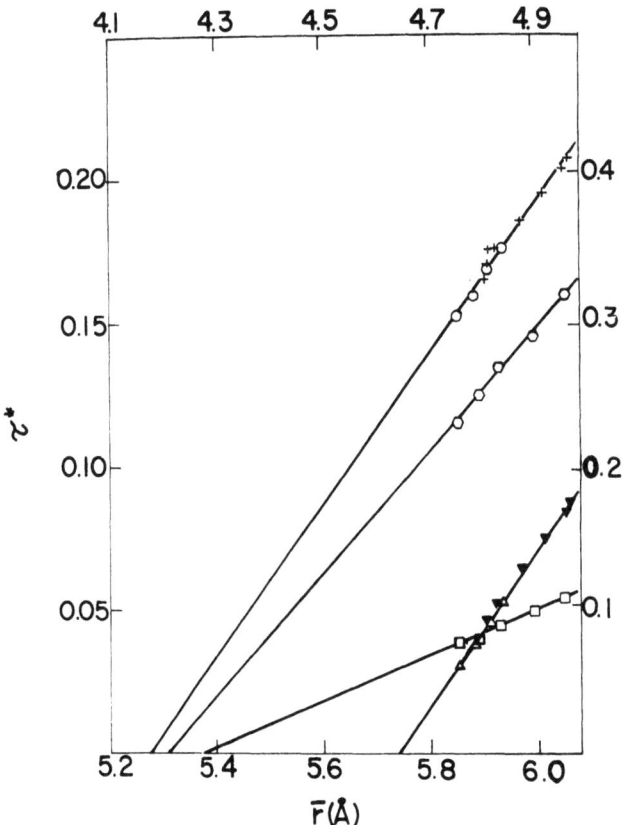

Fig. 5. Cell model plot of τ_{x0} and τ_J vs $\bar{r}$, the intermolecular distance. $\langle \ \rangle$ - τ_{x0} acetone; $\square$ - τ_J acetone; $+$ - τ_{x0} temperature data benzene; $\bigcirc$ - τ_{x0} pressure data benzene; $\blacktriangledown$ - τ_J temperature data benzene; $\triangle$ - τ_J pressure data benzene.

(4.27 Å) is in reasonable agreement with the value estimated from the molecular bond lengths and constituent diameters (between 4.0 and 4.5 Å).

Because τ_J has been commonly identified with τ_{BC} it too is plotted in Figure 5. Again the times seem to follow the $\varrho^{1/3}$ density dependence. In acetone, the intercept is close to the value obtained from τ_{x0}^* indicating that $\tau_J \propto \tau_{x0}$.

The proportionality implies that a Hubbard-like relation exists between $\langle \tau_2^* \rangle$ and τ_{x0}^* in acetone. In benzene on the other hand, τ_J^* has neither the same slope nor the same intercept as τ_{x0} implying that such a Hubbard-like relationship does not hold.

3.3. THE RELATION OF $\langle \tau_2 \rangle$ TO τ_{BC}

In a simple diffusional model such as the Hubbard model, $\langle \tau_2 \rangle$ should be inversely proportional to $\langle \tau_J \rangle$ or equivalently in that case τ_{BC}. To see if such a relation exists here between $\langle \tau_2^* \rangle$ and τ_{x0} the product $\langle \tau_2^* \rangle \tau_{x0}^*$ is plotted in Figure 6. In acetone the product is a constant indicating that the sole cause for the increase in $\langle \tau_2^* \rangle$ with density is the decrease in angular step size due to the decrease in τ_{BC}.

In benzene instead of being a constant, $\langle \tau_2^* \rangle \tau_{x0}^*$ increases by over 100% over the

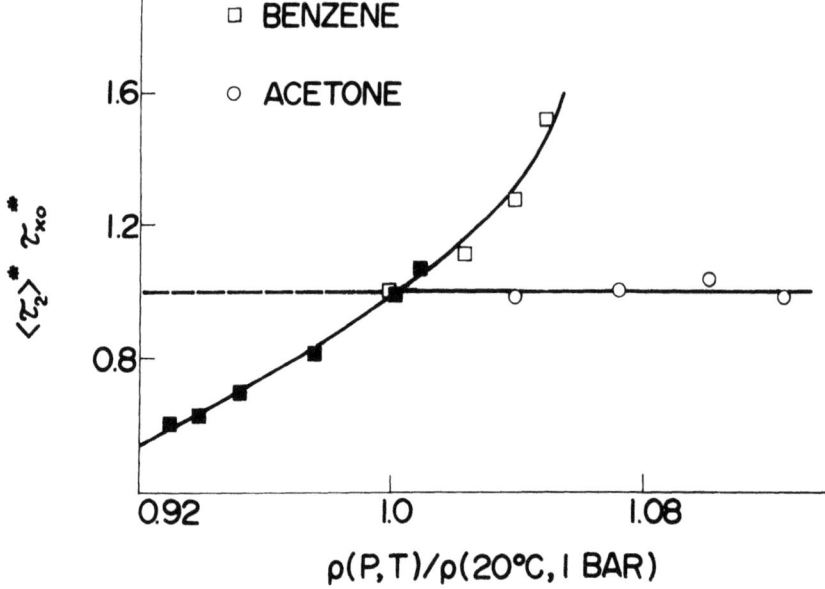

Fig. 6. Density dependence $\langle \tau^*_2 \rangle \tau^*_{x0}(P)/\langle \tau^*_2 \rangle \tau^*_{x0}$ (20°C, 1 atm).

density and temperature ranges studied. This indicates that in contrast to the behavior observed in acetone, the decrease in τ_{BC} is not the only factor determining the density dependence of $\langle \tau_2 \rangle$, in benzene.

3.4. Density dependence of τ_E and τ_E/τ_{BC}

To understand the difference in the behavior of the product $\langle \tau_2 \rangle * \tau^*_{BC}$ in acetone and benzene we must first consider τ_E, the time for the APCF to become exponential. From analysis of Figures 3 and 4 where τ_E is denoted by τ_4 it is apparent that τ_E is just the time required for the AVCF to essentially become zero.

Consider now the ratio $\eta_E = \tau_E/\tau_{BC}$ which is the number of collisions required for angular velocity correlation to be lost.

For acetone, η_E is approximately 2 indicating in a relatively small number of collisions the molecules lose memory of their initial angular velocity. As the density is increased, η_E remains essentially constant showing that the efficiency of a collision for randomizing the angular velocity is independent of density.

In benzene on the other hand instead of being constant η_E increases as the density increases. This directly related to the increase in the oscillatory character of the AVCF. This implies that the efficiency of a collision for randomizing the angular velocity decreases as the density increases. Apparently in benzene, the influence of molecular asymmetry becomes more important as density increases causing this reduction in the collision efficiency.

3.5. Conclusion

This observed difference in the density dependence of the collision efficiency for randomizing the angular velocity can be used to understand the density dependence of $\langle \tau_2 \rangle$ in our two liquids. In acetone, $\langle \tau_2 \rangle$ increases with increasing density only because of the decrease in the angular step size which is directly related to the decrease in time between collisions. In benzene on the other hand, $\langle \tau_2 \rangle$ increases for two reasons. First just as in acetone, the decrease in τ_{BC} with increasing density will cause $\langle \tau_2 \rangle$ to increase. But in addition $\langle \tau_2 \rangle$ increases because of an increase in oscillatory behavior.

References

1. Kivelson, D. and Keyes, T.: *J. Chem. Phys.* **54**, 1786 (1971).
2. Dardy, Volterra, V., and Litovitz, T. A.: *Faraday Symposia of Chemical Society* **6**, 71 (1972).
3. Coakley, J. M.: 'Rayleigh Scattering: Molecular Motions in Liquids Composed of Nearly Spherical Molecules, Ph.D. Thesis, Catholic University of America, 1972.
4. Stewart, H. A.: *Die Struktur Des Frein Molekuls*, Springer-Verlag, 1952, pp. 346–352.
5. Kushick, J. and Berne, B.: to be published in *J. Chem. Phys.*
6. Anderson, J. F. and Uhlman, R.: *J. Chem. Phys.* **55**, 4406 (1971).
7. Steel, W. A.: *J. Chem. Phys.* **38**, 2411 (1963).
8. Hubbard, P. S.: *Phys. Rev.* **131**, 1155 (1963).
9. Bartoli, F. and Litovitz, T. A.: *J. Chem. Phys.* **56**, 413 (1971).
10. Madigoski, W. and Litovitz, T. A.: *J. Chem. Phys.* **34**, 489 (1961).
11. Berne, B. J. and Harp, G.: *Adv. Chem. Phys.* **17**, 63 (1970).
12. Stillinger, J. and Rahman, A.: *J. Chem. Phys.* **57**, 1281 (1972).

DISCUSSION

Jonas: (1) Did you use our NMR data from 1970 (Bull, Jonas: *J. Chem. Phys.*, 1970) for acetone and benzene?

(2) We interpreted our results in terms of relatively free rotations about the six fold axis in benzene and the C_3 rotation in acetone. Since this rotation persists even in the solid state one should not expect dependence on viscosity.

Hertz: I would recommend the measurement of the ^{17}O relaxation rate in acetone as a function of pressure. In the data given by you, you essentially see the internal rotation of the methyl group.

Litovitz: We used both your results and conclusions concerning the relatively free rotations about the C_6 and C_3 symmetry axes in benzene and acetone respectively. Our contribution concerned the close connection between viscosity and reorientation about axes perpendicular to the above axes of symmetry.

Litovitz: This should be quite useful.

Rahman: In the case of water we know the extent to which $\int_0^\infty \langle \omega_\alpha(0) \omega_\alpha(t) \rangle \, dt$ is not zero. Does this information help in any way?

Litovitz: I do not see how this would help in analyzing the two liquids reported here.

Brot: Now you seem convinced that, in dense liquids, many types of molecules 'librate' in their environnment, what is your favorite vocabulary: libration, back correlated collisions, etc.?

Litovitz: My early objection to the word 'liberate' was based on the fact that most investigators upon observing a peak in the Raman wings or in the far infra-red assigned its origin to librational motion totally independent of the orientational process causing the low frequency Rayleigh scattering. If we now agree that these peaks or shoulders are manifestations of the individual steps in the reorientational process then I don't care what vocabulary is used.

Bratos: In majority of cases the separation of the correlation function $\langle \cos\theta(t) \cos\theta(0) \dot\theta(t)^2 \rangle$ in two separate factors $\langle \cos\theta(t) \cos\theta(0) \rangle \langle \dot\theta(t)^2 \rangle$ seems to be justified by the difference of the correlation times associated with these factors. This type of approximation has often been used in the past.

Litovitz: I was not aware that this assumption had been used to analyze Rayleigh data to obtain estimates of the angular velocity autocorrelation function.

FAR INFRA-RED ABSORPTIONS IN NON-DIPOLAR LIQUIDS

MANSEL DAVIES

Edward Davies Chemical Laboratories, University College of Wales, Aberystwyth, U.K.

Abstract. The absorptions usually centred between 50 cm^{-1} and 100 cm^{-1} in all non-dipolar liquids were first recognized from microwave measurements. Studies in the gaseous state establish with certainty their origin in collisionally induced dipoles: but there are often significant changes in going from gas to liquid. Discussion is given of the evaluation of 'effective' quadrupole (or other multipole) moments from the integrated intensities: of possible molecular 'rattling' frequencies: of the preference for distributed charge rather than point-multiple representations: and of the assessment of specific ($A \cdots B$) interactions in mixtures of non-dipolar liquids. Major limitations in our present appraisals are also indicated.

Résumé. Les absorptions dans tous les liquides non-dipolaires souvent centées entre 50 cm^{-1} et 100 cm^{-1} ont été tout d'abord découvertes dans la région micro-onde. Des études à l'état gazeux ont etabli ces origines, dans les dipoles induits, par chocs moléculaires: mais fréquemment il existe des différences importantes entre le gaz et le liquide. L'évaluation des quadrupoles (au autres multipoles) 'effectifs' est donnée au cours de la discussion; ainsi que la fréquence du tremblement ('rattling') moléculaire; les avantages d'un modèle aux charges distribuées (ou lieu d'un multipole de point): et la détermination des interactions spécifiques ($A \cdots B$) dans des liquides non-dipolaires mixtes. Nous avons indiqué quelques limites pour ces évaluations.

This account is aimed at presenting some of the typical features of these absorptions as seen by the chemist: other accounts, some in the present conference [1] present different aspects and their evaluation.

It is relevant to recall that the unexpected presence of these absorptions was clearly established by Whiffen [2] from measurements on pure non-dipolar solvents in the microwave region: small real values of ε'' were found to increase essentially linearly with frequency up to 36 GHz ($\bar{\nu} = 1.2$ cm^{-1}) and their collisionally induced origin was discussed. Welsh and co-workers showed how general this type of absorption was when they established its presence in mixtures of inert gases [3]. The systematic evaluation of the liquid absorptions became possible with the introduction of the interferometric spectrometer which with a Golay detector can produce reliable absorption values ($\pm 5\%$: and often $\pm 2\%$) from 10 cm^{-1} (300 GHz) to higher wavenumbers: whilst with the Putley or Rollin detectors much improved significance is available and reliable absorptions can be obtained down to ca. 2 cm^{-1}. In most cases the collision-induced absorptions in organic liquids do not extend far beyond 150 cm^{-1} but are then replaced by proper-mode absorptions (torsional; low frequency skeletal: or difference tones) whose overlap with the collisional mode provides difficulties in assessing their integrated intensities.

An important feature of the interferometric spectrometer is its ability to produce a refractive index as well as an absorption spectrum: Chamberlain has significantly improved the quality of these measurements by the use of phase-sensitive detection systems [4] and the development of the Mach-Zehnder refractive index method [5].

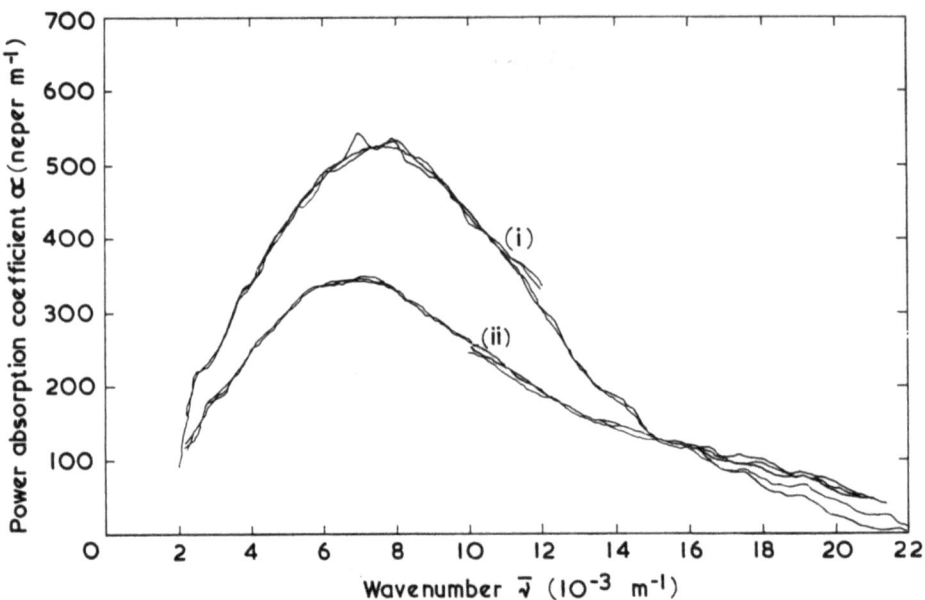

Fig. 1. Actual absorption curves derived from individual interferograms: the overlap of curves near abscissa 10 ($=100\,\text{cm}^{-1}$) arises from the use of different beam splitters: (i) benzene; (ii) carbon disulphide: at 300 K.

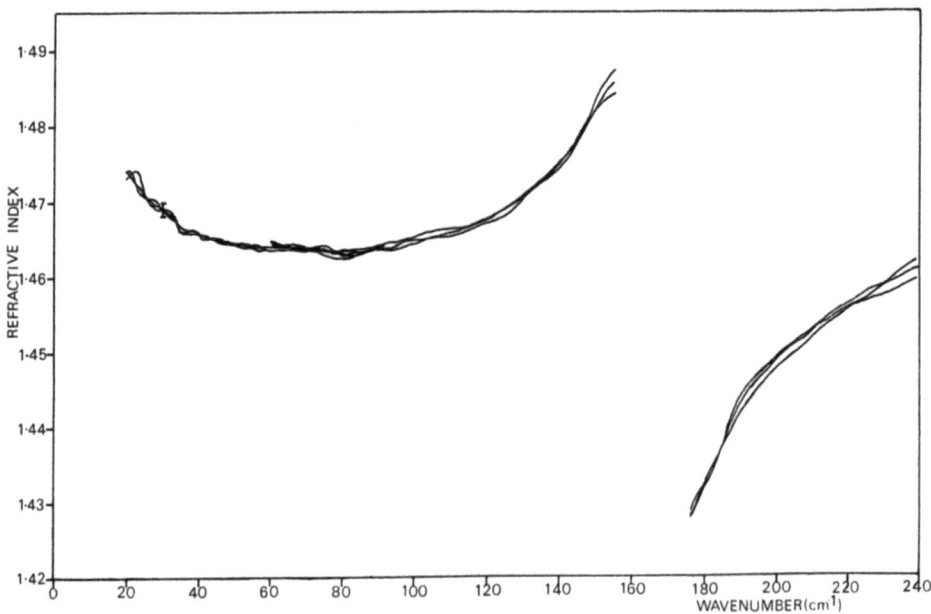

Fig. 2. Refractive index dispersion for p-difluorobenzene at 300 K: compare the absorption curve in Figure 3. (Davies, Graham J.: Ph.D. Thesis, University of Wales, 1972).

Some results typical of those now obtainable are shown: Figure 1 gives the power absorption coefficient [6] from $I=I_0\exp(-\alpha d)$: Figure 2 shows the refractive index dispersion [7]: Figure 3 shows mixtures of two non-dipolar liquids [8] and the presence of proper-mode absorptions near 160 cm^{-1} and 220 cm^{-1}. The order of magnitude of the absorption factors is significant: for benzene ε''(max) is 20×10^{-3} near

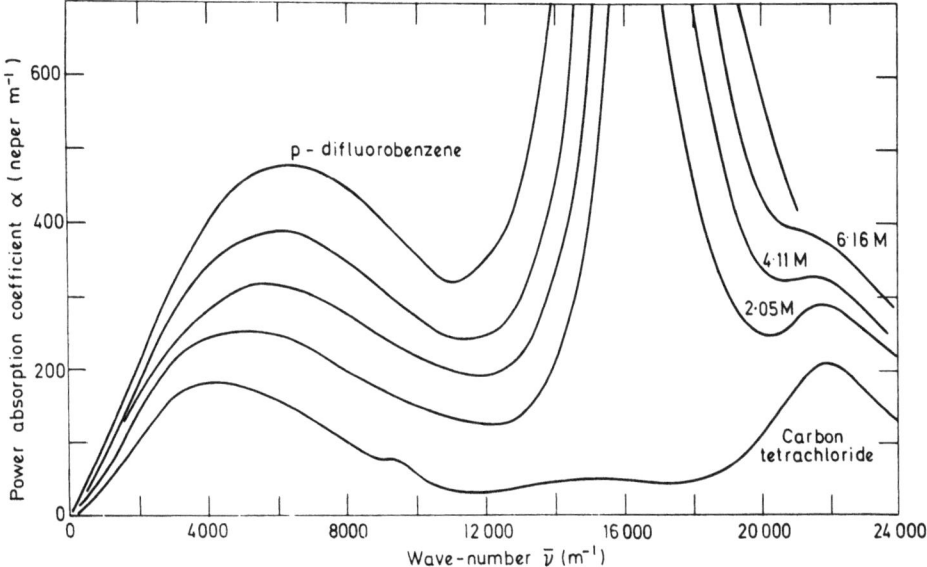

Fig. 3. Absorption spectra of solutions of p-difluorobenzene in carbon tetrachloride at 300 K. Note the units plotted are m^{-1}: i.e. $\bar{v}=4000$ m$^{-1}\equiv 40$ cm^{-1}: $\alpha=200$ neper m$^{-1}\equiv 2$ neper cm^{-1}.

37 cm^{-1}: α(max) being 5 neper cm^{-1} at 73 cm^{-1} or 63% absorption in 0.2 cm. These maxima are displaced, as their different absorption factors are inter-related by:

$$\alpha = \frac{2\pi\bar{v}\varepsilon''}{n},$$

where $n=$ refractive index of the medium.

1. General Aspects

The principal phenomenological aspects of the absorption which have all eventually to be explained, may first be summarized:

(i) the absorption is often centred between 50 cm^{-1} and 100 cm^{-1} and is very broad with width at half-peak height ca. $\Delta\bar{v}=50$ cm^{-1}.

(ii) temperature variation has no pronounced influence. Some initial observations of Pardoe's [9] have been essentially confirmed in recent careful measurements [8]: Figure 4.

(iii) pressure has a pronounced influence although only some early data of Bradley's can be quoted for this conclusion [10]: see Table I.

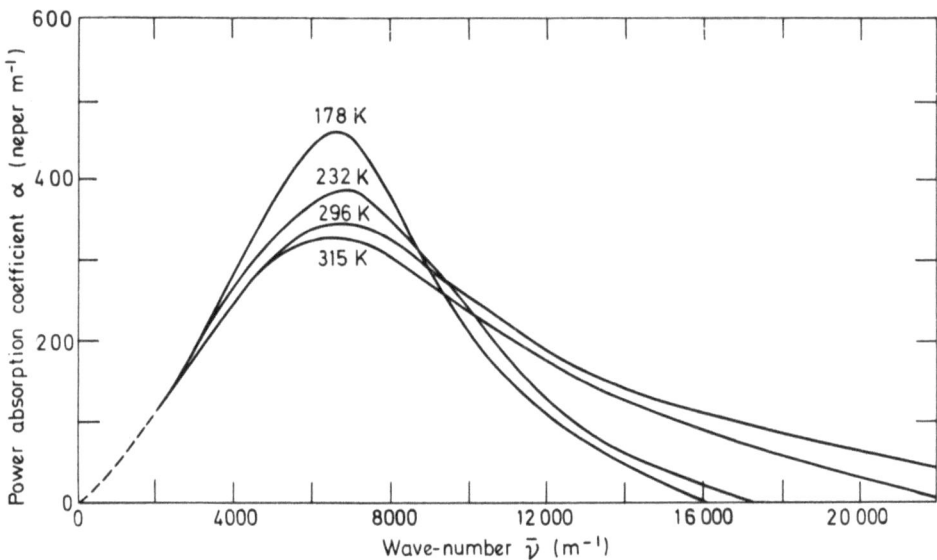

Fig. 4. The absorption coefficient of pure liquid carbon disulphide recorded at four temperatures.

TABLE I

Effect of pressure on absorption maximum for liquid CS_2 at 293 K

Pressure	$\bar{\nu}$ (max) cm^{-1}	α (max) neper cm^{-1}
1 bar	70	2
3 kbar	80	5
11.6 kbar	100	7

(iv) the refractive index spectrum shows a shallow minimum where $\alpha(\bar{\nu})$ is a maximum.

(v) concentration changes in mixed non-dipolar liquids show that the integrated absorption in two-component systems is well represented by: A (mixture) $= A_A N_A^2 + A_B N_B^2 + 2 A_{AB} N_A N_B$. Here $N_A =$ mole fraction of A; A_A is the integrated absorption of pure A; A_B that of B; and A_{AB} measures the absorption specific to the $(A \cdots B)$ collisional interaction.

2. Studies in the Gaseous State

As already suggested by the reference *inter alia* to Welsh's work, the origin of these absorptions is most clearly established in the gaseous state. The non-dipolar molecules do not, in isolation, give any detectable absorptions from their translational or rotational motions. However, as the pressure is raised an absorption appears (one can confidently say for all non-spherically-symmetric molecules) whose strength is, in the first instance, proportional to the square of the pressure. This basic fact establishes its collisional origin. Studies of gaseous non-dipolar polyatomic molecules have now been made in several instances, particularly by Birnbaum [11] and Harries [12]. Birnbaum and Maryott [13] extended the appreciation of Colpa and Ketelaar [14] of the pressure-induced rotational absorption spectrum of hydrogen. The collisionally induced electric moment is then restricted to rotational energy absorptions by the selection rule $\Delta J = +2$. Such features will be broadened *inter alia* by the brief life-time of the collisional state. In addition, a non-resonant absorption will occur corresponding to $\Delta J = 0$ which is possible for each colliding molecule when energy is absorbed directly into the translational mode. [This has analogies to the non-resonant Debye-type absorption present in gaseous symmetric-top molecules where it arises from the non-rotating component of the dipole in a given (JK) rotational state].

Harries studied carbon dioxide and determined the total intensity in two forms: he made alternative derivations of the quadrupole moment, $Q(CO_2)$, whose field gives rise to the induced moments. The first is the integrated optical absorption strength:

$$A = \int_{\text{band}} \alpha (\text{neper cm}^{-1} \text{ bar}^{-2}) \, d\bar{\nu}$$

$$= \sum_{J=0}^{J=\infty} S(J, J+2)$$

which can be equated to the sum of the line strengths of the $(J+2) \leftarrow J$ transitions given for linear molecules by Colpa and Ketelaar:

$$S(J, J+2) = \frac{48\pi^4}{hc} \bar{\nu} N^2 \alpha_p^2 Q^2 \left[\frac{F(J)}{2J+1} - \frac{F(J+2)}{2J+5} \right] \times$$

$$\times \frac{(J+1)(J+2)}{2J+3} \int_0^\infty r^{-6} \exp\left(-\frac{U(r)}{kT}\right) dr$$

N = number of molecules per unit volume: $\bar{\nu}$ = wavenumber of $(J+2) \leftarrow J$ transition: Q = quadrupole moment of CO_2: α_p = mean molecular polarizability: $F(J)$ = fraction of molecules in rotational state J: $U(r)$ = intermolecular (Lennard-Jones) potential. Here, no account is taken of any translational mode absorption.

The second intensity measurement is Δ:

$$\Delta = \int_{\text{band}} \frac{\alpha(\bar{\nu})}{(\bar{\nu})^2} \, d\bar{\nu} = \frac{2\pi}{n} \int_{\text{band}} \frac{\bar{\nu}\varepsilon''(\bar{\nu})}{(\bar{\nu})^2} \, d\bar{\nu} = \frac{\pi^2}{n} (\varepsilon_0 - \varepsilon_\infty)$$

and for this total dispersion amplitude Maryott and Birnbaum deduced classically

$$(\varepsilon_0 - \varepsilon_\infty) = \frac{16\pi^2}{kT} N^2 \alpha_p^2 Q^2 \int_0^\infty r^{-6} \exp(-U(r)/kT) \, dr .$$

Difficulties arise in expressing $U(r)$ sufficiently precisely for collisional interaction with its very close intermolecular approach. Bose and Cole [15] emphasized the inadequacy of the energy representation which assumes the CO_2 molecules to have spherical symmetry (e.g. the Lennard-Jones expression). The quadrupole field and polarizability of CO_2 is distinctly anisotropic and a quadrupole-quadrupole interaction energy of the form $U(Q,Q) = Q_1 Q_2/r_{12}^5 [f(\theta_1 \theta_2)]$ must be included in $U(r)$. Harries's account showed the major significance of this anisotropic term. It is relevant to illustrate the order of magnitude of the agreement provided by this analysis.

TABLE II

Absorption data for CO_2 up to 15 bar and derived quadrupole moments [12]

T (K)	$10^5 \alpha$ (max) neper cm^{-1} bar^{-2}	$\bar{\nu}$ cm^{-1} (obs α_{max})	$\bar{\nu}$ cm^{-1} (calc α_{max})	$\|Q_A\| \times 10^{40}$ Cm^2	$\|Q_A\| \times 10^{40}$ Cm^2
200	12.4	47.5	37	13.9 ± 1.0	12.2 ± 0.5
293	6.5	54	45	14.8 ± 1.0	13.9 ± 0.6
323	5.6	57	46	15.2 ± 1.0	14.2 ± 0.6
373	4.6	67	50	15.5 ± 1.0	14.5 ± 0.6

Ho, Birnbaum and Rosenberg [16] in an extended study of CO_2 found $|Q| \times 10^{40} = = 14.5$ Cm^2 whilst Buckingham and Disch [17] in a different method, of maximum significance, established the value as 13.8. This degree of agreement leaves no doubt that the origin of the absorption is correctly assigned. But there are further indications. Harries's data could well be proved correct in suggesting that $|Q|$ deduced from the collisionally induced absorptions increases with temperature – i.e. the rigid multipole model is inadequate: and Birnbaum has found that above 15 bar a (pressure)^3 term must be invoked [better, a (number density)^3 term] and that its presence is effectively to reduce the absorption strength. This tendency was shown emphatically in the liquid state, where the integrated absorption for CO_2 was far below that predicted by extrapolation of the (molecular number density)^2 term of the gas phase. This observation is interpreted as indicating that intermolecular collisions in CO_2 produce effectively smaller induced dipoles than bimolecular collisions. This is probably related to the pronounced anisotropy of O=C=O and the greater electron-cloud symmetry present in a triple collision. All these aspects are of immediate significance in relation to the absorptions in non-dipolar liquids.

3. The Gas-Liquid Transition

As examination of the gas → liquid transition provides insights into the many complications which arise in the condensed phase, we can consider the further instance of cyanogen (N≡C–C≡N). Evans has recently reported on the far infra-red absorptions of this very anisotropic molecule in the vapour at pressures up to 33 bar and in the liquid [18]. The general pattern resembles that of the O=C=O results but some features are revealing. The contour in the vapour is scarcely influenced by pressure (Figure 5) but whereas the $(J+2) \leftarrow J$ quadrupole induced transitions at 380 K should lead to α(max) at ca. 32 cm^{-1} and a halfpeak height width of 31 cm^{-1}, the observed α(max) is at 47 ± 3 cm^{-1} and the half-width (at 33 bar) is 60 cm^{-1}. A substantial high frequency 'tail' is being built on to the simple $\Delta J = +2$ absorption pattern.

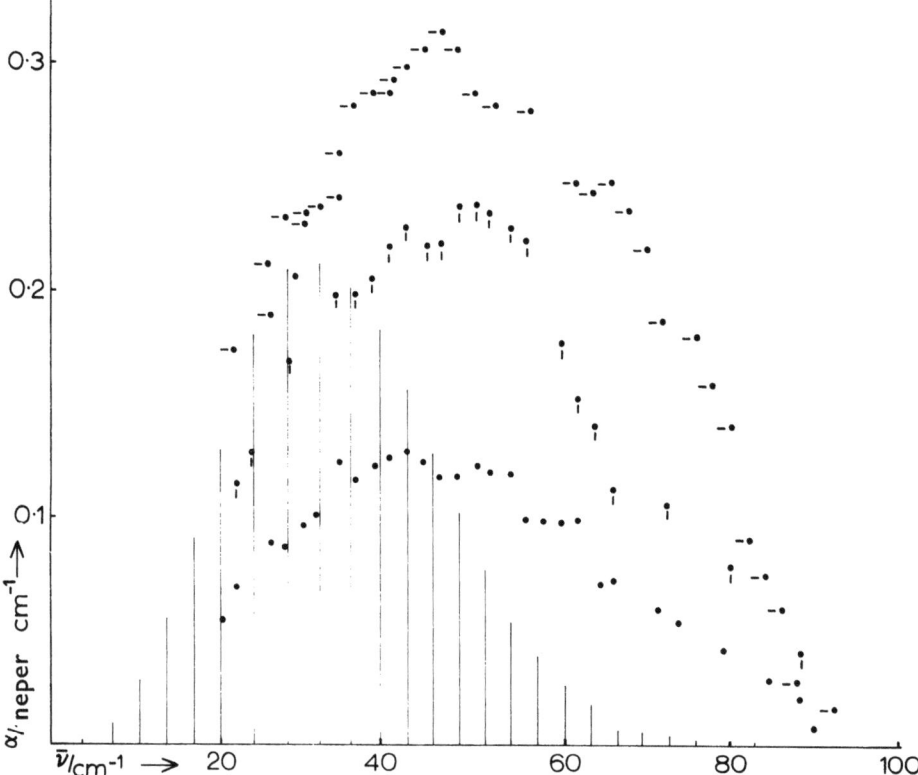

Fig. 5. Actual vapour absorptions recorded for cyanogen at various pressures and temperatures: the mean of several interferograms is plotted. __. 33.5 bar at 383 K; ⎮ 28.4 bar at 381 K; . 24.0 bar at 348 K. The line spectrum drawn gives the positions and *relative* intensities of the $(J+2 \leftarrow J)$ quadrupole induced transitions.

Birnbaum et al. have already commented on this feature in CO_2. Several short-range binary interactions could be contributing, including hexadecapole-induced absorptions, or overlap interactions (cf. charge-transfer species). In quasi-chemical terms the role of the latter could be represented as the formation of a transient polar dimer:

$$M_1 + M_1 \underset{k_{21}}{\overset{k_{12}}{\rightleftharpoons}} M_2.$$

Such systems provide a relaxation time of the order $\tau \sim (k_{12}[M_1] + k_{21})^{-1}$ so that any associated absorption has its maximum at a frequency near $(k_{12}[M_1] + k_{21})$ rad s^{-1}. This factor could quite plausibly be 2×10^{12} s^{-1}, i.e. $\bar{\nu} \sim 66$ cm^{-1}.

In liquid cyanogen the absorption centre has moved to 75 ± 3 cm^{-1}, implying that the process leading to the high frequency tail in the vapour is now a dominant factor: the half-width at 55 cm^{-1} is actually less than in the vapour. This almost certainly means that the overlapping processes of the vapour absorption are less equally balanced in their intensities, i.e. the gas-rotational mode has been suppressed in the liquid. This is not surprising for such an elongated molecule. That the anisotropy of the repulsive field, i.e. the shape of the molecule, is responsible for this is seen in Baise's results [19]. He found that the very slightly polar nitrous oxide (N=N=O) molecule shifted its absorption markedly in going from gas to liquid, whereas the distinctly more polar fluoroform (HCF$_3$), which approaches sphericity, showed a far smaller displacement and probably suffers little hindrance to rotation in the liquid. If interactions in cyanogen liquid produced significant distortion of the linear molecule for times longer than 10^{-10} s then a dipole absorption of the Poley character [20] could be expected near 70 cm^{-1}: but this would seem much less likely in the vapour.

4. Recent Data and the Frequency Centre of the Absorption

Table III summarizes recent data on typical non-dipolar organic liquids [6]. In it A is the experimental integrated band intensity: A_m is the corresponding molecular absorption factor:

$$A_m = \left[n \left(\frac{3}{n^2 + 2} \right)^2 \right] \frac{m}{\varrho} A$$

m = molecular mass; ϱ = density of the liquid; and the expression in square brackets is the Polo-Wilson correction for the influence of the refractive index (internal field) on the absorption intensity.

Qualitative aspects of these results are that the lowest absorption intensities are found in the saturated hydrocarbons whose component bond (or group) moments are minimal: that *trans*-decalin (C$_{10}$H$_{18}$) has only one half the absorption of *cyclo*-hexane (C$_6$H$_{12}$) reflects the greater structural rigidity of the former molecule. Although the component moments $\overrightarrow{\mu(C-Cl)}$ in carbon tetrachloride are very much higher than in the hydrocarbons the almost spherical symmetry and the relative rigidity of the molecule gives it much lower an absorption than *trans*-1,2-dichloroethylene whose polarizability is also enhanced by its π electron system. Similarly, it seems the higher $\overleftarrow{\mu(C-H: \text{aromatic})}$ than $\overleftarrow{\mu(C-H: \text{aliphatic})}$ bond moments result in the enhanced absorption of benzene over *cyclo*-hexane. 1,4-dioxan is interesting as its dipole moment is zero in its symmetric equilibrium configuration but intramolecular distortion (as the result of intermolecular collision) could produce large transient moments.

How does one account for the position of these absorptions? The rotational mode absorptions (e.g. for $\Delta J = 2$) for the molecules in Table III would rarely be centred above 20 cm^{-1} and the displacement of $\bar{\nu}$(max) from such values appears to imply a different origin. The absorption is certainly associated with the dipole formed on

TABLE III

Absorption factors for some non-dipolar liquids in the far infra-red at 298 K

Liquid	α(max) cm^{-1}	$\bar{\nu}$(max) cm^{-1}	$A \times 10^{-6}$ m^{-2}	Polo-Wilson correction	$A_m \times 10^{22}$ m	Δ	$Q \times 10^{40}$ Cm^2 Ref. 8	Lit.
trans-decalin	0.41	67	0.32	0.75	0.60	0.024	5.9	—
cyclohexane	0.70	80	0.74	0.79	1.1	0.028	5.6	3.3 (b)
carbon tetrachloride	1.82	44	1.21	0.79	1.5	0.178	[12.5]^a	[18.2]^a (b)
p-difluorobenzene	4.70	63	3.60	0.76	4.4	0.283	12.9	—
carbon disulphide	3.45	69	3.86	0.69	2.7	0.186	4.0	6.6 (a):5.9 (b)
benzene	5.20	75	5.15	0.75	5.6	0.157	12.5	11.9 (a):11.9 (b)
trans-1,2-dichloroethylene	5.10	86	5.70	0.77	5.6	0.299	11.2	—
1,4-dioxan	8.80	70	6.37	0.79	7.1	0.454	14.9	—

^{a} $\Omega \times 10^{50}$: i.e. octopole moment in Cm^3.

collision: and the liquid state simulates 'the molecule in a box' condition where each molecule rattles within the cage formed by its neighbours.

Accordingly, one representation of the absorption frequency is that given by the frequency of the rattling motion. Various estimates can be offered of this frequency [20]: the absolute values rarely agree to better than a factor of two or three. One relevant calculation is that of Mie who deduced the rattling frequency for molecules whose pair interaction-energy (defining the energy-well of the box) is represented as:

$$U(r) = \frac{B}{r^b} - \frac{C}{r^c}.$$

If l = latent heat of evaporation per molecule; $v_0 = N^{-1}$ (molar volume at OK); m = molecular mass;

$$[v(\text{Mie})]^3 = \frac{1}{4\pi^3 v_0} \left(\frac{bcl}{3m}\right).$$

For CCl_4: CS_2: C_6H_6: $C_4H_8O_2$ (accepting $bc=60$) the ratios of $\bar{v}(\max)/\bar{v}(\text{Mie})$ are 1.48: 1.59: 1.69: 1.89. This agreement is as good as any such model can provide for there is, in any case, a broad spectrum in the rattling mode frequencies.

The variation for a reasonably wide temperature range (178-315 K) has been studied for carbon disulphide. The results agree with earlier observations in showing a slight lowering of $\bar{v}(\max)$ as the temperature falls. The Mie relation implies no temperature dependence of the rattling frequency but, if, following a suggestion of Haigh [21], we consider the bimolecular (gas) collision frequency $\bar{v}(\text{coll})$, the observations can be matched.

$$v(\text{coll}) = \left[1.13\pi \frac{k^{1/2}\sigma^2}{m^{3/2}}\right] \varrho T^{1/2} = K\varrho T^{1/2}$$

σ is the molecular collision diameter: ϱ the density: K is a constant for each molecular species. Careful examination [8] appears to establish that the variation in $\bar{v}(\max)$ is linearly related to $\varrho T^{1/2}$ and the data provide a value for K (in appropriate units) of 46 ± 6, whereas the kinetic theory value is 59. Accordingly, there is a strong presumption that the collisional process does determine $\bar{v}(\max)$ although a consistent quantitative relation is not apparently yet available.

5. The Absorption Intensities and Molecular Multipole Moments

To account for the intensities in Table III clearly needs reference to the dipoles induced by the quadrupole field interactions and by other mechanisms. The molecular appraisal of the variety of contributing factors in the liquid state provides a problem as yet approached theoretically only in very general terms. We have attempted an initial *ad hoc* evaluation on the assumption that the quadrupole (or higher multipole) field is one certain contributor to the induced moments. The geometry of the interactions in the liquid is a complex pattern but the quadrupole field falls off rapidly with distance (r^{-3}) so there is some possibility that its influence is principally confined to bimolecular encounters. If correct, these assumptions would justify the evaluation of 'effective quadrupole moments' in terms of the gas collisional representation. This procedure is at least doubly dubious as (except in the case of O=C=O) the available relations imply spherical symmetry of the molecules: and they represent the molecular electric fields in terms of *point multipoles*. (We shall return to this assumption later). In any case, it is justifiable to evaluate the absorptions in terms of an 'apparently effective quadrupole moment' to find whether its order of magnitude is acceptable.

Accordingly, via the appropriate plot of the experimental data, corrected by the Polo-Wilson factor, we have determined the integral value Δ and used Birnbaum's expression for the dielectric amplitude of the polarization in a compressed gas due to the quadrupole-induced dipole moments:

$$(\varepsilon_0 - \varepsilon_\infty) = \frac{64\pi^3}{\varepsilon_0 kT} N^2 \alpha_p^2 Q^2 \int_0^\infty \frac{1}{r^6} \exp(-U(r)/kT) \, dr,$$

where N=number of molecules per unit volume: α_p=mean molecular polarizability. Combining this relation with Buckingham and Pople's tabulation of the functions $H_n(Y)$ where $Y=2(E/kT)$ and (E/k) and r_0 are standard terms in the Lennard-Jones function assumed to give $U(r)$, the required relation is

$$Q^2 = \frac{3\varepsilon_0}{16\pi^5} \cdot \frac{n^2 kT}{N^2 \alpha_p^2} \left(\frac{3}{n^2+2}\right)^2 \frac{r_0^5 Y^4}{H_8(Y)} \Delta.$$

By the same path, the data for carbon tetrachloride, for which $Q=0$, can be evaluated in terms of its octopole moment Ω. Birnbaum [11] has related the Kramers-Krönig integral to Kielich's expression for the corresponding permittivity term and, incorporating the Polo-Wilson factor, we find:

$$\Omega^2 = \frac{15\varepsilon_0}{64\pi^5} \frac{n}{N^2 \alpha_p^2} \left(\frac{3}{n^2+2}\right)^2 \frac{r_0^7 Y^2 E}{H_{10}(Y)} \Delta.$$

The values of Q and Ω so deduced are given in Table III. Previous values of Q quoted in this table are (a) from pressure-broadening of microwave absorptions, whilst (b) are derived, similarly to the present figures, from liquid state absorptions. In the gas phase Harries found Q evaluated from the far infra-red absorptions was reduced in the ratio 5:3 when correction for the non-spherical character of carbon dioxide was made. There is some reason for believing that this correction may be particularly large for O=C=O. When it is remembered that apparently good estimates of $Q(CO_2)$ varied in recent years between 14 and 30 in units of 10^{-40} Cm^2 and $\Omega(CH_4)$ between 3 and 20 in units of 10^{-50} Cm^3, then the concurrence of the values in Table III is surprisingly good. Undoubtedly it is sufficient to show that acceptable values of the quadrupole moments amply account for the observed intensities. The consistency in this respect tends to support the general terms of the assumptions involved in these evaluations.

Whether the octopole moment for carbon tetrachloride is an acceptable estimate cannot be decided until a reliable gas-phase value is known. Interactions between such molecules would seem to be more determined by the localized features of $(C-Cl)\cdots(Cl-C)$ encounters than by the field of a point-octopole charge representation. Baise (23) has evaluated the octopole moment of neopentane, $C(CH_4)_4$, from its liquid phase absorption. Some related values are shown in Table IV.

TABLE IV

Octopole moments in units of 10^{-50} Cm^3

CH_4	CF_4	CCl_4	$C(CH_3)_4$
3.1 (a)	45 (c)	18 (e)	13 (g)
6 (b)	15 (d)	12 (f)	

(a) (b) (c) Spurling, T. H., et al.: J. Chem. Phys. **48**, 1006 (1968).
(a) M.O. calc: (b) (c) Dielectric second virial coefficient:
(d) Rosenberg, A. and Birnbaum, G.: J. Chem. Phys. **48**, 1396 (1968), far infra-red absorption.
(e) Garg, S. K., et al.: J. Chem. Phys. **49**, 2551 (1968), liquid f.i.r absorption.
(f) Table III: (g) Reference 22.

With the variety of methods involved, these figures serve to establish the order of magnitude of the octopole values. The use of the integrated intensities of the liquid absorptions may, in practice, be far less dubious an approximation for these tetrahedral molecules than in many other cases. The evidence from a distinctly polar but quasi-spherical molecule (HCF_3) and a slightly polar but not too anisotropic a molecule (CO)* demonstrates how little the molecular absorption intensity can change from the gas to the liquid.

6. The Limitations of the Point-Multipole Field

If one accepts that a stepwise unravelling of the contributions to the non-dipolar absorptions can be a profitable approach and that the quadrupole-induced dipole is a major factor, then the validity of its evaluation must be considered. There is no need to repeat the approximations involved in the numerical treatment above, but the question must also be raised whether the point-multipole representation is the most meaningful for polyatomic molecules in the continuous close contacts of the liquid phase. In the gaseous state and insofar as collisions are elastic, the molecules spend only a small fraction of the time at separations less than a few ångströms. In these circumstances an acceptable representation is achieved statistically by expressing the attrac-

---

* Dr Ralph Amey: personal communication.

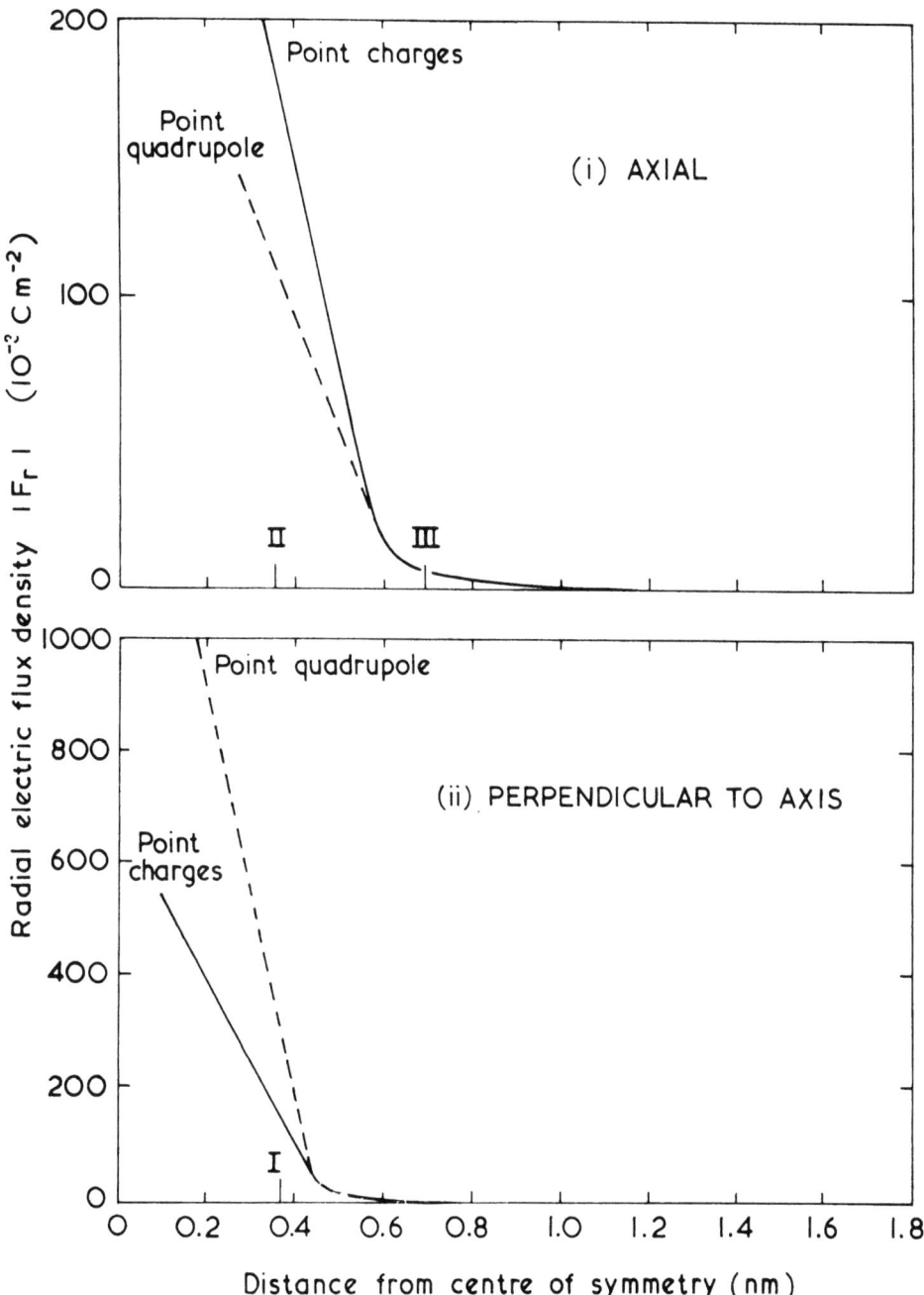

Fig. 6. Variation of the electric field intensity of carbon disulphide (i) along its linear axis, (ii) perpendicular to its linear axis, for alternative models of the quadrupole representation.

tive energies in terms of point-multipole polarizable molecule terms. In the liquid the molecules are in almost continuous contact at their van der Waals envelopes and it becomes clear that the energies can only be adequately approximated if the point-multipole field is expanded to include many terms.

In the liquid state the point-multipole model suffers from two major limitations: the diameters of the molecules, i.e. of their charge distributions, are comparable to

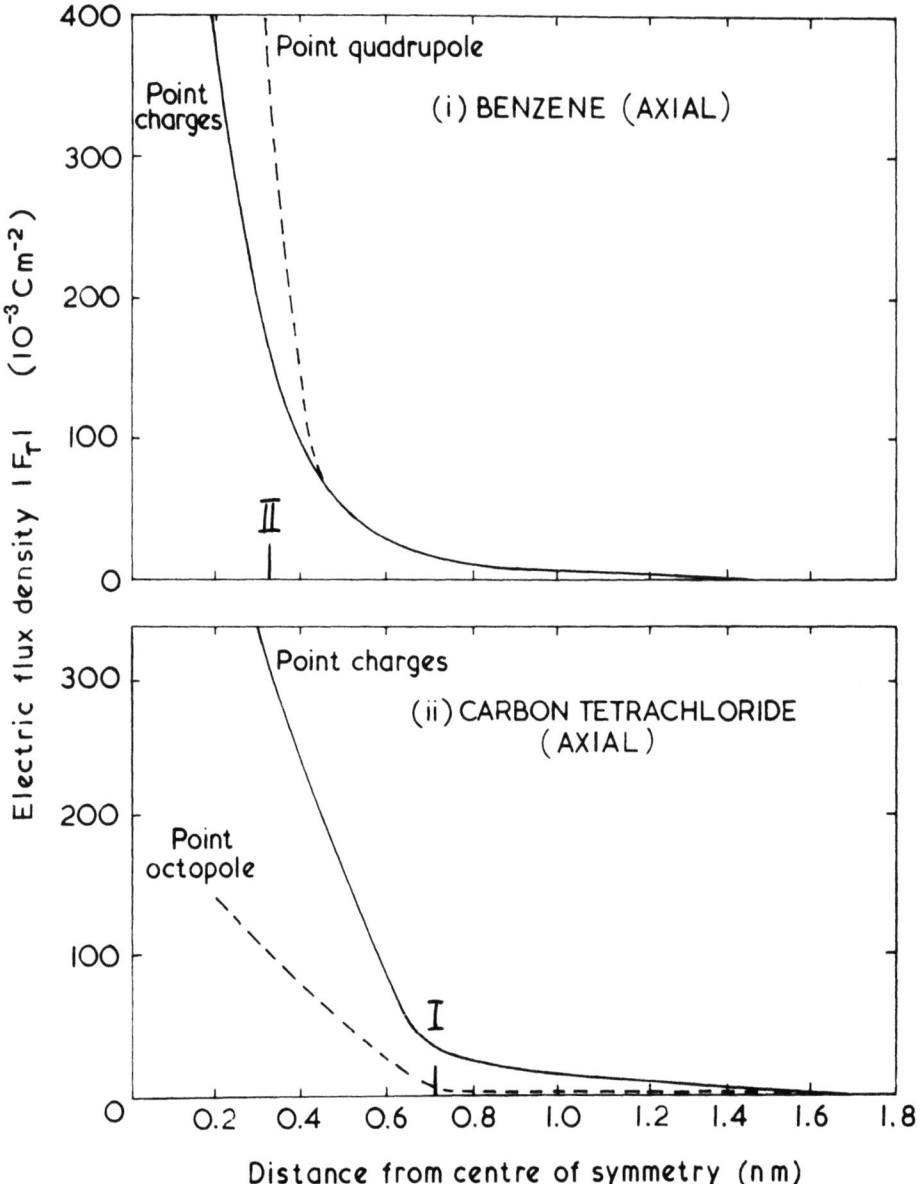

Fig. 7. Variation of the electric field intensity along the axis of symmetric charge distributions for (i) the benzene molecule, (ii) the carbon tetrachloride molecule.

their intermolecular separations and their local structures (e.g. in p-FC_6H_4F) impose considerable anisotropy at their van der Waals peripheries. In assessing these limitations some simple calculations can be made comparing (i) the point-multipole field with (ii) the field given by a pattern of charges located on the atomic centres which gives the same field as (i) at 10 Å or more from the centre of the molecule. The details of these calculations have been given elsewhere [8]: only the order of magnitude of the differences need be considered here. The divergence in the anticipated electric fields is represented for particular instances in Figures 6 and 7.

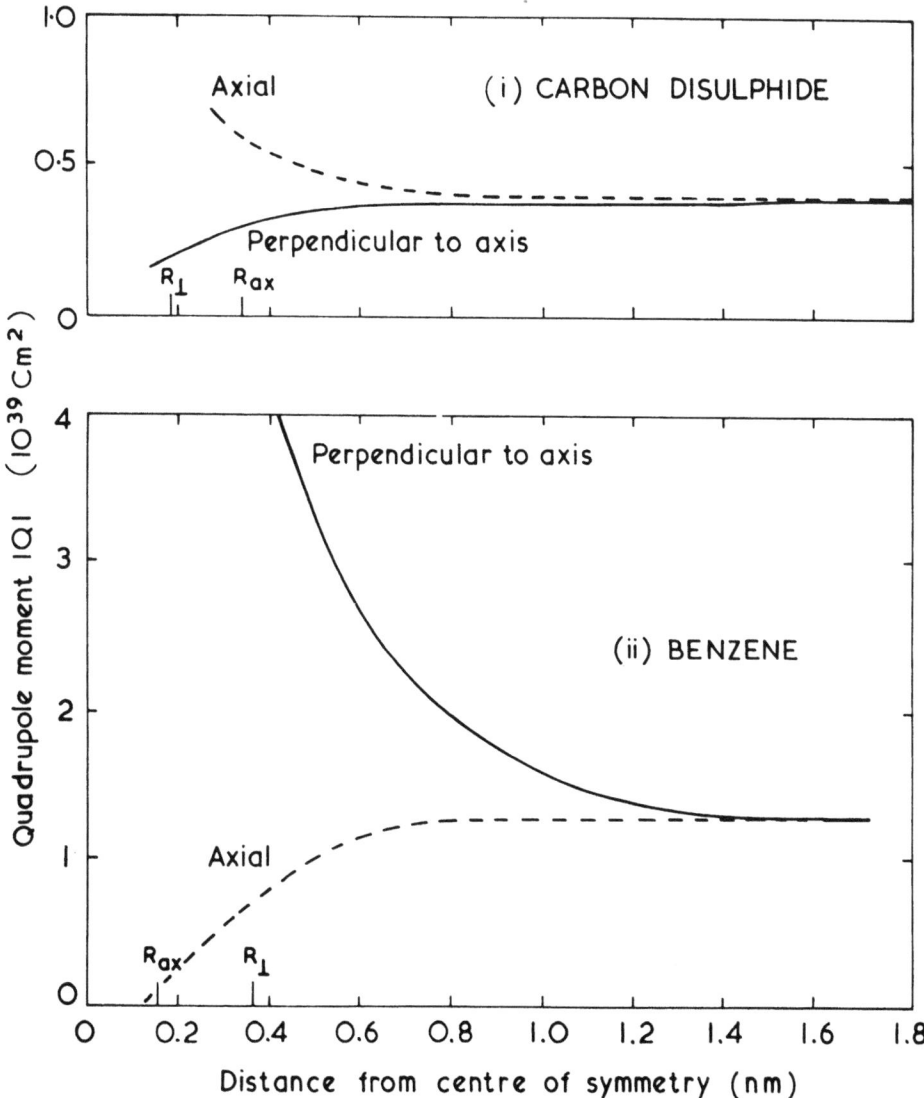

Fig. 8. Variation of the effective quadrupole moment with distance from the centre of symmetry of (i) the carbon disulphide molecule, (ii) the benzene molecule. The van der Waals *radii* of the molecules along and perpendicular to the axis are shown.

An alternative expression of these factors is given in Figure 8. Here the accepted quadrupole values for carbon disulphide and benzene are resolved into local charges on the atoms and from the fields given by those rigid charge distributions, the apparent quadrupoles which would be deduced at various locations near the molecular centres are plotted. The discrepancies are large for separations corresponding to accepted van der Waals diameters – i.e. at distances close to which the molecules spend much of their time in the liquid state: Figure 9. Thus benzene molecules in the superposed-disc alignment (i) each experience a field equivalent to 0.5 of the established Q-value; whilst in the side-by-side alignment (ii) the effective quadrupole is 1.7 times its 'normal' value. Formally, these variations can be incorporated in the multipole

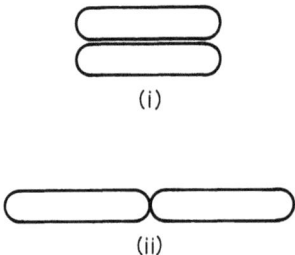

Fig. 9. Alternative modes of closest contact of two benzene molecules: (i) superposed discs; (ii) side-by-side.

model by introducing several higher point-multipoles whose fields are significant only at short distances. But this appears an unrealistic approach to the chemist who is inclined to accept for benzene a localized charge distribution on the twelve atomic centres. At such short distances the point-multipole model is unsatisfactory as molecules unquestionably have an extended spatial distribution of charges: a conclusion well-known to chemists in the form that the local charge model is far preferable to the point dipole.

Further difficulties arise when the non-rigidity of the charge distribution is envisaged. These higher terms may be given alternative representations as hyper-polarizability terms: electron overlap functions: 'charge transfer': etc. They are mentioned here to emphasize the necessity of first achieving some confidence in the characterization of rigid molecule collisions: on that basis only should it subsequently be possible to assess the extent of such quasi-chemical interactions as occur in molecular encounters.

7. The Effective Induced Electric Moment

One further molecular variable not emphasized above is the anisotropy of polarizability. This can lead to preferred directions for the molecular collisions – especially when coupled with the anisotropic fields of localized charge molecules. The chemist is interested to estimate what order of dipole moment is engendered on collision of, e.g. non-dipolar but rigid quadrupole molecules. The mean moment can be evaluated from the integrated intensity of the far infra-red absorption expressed as its equivalent dielectric increment $(\varepsilon_0 - \varepsilon_\infty)$ and, say, the Onsager equation to convert the latter to an effective dipole strength.

$$\mu_{\text{eff}}^2 = \frac{27\varepsilon_0 kT}{8\pi^3} \frac{M}{\varrho N} \frac{1}{(n_2+2)^2} \int_{\text{band}} \frac{\varepsilon''(\bar{\nu}) \, d\bar{\nu}}{\bar{\nu}}.$$

Simple calculations for localized charge models conforming to the accepted multipoles and in which the anisotropic polarizability is considered and alternative collisional orientations are weighted by a Boltzmann factor lead to mean $\mu(\text{calc})$ values. These are compared with the corresponding $\mu(\text{eff})$ values in Table V.

TABLE V

Estimates of collision induced dipole moments

Liquid	$\mu(\text{eff})/10^{-30}$ Cm	$\mu(\text{calc})/10^{-30}$ Cm
Carbon disulphide	0.37	1.27
Benzene	0.50	0.57
Carbon tetrachloride	0.43	0.30

The disagreement for carbon disulphide might be taken to suggest that the difference between bimolecular encounters, i.e. what is envisaged in $\mu(\text{calc})$ and the collisional interactions actually occurring in the liquid, is particularly large. However, this is not supported by the previous finding that $Q(CS_2)$ from the liquid phase absorption is 60% of the value from a gas-phase measurement. Further careful assessment is needed. *Per contra* it would seem that we can already assume with some confidence that no disturbance can be detected in benzene-benzene collisions in the liquid departing from anticipated behaviour in such bimolecular collisions.

8. Absorption Intensities in Mixtures

Physical interaction between molecules in mixtures has been a persistent claim in the chemical literature: complex formation, solvation, charge-transfer, etc., have been adduced for molecular mixtures, $A + B$. The far infra-red absorptions provide (now that they can be measured with appropriate accuracy) an excellent means of assessing any such special features. We are concerned here merely with non-dipolar compounds and, in particular, a summary of these observations for various solvent mixtures: *trans*-1,2-dichloroethylene/cyclohexane: benzene/carbon tetrachloride: *p*-difluorobenzene/cyclohexane: *p*-difluorobenzene/carbon disulphide: *p*-difluorobenzene/benzene. Figure 3: Figure 10: Figure 11.

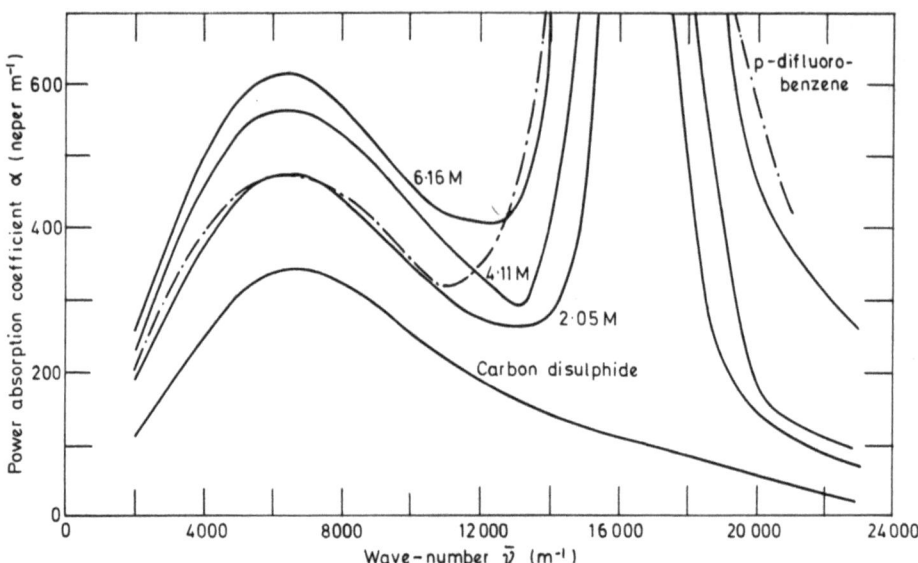

Fig. 10. Absorption spectra of carbon disulphide and of *p*-difluorobenzene, together with solutions of the latter in the former.

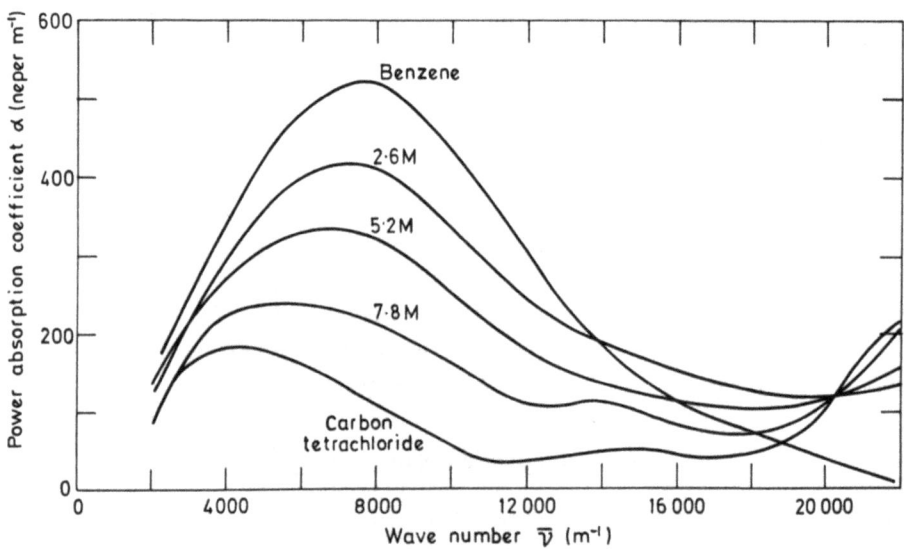

Fig. 11. Absorption spectra of benzene and of carbon tetrachloride together with solutions of the latter in the former.

Firstly, a norm has to be established for such two-component systems. If the mixtures were gaseous then a bimolecular collisional origin of the absorptions would be acceptable, i.e.

$$A(\text{mixture}) = A_A N_A^2 + A_B N_B^2 + 2A_{AB} N_A N_B, \tag{1}$$

where N_A, N_B are mole fractions and each of the two particle collisions has its own induced dipole absorption factor. The latter is given as the integrated value $A = \int_{band} \alpha(\bar{\nu}) d\bar{\nu}$ as this, rather than the absorption at some one wavenumber $\alpha(\bar{\nu})$ is the factor which is correctly related to the total collisional absorption. Especially in liquids, any peculiarities in the absorption contours could lead to anomalies in $\alpha(\bar{\nu})$ (mixture). Transferring the above relation to mixtures of liquids clearly implies that even there bimolecular 'collisions' are predominantly important – an approximation open to question. In this relation the factor A_{AB} will clearly be specific to the $(A \cdots B)$ interaction. However, as an order of magnitude anticipation one might write $A_{AB} = (A_A \cdot A_B)^{1/2}$: if some appreciably different value of A_{AB} is needed this will not be particularly significant until its value differs greatly from $(A_A \cdot A_B)^{1/2}$.

Of the liquid pairs carefully measured here [8] three, *trans*-1,2-dichloroethylene/ *cyclo*hexane: p-difluorobenzene/*cyclo*hexane: p-difluorobenzene/benzene: are adequately represented by Equation (1) even with $A_{AB} = (A_A \cdot A_B)^{1/2}$. For carbon tetrachloride/benzene this assumption leaves some just significant departure at the $N_A = N_B$ composition, suggesting that A_{AB} is here slightly larger than $(A_A \cdot A_B)^{1/2}$; in fact $A_{AB} = 1.25(A_A A_B)^{1/2}$ provides a complete fit (Figure 12b). For p-difluorobenzene and carbon tetrachloride the curve requires $A_{AB} = 1.52(A_A A_B)$: and for p-difluorobenzene and carbon disulphide this numerical factor is 1.76. Figure 12a. It is an open question at which, if any, of these values it can be justifiably claimed that a novel feature appears in the $(A \cdots B)$ interaction which cannot be represented by the normal charge distributions colliding with other anisotropically polarizable partners.

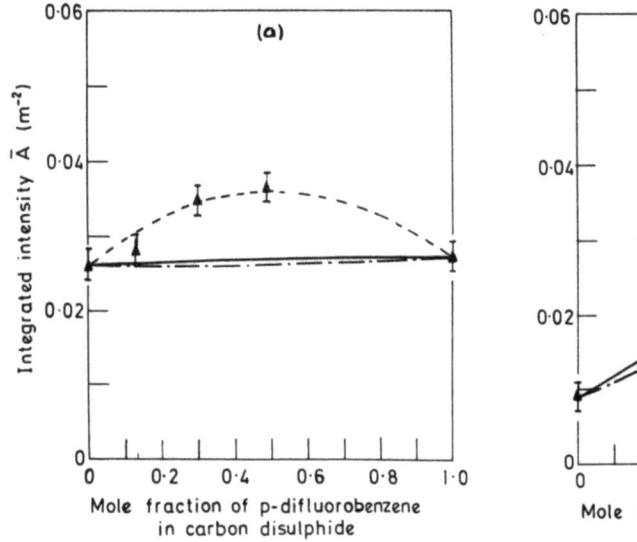
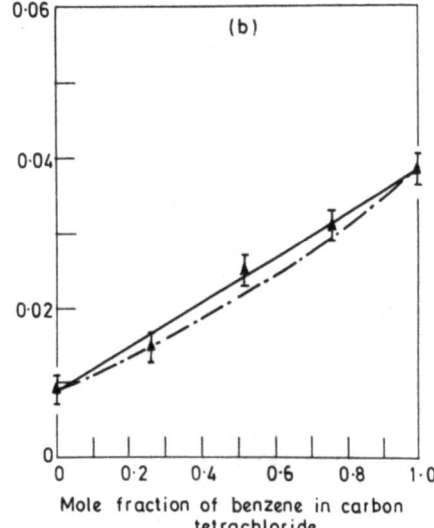

Fig. 12. Integrated intensities for solutions of p-difluorobenzene in carbon disulphide. Experimental points and their uncertainties: $-\cdot-$ calculated for $A_{AB} = (A_A \cdot A_B)^{1/2}$; $--$ calculated for $A_{AB} = 1.76(A_A \cdot A_B)^{1/2}$. (b) Integrated intensities for solutions of benzene in carbon tetrachloride. Experimental points and their uncertainties: $-\cdot-$ calculated for $A_{AB} = (A_A \cdot A_B)^{1/2}$; $-$ calculated for $A_{AB} = 1.25(A_A \cdot A_B)^{1/2}$.

9. Conclusion

The general character of the non-dipolar absorptions is established. Significant deductions can be made, e.g. the values of molecular quadrupoles and other multipoles: the degree to which molecular rotation is unhindered in the liquid can be assessed: possible rattling (collisional) frequencies can be estimated: and the question of specific interactions between non-dipolar molecules can be explored.

Nevertheless we still lack an adequate model to correlate the absorption contour and its frequency centre with the molecular state in the liquid. Evaluation in terms of molecular point-charge models needs to be developed, preferably allowing for anisotropy of polarizability in the molecules. On the basis of an acceptable representation of such first-principle features in molecular collisional interaction it should become possible to assess the extent and character of further specific (electron transfer, etc) features occurring between the molecular partners.

References

1. Price, A. H.: in N. E. Hill, W. Vaughan, A. H. Price, and Davies, M.: *Dielectric Properties and Molecular Behaviour*, Van Nostrand, London, 1969, pp. 223–228 and references therein. Buontempo, U.; Stumper, U.; Kasprowitz-Kielich, B. in this conference.
2. Whiffen, D. H.: *Trans. Faraday Soc.* **46**, 124 (1950).
3. Kiss, Z. J. and Welsh, H. L: *Phys. Rev. Letters* **2**, 166 (1959); Welsh, H. L. and Kriegler, R. J.: *J. Chem. Phys.* **50**, 1043 (1969) and references therein.
4. Chamberlain, J.: *Infrared Physics* **11**, 25 and 57 (1971).
5. Chamberlain, J., Haigh, J., and Hine, M. J.: *Infrared Physics* **11**, 75 (1971).
6. Davies, Graham J., Chamberlain, John, and Davies, Mansel: *J. Chem. Soc. Faraday Trans. II*, **69**, 1223 (1973).
7. Chamberlain, J., Gibbs, J. E., and Gebbie, H. A.: *Infrared Physics* **9**, 185 (1969); Davies, G. J.: Ph.D. Thesis, Univ. of. Wales, 1972.
8. Davies, Graham J. and Chamberlain, John: *J. Chem. Soc. Faraday Trans. II* **69**, 1739 (1973).
9. Pardoe, G. W. F.: *Trans. Faraday Soc.* **6**, 2699 (1970).
10. Bradley, C. C., Gebbie, H. A., Gilby, A. C., Kechin, V. V., and King, J. H.: *Nature* **211**, 839 (1966).
11. Rosenberg, A. and Birnbaum, G.: *J. Chem. Phys.* **52**, 683 (1970); **48**, 1396 (1968).
12. Harries, J. E.: *J. Phys.* **B3**, 704 (1970); **3**, L150 (1971).
13. Birnbaum, G. and Maryott, A. A.: *J. Chem. Phys.* **36**, 2026 (1962).
14. Colpa, J. P. and Ketelaar, J. A. A.: *Mol. Phys.* **1**, 14 and 343 (1958).
15. Bose, T. K. and Cole, R. H.: *J. Chem. Phys.* **52**, 140 (1970).
16. Ho, W., Birnbaum, G., and Rosenberg, A.: *J. Chem. Phys.* **55**, 1028 (1971).
17. Buckingham, A. D. and Disch, R. L.: *Proc. Roy. Soc. (London)* **A273**, 295 (1963).
18. Evans, Myron: *J. Chem. Soc. Faraday Trans. II* **69**, 763 (1973).
19. Baise, A.: *J. Chem. Soc. Faraday Trans. II* **68**, 1904 (1972).
20. Davies, M. Pardoe, G. W. F., Chamberlain, J., and Gebbie, H. A.: *Trans. Faraday Soc.* **64**, 847 (1968); **66**, 273 (1970).
21. Moelwyn-Hughes, E. A.: *Reaction Kinetics in Solution*, Oxford Univ. Press, London, 1947, Chapter VIII.
22. Haigh, J.: D.Phil. Thesis, Oxford Univ., 1970.
23. Baise, A. I.: Ph.D. Thesis, Univ. of Wales, 1971.

DISCUSSION

Cole: A comment on the determination of quadrupole moments from the infrared absorption. Measurements on a pure quadrupolar gas are complicated by quadrupole-quadrupole interaction. This can be avoided by using mixtures with an atomic solvent; this technique has worked well for CO_2–Ar mixtures. A question is whether far infrared absorption intensities have been measured in the solid phases of the molecules studied. This could throw light on the origin of the absorption, as the quadrupole process depends on fluctuations, which one would expect to be smaller in the solid phase and so give smaller absorption.

Davies: Your points serve to emphasize the further significant studies which should be made, as these aspects have not yet been pursued.

Hertz: How do the frequencies you observed compare with those obtained from *inelastic* neutron scattering?

Davies: We have not made such a comparison. One aspect of the infrared data which will limit its significance is the half-width of the generally featureless absorption: this can amount to 50 cm^{-1}.

Rivail: Did you try to look for a connection between the 'excess absorption' (i.e. the difference between the actual absorption and the calculated one assuming an additivity) in systems such as p-difluorobenzene-carbon tetrachloride and (i) the quadrupole moment of the solute; (ii) the molecular polarizability of the solution (i.e. the average polarizability of the solvent and the solute for 1 − 1 mixtures)?

Davies: No, we have not yet made any such analysis. At present the problem appears to be to establish, if possible, the criteria governing the absorption intensities in the absence of specific molecular interactions. Such criteria await a fuller theoretical appraisal of the intensities and the deduction of an 'excess absorption' will be in some appreciable degree an arbitrary process but, as you imply, this is an approach which the chemist will be anxious to pursue.

ROTOTRANSLATIONAL SPECTRUM OF H_2 IN LIQUID A AND COMPARISON BETWEEN INTERCOLLISIONAL INTERFERENCE IN PARA-H_2 AND RARE GASES

U. BUONTEMPO, S. CUNSOLO, P. DORE, and G. JACUCCI

Istituto di Fisica 'G. Marconi', Università di Roma, Roma, Italy

Abstract. Absorption coefficients have been measured in solutions of normal H_2 and pure para-H_2 in liquid A in the spectral range 14–700 cm^{-1}. Such measurements have permitted to study the purely translational contribution to the reduced line shape. The low frequency part of the reduced line shape relative to para-H_2 in liquid A is dominated by intercollisional interference effects, similar to the ones observed in solutions of He and Ne in liquid A.

Résumé. On a mesuré, dans le domaine spectral 14–700 cm^{-1}, les coefficients d'absorption de l'hydrogène normal et du para-hydrogène pur en solution dans l'argon liquide. Ces mesures permettent d'étudier la contribution purement translationnelle au profil réduit de bande. La région de basse fréquence du profil réduit pour para-H_2 dans l'argon liquide est dominée par des effets d'interférence collisionnels, semblables à ceux que présentent He et Ne en solution dans l'argon liquide.

1. Introduction

The collision induced spectra are generally interpreted and discussed with the idea that in a cluster of interacting molecules the two major types of dipole moment induction mechanisms are quadrupole and overlap induction. The first one is present when at least one molecule in the cluster possesses a permanent quadrupole moment, the associated electric field polarizing the other molecules. The second one arises from the deformation of the electronic clouds of the interacting molecules. Because of the different spatial range of the two induction mechanisms, their respective contributions to the absorption spectrum in a given system consist of bands of different width. The overlap induced dipole moment varies with intermolecular separation much more rapidly than the r^{-4} dependence of the quadrupolar electric field. Thus the molecular dynamics of the liquid produces more rapid time variation of the dipole moment induced by the first mechanism than the second one. The duration τ_0 of an overlap optical collision is shorter than the mean correlation time of the quadrupole induced dipole moment. Remembering that the experimental absorption coefficient is related to the Fourier transform $G(\sigma)$ of the dipole moment time autocorrelation function $C(t)$ through the simple formula

$$G(\sigma) \propto [\sigma \tanh(hC\sigma/2KT)]^{-1} A(\sigma), \tag{1}$$

it is then evident that in the frequency space the overlap induced bands be wider than the quadrupole induced one.

An interesting feature that characterizes overlap induced bands is the 'intercollisional interference effect', which is present in the $G(\sigma)$ as a zero frequency dip. This effect has been observed and studied in the translational bands induced in liquid argon by He and Ne impurities [1].

The intercollisional interference is readily discussed by means of $C(t)$. The impurities dissolved in the liquid undergo a motion that is both diffusive and rattling in character. The overlap induced dipole moment associated with each impurity changes sign in the subsequent repulsive interactions with the medium, due to the 'rattling' part of the motion. Such sign alternation produces a negative contribution in the dipole moment autocorrelation function about the time τ_0 which represents the mean rattling half period. This feature in turn results in a low frequency dip in $G(\sigma)$. Such dip is confined to a smaller low frequency portion of the translational band as the time τ_c increases with respect to the duration τ_0 of an overlap optical collision.

We report here the results of an extension of such investigation to H_2 in liquid A. In the absorption spectrum of this system contributions from both the overlap and quadrupole induction mechanism are present. The intercollisional interference effects are featured by the overlap induced contribution only and in order to observe them this contribution has to be extracted from the spectrum. In the translational band of pure para-H_2 in liquid A, this contribution only is present because in para-H_2 fundamental state $(J=0)$ which is the only one really populated, the quadrupolar electric field of the H_2 molecule vanishes.

We have measured the rototranslational spectrum of H_2 in A at various ortho-para concentrations. The translational band has been obtained from the spectrum of pure para-H_2 in liquid A, and intercollisional interference effects have been found in the translational reduced line shape $G(\sigma)$. These effects are discussed and compared to the analogous ones previously measured in He and Ne solutions in liquid A.

The translational band has been measured also in a normal-H_2 solution in liquid A. Here the intercollisional interference effects are hidden by the ortho-H_2 quadrupole induced contribution superimposed to the overlap induced one.

2. Experimental

The reported measurements have been carried out with an RIIC 720 Michelson interferometer. The interferograms have been digitally recorded and the data has been elaborated with the help of a digital computer. The frequency range of the instrument is 10–400 cm^{-1}. This range has been extended to higher frequencies by a partial modification of the electronic apparatus of the instrument that has permitted to divide by two the sampling interval, thus doubling the maximum detectable frequency. This extension was needed in order to measure the entire rototranslational spectrum of the H_2–A mixtures, which extends over 700 cm^{-1}.

The absorption cell consists of a one meter long light pipe cooled with liquid oxygen to the temperature of 91.0 K. Polyethylene windows and lenses have been employed for the region below 350 cm^{-1}, while C_sI has been used from 250 cm^{-1} to 700 cm^{-1}. The absorption coefficient $A(\sigma)$ is obtained from the ratio of the sample and background transmission the background being taken with the cell filled with liquid argon in order to reduce the error due to refractive index mismatich at the liquid window interface.

High purity (5 ppm) standard gases have been employed. The residual water has been eliminated with the help of a liquid nitrogen cooled zeolite trap. This purification procedure causes a partial ortho-para H_2 conversion. For this reason the normal-H_2 mixture (25% para, 75% ortho corresponding to room temperature equilibrium concentration) has been prepared 3 months ahead. The para-H_2-A mixture is prepared by converting the ortho-H_2 present in the normal-H_2 mixture to para-H_2. This task is accomplished by liquifying the hydrogen gas in a special glass container, cooled with help of liquid helium to the liquid hydrogen temperature, in presence of a suitable catalyst (Cr_2O_3-γAl_2O_3).

Great care has been taken to avoid para-ortho back conversion upon raising the temperature. It has been experimentally verified that during the short time needed to take the measurements no appreciable reconversion of para hydrogen occurred.

The low frequency measurements have required an H_2 concentration of the order of one percent, in order to obtain a reasonable absorption of the translational band. One order of magnitude less in largely sufficient at high frequency. The two frequency region have been separately investigated, as already noted and the two series of spectra have been lined by normalizing them in the superposition region.

3. Results

In the induced spectrum of H_2 liquid A the pure translational band is rather separated from the rotational lines, as it has already been noted [2], because of the low value of the moment of inertia of H_2, that places the rotational lines at rather high frequencies. At the temperature of 90 K the only rotational states appreciably populated are the lower $J=0$ and $J=1$ energy states, accessible to para and ortho H_2 respectively. Thus the induced rotational spectrum consists of the lines $S(0)$ and $S(1)$ only, corresponding to the J transitions 0–2 and 1–3 respectively.

Figure 1 shows the rotational spectra of pure para-H_2 and of normal-H_2 both dissolved in liquid A. In its spectrum the $S(0)$ line peaked at $\sigma_0 = 355$ cm^{-1} and $S(1)$ line peaked at $\sigma_1 = 587$ cm^{-1} are both clearly visible. In the first spectrum, on the other hand, only the $S(0)$ line is present. The absence of any appreciable absorption at the $S(1)$ line proves that the residual ortho-H_2 concentration is less than one percent.

In Figure 2 a typical low frequency absorption spectrum is reported, relative to a mixture of normal-H_2 in liquid A. The translational band Q_T is present at low frequencies besides the $S(0)$ line.

Figure 3 shows the same type of spectrum relative to a pure para-H_2-A mixture. Here the $S(0)$ line is four times more intense with respect to the translational band than in the previous case.

In order to observe the shape of the translational contribution to the absorption spectrum, the rotational contribution has to be subtracted.

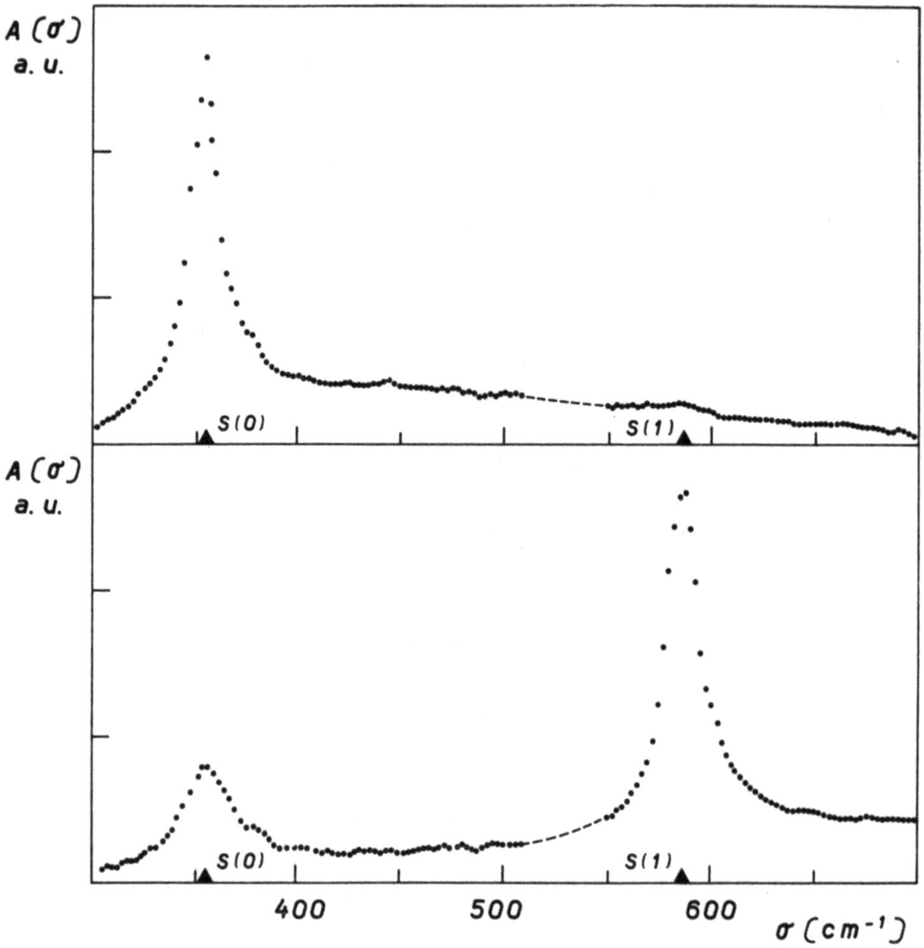

Fig. 1. Induced rotational spectra for para-H_2 (above) and normal-H_2 (below) in liquid A.

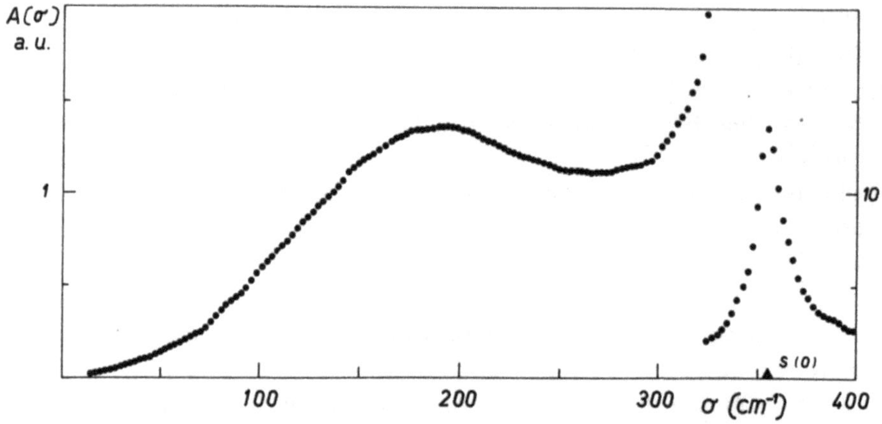

Fig. 2. Low frequency portion of the absorption spectrum in normal-H_2 in liquid A featuring the pure translational band.

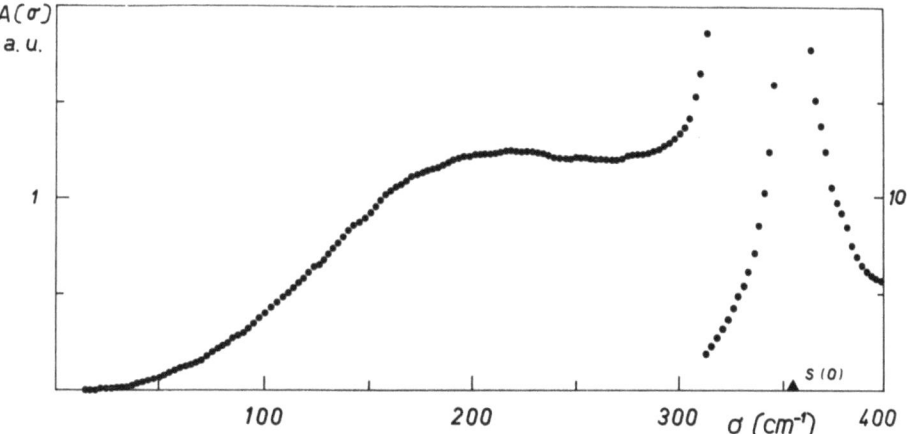

Fig. 3. Low frequency portion of the absorption spectrum in pure para-H_2 in liquid A featuring the pure translational band.

The profile of a rotational component of N_2 in liquid A centered at σ_r is known to be rather well described for $\sigma > \sigma_r$ by the expression (3)

$$A_{r_+}(\sigma) = K\sigma\{\{[(\sigma-\sigma_2)/\delta]^2 + 1\} + A[(\sigma-\sigma_2)/KT]^B \times \\ \times \exp\{-C[(\sigma-\sigma_r)/KT]^D\}\}, \quad (2)$$

where $A = 3.9 \times 10^{-4}$; $B = 1.5$; $C = 0.9$; $D = 1.0$; $\delta = 14.5$ cm^{-1} and K is a normalization constant.

The agreement between the curve represented by this expression and the experimental data has been controlled by fitting with (2) the high frequency side of $S(1)$ in the normal-H_2-A spectrum. Here the contribution of both the Q_T and $S(0)$ tails are small. We find that expression (2) describes rather well the experimental data with a residual discrepancy of 10% at most. The detailed balance principle relates the low and high frequency side of any H_2 rotational component by:

$$A_{r_-}(\sigma_r - \Delta\sigma)/(\sigma_r - \Delta\sigma) \simeq [A_{r_+}(\sigma_r + \Delta\sigma)/(\sigma_r + \Delta\sigma)] \times \\ \times \exp(-2hc\Delta\sigma/kT), \quad (3)$$

where $\Delta\sigma = |\sigma - \sigma_r|$.

Hence one can derive the shape of the low frequency side of the rotational line. With the help of this expression the translational spectrum may be obtained by subtracting from the measured absorption coefficient the rotational contributions. For the pure para-H_2 spectrum the normalization constant is chosen in such a way that expression (2) coincides with the spectrum above 500 cm^{-1}. It is reasonable to expect that in this region the contribution of the Q_T band be small. In the ortho-para-H_2 mixture the $S(1)$ is subtracted first normalizing the curve of expression (2) on the high frequency side. The procedure has then been repeated as for para-H_2. Care has been taken in the subtractions to take into account the slight discrepancies observed between the experimental profile and expression (2).

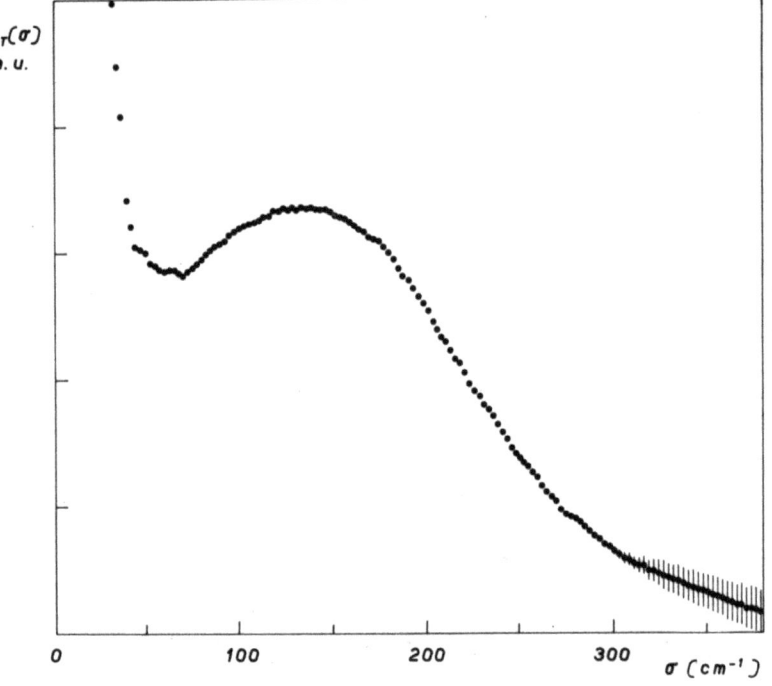

Fig. 4. The reduced line shape $G_T(\sigma)$ for the pure translational band in normal-H_2 in liquid A.

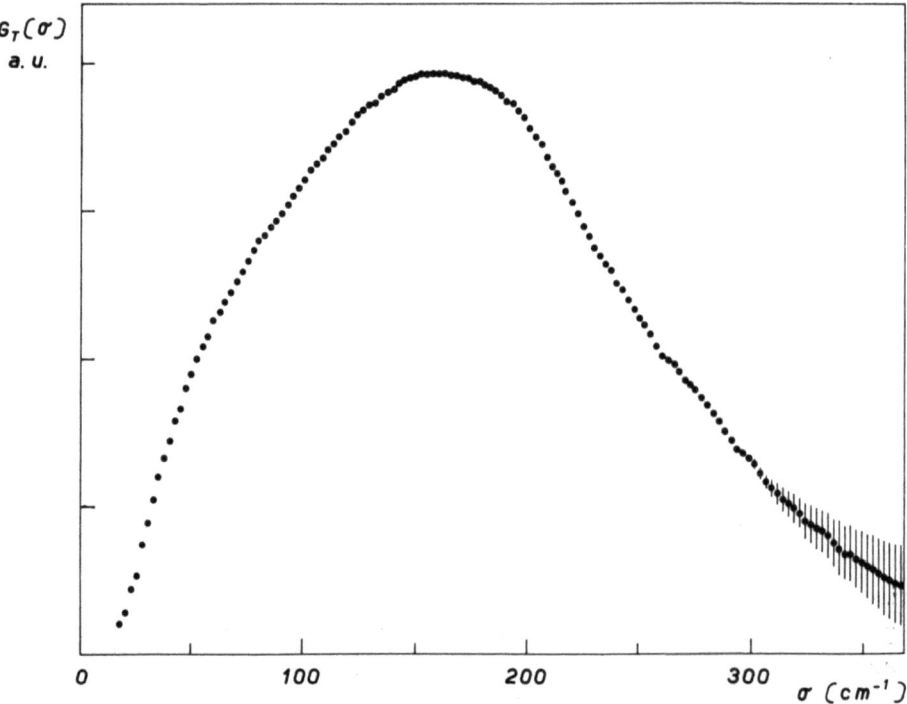

Fig. 5. The reduced line shape $G_T(\sigma)$ for the pure translational band in para-H_2 in liquid A.

The pure translational band has been obtained from a mean of several measurement by the described subtraction procedure. The reduced line shape $G(\sigma)$ of Q_T is shown in Figures 4 and 5 for normal and para-H_2 respectively.

The $G(\sigma)$ of para-H_2 shows a low frequency dip that is, of course, clearly observable independently of the subtraction of the high frequency rotational components. Such a dip is analogous to the ones observed in the overlap induced translational bands in mixtures of noble gases [1]. As a matter of fact the translational band in para-H_2 is due to overlap induction only, the quadrupole moment being here equal to zero. Both contributions are present, on the contrary, in the ortho-H_2 spectrum.

4. Discussion

An interesting feature evidenced in this work is the different behaviour of the reduced line shape of the translational band $G_T(\sigma)$ in pure para-H_2 and in normal H_2.

In para-H_2 $G_T(\sigma)$ vanishes towards zero frequency, while in normal H_2 it rises in the same region and shows a zero frequency peak, which we have assigned to quadrupolar induction by ortho-H_2 molecules. No zero frequency dip is to be expected for the quadrupole induced bands. This can be understood by considering that the quadrupolar electric field is strongly angle dependent and that the hydrogen molecule undergoes a rotational motion which destroys the correlation in the induced dipole moment deriving by memory in the translational motion.

The negative contribution to the dipolar correlation function extending at times of the order of τ_c, characteristic of the overlap translational band, is thus absent here, because the direction of the quadrupole induced dipole moment is randomized by the rotation of the hydrogen molecule. As a consequence no zero frequency dip is to be expected in the Fourier transform $G(\sigma)$ of the quadrupole induced dipolar correlation function.

The reduced line shape of the translational band in pure para-H_2 is of the same type of the ones already measured in mixtures of noble gases.

Thus one expects also in this case that the high frequency tail of the reduced line shape be similar to the gas behaviour, where the binary picture holds, and be described with a curve of expression [4, 5]

$$G(\sigma) \propto \exp(-C\tau_0\sigma/2)^2 \, [1 + \varepsilon_4 H_4(C\tau_0\sigma/2)].$$

With this curve we have fitted the $G_T(\sigma)$ in para-H_2 for $\sigma > 240$ cm^{-1}.

The obtained value of τ_0 is 0.52×10^{-13} s. This characteristic time may then be interpreted as duration of the optical collision, with the help of the exponential model for the overlap dipole moment induced in a couple $\mu(R) = \mu_0 \exp(-R/\varrho)$, where R is the intermolecular distance. τ_0 is related to the range ϱ by the simple relation [4] $\tau_0 = \varrho(m/KT)^{1/2}$, where m is the reduced mass of the colliding couple.

By means of τ_0 a scaled variable may be introduced for the frequency axis $X = C\tau_0\sigma$. Figure 6 shows $g(x) \propto G(X/C\tau_0)$ for para-H_2-A in comparison with the same function

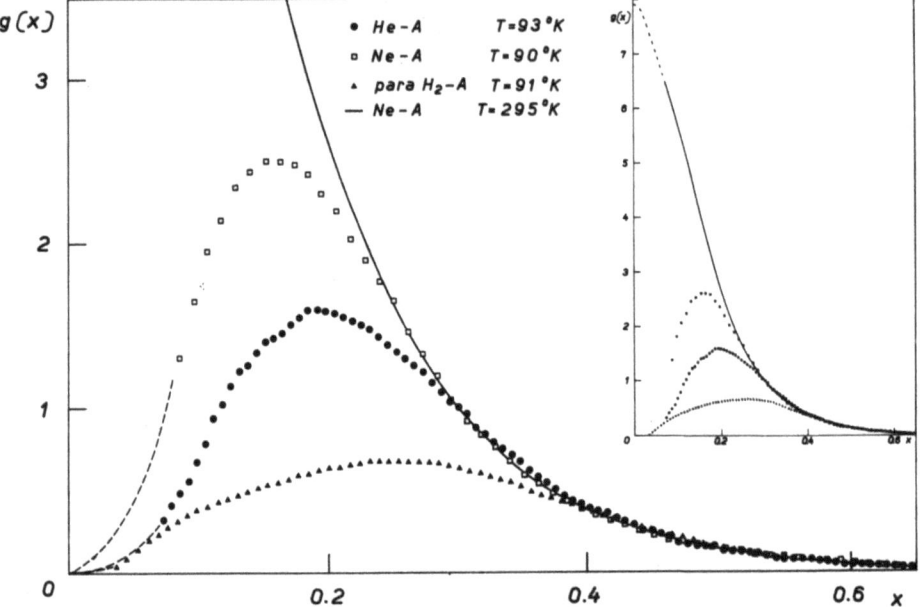

Fig. 6. The reduced line shape $g(x)$ vs the scaled wavenumber. The value $x = 0.20$ corresponds to 130.7 for H$_2$–A, to 90.4 cm^{-1} for He–A, to 54.5 cm^{-1} for liquid Ne–A, and to 94.6 for gaseous Ne–A. The gas behaviour has been calculated from the data of Bosomworth and Gush [6]. The dashed portions of the curves are extrapolations.

obtained for mixtures of noble gases both in the liquid phase and at room temperature. The various curves are normalized together at about $X = 0.4$.

It should be noted that, while the high frequency tails coincide, the various curves for the liquid mixtures depart at low frequencies from the gas behaviour because of the intercollisional interference, the departure being more pronounced for lighter impurities.

We may introduce a characteristic time τ_c, defined as the inverse of the frequency at which $g(x)$ reduces to one half of its value in gas. We obtain the following values for this τ_c

$$\tau_{c_{H2-A}} = 1.89 \times 10^{-13} \text{ s}, \qquad \tau_{c_{He-A}} = 4.04 \times 10^{-13} \text{ s},$$
$$\tau_{c_{Ne-A}} = 9.21 \times 10^{-13} \text{ s}.$$

These times should be proportional to the inverse of the mean rattling frequency of the impurities in the liquid. Assuming that the impurity moves inside a potential well of dimension L, the times τ_c are related to temperature and impurity mass m_0 by the simple relation

$$\tau_c \propto L\sqrt{m_0/KT}.$$

We note that the ratio of τ_c for He and Ne is in good agreement with this model taking the same value for L, while this is not true in the case of H$_2$. We think that such result be due to quantum effects more than to a difference in values of L.

References

1. Buontempo, U., Cunsolo, S., and Jacucci, G.: *Can. J. Phys.* **49**, 2870 (1971).
2. Bulanin, M. O. and Tonkov, M. V.: *Phys. Letters* **3**, 120 (1968).
3. Holleman, G. W. and Ewing, E.: *J. Chem. Phys.* **47**, 571 (1967).
4. Sears, V. F.: *Can. J. Phys.* **46**, 1163 (1968).
5. Brenner, S. L. and McQuarrie, D. A.: *Can. J. Phys.* **49**, 837 (1971).
6. Bosomworth, D. A. and Gush, H. P.: *Can. J. Phys.* **43**, 751 (1965).

DISCUSSION

Marteau: May I know why you use a rather pratical expression for the Boltzman factor to define the transition probability function, namely $\tanh(hc\sigma/2KT)$, instead of $(1 - \exp(-hc\sigma/KT))$?

Buontempo: We use the reduced line shape $G(\sigma)$, previously defined, in order to obtain a symmetric induced dipole moment autocorrelation function.

Carneiro: On your comparison between para-H_2, He and Ne in Ar, H_2 is the 'extreme quantum case'. Should it not be He, according to the De Boer parameter Λ^*, or is the effect a simple mass effect?

Buontempo: Yes, our opinion is that it is mainly a simple mass effect.

Rothschild: How do you extrapolate your data to zero frequency in order to obtain the correlation functions?

Buontempo: We have used a simple parabolic extrapolation from the low frequency part of the experimental $G_T(\sigma)$, in order to obtain from the minimum of the correlation function a characteristic time τ_M to be compared with τ_C previously defined. We have verified that upon varying the extrapolation the obtained value for τ_M does not change.

INTERPRETATION OF FAR INFRARED SPECTRA IN TERMS OF A COLLISION DISTRIBUTION

D. FRENKEL, R. M. VAN AALST, and J. VAN DER ELSKEN

Laboratory for Physical Chemistry, University of Amsterdam, Amsterdam, The Netherlands

Abstract. A generalized model for the description of molecular motions is discussed. Computer simulations are compared with experimental data, which are thereupon analysed in terms of model parameters. Some general trends are indicated and an interpretation in terms of molecular properties is proposed.

Résumé. L'extension d'un modèle collisionnel pour la description des mouvements moleculaires est discutée. Les simulations du calculateur sont comparées avec les données expérimentales, qui sont en conséquent analysées en fonction du modèle. Ainsi on peut dégager certaines lignes générales. On propose une interprétation en termes de propriétés moléculaires.

It is well known that the description of liquids differs greatly from the description of gases on the one hand and solids on the other. In the latter two systems it is possible to express measurable properties in terms of coordinates that have, at least in zero'th order, a simple time dependence. In liquids however this simple zero'th order starting point is lacking and therefore, although it may still be possible to specify the coordinates that are relevant for a certain property, these coordinates will be coupled so strongly to all others that their time dependence will be far from simple. Hence anyone wishing to analyse experimental data on microscopic properties of liquids will have to choose between two approaches: either take all coordinates into explicit consideration or restrict the description to one or a few coordinates and lump the coupling with all others in a coupling-(friction-)parameter which will in general be frequency dependent.

The first approach leads to molecular dynamics calculations, the second to phenomenological descriptions that find their theoretical justification in the generalized Langevin equation.

The advantage of model descriptions lies in the fact that they combine the simplicity obtained by taking only few coordinates into explicit consideration with an intuitively clear picture of the coupling of these coordinates with all others; the main disadvantage is of course the loss of rigour.

In the following we will discuss the description of some dynamical properties of liquids in terms of a molecular model in which the molecules perform rotational or translational motion interrupted by collisions.

The model description of rotational and translational diffusion is as follows: molecules are assumed to undergo infinitely short collisions while rotating or translating freely in between. In contrast to earlier models [1, 2, 3] successive collisions may be correlated i.e. the distribution of times between two consecutive collisions is not necessarily of the Poisson type. The assumption that collisions are Poisson distributed seems justified for a dilute gas but need not necessarily hold for dense systems. The reason for the incorporation of correlated collisions in the model was the hope that it might be possible to interpret certain experimental properties characteristic of liquids in these terms.

In the model the effect of a collision may be twofold: firstly angular or linear momentum may be transferred and secondly energy transfer may take place. Two limiting cases are usually discerned:

(1) No energy is transferred during a collision.
(For rotation this is known as m-diffusion.)

(2) The size of the momentum vector is thermally averaged by every collision (in this case large energy transfer occurs).
(For rotation this is known as J-diffusion.)

If it is assumed that only successive collisions may be directly correlated and that the effect of successive collisions on the molecular motion is uncorrelated, the mathematical formulation of the diffusion problem may be cast in a form that has agreeable analytical properties. In this form all n-collision contributions can be summed explicitly and closed expressions are obtained for the autocorrelation function of the dipole moment and the angular momentum in the case of rotational diffusion and of the velocity in the case of translational diffusion [4]. These correlation functions provide a link with experiment and hence information could be obtained from comparison of model calculations with experimental data.

Experimentally obtained and computer simulated far infrared spectra due to the rotational motion of linear molecules or the translational motion of spherical molecules were thus compared to extract information from model calculations. To this end we separately studied the effect of:

(1) Changing the ratio between energy transfer and momentum transfer (i.e. the limiting elastic and inelastic cases were compared).

(2) Introducing a correlation between successive collisions.

(3) Changing the net effect of one collision on the amount of momentum transferred.

Conclusions based on this procedure can only be qualitative. For a phenomenological description of physical systems this model would actually be rather inefficient due to its flexibility. More importance should be attached to conclusions stating that under certain restrictions a particular type of experimental spectrum can or cannot be simulated than to the actual values of the parameters which give the best fit.

As an example we consider the far infrared rotation spectra of linear molecules in liquids. Qualitative features that can be recognized are:

(a) The presence or absence of rotational fine structure over the entire absorption region or part thereof [5, 6, 7].

(b) The overall intensity distribution. (In fact a wide variety of band shapes can be found in the literature; e.g. [8, 9, 10].)

It turns out that qualitatively all these different shapes can be accounted for by adjusting the variables within the model. A systematic study of the effect of the variation of the different parameters on the simulated spectra led us to the following generalizations:

For liquids showing spectra with rotational fine structure rotational energy transfer is relatively slow whereas angular momentum is readily exchanged. These systems are best described by an m-diffusion model; it seems unnecessary to assume any correlation between successive collisions.

Liquids giving rise to strong absorption at higher than pure rotational frequencies (Poley absorption) have considerable coupling of their rotational coordinates to translational coordinates in the fluid. Because of the large torques acting on the molecules the rotational energy is not even approximately a constant of motion and rotation tends to libration which is of course strongly coupled to translational modes of about the same frequency. In terms of model parameters this strongly coupled motion is best described by a J-diffusion model with correlated collisions that tend to reverse the direction of the angular momentum (thus giving rise to a librational type of diffusive motion).

Figure 1 shows a series of simulated far infrared spectra with rotational fine structure. (If the moment of inertia that appears in the reduced frequency units is chosen to be the moment of inertia of hydrogen chloride, the temperature corresponds to 100 K; by appropriate scaling these simulations may be made to apply to other systems.) For each set of parameters the spectrum is calculated twice; one simulation corresponding to the elastic (m-diffusion) limit, the other to the inelastic (J-diffusion) limit. Two parameters are varied independently:

(1) The average time between collisions (t_{BC}).

(2) The average cosine of the angle through which the angular momentum is rotated during a collision ($\cos \gamma$).

Successive collisions are assumed to be uncorrelated. Clearly the dependence of the rotational fine structure on $\cos \gamma$ is only slight in the J-diffusion case compared to m-diffusion. In contrast the dependence on t_{BC} is much stronger; so much in fact that if t_{BC} (in units $I/\hbar$) is less then one all rotational fine structure disappears in the J-diffusion case. Furthermore the overall spectral intensity distribution for J-diffusion tends to lower frequencies then the corresponding m-diffusion calculations.

Figure 2 shows a series of simulated far infrared spectra without rotational fine structure. Again for each set of parameters an m- and J-diffusion calculation was done. To obtain a spectral intensity distribution resembling Poley absorption it is necessary to assume that $\cos \gamma < 0$ (i.e. the sense of rotation is on the average reversed during a collision). This condition is necessary but not sufficient; a distinct correlation between successive collisions is required to shift the maximum of absorption to higher frequencies. This latter effect is shown in the figure in which corresponding simulations with uncorrelated collisions (A) and strongly correlated collisions (B) are plotted. The position of the maximum of absorption is mainly determined by the time at which the deviation from uncorrelated behaviour in the collision distribution is largest. Clearly the behaviour of m- and J-diffusion simulations is similar; however the shape of the spectra simulated with the J-diffusion model fits experimental spectra better, particularly in the low frequency region.

For the description of translational diffusion in liquids a similar analysis can be

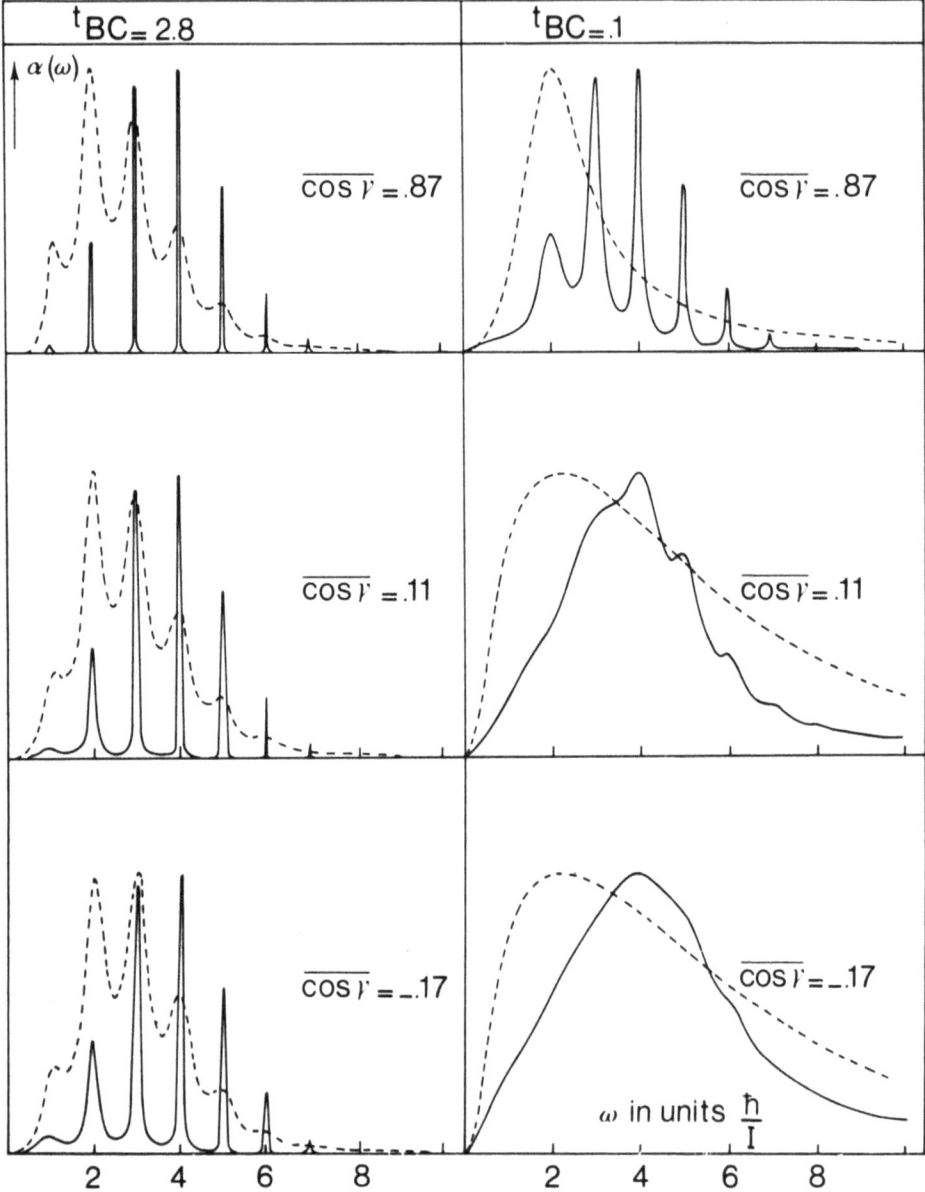

Fig. 1. Dependence of simulated far infrared spectra with rotational fine structure on t_{BC} and $\overline{\cos\gamma}$. Collisions are assumed to be Poisson distributed. Reduced temperature $\theta_R = (2IkT/\hbar^2) = 6.7$; extinction in arbitrary units. m-diffusion simulation = (———); J-diffusion simulation = (-----).

made. The appearance of a hump in the power spectrum of the velocity autocorrelation function of simple liquids at a characteristic frequency is well known [11]. Translational energy transfer depends on the mass ratio of the colliding particles (in a Lorentz gas collisions are completely elastic, for a system of identical particles the

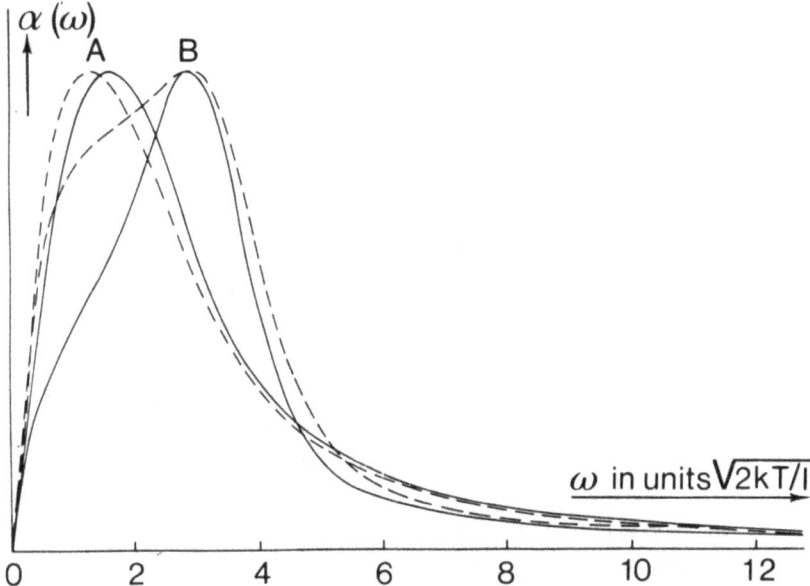

Fig. 2. Simulated far infrared spectra calculated with uncorrelated (A) and strongly correlated (B) collisions. Reduced temperature $\theta_R = (2IkT/\hbar^2) = 680$, $\overline{\cos\gamma} = -0.88$; extinction in arbitrary units. m-diffusion simulation = (———); J-diffusion simulation = (------).

thermalization of translational energy is effected in a few collisions). The completely elastic and strongly inelastic models therefore describe limiting cases. To obtain the characteristic frequency in the power spectrum of the velocity autocorrelation function one must assume that successive collisions are correlated and reverse on the average the direction of the particle velocity. Earlier models for the description of translational diffusion in liquids have been given [12, 13] in which the diffusion is described as a combination of translatory and oscillatory motions. To fit known power spectra of velocity autocorrelation functions of liquids one has to attribute considerable oscillatory character to the diffusional motion; this is commonly interpreted in terms of collective motions that are present in the liquid. Increasing the oscillatory character of the motion in such a description has much the same effect on the power spectrum of the velocity autocorrelation function as the introduction of correlated collisions in the collision model. This indicates a relation between correlated collisions in the model and collective motions in the system described.

Conclusions

Using a collision model for the description of rotational and translational diffusion of molecules certain qualitative conclusions may be drawn about the dynamical behaviour of systems studied experimentally. The relevant parameters in terms of which the experimental data may be interpreted are the amount of energy and momentum transfer during collisions and the possible correlation between successive collisions. The interpretation of experimental far infrared spectra of linear dipole molecules suggests that correlated collisions and large energy transfer appear together for rotation. As it is known that translational energy transfer depends strongly on the mass ratio of the interacting particles it is tempting to offer an interpretation for the rotational case that makes use of an analogous concept.

Molecules that give rise to spectra with rotational fine structure in the liquid phase mainly interact with the surrounding medium through relatively long range forces. These forces couple the rotation to a large number of molecules and this can be considered as an interaction with a pseudoparticle with large effective moment of inertia (this would favor elastic angular momentum transfer, in particular if energy transfer to the relative center of mass motion is small). The precise time dependence of the torques is determined by a large number of coordinates which will all have different time dependences; the coupling to any one of these coordinates separately is weak. Under these circumstances the autocorrelation function of the random torque will tend to δ-function behaviour, hence the autocorrelation function of the angular momentum will become exponential which corresponds to rotational motion modulated by uncorrelated collisions.

Molecules that give rise to Poley absorption mainly couple their rotational coordinates to the fluid through the short range non-central part of the potential. This short range interaction couples the rotation to only a few particles; therefore, using once more the analogy with translation, rotational energy is readily transferred. Furthermore the torques depend mainly on the coordinates of the nearest neighbours. These coordinates are strongly coupled to the coordinates of the rotating molecule itself and therefore the autocorrelation function of the random torque will persist relatively long; the angular momentum will not decay exponentially but exhibit the more or less oscillatory behaviour characteristic of the presence of correlated collisions in the model.

References

1. Gordon, R.: *J. Chem. Phys.* **44**, 1830 (1966).
2. McClung, R. E. D.: *J. Chem. Phys.* **55**, 3459 (1971), **51**, 3842 (1969).
3. St. Pierre, A. G. and Steele, W. A.: *J. Chem. Phys.* **57**, 4638 (1972).
4. Frenkel, D., Wegdam, G. H., and van der Elsken, J.: *J. Chem. Phys.* **57**, 2691 (1972).
5. Birnbaum, G. and Ho, W.: *Chem. Phys. Letters* **5**, 334 (1970).
6. van Aalst, R. M. and van der Elsken, J.: *Chem. Phys. Letters* **13**, 631 (1972).
7. Birnbaum, G.: *Mol. Phys.* **25**, 241, (1973).
8. Kroon, S. G. and van der Elsken, J.: *Chem. Phys. Letters* **1**, 285 (1967).
9. Darmon, I., Gerschel, A., and Brot, C.: *Chem. Phys. Letters* **8**, 454 (1971).
10. Boittiaux, B., Fauquembergue, R., and Desplanques, P.: *Compt. Rend. Acad. Sci. Paris* **271**, 1409 (1970).
11. Egelstaff, P. A.: *An Introduction to the Liquid State*, London, New-York, Acad. Press, 1967.
12. Sears, V. *Proc. Phys. Soc.* **86**, 953 (1965).
13. Caroli, B., Jannink, G., and Saint James, P.: *J. Phys. Chem.* **4**, 545 (1971).

DISCUSSION

Laulicht: Could you suggest what will be the order of magnitude of τ_{BC} for real molecules in liquids?

Frenkel: Calculated τ_{BC} values are of the order of 1×10^{-13} to 5×10^{-13} s, for not too large molecules.

Friedman: In your collision correlations, do you only have a correlation of the interval between successive collisions or do you also have correlations in the energy and momentum transfers of successive collisions?

Frenkel: In this model only correlation between successive collisions has been taken into account.

Comment by M. R. Hoare

I agree entirely, in the spirit of previous remarks, that one must look for tractable models which go beyond the oversimplified J and M-diffusion cases.

In considering these, however, I think one must draw certain distinctions in the way in which the term 'correlation' is used, bearing in mind particularly the subtle difference between the idea of chains of correlation in the probabilities of, say, rotational states before and after collisions and the idea that 'collisions' themselves may be correlated. If there is (first-order) Markovian behaviour, all 'memory-effects' will be statistically determined by the immediately-previous collision, nevertheless correlations in states – in this case the *persistence of angular velocity* – may still propagate over many collision-times.

More immediately, though, I must object to the assertion that the distribution of waiting-times for collisions in a freely-translating gas is Poisson in form. The collision-number function for hard-spheres is naturally velocity-dependent and contains a Gaussian and an error-function term. (See Chapman and Cowling.) A Poisson distribution of waiting-times only applies if one selects a sub-set of spheres with fixed translational velocity. In view of this it seems difficult to understand your statement that deviation from simple Poisson behaviour implies a correlation between successive collisions.

DIELECTRIC ABSORPTION OF NON-POLAR LIQUIDS IN THE MICROWAVE AND FAR INFRARED REGION

ULRICH STUMPER

Drittes Physikalisches Institut der Universität D-34 Göttingen, F.R.G.

Abstract. Measurements of the complex permittivity of seven n-alkanes have been performed in the frequency range from 26 to 1500 GHz. In these liquids of zero permanent electric dipole moment small dielectric losses occur which are probably caused by temporary collision-induced moments. From the loss factor maxima of the dielectric relaxation at microwave frequencies an effective value of the collision-induced dipole moment of 0.063 D is obtained for all alkanes. This and the smallness of the appropriate relaxation times (2–5 ps) give rise to suppose that the dielectric relaxation of the alkanes at microwave frequencies is an effect of the motions of the ends of the molecules.

Résumé. On a déterminé la permittivité complexe de sept n-alcanes liquides dans la gamme des fréquences 26 à 1500 GHz. Dans ces liquides, qui n'ont pas de moments dipolaires électriques permanents, il y a des petites pertes par relaxation diélectrique, qui sont causées par l'action des moments dipolaires temporaires induits par chocs. En bande de micro-ondes le facteur de pertes diélectriques a un maximum de relaxation distinct ($\tan\delta = 4$–13×10^{-4}). De la hauteur du maximum on calcule, que tous les alcanes ont des valeurs égales de 0.063 D du moment dipolaire temporaire. En considération de ce résultat et de la petitesse des temps de relaxation diélectrique (2–5 ps), on peut supposer que la relaxation des alcanes en bande de micro-ondes est probablement un effet du mouvement des bouts des molécules.

In pure liquids without permanent electric dipole moment small dielectric losses are observed in the microwave range as well as in the far infrared, which are probably caused by temporary dipole moments. These dipole moments of the order of 0.1 D arise as a result of mutual interactions of colliding molecules, e.g. by collision induced distortion of the molecular framework or by induction by the molecular multipole fields.

Besides the well-known far infrared absorption at frequencies 10^3 to 10^4 GHz, as resulting from molecular vibrations [1], some non-polar liquids such as CCl_4, CS_2, C_6H_6 show – not very distinctly – additional absorption in the microwave range around 10^2 GHz, which is interpreted as to be correlated with slower diffusive – probably translational – motions of the molecules [2], but detailed explanation of which has not yet been given.

As n-heptane showed a distinct loss factor maximum at microwave frequencies [3, 4], our special interest applied to the homologous series of liquid n-alkanes. These low-loss solvents are so far regarded to be non-polar, as, after a model of Fröhlich [5], their bond moments compensate to zero.

Measurements of the complex permittivity

$$\varepsilon = \varepsilon'(1 - j\tan\delta)$$

of seven n-alkanes (n-pentane to n-hexadecane) have been carried out in the microwave range (26 to 140 GHz) by means of two oversized TE_{01n} – cavity resonator systems in which ε' is determined from the change of the wavelengths and $\tan\delta$ from the half-widths of the resonance curves, both measured in the cases of the resonator being unfilled or filled with the liquid to be tested [6].

In the frequency range 360 to 1500 GHz, the loss factor $\tan\delta$ was determined from the transmission of liquid layers of different thicknesses measured in a Beckman-RIIC Fourier spectrophotometer LR 100 (lamellar grating type). The cells containing the liquids (lengths 1.2 to 7.2 cm, inner diam. 5 cm) were terminated at their faces by polyethylene windows the thickness of which was 0.75 cm to avoid alteration of the thickness of the liquid layer by distortion when the measurement chamber of the spectrophotometer was evacuated. For all cells only one pair of windows was used.

The influence of small polar contaminations of the alkanes on the measured loss was carefully tested. The purity of n-pentane C_5H_{12} and n-heptane C_7H_{16} (Merck AG, Darmstadt), n-hexane C_6H_{14} (Ega-Chemie, Steinheim) and of n-nonane C_9H_{20}, n-decane $C_{10}H_{22}$, n-dodecane $C_{12}H_{26}$ and n-hexadecane $C_{16}H_{34}$ (Fluka AG, Buchs) was greater than 99 wt-%, as stated by vapour phase chromatography. As an example, n-decane (purity 99.82 wt-%) had a small content of 0.15 wt-% non-polar iso-alkenes and of 0.03 wt-% weak polar olefins (e.g. 1-decene). To test the influence of alkanes on the loss, n-decane and n-heptane have been mixed with 0.6 wt-% and 1.4 wt-% 1-decene, resp., and additional loss has been found only of the higher contamination in n-heptane. In n-hexadecane (purity 99.59 wt-%) only iso-alkane contaminations have been determined.

The water content was smaller than 3×10^{-3} wt-% and should give no contribution to the microwave loss factors at the measured concentrations. This was additionally tested by shaking one part of a n-heptane sample with bidistilled water for a week and by comparison of its dielectric loss with the loss of the rest. The loss factor increased significantly only at the highest microwave frequency (138 GHz) by an amount of $\tan\delta = 0.9 \times 10^{-4}$. Samples of n-heptane, n-nonane and n-decane purified in glass columns filled with active Al_2O_3 and Silica-Gel 'Woelm' gave no difference in loss compared with unpurified samples.

The influence of dissolved oxygen was tested by either purifying and treating samples of n-heptane and n-nonane under dried nitrogen of extreme purity and either shaking samples of these alkanes with air for 10 h. No differences in the losses have been found.

In the microwave frequency range, the measured loss factors of all alkanes show significant peaks at frequencies from 40 to 120 GHz; at higher frequencies the loss decreases and then increases again rising presumably to a broad absorption in the far infrared with maxima above 1500 GHz (cf. Figure 1). In Figure 2 the real part ε' of the permittivity of n-nonane is plotted vs microwave frequency as an example. A relaxation dispersion step cannot be determined precisely as it is of the same order of magnitude as the very small error in ε' ($\Delta\varepsilon' = \pm 0.002$).

To obtain the pure loss $\tan\delta_{\text{MICR}}$ only due to relaxation processes at microwave frequencies, the part of the loss $\tan\delta_{\text{FIR}}$ evoked by the far infrared absorption has to be subtracted from the total loss $\tan\delta_M$:

$$\tan\delta_{\text{MICR}} = \tan\delta_M - \tan\delta_{\text{FIR}}.$$

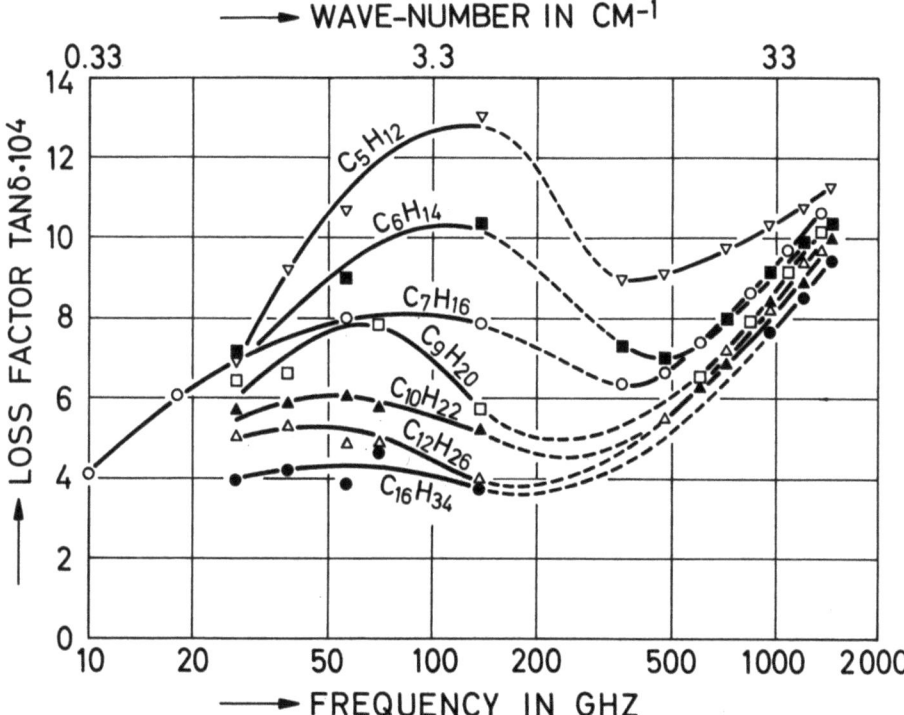

Fig. 1. Plot of the dielectric loss factor $\tan\delta_M$ vs frequency for seven n-alkanes (temperature 20°C, n-hexadecane 25°C). Values for n-heptane are taken from [10] (10 and 18 GHz) and from [3] (55 GHz).

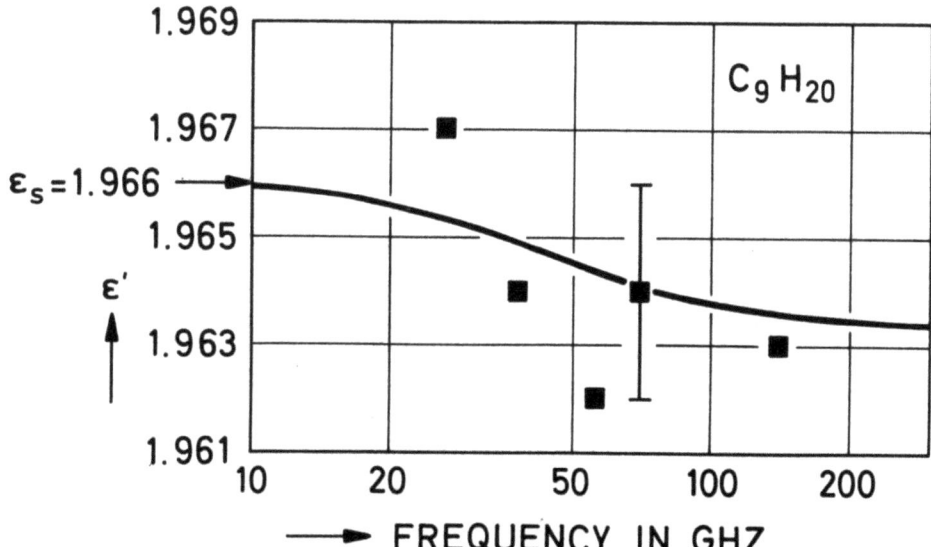

Fig. 2. Plot of the real part ε' of the permittivity vs frequency for n-nonane (20°C). Additionally are given the static permittivity ε_S [9], the dispersion curve of a Debye relaxation calculated under consideration of ε_S, τ_{RELAX} and $\tan\delta_{max}$ of n-nonane, and the error in ε' at 70 GHz.

This has been done in a preliminary state of the measurements, as the course of the loss factor was unknown above 1500 GHz, by supposing the low frequency wing of the far infrared loss $\tan\delta_{FIR}$ to vary linearly with frequency at frequencies f much less (<500 GHz) than the frequency of the far infrared maximum:

$$\tan\delta_{FIR} \propto f.$$

From the maxima $\tan\delta_{max}$ of the loss $\tan\delta_{MICR}$ an effective dipole moment μ_{eff} per molecule (as if in vapour state) has been calculated using the expressions:

$$\varepsilon''_{max} \approx \varepsilon_s \tan\delta_{max}, \qquad \varepsilon''_{max} = \frac{2\pi N}{3kT}\mu_{eff}^2\left(\frac{n^2+2}{3}\right)^2 \qquad (5)$$

with regard to $\varepsilon_s \approx n^2$, $\tan\delta$ being very small. ε_s is the static permittivity, n the refractive index, N the number of molecules per cm^3, k the Boltzmann constant, T the absolute temperature (293 K). Results are given in Table I.

TABLE I

Effective dipole moments and dielectric relaxation times of n-alkanes

Sample	μ_{eff} (D)	τ_{RELAX} (ps)
n-hexane	0.061	2.3
n-heptane	0.061	2.9
n-nonane	0.066	3.5
n-decane	0.061	3.8
n-dodecane	0.063	4.5
n-hexadecane	0.063	3.7

The existence of a value of μ_{eff} (mean value 0.063 D) equal for all alkanes investigated here gives rise to suppose the temporary dipole moment to be localised mainly at the ends of the molecules. Replacing the number N of molecules per cm^3 in the formula given above by the number $2N$ of molecule ends per cm^3, an effective mean dipole moment of 0.044 D per molecule end is obtained.

From the frequencies f_{max} of the microwave loss maxima $\tan\delta_{max}$ the dielectric relaxation times $\tau_{RELAX} = 1/2\pi f_{max}$ have been calculated (cf. Table I).

In Figure 3, dielectric (τ_{RELAX}) and orientational (τ_{OR}) times obtained by Rayleigh scattering experiments are plotted vs the number n_C of carbon atoms per molecule. There τ_{RELAX} obtained for the alkanes are compared with τ_{RELAX} of n-alkylbromides $C_nH_{2n+1}Br$ [7] and with τ_{OR} of n-alkanes [8] and of n-alkylbromides [7]. We observe that τ_{RELAX} of alkanes are about an order of magnitude smaller than τ_{RELAX} of alkylbromides with the same n_C, although τ_{OR} which are a measure for the reorientation time of the whole molecule differ only by a factor $\frac{1}{2}$. This is a second argument for the dielectric relaxation in alkanes at microwave frequencies to be an effect of the molecule ends, i.e. the dielectric relaxation time of alkanes to be a characteristical time for the motions of the ends which are quicker than the orientation of the whole molecule. The strong saturation effect in τ_{RELAX} of alkanes with increasing number n_C of carbon atoms is characteristic for a molecular end effect, too. Thus, the measured

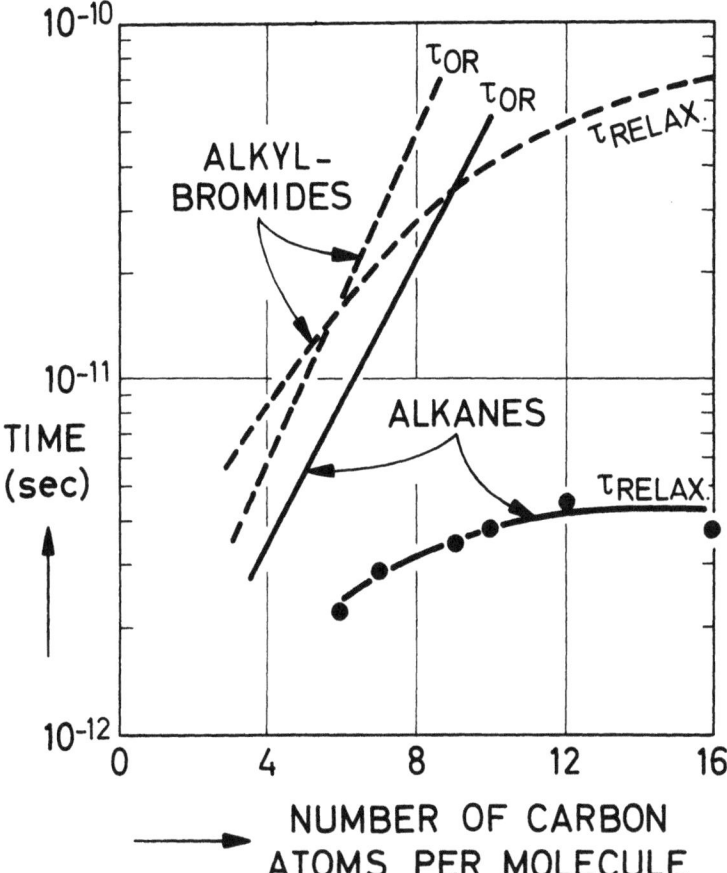

Fig. 3. Plot of the dielectric (τ_{RELAX}) and orientational (τ_{OR}) relaxation times for alkanes and alkylbromides vs number of carbon atoms per molecule.

dielectric relaxation of alkanes seems to occur from fluctuating dipoles, possibly generated by molecular interactions and appearing mainly at the chain ends.

Acknowledgements

The author wishes to thank Prof. Dr R. Pottel for helpful discussions and Dr H. Schulz, Karlsruhe for carrying out vapour phase chromatographic measurements. Fourier spectroscopy was performed in the Institut für Physikalische Chemie der Universität Göttingen. The project was financially supported by the Deutsche Forschungsgemeinschaft, Bad Godesberg.

References

1. Davies, M., Pardoe, G. W. F., Chamberlain, J., and Gebbie, H. A.: *Trans. Faraday Soc.* **66**, 273 (1970).
2. Garg, S. K., Bertie, J. E., Kilp, H., and Smyth, C. P.: *J. Chem. Phys.* **49**, 2551 (1968).

3. Güsewell, D.: *Z. angew. Physik* **22**, 461 (1967).
4. Kilp, H.: *Z. angew. Physik* **30**, 288 (1970).
5. Fröhlich, H.: *Theory of Dielectrics*, 2nd edition, At the Clarendon Press, Oxford 1958, p. 113f.
6. Stumper, U.: *Proc. Conf. on Measurement of High Frequency Dielectric Properties of Materials*, Teddington, Middlesex, England, 1972.
   Stumper, U.: *Rev. Sci. Instr.* **44**, 165 (1973).
   Göttmann, O.: Diplomarbeit, Göttingen, 1972.
   Göttmann, O. and Stumper, U.: to be published.
7. Pinnow, D. A., Candau, S. J., and Litovitz, T. A.: *J. Chem. Phys.* **49**, 347 (1968).
8. Volterra, V., Bucaro, J. A., and Litovitz, T. A.: *Ber. Bunsenges. physik. Chem.* **75**, 309 (1971).
9. Champion, J. V.: *Trans. Faraday Soc.* **66**, 2671 (1970).
10. Dagg, I. R. and Reesor, G. E.: *Can. J. Phys.* **43**, 1552 (1965) and **41**, 1314 (1963).

DISCUSSION

Friedman: (1) It may be interesting to note that several people, perhaps Allerhand first, have found direct evidence from ^{13}C NMR for the changing motions along alkyl chains in the liquid.

(2) I also may raise the question whether one small dipole moment you need may be the permanent moment of the group at the end?

Stumper: In Fröhlich's model, the *n*-alkanes are nondipolar as the sum of all C—H-bond moment vectors in an alkane molecule is zero. This will be, however, only the case if the dipole moments of C—H-bonds in a CH_2-group and in a CH_3-group are of equal values.

Rothschild: If I may make a comment; during a seminar we heard by Dr Zerbi a few weeks ago at our laboratory at Dearborn, and I hope that my memory is correct, he presented some evidence that the motion of the end groups of flexible chain molecules contributes quite appreciably to the overall motion in even rather long chains.

MIX
Papier aus verantwortungsvollen Quellen
Paper from responsible sources
FSC® C105338

If you have any concerns about our products,
you can contact us on
ProductSafety@springernature.com

In case Publisher is established outside the EU,
the EU authorized representative is:
**Springer Nature Customer Service Center GmbH
Europaplatz 3, 69115 Heidelberg, Germany**

Printed by Libri Plureos GmbH
in Hamburg, Germany